Outstanding Contributions to Logic

Volume 22

Outstanding Contributions to Logic puts focus on important advances in modern logical research. Each volume is devoted to a major contribution by an eminent logician. The series will cover contributions to logic broadly conceived, including philosophical and mathematical logic, logic in computer science, and the application of logic in linguistics, economics, psychology, and other specialized areas of study.

A typical volume of Outstanding Contributions to Logic contains:

- A short scientific autobiography by the logician to whom the volume is devoted
- The volume editor's introduction. This is a survey that puts the logician's contributions in context, discusses its importance and shows how it connects with related work by other scholars
- The main part of the book will consist of a series of chapters by different scholars that analyze, develop or constructively criticize the logician's work
- Response to the comments, by the logician to whom the volume is devoted
- A bibliography of the logician's publications Outstanding Contributions to Logic is published by Springer as part of the Studia Logica Library.

This book series, is also a sister series to Trends in Logic and Logic in Asia: Studia Logica Library. All books are published simultaneously in print and online. This book series is indexed in SCOPUS.

Proposals for new volumes are welcome. They should be sent to the editor-in-chief sven-ove.hansson@abe.kth.se

More information about this series at http://www.springer.com/series/10033

Ivo Düntsch · Edwin Mares
Editors

Alasdair Urquhart on Nonclassical and Algebraic Logic and Complexity of Proofs

Editors
Ivo Düntsch
Department of Computer Science
Brock University
St. Catharines, ON, Canada

College of Mathematics and Computer Science
Fujian Normal University
Fuzhou, China

Edwin Mares
School of History, Philosophy, Political Science and International Relations
Victoria University of Wellington
Wellington, New Zealand

ISSN 2211-2758 ISSN 2211-2766 (electronic)
Outstanding Contributions to Logic
ISBN 978-3-030-71429-1 ISBN 978-3-030-71430-7 (eBook)
https://doi.org/10.1007/978-3-030-71430-7

This Springer imprint is published by the registered company Springer Nature Switzerland AG
The registered company address is: Gewerbestrasse 11, 6330 Cham, Switzerland

Preface

It is customary to state a rationale for a book in its preface. This is a series of logicians who have made an outstanding contribution to their field. Alasdair Urquhart's contributions have been truly outstanding, and not to just one isolated field in logic. His work is widely influential on mathematicians, philosophers, computer scientists, and historians of logic. It is only surprising that no collection has been dedicated to his work before this.

It is also customary in books in this series for the editors to talk about their personal connections with the subject. We do this in turn.

I. Düntsch: My first contact with Alasdair goes back to the late 1970s when I asked him for a reprint of his article *A topological representation theorem for lattices* which had appeared in Algebra Universalis in 1978. Owing to our common interest in varieties of projective p-algebras further correspondence ensued. Our exchanges were not restricted to mathematical topics, but also included personal matters and lasted for over 20 years. After all this time we first met in person at a workshop *Relation Days at Brock* in 2002 after I had moved to the Niagara Region in Ontario, Canada. Since then we have become close friends and visit frequently. Out of these visits arose in 2006 our (only) joint publication *Betweenness and comparability obtained from binary relation* which reflects Alasdair's interest in geometry and my interest in relations and their algebras. Balancing art and science and for added pleasure, for many years Alasdair and I joined the bass section in the performance of the Sing-along Messiah by Toronto's Tafelmusik shortly before Alasdair's birthday.

E. Mares: I recall having first seen Alasdair give a talk about Russell's solutions to the paradox prior to ramified type theory. This was at a Russell conference in Toronto in 1984. I had already heard of Alasdair's proof that E and R were undecidable, but I was still an undergraduate and didn't really know what E and R were, although I had an inkling about what undecidability was. When I went to visit Indiana in 1984 before choosing it as my graduate school, Mike Dunn tried to convince me to stay in Toronto to work with Alasdair. This was an expression of Mike's regard for Alasdair and not (I think) an attempt to avoid having me as a graduate student. I first met Alasdair when he gave two talks on relevance logic at Indiana. I recall spending hours talking to him in Mike and Sally's kitchen about all sorts of topics at a party they threw for Alasdair. I found Alasdair, then as now, always approachable, and always kind.

I have always found Alasdair completely free of any sort of intellectual snobbery (that might be warranted, given his achievements) and petty academic jealousy. I am extremely happy to be involved in this project. Alasdair's accomplishments in logic have been truly groundbreaking and he is a great guy.

In putting together this collection, we decided to use an open style of refereeing. We asked people to be second readers for chapters. Then they read the chapters, wrote reports, and in many cases communicated directly with the authors. In every case, this process was undertaken by both second readers and authors in a friendly spirit of collaboration. It was a real pleasure to work with these second readers and they have our heartfelt gratitude. Their names are given on the papers that they read. In addition, we thank the editor of the series, Sven Ove Hansson, for his encouragement and wise guidance, and Ram Prasad Chandrasekar and his team at Springer for their technical assistance.

St. Catharines, Canada/Fuzhou, China — Ivo Düntsch
Wellington, New Zealand — Edwin Mares

Second Readers

A. Avron School of Computer Science, Tel-Aviv University, Tel-Aviv, Email: aa@tauex.tau.ac.il
P. Balbiani Institut de Recherche en Informatique de Toulouse—IRIT, Email: Phi lippe.Balbiani@irit.fr
P. Beame Paul G. Allen School of Computer Science & Engineering, University of Washington, Seattle, Email: beame@cs.washington.edu
G. Bezhanishvili Department of Mathematical Sciences, New Mexico State University, Las Cruces, Email: guram@nmsu.edu
I. Düntsch College of Mathematics and Informatics, Fujian Normal University, Fuzhou and Dept. of Computer Science, Brock University, St Catharines, Email: duentsch@brocku.ca
W. Dzik Institute of Mathematics, University of Silesia, Katowice, Email: woj ciech.dzik@us.edu.pl
R. Goldblatt School of Mathematics and Statistics, Victoria University of Wellington, Wellington, Email: Rob.Goldblatt@vuw.ac.nz
Z. Gyenis Department of Logic, Jagiellonian University, Kraków, Email: zalan. gyenis@uj.edu.pl
I. Hodkinson Department of Computing, Imperial College, London, Email: i. hodkinson@imperial.ac.uk
L. Humberstone Philosophy, Monash University, Melbourne, Email: Lloyd. Humberstone@monash.edu
P. Jipsen Faculty of Mathematics, Chapman University, Orange, Email: jipsen @chapman.edu
J. Madarász Alfréd Rényi Institute of Mathematics, Budapest, Email: madara sz@renyi.hu
E. Mares School of History, Philosophy, Political Science and International Relations, Victoria University of Wellington, Wellington, Email: Edwin.Mares@vuw. ac.nz
G. Payette Department of Philosophy, University of Calgary, Email: ggpaye tt@ucalgary.ca
I. Pratt-Hartmann Department of Computer Science, Manchester University, Manchester, Email: ipratt@cs.man.ac.uk

I. Rewitzky Department of Mathematical Sciences, University of Stellenbosch, Stellenbosch, South Africa, Email: rewitzky@sun.ac.za
H. P. Sankappanavar Department of Mathematics, SUNY, New Paltz, Email: sankapph@newpaltz.edu
I. Sedlár Institute of Computer Science, The Czech Academy of Sciences, Prague, Email: sedlar@cs.cas.cz
F. Wolter Department of Computer Science, University of Liverpool, Email: Wolter@liverpool.ac.uk

Contents

Editors and Contributors

About the Editors

Ivo Düntsch's research areas include lattice theory and universal algebra, modal logic and its application to data analysis and machine learning, statistical foundations of rough set theory, and region-based topology. He obtained his Ph.D. under the guidance of S. Koppelberg and his Habilitation at the Freie Universität Berlin in 1989, referees being R. McKenzie, J. D. Monk, and H.A. Priestley. He has worked in five continents in various academic and administrative capacities. Currently, he serves as visiting professor at Fujian Normal University, Fuzhou, China and is an emeritus professor at Brock University, St Catharines, Canada. He is also a senior member of the International Rough Set Society.

Edwin Mares is a professor of philosophy at Victoria University of Wellington. He has written extensively about non-classical logics (especially relevance logic) as well as about epistemology. He is the author of three books (including Relevant Logic: A Philosophical Interpretation, Cambridge University Press, 2004) and more than 60 journal articles and book chapters. He is also the managing editor of The Australasian Journal of Logic.

Contributors

Shay Allen Logan Department of Philosophy, Kansas State University, Manhattan, KS, USA

Katalin Bimbó Department of Philosophy, University of Alberta, Edmonton, AB, Canada

T. S. Blyth Mathematical Institute, University of St Andrews, St Andrews, Scotland

Sam Buss Department of Mathematics, University of California, La Jolla, CA, USA

Willem Conradie School of Mathematics, University of the Witwatersrand, Witwatersrand, South Africa

J. Michael Dunn Department of Philosophy and Luddy School of Informatics, Computing and Engineering, Indiana University, Bloomington, IN, USA

Ivo Düntsch College of Mathematics and Computer Science, Fujian Normal University, Fuzhou, Fujian, China;
Department of Computer Science, Brock University, Ontario, Canada

Noah Fleming University of Toronto, Toronto, Canada

Jacob Garber University of Alberta, Edmonton, Canada

Robert Goldblatt School of Mathematics and Statistics, Victoria University of Wellington, Kelburn, New Zealand

Valentin Goranko Department of Philosophy, Stockholm University, Stockholm, Sweden;
School of Mathematics (Visiting professorship), University of the Witwatersrand, Witwatersrand, South Africa

Allen P. Hazen Department of Philosophy, University of Alberta, Edmonton, Alberta, Canada

Robin Hirsch Department of Computer Science, University College, London, UK

Marcus Kracht Fakultät Linguistik und Literaturwissenschaft, Universität Bielefeld, Bielefeld, Germany

Philip Kremer Department of Philosophy, University of Toronto, Toronto, Canada

Bernard Linsky Department of Philosophy, University of Alberta, Edmonton, Alberta, Canada

Roger D. Maddux Department of Mathematics, Iowa State University, Ames, IA, USA

David Makinson London School of Economics, London, UK

Brett McLean Laboratoire J. A. Dieudonné UMR CNRS 7351, Université Nice Sophia Antipolis, Nice, France

Ewa Orłowska National Institute of Telecommunications, Warsaw, Poland

Francis Jeffry Pelletier Department of Philosophy, University of Alberta, Edmonton, Alberta, Canada

Toniann Pitassi University of Toronto, Toronto, Canada

Greg Restall School of Historical and Philosophical Studies, The University of Melbourne, Melbourne, Australia

H. J. Silva Centro de Matemática e Aplicaões and Departamento de Matemática, Faculdade de Ciências e Tecnologia, Universidade Nova de Lisboa, Lisboa, Portugal

Shawn Standefer Institute of Philosophy, Slovak Academy of Sciences, Bratislava, Slovakia

Alasdair Urquhart Department of Philosophy, University of Toronto, Toronto, Canada

Dimiter Vakarelov Faculty of Mathematics and Informatics, Department of Mathematical Logic and Applications, Sofia University "St. Kliment Ohridski", Sofia, Bulgaria

André Vellino School of Information Studies, University of Ottawa, Ottawa, Canada

Chapter 1
A Logical Autobiography

Alasdair Urquhart

1.1 Early Years

I was born in Auchtermuchty, a small village in Fife, Scotland, the son of a lowland mother, Meta Mowat, daughter of a country headmaster, and a highland father, William Urquhart from Lairg, Sutherland. (Auchtermuchty appears as the fictional village of Tannochbrae in the 1990s television series *Doctor Finlay*.) The family business in Lairg was a butcher's shop, known as *Urquhart the Butcher*, even for some years after my father's family had given it up.

My father worked as a bank teller in the Bank of Scotland in Auchtermuchty but was called up into the army in the 1940s. He was in training in the south of England for service in Europe, but was invalided out as a consequence of a recurrence of tuberculosis. He died in early 1946, so I do not have any memories of him. In the days before penicillin, tuberculosis was a killer—my father's sister, Mary, died of it as well.

My parents married during the war, and had two children, my sister Gillian, born January 16, 1943, and myself, born December 20, 1945. Gillian trained in physics but eventually became a prominent medical statistician, Gillian M. Raab.

After my father's death, the family moved to St. Ninians, a small town near Stirling. St. Ninians is close to the site of the battle of Bannockburn, a famous Scottish victory in the Wars of Independence. Naturally, growing up where these stirring events took place, as a small boy I was an ardent Scottish nationalist, and William Wallace and King Robert the Bruce were my great heroes. I attended the local primary school down the road, where I didn't need to work hard, as most of the pupils were children of St. Ninians' working folks—as a middle-class child, I was favored.

My experience of school changed for the worse, with a move to Edinburgh when I was in my early teens. I attended Daniel Stewart's College, a boys-only school with

A. Urquhart (✉)
Department of Philosophy, University of Toronto, Toronto, Canada
e-mail: urquhart@cs.toronto.edu

I. Düntsch and E. Mares (eds.), *Alasdair Urquhart on Nonclassical and Algebraic Logic and Complexity of Proofs*, Outstanding Contributions to Logic 22,
https://doi.org/10.1007/978-3-030-71430-7_1

good academic standards, housed in a large Victorian building in the Jacobethan style on the Queensferry Road. This was not a happy time for me, on the whole, and my academic work showed it. I did the minimum of work, normally, though I managed to acquire a good knowledge of French and some mathematics. Our geometry text was based on Euclid, though in a somewhat digested form. I enjoyed geometrical constructions and proofs, a taste that has remained with me.

Scottish teachers in those days disciplined pupils by beating them on the hands with thick leather straps. My indolence and foolish behavior led me to fairly frequent doses of "six of the best" from the Lochgelly tawse. This practice came to an end in the 1980s, after my time.

1.2 Undergraduate Days

After my time at secondary school, I experienced new-found liberation for 4 years at Edinburgh University. I was free to pursue my own interests and spent many happy hours in the University Library or the numerous bookstores in the city. I began in English (my mother's subject) but switched to Philosophy in my second year. I was partly influenced in this shift by tutorials with George Elder Davie, author of *The Democratic Intellect*, a study of nineteenth-century Scottish philosophy. Some years ago I was happy to see a quote from George Davie's book in an article by Noam Chomsky.

Initially, my impulses in philosophy drew me toward a somewhat vague mysticism. I was attracted to the dark sayings of Hegel and F.H. Bradley, and felt there must be some deep meaning to Hegel's idea that pure being and pure nothingness are identical.

I was in the habit of haunting second-hand bookstores, of which there were many at the time. One of my favorites was Grant's, on George IV bridge, a marvelous pile of dusty old books. There I discovered Bertrand Russell's *Introduction to Mathematical Philosophy* (Russell 1920) written while Russell was in prison for anti-war activities. In the second chapter, Russell reveals to the reader what he claims is the first correct definition of number, given by Gottlob Frege. This book blew all the musty cobwebs of Absolute Idealism out of my brain and pointed me toward logic.

The logician in the Edinburgh department at the time was Robert Stoothoff (a Frege specialist), from whom I took a tutorial in my last year on Gödel's theorem. I also studied a bit of axiomatic set theory. But most of my energies went into self-education projects in logic. I worked through most of the first volume of Whitehead and Russell's *Principia Mathematica*, covering up the proofs and trying to rediscover them myself, a very useful training in logical manipulations. I had also heard about Gödel's incompleteness theorem (I think through Nagel and Newman's article in *Scientific American*) and resolved to understand this completely—a project of several years, as it turned out.

Nowadays, there are many books on logic, as well as journals. But in the 1960s, they were quite sparse. The first English translation (Gödel 1962) of the great 1931 incompleteness paper had appeared in 1962, so I studied that assiduously; later I was able to order a copy of Martin Davis's wonderful anthology *The Undecidable* (Davis 1965), containing some of the most important logical papers from the great period of the 1930s. I also studied Paul Rosenbloom's unusual textbook, *The Elements of Mathematical Logic* (Rosenbloom 1950), leading to my interest in Emil Post and his work.

I became aware of Alonzo Church and *The Journal of Symbolic Logic*. I used to lug home bound volumes of the *JSL* from the library and marveled at the reviews of Alonzo Church and W.V. Quine. Some of these were quite brutal. In Volume 6 of the *JSL*, I was amazed to find a devastating review by Quine (1941) of the book *An Inquiry into Meaning and Truth* by my new hero, Bertrand Russell. Quine describes the book as "a rambling discussion of traditional epistemological problems and certain adjacent metaphysical and semantic ones," and at one point describes a move of Russell as a lame performance. The last sentence of the review reads: "For *canoid* ('basket-shaped'), which recurs persistently through the book, read *cynoid* ('dog-shaped')".

I loved the image that Quine projected of himself as a gun-slinging rebel, destroying vague and mushy arguments with his logical six-shooter. I also loved Alonzo Church's combination of logical power with historical sensitivity. Historians of logic owe him a great debt; he is the source of my own interest in the history of logic.

In my fourth year at Edinburgh, I thought of going on to graduate work in logic. At the time, the action in logic seemed to be in the United States—Cohen's independence proof for the Continuum Hypothesis was then only a few years old. On the advice of Robert Stoothoff, I consulted Keith Lehrer, who was visiting the department at the time.

Keith Lehrer recommended the following Departments of Philosophy: The University of Illinois at Chicago Circle, Stanford, UC Berkeley, and Pittsburgh, and I applied to all four. Leon Henkin in Berkeley lost my letter and wrote me a very apologetic response. He offered to expedite my application, but in the end I didn't pursue Berkeley (if I had, my life would have been quite different). I would have very much liked to have studied at Stanford, as Feferman, Hintikka, Kreisel, and Cohen were all there, and they had a great list of logic courses. However, they turned me down, so I ended up at Pittsburgh in 1967.

1.3 Graduate Work in Pittsburgh

When I arrived in Pittsburgh, work on entailment and relevant implication was in full swing; Alan Anderson and Nuel Belnap were writing the first volume of *Entailment* (Anderson and Belnap 1975). My orientation at the time was toward classical logic (it still is), so it is odd how many papers I have published on non-classical logics. However, the problems of relevant implication were fascinating and challenging, so of course I took them up.

In the 1960s, there was a lot of excitement surrounding the development of semantics for non-classical logics, particularly modal logics. I studied the brilliant early papers of Saul Kripke, as well as the first textbook of modal logic by Hughes and Cresswell (1968). It was natural that I would form the ambition of doing for entailment and relevant implication something similar to the work of Kripke.

In my third year at Pittsburgh, I was very happy to discover the semantics for first-degree entailment in which the key condition for negation is

$$x \models \neg A \Leftrightarrow x^* \not\models A,$$

where the operation * satisfies the condition $x^{**} = x$. I told Alan Anderson about the idea immediately, but my hopes were dashed when he told me that I had been anticipated by Andrzej Białynicki-Birula and Helena Rasiowa (1957) in 1957. Looking back, perhaps I should have persisted with my idea, given the enormous subsequent literature on the topic.

Shortly afterward, I found the semilattice semantics for relevant implication (Urquhart 1972); the key condition is

$$x \models A \to B \Leftrightarrow \forall y(y \models A \Rightarrow x \vee y \models B),$$

where the underlying structure is a semilattice $\langle S, \vee, 0\rangle$ with least element 0. This is a very natural condition for implication; its first appearance may have been in Joachim Lambek's calculus inspired by categorial grammar, and it has been discovered in several contexts, such as linear logic, where it forms part of Jean-Yves Girard's "phase semantics." The underlying structure for Lambek is a monoid, a commutative monoid for Girard. I interpreted the elements of the semilattice as "pieces of information."

Naturally, I thought that merging my two ideas would be simple, but this was not so. The * semantics does not combine naturally with the semilattice condition. It is quite easy to prove completeness for the conjunction/implication fragment of relevant implication with respect to the semilattice semantics, but in spite of considerable effort, I did not succeed in adding disjunction. The reason for this emerged shortly when J. Michael Dunn and Bob Meyer discovered that the semilattice system is stronger than **R** if you add disjunction. The formula

$$[(A \to B \vee C) \wedge (B \to D)] \to (A \to D \vee C)$$

is valid in the semilattice semantics with the natural conditions for conjunction and disjunction but is not provable in **R**. Some further formulas satisfying this condition can be found in the contribution by David Makinson.

Dunn and Meyer had left Pittsburgh when I arrived in 1967, as had Bas van Fraassen, but I soon came to count them among my friends and logical colleagues. It was Bob Meyer who discovered how to combine the * semantics with a generalized condition for implication. If you replace the equation $x \vee y = z$ from the semilattice formulation by the ternary relation $Rxyz$, then you can set down conditions that validate all of **R**, including negation with the * semantics. This is the framework developed by Routley and Meyer in a groundbreaking series of papers beginning in 1973 (Routley and Meyer 1973).

I was quite enamored of the semilattice system at the time, and found the Routley–Meyer approach less elegant. I still think it interesting, though it has not been investigated as much as the standard Anderson-Belnap systems. In particular, the decision problem for it appears to be still open. However, I changed my mind on the Routley–Meyer semantics in the 1980s when I discovered the geometrical models discussed below.

While at Pittsburgh, I served as a research assistant for Nicholas Rescher. I took a course with him based on Arthur Prior's books on tense logic, and an essay I wrote giving a completeness proof in tense logic impressed him sufficiently that he invited me to co-author a book with him under the title *Temporal Logic*.

Although the book with Rescher appeared in a handsome volume (Rescher and Urquhart 1971) in the *Library of Exact Philosophy*, I felt that my own contributions to the volume were rather carelessly written, and I did not work in the area of temporal logic after its appearance. It was a surprise in the following years to find that it was my most cited publication for about a decade.

I only recently discovered the story behind this during a talk by Moshe Vardi at the Fields Institute in Toronto. The computer scientist Amir Pnueli around 1970 had somehow become convinced that non-classical logics might provide a useful language for program verification. One of his colleagues suggested deontic logic as a possibility. Pnueli went to the library, but the books on deontic logic did not seem very helpful. However, on a neighboring shelf, he found my book with Nicholas Rescher. This was the answer to his prayers, and the subsequent work developed into an important area in computer science; Pnueli received the Turing award in 1996 for the introduction of temporal logic into computer science. *Temporal Logic* was cited by many of the early papers in program verification, though of course the area has moved on far beyond my book with Rescher.

In spite of the success of my book with Rescher, I continued to have ambivalent feelings about non-classical logics, or "funny logics" as I would sometimes describe them to myself. In the 1960s and 1970s there were rather strong prejudices against non-classical systems in the logic community of North America. This was a time when, in the words of Bob Meyer, if you wrote a fishhook $\multimap$ for strict implication, you had to hide it behind your hand to avoid being hauled before the Use-Mention Committee of the Association for Symbolic Logic.

There were several reasons for this. One was the fact that many logicians worked in mathematics departments, and wanted to be accepted by their colleagues for proving hard mathematical theorems, not results on odd logical systems inspired by philosophical ideas. Another was Quine's vigorous denunciations of modal logic and the "futile subject" (Quine's phrase) of relevance logic and entailment. Quine dominated American philosophy at that time, particularly in areas close to mathematical logic.

The example of Pnueli cited above gives a clue to what happened next. The computer scientists discovered non-classical logics and found they were quite useful. The mathematicians took notice, and some started working in the area. The transformation since that time has been remarkable, and non-classical logic is now an accepted member of the logical community.

1.4 From Pittsburgh to Canada

In 1970 I was offered a job in the Philosophy department at Erindale College, a suburban campus of the University of Toronto. I had met and married my wife Patricia when in Pittsburgh (she was studying philosophy of science at the time). So, in the late summer of that year, Patricia and I drove up to Mississauga, Ontario with our three-month-old daughter Alison and almost all our worldly goods in the trunk of the car.

Canada seemed rather provincial and sleepy after the electric atmosphere of the late 60s and early 70s in the U.S.A. In October 1970, parliament, under the leadership of Prime Minister Pierre Trudeau, in response to FLQ kidnappings, passed the War Measures Act, essentially a declaration of martial law. I was amazed by the reaction, or rather lack of reaction, in English-speaking Canada. If Richard Nixon had declared martial law at that time, there would have been an immediate uproar. But not in Canada!

At the time, I had the ambition to move back to the U.S.A., where things seemed to be happening, both academically and politically. However, the job market was not great, so I stayed in Canada, and I am now very content that I did.

The next few years (logically speaking) were devoted to writing my doctoral thesis (Urquhart 1973b), an exposition of the semilattice semantics and its applications. I also continued various self-education projects, particularly learning basic lattice theory. While on a visit to Edinburgh, I bought a copy of George Grätzer's first lattice theory book (Grätzer 1971). I solved an open problem in this book, and this led to my first lattice theory paper (Urquhart 1973a). My approach to lattice theory at the time was to adapt ideas from the model theory of non-classical logics to solve problems in the area.

As an example of this, my first paper in the area is a description of a certain family of free algebras. These are the algebras of conjunction, disjunction and negation in intuitionistic logic, so you can use a family of simple Kripke models to prove a kind of disjunctive normal form theorem. The papers in lattice theory I wrote at the time are a little hard to read, because I tended to disguise my use of non-classical models.

In the 1970s, academic departments still had money available for visiting professorships. During the 1972–73 year, Bob Meyer visited Toronto, teaching on the downtown campus, and we became good friends. Bob had more technical skill in logic than I did at the time, and I learnt a great deal from him. We also played a lot of chess, since the Fischer–Spassky match had just taken place.

During 1973–74, Kit Fine visited the Toronto department; I had known him from a visit to Edinburgh earlier. Since we both had small children, we would get together on the weekends quite often, and I came to admire Kit's logical insight and power enormously.

Reading the literature of lattice theory, I discovered that there were two famous open problems. The first, the word problem for (free) modular lattices, the second the question of whether every distributive algebraic lattice is isomorphic to the congruence lattice of a lattice. I resolved to attack these questions using my favorite techniques from non-classical logic.

I had recently learned Hilary Priestley's results (Priestley 1970) on representing distributive lattices using ordered Stone spaces. I decided to see if I could find a similar representation for general lattices. The result was my topological representation theory for lattices (Urquhart 1978), a result that is now quite well known in the non-classical logic community.

In 1976, I moved to Oxford for a year with Patricia, Alison, and our son Adam, born in Oakville in 1974, to spend a sabbatical year. Dana Scott, Michael Dummett, and Robin Gandy were all there at the time. I spent a lot of time working on my ideas in lattice theory, but did not succeed in solving either of the two famous problems, though I solved a few minor problems using my approach.

In the 1970s, I participated in the community in lattice theory and universal algebra and got to know a lot of fine researchers such as Ralph Freese, Joel Berman, Bill Lampe, Don Pigozzi, Bjarni Jónsson, Steve Comer and others. I met George Grätzer for the first time at a 1976 meeting in Oberwolfach, and also encountered many of the talented people in the Manitoba group, such as Ivan Rival. They were very welcoming to me, and I was happy to take part in the community at a time when I was feeling rather isolated on the suburban campus in Mississauga.

In October 1978, I attended a meeting devoted to lattice theory and universal algebra in Claremont, California. Many of the top people in the field were there, including Ralph McKenzie, regarded with awe by most participants for his raw mathematical power. Ralph Freese presented a paper (Freese 1980) proving one of the great theorems of lattice theory: the free modular lattice in five generators (variables) has an unsolvable word problem. This talk received the only standing ovation I have witnessed at a mathematics meeting.

Freese's talk made me realize that the simple tricks of non-classical logic in this area were not going to work, and that I had to learn the basic techniques of modular lattice theory. I finally did this in the summer of 1982.

1.5 Australia and Undecidability

On an invitation from J.J.C. Smart, I spent a two month visit in 1982 at the Research School of Social Sciences of the ANU in Canberra. Bob Meyer had gone there after Toronto and had a small cadre of graduate students there, including Steve Giambrone, Errol Martin, Paul Thistlewaite, and Michael McRobbie. My old friend

Chris Mortensen, a fellow graduate student in Pittsburgh, was also there, so it was a most stimulating environment for logic.

Although Freese had proved the word problem for the free modular lattice unsolvable, the word problem for the free orthomodular lattice seemed still open, so I resolved to tackle that by learning the techniques of Freese and others. I soon found out that these techniques have their origins in the basic coordinatization theorem for continuous geometries (von Neumann 1960) due to John von Neumann. The methods of von Neumann in turn are adaptations of classical constructions from projective geometry, going back to von Staudt. This connected up with an old love of mine, the beautiful field of classical projective geometry.

Although I found a lot of pleasure in understanding the geometrical techniques, I still couldn't see how to adapt them for orthomodular lattices. In fact, it appears that the word problem for orthomodular lattices is still open.

When I went to Canberra in 1982, I was surprised to find that Bob was engaged in a project with his students and collaborators, including Paul Thistlewaite, Michael McRobbie, John Slaney, Steve Giambrone, Paul Pritchard, Chris Mortensen, and Adrian Abraham, to prove **R** undecidable. The program of research was strange, but very original, as most of Bob's ideas were. The idea was to *guess* a two-place connective definable in **R** that you can prove is associative, and also *freely* associative (that is, not satisfying any identities except those of the free monoid), allowing the coding of undecidable word problems for semigroups (Thistlewaite et al. 1988).

In pursuing this objective, they had written a number of very impressive programs, including theorem provers such as KRIPKE, and also programs for generating finite matrices for formulas. This attack on the decidability problem is described in the monograph (Thistlewaite et al. 1988) by Thistlewaite, McRobbie, and Meyer. This was my first encounter with computers, and the first time I had seen large-scale programs for research in logic.

Bob Meyer's group had access to a PDP-10 (DECsystem-10) computer and I was amazed to see big logical matrices appearing on the printer in Bob's office. I was used to laboriously constructing such things by hand, and this seemed quite futuristic to me. I didn't do any programming at that time (Paul Thistlewaite did most of the code hacking for Bob when I was there) but I spent many happy hours exploring the Great Underground Empire in the text-based adventure game **Dungeon** or **Zork** as it came to be known in its PC incarnation. It was only when I returned to Toronto that I started programming, and became a devotee of Lisp for some years, writing some simple theorem proving programs for the classical propositional calculus.

Up to that point, I had believed **R** decidable, or at least the positive fragment $\mathbf{R}^+$, based on my knowledge of Dunn's sequent calculus Anderson and Belnap (1975, Sect. 28.5). Now I reconsidered this belief. One day in the cafeteria of ANU, Bob mentioned to me that their programs had turned up a lot of non-trivial matrices for the logic **KR**, the resulting of adding *ex falso quodlibet* to **R**, that is to say, $(A \wedge \neg A) \to B$. This surprised me, because relevance logics were invented precisely for the purpose of avoiding the "paradoxes of material implication." The corresponding semantic condition is $\forall x(x^* = x)$; this in turn forces the ternary relation in a Routley–Meyer model to be *totally symmetric*. Since I was reading some of the classical literature on projective geometry at the time, this enabled me to form the right connections.

The collinearity relation in a projective space is also totally symmetric, so you can build models for **KR** from projective spaces. Relative to a suitably chosen coordinate frame, you can use von Staudt's method to define multiplication on a line; the construction is general enough (free enough) that you can embed any countable semigroup in the defined multiplicative semigroup. This yields undecidability (Urquhart 1984) for a large family of relevance logics, including **E**, **R** and **KR**.

Some years later, I realized that the geometrical ideas used for undecidability could also be used to refute the interpolation theorem for many relevance logics. My construction was adapted from a 1973 paper (Grätzer et al. 1973) by Grätzer, Jónsson and Lakser showing that the amalgamation property fails in any variety of modular lattices containing a nondistributive lattice. Their argument is based on the fact that a non-Arguesian projective plane cannot be embedded in a three-dimensional projective space. The construction for relevance logics is somewhat more complicated because of the presence of additional connectives such as negation. I was pleased with the resulting paper (Urquhart 1993), since I believe it was the first time that a non-Arguesian projective plane had been used to prove a result in pure logic.

After proving the undecidability results, I decided to move on from non-classical logics and duality theory to a new area. I still retained a great love for classical propositional logic, going back to my early study of *Principia Mathematica*, and I was aware that the central problem in theoretical computer science was the question whether there is a feasible decision procedure for satisfiability, the $\mathscr{P} =?\mathscr{N}\mathscr{P}$ problem. I devoted a sabbatical year, 1982–1983, to learning basic complexity theory.

1.6 Moving to Computer Science

The move into computer science was a fortunate choice. Although I enjoyed interacting with colleagues in the lattice theory community, who were very welcoming, this only happened now and then at conferences. I was also an active member of the logic community, but was largely cut off from face-to-face interactions except at logic meetings. However, the University of Toronto was home to one of the leading departments for theoretical computer science; the faculty included Steve Cook, Allan Borodin, Charles Rackoff, Joachim von zur Gathen, Faith Fich, Russell Impagliazzo, as well as a large contingent of talented graduate students.

For the first time in my academic career since graduate school, I could interact on a daily basis with a group of first-rate mathematical researchers, attending seminars, and hearing all the latest news in the area of theory. This was made all the easier because I had moved from Erindale College to the central campus, after Bas van Fraassen had left for Princeton. On the downtown campus, I encountered some first-rate undergraduates. Three students, Jamie Tappenden, Peter Koellner, and Philip Kremer, to whom I taught logic on the undergraduate level, have gone on to distinguished careers in logic and philosophy.

I started work on the complexity of proofs in classical logic. I was lucky in my choice of this area, since the literature on the subject in the early 1980s was very small, and I could master all of it very quickly, particularly since Steve Cook had supervised two doctoral theses, those of Robert Reckhow and Martin Dowd, containing basic results in the area of propositional proof complexity.

The most important early paper on the subject is by Grigori Tseitin (1970), presented to the Leningrad Seminar on Mathematical Logic in September 1966. I studied his paper very closely, together with by Zvi Galil's exposition and continuation (Galil 1977) of its results. Tseitin had proved the first really significant lower bounds in the area, showing an exponential lower bound for regular resolution refutations of graph-based formulas.

A regular refutation is one in which the same variable is never resolved on twice in a given proof thread. In spite of great efforts, I was never able to generalize Tseitin's lower bound to unrestricted resolution refutations. Tseitin had proved his result by representing resolution proofs as deletion processes on graphs. Try as I might, I couldn't get a more general version to work.

Shortly afterward, Armin Haken (son of Wolfgang Haken, who together with Kenneth Appel proved the four-color theorem), proved an exponential lower bound (Haken 1985) for general resolution proofs. Armin's paper was a revelation to me. The key to his result was a simple but elegant counting argument. Studying his work, I realized that my thinking had been along incorrect lines. Following traditional proof theory, I had attempted to prove a lower bound by constructive methods; Armin's argument cut through this by simple counting.

I began to educate myself in the probabilistic method, a technique that has stood me in good stead since. The contribution by Sam Buss in this volume discusses one of my papers where it plays an essential role. Armin Haken's lower bound had been for the pigeonhole principle; I was able to adapt it for Tseitin's graphical formulas. The resulting paper (Urquhart 1987) is one of my most cited publications. I refer the reader to the contribution of Noah Fleming and Toniann Pitassi for an excellent survey of Tseitin's graph-based clauses and their uses in proof complexity

Soon after my paper appeared, Vašek Chvátal and Endre Szemerédi proved one of the great results in the area of proof complexity, by building on my result for the graphical formulas. They show that randomly chosen formulas in 3CNF, within a particular range of clause density, require exponentially long resolution refutations. I had been trying to prove something similar myself, but did not have the expertise required in probabilistic combinatorics. Vašek Chvátal was visiting at the University of Waterloo at that time, and called me up to talk about their result. He paid me what I took to be a sincere, if unintended, compliment. He asked, "What are you doing in a philosophy department?"

My paper proving exponential lower bounds for the graph-based formulas led to an involvement with the community of researchers investigating algorithms for satisfiability. In this way, I became friendly with Armin Biere, John Franco, Hans van Maaren, Karem Sakallah, Hans Kleine Büning, Oliver Kullmann and other members of the SAT community. There is an overlap with the theory community, but the main thrust of SAT research is to develop fast algorithms for satisfiability.

The progress in this area has been astonishing. When I first attended conferences in this area, the existing programs could only deal with relatively small instances of SAT. The introduction of Conflict Driven Clause Learning (CDCL) algorithms revolutionized the field. Such algorithms can now deal with very large instances for applications such as circuit verification and model checking, and SAT solvers are used routinely in industry. Donald Knuth has given an excellent exposition (Knuth 2015) of the state of the art in 2015. Our theoretical understanding of these algorithms and their excellent performance in some cases (though not in all) lags considerably behind the amazing advances in practice.

I was very happy to find in the Toronto computer science department a culture of cooperation with a focus on making progress on the many open problems of theoretical computer science. This was a fertile period for my research, as I collaborated with several members of the Toronto department.

I wrote a very long paper with Steve Cook (Cook and Urquhart 1993) on the intuitionistic version of feasibly constructive arithmetic, and learned Kurt Gödel's elegant *Dialectica* interpretation of intuitionistic number theory as part of this research. This paper also contains a kind of independence result, showing that there is no feasibly constructive proof that the extended Frege system has super-polynomial proof complexity.

I also collaborated with graduate students in the department, including Michael Soltys and my student Alex Hertel. My most extended collaboration was with Stephen Bellantoni and Toni Pitassi, who were both graduate students at the time (Toni is now a star of the Toronto department, while Stephen eventually moved into the financial sector).

In 1988, Miklós Ajtai made a remarkable breakthrough with a paper (Ajtai 1988) proving super-polynomial lower bounds for the pigeonhole principle in Frege systems where the proofs are required to have bounded depth. His proof used nonstandard models for first-order number theory. I set to work with Stephen and Toni with two goals in mind: first, to give a standard proof of Ajtai's result, second, to improve the lower bound to exponential.

We succeeded in the first goal, but not the second (Bellantoni et al. 1992). However, Toni persisted and eventually succeeded in the second goal in a paper with Paul Beame and Russell Impagliazzo—Jan Krajíček, Pavel Pudlák and Alan Woods proved the same result independently (Beame et al. 1992). This was one of my most intense collaborations, from which I learned a great deal.

I was pleased to discover that my new-found expertise in complexity theory could be adapted to my former interests in relevance logic. In the 1970s, I had tried, but failed, to simplify the proof by Saul Kripke that the implicational fragment $\mathbf{R}_{\rightarrow}$ of $\mathbf{R}$, is decidable. Bob Meyer had extended Kripke's proof to include conjunction in his doctoral thesis.

In the early 1970s, I thought that it might be possible to simplify Kripke's argument by proving that $\mathbf{R}_{\rightarrow}$ has the finite model property in the semilattice semantics, and encouraged Bob Meyer to show this when he was in Toronto. He carried out my plan, but I was disappointed that the argument was basically an adaptation of the old Kripke method. The true picture emerged much later, when I showed that the implication-conjunction fragment of **R** is Ackermann-complete (Urquhart 1999). This means that the Kripke-Meyer decision procedure is essentially optimal.

I had earlier proved that the pure implication fragment $\mathbf{R}_{\rightarrow}$ is EXPSPACE-hard (Urquhart 1990), and so very intractable, but this left a gap between my lower bound and the Ackermann upper bound. This problem was eventually resolved in a brilliant contribution by the young French logician Sylvain Schmitz, who showed (Schmitz 2016) that the decision problem for $\mathbf{R}_{\rightarrow}$ is complete for doubly exponential time.

1.7 History of Logic and Philosophy

The preceding sections follow my work on logical problems. However, I have always kept up an interest in the history of logic, starting from my reading of the early issues of *The Journal of Symbolic Logic* and the painstaking work of Church.

Most of my historical work is connected with Bertrand Russell, an old hero from my undergraduate days. I spent several years editing Volume 4 of the *Collected Papers* of Russell (1994) The original manuscripts are in the Russell archives in McMaster University, Hamilton. I would spend a few days each week driving back and forth between Toronto and Hamilton, sometimes thinking about the problem of the complexity of bounded depth Frege proof systems.

The most interesting manuscript for me was a kind of logical diary that Russell wrote around 1904. This is "Russell thinking aloud on paper," and is quite fascinating to read, as he tries one idea after another in an attempt to avoid the paradoxes. He was aiming to construct a consistent type-free logic including classical logic. It was natural that (like Frege) he failed in this attempt, and eventually was forced to adopt type theory, though rather against his (and Whitehead's) will. In addition to my work on Russell, I have written substantial articles on Kurt Gödel, John von Neumann, Emil Post and their interactions (Urquhart 2009a, 2010, 2016).

While still an undergraduate, I became fascinated by the fact that Boolean functions can be transformed into one another by permuting and complementing variables. I investigated this in a very primitive way, since I was unaware of the techniques needed for this problem. In 2009, I returned to this area, and published a historical survey giving the solution to the problem of enumerating types of Boolean functions, using George Pólya's theory of enumeration under group action. I found a lot of satisfaction in writing this article (Urquhart 2009b), based on a return to my very first logical research.

I also wrote an article (Urquhart 2012) on the work of Henry M. Sheffer, whose mysterious monograph, *The General Theory of Notational Relativity* is referenced in the second edition of *Principia Mathematica*; this was the result of Bernard Linsky giving me a copy during a meeting of the *Society for Exact Philosophy* in Edmonton.

I enjoy historical research. If you are trying to solve a difficult logical problem, you may work away for months, and then discover your whole approach is wrong (the earlier sections of this essay have several examples of this). However, if you work on history, you will usually be rewarded by something worth publishing.

Although I taught in a philosophy department, I have not published much in philosophy. However, I have ventured to express philosophical opinions in the numerous reviews I have written over my career. I like writing reviews for the same reason I like writing history—something almost always results.

1.8 Retirement

I retired from teaching in the philosophy department in 2007, but continued my work in logic, and supervised graduate students for a few more years. I was involved with the Association for Symbolic Logic, and did a good deal of editorial work for their journals.

When Herb Enderton retired as managing editor for reviews for the *Bulletin of Symbolic Logic*, I took over from him, 2002–2008. Although this was no longer a paid position, I did my best to keep up the tradition of Alonzo Church, though I could not hope to do as thorough a job as Herb. I served as an editor of the *Journal of Symbolic Logic* and of the *Review of Symbolic Logic* from 2008 to 2015, and served as the President of the Association from 2013 to 2016.

References

Ajtai, M. (1988). The complexity of the pigeonhole principle. In *Proceedings of the 29th Annual IEEE Symposium on the Foundations of Computer Science* (pp. 346–355).

Anderson, A. R., & Belnap, N. D. (1975). *Entailment* (Vol. 1). Princeton, NJ: Princeton University Press.

Beame, P., Impagliazzo, R., Krajíček, J., Pitassi, T., Pudlák, P., & Woods, A. (1992). Exponential lower bounds for the pigeonhole principle. In *Proceedings of the 24th Annual ACM Symposium on theory of computing* (pp. 200–220).

Bellantoni, S., Pitassi, T., & Urquhart, A. (1992). Approximation and small-depth Frege proofs. SIAM Journal of Computing, 21, 1161–1179.

Białynicki-Birula, A., & Rasiowa, H. (1957). On the representation of quasi-Boolean algebras. Bulletin de l'académie polonaise des sciences, 5, 259–261.

Cook, S. A., & Urquhart, A. (1993). Functional interpretations of feasibly constructive arithmetic. Annals of Pure and Applied Logic, 63, 103–200.

Davis, M. (Ed.). (1965). *The Undecidable: Basic Papers on Undecidable Propositions, Unsolvable Problems and Computable Functions*. Hewlett, New York: Raven Press.

Freese, R. (1980). Free modular lattices. *Trans. Amer. Math. Soc.*, *261*, 81–91.

Galil, Z. (1977). On the complexity of regular resolution and the Davis-Putnam procedure. Theoretical Computer Science, 4, 23–46.

Gödel, K. (1962). *On formally undecidable propositions of Principia Mathematica and related systems*. Edinburgh: Oliver and Boyd. Translated by B. Meltzer, with an Introduction by R.B. Braithwaite.

Grätzer, G. (1971). *Lattice Theory: First Concepts and Distributive Lattices*. San Francisco: Freeman.

Grätzer, G., Jónsson, B., & Lakser, H. (1973). The amalgamation property in equational classes of modular lattices. Pacific Journal of Mathematics, 45, 507–524.

Haken, A. (1985). The intractability of resolution. Theoretical Computer Science, 39, 297–308.
Hughes, G., & Cresswell, M. (1968). *An Introduction to Modal Logic*. London: Methuen.
Knuth, D. E. (2015). *The art of computer programming, Volume 4, Fascicle 6: Satisfiability*. Boston: Addison-Wesley.
Priestley, H. (1970). Representation of distributive lattices by means of ordered Stone spaces. *Bull. London Math. Soc.*, *2*, 186–190.
Quine, W. (1941). Review of Bertrand Russell: An inquiry into meaning and truth. The Journal of Symbolic Logic, 6, 29–30.
Rescher, N., & Urquhart, A. (1971). *Temporal logic*. Berlin: Springer.
Rosenbloom, P. C. (1950). *The elements of mathematical logic*. Mineola: Dover Publications, Inc.
Routley, R., & Meyer, R. K. (1973). Semantics of entailment. In H. Leblanc (Ed.), *Truth syntax and modality* (pp. 199–243). North-Holland Publishing Company. *Proceedings of the Temple University Conference on Alternative Semantics*.
Russell, B. (1920). *Introduction to mathematical philosophy* (2nd ed.). Crows Nest: George Allen and Unwin.
Russell, B. (1994). *Collected papers, Volume 4: Foundations of logic 1903–05*. London: Routledge. Edited by Alasdair Urquhart with the assistance of Albert C. Lewis.
Schmitz, S. (2016). Implicational relevance logic is 2-Exptime-complete. *Journal of Symbolic Logic*, *81*, 641–661.
Siekmann, J., & Wrightson, G. (Eds.). (1983). *Automation of Reasoning*. New York: Springer-Verlag.
Thistlewaite, P., McRobbie, M., & Meyer, R. (1988). Automated theorem-proving in non-classical logics. London: Pitman.
Tseitin, G. (1970). On the complexity of derivation in propositional calculus. In A. O. Slisenko (Ed.), *Studies in constructive mathematics and mathematical logic, Part 2* (pp. 115–125). New York: Consultants Bureau. Reprinted in [24], Vol. 2, pp. 466–483.
Urquhart, A. (1972). Semantics for relevant logics. Journal of Symbolic Logic, 37, 159–169.
Urquhart, A. (1973a). Free distributive pseudocomplemented lattices. *Algebra Universalis*, *3*, 13–15.
Urquhart, A. (1973b). *The semantics of entailment*. PhD thesis, University of Pittsburgh.
Urquhart, A. (1978). A topological representation theory for lattices. Algebra Universalis, 8, 45–58.
Urquhart, A. (1984). The undecidability of entailment and relevant implication. Journal of Symbolic Logic, 49, 1059–1073.
Urquhart, A. (1987). Hard examples for resolution. Journal of the Association for Computing Machinery, 34, 209–219.
Urquhart, A. (1990). The complexity of decision procedures in relevance logic. In J. M. Dunn & A. Gupta (Eds.), *Truth or Consequences: Essays in honour of Nuel Belnap* (pp. 61–76). Dordrecht: Kluwer.
Urquhart, A. (1993). Failure of interpolation in relevant logics. *J. Philosophical Logic*, *22*, 449–479.
Urquhart, A. (1999). The complexity of decision procedures in relevance logic II. Journal of Symbolic Logic, 64, 1774–1802.
Urquhart, A. (2009a). Emil post. In D. M. Gabbay & J. Woods (Eds.), *Handbook of the history of logic, Volume 5: Logic from Russell to Church*. Amsterdam: North-Holland Elsevier.
Urquhart, A. (2009b). Enumerating types of Boolean functions. Bulletin of Symbolic Logic, 15, 273–299.
Urquhart, A. (2010). Von Neumann, Gödel and complexity theory. Bulletin of Symbolic Logic, 16, 516–530.
Urquhart, A. (2012). Henry M. Sheffer and notational relativity. History and Philosophy of Logic, 33, 33–47.
Urquhart, A. (2016). Russell and Gödel. Bulletin of Symbolic Logic, 22, 504–520.
von Neumann, J. (1960). *Continuous Geometry*. Princeton, New Jersey: Princeton University Press.

Alasdair Urquhart's Publications—Doctoral Thesis, Books and Articles

Alekhnovich, M., Johannsen, J., Pitassi, T., & Urquhart, A. (2007). An exponential separation between regular and general resolution. *Theory of Computing, 3*, 81–102. In *Preliminary Version in Proceedings of the 34th Annual ACM Symposium on Theory of Computing*, 19–21 May 2002, Montréal, Québec, Canada.

Arai, N. H., Pitassi, T., & Urquhart, A. (2006). The complexity of analytic tableaux. *Journal of Symbolic Logic, 71*, 777–790. In *Preliminary Version: 33rd Annual ACM Symposium on Theory of Computing*, Hersonissos, Greece (pp. 356–363). Association for Computing Machinery, 2001.

Arai, N. H., & Urquhart, A. (2000). Local symmetries in propositional logic. In *Automated reasoning with analytic tableaux and related methods* (Vol. 1847, pp. 40–51). Berlin: Springer. Lecture notes in computer science.

Bellantoni, S., Pitassi, T., & Urquhart, A. (1992). Approximation and small-depth Frege proofs. *SIAM Journal of Computing, 21*, 1161–1179.

Brown, J. R., & Urquhart, A. (1998). Review of Benacerraf and his critics, edited by Adam Morton and Stephen Stich. *Dialogue, 37*, 633–637.

Čačić, V., Pudlák, P., Restall, G., Urquhart, A., & Visser, A. (2007). Decorated linear order types and the theory of concatenation. In F. Delon, U. Kohlenbach, P. Maddy, & F. Stephan (Eds.), *Logic Colloquium 2007* (pp. 1–13). Association for Symbolic Logic and Cambridge University Press.

Cook, S. A., & Urquhart, A. (1993). Functional interpretations of feasibly constructive arithmetic. *Annals of Pure and Applied Logic, 63*, 103–200.

Düntsch, I., & Urquhart, A. (2006). Betweenness and comparability obtained from binary relations. In *Relations and Kleene algebras in computer science* (Vol. 4136). Lecture notes in computer science. Berlin: Springer.

Giambrone, S., Meyer, R. K., & Urquhart, A. (1987). A contractionless semilattice semantics. *Journal of Symbolic Logic, 52*, 526–529.

Giambrone, S., & Urquhart, A. (1987). Proof theories for semilattice logics. Zeitschrift für mathematische Logik und Grundlagen der Mathematik, 33, 433–439.

Hertel, A., Hertel, P., & Urquhart, A. (2007). Formalizing dangerous SAT encodings. In *Theory and Applications of Satisfiability Testing – SAT 2007 Proceedings* (Vol. 4501, pp. 159–172). Lecture notes in computer science. Springer.

Hertel, A., & Urquhart, A. (2007). Game characterizations and the PSPACE-completeness of tree resolution space. In *Computer Science Logic 2007* (Vol. 4646, pp. 527–541). Lecture notes in computer science. Springer.

Hertel, A., & Urquhart, A. (2009). Algorithms and complexity results for input and unit resolution. Journal on Satisfiability, Boolean Modeling and Computation, 6, 141–164.

Impagliazzo, R., Pitassi, T., & Urquhart, A. (1994). Upper and lower bounds for tree-like cutting-plane proofs. In *9th IEEE Symposium on Logic in Computer Science* (pp. 220–228).

Kremer, P., & Urquhart, A. (2008). Supervaluation fixed-point logics of truth. *Journal of Philosophical Logic*, 407–440.

Meyer, R. K., Giambrone, S., Urquhart, A., & Martin, E. P. (1988). Further results on proof theories for semilattice logics. Zeitschrift für mathematische Logik und Grundlagen der Mathematik, 34, 301–304.

Mints, G., Olkhovikov, G., & Urquhart, A. (2010). Failure of interpolation in constant domain intuitionistic logic. Journal of Symbolic Logic, 78, 937–950.

Pelham, J., & Urquhart, A. (1994). Russellian propositions. In *Logic, methodology and philosophy of science IX: Proceedings of the 9th International Congress of Logic, Methodology, and Philosophy of Science*, Uppsala, Sweden, August 7–14, 1991 (pp. 307–326). North Holland.

Pelletier, F. J., & Urquhart, A. (2003). Synonymous logics. Journal of Philosophical Logic, 32, 259–285.

Pelletier, F. J., & Urquhart, A. (2008). Synonymous logics: A correction. Journal of Philosophical Logic, 37, 95–100.
Pitassi, T., & Urquhart, A. (1995). The Complexity of the Hajós calculus. *SIAM Journal on Discrete Mathematics*, *8*, 464–483.
Rescher, N., & Urquhart, A. (1971). *Temporal logic*. Berlin: Springer.
Russell, B. (1994). *Collected Papers Volume 4; Foundations of Logic 1903–05*. Routledge. Edited by Alasdair Urquhart with the assistance of Albert C. Lewis.
Soltys, M., & Urquhart, A. (2004). Matrix identities and the pigeonhole principle. Archive for Mathematical Logic, 43, 351–357.
Urquhart, A. (1971). Completeness of weak implication. Theoria, 37, 274–282.
Urquhart, A. (1972). Semantics for relevant logics. *Journal of Symbolic Logic, 37*, 159–169.
Urquhart, A. (1973a). Free distributive pseudocomplemented lattices. *Algebra Universalis, 3*, 13–15.
Urquhart, A. (1973b). Free Heyting algebras. Algebra Universalis, 3, 94–97.
Urquhart, A. (1973c). An interpretation of many-valued logic. Zeitschrift für Mathematische Logik und Grundlagen der Mathematik, 2, 212–219.
Urquhart, A. (1973d). A semantical theory of analytic implication. Journal of Philosophical Logic, 2, 212–219.
Urquhart, A. (1973e). *The semantics of entailment*. PhD thesis, University of Pittsburgh. https://utoronto.academia.edu/AlasdairUrquhart/Thesis.
Urquhart, A. (1974a). Implicational formulas in intuitionistic logic. Journal of Symbolic Logic, 39, 661–664.
Urquhart, A. (1974b). Proofs, snakes and ladders. Dialogue, 13, 723–731.
Urquhart, A. (1975). Popper's logical conceptions. *Communication and Cognition, 8*, 237–242.
Urquhart, A. (1977). A finite matrix whose consequence relation is not finitely axiomatizable. Reports on Mathematical Logic, 9, 71–73.
Urquhart, A. (1978). A topological representation theory for lattices. *Algebra Universalis, 8*, 45–58.
Urquhart, A. (1979). Distributive lattices with a dual homomorphic operation. Studia Logica, 38, 201–209.
Urquhart, A. (1981a). Decidability and the finite model property. Journal of Philosophical Logic, 10, 367–370.
Urquhart, A. (1981b). The decision problem for equational theories. Houston Journal of Mathematics, 7, 587–589.
Urquhart, A. (1981c). Distributive lattices with a dual homomorphic operation II. Studia Logica, 40, 391–404.
Urquhart, A. (1981d). Projective distributive p-algebras. Bulletin of the Australian Mathematical Society, 24, 269–275.
Urquhart, A. (1982a). Equational classes of distributive double p-algebras. Algebra Universalis, 14, 235–243.
Urquhart, A. (1982b). Intensional languages via nominalization. Pacific Philosophical Quarterly, 63, 186–192.
Urquhart, A. (1983). Relevant implication and projective geometry. *Logique et Analyse*, 103–104, 345–357. Special issue on Canadian logic.
Urquhart, A. (1984a). Many-valued logic. In D. Gabbay & F. Guenthner (Eds.), *Handbook of philosophical logic* (Vol. III). Dordrecht: D. Reidel Publishing Company.
Urquhart, A. (1984b). The undecidability of entailment and relevant implication. Journal of Symbolic Logic, 49, 1059–1073.
Urquhart, A. (1987). Hard examples for resolution. *Journal of the Association for Computing Machinery, 34*, 209–219.
Urquhart, A. (1988). Russell's zigzag path to the ramified theory of types. *Russell, N.S., 8*, 82–91.
Urquhart, A. (1989a). The complexity of Gentzen systems for propositional logic. Theoretical Computer Science, 66, 87–97.

Urquhart, A. (1989b). What is relevant implication? In J. Norman, & R. Sylvan (Eds.), *Directions in relevant logic* (pp. 167–74). Amsterdam: Kluwer.
Urquhart, A. (1990a). The complexity of decision procedures in relevance logic. In J. M. Dunn & A. Gupta (Eds.), *Truth or Consequences: Essays in Honour of Nuel Belnap* (pp. 61–76). Dordrecht: Kluwer.
Urquhart, A. (1990b). The logic of physical theory. In A. D. Irvine (Ed.), *Physicalism in mathematics* (pp. 145–154). Amsterdam: Kluwer.
Urquhart, A. (1992a). Complexity of proofs in classical propositional logic. In Y. N. Moschovakis (Ed.), *Logic from computer science* (pp. 597–608). Berlin: Springer.
Urquhart, A. (1992b). The relative complexity of resolution and cut-free Gentzen systems. *Annals of mathematics and artificial intelligence, 6*, 157–168.
Urquhart, A. (1992c). Semilattice semantics for relevance logics. In A. R. Anderson, N. D. Belnap Jr., & J. M. Dunn (Eds.), *Entailment* (Vol. II, pp. 157–168). Princeton: Princeton University Press.
Urquhart, A. (1992d). The undecidability of all principal relevance logics. In A. R. Anderson, N. D. Belnap Jr., & J. M. Dunn (Eds.), *Entailment* (Vol. II, pp. 348–374). Princeton: Princeton University Press.
Urquhart, A. (1993). Failure of interpolation in relevant logics. *Journal Philosophica Logic, 22*, 449–479.
Urquhart, A. (1995a). The complexity of propositional proofs. The Bulletin of Symbolic Logic, 1, 425–467.
Urquhart, A. (1995b). Decision problems for distributive lattice-ordered semigroups. Algebra Universalis, 33, 399–418.
Urquhart, A. (1995c). G.F. Stout and the theory of descriptions. Russell (New Series), 14, 163–171.
Urquhart, A. (1996). Duality for algebras of relevant logics. Studia Logica, 56, 263–276.
Urquhart, A. (1997a). The graph constructions of Hajós and Ore. Journal of Graph Theory, 26, 211–215.
Urquhart, A. (1997b). The number of lines in Frege proofs with substitution. Archive for Mathematical Logic, 37, 15–19.
Urquhart, A. (1998a). Complexity, computational. In Craig, E. (Ed.), *Routledge encyclopedia of philosophy* (Vol. 2, pp. 471–476). London: Routledge.
Urquhart, A. (1998b). The complexity of propositional proofs. Canadian Journal of Artificial Intelligence, 42, 8–18.
Urquhart, A. (1998c). The complexity of propositional proofs. Bulletin of the EATCS, 64, 128–138.
Urquhart, A. (1999a). Beth's definability theorem in relevant logics. In E. Orłowska (Ed.), *Logic at work. Essays dedicated to the memory of Helena Rasiowa*. Berlin: Physica-Verlag.
Urquhart, A. (1999b). The complexity of decision procedures in relevance logics II. Journal of Symbolic Logic, 64, 1774–1802.
Urquhart, A. (1999c). The symmetry rule in propositional logic. *Discrete Applied Mathematics*, 96–97, 177–193.
Urquhart, A. (2000). The complexity of linear logic with weakening. In S. R. Buss, P. Hájek, & P. Pudlák (Eds.), *Logic Colloquium'98* (pp. 500–515). The Association for Symbolic Logic.
Urquhart, A. (2001a). Basic many-valued logic. In Gabbay, D., & Guenthner, F. (Eds.), *Handbook of philosophical logic* (2nd ed., Vol. 2, pp. 249–295). Amsterdam: Kluwer.
Urquhart, A. (2001b). The complexity of propositional proofs. In G. Päun, G. Rosenberg, & A. Salomaa (Eds.), *Current trends in theoretical computer science*. Singapore: World Scientific.
Urquhart, A. (2002). Metatheory. In D. Jacquette (Ed.), *A companion to philosophical logic* (pp. 307–318). Hoboken: Blackwell.
Urquhart, A. (2003a). Resolution proofs of matching principles. Annals of Mathematics and Artificial Intelligence, 37, 231–250.
Urquhart, A. (2003b). Sections 11.5 and 11.6. In R. Sylvan, & R. Brady (Eds.), *Relevant logics and their rivals II* (pp. 217–230). Farnham: Ashgate Publishing Company.
Urquhart, A. (2003c). The theory of types. In N. Griffin (Ed.), *The Cambridge companion to Russell* (pp. 286–309). Cambridge: Cambridge University Press.

Urquhart, A. (2004). Complexity. In L. Floridi (Ed.), *The Blackwell guide to the philosophy of computing and information*. Hoboken: Blackwell.
Urquhart, A. (2005a). The complexity of propositional proofs with the substitution rule. Logic Journal of the IGPL, 13, 287–291.
Urquhart, A. (2005b). Russell on meaning and denotation. In B. Linsky, & G. Imaguire (Eds.), *On denoting: 1905–2005* (pp. 99–120). Freiburg im Breisgau: Philosophia Verlag.
Urquhart, A. (2006a). Duality theory for projective algebras. In W. MacCaull, M. Winter, & I. Düntsch (Eds.), *Relational methods in computer science* (Vol. 3929). Lecture notes in computer science. Berlin: Springer.
Urquhart, A. (2006b). Width versus size in resolution proofs. In J.-Y. Cai, S. B. Cooper, & A. Li (Eds.), *3rd International Conference on Theory and Applications of Models of Computation, TAMC 2006, Beijing, China, May 2006 Proceedings* (Vol. 3959, pp. 79–88). Lecture notes in computer science. Springer.
Urquhart, A. (2007). Four variables suffice. Australasian Journal of Logic, 5, 66–73.
Urquhart, A. (2008a). Mathematics and physics: Strategies of assimilation. In P. Mancosu (Ed.), *The philosophy of mathematical practice* (pp. 417–440). Oxford: Oxford University Press.
Urquhart, A. (2008b). The boundary between mathematics and physics. In P. Mancosu (Ed.), *The philosophy of mathematical practice* (pp. 407–416). Oxford: Oxford University Press.
Urquhart, A. (2008c). The unnameable. In J. Tappenden, A. Varzi, & W. Seager (Eds.), *Truth and values: Essays for Hans Herzberger* (pp. 119–135). Calgary: University of Calgary Press. *Canadian Journal of Philosophy*, Supplementary Volume 34.
Urquhart, A. (2009a). Emil post. In D. Gabbay, & J. Woods (Eds.), *Handbook of the history of logic. Volume 5: Logic from Russell to Church* (pp. 617–666). Amsterdam: North-Holland.
Urquhart, A. (2009b). Enumerating types of Boolean functions. *Bulletin of Symbolic Logic, 15*, 273–299.
Urquhart, A. (2009c). Logic and denotation. In N. Griffin, & D. Jacquette (Eds.), *Russell vs. Meinong: The legacy of on denoting* (pp. 10–25). London: Routledge.
Urquhart, A. (2009d). Weakly additive algebras and a completeness problem. In P. Schotch, B. Brown, & R. Jennings (Eds.), *On preserving: Essays on preservationism and paraconsistent logic* (pp. 33–47). Toronto: University of Toronto Press.
Urquhart, A. (2010a). Anderson and Belnap's Invitation to Sin. *Journal of Philosophical Logic, 39*, 453–472.
Urquhart, A. (2010b). Proof theory. In Y. Crama, & P. L. Hammer (Eds.), *Boolean models and methods in mathematics, computer science, and engineering* (pp. 79–98, Vol. 134). Encyclopedia of mathematics and its applications. Cambridge: Cambridge University Press.
Urquhart, A. (2010c). Von Neumann, Gödel and complexity theory. Bulletin of Symbolic Logic, 16, 516–530.
Urquhart, A. (2011a). The depth of resolution proofs. Studia Logica, 99, 349–364.
Urquhart, A. (2011b). A near-optimal separation of regular and general resolution. SIAM Journal on Computing, 40, 107–121.
Urquhart, A. (2012a). Henry M. Sheffer and notational relativity. History and Philosophy of Logic, 33, 33–47.
Urquhart, A. (2012b). Width and size of regular resolution proofs. Logical Methods in Computer Science, 8, 1–15.
Urquhart, A. (2013). Principia Mathematica: The first 100 years. In N. Griffin, & B. Linsky (Eds.), *The Palgrave centenary companion to Principia Mathematica* (pp. 3–20). London: Palgrave Macmillan.
Urquhart, A. (2015a). First degree formulas in quantified S5. Australasian Journal of Logic, 12, 204–210.
Urquhart, A. (2015b). Mathematical depth. Philosophia Mathematica, 23, 233–241.
Urquhart, A. (2016a). From Jónsson and Tarski to Schotch and Jennings. In G. Payette (Ed.), *"Shut up," he explained: Essays in honour of Peter K. Schotch*. London: Palgrave Macmillan.

Urquhart, A. (2016b). Relevance Logic: Problems Open and Closed. *Australasian Journal of Logic, 13*, 11–20.
Urquhart, A. (2016c). Russell and Gödel. Bulletin of Symbolic Logic, 22, 504–520.
Urquhart, A. (2016d). The story of γ. In K. Bimbó (Ed.), *J. Michael Dunn on information based logics* (pp. 93–106). Berlin: Springer.
Urquhart, A. (2017). The geometry of relevant implication. IFCoLog Journal of Logics and their Applications, 4, 591–604.
Urquhart, A. (2019). Relevant Implication and Ordered Geometry. *Australasian Journal of Logic, 16*, 342–354.
Urquhart, A., & Apostoli, P. (2001). Weakly additive operators on distributive lattices. In J. Woods, & B. Brown (Eds.), *Logical consequences: Rival approaches*. Paris: Hermes Science Publishing.
Urquhart, A., & Fu, X. (1996). Simplified lower bounds for propositional proofs. Notre Dame Journal of Formal Logic, 37, 523–544.
Urquhart, A. (1971). Review of the paradox of the Liar, edited by Robert Martin. *Dialogue, 10*, 823–825.
Urquhart, A. (1974). Review of The Development of Mathematical Logic, by R.L. *Goodstein. Historia Mathematica, 1*, 212–214.
Urquhart, A. (1975). Critical Notice of Formal Philosophy, by Richard Montague. *Canadian Journal of Philosophy, 4*, 573–578.
Urquhart, A. (1977). Review of distributive lattices, by Balbes and Dwinger. *Journal of Symbolic Logic, 42*, 587–588.
Urquhart, A. (1978). Review of Meaning and Modality by Casimir Lewy. *The Journal of Philosophy, 75*, 438–446.
Urquhart, A. (1982). Review of sceptical essays, by Benson Mates. *Canadian Philosophical Reviews, 2*, 29–31.
Urquhart, A. (1983a). Review of classical propositional operators, by Krister Segerberg. *Canadian Philosophical Reviews, 3*, 306–308.
Urquhart, A. (1983b). Review of realism, mathematics and modality, by Hartry Field. *History and Philosophy of Logic, 14*, 117–119.
Urquhart, A. (1984). Critical Notice of Handbook of Mathematical Logic, edited by Jon Barwise. *Canadian Journal of Philosophy, 14*, 675–682.
Urquhart, A. (1985a). Review of mathematics in philosophy, by Charles Parsons. *History and Philosophy of Logic, 6*, 239–241.
Urquhart, A. (1985b). Review of probabilistic metaphysics, by Patrick Suppes. *Canadian Philosophical Reviews, 5*, 478–480.
Urquhart, A. (1985c). Review of papers by Šehtman and Popov. Journal of Symbolic Logic, 50, 1081–1083.
Urquhart, A. (1987). Review of Beyond Analytic Philosophy by Hao Wang. *Canadian Journal of Philosophy, 17*, 477–482.
Urquhart, A. (1988a). Review of logic, bivalence and denotation, by Ermanno Bencivenga, Karel Lambert and Bas C. van Fraassen. *Canadian Philosophical Reviews, 8*, 121–123.
Urquhart, A. (1988b). Review of modern logic and quantum mechanics, by Rachel Wallace Garden. *Journal of Symbolic Logic, 53*, 648–649.
Urquhart, A. (1989). Review of relevant predication I: The formal theory, by J. Michael Dunn. *Journal of Symbolic Logic, 54*, 347–381.
Urquhart, A. (1990a). Review of constructivism in mathematics, Vol. 1, by Troelstra and van Dalen. *Studia Logica, 49*, 151–152.
Urquhart, A. (1990b). Review of quantum theory, the Church-Turing principle and the universal quantum computer, by D. Deutsch. *Journal of Symbolic Logic, 55*, 1309–1310.
Urquhart, A. (1990c). Review of relevant logic: A philosophical examination of inference, by Stephen Read. *History and Philosophy of Logic, 11*, 98–99.
Urquhart, A. (1990d). Review of the situation in logic, by Jon Barwise. *Canadian Philosophical Reviews, 10*, 96–98.

Urquhart, A. (1991a). Review of constructivism in mathematics, Vol. 2, by Troelstra and van Dalen. *Studia Logica, 50*, 355–356.
Urquhart, A. (1991b). Review of Hermes number 7. *Russell, N.S., 11*, 103–105.
Urquhart, A. (1991c). Review of Intensional Mathematics, edited by Stuart Shapiro. *Studia Logica, 50*, 161–162.
Urquhart, A. (1992a). Review of Proof and Knowledge in Mathematics, edited by Michael Detlefsen. *Canadian Philosophical Reviews, 12*, 237–238.
Urquhart, A. (1992b). Review of two papers on relevance and paraconsistency by Arnon Avron. Journal of Symbolic Logic, 57, 1481–1482.
Urquhart, A. (1993a). Review of lectures on linear logic, by Anne Troelstra. *Canadian Philosophical Reviews, 13*(3), 126–128.
Urquhart, A. (1993b). Review of Russell's idealist apprenticeship by Nicholas Griffin. *Russell, N.S., 13*, 104–108.
Urquhart, A. (1993c). Review of the mathematical philosophy of Bertrand Russell, by Rodriguez-Consuegra. *Philosophia Mathematica, 1*, 90–93.
Urquhart, A. (1995). Review of Substructural Logics, edited by Peter Schroeder-Heister and Kosta Došen. *History and Philosophy of Logic, 16*, 138–139.
Urquhart, A. (1996). Review of Russell and analytic philosophy, edited by Andrew Irvine and G.A. *Wedeking. Journal of Symbolic Logic, 61*, 1391–1392.
Urquhart, A. (1997a). Review of Feasible Mathematics II, edited by Peter Clote and Jeffrey Remmel. *Logic Journal of the IGPL, 5*, 301–302.
Urquhart, A. (1997b). Review of the beat of a different drummer, by Jagdish Mehra. *International Studies in the Philosophy of Science, 11*, 311–313.
Urquhart, A. (1999a). Critical Notice of From Kant to Hilbert: A Source Book in the Foundations of Mathematics, edited by William B. *Ewald. Dialogue, 38*, 587–592.
Urquhart, A. (1999b). Review of countable Boolean algebras and decidability, by Sergei Goncharov. *Studia Logica, 63*, 443–445.
Urquhart, A. (1999c). Review of logic, logic and logic, by George Boolos. *Philosophy in Review, 19*, 244–246.
Urquhart, A. (1999d). Review of Russell's hidden substitional theory, by Gregory Landini. *Journal of Symbolic Logic, 614*, 1370–1371.
Urquhart, A. (2000). Review of Hugh MacColl and the Tradition of Logic, edited by Astroh and Read. *History and Philosophy of Logic, 21*, 308–314.
Urquhart, A. (2001a). Review of an introduction to substructural logics, by Greg Restall. *International Studies in the Philosophy of Science, 15*, 108–110.
Urquhart, A. (2001b). Review of Ludwig Boltzmann: The man who trusted atoms by Carlo Cercignanni. *International Studies in the Philosophy of Science, 15.*
Urquhart, A. (2001d). Review of the search for mathematical roots 1870–1940 by Ivor Grattan-Guiness. *Russell N.S., 21*, 91–94.
Urquhart, A. (2002). Review of the Couturat-Russell correspondence, edited by Anne-Françoise Schmid. *Russell, N.S., 22*, 188–193.
Urquhart, A. (2004a). Review of model theory of stochastic processes, by Sergio Fajardo and H. Jerome Keisler. *Bulletin of Symbolic Logic, 14*, 110–112.
Urquhart, A. (2004b). Review of the limits of abstraction, by Kit Fine. *The Journal of Philosophy, 101*, 594–598.
Urquhart, A. (2004c). Review of towards a philosophy of real mathematics, by David Corfield. *Canadian Philosophical Reviews, 24*, 175–177.
Urquhart, A. (2005a). Review of Algebraic Methods in Philosophical Logic by J. Michael Dunn and Gary Hardegree. *Studia Logica, 79*, 305–306.
Urquhart, A. (2005b). Review of Collected Works Volumes IV and V by Kurt Gödel. *Review of Modern Logic, 10*, 191–200.
Urquhart, A. (2005c). Review of Collected Works Volumes IV and V by Kurt Gödel. *Review of Modern Logic, 10*, 191–200.

Urquhart, A. (2005d). Review of the Couturat-Russell Correspondence edited by Anne-Françoise Schmid. *Bulletin of Symbolic Logic*, *11*, 442–444.
Urquhart, A. (2007a). Review of A First Course in Logic by Shawn Hedman. *Bulletin of Symbolic Logic*, *13*, 538–540.
Urquhart, A. (2007b). Review of Nonstandard Methods and Applications in Mathematics edited by Cutland, Di Nasso and Ross. *Bulletin of Symbolic Logic*, *13*, 372–374.
Urquhart, A. (2007c). Review of Truth and Games: Essays in Honour of Gabriel Sandu edited by Tuomo Aho and Ahti-Veikko Pietarinen. *Bulletin of Symbolic Logic*, *13*, 119–121.
Urquhart, A. (2008). Review of Current topics in logic and analytic philosophy edited by Martínez, Falguera and Sagüillo. *Bulletin of Symbolic Logic*, *14*, 271–272.
Urquhart, A. (2010). Review of Relational Semantics of Nonclassical Logical Calculi by Katalin Bimbó and. *J. Michael Dunn. Bulletin of Symbolic Logic*, *16*, 277–278.
Urquhart, A. (2012a). Review of The Development of Modern Logic edited by Leila Haaparanta. *Bulletin of Symbolic Logic*, *18*, 268–270.
Urquhart, A. (2013). Review of Russell's Unknown Logicism: A study in the history and philosophy of mathematics by Sébastien Gandon. *Philosophia Mathematica*, *21*, 399–402.
Urquhart, A. (2015). Review of Logical Foundations of Mathematics and Computational Complexity: A Gentle Introduction by Pavel Pudlák. *Philosophia Mathematica*, *23*, 435–438.
Urquhart, A. (2016). Review of The Once and Future Turing edited by S. Barry Cooper and Andrew Hodges. *Bulletin of Symbolic Logic*, *22*, 354–356.
Urquhart, A. (2012b). Review of the evolution of Principia Mathematica: Bertrand Russell's manuscripts and notes for the second edition by Bernard Linsky. *Notre Dame Philosophical Reviews*.
Urquhart, A. (2018). Review of philosophical explorations of the legacy of Alan Turing: Turing 100, edited by Juliet Floyd and Lisa Bokulich. *Notre Dame Philosophical Reviews*.
Urquhart, A. (2009). Review of Philosophical Logic by John P. Burgess. *Notre Dame Philosophical Reviews*.
Urquhart, A. (2000–2001c). Review of Russell's metaphysical logic, by Bernard Linsky. *University of Toronto Quarterly*, *70*, 455–457.

Chapter 2
Relevance-Sensitive Truth-Trees

David Makinson

Second Reader—*A. Avron, Tel-Aviv University*

Abstract Our goal is to articulate a clear rationale for relevance-sensitive propositional logic. The method: truth-trees. Familiar decomposition rules for truth-functional connectives, accompanied by novel ones for the for the arrow, together with a recursive rule, generate a set of 'acceptable' formulae that properly contains all theorems of the well-known system R and is closed under substitution, conjunction, and detachment. We conjecture that it satisfies the crucial letter-sharing condition.

Keywords Relevance-sensitivity · Relevance logic · Truth-trees · Decomposition trees · Semantic tableaux · Explosion · Variable-sharing · Parity · Natural deduction

2.1 Introduction

The chapter is an extended version of 'Relevance via decomposition: a project, some results, an open question' (Makinson 2017). Sections 2.8–2.10 and Appendices 4 and 5 are new; Appendices 2 and 3 have been entirely rewritten; new results and explanatory remarks have been added throughout, the most important new result being Observation 2.8.

In its standard Hilbertian axiomatization, the well-known relevance logic R has many axiom schemes and two derivation rules. The derivation rules are straightforward. One of them is detachment $\phi, \phi \rightarrow \psi \ / \ \psi$, where $\rightarrow$ is the non-classical

D. Makinson (✉)
London School of Economics, London, UK
e-mail: david.makinson@gmail.com

I. Düntsch and E. Mares (eds.), *Alasdair Urquhart on Nonclassical and Algebraic Logic and Complexity of Proofs*, Outstanding Contributions to Logic 22,
https://doi.org/10.1007/978-3-030-71430-7_2

connective intuitively representing 'relevant implication'; the other is conjunction (also called adjunction in this context) $\phi, \psi/\phi \wedge \psi$, which is needed because of the absence of the formula $\phi \rightarrow (\psi \rightarrow (\phi \wedge \psi))$ among the theorems. But the axiom schemes form a rather motley and untidy crew of about a dozen—the exact number depending on how we count, for example, the two forms of $\wedge$-elimination and $\vee$-introduction.

The axiom system for R has also been characterized, along with several of its neighbors, by a Routley–Meyer possible-worlds semantics with three-place relations satisfying various constraints. However, notwithstanding its versatility and technical usefulness, the semantics can hardly be said to provide a satisfying rationale. Some intuitive appreciation of the three-place relations may be obtained by thinking of them as indexed families of two-place ones; but there is still no rationale for the complex constraints that one needs to place on those families to ensure the validity of favored formulae.

Other semantic approaches have their drawbacks. In particular, the semi-lattice semantics of Urquhart (1972, 1989) covers only the $\wedge$, $\vee$, $\rightarrow$ fragment of the language, not handling negation in a classically acceptable manner; it can only be treated intuitionistically, by defining $\neg\phi$ as $\phi \rightarrow \bot$, where $\rightarrow$ is relevance-sensitive implication and $\bot$ is a zero-ary falsum connective added to the language. The relational algebra semantics of Maddux (2010) is even more difficult than the Routley–Meyer semantics to connect with any intuitive ideas behind relevance logic.

A more convincing perspective is provided by one of the earliest approaches to relevance logic, natural deduction, which is centered on a well-motivated restriction of the rule of arrow introduction. However, the systems obtained are still not entirely transparent. To block deductive detours that get around the constraint on arrow introduction, one needs to impose certain restrictions on the rules of conjunction introduction and disjunction elimination. These, in turn, lead to overkill, for they have the effect of blocking derivation of the generally accepted classical principle of distribution of conjunction over disjunction. One is thus forced into the ad hoc step of adding that principle as a separate inference rule.

It is thus natural to seek a clearer picture from a different direction. Our project is to do so by tweaking the classical procedure of truth-trees, also known as semantic decomposition trees or semantic tableaux.

To that end, we begin by defining 'directly acceptable' truth-trees. The set of formulae that are validated by such trees, likewise called directly acceptable, turns out to be closed under substitution and conjunction, and to contain all the standard axioms of R together with certain interesting formulae that are not theorems of R. It also satisfies the well-known letter-sharing condition, and successfully excludes other formulae which, whilst satisfying that condition, are notoriously relevance-insensitive, such as $p \rightarrow (q \rightarrow p)$ and its instance $p \rightarrow (p \rightarrow p)$, as well as others for which the condition does not arise since they do not have the arrow as principal connective, for example, $(p \rightarrow q) \vee (q \rightarrow p)$ and its instance $(p \rightarrow \neg p) \vee (\neg p \rightarrow p)$.

However, it also turns out that the set of directly acceptable formulae is not closed under detachment, and inspection of examples suggests that there is a real need to ensure such closure. We do so by introducing a natural recursion into the construction of the trees themselves. The resulting set of 'acceptable' formulae may equivalently be characterized by closing directly acceptability under detachment and conjunction. It properly includes the set of theorems of R; but it is still an open question whether it satisfies the letter-sharing condition, which is usually seen as a *sine qua non* for any relevance logic worthy of the name. We conjecture that it does so.

Although often used to introduce students to classical and first-order logic, truth-trees are not always regarded with great favor; Humberstone (2011, page 189) expressed a widespread sentiment when he remarked that they 'are not conducive to a clear-headed separation of proof-theoretic from semantic considerations'. We beg to differ: even classical truth-trees may be read in both semantic and syntactic ways and we need not be Manichean about the matter. One may allow both perspectives while being none the less clear-headed about their roles. The constraints that we impose on classical trees in this paper may be understood as effecting syntactic controls on a semantically powered procedure.

In general terms, we tend to agree with Tennant (1979), van Benthem (1983) that relevance is not itself a semantic notion. There is no philosophical need to give its logic a semantics beyond that which is already provided by classical two-valued logic, nor does the present text propose one. The paper has no metaphysical agenda. Its purpose is not to reject classical logic as incorrect, as has often been done by those working on relevance, but to examine one way of controlling its application. It is logic at play rather than logic with an evangelical mission.

2.2 Recalling Classical Truth-Trees

In this section, we recall briefly the use of truth-trees in classical propositional logic for formulae built using the connectives $\wedge, \vee, \neg$. Given such a formula ϕ, one can test whether it is a tautology by building a tree whose root-node r is labeled by $\neg\phi$ and whose construction is continued by means of the decomposition rules in Table 2.1. All of them act on a single input. The number of formulae in the output may be one (rule for negation) or two (all the others) and, in the latter case, with or without forking according to the rule under consideration.

A completed tree is said to be *successful* iff every branch contains a *crash-pair*, i.e., a pair of nodes z_1: ζ, z_2: $\neg\zeta$ for some formula ζ. Success is independent of the order of decomposition, although that can influence the shape and size of the tree. It is known that the classical tautologies are just those formulae ϕ such that there is a successful decomposition tree with root labeled by $\neg\phi$.

2.3 Decomposing Arrows

We add decomposition rules for a 'relevance-sensitive' conditional connective $\rightarrow$ in both plain and negated forms, in the manner described by Table 2.2.

While perfectly familiar as a rule of inference, modus ponens is novel as a rule of decomposition. In all previous decomposition systems for relevance logic (or others) of which the present author is aware (see Appendix 3), a rule of 'implicative forking' is used: the branch forks with the consequent on one arm and the negation of the antecedent on the other, reproducing what is done in classical logic for material implication. This difference is important, for reasons that will be explained in Sect. 2.4.

The rule for decomposing negated arrows, which we call counter-case, is exactly the same as for truth-functional implication, but with a global 'parity' constraint on what is subsequently done with the outputs. We will articulate that constraint formally in Sect. 2.4.1 but, before doing so, we explain its rationale.

Counter-case differs conceptually from the decomposition rules in Table 2.1 for truth-functional connectives, as well as from modus ponens for unnegated arrow. For those rules, we have the following, where 'implies' may be read as logical consequence understood indifferently in a classical manner or in terms of a relevance-sensitive logic such as the one being articulated here.

- For the three rules with a single input that don't fork (decomposing $\phi \wedge \psi$, $\neg(\phi \vee \psi)$, $\neg\neg\phi$), the input implies each output.

Table 2.1 Classical decomposition rules

Given a node m on a branch, labeled by				
$\varphi \wedge \psi$	$\neg(\varphi \wedge \psi)$	$\varphi \vee \psi$	$\neg(\varphi \vee \psi)$	$\neg\neg\varphi$
We can add to that branch				
Two	Two	Two	Two	One node n
Further nodes $n1, n2$ with				
No fork	A fork	A fork	No fork	
Where n_1, n_2 are labeled by the formulae				Labeled
φ, ψ	$\neg\varphi, \neg\psi$	φ, ψ	$\neg\varphi, \neg\psi$	φ

Table 2.2 Decomposition rules for arrow

Modus ponens	Counter-case
Given modes m, m' on a branch, labeled by	Given a node m on a branch labeled by
$\varphi, \varphi \rightarrow \psi$	$\neg(\varphi \rightarrow \psi)$
We can add to that branch, without forking	
One further node n labeled by	Two nodes n_1, n_2 labeled respectively by
ψ	φ, ψ

- For the two rules with a single input that do fork (decomposing $\phi \vee \psi$, $\neg(\phi \wedge \psi)$), the input implies the disjunction of the two outputs.
- For the sole rule with two inputs (modus ponens), the two inputs together imply the single output. As the notion of 'together' is rather delicate in relevance-sensitive contexts, we add that not only the inference from $\phi \wedge (\phi \rightarrow \psi)$ to ψ, but also that from ϕ to $(\phi \rightarrow \psi) \rightarrow \psi$, is valid in, say, the usual natural deduction systems for R and is acceptable in the sense to be defined in the present text.

In brief, each of those decomposition rules corresponds to an uncontested form of inference. But for counter-case the situation is quite different. While the input $\neg(\phi \rightarrow \psi)$, with the arrow read as material implication, does classically imply each of the two outputs ϕ, $\neg\psi$, these implications fail when the arrow is taken to be relevance-sensitive, since the falsehood of $\phi \rightarrow \psi$ can arise not only from the truth of ϕ accompanied by the falsehood of ψ, but also from a lack of relevance between them. Nevertheless, the rule retains an *indirect* rationale: its two output formulae ϕ, $\neg\psi$ may serve as without loss of generality (wlog) assumptions for getting further items. The idea is that if a formula θ can be obtained by continuing the decomposition procedure with the help of those two items, then θ follows from $\neg(\phi \rightarrow \psi)$, *provided* both ϕ and $\neg\psi$ are actually called upon to get there. That proviso is what the parity condition in the following section is designed to express.

The status of counter-case in a relevance-sensitive context is thus reminiscent of that of existential instantiation in classical first-order logic. There, truth-trees and natural deduction systems alike decompose an existential quantification $\exists x(\varphi)$ to $\varphi_{x:=a}$ where a is a 'fresh' constant, that is, one that does not occur anywhere up to that point. Clearly, the output formula $\varphi_{x:=a}$ is not in general a consequence of the input formula $\exists x(\phi)$ (although the two are 'friendly' in a sense defined in Makinson 2005). But whilst $\exists x(\varphi) \not\models \varphi_{x:=a}$, we do have that $A, \exists x(\varphi) \models \psi$ whenever $A, \exists x(\varphi), \varphi_{x:=a} \models \psi$ and a does not occur in any formula in $A \cup \{\varphi, \psi\}$. In other words, the rule of existential instantiation serves to allow assumption of $\varphi_{x:=a}$ without loss of generality in appropriate circumstances. For a general discussion of wlog procedures see Harrison (2009).

Finally, we comment on a matter that might lead some readers to doubt the legitimacy of using truth-trees at all in the present context. Evidently, by labeling the root of such a tree with the negation of the target formula and working toward crash-pairs, we are implementing a form of reductio ad absurdum. Now, it may be imagined from the rejection of the classical principle of 'right explosion' $(p \wedge \neg p) \rightarrow q$ and its dual $q \rightarrow (p \vee \neg p)$ that the spirit of relevance logic is intrinsically hostile to reductio. But this is not so. Indeed, formulae reflecting that principle, such as $((\neg\alpha \rightarrow \zeta) \wedge (\neg\alpha \rightarrow \neg\zeta)) \rightarrow \alpha$ and its 'unpacked' version $(\neg\alpha \rightarrow \zeta) \rightarrow ((\neg\alpha \rightarrow \neg\zeta) \rightarrow \alpha)$ are derivable from the axioms of the most widely known relevance logic R (see, e.g., Anderson and Belnap (1975) Sect. 9, especially pages 109–110), and a derivable rule of reductio ad absurdum is likewise obtainable in the natural deduction version of R (see, e.g., Loveland et al. 2014, Sect. 8.4, page 272). Those formulae are also directly acceptable in the sense shortly to be defined.

In short, there is nothing incoherent about the idea of giving reductio a central role in our analysis.

2.4 Directly Acceptability—The Concept

In this section, we define a notion of directly acceptability for trees and formulae; in Sect. 2.7 it will be extended recursively to one of acceptability *tout court*. We need the preliminary notions of a *crash-pair*, *trace*, and *critical node*, leading to the central *parity constraint.*

2.4.1 Definition

Crash-pairs are defined as in the classical context, where they are often given the rather long name of 'explicit contradictions'. They are pairs $\{z_1 : \zeta, z_2 : \neg\zeta\}$ of nodes on the same branch, for some formula ζ.

The *trace* T(n) of a node n in a tree is defined in essentially the same way as its classical counterpart. Conceptually, it is the set of all nodes in the tree on which n depends, that is, the nodes that are used in 'getting from the root to' n. Technically, it is best defined by a recursion in the reverse direction, as the least set of nodes of the tree that contains n and is such that whenever it contains a node then it also contains the node (both nodes, for modus ponens, no nodes for the root) from which it was obtained by application of a decomposition rule. Note, in particular, that each output of a forking rule contains the input of that rule in its trace.

For this notion of trace to be well defined as a function on nodes to sets of nodes, we must understand trees to be labeled not only by formulae but also by a record, for each node, of the preceding node(s) from which it was obtained. Just as two trees can agree in their 'naked geometry' but differ in the formulae attached to nodes, so too they may agree in both those respects but not in their justificational structure; there may be two different ways in which a given node could be obtained from its predecessors. An example will come up in Appendix 1, in a truth-tree for $(\alpha \to \alpha) \to ((\alpha \to \alpha) \to (\alpha \to \alpha))$. However, this kind of situation is comparatively infrequent and, to reduce clutter, we will omit the justificational part of a label whenever the structure of a tree is unambiguous without it.

The trace of a *set of nodes* is understood to be the union of their separate traces: $\mathrm{T}(X) = \cup\{\mathrm{T}(n){:}\ n \in X\}$. The only non-singleton sets for which we will actually need this function are pairs, specifically crash-pairs for which, therefore, $\mathrm{T}(\{n, m\}) = \mathrm{T}(n)\cup\mathrm{T}(m)$.

In contrast to crash-pair and trace, the notions of critical pair and the parity constraint are new to the present context. A *critical pair* is a pair $\{c_1 : \phi, c_2 : \neg\psi\}$ of nodes that are introduced by an application of the counter-case rule to a node m:

$\neg(\phi \rightarrow \psi)$; the individual nodes c_1, c_2 are called *critical nodes* and are *partners* of each other.

A branch B of a truth-tree is said to *crash with parity* iff it contains some crash-pair $Z = \{z_1 : \zeta, z_2 : \neg\zeta\}$ such that for every critical node c on B, if c is in the trace of Z then its partner is also (on B and) in the trace of Z. The words in parentheses are redundant, since all nodes in the trace of a node on B must also be on B, but it is sometimes helpful to make them explicit. Another way of expressing the definition: for every critical node on B, either both it and its partner, or neither of them, is in the trace of Z.

A *directly acceptable truth-tree* is one such that every branch B crashes with parity. Thus, a directly acceptable tree is one where every branch B contains some crash-pair Z such that for every critical node c on the branch, either both it and its partner are in the trace of Z, or neither of them are.

Directly acceptable formulae are those with some directly acceptable tree, that is, such that there is a directly acceptable truth-tree with root labeled by $\neg\phi$.

2.4.2 Comments and Examples

The following remarks and examples may help appreciate the contours of the above definitions.

Choice of primitive connectives. We are assuming that the primitive connectives of the propositional language are $\neg$, $\wedge$, $\vee$, $\rightarrow$. Further truth-functional connectives may be added with appropriate decomposition rules, so long as they are not zero-ary (the falsum and the verum); it is essential for Lemma 2.1 below that every formula contains at least one sentence letter.

Formulae in $\neg$, $\wedge$, $\vee$ *alone.* Clearly, a formula using at most the connectives $\neg$, $\wedge$, $\vee$ is directly acceptable iff it is a classical tautology, since the decomposition rules for those connectives are unchanged and, as there are no applications of counter-case, the parity condition 'does not arise', in other words, is vacuously satisfied.

Parity. The requirement of crashing with parity is reminiscent of the proviso in natural deduction systems for relevance logic that a supposition should *actually be used* before being discharged by the rule $\rightarrow^+$ of conditional proof (see, e.g., Loveland et al. 2014, Sect. 3). As remarked in Sect. 2.3, the decomposition rule for negated arrows treats a pair of critical nodes, obtained by an application of counter-case, as wlog assumptions that may be made without loss of generality when seeking a contradiction by further steps of decomposition. The parity constraint requires that if either one of those two suppositions is used to get the designated crash-pair in a given branch, then so is the other. One may thus see the constraint as transposing the demand for 'actual use' from conditional proof in natural deduction to counter-case in a truth-tree where, as we shall see, it turns out to work more transparently.

This prompts the question whether it would make any difference to the set of directly (or indirectly) acceptable formulae if we were to require that *both* critical nodes, rather than *either both or none*, are in the trace of the crash-pair; to be more

Fig. 2.1 Disjunctive Syllogism

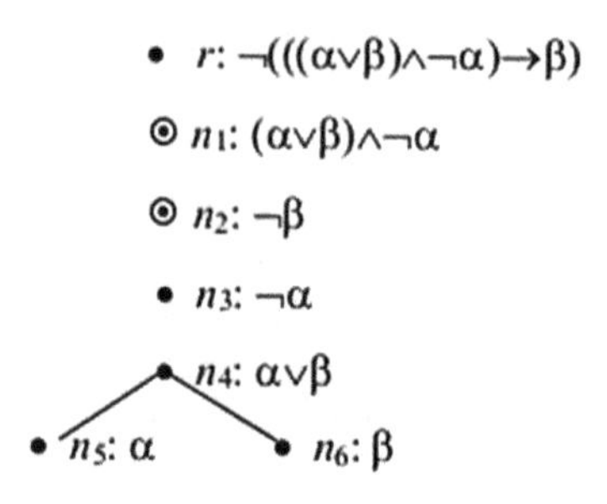

precise, that for every branch B and every critical node c on B, both it and its partner are in the trace of Z_B. All proofs in this paper go through without change for both formulations, but it is an open question whether they are equivalent. There is a partial result in this direction, and further discussion, in Appendix 4.

Quantificational structure. The quantificational prefix of the definition of a directly acceptable tree is $\forall B \exists Z \forall c$, the variables ranging over branches, crash-pairs, and critical nodes, respectively. Its delicacy can be illustrated by trees for 'symmetric explosion' and mingle, which are not directly acceptable but do satisfy a weaker $\forall B \forall c \exists Z$ condition (details in Sect. 2.5.2 below). It is often convenient to replace the existential quantifier sandwiched between universals in $\forall B \exists Z \forall c$, by a choice function. In such terms, the definition of direct acceptability requires that there is a choice function $B \mapsto Z_B$ taking branches to crash-pairs lying on them such that for every branch B and every critical node c on B, either both c and its partner are in the trace of Z_B or neither of them are. We will often refer implicitly to such choice functions by speaking of 'the designated crash-pair' on a branch.

Example: right explosion. The effect of the parity constraint is illustrated quite trivially by the example of right explosion $(\alpha \wedge \neg\alpha) \to \beta$: the root of its truth-tree is $r\colon \neg((\alpha \wedge \neg\alpha) \to \beta)$, decomposed by counter-case to the critical nodes $n_1 : \alpha \wedge \neg\alpha$ and $n_2 : \neg\beta$, with n_1 decomposed to the crash-pair n_3: α and n_4: $\neg\,\alpha$, which has in its trace the critical node n_1 but not n_1's partner n_2.

Example: disjunctive syllogism. A more instructive example is disjunctive syllogism, $((\alpha \vee \beta) \wedge \neg\alpha) \to \beta$, which is notorious from a relevance-sensitive perspective because it so easily yields right explosion by the celebrated Lewis derivation. When we construct a truth-tree for disjunctive syllogism we get Fig. 2.1, where non-critical nodes are indicated by •, while for critical nodes the dot is circled.

The tree in Fig. 2.1 has two branches. While both crash (which is all one needs in the classical context) the left branch does not do so with parity, since the critical node n_1 is in the trace of the unique crash-pair $\{n_3, n_5\}$on that branch but its partner n_2 is not.

Reading arrow as material implication. From these examples, we know that there are formulae that are not directly acceptable although they are tautologies when $\to$ is read as material implication (in more formal terms, when sub-formulae $\psi \to \theta$ are systematically replaced by $\psi \supset \theta$ abbreviating $\neg\psi \vee \theta$). On the other hand, the converse does hold: if a formula ϕ is directly acceptable then it is a tautology when

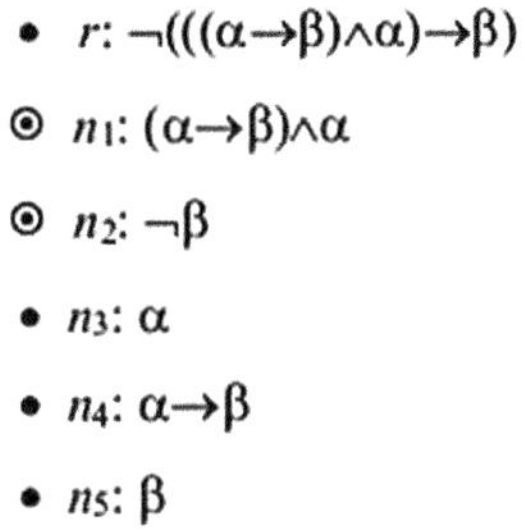

Fig. 2.2 Modus ponens

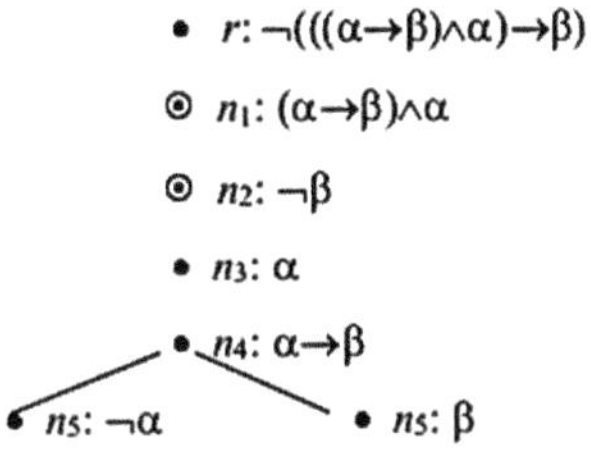

$\bullet$ r: $\neg(((\alpha\to\beta)\wedge\alpha)\to\beta)$
$\odot$ n_1: $(\alpha\to\beta)\wedge\alpha$
$\odot$ n_2: $\neg\beta$
$\bullet$ n_3: α
$\bullet$ n_4: $\alpha\to\beta$
$\bullet$ n_5: β

Fig. 2.3 Implicative forking can spoil parity

$\bullet$ r: $\neg(((\alpha\to\beta)\wedge\alpha)\to\beta)$
$\odot$ n_1: $(\alpha\to\beta)\wedge\alpha$
$\odot$ n_2: $\neg\beta$
$\bullet$ n_3: α
$\bullet$ n_4: $\alpha\to\beta$
$\bullet$ n_5: $\neg\alpha$ $\quad$ $\bullet$ n_5: β

$\to$ is read as material implication. For, suppose that ϕ is directly acceptable. Then it has a directly acceptable truth-tree with root r: $\neg\,\phi$; choose any one of them. Every branch of that tree contains a crash-pair. Since both modus ponens and counter-case are sound for material implication (and all the other decomposition rules used are also sound for their connectives) it follows that $\neg\,\phi$ is classically unsatisfiable, so that ϕ is a tautology.

Modus ponens versus implicative forking as decomposition rules. We can now explain, as promised at the beginning of Sect. 2.3, why unnegated arrows are decomposed by modus ponens rather than by implicative forking as is standard for classical logic. It is because unnecessary forking can provoke undesired failures of the parity condition.

This can be illustrated by the scheme $((\alpha \to \beta) \wedge \alpha) \to \beta$, which mirrors modus ponens as a formula in the object language. Intuitively, it is highly desirable for relevance-sensitive arrow; it is also directly acceptable, as witnessed by the truth-tree in Fig. 2.2, whose unique branch contains a crash-pair $\{n_2, n_5\}$with both critical nodes $\{n_1, n_2\}$in its trace.

But, if we handle unnegated arrows by implicative forking, we cannot make the step to n_5. The formula $\alpha \to \beta$ at n_4 is decomposed in the same way as is $\neg\,\alpha \vee \beta$ and so $((\alpha \to \beta) \wedge \alpha) \to \beta$ gets the same tree structure as disjunctive syllogism, failing parity in the same way: in Fig. 2.3, the crash-pair of the left branch is $\{n_3, n_5\}$with critical node n_1 in its trace but not its partner n_2.

Decidability. Clearly the set of directly acceptable formulae is decidable. Given the undecidability of R (Urquhart 1984), this already tells us that the two cannot coincide.

2.5 Direct Acceptability—Exclusions

This section establishes some outer limits to direct acceptability, showing that certain formulae (and kinds of formula) are not directly acceptable. Among them are some formulae that are 'not far' from that status, in the sense that their truth-trees fail the parity condition but would satisfy it if in the definition of parity the quantificational prefix was to be weakened by moving its existential quantifier inwards.

2.5.1 *Letter-Sharing*

Our main goal in this subsection is to establish letter-sharing for direct acceptability. We do so via a lemma that makes no appeal to the parity constraint nor even to the presence of crash-pairs, although it does make essential use (in subcase 2.2 of its proof) of our background assumption that every formula of the language contains at least one sentence letter. It is thus a fact of universal logic rather than specific to relevance logic. To motivate the rather intricate formulation, readers are invited to begin by seeing how the lemma is deployed in the proof of Observation 2.2.

Lemma 2.1 *Let ϕ, ψ be formulae with no sentence letters in common and consider any truth-tree T for $\phi \to \psi$ (with or without crash-pairs) that is built using the decomposition rules in Tables 2.1 and 2.2 (with or without the parity constraint). For every node m: θ in T after the root r: $\neg (\phi \to \psi)$ we have the following: (i) θ is shorter than the root formula (under any reasonable definition of length), (ii) m has in its trace exactly one of the critical nodes n_1: ϕ and n_2: $\neg \psi$ obtained from the root by the counter-case rule, (iii) every sentence letter occurring in θ occurs in the formula, ϕ or $\neg \psi$, attached to that critical node.*

Proof We induce the construction of T from its root. The base concerns the cases that m: $\theta = n_1$: ϕ or m: $\theta = n_2$: $\neg \psi$, since these are the only items immediately obtainable from root r: $\neg(\phi \to \psi)$ by a single application of the decomposition rules available; in both cases the three desired properties are immediate. For the induction step, there are two cases to consider.

Case 1. Suppose that m is obtained from a single node m': θ' other than the root by one of the decomposition rules other than modus ponens, so that $\mathrm{T}(m) = \{m\} \cup \mathrm{T}(m')$. Since the decomposition shortens its input formula, θ is shorter than θ' and so, by the induction hypothesis for (i), remains shorter than the root formula. Since $\mathrm{T}(m) = \{m\} \cup \mathrm{T}(m')$ it follows by the induction hypothesis for (ii) that $\mathrm{T}(m)$ contains the same node from $\{n_1, n_2\}$as does $\mathrm{T}(m')$. By the induction hypothesis for (iii), every letter in θ' occurs in whichever of n_1, n_2 is in $\mathrm{T}(m')$ so, since the decomposition does not introduce any new letters, every letter in θ occurs in whichever of n_1, n_2 is in $\mathrm{T}(m) = \{m\} \cup \mathrm{T}(m')$.

Case 2. Suppose that m: θ is obtained from a pair of nodes m_1: α and m_2: $\alpha \to \theta$ by modus ponens, so that $\mathrm{T}(m) = \{m\} \cup \mathrm{T}(m_1) \cup \mathrm{T}(m_2)$. Neither of m_1, m_2 can be the root-node—the latter because $\alpha \to \theta$ has arrow rather than negation as principal connective, the former because α is shorter than $\alpha \to \theta$ and so, by the induction hypothesis, shorter than the root-node formula. So, by the induction hypothesis, exactly one of n_1, n_2 is in T(m_1) and exactly one of n_1, n_2 is in T(m_2). This gives rise to subcases according to which of the n_i is in the trace of which of the m_j.

Subcase 2.1. Suppose that $n_i \in \mathrm{T}(m_1) \cap \mathrm{T}(m_2)$ for some $i \in \{1,2\}$. Consider the case that $i = 1$, the other is similar. By the induction hypothesis, $n_2 \notin T(m_1) \cup T(m_2)$. Since $T(m) = \{m\} \cup T(m_1) \cup T(m_2)$ we have $n_1 \in \mathrm{T}(m)$ while $n_2 \notin T(m)$. Moreover, every letter in θ occurs in $\alpha \to \theta$, and so by the induction hypothesis occurs in n_1 as desired.

Subcase 2.2. Suppose that $n_i \in \mathrm{T}(m_1)$ while $n_j \in \mathrm{T}(m_2)$ for some $i \neq j \in \{1,2\}$; we derive a contradiction. Consider the case that $i = 1$, the other is similar. By the induction hypothesis, every letter in α occurs in ϕ and every letter in $\alpha \to \theta$ occurs in $\neg\psi$. From the latter, every letter in α occurs in ψ. Since α contains at least one letter, this implies that ϕ and ψ share a letter, contrary to the initial supposition of the lemma. □

Observation 2.2 Direct acceptability satisfies the letter-sharing condition. That is, whenever an arrow formula $\phi \to \psi$ is directly acceptable, its antecedent ϕ and consequent ψ share at least one sentence letter.

Proof Suppose for reductio that T is a directly acceptable truth-tree with root r: $\neg(\phi \to \psi)$ where ϕ shares no sentence letters with ψ. Let B be any branch of T, with designated crash-pair $Z_B = \{z : \zeta, z' : \neg\zeta\}$. By Lemma 2.1, exactly one of n_1: ϕ and n_2: $\neg\,\psi$ is in the trace of z, and exactly one of them is in the trace of z'. If n_1 (resp. n_2) is in the trace of both of z, z', then n_2 (resp. n_1) is in the trace of neither of z, z', violating the parity condition. Thus, n_1 is in the trace of z while n_2 is in the trace of z', or inversely. Consider the first case, the second is similar. By Lemma 2.1 again, every letter in ζ occurs in ϕ and every letter in $\neg\,\zeta$ occurs in ψ. Since ζ has at least one letter, it follows that ϕ, ψ share at least one letter, contradicting the initial supposition. □

We note in passing that this proof yields rather more information than actually needed for Observation 2.2. As pointed out by Karl Schlechta (personal communication), it does not require that *all* branches contain a crash-pair—it goes through so long as *at least one* branch crashes and does so with parity.

Observation 2.2 has as corollary that whenever $\alpha \to (\beta \to \gamma)$ is directly acceptable then each of α, β, γ shares a letter with at least one of the other two. The verification consists in first showing that when $\alpha \to (\beta \to \gamma)$ is directly acceptable then so too are $\beta \to (\alpha \to \gamma)$ and $\neg\gamma \to (\alpha \to \neg\beta)$, then applying Observation 2.2 to each of them.

By re-running the proofs for Lemma 2.1 and Observation 2.2 with suitable editing, we can obtain an interesting 'omissibility' property: whenever $(\alpha \wedge \beta) \to \gamma$ or $\alpha \to (\beta \vee \gamma)$ is directly acceptable and β shares no letters with either α or γ, then $\alpha \to \gamma$ is directly acceptable. This property is related to what Avron et al. (2018), Sect. 11.1.1 call 'the basic relevance criterion', but exact comparison is difficult since the latter is defined in terms of consequence relations and the coordination of such relations and theorem sets is not univocal as it is in classical logic (see Sect. 2.9). To re-run the proof of Lemma 2.1 for $(\alpha \wedge \beta) \rightarrow \gamma$, say, we may assume wlog that the tree begins with nodes labeled r: $\neg((\alpha \wedge \beta) \rightarrow \gamma)$, n_1: $\alpha \wedge \beta$, n_2: $\neg \gamma$, n_3: α, n_4: β and show that for any node m: θ coming after n_1: (i) θ is shorter than the root formula; (ii) if n_4 is in the trace of m then neither n_2 or n_3 is in its trace; (iii) every sentence letter in θ occurs β or in at least one of α, γ.

The proofs can also be re-run with editing to show that direct acceptability satisfies the Halldén condition: if $\phi \vee \psi$ is directly acceptable and ϕ, ψ share no sentence letters then at least one of ϕ, ψ is directly acceptable. In the re-run of the derivation of the observation from the lemma, we no longer use reductio and, instead of using the parity constraint to get a contradiction, we infer directly that one or other of the separate disjuncts is directly acceptable.

It is an open question whether direct acceptability satisfies interpolation, that is, whenever $\alpha \rightarrow \beta$ is directly acceptable there is a formula γ, all of whose letters are common to α and β, such that both $\alpha \rightarrow \gamma$ and $\gamma \rightarrow \beta$. The limiting case of interpolation that arises when antecedent and consequent share no letters is given vacuously by Observation 2.2; we conjecture that the principal case holds too.

2.5.2 Examples of Quantifier Permutation

It is instructive to consider the formula $(p \wedge \neg p) \rightarrow (q \vee \neg q)$ (symmetric explosion), which notoriously fails the letter-sharing condition. Observation 2.2 tells us that it is not directly acceptable. This may at first be puzzling, since we can construct the tree in Fig. 2.4.

The tree in Fig. 2.4 has a unique branch, with a single critical pair $\{n_1, n_2\}$and two crash-pairs $\{n_3, n_4\}$and $\{n_5, n_6\}$. But $\{n_3, n_4\}$fails the parity constraint because n_1 is in its trace while its partner n_2 is not and, similarly, $\{n_5, n_6\}$fails the constraint since n_2 is in its trace while n_1 is not.

Fig. 2.4 Symmetric explosion: tree failing parity

- r: $\neg((p\wedge\neg p)\rightarrow(q\vee\neg q))$
- ⊙ n_1: $p\wedge\neg p$
- ⊙ n_2: $\neg(q\vee\neg q)$
- n_3: p
- n_4: $\neg p$
- n_5: $\neg q$
- n_6: $\neg\neg q$

Fig. 2.5 Mingle: tree failing parity

- r: $\neg(p \to (p \to p))$
- ⊙ n_1: p
- ⊙ n_2: $\neg(p \to p)$
- ⊙ n_3: p
- ⊙ n_4: $\neg p$

This example brings out the importance, remarked in Sect. 2.4.2, of the order of the second and third quantifiers $\exists Z \forall c$ in the definition of parity, contrasting with $\forall c \exists Z$. For symmetric explosion, the latter prefix is satisfied since for every critical node c (on the unique branch) there is a crash-pair Z such that c is not in the trace of Z so that, vacuously, if c is in the trace of Z then so is its partner.

A similar pattern emerges in the truth-tree for mingle, $p \to (p \to p)$ which, unlike symmetric explosion, satisfies the letter-sharing condition; see Fig. 2.5.

The tree in Fig. 2.5 has a single branch, with two crash-pairs $\{n_1, n_4\}$, $\{n_3, n_4\}$sharing the node n_4. But critical node n_3 is not in the trace of the first crash-pair although its partner n_4 is, while critical node n_1 is not in the trace of the second one although its partner n_2 is. So the unique branch fails the $\exists Z \forall c$ condition for parity although it satisfies the weaker $\forall c \exists Z$ one.

Mingle and symmetric explosion thus get 'quite close' to direct acceptability, missing out only on the order of the quantifier prefix in the parity condition. This formal fact resonates with a vague but widely shared intuition that they are, in some sense, two of the 'least irrelevant' among the formulae rejected by relevance-sensitive logicians. In contrast, the formula $p \to (q \to p)$—which we like to call 'mangle' in preference to the more common (and rather bland) name 'positive paradox'—is aberrant to a more serious extent than its substitution instance $p \to (p \to p)$, since the unique branch of its tree (write q in place of p at appropriate places in the tree for mingle) fails even the weaker $\forall c \exists Z$ condition.

This affinity between mingle and symmetric explosion wrt parity in their truth-trees is reminiscent of connections between those formulae in an axiomatic setting. Meyer has shown that the system RM defined by adding mingle to the standard axiomatization of R contains $\alpha \to \beta$ whenever it contains both $\neg\alpha$ and β, even when those two formulae share no letters. Thus, in particular, RM contains symmetric explosion as well as, for example, $\neg(p \to p) \to (q \to q)$. Derivations can be found in Anderson and Belnap (1975), Sect. 29.5 and in Priest (2008), Sect. 10.11, question 6.

2.6 Direct Acceptability—Inclusions

We now describe the surprisingly broad reach of direct acceptability as well as its closure under certain operations.

Observation 2.3 All axiom schemes of a standard axiomatization of the relevance logic R are directly acceptable.

Proof It suffices to check the axiom schemes one by one, finding a directly acceptable truth-tree for each. The verifications are routine but often interesting; see Appendix 1.

Observation 2.4 The set of all directly acceptable formulae is closed under substitution and conjunction.

Proof For conjunction, simply put the trees together with a new root and apply the decomposition rule for negated conjunctions. For substitution, substitute throughout the tree. In both cases, parity is preserved. □

Observation 2.5 There are directly acceptable formulae that are not in R.

Proof Conceptually, the simplest example is the two-letter formula $\{p \rightarrow (q \vee (p \rightarrow q))\} \rightarrow (p \rightarrow q)$ that one might call *skewed cases*. The tree in Fig. 2.6 satisfies parity; indeed, all critical nodes are in the trace of each crash-pair.

To verify that the skewed cases formula is not in R, one can use the matrix M_0 of Anderson and Belnap (1975), pp. 252–253, which is known to validate R. Assign the value -0 to each of p, q and calculate according to the tables for M_0 to obtain the undesignated value -3. □

Some comments on Fig. 2.6 may be helpful. (1) The right branch contains another crash-pair, namely, $\{n_2, n_8\}$, which would permit shortening that branch by one node, but with that crash-pair, parity fails. (2) The node n_3 is used twice, once in the common part of the two branches and then in the right branch alone. This kind of asymmetry between branches would not be allowed in their translation into subproofs in the usual natural deduction system for R (Appendix 2). (3) The substitution

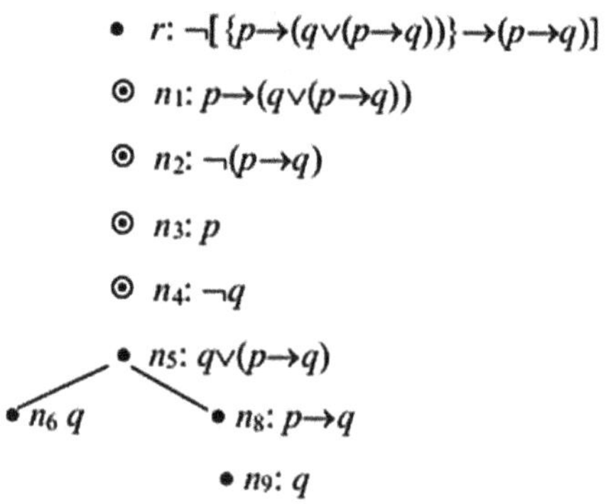

Fig. 2.6 Skewed cases

• r: $\neg[\{p \rightarrow (q \vee (p \rightarrow q))\} \rightarrow (p \rightarrow q)]$
⊙ n_1: $p \rightarrow (q \vee (p \rightarrow q))$
⊙ n_2: $\neg(p \rightarrow q)$
⊙ n_3: p
⊙ n_4: $\neg q$
• n_5: $q \vee (p \rightarrow q)$
• n_6 q • n_8: $p \rightarrow q$
• n_9: q

instance $\{p \to (p \vee (p \to p))\} \to (p \to p)$ of skewed cases, with just one letter, also serves as a witness for Observation 2.5. It is directly acceptable (substituting in the same tree) but not in R (same matrix, same value for p).

Both skewed cases and its substitution instance are theorems of the system RMI of Avron et al. (2018), Sect. 14.3, although that system differs from direct acceptability in that its theorem-set is not closed under conjunction. They are also valid in Urquhart's 'semi-lattice semantics' for relevance logic without classical negation. In the next section, following Corollary 2.7.2, we will see a formula that, in contrast, is acceptable (though not directly so) while not valid in Urquhart's semantics.

Given Observation 2.5, philosophically inclined readers may ask how its witness formulae should be seen from a relevance-sensitive perspective. In the author's view, little guidance as to their desirability may be extracted from bare intuition, which may legitimately be educated by the outcome of a satisfying formal account.

For the record, we note that the literature contains several more formulae that are valid in Urquhart's semantics yet absent from R (likewise failing the matrix M_0); all those that the author has seen are also directly acceptable. However, they all have at least three different connectives, three sentence letters, and seven letter occurrences (compared to two, one or two, six for our witnesses), the simplest among them being $\{(p \to (q \vee r)) \wedge (q \to r)\} \to (p \to r)$ (three letters) and $\{(p \to (q \vee r)) \wedge (q \to s)\} \to (p \to (s \vee r))$ (four letters). For more on such formulae, see Dunn (1986), Sect. 4.6, Urquhart (1989), page 169, Anderson et al. (1992), page 151, Humberstone (2011), page 1210, Example 8.13.21, Bimbó and Dunn (2018), Sect. 4, Bimbó et al. (2018), introduction and Sect. 2 (be aware, however, of a misprint in the last of these references, page 179, the formula $[(p \to p) \wedge ((p \wedge q) \to r) \wedge (p \to (q \vee r))] \to (p \to r)$ is printed with an arrow in place of its third conjunction sign).

2.7 Going Recursive

Despite its strengths, highlighted in the preceding section, direct acceptability has serious limitations. In this section we identify the shortcomings and introduce a recursive step into the tree construction to transcend them.

2.7.1 Undesirable Limits to Direct Acceptability

There are formulae that seem quite innocuous from a relevance-sensitive standpoint, but which are not directly acceptable. For example, $(p \to q) \to (\neg\neg p \to q)$ and its converse $(\neg\neg p \to q) \to (p \to q)$ appear equally unobjectionable but, while the former is directly acceptable, the latter is not. Any truth-tree for it would have to begin as in Fig. 2.7. On reaching n_4 in Fig. 2.7, one would like to apply modus ponens to $\neg\neg p \to q$ and $\neg\neg\ p$ to get q and finish, but the minor premise $\neg\neg\ p$ is missing: while p is available at n_3, its double negation is not. We cannot obtain that premise by

Fig. 2.7 A desirable decomposition blocked

- r: $\neg((\neg\neg p \to q) \to (p \to q))$
- ⊙ n_1: $\neg\neg p \to q$
- ⊙ n_2: $\neg(p \to q)$
- ⊙ n_3: p
- ⊙ n_4: $\neg q$

decomposing p, for decomposition always *eliminates* a principal connective, never introduces one. Nor can we apply modus tollens to n_1: $\neg\neg\ p \to q$ using n_4: $\neg\ q$, since modus tollens is not one of our direct decomposition rules. The truth-tree is blocked at n_4 and can go no further with the resources of direct decomposition.

The same example also reveals the failure of a general closure condition for the set of directly acceptable formulae.

Observation 2.6 The set of directly acceptable formulae is not closed under detachment wrt the arrow.

Proof Put $\psi = (\neg\neg p \to q) \to (p \to q)$ and $\phi = p \to \neg\neg p$. We have noted that ψ is not directly acceptable. But it is easy to check that both ϕ and $\phi \to \psi$ are directly acceptable. □

To prevent misunderstanding, we emphasize that closure under detachment of the set of directly acceptable formulae is a quite different matter from the availability of modus ponens as one of the decomposition rules for building directly acceptable trees. Observation 2.6 tells us, in effect, that the latter does not imply the former. To emphasize the distinction, we will always use the word 'detachment' in the context of closure of the output set of formulae, and 'modus ponens' in the context of decomposing within a tree.

It is easy to multiply instances of formulae that appear quite harmless from a relevance-sensitive perspective, but which are not directly acceptable and reflect failures of detachment. An example with conjunction in place of double negation is $((p \wedge p) \to q) \to (p \to q)$. Examples with arrow as sole connective also exist; one pointed out by Lloyd Humberstone (personal communication) is $((p \to p) \to q) \to q$. An interesting example of a quite different kind is the negated conditional $\neg\,((p \vee \neg p) \to (p \wedge \neg p))$. Neither it, nor any other negated arrow formula $\neg(\alpha \to \beta)$ is directly acceptable. For, given a root labeled $\neg\neg(\alpha \to \beta)$ the only decomposition rule we can apply is double negation elimination to get $\alpha \to \beta$ and the only rule that might then be applied to $\alpha \to \beta$ is modus ponens, but no minor premise is available for such a step. Here too, detachment fails, since both $p \vee \neg p$ and $(p \vee \neg p) \to \neg((p \vee \neg p) \to (p \wedge \neg p))$ are directly acceptable.

It should be understood that the snags in the attempted truth-trees for these formulae have nothing to do with the parity condition and everything to do with modus ponens, for they persist when one uses the decomposition rules of Tables 2.1 and 2.2

without requiring parity. Now, the only difference between such rules and those that are standard in truth-trees for classical logic is that modus ponens replaces implicative forking for unnegated arrow. One can therefore say, roughly, that while implicative forking undesirably magnifies the severity of the parity constraint, its replacement by modus ponens blocks steps where parity is not an issue.

The question thus arises: Is there a principled way of extending our decomposition procedure to validate formulae such as those we have mentioned and ensure closure under detachment—without losing the vital letter-sharing property enjoyed by direct acceptability?

2.7.2 Recursion to Acceptability

To achieve that result, it is natural to introduce a recursive step into the decomposition procedure, generating a notion of *acceptable* trees and formulae. The basis of the recursion takes directly acceptable trees and formulae (as defined in Sect. 2.4.1) to be acceptable. For the recursion step, when constructing an acceptable tree, we allow passage from a node m: ϕ to a node n: ψ in the tree whenever the formula $\phi \rightarrow \psi$ is acceptable.

Note that this is not the same as the decomposition rule of modus ponens, which proceeded from *two* nodes labeled ϕ, $\phi \rightarrow \psi$ to a node labeled ψ (whether or not the formula $\phi \rightarrow \psi$ is acceptable). The recursive rule proceeds from a *single* node labeled ϕ to one labeled ψ provided that $\phi \rightarrow \psi$ is acceptable (whether or not the formula labels a node in the tree).

Strictly speaking, the recursive step does not always decompose, since there is no limit on the letters or complexity of the formula ψ as compared to ϕ, but we will call it a 'decomposition' rule in a derivative sense of the term. Because of the rule's broad sweep, it is not immediately clear whether the sets of acceptable trees and formulae are decidable as are their counterparts for direct acceptability; but semi-decidability, i.e., recursive enumerability, is retained.

All the examples of intuitively agreeable but not directly acceptable formulae in Sect. 2.7.1 are evidently acceptable. Moreover, acceptability has the following general properties.

Observation 2.7 All directly acceptable formulae are acceptable. Modus tollens and implicative forking are admissible as decomposition rules in the construction of acceptable truth-trees. The set of acceptable formulae is closed under substitution, conjunction, contraposition, and detachment.

Proof Modus tollens (decomposing $\phi \rightarrow \psi$, $\neg\, \psi$ to $\neg\phi$) is admissible, since it may be taken as abbreviating an application of the recursive rule using the directly acceptable $(\phi \rightarrow \psi) \rightarrow (\neg\psi \rightarrow \neg\phi)$ followed by modus ponens. Implicative forking (decomposing $\phi \rightarrow \psi$ by forking to $\neg\phi$ on one branch and ψ on the other) is also admissible, using the directly acceptable $(\phi \rightarrow \psi) \rightarrow (\neg\phi \vee \psi)$ followed by the rule for decomposing disjunctions. Closure under substitution and conjunction are checked in the same way as in the context of direct acceptability. For closure under contraposition, suppose that $\phi \rightarrow \psi$ is acceptable. We can build an acceptable tree with root labeled $\neg(\neg\psi \rightarrow \neg\phi)$ by checking that $\neg(\neg\psi \rightarrow \neg\phi) \rightarrow \neg(\phi \rightarrow \psi)$ is directly acceptable, applying the recursive rule to get a node labeled $\neg(\phi \rightarrow \psi)$, and pasting in any acceptable tree for $\phi \rightarrow \psi$. Clearly, parity is preserved. Closure under detachment is verified as follows. Suppose ϕ and $\phi \rightarrow \psi$ are both acceptable. From the latter, we know from the above that $\neg\psi \rightarrow \neg\phi$ is acceptable. We can build an acceptable tree with $\neg\psi$ as root by using $\neg\psi \rightarrow \neg\phi$ in the recursive rule to get $\neg\phi$, then pasting in the acceptable tree for ϕ. Again, parity is clearly preserved. □

Corollary 2.7.1 *All formulae derivable in the system R are acceptable, but not conversely.*

Proof By Observation 2.3, all axiom schemes of R are directly acceptable and so, by Observation 2.7, are acceptable. Since, by Observation 2.7 again, the set of acceptable formula is closed under the derivation rules of R (conjunction and detachment), we are done. □

Corollary 2.7.2 *Acceptability supports a straightforward construction of a characteristic Lindenbaum algebra.*

Proof Given Corollary 2.7.1, this follows from the fact that R supports such a construction. It can also be verified directly. □

In Observation 2.5 and its ensuing discussion, we gave some examples of formulae that are not in R but are both directly acceptable and valid in Urquhart's semi-lattice semantics. The question arises whether among formulae without negation (i.e., those to which that semantics is applicable), there are any that are acceptable but *not* Urquhart valid. We have not found any that are directly acceptable, but we do have an acceptable one, $((p \rightarrow q)\wedge((p \wedge q) \rightarrow r))\rightarrow (p \rightarrow r)$, which expresses as a formula the familiar Tarski condition of cumulative transitivity for consequence relations. On the one hand, it is easily checked to be acceptable. On the other hand, it is known that formulae using at most the connectives $\rightarrow$, $\wedge$ are valid in the Urquhart semantics iff they are in R (see Humberstone (2011), Sect. 6.45, page 906). But the above formula is not in R since, using contraposition for $\rightarrow$, commutation for $\wedge$ and de Morgan, we can show that it is inter-derivable in R with its dual $((p \rightarrow (q \vee r))\wedge(q \rightarrow r))\rightarrow (p \rightarrow r)$ which, as we saw in Sect. 2.6, is not in R although it is valid in Urquhart's semantics.

This, in turn, reveals an interesting asymmetry in the set of formulae that are valid in that semantics. While the last-mentioned formula is valid, the first-mentioned formula of that paragraph is not although, in a natural sense, they are duals. The above verification of that duality deploys contraposition and de Morgan, thus using negation, which is not available in the Urquhart semantics in more than an intuitionistic sense.

Before leaving the section, we comment on a feature of the recursive rule that may, at first, be puzzling. As already remarked, the rule allows passage from a *single* node

Fig. 2.8 Two-premise recursive rule yields mangle

- r: $\neg(p\to(q\to p))$
- ⊙ n_1: p
- ⊙ n_2: $\neg(q\to p)$
- ⊙ n_3: q
- ⊙ n_4: $\neg p$
- n_5: $p\wedge q$
- n_6: p

m: ϕ to a node n: ψ whenever the formula $\phi \to \psi$ is acceptable; why not also two (or more) nodes working together as inputs? To see the reason, consider the case of two inputs. Such a rule would tell us that whenever we have nodes m: ϕ, n: ψ on a branch we may pass to a node labeled θ provided a certain conditional formula involving ϕ, ψ, θ is acceptable. But what is that formula? Two options suggest themselves: $\phi \to (\psi \to \theta)$ and $(\phi \wedge \psi) \to \theta$.

Now, the former option adds nothing to the one-premise rule, since we may pass from m: ϕ to m': $\psi \to \theta$ by an application of the one-premise version, then apply modus ponens to n: ψ, m': $\psi \to \theta$ to get θ as desired. On the other hand, the latter option is catastrophic, since it allows a notorious classical derivation of the undesirable 'mangle' formula $p \to (q \to p)$ (Sect. 2.5.2), by justifying the step from nodes n_1, n_3 to n_5 in the tree of Fig. 2.8, which satisfies parity.

2.7.3 The Main Open Problem

However, we have not been able to determine *whether the letter-sharing property continues to hold for acceptability*. Worse, we have not even been able to show non-triviality, in the sense that there is at least one formula in the language of $\neg$, $\wedge$, $\vee$, $\to$ that is a tautology when $\to$ is read as material implication, but which is not acceptable. We have only a partial result in this direction: letter-sharing, and hence non-triviality, hold for acceptability when it is formulated in the $\neg$, $\to$ fragment of the full $\neg$, $\wedge$, $\vee$, $\to$ language (see Appendix 4).

If letter-sharing does survive, as we conjecture, then all forms of explosion will be rejected, as too will mingle and mangle (Sect. 2.5.2). In that situation, it would be reasonable to see decomposition-with-parity as supplying a clear rationale for relevance-sensitive logic. On the other hand, if letter-sharing is lost then, since it is a *sine qua non* for any relevance logic worthy of the name, that title cannot be awarded to acceptability. And if the non-triviality property fails, then acceptability collapses into classical tautologicality.

It is natural to seek a proof of letter-sharing for acceptability by extending the one given in Observation 2.2 for direct acceptability. However, this idea faces two difficulties. An obvious one is that the recursive rule allows passage from n: ϕ to m: ψ if $\phi \rightarrow \psi$ is already known to be acceptable, even when ψ contains letters not already in ϕ. In Sect. 2.8.2, after having established an equivalent characterization of acceptability, we will see how to get around that obstacle.

But a more recalcitrant obstacle is that in a tree for an arrow formula $\alpha \rightarrow \beta$, the recursive rule may be applied directly to the root r: $\neg (\alpha \rightarrow \beta)$, bypassing the nodes n_1: α and n_2: $\neg \beta$ that would otherwise be obtained from it by counter-case, thus preventing us from transposing the proof of Lemma 2.1 to the context of acceptability. Attempts to get around this obstacle by properly weakening the definition of acceptability create unfortunate losses, as we will also show in Sect. 2.8.2.

2.8 An Equivalent Definition of Acceptability

In this section we show that the set of acceptable formulae is precisely the closure, under conjunction and detachment with respect to arrow, of the set of directly acceptable formulae, and comment on some implications of this result.

2.8.1 *Characterization*

For brevity, write A for the set of acceptable formulae, DA for the set of directly acceptable ones, and (DA)* for the closure of DA under conjunction and detachment with respect to arrow.

Observation 2.8 A = (DA)*.

Proof The inclusion (DA)* $\subseteq$ A is immediate from Observation 2.7. For the converse, let $\alpha \in$ A. Then there is an acceptable truth-tree T with root r: $\neg \alpha$. We argue by induction. As the induction hypothesis, we assume that for every application of the recursive rule in T, its justifying formula $\phi \rightarrow \psi$ is in (DA)*; we need to show that $\alpha \in$ (DA)*. Since (DA)* is by definition closed under detachment, it will suffice to construct a formula θ such that both $\theta, \theta \rightarrow \alpha \in$ (DA)*.

Without loss of generality, we may assume that every designated crash-pair in T has in its trace both the input and output of some application of the recursive rule. For, suppose B is a branch of T whose designated crash-pair $\{z$: ζ, z': $\neg \zeta\}$does not have some such formulae in its trace. We can create a detour that applies the recursive rule to z: ζ to output a new node z'': ζ, with justifying formula the directly acceptable formula $\zeta \rightarrow \zeta$ and treat $\{z''$: ζ, z': $\neg \zeta\}$as the crash-pair for that branch. Clearly the tree so formed remains acceptable and has one less violation of the wlog property.

Since there are only finitely many applications of the recursive rule in T, we may legitimately put θ to be the conjunction $\wedge\{\phi_i \rightarrow \psi_i\}$of all their justifying formulae $\phi_i \rightarrow \psi_i$. By the induction hypothesis, each conjunct of θ is in (DA)*. Since (DA)* is, by definition, closed under conjunction, we thus have $\theta \in$ (DA)*. It remains to show that $\theta \rightarrow \alpha \in$ (DA)*; we show that in fact $\theta \rightarrow \alpha \in$ DA $\subseteq$ (DA)* by building a directly acceptable tree S for it.

Put the root of S to be s: $\neg(\theta \rightarrow \alpha)$; apply counter-case to get n_1: θ and n_2: $\neg\alpha$; then continue as in T from its root labeled $\neg\alpha$, but replacing each application of the recursive rule, justified by a formula $\phi_i \rightarrow \psi_i$, by sufficiently many applications of the decomposition rule $\wedge-$ to n_1: $\theta = \wedge\{\phi_i \rightarrow \psi_i\}$to get $\phi_i \rightarrow \psi_i$ and then applying modus ponens to ϕ_i, $\phi_i \rightarrow \psi_i$ to get ψ_i. Since in T the root r: $\neg\alpha$ is in the trace of every crash-pair, the same holds in S for the node n_2: $\neg\alpha$ so, for parity, we need to check that the same holds in S for its partner n_1: $\theta = \wedge\{\phi_i \rightarrow \psi_i\}$. But that is ensured by the wlog assumption above. For all other critical nodes in S, parity is clearly preserved from parity in T. □

2.8.2 *Some General Lessons from Observation 2.8*

Observation 2.8 is technical, but it has several broad implications for acceptability. Three of them are 'good news' for the concept, while one sounds a warning bell.

The first point on the positive side is that the observation reveals some robustness in the notion of acceptability, since there are at least two natural ways of widening direct acceptability to get it. Second, it delivers a Hilbertian axiomatization of the acceptable formulae, since we may simply declare the decidable set DA as the axioms and take the derivation rules to be detachment and conjunction.

The third agreeable feature is that the result allows us to bypass, without cost, the first of two obstacles facing a proof of letter-sharing for acceptability that were mentioned in Sect. 2.7.3, namely that the recursive rule allows passage from n: ϕ to m: ψ even when ψ contains letters not already in ϕ. To do that, we can simply restrict the rule to those instances where every letter in ψ already occurs in ϕ and shows that the restriction leaves the set of acceptable formulae unchanged. Call the constrained version *timid acceptability*. By Observation 2.8, to show that timid acceptability coincides with plain acceptability, it suffices to check that timid acceptability (1) includes direct acceptability and is closed under both (2) conjunction and (3) detachment wrt the arrow.

Property (1) is immediate from the definitions. Property (2) has the same verification as before, without change. For (3), first note that timid acceptability is closed under both substitution and contraposition, with verifications unchanged from those in the proof of Observation 2.7, noting also that every letter in the consequent of $\neg(\neg\psi \rightarrow \neg\phi) \rightarrow \neg(\phi \rightarrow \psi)$, used there to support an application of the recursive rule, already occurs in the antecedent. Now suppose α, $\alpha \rightarrow \beta$ are both timidly acceptable; we need to show that β is too. Let p be any one of the sentence letters in β, and let σ be the substitution that is the identity on letters occurring in β but sends all other letters to p. Since timid acceptability is closed under substitution, both $\sigma(\alpha)$ and $\sigma(\alpha \to \beta) = \sigma(\alpha) \to \sigma(\beta) = \sigma(\alpha) \to \beta$ are timidly acceptable and thus, by closure under contraposition, so too is $\neg\beta \to \neg\sigma(\alpha)$. Since all letters in $\sigma(\alpha)$ occur in β, we can build a timidly acceptable tree with root $r\colon \neg\, \beta$ by applying the timidly recursive rule to that root to get $n\colon \neg\, \sigma(\alpha)$, where we can paste in a timidly acceptable tree for $\sigma(\alpha)$.

On the negative side, Observation 2.8 shows that if one seeks to guarantee letter-sharing by *properly* weakening the notion of acceptability, there is limited room for maneuver. One must either (i) cut into the set of directly acceptable formulae, or (ii) give up (or at least restrict) closure under conjunction, or (iii) do the same for detachment. All three options seem, to the present author, to rather unpalatable but there are systems in the literature exemplifying at least the first two: R itself takes option (i), while option (ii) is followed by the system RMI of Avron et al. (2018), Sect. 14.3.

Suppose, for example, that one tries to overcome the second obstacle facing a proof of letter-sharing mentioned in Sect. 2.7.3 by prohibiting applications of the recursive rule to the root $r\colon \neg\,(\alpha \to \beta)$. It is not clear whether this weakening is proper, but it creates serious difficulties for the proof in Observation 2.7 that acceptability is closed under detachment. The difficulties can be overcome in the special case that one of the formulae α, β contains all the sentence letters that occur in the other, but the author has not been able to find a way around them in general.

Nevertheless, we conjecture that letter-sharing does hold for acceptability. Some encouragement for this conjecture may perhaps be found in a certain six-element model for relevance logics that was defined in Routley et al. (1982) and further studied in Swirydowicz (1999), Brady (2003), Theorem 9.8.3; see also the textbook Schechter (2005) where it is called the 'crystal' model and, most recently, Øgaard (2019), footnote 12. It is known that (i) all arrow formulae that are valid in the crystal model satisfy the letter-sharing condition, (ii) the crystal model validates all axioms of R, also preserving validity under conjunction and detachment for arrow. Moreover, as communicated to the author by Tore Øgaard on 17 June 2019, the MaGIC software of Slaney (1995) verifies that the crystal model validates our skewed cases formula $(p \to (q \vee (p \to q))) \to (p \to q)$. It is thus conceivable that it may validate all directly acceptable formulae—in which case, by Observation 2.8 above, it will validate all acceptable formulae and so acceptability will satisfy letter-sharing.

A semantic strategy for establishing letter-sharing for acceptability could thus be to prove that all directly acceptable formulae are valid in the crystal model. One way of approaching such a proof might be to transform the system of directly acceptable trees into a system of sequents and check for that.

2.9 From Acceptable Formulae to Acceptable Consequence

What are the options for defining an interesting consequence relation $\vdash$ in terms of the set of acceptable formulae? Presumably, we would like the relation to be closely coordinated with a conditional connective of the object language, so an important conceptual choice immediately presents itself: which conditional connective should we use—relevance-sensitive $\rightarrow$ or classical $\supset$? The decision makes a serious difference. Suppose, on the one hand, we take $\alpha \vdash \beta$ to hold just if the formula $\alpha \supset \beta$, abbreviating $\neg \alpha \vee \beta$, is acceptable. Then, since $(p \wedge \neg p) \supset q$ is acceptable, as is every tautology in the connectives $\neg$, $\wedge$, $\vee$ alone, we have $p \wedge \neg p \vdash q$. Suppose, on the other hand, we put $\alpha \vdash \beta$ iff $\alpha \rightarrow \beta$ is acceptable. Then $p \wedge \neg p \nvdash q$, so long as acceptability satisfies letter-sharing.

Behind the question of choosing which connective to align with $\vdash$, lies one of purpose. Do we have ambitions to apply our relevance-sensitive logic as the working engine for some substantive theory? For example, one might hope to resuscitate set theory with naïve comprehension, which is notoriously inconsistent, by articulating it within the context of such a logic. Again, one might seek a workaday version of Peano arithmetic that can be shown to be consistent by finitistic reasoning, formulating it within such a logic. Such tasks would presumably call for a relevance-sensitive consequence relation and so should coordinate $\vdash$ with $\rightarrow$. But if, like the present author, we are sceptical about the potential for such attempts at application (see, for example, Incurvati (2020), Chap. 4 on the case of set theory), then it is more pertinent to ask which option may be more helpful in the exploration of relevance-sensitive logic itself.

Both options seem possible, as well as various compromises between them, each with its attractions and limitations. Nor is there anything to prevent us from considering two different consequence relations together, which would raise interesting questions about their interaction. But in this section, we will merely see how each of the two options, considered separately, plays out for acceptability.

We begin, in Sect. 2.9.1, by looking at the coordination of consequence with material implication. Although very few authors have gone that way, it has some attractive formal features; it is regularly behaved and it makes the relation a conservative extension of classical consequence, thus permitting a vision of relevance-sensitive logic as an extension rather than a rival or subsystem of classical logic. In Sect. 2.9.2 we consider coordination with relevance-sensitive arrow, much more frequently envisaged, where the situation is considerably more complex.

2.9.1 $\vdash$ Coordinated with $\supset$

The following definition of $\Gamma \vdash \beta$, where Γ is any set of formulae and β is an individual formula, was suggested in Makinson (2014) for the relevance logic R and may equally be articulated for A. Put $\Gamma \vdash \beta$ iff there are $\alpha_1, \ldots, \alpha_n \in \Gamma$ $(n \geq 0)$ such that $(\alpha_1 \supset (\alpha_2 \supset (\ldots(\alpha_n \supset \beta)\ldots) \in$ A, with the α_i in arbitrary order and with the iterated conditional formula understood to be β in the limiting case that $n = 0$.

This definition can be put in several equivalent forms. Clearly, since A contains all classical tautologies and is closed under substitution and detachment, we may replace $(\alpha_1 \supset (\alpha_2 \supset (\ldots(\alpha_n \supset \beta)\ldots)$ by $(\alpha_1 \wedge \ldots \wedge \alpha_n) \supset \beta$. We may also break the definition in two parts, separating the principal case where $n \geq 1$ from the limiting case that $n = 0$. The former is a 'deduction condition' that $\Gamma \cup \{\alpha\} \vdash \beta$ iff $\Gamma \vdash \alpha \supset \beta$, which is a property internal to $\vdash$, not mentioning A; the latter is the 'output condition' that $\emptyset \vdash \beta$ iff $\beta \in$ A, making the bridge to acceptability.

So defined, $\vdash$ has many attractive properties. It is a closure relation, compact, closed under substitution, and outputs the set A of acceptable formulae as the set of consequences of the empty set. It also allows one to see relevance logic as an extension of classical logic in $\neg$, $\wedge$, $\vee$ by a non-classical connective $\rightarrow$ rather than, as is customary, as a subsystem of classical logic in $\neg$, $\wedge$, $\vee$, $\rightarrow$, for it can be shown that the relation conservatively extends classical consequence in the connectives $\neg$, $\wedge$, $\vee$. Nothing classical is lost, some expressivity is gained.

The verifications of these properties for A are the same as were carried out for R in Makinson (2014), with the following addition for conservativeness. Clearly, by the definition of $\vdash$, it suffices to show that if a formula ϕ in the language of $\neg$, $\wedge$, $\vee$ is in A, then it is a classical tautology. We know that this is the case for direct acceptability DA (Sect. 2.4) but we have not actually verified it for A, where the recursive rule may introduce into a truth-tree occurrences of the arrow that are not in ϕ. It suffices to show the following lemma: whenever a formula ϕ in the full language $\neg$, $\wedge$, $\vee$, $\rightarrow$ has an acceptable truth-tree T, then the truth-tree T^* obtained from T by replacing all arrows with material implications is a classically correct tree for the formula ϕ^* obtained from ϕ by the same replacements. This can be shown by an easy induction. The desired conservation result then holds as the limiting case where ϕ has no arrows so that $\phi^* = \phi$.

As noted when defining $\vdash$, the relation satisfies the deduction condition wrt $\supset$, that is, $\Gamma \cup \{\alpha\} \vdash \beta$ iff $\Gamma \vdash \alpha \supset \beta$. In contrast, the corresponding condition wrt $\rightarrow$ fails: we have $\{p \wedge \neg p\} \vdash q$ since $(p \wedge \neg p) \supset q \in$ A, but $\emptyset \nvdash (p \wedge \neg p) \rightarrow q$ since, on the assumption that A satisfies letter-sharing, $(p \wedge \neg p) \rightarrow q \notin$ A.

In summary: consequence defined from A by coordination with $\supset$ is internally well-behaved and neatly related, as a conservative extension, to classical consequence. However, failure of the deduction property wrt $\rightarrow$ means that this relation would hardly be appropriate in a program for replacing classical logic by a relevance-sensitive one as the inferential motor of a substantive mathematical theory.

2.9.2 $\vdash$ Coordinated with $\rightarrow$

The subtleties of coordinating $\vdash$ with a connective $\rightarrow$ in substructural logics have been studied by Avron (1992, 2014), Avron et al. (2018), as well as other investigators. In contrast to the procedure followed in the present paper, Avron and co-authors prefer to put the consequence relation center-stage, leaving the set of logically true formulae in the background; in particular, the former is not defined from the latter but articulated concurrently. But much of what they observe carries over, *mutatis mutandis*, to our procedure, and this section essentially transposes their analysis to the case of A.

It is tempting to edit the definition in Sect. 2.9.1 by putting $\Gamma \vdash \beta$ iff there are $\alpha_1,\ldots,\alpha_n \in \Gamma$ $(n \geq 0)$ such that $(\alpha_1 \rightarrow (\alpha_2 \rightarrow (\ldots(\alpha_n \rightarrow \beta)\ldots) \in A$, with the iterated conditional formula again understood to be β in the limiting case that $n = 0$. However, this relation is not-well defined unless both the ordering of antecedents $\alpha_1,\ldots\alpha_n$ in the formula $(\alpha_1 \rightarrow (\alpha_2 \rightarrow (\ldots(\alpha_n \rightarrow \beta)\ldots)$, and the introduction or suppression of repetitions among those antecedents, make no difference to the status of the formula as an element of A. This is indeed the case for the ordering, since we know that $(\alpha_1 \rightarrow (\alpha_2 \rightarrow (\ldots(\alpha_n \rightarrow \beta)\ldots) \in A$ iff $(\alpha'_1 \rightarrow (\alpha'_2 \rightarrow (\ldots(\alpha'_n \rightarrow \beta)\ldots) \in A$ where the primes indicate any reordering of the α_i. But it is a problem for repetition. While $p \rightarrow p \in A$, we hope and conjecture that $p \rightarrow (p \rightarrow p) \notin A$ since, as remarked in Sect. 2.5.2, its presence would create failure of letter-sharing and thereby the collapse of our enterprise.

In effect, the definition wishes to put $f(\gamma) \vdash \beta$ iff $\gamma \in A$ where f is a surjective function f taking iterated arrow formulae to finite sets Γ of formulae, namely, $f((\alpha_1 \rightarrow (\alpha_2 \rightarrow (\ldots(\alpha_n \rightarrow \beta)\ldots)) = \{\alpha_1, \alpha_2,\ldots, \alpha_n\}$; although there are formulae γ, γ' with $f(\gamma) = f(\gamma')$ but $\gamma \in A$ while $\gamma' \notin A$. It is thus an example of the familiar phenomenon of a notion failing to be well defined because of the non-injectivity, up to an appropriate equivalence relation, of a function implicit in the definition.

In the case of R, where these considerations are well-known, it is customary to repair the situation by taking Γ to be a *multiset* of formulae, and that option works for A as well. For weaker systems that do not allow permutation of antecedents in an iterated arrow formula, one needs to tighten even further the identity criterion for assemblies on the left of the turnstile, taking them as, say, sequences. But that is not our concern here, so we continue the discussion by re-casting the definition of $\vdash$ in terms of multisets, minimizing notational clutter by simply re-reading set-theoretic signs such as $\emptyset$, $\cup$ and $\{\ldots\}$in a manner appropriate for such items.

Assuming that A satisfies letter-sharing, we can say that certain structural properties of classical consequence fail for $\vdash$. Specifically, strong reflexivity, the condition that $\Gamma \vdash \beta$ whenever $\beta \in \Gamma$, fails. For example, $\{p, q\} \nvdash q$ since $p \rightarrow (q \rightarrow q) \notin$ A, assuming letter-sharing for A. By the same token, conjunction of premises fails: $\{p, q\} \nvdash p \wedge q$ since $p \rightarrow (q \rightarrow (p \wedge q)) \notin A$, under the same assumption. Also, monotony fails, even in its limiting form that $\{\alpha\} \vdash \beta$ whenever $\emptyset \vdash \beta$. On the one hand, $\emptyset \vdash q \rightarrow q$ since $q \rightarrow q \in A$; on the other hand, $\{p\} \nvdash q \rightarrow q$ since $p \rightarrow (q \rightarrow q) \notin A$, again under the assumption that A satisfies letter-sharing.

In brief, assuming that the set of acceptable formulae satisfies the letter-sharing condition, defining a consequence relation in a way that coordinates it with the arrow forces a reconfiguration of its left arguments as items with a tighter identity criterion than sets, say as multisets, and leads to failure of both strong reflexivity and monotony for the relation.

It is possible to block at least some of the above outcomes by tweaking the definition, and various ways of doing this have been suggested in the literature (see, e.g., Avron 1992; Avron et al. 2018). One such definition, suggested by Avron in correspondence with the author, would put $\Gamma \vdash \beta$ iff there is a $\gamma \in A$ such that for some $\alpha_1,\ldots, \alpha_n \in \Gamma$ $(n \geq 0)$, we have $(\alpha_1 \wedge \ldots \wedge \alpha_n \wedge \gamma) \rightarrow \beta \in A$. Conceptually, this can be seen as a compromise between coordinating $\vdash$ with $\rightarrow$ and doing so with

$\supset$, for the principal connective is $\rightarrow$ while the antecedent is a conjunction rather than iterating arrows. It has some of the attractions of each of the two definitions discussed. Like the first one, it does not force us to leave the familiar world of sets on the left of the turnstile, and it satisfies strong reflexivity and monotony. Like the second one, we do not have $\{p \wedge \neg p\} \vdash q$. However, by taking γ in the definition as $q \vee \neg q$, we do have its dual $p \vdash q \vee \neg q$, which many would see as contrary to the spirit of a relevance-sensitive consequence relation.

In summary: Whether we wish to link $\vdash$ with $\rightarrow$, with $\supset$, or seek a compromise between the two, all candidate definitions or characterizations of the consequence relation in terms of the theorem-set of a relevance-sensitive logic have their limitations, whether that logic be our A, R, or another. Choosing one link over another should be guided by convenience in executing whatever project one has underway—external application or internal investigation—rather than by doctrinal considerations.

2.10 Recapitulation of Open Questions

We bring together the various open questions that are mentioned in the text and appendices.

2.10.1 *Open Problems About Acceptability*

- Are there any classical tautologies in the language of $\neg$, $\wedge$, $\vee$, $\rightarrow$ (with $\rightarrow$ read as material implication) that are not acceptable? (Sect. 2.7.3)
- Does acceptability have the letter-sharing property? (Sect. 2.7.3)
- Does it have the 'omissibility property'? (Sect. 2.5.1)
- Does it satisfy the Halldén condition? (Sect. 2.5.1)
- Does acceptability (or even direct acceptability) satisfy interpolation? (Sect. 2.5.1)
- Are all directly acceptable formulae (and hence all acceptable ones) valid in the six-element crystal model? (Sect. 2.5.2)
- Does it make a difference if the definition of parity is strengthened in the manner described in Sect. 2.4.2 and Appendix 4?
- Is acceptability decidable? (Sect. 2.7.2)
- Does the variant natural deduction system $\vee E^s$, at one point mooted by Anderson and Belnap, give the same output as acceptability? (Appendix 2)
- In Sect. 2.5.1 we saw that the proof of Observation 2.2 goes through so long at least one branch has a crash-pair satisfying parity. Can this excess capacity be exploited further?

2.10.2 Some Directions of Exploration

- What would a Gentzen-style account of acceptability look like? Is there a version satisfying cut-elimination?
- Can we characterize acceptability with a Frege–Hilbert-style axiomatization?
- Does anything new or interesting arise on the first-order level?

2.10.3 Potential Adaptations

- In Sect. 2.4.2 we saw that if one permutes the quantifier prefix in the definition of acceptability from $\forall B \exists Z \forall c$ to $\forall B \forall c \exists Z$, one obtains mingle and symmetric explosion. How interesting is this system and how close is it to RM? (see also Sect. 2.5.2, Appendix 2)
- Can one elegantly adapt the ideas of truth-trees with parity to analyze an arrow connective reflecting necessity as well as relevance-sensitivity (Appendices 1 and 5)?
- Can we adapt the same ideas to obtain interesting systems for resource-sensitive arrows?

Acknowledgements Thanks to Marcello d'Agostino, Lloyd Humberstone, Paul McNamara, Karl Schlechta and Alasdair Urquhart for comments on various drafts over several years; Michael McRobbie for kindly providing a copy of his dissertation; Tore Øgaard for checking some formulae with the programs MaGIC and Prover9; and students of LSE's PH217/419 for challenging questions in the classroom. Special thanks to Arnon Avron, whose penetrating and constructive comments as a reader for this publication led to considerable improvements.

Appendices

The appendices enlarge on matters arising in the main text. They deal with the following five matters: (1) Verification of direct acceptability for the axioms of R; (2) Comparison of our treatment of disjunction and conjunction with that of familiar natural deduction systems; (3) Earlier attempts to develop tree/tableau procedures for relevance logic, notably by Belnap & McRobbie; (4) Some properties of acceptability when restricted to the language of negation and arrow; (5) Experience teaching the material to students.

Appendix 1: Direct Acceptability for Axioms of R

In this appendix we verify the acceptability of the axiom schemes of the relevance logic R, noted in Observation 2.3, with comments on those axioms as we go. The axiomatization considered is the standard one found in Anderson et al. (1992), page xxiv (also in Mares 2012, Appendix A), using the 'basic' connectives $\neg, \wedge, \vee, \rightarrow$.

Some formulations of R in the literature also contain auxiliary primitives, notably non-classical two-place connectives of 'fusion' and 'fission' (known in the linear logic literature as multiplicative conjunction and disjunction) and/or a zero-ary connective (propositional constant) t, in an effort to smooth some of the wrinkles in the Routley–Meyer semantics. The present account in terms of acceptability has no need for auxiliary connectives: we treat fusion and fission of ϕ with ψ as no more than abbreviations for $\neg(\phi \rightarrow \neg\psi)$ and $\neg\phi \rightarrow \psi$ respectively, and we dispense entirely with the constant t which, in the view of the author (and, e.g., of Avron et al. 2018, Sect. 11.1.1) is contrary to the spirit of relevance logic.

For the first-degree axiom schemes of R, that is, those of the form $\alpha \rightarrow \beta$ where neither antecedent or consequent contains arrows, it suffices to take a classical truth-tree with root r: $\neg(\alpha \rightarrow \beta)$, apply counter-case to get $\alpha, \neg\beta$ then use the rules for the classical connectives $\neg, \wedge, \vee$ only, finally checking by inspection that every branch satisfies parity. That covers axiom schemes 1 (identity) $\alpha \rightarrow \alpha$; 5 ($\wedge$-elimination) $(\alpha \wedge \beta) \rightarrow \alpha, (\alpha \wedge \beta) \rightarrow \beta$; 6 ($\vee$-introduction) $\alpha \rightarrow (\alpha \vee \beta), \beta \rightarrow (\alpha \vee \beta)$; 9 (distribution) $(\alpha \wedge (\beta \vee \gamma)) \rightarrow ((\alpha \wedge \beta) \vee (\alpha \wedge \gamma))$; and 11 (double negation elimination) $\neg\neg\alpha \rightarrow \alpha$.

Of these first-degree schemes, the only one whose tree we write out in full is distribution. It deserves explicit attention because it fails under some other approaches to relevance logic (notably that of McRobbie and Belnap 1984, see Appendix 3) as well as causing headaches for the standard natural deduction approach (as discussed in Appendix 2). The tree for distribution in Fig. 2.9 is a familiar one. It has four branches, each with its crash-pair. There are just two critical nodes, and they are common to all four branches. Both critical nodes are in the trace of each of the four crash-pairs, so parity is satisfied.

The remaining schemes of R are of higher degree. For each of them we exhibit a directly acceptable truth-tree, with comments.

Scheme 2 (suffixing): $(\alpha \rightarrow \beta) \rightarrow ((\beta \rightarrow \gamma) \rightarrow (\alpha \rightarrow \gamma))$. This may be understood as an 'exported' or 'unpacked' (and hence strengthened) form of transitivity $((\alpha \rightarrow \beta) \wedge (\beta \rightarrow \gamma)) \rightarrow (\alpha \rightarrow \gamma)$ for the arrow. The unique branch contains a single crash-pair, with all six critical nodes in its trace, so parity is satisfied. The linearity of the tree is a consequence of the fact that the only connectives involved are $\rightarrow, \neg$, (cf. Appendix 4).

We can use Fig. 2.10 to illustrate a point made in Sect. 2.4.1. Substituting α for β, γ gives $(\alpha \rightarrow \alpha) \rightarrow ((\alpha \rightarrow \alpha) \rightarrow (\alpha \rightarrow \alpha))$, with an acceptable tree likewise obtainable by the same substitution in the tree. Now, for suffixing, there is no ambiguity about how each node is obtained. For example, n_8: γ is obtained from n_7: β and n_3: $\beta \rightarrow \gamma$ by modus ponens. But in the tree for the substitution instance, n_8: α may be

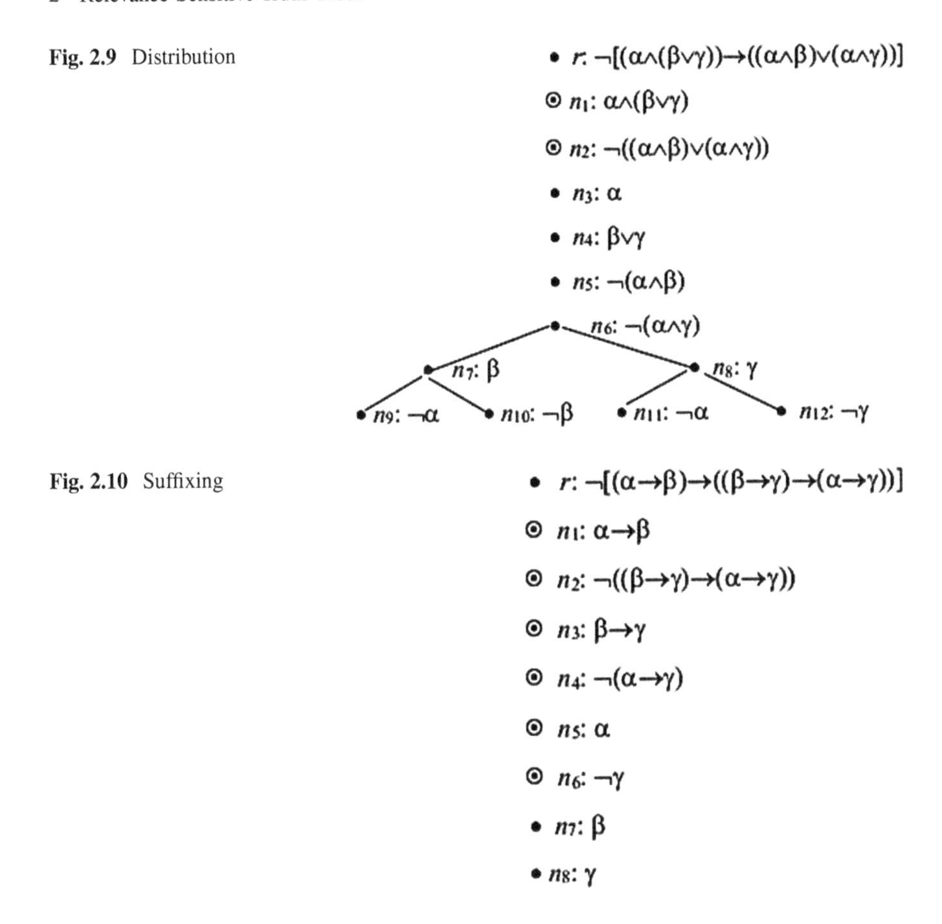

Fig. 2.9 Distribution

Fig. 2.10 Suffixing

obtained in various ways—from n_7: α and n_3: $\alpha \to \alpha$ as before, but also from, say, n_5: α and n_1: $\alpha \to \alpha$. The former pattern of justification satisfies parity just as it did in the tree for suffixing, but the latter does not: the critical node n_3: $\alpha \to \alpha$ is no longer in the trace of the designated crash-pair $\{n_6{:}\ \neg\,\alpha,\ n_8{:}\ \alpha\}$although its partner n_4: $\neg\,(\alpha \to \alpha)$ is. So, in this example, the identification of trace, satisfaction of parity, and status of the tree as directly acceptable or not, all depend on its justificational pattern, going beyond its bare tree structure, the formulae attached to nodes and the choice of crash-pairs.

Scheme 3 (assertion): $\alpha \to ((\alpha \to \beta) \to \beta)$. As for suffixing (above) and contraction (below), its only connectives are $\neg$, $\to$ so it has only one branch. In the directly acceptable tree of Fig. 2.11, the crash-pair $\{n_4,\ n_5\}$has all four critical nodes in its trace so parity is satisfied.

Assertion may be seen as an 'exported' or 'unpacked' version of the formula $(\alpha \wedge (\alpha \to \beta)) \to \beta$ expressing modus ponens (see Sect. 2.4.1). Another way of understanding it intuitively is as the result of permuting the antecedents of the trivial $(\alpha \to \beta) \to (\alpha \to \beta)$. It is the only axiom from the list that is unprovable in the

Fig. 2.11 Assertion

- r: $\neg[\alpha \to ((\alpha \to \beta) \to \beta)]$
- ⊙ n_1: α
- ⊙ n_2: $\neg((\alpha \to \beta) \to \beta)$
- ⊙ n_3: $\alpha \to \beta$
- ⊙ n_4: $\neg\beta$
- n_5: β

Fig. 2.12 Contraction

- r: $\neg[(\alpha \to (\alpha \to \beta)) \to (\alpha \to \beta)]$
- ⊙ n_1: $\alpha \to (\alpha \to \beta)$
- ⊙ n_2: $\neg(\alpha \to \beta)$
- ⊙ n_3: α
- ⊙ n_4: $\neg\beta$
- n_5: $\alpha \to \beta$
- n_6: β

weaker axiom systems NR and E that seek to embody a composite requirement of relevance-and-necessity for the arrow (see, e.g., Sect. 28.1 of Anderson and Belnap 1975, Sect. 2.4 of Mares 2012). It is also the only one that is of the form $\phi \to \psi$ where ϕ contains no arrows but ψ has arrow as principal connective.

Scheme 4 (contraction): $(\alpha \to (\alpha \to \beta)) \to (\alpha \to \beta)$. The crash-pair $\{n_4, n_6\}$has all four critical nodes in its trace, so parity is satisfied (Fig. 2.12).

An interesting feature of this tree, reflecting repetition of α in the antecedent of the scheme itself, is that node n_3: α is used twice: first to get n_5: $\alpha \to \beta$ and then to get n_6: β. Another feature is that the unique branch contains a second crash-pair $\{n_2, n_5\}$which, however, does not satisfy the parity condition since the critical node n_4 is not in its trace although its partner n_3 is.

Scheme 7 ($\wedge$-introduction, $\wedge^+$): $\{(\alpha \to \beta) \wedge (\alpha \to \gamma)\} \to \{\alpha \to (\beta \wedge \gamma)\}$. Two branches each with its crash-pair, every critical node in the trace of each crash-pair.

This is a second-degree scheme, where degree measures the maximum embedding of arrows within arrows, defined recursively in the obvious way. If we consider its unpacked third-degree version $(\alpha \to \beta) \to ((\alpha \to \gamma) \to (\alpha \to (\beta \wedge \gamma))$, we find that while both branches in its truth-tree still crash, they do so without parity. The same happens for scheme 8 ($\vee$-elimination) below. This illustrates the well-known difference of power, for A as for many relevance-sensitive logics, between the classically equivalent formulae $(\phi \wedge \psi) \to \theta$ and $\phi \to (\psi \to \theta)$ (Fig. 2.13).

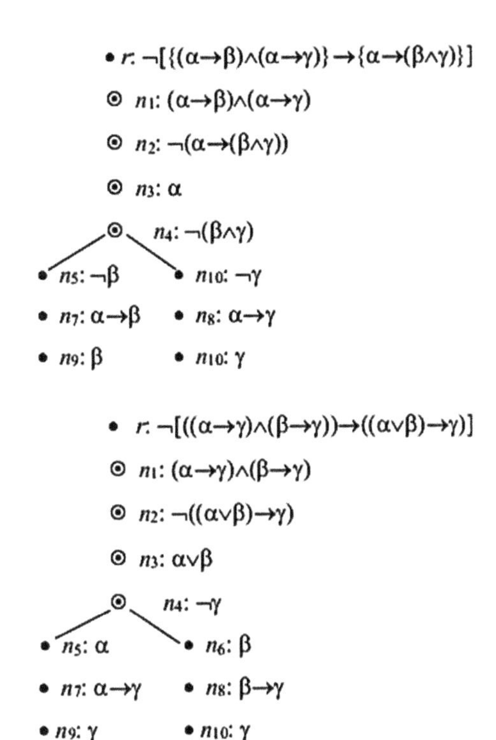

Fig. 2.13 $\wedge$-introduction

Fig. 2.14 $\vee$-elimination

From our perspective, two factors appear to lie behind this difference of power. On the one hand, decomposing $\neg\,(\phi \rightarrow (\psi \rightarrow \theta))$ gives rise to two applications of counter-case, thus two critical pairs risking failure of parity, while $\neg\,((\phi \wedge \psi) \rightarrow \theta)$ produces only one critical pair. But that is not the whole story for, while the third-degree versions of $\wedge$-introduction and $\vee$-elimination are not directly acceptable (and, apparently, unacceptable), we have already seen that there are other unpacked third-degree schemes, namely, suffixing and assertion, that are directly acceptable just like their packed second-degree counterparts. The reason for this contrast seems to be that, for each of $\wedge+$, $\vee+$, decomposition creates a fork with a critical node used on one branch while its partner is used on the other, while for suffixing and assertion there are no forks to create such problems.

This example illustrates the way in which the decompositional approach can help explain similarities and differences between formulae that otherwise can be difficult to understand. Explanations can also be given in terms of relevance-sensitive natural deduction, but less transparently.

Scheme 8 ($\vee$-elimination): $((\alpha \rightarrow \gamma) \wedge (\beta \rightarrow \gamma)) \rightarrow ((\alpha \vee \beta) \rightarrow \gamma)$. A directly acceptable tree for it is given in Fig. 2.14. The same comments may be made as for $\wedge$-introduction.

Scheme 10 (contraposition): $(\alpha \rightarrow \neg\beta) \rightarrow (\beta \rightarrow \neg\alpha)$. This is the form used in the standard axiomatization of R, with other familiar forms of contraposition deriv-

Fig. 2.15 Contraposition

- r: $\neg[(\alpha \rightarrow \neg\beta) \rightarrow (\beta \rightarrow \neg\alpha)]$
- ⊙ n_1: $\alpha \rightarrow \neg\beta$
- ⊙ n_2: $\neg(\beta \rightarrow \neg\alpha)$
- ⊙ n_3: β
- ⊙ n_4: $\neg\neg\alpha$
- n_5: α
- n_6: $\neg\beta$

able there as theorems. In the tree of Fig. 2.15, the unique branch has crash-pair $\{n_3, n_6\}$ and each of the four critical nodes is in its trace.

Appendix 2: Disjunction, Conjunction

This appendix compares the treatment of disjunction and conjunction in our truth-trees-with-parity, with the way they are handled in the usual natural deduction system for the relevance logic R. We assume some familiarity with the latter; for background see Anderson and Belnap (1975) or, for a textbook presentation, part 3 of Loveland et al. (2014).

It is convenient to begin with disjunction. Consider the formula $(p \rightarrow (q \vee (p \rightarrow q))) \rightarrow (p \rightarrow q)$ (skewed cases) which, as noted in Observation 2.5, is directly acceptable but not in R. We recall how an attempt to derive it by natural deduction for R fails. From the suppositions $p \rightarrow (q \vee (p \rightarrow q))$ and p one infers $q \vee (p \rightarrow q)$; one then makes two sub-proofs, the first supposing q to obtain q *ipso facto*, the second supposing $p \rightarrow q$ and applying modus ponens to that with p, to get q again. Thus one of the sub-derivations appeals to the supposition p while the other does not, violating the special proviso on $\vee^-$ in natural deduction for R, that these sub-derivations must appeal to exactly the same suppositions (other than the two disjuncts themselves).

In contrast, while the truth-tree for $(p \rightarrow (q \vee (p \rightarrow q))) \rightarrow (p \rightarrow q)$ forks at its node labeled $q \vee (p \rightarrow q)$, the parity constraint acts on each branch separately without comparison with the other branch and, as we saw in Observation 2.5, it is satisfied. We might say, roughly, that the parity condition is more generous toward disjunctive reasoning than is the 'same suppositions' constraint.

But, one may ask, isn't it *too* generous? Can't we build acceptable trees for mangle $\alpha \rightarrow (\beta \rightarrow \alpha)$ and its instance mingle $\alpha \rightarrow (\alpha \rightarrow \alpha)$ by simulating the notorious trick, legitimate in classical natural deduction, that the 'same suppositions' proviso of relevant natural deduction was designed to block? Recall that classically, for mangle, one may first suppose α, then suppose β, use $\vee^+$ on the former to get $\alpha \vee (\beta \rightarrow \alpha)$, then carry out sub-derivations with the two disjuncts as suppositions, that both get α,

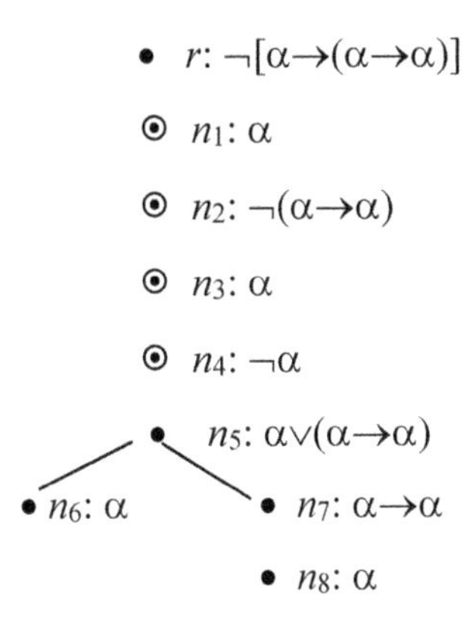

Fig. 2.16 Mingle: a devious tree failing parity

discharge those two suppositions by $\vee^-$, and finally apply conditional proof twice. For mingle, one does the same with β instantiated to α throughout. In relevantized natural deduction, the derivation is blocked by the 'same suppositions' proviso on $\vee^-$ mentioned above, but the question is: can't we imitate the classical procedure in an acceptable truth-tree?

For direct acceptability, the answer is simple: we cannot transcribe the application of $\vee^+$ since our trees can only decompose, never compose. No disjunctions can be introduced into the direct decomposition trees for mangle or mingle, the trees do not fork, and the unique branch, exhibited in Fig. 2.5 of Sect. 2.5.2, fails parity.

For the more general notion of acceptability, the answer is a little more complex. With the recursive rule available, we can indeed simulate the natural deduction step $\vee^+$, but the crash-pair in one of the ensuing branches fails parity. In detail, the truth-tree in Fig. 2.16 for mingle is the same, up to node n_4, as the one in Fig. 2.5 that failed for direct acceptability.

Turning now to conjunction, we can say that it too is treated more generously by truth-trees-with-parity than by the standard natural deduction system for R. For, on the one hand as we saw in Appendix 1, a familiar classical truth-tree for the distribution principle $(\alpha \wedge (\beta \vee \gamma)) \rightarrow ((\alpha \wedge \beta) \vee (\alpha \wedge \gamma))$ happily satisfies parity. On the other hand, notoriously, distribution faces a difficulty in the natural deduction system for R because of a constraint that the system places on $\wedge^+$, which we briefly recall.

Just as unbridled $\vee^-$ can be used to 'cheat' its way around the 'actual use' constraint on conditional proof to establish mangle, so too can unrestrained $\wedge^+$. One may first suppose α, then suppose β, use $\wedge^+$ on these to get $\alpha \wedge \beta$, then $\wedge^-$ back to α and finally apply conditional proof twice. The supposition β is used in the $\wedge^+/\wedge^-$ detour and so the 'actual use' constraint on CP is satisfied. The trick cannot be simulated in our truth-trees. For direct acceptability, all rules decompose so that $\wedge^+$, like $\vee^+$, has no role. For indirect acceptability, the recursive rule takes only one node as input, never two (see the discussion in Sect. 2.7.2) so that $\wedge^+$ is not available.

To block this $\wedge^+/\wedge^-$ 'funny business', Anderson & Belnap introduced a 'same suppositions' proviso on $\wedge^+$ echoing the one on $\vee^-$: one can infer a conjunction from its two conjuncts only if the two conjuncts depend on exactly the same suppositions. However, this proviso has the side effect of also blocking derivations of distribution, which is recuperated by postulating it as a primitive rule (and as an axiom in the Hilbertian presentation of R, as seen in Appendix 1). This rather ad hoc move has long been a source of unease; as remarked by D'Agostino et al. (1999), page 416, 'the integration of this axiom into the proof-theory of R has always been a source of considerable difficulty'.

It is seldom noted that, in response to the problem, Anderson and Belnap (1975) (Sect. 27.2, page 348) mooted a modification of their definition of dependence in natural deduction. The essential idea behind the change is that when, in a derivation, one reaches $\phi \vee \psi$ and creates auxiliary derivations headed, respectively, by ϕ, ψ, those two formulae are not given fresh dependency labels but are taken to depend on the same suppositions as did $\phi \vee \psi$. This has the effect of increasing the set of earlier suppositions that are treated as being used in each of its two auxiliary derivations, thus softening the bite of the provisos on $\vee^-$ and $\wedge^+$, whose formulations are left unchanged. Essentially the same proposal has been made by Urquhart (1989) (see also the brief remark in his Urquhart 2016), Dunn and Restall (2002), Brady (2006).

Anderson & Belnap observe that the suggested change rehabilitates the classical derivation of distribution, by allowing its applications of $\wedge^+$ in its two subordinate derivations. They also note that it yields the formula $\{(p \rightarrow (q \vee r)) \wedge (q \rightarrow s)\} \rightarrow (p \rightarrow (s \vee r))$, which we mentioned in Sect. 2.6 as being one of several that are directly acceptable but not in R. Nevertheless, not having at hand a matrix-like M_0 (see Sect. 2.6) to validate all formulae that the change renders derivable while still invalidating explosion, they seem to have feared that it might yield too much and refrained from explicitly recommending it, unlike Urquhart (1989), page 169, who greeted it enthusiastically.

Although quite technical, the mooted revision raises two terminological issues, each with an interesting conceptual resonance. For Anderson & Belnap, the formulae ϕ, ψ heading the subordinate derivations are still suppositions (in their wording, hypotheses) although of a rather ghostly kind; on the other hand, Urquhart (1989) prefers not to see them as suppositions (in his wording, assumptions) at all, but as marking a split of the argument into two parts (we might say, into two 'cases'). Again, Anderson & Belnap saw the revision as replacing one rule for disjunction elimination (in their notation, $\vee$E, where 'E' is for elimination) by another ($\vee$E^s, where 's' is for strong); for the present author, it is perhaps more transparent to see the move as keeping the same rule but with a new definition of dependence that weakens the impact of its proviso.

The suggested revision does not quite correspond to what is going on in our truth-trees, because the proviso on disjunctive roof in the natural deduction system continues to compare its subordinate derivations, whereas the parity condition is *internal to each branch* of a truth-tree, without comparing them. But the output does come closer to acceptability, at least in so far as the status of distribution is concerned. Perhaps the two outputs coincide, although the author suspects that there are still some subtle differences.

Appendix 3: Earlier Work on Truth-Trees for Relevance Logic

Perhaps the first attempt at articulating truth-trees (aka semantic tableaux) for relevance logics is contained in Sects. 6 and 7 of Routley (2018), a manuscript that was circulated privately to Meyer and some others in 1970/71, and not published until 2018.

In 1976 Dunn (1976) gave what appears to be the first published construction. Using pairs of trees, it works beautifully—but it covers only first-degree conditionals, i.e., formulae of the form $\phi \rightarrow \psi$ where ϕ, ψ contain no arrows, and has resisted attempts to extend coverage much further.

The most intensive work on the topic was carried out by in his doctoral thesis McRobbie (McRobbie, 1979) (with parts in the abstracts McRobbie 1977, McRobbie and Belnap 1977), followed by the papers of McRobbie and Belnap (1979, 1984). Much of this is brought together in Anderson et al. (1992), Sect. 60; there is also a very clear exposition, with insightful discussion, in D'Agostino et al. (1999), Sect. 2.

Because of his limited coverage, Dunn (1976) did not consider the decomposition of unnegated arrows. They are treated by Routley, and by McRobbie & Belnap in the texts mentioned above; however, they are not decomposed by modus ponens, but by implicative forking: the tree forks into the consequent and the negation of the antecedent. These authors decompose negated arrows by the counter-case rule; but there is no recognition of the suppositional nature of its outputs when we leave the classical context for one that is sensitive to relevance. None of the authors bring recursion into the decomposition procedure.

Examining the publications of McRobbie and Belnap in more detail, we can note the following features.

- The language of McRobbie's 1977 abstract lacks negation, covering only the 'positive' connectives $\rightarrow$, $\wedge$, $\vee$, $\circ$ (fusion) and a truth-constant t.
- In McRobbie and Belnap (1977, 1979), negation is present, but the language is still severely restricted, since arrow is the only other connective allowed. Of course, as remarked at the beginning of Appendix 1, intensional versions of conjunction and disjunction (aka fusion and fission) are definable from arrow with negation, but their ordinary (extensional) counterparts are not, so the restriction is a serious one. Despite the absence of ordinary conjunction and disjunction, the trees of McRobbie & Belnap can still have multiple branches, since unnegated arrows are decomposed by forking. The trees are subject to a dependency condition, requiring that every node is in the trace of some crash-pair on some branch. This contrasts with our parity constraint in several respects, highlighted by italics in the following statement of the latter: *every* branch contains *some* crash-pair such that for every *critical* node in the tree, *if* it is in the trace of the crash-pair *then* so is its partner.
- Ordinary conjunction and disjunction are finally tackled by McRobbie and Belnap (1984). Standard classical decomposition rules for those connectives are accompanied by a rather cumbersome procedure of copying unused items from above a fork into each of the branches issuing from the fork. We do not give the details of the copying procedure; apart from the paper itself, there is a clear account in

D'Agostino et al. (1999), Sect. 2.1 pp. 414–416. The important point to note is that the system goes into overkill, failing to validate distribution of $\wedge$ over $\vee$, just as did Anderson & Belnap's system of natural deduction (see Appendix 2). In both contexts, repair can be carried out by hacking—just add distribution as an additional decomposition or deduction rule—but the patch is hardly satisfying.

From our perspective, the difficulties faced by McRobbie and Belnap have two main sources. One is in the decomposition of negated arrows, where there is no recognition of the special role, and suppositional status, of *critical nodes*, thus blocking articulation of a notion of branches *crashing with parity*. The other source is in the decomposition of unnegated arrows, where the authors stick with the classical step of *implicative forking*, rather than modus ponens.

Both these features appear to stem from a failure to take sufficiently seriously, in the context of truth-trees, the different logical powers of $\phi \rightarrow \psi$ and $\neg \phi \vee \psi$. For negated arrows, counter-case does not track a relevantly admissible inference, but rather sets up a pair of wlog suppositions within the framework of an overall proof by reductio ad absurdum. For unnegated arrows, decomposition by implicative forking misses out on part of the force of the input.

Why were these obstacles not overcome in the years immediately following 1984? The author suspects that work on them was cut short by a tsunami in the relevance logic community, brought about by the Routley–Meyer possible-worlds semantics. Briefly suggested in the manuscript of Routley (2018) of 1970/1 and immediately developed by Routley and Meyer (see Bimbó et al. 2018) that semantics provided an exciting new technique to explore. Work on truth-trees for relevance logic went backstage, to appear from time to time in a minor role. Inspired by Kripke's use of tableaux in his completeness proofs for modal logic in the 1960s, the following decades saw some attempts to render versions of the Routley–Meyer semantics for substructural logics computationally manageable by expressing some of the machinery of the models in terms of semantic tableaux; see, for example, papers of Pabion (1979), Bloesch (1993), Priest (2008).

The Routley–Meyer semantics also created an unfortunate methodological orientation. Its rules for evaluating formulae in a given world gave the impression that for relevance-sensitive logic not only the arrow but also negation, and perhaps even conjunction and disjunction, need to be treated non-classically. For negation, this vision was also encouraged by Dunn's 1976 work on truth-tree pairs for first-degree arrow formulae, where the connective is treated in a four-valued way; while for conjunction and disjunction it was comforted by the 'anti-cheating' constraints placed on rules for those connectives in Anderson & Belnap's natural deduction system. The general perspective was expressed in an influential textbook on modal logic by Hughes and Cresswell (1996), where it is said in passing that '... in fact relevance logics differ from all the logics we have so far considered in that they require a non-standard interpretation of the PC symbols, in particular of negation' (page 205).

Appendix 4: Some Properties of the $\neg$, $\rightarrow$ Fragment

In this appendix we establish two desirable properties that hold for acceptability when it is defined on the $\neg$, $\rightarrow$ fragment of our $\neg$, $\wedge$, $\vee$, $\rightarrow$ language. Note that we are not just looking at those $\neg$, $\rightarrow$ formulae that are acceptable in the original sense; such formulae may have acceptable trees in which the recursive rule is applied to introduce the connectives $\wedge$, $\vee$. We are considering a situation in which the recursive rule is also limited to the $\neg$, $\rightarrow$ language. The arguments below go through essentially because in that language all truth-trees have a single branch, without forks.

The first result concerns the notion of parity. Recall from Sect. 2.4.1 that its definition requires that for every branch B and every critical node c on B, *if* c *is in the trace of* Z_B *then its partner is also on* B *and in the trace of* Z_B. This prompts the question, raised in Sect. 2.4.2, whether it makes a difference if we impose the apparently stronger condition that for every branch B and every critical node c on B, *both it and its partner* are in the trace of Z_B.

Observation 2.9 For the language of $\neg$, $\rightarrow$: whenever a formula has an acceptable truth-tree then it has one in which its unique branch satisfies the strengthened parity condition.

Outline of Proof Let α be a formula in that language and let T be a truth-tree with root $r\colon \neg\,\alpha$ that is acceptable (with the recursive rule also restricted to that language). Then T has a unique branch B that may be identified with T itself. Form $T^* = B^*$ by deleting all nodes in B that are not in the trace of the designated crash-pair Z_B. Then T^* is also a single branch decomposition tree, with the same root as T and the same designated crash-pair $Z_{B^*} = Z_B$. It remains to check that T^* satisfies the stronger version of parity. Suppose c is a critical node on B^*. Then c was already on B and was in the trace of Z_B. So, by the original definition of parity, applied to T, the partner c' of c is also on B and is in the trace of Z_B. Hence c' is on B^* where it is in the trace of Z_{B^*}. □

The author has not been able to settle the question whether Observation 2.9 continues to hold for the full language using $\neg$, $\wedge$, $\vee$, $\rightarrow$. In that context, a tree may have more than one branch and the operation, for a given branch B, of simply deleting all nodes that are not in the trace of Z_B, can create havoc, as may be appreciated by considering the example of distribution Appendix 1, Fig. 2.9. If we apply that deletion to, say, the leftmost branch of that figure, then we eliminate the nodes n_4, n_6 which produced forks, thus marooning nodes on the other side of each fork. Deleting the marooned nodes as well does not get one out of the difficulty, as it can destroy needed crash-pairs. In general, it does not seem possible to transform every acceptable tree into one in which every node on an arbitrary branch is in the trace of its designated crash-pair.

The second result of this appendix is that in the restricted language, acceptability coincides with being a theorem of R.

Observation 2.10 For the language of $\neg$, $\rightarrow$: A formula is acceptable iff it is a theorem of R.

Proof Sketch For the right-to-left direction, it is known that when a $(\neg, \rightarrow)$-formula is a theorem of R, then it has a derivation using only axioms that are $(\neg, \rightarrow)$-formulae with detachment the sole derivation rule (Anderson and Belnap 1975, Sect. 28.3.2, page 375). Those axioms continue to be directly acceptable in the restricted language (with the same verifications as in Observation 2.3), as does closure of acceptability under detachment (with the same verification as in Observation 2.7).

Left-to-right: Let α be a formula in the language of $\neg$, $\rightarrow$ with an acceptable truth-tree (where the recursive rule is restricted to such formulae), with root $r\colon \neg\, \alpha$. It has a unique branch B. We need to work delicately with *sets* X of nodes and *multisets* Γ of formulae.

Call a finite multiset Γ of formulae, with elements $\alpha_1,\ldots\alpha_n$, R-*inconsistent* iff $!(\Gamma) \in$ R, where $!(\Gamma) = \alpha_1 \rightarrow (\alpha_2 \rightarrow (\ldots(\alpha_{n-1} \rightarrow \neg\alpha_n)\ldots)$; the order of the α_i does not matter, since all permutations (with an appropriate contraposition when the last item is moved) are equivalent in R. Clearly, for the designated crash-pair $\{z : \zeta, z' : \neg\zeta\}$ of B, the formula $!\{\zeta, \neg\zeta\}$ is in R. Also, R satisfies each of the following closure conditions corresponding to decomposition rules:

When $\phi \rightarrow !(\Gamma) \in$ R then $\neg\neg\, \phi \rightarrow !(\Gamma) \in$ R (for double negation elimination)
When $\psi \rightarrow !(\Gamma) \in$ R then $\phi \rightarrow ((\phi \rightarrow \psi) \rightarrow !(\Gamma)) \in$ R (for modus ponens)
When $\phi \rightarrow (\neg\psi \rightarrow !(\Gamma) \in$ R then $\neg(\phi \rightarrow \psi) \rightarrow !(\Gamma) \in$ R (for counter-case)
When $\psi \rightarrow !(\Gamma) \in$ R and $\phi \rightarrow \psi \in$ R then $\phi \rightarrow !(\Gamma) \in$ R (for the recursive rule).

Using these closure conditions, one can follow R-inconsistency backwards from the crash-pair to sets X of nodes in its trace that are ever-closer to r, as measured by the mean of the set of integer distances of the elements of X from r. Note that to activate the closure condition corresponding to counter-case, the critical nodes ϕ, $\neg\psi$ must *both* be in the trace of the crash-pair, as assured by strengthened parity using Observation 2.9. This progression stops with $X = \{r\colon \neg\, \alpha\}$, so $\neg\neg\, \alpha \in$ R and thus $\alpha \in$ R as desired. □

Corollary 2.10.1 *Acceptability for the* $\neg$, $\rightarrow$ *language is non-trivial (in the sense defined in Sect. 2.7) and satisfies letter-sharing.*

Proof By Observation 2.10, since R has those properties. □

Appendix 5: Pedagogical Remarks

Teaching this material to students of philosophy at LSE in 2017 through 2019 has provided the author with some classroom experience, shared in this appendix.

Honesty toward the students requires one to begin by frankly reviewing doubts about the whole enterprise of relevance logic (cf., e.g., Veltman 1985, Sect. I.2.1.4, Burgess 2009, Chap. 5, Makinson 2014, Sect. 6). There are three main grounds for scepticism.

In general terms, one should not confuse the *logical* question of whether an inference is valid, with the *pragmatic* one of whether it is wise to insist on pursuing the process rather than, say, going back to assumptions to revise or abandon some among them. Perhaps the very desire for relevance-sensitivity stems from the pragmatic concern rather than the logical one.

Second, the relevance or irrelevance of one proposition to another is largely a question of *subject matter*. As observed by van Benthem (1983), in ordinary discourse relevance is typically supplied by some background of assumptions that are taken for granted. Suppose, for example, that you are told: 'If your cat has white, ginger, and black fur, then you can soon expect kittens'. Without any background information, the antecedent may appear quite irrelevant to the consequent. But given the information that, with rare exceptions, only female cats can have fur of three colors, plus the more widely known fact that, so long as they have not been neutered, female cats are usually quite fertile, there is indeed a substantive connection between the two parts of the conditional. Given this dependence on whatever subject matter happens to be in the speaker's implicit assumptions, attempts to specify formal criteria for the truth of a relevant conditional may face major difficulties, and they will be inherited by any relevance-sensitive logic that is defined in terms of truth.

Finally, the history of the subject has revealed serious shortcomings in each of the main approaches in the literature—whether inelegance in formal implementation of a promising idea (the natural deduction account), poorly motivated and heteroclite axiom sets (the Hilbertian axiomatic approach) or difficulties in finding convincing intuitive motivation for formal devices (such as for the constraints imposed on three-place relations in the Meyer–Routley semantics). The area has acquired the reputation of a failed research project.

Students should also be warned that the subject has served as a trampoline for some quite startling positions in the philosophy of logic. But they can also be reassured that one may perfectly well be interested in relevance-sensitive logics as objects of study, without preaching that any of them should sweep classical logic aside, nor taking seriously doctrines of the existence of true contradictions, subsisting impossible worlds and so on, associated with the philosophy known as dialetheism. One may thus, with a free conscience, make unconstrained use of the resources of classical logic and pertinent parts of mathematics when reasoning about relevance-sensitivity, without constantly looking over one's shoulder.

It is important that, from the beginning, students distinguish the requirement of relevance from the composite one of relevance-and-necessity. Following Descartes' dictum, the present author's personal view is that the composite condition should be broken down into its two parts, to be investigated separately until each is perfectly clear. Only then should their combination be attempted.

Optimally, to be able to appreciate what is going on, students should already have been exposed to truth-trees for classical propositional logic. The tweaks needed to incorporate a control for relevance then become understandable and even natural. Those who have spent time on natural deduction will need frequent reminding that the rules for direct acceptability always proceed by decomposition, never by composition—for example, we are not permitted to proceed from nodes labeled ϕ

and ψ to one labeled $\phi \wedge \psi$, nor from one labeled ϕ to another for $\phi \vee \psi$. The only decomposition rule with more than one input formula is modus ponens; even for the recursive rule, we allow only one input, never more (Sect. 2.7.2, Fig. 2.8).

Once students have understood the limitations of direct acceptability and have been introduced to the recursive decomposition rule, another admission should frankly be made: acceptability currently faces the open problem of letter-sharing, for which a negative answer could well spell ruin. The author's experience is that students appreciate being brought on stage in this way and are agreeably surprised to learn that there are still formal questions about classical propositional logic that have not been resolved.

In the classroom, the material is best presented through an analysis of examples, with definitions articulated formally only after their illustration in specific instances; this policy also guides the presentation in Chap. 11 of the textbook Makinson (2020). At the same time, it is helpful to encourage an ability to 'smell' examples—to articulate intuitions about them and conjecture whether they are acceptable before beginning the formal work of checking. At the beginning, students throw up their hands in bewilderment, unable to express suspicions or place confidence in guesses but, with sufficient practice, they can develop a fine-tuned sensitivity just as for natural deductions in classical logic and the undecidable system R (cf. remarks of Anderson and Belnap 1975, Sect. 28.1, page 350).

In any particular example, there are quite different two jobs to be done: construct a candidate tree and check it for parity. What is the best way of scheduling the two tasks? Two basic options arise. One can annotate the tree as one builds it, in such a way that satisfaction or failure of parity can be read directly from the completed tree; or one can build the tree without worrying too much about parity and then check when the construction is finished.

The construct-then-check order frees up the mind and so is less boring. The main job of the teacher may be to stop students from then doing the check by merely eyeballing the trees they have constructed; it should be carried out systematically. An algorithm that corresponds closely to the definition of parity iterates the following steps. Choose a branch and follow the trace of its designated crash-pair, from the crash-pair itself upwards toward the root, marking the nodes that are in the trace. When that has been done, inspect the critical nodes on that branch to see whether any one of them has a mark while its partner does not. If so, parity fails, else it succeeds for that branch and we can go on to check for the next branch. Evidently, care is needed in that transition, since a node that is common to two branches may be in the trace of the crash-pair of one of them but not in that of the other. When there are only two branches in the tree, one can place the marks recording the trace of the crash-pair on the left branch to the left of the nodes, and those for the right branch to the right. But when there are more than two branches, one needs either to erase marks made for previous branches or to use more elaborate annotations.

The check-as-you-go order of verification also involves marking nodes, but in reverse direction as one builds the tree from its root. To reduce clutter, the following 'minimalist' labeling can be used. Whenever one applies the counter-case rule to a node $n\colon \neg(\phi \rightarrow \psi)$, take a natural number i (say, the first one not yet used) and label the output nodes m: ϕ and m': $\neg\ \psi$ by, say, i_a, i_c where 'a' is for antecedent and 'c' is for consequent (don't bother to annotate the input node n). Propagate those labels through all of the single-input decomposition rules (including the recursive rule and further applications of counter-case). For the sole double-input rule, namely, modus ponens, going from m: ϕ, m': $\phi \rightarrow \psi$ to n: ψ, consider the union $L \cup L'$ of the label sets L, L' on m, m', respectively. One could propagate all labels in $L \cup L'$ to label the output node, but it is (equivalent and) more parsimonious to proceed as follows: (i) for each i, if both marks i_a, i_c are in $L \cup L'$ then neither of them goes into the label for the output node n; (ii) otherwise whichever, if any, of i_a, i_c is in $L \cup L'$ is put into that label. To check the completed tree for parity, one inspects in turn each designated crash-pair $\{z_1\colon \zeta, z_2\colon \neg\ \zeta\}$with label sets L_1, L_2. If $L_1 \cup L_2$ contains i_a but not i_c, or conversely, for some integer i, then parity fails; else it succeeds for that branch and we can pass to the next one. Evidently, this downwards-propagation labeling can also be effected *after* having completed the tree, as an alternative to the upwards-moving 'tracing the trace'.

To summarize, we have two work schedules: construct-then-check and check-as-you-go. The latter is carried out by labeling in the direction from root to crash-pairs while the former can be done either in the same way, but after tree construction has finished, or by working backwards from crash-pairs. The author hesitates to recommend one of these options over another but feels that the check-as-you-go procedure can cramp the student's style and more easily become centered on boring bookkeeping. It would be nice to have a software in which the user can write a candidate tree, leaving it to the computer to then check it for parity.

To the author's surprise, students did not appear to have much difficulty handling the triple quantification $\forall B \exists Z \forall c$ in the definition of parity (and thus the quadruple quantification $\exists T \forall B \exists Z \forall c$ in the definition of an acceptable formula). This is perhaps because once the idea is understood it becomes quite natural, especially when expressed using an implicit choice function $B \mapsto Z_B$.

When looking at specific formula, it was often convenient to make use of the fact that although modus tollens and implicative forking are not allowed as decomposition rules in the construction of directly acceptable truth-trees, they become admissible when the recursive rule is available (Observation 2.7). Modus tollens is handy when verifying the acceptability of a scheme that rather neatly reflects the roles of modus ponens and counter-case in decomposition, namely $(\neg(\phi \rightarrow \psi) \rightarrow \theta) \leftrightarrow (\phi \rightarrow (\neg\psi \rightarrow \theta))$ where the biconditional abbreviates the conjunction of an arrow and its converse. Implicative forking gives a straightforward way of establishing the acceptability of formulae of the kind $\neg\ (\phi \rightarrow \psi)$, where ϕ, $\neg\ \psi$ are both acceptable. Contraposing $\phi \rightarrow \psi$ provides a route for formulae of the kind $(\phi \rightarrow \psi) \rightarrow (\phi' \rightarrow \psi)$ where $\neg\ \phi \rightarrow \neg\phi'$ is known to be acceptable.

Finally, work on acceptability provides students with simple examples of some relatively sophisticated concepts of 'universal logic'. It illustrates the difference between a set being computable (as is the set of directly acceptable ones) and merely semi-computable (as is, for all we know at present, the set of acceptable ones). Counter-case decomposition puts into the spotlight the difference between inferential and procedural steps in a train of reasoning. Of course, that distinction already arises in classical natural deduction, where all acts of supposition and discharge are procedural (as noted for $\exists^-$ in Sect. 2.3). To be sure, there are inferential principles underlying such procedural steps, but they are subtler than simply saying that one formula implies another; they state that if one or more inferences are valid then so is another (see, e.g., Makinson 2020, Chap. 10 for a student-oriented explanation).

References

Anderson, A. R., & Belnap, Jr., N. D. (1975). *Entailment. The logic of relevance and necessity* (Vol. 1). Princeton: Princeton University Press.

Anderson, A. R., Belnap, Jr., N. D., & Dunn, J. M. (1992). *Entailment. The logic of relevance and necessity* (Vol. 2). Princeton: Princeton University Press.

Avron, A. (1992). Whither relevance logic? *J. Philos. Logic*, *21*(3), 243–281.

Avron, A. (2014). What is relevance logic? *Ann. Pure Appl. Logic*, *165*(1), 26–48.

Avron, A., Arieli, O., & Zamansky, A. (2018). *Theory of effective propositional paraconsistent logics* (Vol. 75). Studies in logic. London: College Publications.

Bimbó, K., & Dunn, J. M. (2018). Larisa Maksimova's early contributions to relevance logic. In *Larisa Maksimova on implication, interpolation, and definability* (Vol. 15, pp. 33–60). Outstanding contributions to logic. Berlin: Springer.

Bimbó, K., Dunn, J. M., & Ferenz, N. (2018). Two manuscripts, one by Routley, one by Meyer: the origins of the Routley-Meyer semantics for relevance logics. *Australas. J. Log.*, *15*(2), 171–209.

Bloesch, A. (1993). A tableau style proof system for two paraconsistent logics. *Notre Dame J. Formal Logic*, *34*(2), 295–301.

Brady, R. (Ed.). (2003). *Relevance logics and their rivals* (Vol. 2). Farnham: Ashgate Publishing.

Brady, R. T. (2006). Normalized natural deduction systems for some relevant logics. I. The logic DW. *The Journal of Symbolic Logic*, *71*(1), 35–66.

Burgess, J. P. (2009). *Philosophical logic*. Princeton foundations of contemporary philosophy. Princeton: Princeton University Press.

D'Agostino, M., Gabbay, D., & Broda, K. (1999). Tableau methods for substructural logics. In *Handbook of tableau methods* (pp. 397–467). Dordrecht: Kluwer Academic Publishers.

Dunn, J. (1986). Relevance logic and entailment. In D. Gabbay, & F. Guenthner (Eds.), *Handbook of philosophical logic* (1st ed., Vol. 3, pp. 117–224). Dordrecht: Reidel

Dunn, J., & Restall, G. (2002). Relevance logic. In D. Gabbay, & F. Guenthner (Eds.), *Handbook of philosophical logic* (2nd ed., Vol. 6, pp. 1–128). Amsterdam: Kluwer.

Dunn, J. M. (1976). Intuitive semantics for first-degree entailments and 'coupled trees'. *Philos. Studies*, *29*(3), 149–168.

Harrison, J. (2009). Without loss of generality. In *Theorem proving in higher order logics* (Vol. 5674, pp. 43–59). Lecture notes in computer science. Berlin: Springer.

Hughes, G., & Cresswell, M. (1996). *A New Introduction to Modal Logic*. London: Routledge.

Humberstone, L. (2011). *The connectives*. Cambridge: MIT Press.

Incurvati, L. (2020). *Conceptions of set and the foundations of mathematics*. Cambridge: Cambridge University Press.

Loveland, D. W., Hodel, R. E., & Sterrett, S. G. (2014). *Three views of logic. Mathematics, philosophy, and computer science*. Princeton: Princeton University Press.

Maddux, R. D. (2010). Relevance logic and the calculus of relations. *Rev. Symb. Log.*, *3*(1), 41–70.

Makinson, D. (2005). Logical friendliness and sympathy. In Logica Universalis, pages 191–205. Birkhäuser, Basel.

Makinson, D. (2014). Relevance logic as a conservative extension of classical logic. In *David Makinson on classical methods for non-classical problems* (Vol. 3, pp. 383–398). Outstanding contributions to logic. Dordrecht: Springer.

Makinson, D. (2017). Relevance via decomposition: a project, some results, an open question. *Australas. J. Log.*, *14*(3), 356–377.
Makinson, D. (2020). *Sets, logic and maths for computing* (3rd ed.). Undergraduate topics in computer science. Berlin: Springer.
Mares, E. (2012). Relevance logic. Stanford encyclopedia of philosophy. https://plato.stanford.edu/entries/logic-relevance/.
McRobbie, M. (1977). A tableau system for positive relevant implication (abstract). *Bulletin of the Section of Logic* (Polish Academy of Sciences, Institute of Philosophy and Sociology), *6*, 99–101. *Relevance Logic Newsletter*, *2*, 99–101. Accessible at http://aal.ltumathstats.com/curios/relevance-logic-newsletter.
McRobbie, M. (1979). *A proof-theoretic investigation of relevant and modal logics*. PhD thesis, Australian National University.
McRobbie, M., & Belnap, N. (1977). Relevant analytic tableaux (abstract). *Relevance Logic Newsletter*, *2*, 46–49. Accessible at http://aal.ltumathstats.com/curios/relevance-logic-newsletter.
McRobbie, M. A. and Belnap, N. D. (1979). Relevant analytic tableaux. Studia Logica, 38(2), 187–200.
McRobbie, M. A., & Belnap, N. D. (1984). Proof tableau formulations of some first-order relevant orthologics. *Bulletin of the Section of Logic* (Polish Academy of Sciences, Institute of Philosophy and Sociology), *13*(4), 233–240.
Øgaard, T. (2019). Non-Boolean classical relevant logics. *Synthese*. https://doi.org/10.1007/s11229-019-02507-z.
Pabion, J.-F. (1979). Beth's tableaux for relevant logic. *Notre Dame J. Formal Logic*, *20*(4), 891–899.
Priest, G. (2008). *An introduction to non-classical logic. From if to is* (2nd ed.). Cambridge introductions to philosophy. Cambridge: Cambridge University Press.
Routley, R. (2018). Semantic analysis of entailment and relevant implication: I. *The Australasian Journal of Logic*, *15*(2), 210–279. Circulated privately 1970/71, transcribed by Nicholas Ferenz.
Routley, R., Plumwood, V., Meyer, R. K., & Brady, R. T. (1982). *Relevant logics and their rivals. Part I*. Atascadero: Ridgeview Publishing Co.
Schechter, E. (2005). *Classical and nonclassical logics*. Princeton, NJ: Princeton University Press.
Slaney, J. (1995). *MaGIC, matrix generator for implication connectives, release 2.1*. Technical report, Australian National University.
Swirydowicz, K. (1999). There exist exactly two maximal strictly relevant extensions of the relevant logic *R*. *J. Symbolic Logic*, *64*(3), 1125–1154.
Tennant, N. (1979). Entailment and proofs. Proceedings of the Aristotelian Society, New Series, 179:167–189.
Urquhart, A. (1972). Semantics for relevance logics. The Journal of Symbolic Logic, 37:159–169.
Urquhart, A. (1984). The undecidability of entailment and relevant implication. *J. Symbolic Logic*, *49*(4), 1059–1073.
Urquhart, A. (1989). What is relevant implication? In J. Norman, & R. Sylvan (Eds.), *Directions in relevant logic* (Vol. 1, pp. 167–174). Reason and argument. Amsterdam: Kluwer.
Urquhart, A. (2016). Relevance logic: problems open and closed. *Australas. J. Log.*, *13*(1), 11–20.
van Benthem, J. (1983). Review of B.J. Copeland "On when a semantics is not a semantics: some reasons for disliking the Routley-Meyer semantics for relevance logic". *Journal of Symbolic Logic*, *49*, 994–995.
Veltman, F. (1985). *Logics for conditionals*. PhD thesis, University of Amsterdam.

Chapter 3
Tarskian Classical Relevant Logic

Roger D. Maddux

Second Reader
I. Hodkinson
Imperial College

Abstract The Tarskian classical relevant logic TR arises from Tarski's work on the foundations of the calculus of relations and on first-order logic restricted to finitely many variables, presented by Tarski and Givant their book, *A Formalization of Set Theory without Variables*, and summarized in the first nine sections. TR is closely related to the well-known logic KR. Every formula of relevance logic has a corresponding sentence in Tarski's extended first-order logic of binary relations with operators on the relation symbols. A formula is in TR (by definition), or in KR (by a theorem), if and only if its corresponding sentence can be proved in first-order logic, using at most four variables, from the assumptions that all binary relations are dense and, for TR, commute under composition, or, for KR, are symmetric. The vocabulary of TR is the same as the classical relevant logic CR* proposed by Meyer and Routley but TR properly contains CR*. The frames characteristic for TR are the ones that are characteristic for CR* and satisfy an extra frame condition. There are formulas in TR (but not in CR*) that correspond to this frame condition and provide a counterexample to a theorem of T. Kowalski. The frames characteristic for TR, or KR, are the ones whose complex algebras are integral dense relation algebras that are commutative, or symmetric, respectively. For both classes, the number of isomorphism types grows like the number of isomorphism types of ternary relations. Asymptotic formulas are obtained for both classes. Similar results apply to a hierarchy of logics defined by the number of variables used in the first-order proofs of their corresponding sentences.

Keywords Relevance logic · Classical relevant logic · Relation algebras · Semi-associative relation algebras · Provability in first-order logic with finitely many variables · Sequent calculus

---

R. D. Maddux (✉)
Department of Mathematics, Iowa State University, Ames, IA 50011-2066, USA
e-mail: maddux@iastate.edu

I. Düntsch and E. Mares (eds.), *Alasdair Urquhart on Nonclassical and Algebraic Logic and Complexity of Proofs*, Outstanding Contributions to Logic 22,
https://doi.org/10.1007/978-3-030-71430-7_3

3.1 Introduction

In 1975, Alfred Tarski delivered a pair of lectures on relation algebras at the University of Campinas. The videotaped lectures were eventually transcribed and published in 2016 (Suguitani et al. 2016). At the end of his second lecture, Tarski said (Suguitani et al. 2016, p. 154),

> And finally, the last question, if it is so, you could ask me a question whether this definition of relation algebra which I have suggested and which I have founded—I suggested it many years ago—is justified in any intrinsic sense. If we know that these are not all equations which are needed to obtain representation theorems, this means, to obtain the algebraic expression of first-order logic with two-place predicate, if we know that this is not an adequate expression of this logic, then why restrict oneself to these equations? Why not to add strictly some other equations which hold in representable relation algebras or maybe all?

Tarski defined relation algebras as those that satisfy the axioms (R_1)–(R_{10}) in Table 3.6. Each axiom is an equation $A \stackrel{\circ}{=} B$ between predicates A, B in Tarski's extended system $\mathcal{L}^+$ of first-order logic (described in detail in Sect. 3.3). In this system, 1' denotes the identity relation, $+$ is an operation on predicates denoting union, $^-$ denotes complementation, ; denotes relative multiplication and $\stackrel{\circ}{=}$ is a symbol denoting the equality of predicates, according to Tarski's definitional axioms for $\mathcal{L}^+$ listed in Table 3.4. Since $\mathcal{L}^+$ is a definitional extension of first-order logic $\mathcal{L}$ (described in Sect. 3.2) every equation $A \stackrel{\circ}{=} B$ in $\mathcal{L}^+$ can be translated into a sentence $\mathsf{G}(A \stackrel{\circ}{=} B)$ of first-order logic $\mathcal{L}$ by eliminating predicate operators according to the elimination mapping G, defined in Table 3.5. The answer to Tarski's question "whether this definition of relation algebra … is justified in any intrinsic sense" is Theorem 3.9 (3) in Sect. 3.9: an equation is derivable from the axioms for relation algebras iff its translation can be proved with no more than four variables.

Half of this answer was known to Tarski already in the early 1940s. The other half was proved 30 years later (Maddux 1978). In a manuscript started in 1942, Tarski created a system $\mathcal{L}_3$ of logic with only three variables (described in Sect. 3.7) that is equipollent with the equational theory $\mathcal{L}^\times$ of relation algebras (described in Sect. 3.5). The equipollence of $\mathcal{L}_3$ with $\mathcal{L}^\times$ is stated as Theorem 3.6 in Sect. 3.8.

By 1953, Tarski had shown that set theory can be formalized in $\mathcal{L}^\times$ as equations between predicates of first-order logic, with proofs based on just the axioms for relation algebras with substitution and modus ponens as the only rules of inference. This result, announced in Tarski (1953), was eventually published in the book by Tarski and Givant, *A Formalization of Set Theory without Variables*, where Theorem 3.9 (3) is mentioned (Tarski and Givant 1987, p. 89, p. 209).

The characterization of the equations true in relation to algebras as the ones whose translations into first-order logic are provable with four variables can be applied to the relevance logic R of Anderson and Belnap (1959, 1975), Anderson et al. (1992), Belnap (1960, 1967) and to the classical relevant logic CR^* of Meyer and Routley (1973, 1974). The connectives of R and CR^* can be interpreted as operations on

Table 3.1 Operations for interpreting formulas as relations on a base set U

Name of connective	Interpretation as an operation
disjunction	$A \vee B = \{\langle x, y\rangle : \langle x, y\rangle \in A \text{ or } \langle x, y\rangle \in B\}$
conjunction	$A \wedge B = \{\langle x, y\rangle : \langle x, y\rangle \in A \text{ and } \langle x, y\rangle \in B\}$
Boolean negation	$\neg A = \{\langle x, y\rangle : x, y \in U \text{ and } \langle x, y\rangle \notin A\}$
De Morgan negation	$\sim A = \{\langle x, y\rangle : x, y \in U \text{ and } \langle y, x\rangle \notin A\}$
implication	$A \rightarrow B = \{\langle x, y\rangle : x, y \in U, \text{and for all } z \in U,$ if $\langle z, x\rangle \in A$ then $\langle z, y\rangle \in B\}$
fusion	$A \circ B = \{\langle x, y\rangle : \text{for some } z,\ \langle x, z\rangle \in B \text{ and } \langle z, y\rangle \in A\}$
Routley star	$A^* = \{\langle x, y\rangle : \langle y, x\rangle \in A\}$
truth	$\mathbf{t} = \{\langle x, x\rangle : x \in U\}$

binary relations according to Table 3.1. Define a predicate A of $\mathcal{L}^+$ to be valid under density and commutativity if A denotes a relation containing the identity relation whenever the connectives in A are interpreted as operations on a set S of dense binary relations, where S is closed under the operations in Table 3.1 that correspond to connectives occurring in A and S is closed and commutative under relative multiplication. When interpreted this way, every predicate in R or CR* is valid under density and commutativity.

For predicates A and B, let $A \leq B$ be the equation $A + B \overset{\circ}{=} B$. The equation $1' \leq A$ asserts that A contains the identity relation. The density of A is expressed by the equation $A \leq A ; A$ and commutativity by $A ; B \overset{\circ}{=} B ; A$. By the completeness theorem for first-order logic, if A is in R or CR* then the translation $\mathsf{G}(1' \leq A)$ is provable from the sets of equations expressing density and commutativity, as defined in (3.26) and (3.27) in Sect. 3.10. Any such proof will involve some finite number of variables but it turns out that if A is a theorem of R or CR* then the translation of $1' \leq A$ into $\mathcal{L}$ can be actually be proved with no more than four variables (Theorem 3.22).

In Sect. 3.20, there are two examples, (3.148) and (3.149), of a predicate A with the property that $1' \leq A$ translates to a logically valid sentence that cannot be proved with four variables. In both cases the translation $\mathsf{G}(1' \leq A)$ can be proved with five variables and requires no appeal to density or commutativity. Theorem 3.27 shows that they are 5-provable but not 4-provable. Predicates in the vocabulary of CR* that are 5-provable but not 4-provable have been known for a long time but Mikulás (2009) was the first to find examples in the vocabulary of R. Mikulás (2009) created infinitely many such predicates. The two examples in Sect. 3.20 were created later (Maddux 2010, Sect. 8).

Tarski's classical relevant logic TR is defined in (3.29) of Sect. 3.10 as the set of predicates A in the vocabulary of CR* such that $\mathsf{G}(1' \leq A)$ is 4-provable from density and commutativity. Consequently CR* $\subseteq$ TR. Although TR does not contain any formulas that are 5-provable but not 4-provable, equality still fails. The frame conditions (3.40)–(3.42) in Sect. 3.11 hold in the frames characteristic for TR (Theorem 3.17(3)) but they do not hold in all the frames that are characteristic for CR^* (Theorem 3.26 (1)). The frame conditions (3.40)–(3.42) correspond to predicates (3.101) and (3.102). These predicates were created using the same device by which (3.148) and (3.149) were obtained from predicates in the vocabulary of CR^* that are 5-provable but not 4-provable. They are confined to the vocabulary of R and belong to TR, but they are not theorems of R (Theorem 3.26(1)). In Theorem 3.26 (2), they are shown to be valid in a frame satisfying (3.43) iff it satisfies (3.41). Furthermore, (3.101) and (3.102) are 3-provable without assuming density or commutativity (Theorem 3.22). These observations show in Sect. 3.19 that Kowalski (2013, Thm 8.1) is incorrect.

Although CR^* and R cannot be characterized as the formulas that are 4-provable from density and commutativity, TR does have that characterization simply because it is defined that way. The logic KR, which figures prominently in the research of Alasdair Urquhart (Anderson et al. 1992, Sect. 65); (Urquhart 1984, 1993, 1999, 2017, 2019) also has such a characterization despite being defined in a completely different way. Both logics TR and KR can be correlated with classes of relation algebras. A predicate A is in TR iff the equation $1' \leq A$ is true in every dense commutative relation algebra iff $\mathsf{G}(1' \leq A)$ is 4-provable from density and commutativity (Theorem 3.17 (3)). Similarly, A is in KR iff the equation $1' \leq A$ is true in every dense symmetric relation algebra iff $\mathsf{G}(1' \leq A)$ is 4-provable from density and symmetry (Theorem 3.17 (4)).

Theorem 3.22 shows that dozens of formulas and rules are provable with one to four variables, with or without additional non-logical assumptions selected from density, commutativity, or symmetry. For example, the permutation axiom $(A \to (B \to C)) \to (B \to (A \to C))$ is 4-provable from commutativity (Theorem 3.22 (3.140)) and the contraction axiom $(A \to (A \to B)) \to (A \to B)$ is 4-provable from density (Theorem 3.22 (3.136)). It was recognized long ago that density and commutativity are optional hypotheses. For example, permutation (3.140) and contraction (3.136) are not taken as axioms of Basic Logic (Routley et al. 1982; Sylvan et al. 2003).

Many formulas of R and CR^* depend on density and commutativity. The number of variables required for a proof of validity is another classificatory principle. For example, Theorem 3.23 shows that permutation (3.140) is not 3-provable from density and symmetry (which implies commutativity by Lemma 3.1). In fact, Theorem 3.24 shows that permutation is not even ω-provable from density alone without commutativity. Similarly, Theorem 3.23 shows that contraction (3.136) is not 3-provable from density and symmetry, while Theorem 3.25 shows that it is not ω-provable from commutativity alone without density.

The remainder of this introduction is a detailed review of the contents of each section. Sections 3.2–3.9 present Tarski's work on logic with finitely many variables. These sections summarize the first 100 pages of (Tarski and Givant 1987). Table 3.2 lists the formalisms treated in Sects. 3.2–3.9. First-order logic $\mathcal{L}$ is presented in Sect. 3.2. Axioms for first-logic are listed in Table 3.3. Section 3.3 presents the definitional extension $\mathcal{L}^+$ of $\mathcal{L}$. Axioms that define the predicate operators are listed

Table 3.2 For a given system $\mathcal{F}$, the axioms for $\mathcal{F}^+$ are those of $\mathcal{F}$ plus (DI)–(DV). Axioms for $\mathcal{L}^\times$ and RA coincide. Axioms for $\mathcal{L}w^\times$ and SA coincide. Formalisms between horizontal lines are equipollent

Formalisms of Tarski and Givant (1987)

Name	Sect.			Axioms	Rules
$\mathcal{L}$	3.2	Σ	$\vdash$	(AI) – (AIX)	MP
$\mathcal{L}^+$	3.3	Σ^+	$\vdash^+$	" plus (DI) – (DV)	MP
$\mathcal{L}^\times$	3.5	$\Sigma^\times$	$\vdash^\times$	(R_1) – (R_3) (R_4) (R_5) – (R_{10})	Repl, Trans
$\mathcal{L}_3$	3.7	Σ_3	$\vdash_3$	(AI) – (AVIII)(AIX′)(AX)	MP
$\mathcal{L}_3^+$	3.7	Σ_3^+	$\vdash_3^+$	" plus (DI) – (DV)	MP
$\mathcal{L}w^\times$	3.9	$\Sigma^\times$	$\vdash_s^\times$	(R_1) – (R_3) (R_4') (R_5)–(R_{10})	Repl, Trans
$\mathcal{L}s_3$	3.9	Σ_3	$\vdash_s$	(AI) – (AVIII)(AIX′)	MP
$\mathcal{L}s_3^+$	3.9	Σ_3^+	$\vdash_s^+$	" plus (DI) – (DV)	MP
$\mathcal{L}_4$	3.9	Σ_4	$\vdash_4$	(AI) – (AVIII)(AIX′)	MP
$\mathcal{L}_4^+$	3.9	Σ_4^+	$\vdash_4^+$	" plus (DI) – (DV)	MP

in Table 3.4. In Sect. 3.4, the translation mapping G, defined in Table 3.5, shows how to translate equations into first-order sentences. Theorem 3.1 states that $\mathcal{L}$ and $\mathcal{L}^+$ are equipollent in means of expression and proof. Sections 3.2–3.4 summarize (Tarski and Givant 1987, Ch. 1–2).

The equational formalism $\mathcal{L}^\times$ of Tarski and Givant (1987, Ch. 3) is defined in Sect. 3.5. Axioms for $\mathcal{L}^\times$ and $\mathcal{L}w^\times$ are listed in Table 3.6. The axioms for $\mathcal{L}^\times$ and $\mathcal{L}w^\times$ also axiomatize the variety RA of relation algebras and the variety SA of semi-associative relation algebras, respectively. If both associative laws (R_4) and (R_4') are excluded, the remaining equations in Table 3.6 axiomatize the variety NA of non-associative relation algebras. Theorem 3.2 says that $\mathcal{L}^\times$ and $\mathcal{L}w^\times$ are subformalisms of $\mathcal{L}^+$. Section 3.6 contains basic definitions and facts about relation algebras, semi-associative relation algebras, non-associative relation algebras, representable relation algebras, and the rules of equational logic. Dense, commutative, symmetric, simple, and integral algebras are defined in Sect. 3.6. Key facts, presented in Lemmas 3.1 and 3.2, are that symmetric semi-associative relation algebras are commutative and simple commutative semi-associative relation algebras are integral. The predicate algebra $\mathfrak{P}$, proper relation algebras, the variety RRA of representable relation algebras, algebras of binary relations, the satisfaction relation, and the denotation function are all defined in Sect. 3.6. Theorems 3.3 and 3.4 relate provability in $\mathcal{L}^\times$, $\mathcal{L}w^\times$, and $\mathcal{L}^+$ to truth in RA, SA, and RRA, respectively. The free RA, SA, and RRA are constructed as quotients of the predicate algebra $\mathfrak{P}$.

$\mathcal{L}^\times$ is compared to $\mathcal{L}^+$ in Sect. 3.7. $\mathcal{L}^\times$ is weaker than $\mathcal{L}^+$ because there is a 5-provable but not 4-provable equation and a sentence not equivalent to any equation. A study of the translation mapping G leads to Tarski's idea for a 3-variable formalism. Tarski's proposal is realized in the construction of the 3-variable formalisms $\mathcal{L}_3$ and $\mathcal{L}_3^+$. The axioms for these formalisms include the associative law (R_4), which requires four variables to prove. Theorems 3.5 and 3.6 in Sect. 3.8 state that $\mathcal{L}_3$, $\mathcal{L}_3^+$, and $\mathcal{L}^\times$ are equipollent in means of expression and proof. An alternative way to express 3-variable sentences as equations is also presented.

Tarski and Givant (1987, Sect. 3.10) introduced the *standardized* 3-variable formalisms $\mathcal{L}_s$ and $\mathcal{L}_s^+$. These formalisms (which should have been called $\mathcal{L}_3$ and $\mathcal{L}_3^+$, but the names were already taken) are shown to be equipollent to $\mathcal{L}w^\times$ in Sect. 3.9. Weakening (R_4) to the semi-associative law (R_4'), which only requires three variables to prove, produces the equational formalism $\mathcal{L}w^\times$ and the standardized 3-variable formalisms $\mathcal{L}s_3$ and $\mathcal{L}s_3^+$. Because it is 4-provable, the associative law (R_4) is part of the standard 4-variable formalism $\mathcal{L}_4^+$. Theorem 3.7 states that the formalisms $\mathcal{L}w^\times$, $\mathcal{L}s_3$, and $\mathcal{L}s_3^+$ are equipollent in means of expression and proof, Theorem 3.8 gives the connections between provability in $\mathcal{L}_4^+$ $\mathcal{L}_3^+$, and $\mathcal{L}^\times$, and Theorem 3.9 links theories in $\mathcal{L}_4^+$, $\mathcal{L}_3^+$, and $\mathcal{L}^\times$.

In Sect. 3.10, the logics T_n^Ψ, CT_n^Ψ, T_n, CT_n for $3 \leq n \leq \omega$, and TR are officially defined, where Ψ is a set of equations that serve as non-logical assumptions. For this purpose, equations of density Ξ^d, commutativity Ξ^c, and symmetry Ξ^s are defined in (3.26), (3.27), and (3.28), respectively, and some predicate operators are defined in (3.14)–(3.21) for use as connectives in relevance logic. The definitions match the interpretations in Table 3.1. For example, the Routley star is converse, truth **t** is the identity predicate 1', $\neg$ is used as Boolean negation (as well as negation in $\mathcal{L}$), and $\circ$ is defined as reversed relative multiplication. Two groups of operators are distinguished, the "relevance logic operators" and "classical relevant logic operators." The difference between logics T_n and CT_n lies only in their vocabularies: T_n uses the relevance logic operators, while CT_n, its "classical" counterpart, uses the classical relevant logic operators.

Section 3.11 contains a review of material on frames, including the definitions of KR-frames, KR itself, the CR^*-frames characteristic for CR^*, complex algebras of frames, the pair-frame on a set, validity in a frame, and 12 frame conditions. Lemmas 3.3 and 3.4 present some basic connections between frame conditions. Theorem 3.10 relates conditions on complex algebras to frame conditions. Theorem 3.11 characterizes frames whose complex algebras are in NA, SA, and RA. Theorem 3.12 is the Representation Theorem for NA, SA, and RA: every algebra in NA, SA, or RA is embeddable in the complex algebra of a frame satisfying the characteristic conditions in Theorem 3.11. Theorem 3.13 says that every group can be viewed as a frame. By Lemma 3.5, all CR^*-frames are commutative. Theorem 3.14 is the key connection between frames and relation algebras: the complex algebra of a frame $\mathfrak{K}$ is a dense commutative relation algebra iff $\mathfrak{K}$ is a CR^*-frame satisfying (3.41) and the complex algebra of $\mathfrak{K}$ is a dense symmetric relation algebra iff $\mathfrak{K}$ is a KR-frame.

The sequent calculus from (Maddux 1983) is presented in Sect. 3.12 with definitions of n-provability in the sequent calculus, n-dimensional relational basis (Definition 3.1), and the variety RA_n of n-dimensional relation algebras. Theorem 3.15 gives the key connections between algebras and provability in the sequent calculus: 3-provability matches up with SA, 4-provability with RA, ω-provability with RRA, and the n-provability of a sequent is characterized as a satisfaction relation on an algebra. Lemma 3.6 has several derived rules of inference for the sequent calculus.

In Sect. 3.13, the results of Tarski and Givant are combined with the sequent calculus and frame characterizations to characterize KR, TR, and Tarski's relevance logics of 3, 4, and ω variables. Lemma 3.7 says that an equation is true in every algebra satisfying the equations in Ψ iff a certain condition on homomorphisms holds. This lemma is used for the major characterization theorems. Theorem 3.16 characterizes CT_3^Ψ and CT_3 in six ways, Theorem 3.17 characterizes KR, TR, CT_4^Ψ, and CT_4 in eight ways, Theorem 3.18 characterizes KR and TR in two more ways, and Theorem 3.19 characterizes CT_ω^Ψ and CT_ω in four ways. Theorem 3.21 extends the characterizations to cover T_3^Ψ, T_3, T_4^Ψ, T_4, T_ω^Ψ, and T_ω.

For Theorem 3.22 in Sect. 3.14, dozens of predicates and rules are provided with n-proofs in the sequent calculus, where n ranges from 1 to 4, sometimes under various non-logical assumptions. Theorem 3.23 in Sect. 3.15 shows that the eleven predicates of Theorem 3.22 that are 4-provable (sometimes from density, commutativity, or both) are not 3-provable from density and symmetry. The proof uses a frame in Table 3.9 whose complex algebra is a dense symmetric semi-associative relation algebra that is not a relation algebra. Theorem 3.24 in Sect. 3.16 shows the five predicates of Theorem 3.22 that use commutativity are not ω-provable from density alone. The proof uses a frame in Table 3.10 whose complex algebra is a non-commutative dense representable relation algebra. Theorem 3.25 in Sect. 3.17 shows six predicates of Theorem 3.22 that rely on density are not ω-provable from symmetry. The proof uses the frame in Table 3.12 of the 2-element group.

Theorem 3.26 of Sect. 3.18 shows that TR exceeds CR^*. The frames characteristic for TR satisfy the frame condition (3.41), which is needed to insure axiom (R_9) holds. However, according to Dunn (2001, p. 104),

> But (56) [$(A \circ B)^{-1} = B^{-1} \circ A^{-1}$] does not correspond to any formula in the primitive vocabulary of R, nor do I know of any such formula that it implies which is not also a theorem of R. So we are left with a nagging question.

It turns out that axiom (R_9), $(A \circ B)^{-1} = B^{-1} \circ A^{-1}$, corresponds to (3.101). The predicates (3.101) and (3.102) are in the vocabulary of R but they are not theorems of R because they fail in the frame $\mathfrak{K}_4$ in Table 3.13 whose complex algebra is a Dunn monoid that cannot be embedded in a relation algebra because it fails to satisfy (3.41). Consequently, predicates (3.101) and (3.102) are in T_3 but not R. Predicate (3.103), which uses the Routley star, is in CT_3 but not CR^*. Section 3.19 points out that (3.101) is a counterexample to Kowalski (2013, Thm 8.1) and the complex algebra of $\mathfrak{K}_4$ is a counterexample to Kowalski (2013, Thm 7.1). Section 3.20 presents the two examples of predicates in the vocabulary of R that are 5-provable but not 4-provable. Asymptotic formulas for the numbers of TR-frames and KR-frames are obtained in

Sect. 3.21. Their numbers grow like c^{n^3} for some $c > 1$. For any fixed $3 \leq n \in \omega$, the probability that a randomly selected TR-frame or KR-frame validates every n-provable predicate approaches 1 as the number of elements in the frame grows. Some questions are raised in Sect. 3.22.

3.2 First-Order Logic $\mathcal{L}$ of Binary Relations

Tarski and Givant let $\mathcal{L}$ be a first-order language with equality symbol $\mathring{\mathbf{1}}$ and exactly one binary relation symbol $\mathbf{E}$ (Tarski and Givant 1987, p. 5), while $\mathcal{M}^{(n)}$ is a first-order language with equality symbol $\mathring{\mathbf{1}}$ and exactly $n + 1$ binary relation symbols, where $n \geq 0$ (Tarski and Givant 1987, p. 191). They also consider formalisms $\mathcal{M}$ and $\mathcal{M}^{\times}$ with any cardinality of binary relation symbols (Tarski and Givant 1987, p. 237). Tarski and Givant used a single relation symbol because they were presenting Tarski's formalization of set theory without variables. For set theory, it is usually sufficient to have just one relation symbol intended to denote the relation of membership. There is no need here for such a restriction. Changing notation and the number of relation symbols, we assume instead that 1' is the **equality symbol of** $\mathcal{L}$ and that $\mathcal{L}$ has a countable infinite set Π of binary relation symbols (including 1'), but no function symbols and no constants. (Tarski and Givant let $\mathbf{\Pi}$ be what we call Π^{+} in Sect. 3.3.)

The relation symbols in Π are called **atomic predicates**, and those that are distinct from 1' are also called **propositional variables** because they will play the rôle of variables in formulas of relevance logic. The **connectives of** $\mathcal{L}$ are **implication** $\Rightarrow$ and **negation** $\neg$, and $\forall$ is the **universal quantifier**. $\mathcal{L}$ has a countable set of **variables** $\Upsilon = \{\mathsf{v}_i : i \in \omega\}$, ordered in the natural way so that v_i precedes v_j if $i < j$. Thus, v_0 and v_1 are the first and second variables. For every $n \in \omega$, let $\Upsilon_n = \{\mathsf{v}_i : i < n\}$. The **atomic formulas of** $\mathcal{L}$ are the ones of the form xAy, where $x, y \in \Upsilon$ are variables and $A \in \Pi$ is an atomic predicate. For example, $x\,1'\,y$ is an atomic formula since $1' \in \Pi$. The set Φ of **formulas of** $\mathcal{L}$ is the intersection of every set that contains the atomic formulas and includes $\varphi \Rightarrow \psi$, $\neg\varphi$, and $\forall_x\varphi$ for every variable $x \in \Upsilon$ whenever it contains φ and ψ. The set of **sentences of** $\mathcal{L}$ (formulas with no free variables) is Σ. The connectives $\vee$, $\wedge$, and $\Leftrightarrow$, and the existential quantifier $\exists$ are defined for all $\varphi, \psi \in \Phi$ by $\varphi \vee \psi = \neg\varphi \Rightarrow \psi$, $\varphi \wedge \psi = \neg(\varphi \Rightarrow \neg\psi)$, $\varphi \Leftrightarrow \psi = \neg((\varphi \Rightarrow \psi) \Rightarrow \neg(\psi \Rightarrow \varphi))$, and $\exists_x\varphi = \neg\forall_x\neg\varphi$ for every $x \in \Upsilon$. When a connective is used more than once without parentheses, we restore them by association to the left. For example, $\varphi \vee \psi \vee \xi = (\varphi \vee \psi) \vee \xi$. When parentheses are omitted from a formula, the unary connective $\neg$ should be applied first, followed by $\wedge$, $\vee$, $\Rightarrow$, and $\Leftrightarrow$, in that order.

In formulating axioms and deductive rules for $\mathcal{L}$, Tarski and Givant (1987, p. 8) adopt the system $\mathcal{S}_1$ of Tarski (1965), which provides axioms for the logically valid sentences and requires only the rule MP of **modus ponens** (to infer B from $A \rightarrow B$ and A). Tarski's system $\mathcal{S}_2$ provides axioms for the logically valid formulas (not just the sentences), and uses the rule of **generalization** (to infer $\forall_x\varphi$ from φ) as well as

Table 3.3 Axioms for first-order logic $\mathcal{L}$, where $x, y \in \Upsilon, \varphi, \psi \in \Phi$

Axiom	
$[(\varphi \Rightarrow \psi) \Rightarrow ((\psi \Rightarrow \xi) \Rightarrow (\varphi \Rightarrow \xi))]$	(AI)
$[(\neg\varphi \Rightarrow \varphi) \Rightarrow \varphi]$	(AII)
$[\varphi \Rightarrow (\neg\varphi \Rightarrow \psi)]$	(AIII)
$[\forall_x \forall_y \varphi \Rightarrow \forall_y \forall_x \varphi]$	(AIV)
$[\forall_x(\varphi \Rightarrow \psi) \Rightarrow (\forall_x \varphi \Rightarrow \forall_x \psi)]$	(AV)
$[\forall_x \varphi \Rightarrow \varphi]$	(AVI)
$[\varphi \Rightarrow \forall_x \varphi]$ where x is not free in φ	(AVII)
$[\exists_x (x1\text{'}y)]$ where $x \neq y$	(AVIII)
$[x1\text{'}y \Rightarrow (\varphi \Rightarrow \psi)]$ where φ is atomic, x occurs in φ, and ψ is obtained from φ by replacing a single occurrence of x by y	(AIX)

MP. The systems $\mathcal{S}_1$ and $\mathcal{S}_2$ in Tarski (1965) were obtained by modifying a system of Quine (1940, 1951, 1962, 1981) which also uses only MP. Tarski's systems avoid the notion of substitution. Henkin (1949, 1996) proved Gödel's completeness theorem for the case in which there are relation symbols of arbitrary finite rank but no constants and no function symbols. He used MP and a restricted form of generalization as rules of inference. Tarski (1965, Thms 1 and 5) proved that the systems $\mathcal{S}_1$ and $\mathcal{S}_2$ are complete by deriving Henkin's axioms and noting that both systems are semantically sound.

For every formula $\varphi \in \Phi$, the **closure** $[\varphi]$ of φ is a sentence obtained by universally quantifying φ with respect to every free variable in φ. The closure operator is determined by the following conditions: $[\varphi] = \varphi$ for every sentence $\varphi \in \Sigma$, and if x is the last variable (in the ordering of the variables) that occurs free in φ, then $[\varphi] = [\forall_x \varphi]$.

The set $\Lambda[\mathcal{L}]$ of **logical axioms for** $\mathcal{L}$, or simply Λ, is the set of sentences that coincide with one of the sentences (AI)–(AIX) shown in Table 3.3, where $\varphi, \psi, \xi \in \Phi$. If $\Psi \subseteq \Sigma$ then a sentence $\varphi \in \Sigma$ is **provable in $\mathcal{L}$ from** Ψ, written $\Psi \vdash \varphi$ or $\vdash \varphi$ if $\Psi = \emptyset$, if φ is in every set that is closed under MP and contains $\Psi \cup \Lambda$. The **theory generated by** Ψ in $\mathcal{L}$ is

$$\Theta\eta\,\Psi = \{\varphi : \varphi \in \Sigma,\ \Psi \vdash \varphi\}.$$

Two formulas $\varphi, \psi \in \Phi$ are **logically equivalent** in $\mathcal{L}$, written $\varphi \equiv \psi$, if $\vdash [\varphi \Leftrightarrow \psi]$.

3.3 Extending $\mathcal{L}$ to $\mathcal{L}^+$

Tarski and Givant extend $\mathcal{L}$ to $\mathcal{L}^+$ by adding a second equality symbol $\mathring{=}$ and four operators $+$, $^-$, ;, and $^\smile$ that act on relation symbols and produce new relation

Table 3.4 Definitional axioms for the extension $\mathcal{L}^+$ of $\mathcal{L}$, where $A, B \in \Pi^+$

$[\,\mathrm{v}_0 A + B \mathrm{v}_1 \Leftrightarrow \mathrm{v}_0 A \mathrm{v}_1 \vee \mathrm{v}_0 B \mathrm{v}_1\,]$	(DI)
$[\,\mathrm{v}_0 \overline{A} \mathrm{v}_1 \Leftrightarrow \neg \mathrm{v}_0 A \mathrm{v}_1\,]$	(DII)
$[\,\mathrm{v}_0 A ; B \mathrm{v}_1 \Leftrightarrow \exists_z (\mathrm{v}_0 A z \wedge z B \mathrm{v}_1)\,]$	(DIII)
$[\,\mathrm{v}_0 A^{\smile} \mathrm{v}_1 \Leftrightarrow \mathrm{v}_1 A \mathrm{v}_0\,]$	(DIV)
$A \mathrel{\mathring{=}} B \Leftrightarrow [\,\mathrm{v}_0 A \mathrm{v}_1 \Leftrightarrow \mathrm{v}_0 B \mathrm{v}_1\,]$	(DV)

symbols. The set Π^+ of **predicates** of $\mathcal{L}^+$ is the intersection of every set containing Π that also contains $A + B$, $\overline{A}$, $A ; B$, and $A^{\smile}$ whenever it contains A and B. Predicates obtained in distinct ways are distinct, so, for example, if $A + B = C + D$ then $A = C$ and $B = D$. Three predicates in Π^+ are defined by

$$1 = 1' + \overline{1'}, \qquad 0 = \overline{1' + \overline{1'}}, \qquad 0' = \overline{1'}, \tag{3.1}$$

and two additional predicate operators are defined for all $A, B \in \Pi^+$ by

$$A \cdot B = \overline{\overline{A} + \overline{B}}, \qquad A \dagger B = \overline{\overline{A} ; \overline{B}}. \tag{3.2}$$

When parentheses are omitted, the unary operators should be evaluated first, followed by $;$, $\cdot$, $\dagger$, and then $+$, in that order. For example, $A \dagger B + C ; D \cdot E = (A \dagger B) + ((C ; D) \cdot E)$. Tarski and Givant add a formula $A \mathrel{\mathring{=}} B$, called an **equation**, for any predicates $A, B \in \Pi^+$. The set of **equations of** $\mathcal{L}^+$ is $\Sigma^\times$. For all $A, B \in \Pi^+$, let $A \leq B$ be the equation $A + B \mathrel{\mathring{=}} B \in \Sigma^\times$ (Tarski and Givant 1987, p. 236), which we call an **inclusion**.

Our notation for the equality symbol is derived from Schröder (1966), who denoted the identity relation on a set by 1'. Schröder obtained his notation from the Boolean unit 1 by adding an apostrophe. Similarly, Schröder added a comma to the symbol $\cdot$ for intersection to obtain his symbol ; for relative product. In (Tarski and Givant 1987), Tarski altered Schröder's system by using a circle instead of an apostrophe or comma, as well as making many symbols boldface. For example, instead of ; he used $\odot$. Tarski and Givant originally used a boldface equality symbol instead of $\mathring{=}$, but $\mathbf{=}$ is not as easily distinguished from the usual equality symbol $=$ as is $\mathring{=}$. The notation for $\mathring{=}$ used here was inspired by Tarski's device of adding circles to Boolean notation.

The **atomic formulas of** $\mathcal{L}^+$ are $x A y$ and $A \mathrel{\mathring{=}} B$, where $x, y \in \Upsilon$ and $A, B \in \Pi^+$. The set Φ^+ of **formulas of** $\mathcal{L}^+$ is the intersection of every set containing the atomic formulas of $\mathcal{L}^+$ that also contains $\varphi \Rightarrow \psi$, $\neg\varphi$, and $\forall_x \varphi$ for every $x \in \Upsilon$ whenever it contains φ and ψ. The set Σ^+ of **sentences of** $\mathcal{L}^+$ is the set of formulas that have no free variables. Equations have no free variables so $\Sigma^\times \subseteq \Sigma^+$.

The set $\Lambda[\mathcal{L}^+]$ of **logical axioms of** $\mathcal{L}^+$, or simply Λ^+ (Tarski and Givant 1987, p. 25), is the union of $\Lambda[\mathcal{L}]$ with the set of sentences that coincide with one of the sentences in Table 3.4 for some $A, B \in \Pi^+$. If $\Psi \subseteq \Sigma^+$, then a sentence $\varphi \in \Sigma^+$ is

provable in $\mathcal{L}^+$ from Ψ, written $\Psi \vdash^+ \varphi$ or $\vdash^+ \varphi$ if $\Psi = \emptyset$, if φ is in every set that contains $\Psi \cup \Lambda^+$ and is closed under MP. The **theory generated by** Ψ in $\mathcal{L}^+$ is

$$\Theta\eta^+\Psi = \{\varphi : \varphi \in \Sigma^+,\ \Psi \vdash \varphi\}.$$

Two formulas $\varphi, \psi \in \Pi^+$ of $\mathcal{L}^+$ are **logically equivalent** in $\mathcal{L}^+$, written $\varphi \equiv^+ \psi$, if $\vdash^+ [\varphi \Leftrightarrow \psi]$. The **calculus of relations** may be defined as $\Theta\eta^+\emptyset$. One may also consider it to be the closure of $\Theta\eta^+\emptyset$ under the connectives $\neg$ and $\Rightarrow$, since Schröder and Tarski showed that every propositional combination of equations is logically equivalent to an equation (Tarski and Givant 1987, 2.2(vi)).

3.4 Equipollence of $\mathcal{L}$ and $\mathcal{L}^+$

$\mathcal{L}$ and $\mathcal{L}^+$ are expressively and deductively equipollent. To prove this, Tarski defined a translation mapping G from formulas of $\mathcal{L}^+$ to formulas of $\mathcal{L}$. See (Tarski and Givant 1987, 2.3(iii)) for the definition of G and (Tarski and Givant 1987, 2.4(iii)) for the definition of **translation mapping** from one formalism to another. G eliminates operators in accordance with the definitional axioms (DI)–(DV). If $\varphi, \psi \in \Phi^+$, $x, y \in \Upsilon$, and $A, B \in \Pi^+$, then the conditions determining G are shown in Table 3.5. From the first four conditions it follows that G leaves formulas of $\mathcal{L}$ unchanged. The next result states that $\mathcal{L}$ is a subformalism of $\mathcal{L}^+$ and $\mathcal{L}$ is expressively and deductively equipollent with $\mathcal{L}^+$. Part (4) is the **main mapping theorem for $\mathcal{L}$ and $\mathcal{L}^+$**.

Theorem 3.1 (Tarski and Givant 1987, Sect. 2.3) *Formalisms $\mathcal{L}$ and $\mathcal{L}^+$ are equipollent.*

1. $\Phi \subseteq \Phi^+$ *and* $\Sigma \subseteq \Sigma^+$ [2.3(i)].
2. G *maps* Φ^+ *onto* Φ *and* Σ^+ *onto* Σ [2.3(iv)(δ)].
3. $\varphi \equiv^+ \mathsf{G}(\varphi)$ *if* $\varphi \in \Phi^+$ [2.3(iv)(ε)].
4. $\Psi \vdash^+ \varphi$ *iff* $\{\mathsf{G}(\psi) : \psi \in \Psi\} \vdash \mathsf{G}(\varphi)$ *if* $\Psi \subseteq \Sigma^+$ *and* $\varphi \in \Sigma^+$ [2.3(v)].
5. $\Psi \vdash^+ \varphi$ *iff* $\Psi \vdash \varphi$, *if* $\Psi \subseteq \Sigma$ *and* $\varphi \in \Sigma$ [2.3(ii)(ix)].
6. $\Theta\eta\,\Psi = \Theta\eta^+\Psi \cap \Sigma$ *if* $\Psi \subseteq \Sigma$ [2.3(x)].

Table 3.5 Definition of translation mapping $\mathsf{G}\colon \Phi^+ \to \Phi$, where $x, y, z \in \Upsilon$, $A, B \in \Pi^+$, and $\varphi, \psi \in \Phi^+$

$$\begin{aligned}
\mathsf{G}(xAy) &= xAy \quad \text{if } A \in \Pi\\
\mathsf{G}(\varphi \Rightarrow \psi) &= \mathsf{G}(\varphi) \Rightarrow \mathsf{G}(\psi)\\
\mathsf{G}(\neg\varphi) &= \neg\mathsf{G}(\varphi)\\
\mathsf{G}(\forall_x\varphi) &= \forall_x\mathsf{G}(\varphi)\\
\mathsf{G}(xA+By) &= \mathsf{G}(xAy) \vee \mathsf{G}(xBy)\\
\mathsf{G}(x\overline{A}y) &= \neg\mathsf{G}(xAy)\\
\mathsf{G}(xA;By) &= \exists_z(\mathsf{G}(xAz) \wedge \mathsf{G}(zBy))\\
&\quad\text{where } z \text{ is the first variable distinct from } x \text{ and } y\\
\mathsf{G}(xA^{\smile}y) &= \mathsf{G}(yAx)\\
\mathsf{G}(A \mathrel{\mathring{=}} B) &= [\,\mathsf{G}(\mathsf{v}_0 A\mathsf{v}_1) \Leftrightarrow \mathsf{G}(\mathsf{v}_0 B\mathsf{v}_1)\,]
\end{aligned}$$

3.5 Equational Formalisms $\mathcal{L}^\times$ and $\mathcal{L}w^\times$

The equational formalisms $\mathcal{L}^\times$ and $\mathcal{L}w^\times$ are defined by Tarski and Givant (1987, Sect. 3.1, p. 89). $\mathcal{L}^\times$ is the primary subject of their book but $\mathcal{L}w^\times$ makes only an incidental appearance as a weakening of $\mathcal{L}^\times$. $\mathcal{L}^\times$ is equipollent with the 3-variable formalisms $\mathcal{L}_3$ and $\mathcal{L}_3^+$ described in Sect. 3.7 while $\mathcal{L}w^\times$ is equipollent with the "*(standardized) formalisms*" $\mathcal{L}s_3$ and $\mathcal{L}s_3^+$ described in Sect. 3.9. Tarski and Givant said, "These standardized formalisms are undoubtedly more natural and more interesting in their own right than $\mathcal{L}_3$ and $\mathcal{L}_3^+$" (Tarski and Givant 1987, p. 89).

The axioms of $\mathcal{L}^\times$ and $\mathcal{L}w^\times$ are certain equations in $\Sigma^\times$ and their deductive rules apply to equations. $\Lambda[\mathcal{L}^\times]$, or simply $\Lambda^\times$, is the set of **axioms of** $\mathcal{L}^\times$ and $\Lambda[\mathcal{L}w^\times]$, or simply $\Lambda^\times_s$, is the set of **axioms of** $\mathcal{L}w^\times$. An equation $\varepsilon \in \Sigma^\times$ belongs to $\Lambda^\times$ if there are predicates $A, B, C \in \Pi^+$ such that ε coincides with one of the equations (R_1)–(R_{10}) listed in Table 3.6, and ε belongs to $\Lambda^\times_s$ if ε coincides with one of the equations (R_1)–(R_3), (R'_4), (R_5)–(R_{10}) in Table 3.6. Note that $\Lambda^\times_s \subseteq \Lambda^\times$ because $1 \in \Pi^+$, hence every instance of (R'_4) is also an instance of (R_4). Deducibility in $\mathcal{L}^\times$ and $\mathcal{L}w^\times$ is defined as it is in equational logic. The **transitivity rule** Trans is to infer $B \mathrel{\mathring{=}} C$ from $A \mathrel{\mathring{=}} B$ and $A \mathrel{\mathring{=}} C$, and the **replacement rule** Repl is to infer $\overline{A} \mathrel{\mathring{=}} \overline{B}$, $A^\smile \mathrel{\mathring{=}} B^\smile$, $A+C \mathrel{\mathring{=}} B+C$, and $A;C \mathrel{\mathring{=}} B;C$ from $A \mathrel{\mathring{=}} B$. For every $\Psi \subseteq \Sigma^\times$, an equation $\varepsilon \in \Sigma^\times$ is **provable in** $\mathcal{L}^\times$ **from** Ψ, written $\Psi \vdash^\times \varepsilon$ or $\vdash^\times \varepsilon$ when $\Psi = \emptyset$, iff ε is in every set that contains $\Psi \cup \Lambda^\times$ and is closed under Trans and Repl. Similarly, ε is **provable in** $\mathcal{L}w^\times$ **from** Ψ, written $\Psi \vdash^\times_s \varepsilon$ or $\vdash^\times_s \varepsilon$ when $\Psi = \emptyset$, iff ε belongs to every set that contains $\Psi \cup \Lambda^\times_s$ and is closed under Trans and Repl. For every $\Psi \subseteq \Sigma^\times$, the **theory generated by** Ψ in $\mathcal{L}^\times$ is

$$\Theta\eta^\times\Psi = \{\varepsilon : \varepsilon \in \Sigma^\times,\ \Psi \vdash^\times \varepsilon\},$$

and the **theory generated by** Ψ in $\mathcal{L}w^\times$ is

$$\Theta\eta_s^\times\Psi = \{\varphi : \varepsilon \in \Sigma^\times,\ \Psi \vdash_s^\times \varepsilon\}.$$

The rules stated here employ simplifications (mentioned but not proved by Tarski and Givant (1987, p. 47)) made possible by the presence of certain equations in $\Lambda_s^\times$. The equation $A \mathrel{\mathring{=}} A$ is deducible in $\mathcal{L}^\times$ and $\mathcal{L}w^\times$ for every predicate $A \in \Pi^+$ because it follows by Trans from two instances of (R_3), (R_5), or (R_7). To derive $B \mathrel{\mathring{=}} A$ from $A \mathrel{\mathring{=}} B$, first derive $A \mathrel{\mathring{=}} A$ using Trans and one of (R_3), (R_6), or (R_7) and then apply Trans to $A \mathrel{\mathring{=}} B$ and $A \mathrel{\mathring{=}} A$. To derive $A \mathrel{\mathring{=}} C$ from $A \mathrel{\mathring{=}} B$ and $B \mathrel{\mathring{=}} C$, first derive $B \mathrel{\mathring{=}} A$ from $A \mathrel{\mathring{=}} B$ and apply Trans to $B \mathrel{\mathring{=}} A$ and $B \mathrel{\mathring{=}} C$.

For equational logic in general the replacement rule would include the equations $C + A \mathrel{\mathring{=}} C + B$ and $C ; A \mathrel{\mathring{=}} C ; B$ as equations derivable from $A \mathrel{\mathring{=}} B$, but they can be derived. To get $C ; A \mathrel{\mathring{=}} C ; B$ from $A \mathrel{\mathring{=}} B$, first derive $A^\smile ; C^\smile \mathrel{\mathring{=}} B^\smile ; C^\smile$ by applying Repl twice. Two instances of (R_9) are $C ; A^\smile \mathrel{\mathring{=}} A^\smile ; C^\smile$ and $C ; B^\smile \mathrel{\mathring{=}} B^\smile ; C^\smile$. Use the laws of equality proved above to get $C ; A^\smile \mathrel{\mathring{=}} C ; B^\smile$ from these last three equations. Next, obtain $(C ; A^\smile)^\smile \mathrel{\mathring{=}} (C ; B^\smile)^\smile$ by Repl and complete the proof using two instances of (R_7) and the laws of equality. Thus, the presence of (R_7) and (R_9) in $\Lambda^\times$ and $\Lambda_s^\times$ is enough to make $C ; A \mathrel{\mathring{=}} C ; B$ derivable from $A \mathrel{\mathring{=}} B$. It is easier to derive $C + A \mathrel{\mathring{=}} C + B$ from $A \mathrel{\mathring{=}} B$ using (R_1) and one of (R_3), (R_6), or (R_7).

Theorem 3.2 (Tarski and Givant 1987, Sect. 3.4) *$\mathcal{L}w^\times$ is a subformalism of $\mathcal{L}^\times$ and $\mathcal{L}^\times$ is a subformalism of $\mathcal{L}^+$.*

1. $\Sigma^\times \subseteq \Sigma^+$ [3.4(i)].
2. *If $\Psi \vdash_s^\times \varepsilon$ then $\Psi \vdash^\times \varepsilon$, for every $\Psi \subseteq \Sigma^\times$ and $\varepsilon \in \Sigma^\times$.*
3. $\Theta\eta_s^\times\Psi \subseteq \Theta\eta^\times\Psi$ *for every $\Psi \subseteq \Sigma^\times$.*
4. *If $\Psi \vdash^\times \varepsilon$ then $\Psi \vdash^+ \varepsilon$, for every $\Psi \subseteq \Sigma^\times$ and $\varepsilon \in \Sigma^\times$* [3.4(ii)].
5. $\Theta\eta^\times\Psi \subseteq \Theta\eta^+\Psi \cap \Sigma^\times$ *for every* $\Psi \subseteq \Sigma^\times$[3.4(vii)].

Proof Part (1) is the observation made in the previous section that equations have no free variables. Part (2) follows from $\Lambda_s^\times \subseteq \Lambda^\times$ and part (3) follows from part (2). Part (4) can be proved by induction on provability in $\mathcal{L}^+$. One shows that the axioms of $\mathcal{L}^\times$ are provable in $\mathcal{L}^+$ and that if the hypotheses of Trans or Repl are provable in $\mathcal{L}^+$ then so are their conclusions. This would be tedious to carry out according to the definitions of the notions involved. For example, if φ were $A + B \mathrel{\mathring{=}} B + A$, an instance of axiom (AI) where $A, B \in \Pi^+$, one would have to provide a sequence of sentences in Σ^+, each of which is either an instance of (AI)–(AIX) or (DI)–(DV) or follows from two previous sentences by MP, ending with $A + B \mathrel{\mathring{=}} B + A$. It is much easier to proceed semantically, taking advantage of Gödel's completeness theorem for $\mathcal{L}$ (Tarski and Givant 1987, Sect. 1.4) and its implications for $\mathcal{L}^+$ (Tarski and Givant 1987, Sect. 2.2). It then becomes clear that (AI) expresses the fact that the operation of forming the union of two binary relations is commutative and that this fact can be proved in $\mathcal{L}^+$. Similarly, all the other axioms of $\mathcal{L}^\times$ can be seen as logically valid (and therefore provable in $\mathcal{L}^+$) when they are interpreted according to (DI)–(DV).

Table 3.6 Axioms for the equational formalisms $\mathcal{L}^{\times}$ and $\mathcal{L}w^{\times}$, where $A, B, C \in \Pi^{+}$

Axiom	
$A + B \stackrel{\circ}{=} B + A$	(R_1)
$A + (B + C) \stackrel{\circ}{=} (A + B) + C$	(R_2)
$\overline{\overline{A} + \overline{B}} + \overline{\overline{A} + B} \stackrel{\circ}{=} A$	(R_3)
$A;(B;C) \stackrel{\circ}{=} (A;B);C$	(R_4)
$A;(B;1) \stackrel{\circ}{=} (A;B);1$	(R_4')
$(A + B);C \stackrel{\circ}{=} A;C + B;C$	(R_5)
$A;1\text{'} \stackrel{\circ}{=} A$	(R_6)
$(A^{\smile})^{\smile} \stackrel{\circ}{=} A$	(R_7)
$(A + B)^{\smile} \stackrel{\circ}{=} A^{\smile} + B^{\smile}$	(R_8)
$(A;B)^{\smile} \stackrel{\circ}{=} B^{\smile};A^{\smile}$	(R_9)
$A^{\smile};\overline{A;B} + \overline{B} \stackrel{\circ}{=} \overline{B}$	(R_{10})

3.6 Relation Algebras, Semi-associative and Representable

Since Π^{+} is closed under the predicate operators $+, ^{-}, ;, ^{\smile}$, and contains 1', we may define the **predicate algebra** $\mathfrak{P}$ by

$$\mathfrak{P} = \langle \Pi^{+}, +, ^{-}, ;, ^{\smile}, 1\text{'} \rangle.$$

Then $\mathfrak{P}$ is an absolutely free algebra that is freely generated by the propositional variables $\{A : 1\text{'} \neq A \in \Pi\}$ (Tarski and Givant 1987, p. 238). This means that any function mapping the propositional variables into an algebra of the same similarity type as $\mathfrak{P}$ has a unique extension to a homomorphism from $\mathfrak{P}$ into that algebra. Consider an algebra $\mathfrak{A}$ having the same similarity type as $\mathfrak{P}$, say

$$\mathfrak{A} = \langle U, +, ^{-}, ;, ^{\smile}, 1\text{'} \rangle,$$

where U is a set called the **universe** of $\mathfrak{A}$, $1\text{'} \in U$, $+$ and $;$ are binary operations on U, and $^{-}$ and $^{\smile}$ are unary operations on U. Define three additional elements of U by $0\text{'} = \overline{1\text{'}}$, $1 = 1\text{'} + 0\text{'}$, and $0 = \overline{1}$. Define the binary operation $\cdot$ on U by $x \cdot y = \overline{\overline{x} + \overline{y}}$ and the binary relation $\leq$ on U by $x \leq y$ if $x + y = y$. The algebra $\mathfrak{A}$ is **dense** if $x \leq x;x$ for every $x \in U$, **commutative** if $x;y = y;x$ for all $x, y \in U$, **symmetric** if $x^{\smile} = x$ for every $x \in U$, and **integral** if $0 \neq 1$ and $x;y = 0$ imply $x = 0$ or $y = 0$, for all $x, y \in U$. An element $x \in U$ is an **atom** of $\mathfrak{A}$ if $x \neq 0$ and for every $y \in U$ either $x \leq y$ or $y = 0$, and $\text{At}(\mathfrak{A})$ is the set of atoms of $\mathfrak{A}$. The algebra $\mathfrak{A}$ **atomic** if $0 \neq y \in U$ implies $x \leq y$ for some atom $x \in \text{At}(\mathfrak{A})$. The algebra $\mathfrak{A}$ is **simple** if $0 \neq 1$ and has no non-trivial homomorphic images, meaning that every homomorphic image of $\mathfrak{A}$ is either a 1-element algebra or is isomorphic to $\mathfrak{A}$. The algebra $\mathfrak{A}$ is **semi-**

simple if it is isomorphic to a subdirect product of simple homomorphic images of $\mathfrak{A}$. It follows from the definition of subdirect product that every semi-simple algebra is isomorphic to a subalgebra of a direct product of simple homomorphic images of $\mathfrak{A}$.

For every homomorphism $h\colon \mathfrak{P} \to \mathfrak{A}$ that maps the predicate algebra $\mathfrak{P}$ into $\mathfrak{A}$, let $\models_h$ be the relation that holds between the algebra $\mathfrak{A}$ and a set of equations $\Psi \subseteq \Sigma^\times$ if $h(A) = h(B)$ whenever $A, B \in \Pi^+$ and $A \mathrel{\mathring{=}} B \in \Psi$. For any $\varepsilon \in \Sigma^\times$ let $\mathfrak{A} \models_h \varepsilon$ mean the same as $\mathfrak{A} \models_h \{\varepsilon\}$. An equation $A \mathrel{\mathring{=}} B \in \Sigma^\times$ is **true in** $\mathfrak{A}$ if $\mathfrak{A} \models_h A \mathrel{\mathring{=}} B$ for every homomorphism $h\colon \mathfrak{P} \to \mathfrak{A}$ from the predicate algebra $\mathfrak{P}$ to $\mathfrak{A}$. For example, if $1' \neq A, B \in \Pi$, then $\mathfrak{A}$ is commutative iff the equation $A\,;B \mathrel{\mathring{=}} B\,;A$ is true in $\mathfrak{A}$, dense iff the inclusion $A \leq A\,;A$ is true in $\mathfrak{A}$, and symmetric iff $A^{\smile} \mathrel{\mathring{=}} A$ is true in $\mathfrak{A}$. The way these equivalences are established is illustrated by the proof of Lemma 3.1 below.

The algebra $\mathfrak{A}$ is a **relation algebra** if the equations (R_1)–(R_{10}) in Table 3.6 are true in $\mathfrak{A}$. $\mathfrak{A}$ is a **semi-associative relation algebra** if the equations (R_1)–(R_3), (R_4'), and (R_5)–(R_{10}) are true in $\mathfrak{A}$. $\mathfrak{A}$ is a **non-associative relation algebra** if the equations (R_1)–(R_3) and (R_5)–(R_{10}) are true in $\mathfrak{A}$. Let RA be the class of relation algebras, let SA be the class of semi-associative relation algebras, and let NA be the class of non-associative relation algebras. It follows immediately from their definitions that $\mathsf{RA} \subseteq \mathsf{SA} \subseteq \mathsf{NA}$.

Lemma 3.1 *Assume* $\mathfrak{A} = \langle U, +, ^-, ;, ^{\smile}, 1'\rangle$ *is a symmetric algebra in which* (R_9) *is true. Then* $\mathfrak{A}$ *is commutative. In particular, every symmetric semi-associative relation algebra is commutative.*

Proof Assume $\mathfrak{A} = \langle U, +, ^-, ;, ^{\smile}, 1'\rangle$, $\mathfrak{A}$ is symmetric, and (R_9) is true in $\mathfrak{A}$. Let $x, y \in U$ and let $1' \neq A, B \in \Pi$ be distinct propositional variables. Since $\mathfrak{P}$ is absolutely freely generated by the propositional variables, there is a homomorphism $h\colon \mathfrak{P} \to \mathfrak{A}$ such that $h(A) = x$ and $h(B) = y$. Since (R_9) is true in $\mathfrak{A}$, $h((A\,;B)^{\smile}) = h(B^{\smile}\,;A^{\smile})$, hence

$$
\begin{aligned}
x\,;y &= (x\,;y)^{\smile} && \mathfrak{A} \text{ is symmetric}\\
&= (h(A)\,;h(B))^{\smile} && \text{choice of } h\\
&= h((A\,;B)^{\smile}) && h \text{ is a homomorphism}\\
&= h(B^{\smile}\,;A^{\smile}) && (R_9) \text{ is true in } \mathfrak{A}\\
&= h(B)^{\smile}\,;h(A)^{\smile} && h \text{ is a homomorphism}\\
&= y^{\smile}\,;x^{\smile} && \text{choice of } h\\
&= y\,;x && \mathfrak{A} \text{ is symmetric.}
\end{aligned}
$$

Jónsson and Tarski (1952, Th. 4.15) proved that every relation algebra is semi-simple; see Givant (2017, Thm 12.10). By (Maddux 1978, Cor 8(7)) or (Maddux 2006, Thm 388) it is also true that every semi-associative relation algebra is semi-simple. By (Maddux 2006, Thm 379(iii)), (Maddux 1978, Thm 7(20)), or (Maddux 1991, Thm 29), a semi-associative relation algebra $\mathfrak{A} = \langle U, +, ^-, ;, ^{\smile}, 1'\rangle$ is simple iff $0 \neq 1$ and for all $x, y \in U$, if $(x\,;1)\,;y = 0$ then $x = 0$ or $y = 0$. Jónsson and Tarski (1952, 4.17) proved that a relation algebra is integral iff its identity element is an atom. This result also extends to SA. By (Maddux 2006, Thm 353) or (Maddux 1990, Thm 4) a semi-associative relation algebra $\mathfrak{A}$ is integral iff 1' is an atom of $\mathfrak{A}$. These facts and some basic observations from universal algebra are used in the proof of the following lemma.

Lemma 3.2

1. *Assume* $\mathfrak{A} = \langle U, +, ^{-}, ;, ^{\smile}, 1'\rangle$ *is a simple commutative semi-associative relation algebra. Then* $\mathfrak{A}$ *is integral and* 1' *is an atom of* $\mathfrak{A}$.
2. *An equation is true in every commutative semi-associative relation algebra iff it is true in every integral commutative semi-associative relation algebra.*
3. *An equation is true in every commutative semi-associative relation algebra iff it is true in every commutative semi-associative relation algebra in which the identity element is an atom.*

Proof For part (1), suppose $\mathfrak{A} \in$ SA is commutative and simple. First note that $0 \neq 1$ since $\mathfrak{A}$ is simple. To show $\mathfrak{A}$ is integral, suppose that $x\,;y = 0$. We have $(x\,;1)\,;y = (1\,;x)\,;y$ since $\mathfrak{A}$ is commutative, $(1\,;x)\,;y = 1\,;(x\,;y) = 1\,;0$ by (Maddux 1991, Thm 13) or (Maddux 2006, Thm 354) and the assumption that $x\,;y = 0$, and $1\,;0 = 0$ by (Maddux 2006, Thm 287), so $(x\,;1)\,;y = 0$. Since $\mathfrak{A}$ is a simple semi-associative relation algebra, it follows, as noted above, that either $x = 0$ or $y = 0$. This shows $\mathfrak{A}$ is integral, so we conclude that 1' is an atom of $\mathfrak{A}$, also noted above.

Parts (2) and (3) are equivalent because a semi-associative relation algebra is integral iff its identity element is an atom. One direction of each part is also trivially true. It suffices therefore to assume that an equation ε is true in every integral commutative semi-associative relation algebra and show that it is true in every commutative semi-associative relation algebra.

Suppose $\mathfrak{A}$ is a commutative semi-associative relation algebra. As noted above, $\mathfrak{A}$ is isomorphic to a subalgebra of a direct product of simple semi-associative relation algebras that are homomorphic images of $\mathfrak{A}$. Homomorphic images of commutative algebras are commutative, so all these simple semi-associative relation algebras are commutative. By part (1), they are also integral, so by hypothesis the equation ε is true in all of them. If an equation is true in a collection of algebras, then it is also true in their direct product, and if it is true in an algebra then it is also true in all the subalgebras of that algebra. These two facts combine to show that ε must therefore be true in $\mathfrak{A}$, as desired.

The next theorem provides a link between relation algebras and deducibility in $\mathcal{L}^{\times}$ and between semi-associative relation algebras and deducibility in $\mathcal{L}w^{\times}$. Following their proof of (Tarski and Givant 1987, 8.2(x)), which is part (5), Tarski and Givant say, "It may be noticed that the theorem just proved could be given a stronger form by using the notion of a free algebra with defining relations. … A precise formulation of the improved Theorem (x) would be rather involved, and we leave it to the reader." This stronger form is part (3).

For every $\Psi \subseteq \Sigma^{\times}$, let $\simeq^{\times}_{\Psi}$ and $\simeq^{s}_{\Psi}$ be the binary relations defined for any $A, B \in \Pi^{+}$ by $A \simeq^{\times}_{\Psi} B$ iff $\Psi \vdash^{\times} A \mathrel{\mathring{=}} B$ (Tarski and Givant 1987, p. 238), and $A \simeq^{s}_{\Psi} B$ iff $\Psi \vdash^{\times}_{s} A \mathrel{\mathring{=}} B$. For every $\Psi \subseteq \Sigma^{+}$, let $\simeq^{+}_{\Psi}$ be the binary relation defined for any $A, B \in \Pi^{+}$ by $A \simeq^{+}_{\Psi} B$ iff $\Psi \vdash^{+} A \mathrel{\mathring{=}} B$ (Tarski and Givant 1987, p. 240). In all three relations, reference to Ψ is omitted when $\Psi = \emptyset$. Theorem 3.3 concerns $\simeq^{\times}_{\Psi}$ and $\simeq^{s}_{\Psi}$, while Theorem 3.4 deals with $\simeq^{+}_{\Psi}$.

Theorem 3.3 *Assume* $\Psi \subseteq \Sigma^{\times}$.

1. $\simeq^{\times}_{\Psi}$ *is a congruence relation on* $\mathfrak{P}$ *and the quotient algebra* $\mathfrak{P}/\simeq^{\times}_{\Psi}$ *is a relation algebra* (Tarski and Givant 1987, 8.2(ix)).
2. $\simeq^{s}_{\Psi}$ *is a congruence relation on* $\mathfrak{P}$ *and the quotient algebra* $\mathfrak{P}/\simeq^{s}_{\Psi}$ *is a semi-associative relation algebra.*
3. *For every* $\varepsilon \in \Sigma^{\times}$, $\Psi \vdash^{\times} \varepsilon$ *iff for every* $\mathfrak{A} \in \mathsf{RA}$ *and every homomorphism* $h\colon \mathfrak{P} \to \mathfrak{A}$, *if* $\mathfrak{A} \models_h \Psi$ *then* $\mathfrak{A} \models_h \varepsilon$.
4. *For every* $\varepsilon \in \Sigma^{\times}$, $\Psi \vdash^{\times}_{s} \varepsilon$ *iff for every* $\mathfrak{A} \in \mathsf{SA}$ *and every homomorphism* $h\colon \mathfrak{P} \to \mathfrak{A}$, *if* $\mathfrak{A} \models_h \Psi$ *then* $\mathfrak{A} \models_h \varepsilon$.
5. $\mathfrak{P}/\simeq^{\times}$ *is a relation algebra that is* RA*-freely generated by* $\{A/\simeq^{\times} : 1' \neq A \in \Pi\}$ (Tarski and Givant 1987, 8.2(x)).
6. $\mathfrak{P}/\simeq^{\times}$ *is a semi-associative relation algebra that is* SA*-freely generated by* $\{A/\simeq^{s} : 1' \neq A \in \Pi\}$.

Proof For part (1), note that $\simeq^{\times}_{\Psi}$ is a congruence relation on $\mathfrak{P}$ because of the rules Trans and Repl and their consequences. It follows that $\simeq^{\times}_{\Psi}$ determines a quotient homomorphism $q\colon \Pi^{+} \to \{A/\simeq^{\times}_{\Psi} : A \in \Pi^{+}\}$ that carries each predicate $A \in \Pi^{+}$ to its equivalence class $A/\simeq^{\times}_{\Psi}$ under $\simeq^{\times}_{\Psi}$. To show that the quotient algebra $\mathfrak{P}/\simeq^{\times}_{\Psi}$ is in RA, we must prove that every axiom in $\Lambda^{\times}$ is true in $\mathfrak{P}/\simeq^{\times}_{\Psi}$. For that we assume $h\colon \mathfrak{P} \to \mathfrak{P}/\simeq^{\times}_{\Psi}$ is a homomorphism. We must show that if $\lambda \in \Lambda^{\times}$ then $\mathfrak{P}/\simeq^{\times}_{\Psi} \models_h \lambda$. We do just one example, say an instance $A + B \mathrel{\mathring{=}} B + A$ of (R_1). Since $h(A)$ and $h(B)$ are equivalence classes of predicates, we may choose $C, D \in \Pi^{+}$ such that $h(A) = C/\simeq^{\times}_{\Psi} = q(C)$ and $h(B) = D/\simeq^{\times}_{\Psi} = q(D)$. Then

$h(A+B) = h(A) + h(B)$	h is a homomorphism
$= q(C) + q(D)$	choice of C, D
$= q(C + D)$	q is a homomorphism
$= q(D + C)$	$\Psi \vdash^{\times} C + D \mathrel{\mathring{=}} D + C$
$= q(D) + q(C)$	q is a homomorphism
$= h(B) + h(A)$	choice of C, D
$= h(B + A)$	h is a homomorphism

Proofs for the other axioms are similar. Thus, $\mathfrak{P}/\simeq^{\times}_{\Psi} \in \mathsf{RA}$. Furthermore, $\mathfrak{P}/\simeq^{\times}_{\Psi} \models_q \Psi$ just by the definitions of $\simeq^{\times}_{\Psi}$ and q. The proof of part (2) is the same, but with $\simeq^{\times}_{\Psi}$, RA, and $\Lambda^{\times}$ replaced by $\simeq^{s}_{\Psi}$, SA, and $\Lambda^{\times}_{s}$, respectively.

For part (3), assume $\Psi \vdash^{\times} \varepsilon$, $\mathfrak{A} \in \mathsf{RA}$, $h\colon \mathfrak{P} \to \mathfrak{A}$ is a homomorphism, and $\mathfrak{A} \models_h \Psi$. We wish to show $\mathfrak{A} \models_h \varepsilon$. Let $\Theta = \{\theta : \theta \in \Sigma^{\times}, \mathfrak{A} \models_h \theta\}$. We will show

$\Theta\eta^\times\Psi \subseteq \Theta$. We have $\Lambda^\times \subseteq \Theta$ by the definition of RA and $\Psi \subseteq \Theta$ by the hypothesis $\mathfrak{A} \models_h \Psi$ and the definition of $\models_h$. Next, we show that Θ is closed under the rules Trans and Repl. To see this for Trans, assume $A \mathrel{\mathring{=}} B, A \mathrel{\mathring{=}} C \in \Theta$. Then $h(A) = h(B)$ and $h(A) = h(C)$ by the definition of Θ. It follows that $h(B) = h(C)$, so $B \mathrel{\mathring{=}} C \in \Theta$. For Repl, we assume $A \mathrel{\mathring{=}} B \in \Theta$ and wish to show the conclusions of Repl are in Θ. We have $h(A) = h(B)$ since $A \mathrel{\mathring{=}} B \in \Theta$. This implies $h(A)^\smile = h(B)^\smile$, $\overline{h(A)} = \overline{h(B)}$, $h(A);h(C) = h(B);h(C)$, and $h(A) + h(C) = h(B) + h(C)$ for every $C \in \Pi^+$. Then $h(A^\smile) = h(B^\smile)$, $h(\overline{A}) = h(\overline{B})$, $h(A;C) = h(B;C)$, and $h(A + C) = h(B + C)$ since h is a homomorphism. Thus, the conclusions of Repl are also in Θ. Since Θ is a set containing $\Lambda^\times \cup \Psi$ and is closed under Trans and Repl, it contains $\Theta\eta^\times\Psi$. By hypothesis, we have $\varepsilon \in \Theta\eta^\times\Psi$, hence $\varepsilon \in \Theta$, as desired. This completes the proof of one direction of part (3).

Now suppose that $\varepsilon = (A \mathrel{\mathring{=}} B)$ for some $A, B \in \Pi^+$ and $\Psi \nvdash^\times \varepsilon$. We wish to show $\mathfrak{A} \not\models_h \varepsilon$ for some $\mathfrak{A} \in \mathsf{RA}$ and some homomorphism $h : \mathfrak{P} \to \mathfrak{A}$ such that $\mathfrak{A} \models_h \Psi$. It suffices to let $\mathfrak{A} = \mathfrak{P}/{\simeq^\times_\Psi}$ and $h = q$. Since $A \mathrel{\mathring{=}} B$ is not provable in $\mathcal{L}^\times$ from Ψ, the equivalence classes $q(A) = A/{\simeq^\times_\Psi}$ and $q(B) = B/{\simeq^\times_\Psi}$ are distinct, hence $q(A) \neq q(B)$. Thus, we have an algebra $\mathfrak{A} \in \mathsf{RA}$ and a homomorphism $h : \mathfrak{P} \to \mathfrak{A}$ such that $\mathfrak{A} \models_h \Psi$ but not $\mathfrak{A} \models_h \varepsilon$, as desired. This completes the proof of part (3). For part (4), repeat the proof of part (3) using SA, $\Lambda^\times_s$, and $\simeq^s_\Psi$ in place of RA, $\Lambda^\times$, and $\simeq^\times_\Psi$.

Parts (5) and (6) follow from parts (3) and (4) when $\Psi = \emptyset$. We show that the quotient algebra $\mathfrak{P}/{\simeq^\times}$ is RA-freely generated by the $\simeq^\times$-equivalence classes of the propositional variables. One starts with an arbitrary map f from $\{A/{\simeq^\times} : 1' \neq A \in \Pi\}$ into a relation algebra $\mathfrak{A}$. Then f determines a map on the propositional variables $\{A : 1' \neq A \in \Pi\}$ that sends A to $f(A/{\simeq^\times})$. Since $\mathfrak{P}$ is absolutely freely generated by the propositional variables, this map has a unique extension to a homomorphism $h : \mathfrak{P} \to \mathfrak{A}$ that vacuously satisfies the condition $\mathfrak{A} \models_h \emptyset$. Now h sends any two equivalent predicates to the same thing, for if $A \simeq^\times B$, then $\vdash^\times A \mathrel{\mathring{=}} B$, hence $\mathfrak{A} \models_h A \mathrel{\mathring{=}} B$ by part (3), i.e., $h(A) = h(B)$. This means that h determines a map g from $\{A/{\simeq^\times} : A \in \Pi^+\}$ into $\mathfrak{A}$ that satisfies the condition $g(A/{\simeq^\times}) = h(A)$ for every $A \in \Pi^+$. This condition implies g is a homomorphism because, for example, if $A, B \in \Pi^+$ then

$$\begin{aligned} g(A/{\simeq^\times} + B/{\simeq^\times}) &= g((A + B)/{\simeq^\times}) = h(A + B) \\ &= h(A) + h(B) = g(A/{\simeq^\times}) + g(B/{\simeq^\times}). \end{aligned}$$

This proves part (5). The proof of part (6) is essentially the same.

An algebra $\mathfrak{A} = \langle U, +, ^-, ;, ^\smile, 1'\rangle$ is **proper relation algebra** if there is an equivalence relation $E \in U$ such that U is a set of binary relations included in E and the following conditions hold.

$$A + B = \{\langle x, y\rangle : \langle x, y\rangle \in A \text{ or } \langle x, y\rangle \in B\} = A \cup B, \tag{3.3}$$
$$\overline{A} = \{\langle x, y\rangle : \langle x, y\rangle \in E \text{ and } \langle x, y\rangle \notin A\} = E \setminus A, \tag{3.4}$$
$$A ; B = \{\langle x, y\rangle : \text{for some } z, \langle x, z\rangle \in A \text{ and } \langle z, y\rangle \in B\}, \tag{3.5}$$
$$A^{\smile} = \{\langle x, y\rangle : \langle y, x\rangle \in A\}, \tag{3.6}$$
$$1' = \{\langle x, x\rangle : \langle x, x\rangle \in E\}. \tag{3.7}$$

For any equivalence relation E, $\mathfrak{Sb}(E)$ is the proper relation algebra whose universe $\wp(E)$ consists of all relations included in E, called the **algebra of subrelations of** E. For any set U, $\mathfrak{A}$ is **proper relation algebra on** U if conditions (3.3)–(3.7) hold when $E = U \times U$. For example, $\mathfrak{Re}(U)$ is defined as the proper relation algebra on U whose universe $\wp(U \times U)$ is the set of all binary relations on U, called the **algebra of binary relations on** U. The algebra $\mathfrak{A}$ is a **representable relation algebra** if it is isomorphic to a proper relation algebra. Let RRA be the class of representable relation algebras. Straightforward computations show that the axioms of $\mathcal{L}^{\times}$ and $\mathcal{L}w^{\times}$ are true in every representable relation algebra. Therefore, $\mathsf{RRA} \subseteq \mathsf{RA} \subseteq \mathsf{SA}$.

Suppose U is a set and $R(A)$ is a binary relation on U whenever $1' \neq A \in \Pi$. This determines a relational structure $\mathfrak{U} = \langle U, R(A)\rangle_{1' \neq A \in \Pi}$. The **denotation function** $\mathsf{De}_{\mathfrak{U}}$ of $\mathfrak{U}$ is defined for all $A, B \in \Pi^{+}$ by the following conditions.

$$\mathsf{De}_{\mathfrak{U}}(A) = R(A) \text{ if } 1' \neq A \in \Pi, \tag{3.8}$$
$$\mathsf{De}_{\mathfrak{U}}(A + B) = \mathsf{De}_{\mathfrak{U}}(A) \cup \mathsf{De}_{\mathfrak{U}}(B), \tag{3.9}$$
$$\mathsf{De}_{\mathfrak{U}}(\overline{A}) = (U \times U) \setminus \mathsf{De}_{\mathfrak{U}}(A), \tag{3.10}$$
$$\mathsf{De}_{\mathfrak{U}}(A ; B) = \mathsf{De}_{\mathfrak{U}}(A) ; \mathsf{De}_{\mathfrak{U}}(B), \tag{3.11}$$
$$\mathsf{De}_{\mathfrak{U}}(A^{\smile}) = \mathsf{De}_{\mathfrak{U}}(A)^{\smile}, \tag{3.12}$$
$$\mathsf{De}_{\mathfrak{U}}(1') = \{\langle u, u\rangle : u \in U\}, \tag{3.13}$$

where the symbols on the right in (3.11) and (3.12) denote the operations on binary relations defined in (3.5) and (3.6). This definition is used by Tarski and Givant (1987, p. 26, p. 47, 6.1(i)). Note that $\mathsf{De}_{\mathfrak{U}}$ is a homomorphism from $\mathfrak{P}$ into $\mathfrak{Re}(U)$. Since $\mathfrak{P}$ is absolutely freely generated by $\{A : 1' \neq A \in \Pi\}$, $\mathsf{De}_{\mathfrak{U}}$ could also have been defined as the unique homomorphism from $\mathfrak{P}$ into $\mathfrak{Re}(U)$ determined by condition (3.8).

For any sequence $s\colon \omega \to U$ of elements of U and any formula $\varphi \in \Phi^{+}$ the **satisfaction relation** is defined by induction on the complexity of formulas in Φ^{+} as follows.

$$\begin{array}{ll}
\mathfrak{U} \models A \mathrel{\mathring{=}} B[s] & \text{iff } \mathsf{De}_{\mathfrak{U}}(A) = \mathsf{De}_{\mathfrak{U}}(B), \text{ for} A, B \in \Pi^+, \\
\mathfrak{U} \models \mathsf{v}_i A \mathsf{v}_j[s] & \text{iff } \langle s_i, s_j \rangle \in \mathsf{De}_{\mathfrak{U}}(A), \text{ for } A \in \Pi^+ \text{and } i, j \in \omega, \\
\mathfrak{U} \models \varphi \Rightarrow \psi[s] & \text{iff } \mathfrak{U} \models \psi[s] \text{ or not } \mathfrak{U} \models \varphi[s], \\
\mathfrak{U} \models \neg\varphi[s] & \text{iff not } \mathfrak{U} \models \varphi[s], \\
\mathfrak{U} \models \forall_{\mathsf{v}_i}\varphi[s] & \text{iff } \mathfrak{U} \models \varphi[s^{i/u}] \text{ for every } u \in U \text{ and } i \in \omega, \text{ where}
\end{array}$$

$$s_j^{i/u} = \begin{cases} s_j & \text{if } i \neq j \in \omega, \\ u & \text{if } j = i. \end{cases}$$

Finally, $\mathfrak{U} \models \varphi$ iff $\mathfrak{U} \models \varphi[s]$ for all $s : \omega \to I$, and for every $\Psi \subseteq \Phi^+$, $\mathfrak{U} \models \Psi$ if $\mathfrak{U} \models \varphi$ for every $\varphi \in \Psi$. When $\mathfrak{U} \models \varphi$ or $\mathfrak{U} \models \Psi$, we say $\mathfrak{U}$ is a **model of** φ or Ψ, respectively.

Theorem 3.4 *Assume* $\Psi \subseteq \Sigma^+$.

1. $\simeq^+_\Psi$ *is a congruence relation on* $\mathfrak{P}$ *and the quotient algebra* $\mathfrak{P}/\simeq^+_\Psi$ *is a representable relation algebra* (Tarski and Givant 1987, 8.3(vii)).
2. *For every* $\varphi \in \Sigma^+$, $\Psi \vdash^+ \varphi$ *iff for every proper relation algebra* $\mathfrak{A}$ *on a set* U *and every homomorphism* $h\colon \mathfrak{P} \to \mathfrak{A}$, *if* $\mathfrak{U} = \langle U, h(A)\rangle_{A\in\Pi}$ *is a model of* Ψ *then* $\mathfrak{U}$ *is a model of* φ.
3. $\mathfrak{P}/\simeq^+$ *is a representable relation algebra that is* RRA*-freely generated by* $\{A/\simeq^+ : 1\text{'} \neq A \in \Pi\}$ (Tarski and Givant 1987, 8.3(viii); Maddux 1978, Thm 11(4); Maddux 2006, Thm 553; Maddux 1999, Thm 4.3).

Proof The proof of Theorem 3.4 makes use of Gödel's completeness theorem for $\mathcal{L}$ and its extension to $\mathcal{L}^+$ (Tarski and Givant 1987, Sects. 1.4, 2.2). For a detailed proof of the completeness theorem for $\mathcal{L}^+$ in a more general setting that allows predicates of arbitrary finite rank, see (Maddux 2006, Thm 170). By the extension of the completeness theorem to $\mathcal{L}^+$, $A \simeq^+_\Psi B$ iff $\mathsf{De}_{\mathfrak{U}}(A) = \mathsf{De}_{\mathfrak{U}}(B)$ whenever $\mathfrak{U}$ is a model of Ψ. It is apparent from this that $\simeq^+_\Psi$ is a congruence relation, but for a detailed proof see (Maddux 2006, Thm 130).

Tarski and Givant present the following proof of part (1). First construct an indexed system $\langle \mathfrak{U}_i : i \in I\rangle$ of structures $\mathfrak{U}_i = \langle U_i, R_i(A)\rangle_{1\text{'}\neq A\in\Pi}$ such that every model of Ψ is elementarily equivalent to (satisfies the same sentences as) one of the indexed structures. Let $\mathfrak{A}$ be the direct product of the system $\langle \mathfrak{Re}(U_i) : i \in I\rangle$. For each $i \in I$ let h_i be the unique homomorphism from $\mathfrak{P}$ into $\mathfrak{Re}(U_i)$ that extends $R_i\colon \{1\text{'} \neq A \in \Pi\} \to \wp(U_i \times U_i)$. Since h_i is a homomorphism that agrees with $\mathsf{De}_{\mathfrak{U}_i}$ on the propositional variables it follows from the definition of $\mathsf{De}_{\mathfrak{U}_i}$ by induction on predicates that $h_i = \mathsf{De}_{\mathfrak{U}_i}$. If $A \simeq^+_\Psi B$, i.e., $\Psi \vdash^+ A \mathrel{\mathring{=}} B$, then $\mathfrak{U}_i \models A \mathrel{\mathring{=}} B$ by the soundness part of the completeness theorem, hence $h_i(A) = \mathsf{De}_{\mathfrak{U}_i} A = \mathsf{De}_{\mathfrak{U}_i} B = h_i(B)$. It follows that h_i determines a map g_i from $\{A/\simeq^+_\Psi : A \in \Pi^+\}$ into $\mathfrak{A}$ that satisfies the condition $g(A/\simeq^+_\Psi) = h(A)$ for every $A \in \Pi^+$. This condition implies g_i is a homomorphism as in the proof of Theorem 3.3 (5). Define k by setting

$$k(A/\simeq^+_\Psi) = \langle g_i(A/\simeq^+_\Psi) : i \in I\rangle \text{ for every } A \in \Pi^+.$$

Then k is a homomorphism from $\mathfrak{P}/\simeq^{+}_{\Psi}$ into $\mathfrak{A}$. Suppose $A/\simeq^{+}_{\Psi} \neq B/\simeq^{+}_{\Psi}$ for some $A, B \in \Pi^{+}$. By the completeness theorem for $\mathcal{L}^{+}$, there is a model of Ψ in which $A \mathrel{\mathring{=}} B$ fails, so by the definition of the system $\langle \mathfrak{U}_i : i \in I \rangle$ there is some $j \in I$ such that $g_j(A/\simeq^{+}_{\Psi}) \neq g_j(B/\simeq^{+}_{\Psi})$ and hence $k(A/\simeq^{+}_{\Psi}) \neq k(B/\simeq^{+}_{\Psi})$. This shows that k is one-to-one and therefore an isomorphism of $\mathfrak{P}/\simeq^{+}_{\Psi}$ into the proper relation algebra $\mathfrak{A}$.

3.7 3-Variable Formalisms $\mathcal{L}_3^{+}$ and $\mathcal{L}_3$

$\mathcal{L}^{\times}$ is weaker than $\mathcal{L}^{+}$ and $\mathcal{L}$. It follows from Theorem 3.3 (3) and Theorem 3.4 (2) that if an equation is provable in $\mathcal{L}^{+}$ but is not true in some $\mathfrak{A} \in \mathsf{RA}$, then it is not provable in $\mathcal{L}^{\times}$ and $\mathcal{L}^{\times}$ is therefore weaker than $\mathcal{L}^{+}$ in means of proof. To get the following equation, which is provable in $\mathcal{L}^{+}$ but not $\mathcal{L}^{\times}$, Tarski and Givant used a non-representable relation algebra found by McKenzie (1970) that is generated by a single element. Givant constructed the equation and it was later simplified by George McNulty and Tarski (Tarski and Givant 1987, 3.4(vi)).

$$1 = 1; (A \dagger A + (A; A + 1' + (\overline{A + A^{\smile}}); (\overline{A + A^{\smile}}) \cdot \overline{A^{\smile}}) \cdot \overline{A} + \overline{A; A^{\smile}}); 1.$$

$\mathcal{L}^{\times}$ is also weaker than $\mathcal{L}$ and $\mathcal{L}^{+}$ in means of expression due to Korselt's result, reported by Löwenheim (1915), that no equation in $\Sigma^{\times}$ is logically equivalent to any sentence asserting the existence of four distinct objects, such as

$$[\,\exists_w \neg(w1'x \vee w1'y \vee w1'z)\,].$$

Tarski greatly generalized Korselt's theorem (Tarski and Givant 1987, 3.5(viii)).

$\mathcal{L}^{\times}$ seems to be correlated with the logic of three variables because Korselt's sentence uses four variables while $\mathsf{G}(A \mathrel{\mathring{=}} B)$ contains at most three. Indeed, it is apparent from the definition of G that if neither A nor B contains an occurrence of ; , then only the first two variables occur in $\mathsf{G}(A \mathrel{\mathring{=}} B)$ but if ; occurs in A or B then $\mathsf{G}(A \mathrel{\mathring{=}} B)$ does have the first three variables in it. This suggests that perhaps *every* sentence containing only the first three variables is logically equivalent to an equation in $\Sigma^{\times}$. Tarski was able to show that this is actually the case. He proposed the construction of 3-variable formalisms $\mathcal{L}_3^{+}$ and $\mathcal{L}_3$ that would be equipollent with $\mathcal{L}^{\times}$ in means of expression and proof.

For every finite $n \geq 3$ let Φ_n^{+} be the set of formulas in Φ^{+} that contain only variables in Υ_n and let

$$\Phi_n = \Phi \cap \Phi_n^{+}, \qquad \Sigma_n = \Sigma \cap \Phi_n^{+}, \qquad \Sigma_n^{+} = \Sigma^{+} \cap \Phi_n^{+}.$$

Tarski's theorem that every sentence in Σ_3^{+} is logically equivalent to an equation in $\Sigma^{\times}$ suggests that Σ_3^{+} and Σ_3 should be the sets of sentences of $\mathcal{L}_3^{+}$ and $\mathcal{L}_3$. The restriction of G to Σ_3^{+} could serve as the translation mapping from $\mathcal{L}_3^{+}$ to $\mathcal{L}_3$.

Tarski's initial proposal came in two parts. First, Tarski proposed restricting the axioms (AI)–(AIX) and the rule MP to those instances that belong to Σ_3^+. Givant found these restricted axioms were too weak and suggested replacing (AIX) with (AIX′), called the **general Leibniz law**, which is formulated in terms of a variant type of substitution defined by Tarski and Givant (1987, pp. 66–67).

$$[\, x\, 1\text{'}\, y \Rightarrow (\varphi \Rightarrow \varphi[x/y])\,]. \tag{AIX′}$$

The variant substitution is complicated so Tarski and Givant borrowed an idea from Maddux (1978) to formulate an alternate axiom (AIX″). If $x, y \in \Upsilon$ and $\varphi \in \Phi^+$ then $\mathsf{S}_{xy}\varphi$ is the result of interchanging x and y throughout the formula φ. The function $\mathsf{S}_{xy} : \Phi^+ \to \Phi^+$ is determined by these rules, in which $\hat{x} = y$, $\hat{y} = x$, and $\hat{v} = v$ if $v \neq x, y$ for every $v \in \Upsilon$.

$$\begin{aligned} \mathsf{S}_{xy}(vAw) &= \hat{v}A\hat{w}, \\ \mathsf{S}_{xy}(\varphi \Rightarrow \psi) &= \mathsf{S}_{xy}(\varphi) \Rightarrow \mathsf{S}_{xy}(\psi), \\ \mathsf{S}_{xy}(\neg\varphi) &= \neg\mathsf{S}_{xy}(\varphi), \\ \mathsf{S}_{xy}(\forall_v \varphi) &= \forall_{\hat{v}}\mathsf{S}_{xy}(\varphi). \end{aligned}$$

Givant proved that (AIX″) can be used instead of (AIX′) in the axiomatization of $\mathcal{L}_3^+$,

$$[\, x\, 1\text{'}\, y \Rightarrow (\varphi \Rightarrow \mathsf{S}_{xy}\varphi)\,]. \tag{AIX″}$$

Tarski knew by the early 1940s that (R_4) can not be proved with only three variables and would have to be included in the axiomatization of $\mathcal{L}_3^+$ by *fiat*. The second part of Tarski's proposal was to include the general associativity axiom

$$[\, \exists_z(\exists_y(\varphi[x,y] \wedge \psi[y,z]) \wedge \xi[z,y]) \Leftrightarrow \exists_z(\varphi[x,z] \wedge \exists_x(\psi[z,x] \wedge \xi[x,y]))\,]. \tag{AX}$$

This axiom involves the complicated substitution but Givant proved it could be replaced by (AX′), in which the free variables of formulas $\varphi, \psi, \xi \in \Phi^+$ are just x and y,

$$[\, \exists_z(\exists_y(\varphi \wedge \mathsf{S}_{xz}\psi) \wedge \mathsf{S}_{xz}\xi) \Leftrightarrow \exists_z(\mathsf{S}_{yz}\varphi \wedge \exists_x(\mathsf{S}_{yz}\psi \wedge \xi)\,)\,]. \tag{AX′}$$

The sets of sentences of $\mathcal{L}_3^+$ and $\mathcal{L}_3$ are Σ_3^+ and Σ_3, their axioms are the sentences in Σ_3^+ and Σ_3 that are instances of (AI)–(AVIII), (AIX′), or (AX), and their rule of inference is MP. For simpler axiom sets use (AIX″) and (AX′) instead of (AIX′) and (AX). For every $\Psi \subseteq \Sigma_3^+$ a sentence $\varphi \in \Sigma_3^+$ is **provable in $\mathcal{L}_3^+$ from** Ψ, written $\Psi \vdash_3^+ \varphi$ or $\vdash_3^+ \varphi$ if $\Psi = \emptyset$, if φ is in every set that contains Ψ and the axioms of $\mathcal{L}_3^+$ and is closed under MP. The **theory generated by** Ψ in $\mathcal{L}_3^+$ is

$$\Theta\eta_3^+\,\Psi = \{\varphi : \varphi \in \Sigma_3^+,\ \Psi \vdash_3^+ \varphi\}.$$

The notions of $\Psi \vdash_3 \varphi$, φ is **provable in** $\mathcal{L}_3$ **from** Ψ, and $\Theta\eta_3\,\Psi$, the **theory generated by** Ψ in $\mathcal{L}_3$, are defined similarly for every $\Psi \subseteq \Sigma_3$ and sentence $\varphi \in \Sigma_3$.

3.8 Equipollence of $\mathcal{L}^\times$, $\mathcal{L}_3$, and $\mathcal{L}_3^+$

With Givant's changes Tarski's proposal worked. It provided a 3-variable restriction $\mathcal{L}_3^+$ of $\mathcal{L}^+$ and a 3-variable restriction $\mathcal{L}_3$ of $\mathcal{L}$. Both are equipollent with $\mathcal{L}^\times$ in means of expression and proof. .f $\mathcal{L}_3$ and $\mathcal{L}_3^+$ the appropriate translation mapping from $\mathcal{L}_3^+$ to $\mathcal{L}_3$ is simply the restriction of G to Σ_3^+. Part (4) of the following theorem is the **main mapping theorem** for $\mathcal{L}_3$ and $\mathcal{L}_3^+$.

Theorem 3.5 (Tarski and Givant 1987, Sect. 3.8) *Formalisms $\mathcal{L}_3$ and $\mathcal{L}_3^+$ are equipollent.*

1. $\Phi_3 \subseteq \Phi_3^+$ *and* $\Sigma_3 \subseteq \Sigma_3^+$ [3.8(viii)(α)].
2. G *maps* Φ_3^+ *onto* Φ_3 *and* Σ_3^+ *onto* Σ_3 [3.8(ix)(δ)].
3. $\varphi \equiv_3^+ \mathsf{G}(\varphi)$ *if* $\varphi \in \Phi_3^+$ [3.8(ix)(ε)].
4. $\Psi \vdash_3^+ \varphi$ *iff* $\{\mathsf{G}(\psi) : \psi \in \Psi\} \vdash_3 \mathsf{G}(\varphi)$, *for* $\Psi \subseteq \Sigma_3^+$ *and* $\varphi \in \Sigma_3^+$ [3.8(xi)].
5. $\Psi \vdash_3^+ \varphi$ *iff* $\Psi \vdash_3 \varphi$, *for* $\Psi \subseteq \Sigma_3$ *and* $\varphi \in \Sigma_3$, [3.8(viii)(β), 3.8(xii)(β)].

For the equipollence of $\mathcal{L}^\times$ and $\mathcal{L}_3^+$, Tarski and Givant construct a function H on Φ_3^+ whose restriction to Σ_3^+ is an appropriate translation mapping from $\mathcal{L}_3^+$ to $\mathcal{L}^\times$ (Tarski and Givant 1987, pp. 77–79). They begin, "Its definition is complicated and must be formulated with care. We give here enough hints for constructing such a definition, without formulating it precisely in all details," and end with, "We hope the above outline gives an adequate idea of the definition of H." After obtaining the main mapping theorem for $\mathcal{L}^\times$ and $\mathcal{L}_3^+$ Tarski and Givant (1987, p. 87) say,

> The construction used here to establish these equipollence results has clearly some serious defects, if only from the point of view of mathematical elegance. Actually, this applies to the proof of the equipollence of $\mathcal{L}_3^+$ and $\mathcal{L}^\times$. The splintered character of the definition of the translation mapping H, with its many cases, is a principal cause of the fragmented nature of certain portions of the argument; the involved notion of substitution (which we have to use because of the restricted number of variables in our formalisms) is another detrimental factor. As a final result, the construction is so cumbersome in some of its parts—culminating in the proofs of (iv) and (v)—that we did not even attempt to present them in full. A different construction that would remove most of the present defects would be very desirable indeed.

Another description of H was given by Givant (2006) and simpler alternative constructions appear in (Maddux 1978, p. 192–3) and (Maddux 2006, p. 543–4, p. 548–9). One of these is presented here as a response to Tarski and Givant. It can be precisely defined in one page instead of outlined in three. We start with an auxiliary map $\mathsf{J}\colon \Phi_3^+ \to \Phi_3^+$. The map J has the property that for every formula $\varphi \in \Phi_3^+$ there are $k \in \omega$ and finite sequences of atomic predicates $R, S, T \in {}^k\Pi$ such that

$$\mathsf{J}(\varphi) = \bigwedge_{i\langle k}(\mathsf{v}_0 R_i \mathsf{v}_2 \vee \mathsf{v}_2 S_i \mathsf{v}_1 \vee \mathsf{v}_0 T_i \mathsf{v}_1) \in \Phi_3^+.$$

J is defined by induction on the complexity of formulas. For the meanings of 0, 1, $\cdot$, and $\dagger$, recall definitions (3.1) and (3.2). If $A, B \in \Pi$ then

$$\begin{aligned}
\mathsf{J}(\mathsf{v}_0 A \mathsf{v}_1) &= \mathsf{v}_0 0 \mathsf{v}_2 \vee \mathsf{v}_2 0 \mathsf{v}_1 \vee \mathsf{v}_0 A \mathsf{v}_1,\\
\mathsf{J}(\mathsf{v}_1 A \mathsf{v}_0) &= \mathsf{v}_0 0 \mathsf{v}_2 \vee \mathsf{v}_2 0 \mathsf{v}_1 \vee \mathsf{v}_0 A^{\smile} \mathsf{v}_1,\\
\mathsf{J}(\mathsf{v}_1 A \mathsf{v}_2) &= \mathsf{v}_0 0 \mathsf{v}_2 \vee \mathsf{v}_2 A^{\smile} \mathsf{v}_1 \vee \mathsf{v}_0 0 \mathsf{v}_1,\\
\mathsf{J}(\mathsf{v}_2 A \mathsf{v}_1) &= \mathsf{v}_0 0 \mathsf{v}_2 \vee \mathsf{v}_2 A \mathsf{v}_1 \vee \mathsf{v}_0 0 \mathsf{v}_1,\\
\mathsf{J}(\mathsf{v}_0 A \mathsf{v}_2) &= \mathsf{v}_0 A \mathsf{v}_2 \vee \mathsf{v}_2 0 \mathsf{v}_1 \vee \mathsf{v}_0 0 \mathsf{v}_1,\\
\mathsf{J}(\mathsf{v}_2 A \mathsf{v}_0) &= \mathsf{v}_0 A^{\smile} \mathsf{v}_2 \vee \mathsf{v}_2 0 \mathsf{v}_1 \vee \mathsf{v}_0 0 \mathsf{v}_1,\\
\mathsf{J}(\mathsf{v}_0 A \mathsf{v}_0) &= \mathsf{v}_0 0 \mathsf{v}_2 \vee \mathsf{v}_2 0 \mathsf{v}_1 \vee \mathsf{v}_0 (A \cdot 1\text{'}) ; 1 \mathsf{v}_1,\\
\mathsf{J}(\mathsf{v}_1 A \mathsf{v}_1) &= \mathsf{v}_0 0 \mathsf{v}_2 \vee \mathsf{v}_2 0 \mathsf{v}_1 \vee \mathsf{v}_0 1 ; (A \cdot 1\text{'}) \mathsf{v}_1,\\
\mathsf{J}(\mathsf{v}_2 A \mathsf{v}_2) &= \mathsf{v}_0 (1 ; (A \cdot 1\text{'})) \mathsf{v}_2 \vee \mathsf{v}_2 0 \mathsf{v}_1 \vee \mathsf{v}_0 0 \mathsf{v}_1,\\
\mathsf{J}(A \mathrel{\mathring{=}} B) &= \mathsf{v}_0 0 \mathsf{v}_2 \vee \mathsf{v}_2 0 \mathsf{v}_1 \vee \mathsf{v}_0 0 \dagger (A \cdot B + \overline{A} \cdot \overline{B}) \dagger 0 \mathsf{v}_1.
\end{aligned}$$

This completes the cases in which φ is atomic. If $\varphi, \varphi' \in \Phi_3^+$ and there are $k, k' \in \omega$, $R, S, T \in {}^k\Pi$, and $R', S', T' \in {}^{k'}\Pi$ such that

$$\begin{aligned}
\mathsf{J}(\varphi) &= \bigwedge_{i \langle k} (\mathsf{v}_0 R_i \mathsf{v}_2 \vee \mathsf{v}_2 S_i \mathsf{v}_1 \vee \mathsf{v}_0 T_i \mathsf{v}_1),\\
\mathsf{J}(\varphi') &= \bigwedge_{j \langle k'} (\mathsf{v}_0 R'_j \mathsf{v}_2 \vee \mathsf{v}_2 S'_j \mathsf{v}_1 \vee \mathsf{v}_0 T'_j \mathsf{v}_1),
\end{aligned}$$

then

$$\begin{aligned}
\mathsf{J}(\neg\varphi) &= \bigwedge_{f \in {}^k 3} \left(\mathsf{v}_0 \Big(\sum_{f(i)=0} \overline{R_i} \Big) \mathsf{v}_2 \vee \mathsf{v}_2 \Big(\sum_{f(i)=1} \overline{S_i} \Big) \mathsf{v}_1 \vee \mathsf{v}_0 \Big(\sum_{f(i)=2} \overline{T_i} \Big) \mathsf{v}_1 \right),\\
\mathsf{J}(\varphi \Rightarrow \varphi') &= \bigwedge_{f \in {}^k 3,\, j \langle k'} \left(\mathsf{v}_0 \Big(\sum_{f(i)=0} \overline{R_i} + R'_j \Big) \mathsf{v}_2 \vee \right.\\
&\qquad \left. \mathsf{v}_2 \Big(\sum_{f(i)=1} \overline{S_i} + S'_j \Big) \mathsf{v}_1 \vee \mathsf{v}_0 \Big(\sum_{f(i)=2} \overline{T_i} + T'_j \Big) \mathsf{v}_1 \right),\\
\mathsf{J}(\forall_{\mathsf{v}_0} \varphi) &= \bigwedge_{i \langle k} \Big(\mathsf{v}_0 0 \mathsf{v}_2 \vee \mathsf{v}_2 (R^{\smile}{}_i \dagger T_i) + S_i \mathsf{v}_1 \vee \mathsf{v}_0 0 \mathsf{v}_1 \Big),\\
\mathsf{J}(\forall_{\mathsf{v}_1} \varphi) &= \bigwedge_{i \langle k} \Big(\mathsf{v}_0 (T_i \dagger S^{\smile}{}_i) + R_i \mathsf{v}_2 \vee \mathsf{v}_2 0 \mathsf{v}_1 \vee \mathsf{v}_0 0 \mathsf{v}_1 \Big),\\
\mathsf{J}(\forall_{\mathsf{v}_2} \varphi) &= \bigwedge_{i \langle k} \Big(\mathsf{v}_0 0 \mathsf{v}_2 \vee \mathsf{v}_2 0 \mathsf{v}_1 \vee \mathsf{v}_0 (R_i \dagger S_i) + T_i \mathsf{v}_1 \Big).
\end{aligned}$$

Now we can define H on sentences $\varphi \in \Sigma_3$. Apply J to $\forall_{\mathsf{v}_2}\varphi$, obtaining $k < \omega$ and finite sequences $R, S, T \in {}^k\Pi$ such that

$$\mathsf{J}(\forall_{\mathsf{v}_2}\varphi) = \bigwedge_{i\langle k} \Big(\mathsf{v}_0 0 \mathsf{v}_2 \vee \mathsf{v}_2 0 \mathsf{v}_1 \vee \mathsf{v}_0(R_i \dagger S_i) + T_i \mathsf{v}_1\Big),$$

and set $\mathsf{H}(\varphi)$ equal to

$$1 \mathrel{\mathring{=}} \big((R_0 \dagger S_0) + T_0\big) \cdot \big((R_1 \dagger S_1) + T_1\big) \cdot \ldots \cdot \big((R_{k-1} \dagger S_{k-1}) + T_{k-1}\big).$$

The original construction of H by Tarski and Givant (1987, Sect. 3.9) has desirable properties not shared by the mapping H defined above. Tarski's H is defined on all formulas but the definition given here is restricted to sentences. By (Tarski and Givant 1987, 3.9(iii)(δ)) the sets of free variables of φ and $\mathsf{H}(\varphi)$ are the same for every formula $\varphi \in \Phi_3^+$ and H maps Σ_3^+ onto $\Sigma^\times$ (not just into, as is the case for the H constructed here). Furthermore, H produces simpler output in certain cases. For example, $\mathsf{H}(\neg A \mathrel{\mathring{=}} B) = 1 ; \overline{A \cdot B + \overline{A} \cdot \overline{B}} ; 1 \mathrel{\mathring{=}} 1$, $\mathsf{H}(\neg A \mathrel{\mathring{=}} 1) = 1 ; \overline{A} ; 1 \mathrel{\mathring{=}} 1$, and $\mathsf{H}(A \mathrel{\mathring{=}} B) = A \mathrel{\mathring{=}} B$ whenever $A, B \in \Pi$. On the other hand, to insure that the output of H has the same set of free variables as the input, it is necessary in the definition of $\mathsf{H}(\neg\varphi)$ and $\mathsf{H}(\varphi \Rightarrow \psi)$ to consider many cases that depend on the free variables of $\mathsf{H}(\varphi)$ and $\mathsf{H}(\psi)$. This causes the very long proof by cases of the main mapping theorem for $\mathcal{L}^\times$ and $\mathcal{L}_3^+$ encountered by Tarski and Givant.

The following theorem summarizes the principal parts of the equipollence of $\mathcal{L}^\times$ and $\mathcal{L}_3^+$. It is stated for the versions of J and H constructed here, so part (2) only says "into" instead of "onto." The other parts are the same as the corresponding versions by Tarski and Givant (1987, Sect. 3.9). Part (4) is the **main mapping theorem** for $\mathcal{L}^\times$ and $\mathcal{L}_3^+$, while part (5), a corollary of part (4), is the equipollence of $\mathcal{L}^\times$ and $\mathcal{L}_3^+$ in means of proof.

Theorem 3.6 (Tarski and Givant 1987, Sect. 3.9) *Formalisms $\mathcal{L}^\times$ and $\mathcal{L}_3^+$ are equipollent in means of expression and proof.*

1. $\Sigma^\times \subseteq \Sigma_3^+$ [3.9(i)].
2. H *maps* Σ_3^+ *into* $\Sigma^\times$.
3. $\varphi \equiv_3^+ \mathsf{J}(\varphi)$ *if* $\varphi \in \Phi_3^+$ [3.9(iii)(ε)] *and* $\mathsf{J}(\varphi) \equiv \mathsf{H}(\varphi)$ *if* $\varphi \in \Sigma_3^+$.
4. $\Psi \vdash_3^+ \varphi$ *iff* $\{\mathsf{H}(\psi) : \psi \in \Psi\} \vdash^\times \mathsf{H}(\varphi)$, *for* $\Psi \subseteq \Sigma_3^+$ *and* $\varphi \in \Sigma_3^+$ [3.9(vii)].
5. $\Psi \vdash_3^+ \varepsilon$ *iff* $\Psi \vdash^\times \varepsilon$, *for* $\Psi \subseteq \Sigma^\times$ *and* $\varepsilon \in \Sigma^\times$ [3.9(ix)].
6. $\Theta\eta_3\, \Psi = \Theta\eta_3^+\, \Psi \cap \Sigma_3$, *for* $\Psi \subseteq \Sigma_3$ [3.9(x)].
7. $\Theta\eta^\times\, \Psi = \Theta\eta_3^+\, \Psi \cap \Sigma^\times$, *for* $\Psi \subseteq \Sigma^\times$ [3.9(xi)].

3.9 Equipollent 3-Variable Formalisms $\mathcal{L}s$, $\mathcal{L}s^+$, $\mathcal{L}w^\times$

Since Tarski and Givant included the axiom (AX) only to achieve the equipollence of $\mathcal{L}_3$ and $\mathcal{L}_3^+$ with $\mathcal{L}^\times$, they defined the "*(standardized) formalisms*" $\mathcal{L}s_3$ and $\mathcal{L}s_3^+$

by deleting (AX) from the axiom sets of $\mathcal{L}_3$ and $\mathcal{L}_3^+$ (Tarski and Givant 1987, p. 89). They did not introduce any special notation for provability in these formalisms. We use $\vdash_s$ and $\vdash_s^+$ for $\mathcal{L}s_3$ and $\mathcal{L}s_3^+$. Since (R_4) is the axiom of $\mathcal{L}^\times$ that corresponds to (AX), Tarski and Givant asked whether simply deleting (R_4) from the axioms of $\mathcal{L}^\times$ would produce a formalism equipollent with the standardized formalisms $\mathcal{L}s_3$ and $\mathcal{L}s_3^+$. The answer is "no." For example, the **semi-associative law** (R_4') in Table 3.6 is provable in $\mathcal{L}s_3^+$ but cannot be derived in $\mathcal{L}^\times$ from just the axioms (R_1)–(R_3) and (R_5)–(R_{10}) when 1' $\neq A, B \in \Pi$. Another equation with the same property is

$$A ; 1 \mathrel{\mathring{=}} (A ; 1) ; 1.$$

Adding either one of these as an axiom produces a formalism equipollent with $\mathcal{L}s_3$ and $\mathcal{L}s_3^+$. Therefore, Tarski and Givant defined a weakened equational formalism $\mathcal{L}w^\times$ by replacing (R_4) with (R_4') in the axiomatization of $\mathcal{L}^\times$. The equipollence of $\mathcal{L}w^\times$ with $\mathcal{L}s_3$ and $\mathcal{L}s_3^+$ is stated in the next theorem and was noted by Tarski and Givant (1987, p. 89, p. 209). Part (4) tells us that the axioms for SA characterize the equations provable in standard first order logic of three variables without associativity.

Theorem 3.7 (Maddux 1978, Thm 11(30); Maddux 1999, Thm 6.3; Maddux 2006, Thm 569(i)(ii)) *Formalisms $\mathcal{L}w^\times$, $\mathcal{L}s_3$, and $\mathcal{L}s_3^+$ are equipollent in means of expression and proof.*

1. $\varphi \equiv_s^+ \mathsf{G}(\varphi) \equiv_s^+ \mathsf{J}(\varphi)$, *for* $\varphi \in \Phi_3^+$.
2. $\Psi \vdash_s^+ \varphi$ *iff* $\{\mathsf{G}(\psi) : \psi \in \Psi\} \vdash_s \mathsf{G}(\varphi)$ *iff* $\{\mathsf{H}(\psi) : \psi \in \Psi\} \vdash_s^\times \mathsf{H}(\varphi)$, *for* $\Psi \subseteq \Sigma_3^+$ *and* $\varphi \in \Sigma_3^+$.
3. $\Psi \vdash_s^+ \varphi$ *iff* $\Psi \vdash_s \varphi$, *for* $\Psi \subseteq \Sigma_3$ *and* $\varphi \in \Sigma_3$.
4. $\Psi \vdash_s^+ \varepsilon$ *iff* $\Psi \vdash_s^\times \varepsilon$, *for* $\Psi \subseteq \Sigma^\times$ *and* $\varepsilon \in \Sigma^\times$.

Tarski and Givant (1987, p. 91) also define formalisms $\mathcal{L}_n$ and $\mathcal{L}_n^+$ for every finite $n \geq 4$, imitating the definitions of $\mathcal{L}s_3$ and $\mathcal{L}s_3^+$, but with n in place of 3. The sets of formulas of $\mathcal{L}_n$ and $\mathcal{L}_n^+$ are Φ_n and Φ_n^+, the sets of sentences are Σ_n and Σ_n^+, the sets of axioms are those instances of axioms (AI)–(AVIII) and (AIX') that lie in Σ_n and Σ_n^+, and the only rule of inference is MP. The next two theorems involve the first of these formalisms, when $n = 4$. They are a precise expression of the fact, noted by Tarski and Givant (1987, p. 92), that a sentence in Σ_3^+ (containing only three variables) can be proved using four variables iff it can be proved using just three variables together with the assumption that relative multiplication is associative. (The notationally peculiar equivalence of $\vdash_3$ with $\vdash_4$ and $\vdash_3^+$ with $\vdash_4^+$ is due to the inclusion of associativity in the definitions of $\vdash_3$ and $\vdash_3^+$.)

Theorem 3.8 (Maddux 1978, Thm 11(31); Maddux 1999, Thm 6.4; Maddux 2006, Thm 569(iii)(iv))

1. $\Psi \vdash_4^+ \varphi$ *iff* $\{\mathsf{G}(\psi) : \psi \in \Psi\} \vdash_4 \mathsf{G}(\varphi)$ *iff* $\{\mathsf{H}(\psi) : \psi \in \Psi\} \vdash^\times \mathsf{H}(\varphi)$, *for* $\Psi \subseteq \Sigma_3^+$ *and* $\varphi \in \Sigma_3^+$.
2. $\Psi \vdash_4^+ \varphi$ *iff* $\Psi \vdash_3^+ \varphi$, *for* $\Psi \subseteq \Sigma_3^+$ *and* $\varphi \in \Sigma_3^+$.
3. $\Psi \vdash_4^+ \varphi$ *iff* $\Psi \vdash_4 \varphi$ *iff* $\Psi \vdash_3 \varphi$, *for* $\Psi \subseteq \Sigma_3$ *and* $\varphi \in \Sigma_3$.

4. $\Psi \vdash_4^+ \varepsilon$ *iff* $\Psi \vdash^\times \varepsilon$, *for* $\Psi \subseteq \Sigma^\times$ *and* $\varepsilon \in \Sigma^\times$.

Theorem 3.8 has the following consequences when $\Psi = \emptyset$.

Theorem 3.9 (Maddux 1989, Thm 24)

1. $\Theta\eta_3\,\emptyset = \Theta\eta_4\,\emptyset \cap \Sigma_3$.
2. $\Theta\eta_3^+\,\emptyset = \Theta\eta_4^+\,\emptyset \cap \Sigma_3^+$.
3. $\Theta\eta^\times\,\emptyset = \Theta\eta_4^+\,\emptyset \cap \Sigma^\times$.

Theorem 3.9 (3) asserts that an equation is true in every relation algebra iff its translation into a sentence containing only three variables can be proved with four variables. Theorem 3.9 (3) provides an answer to Tarski's question, "whether this definition of relation algebra ... is justified in any intrinsic sense." The answer is that Tarski's definition characterizes the equations provable with four variables. Theorem 3.9 says,

true in relation algebras $\Leftrightarrow$ 3-provable with associativity $\Leftrightarrow$ 4-provable.

3.10 Relevance Logics T_n, CT_n, $3 \le n \le \omega$, and TR

The connectives of relevance logic that are interpreted in Table 3.1 as operations on binary relations are $\vee$, $\wedge$, $\neg$, $\sim$, $\to$, $\circ$, *, and $\mathbf{t}$. To connect these interpretations with the formalisms of Tarski and Givant we use the symbols for these connectives to denote operators on Π^+. Although the symbols $\vee$, $\wedge$, and $\neg$ have already appeared as connectives in $\mathcal{L}$, they will also denote operators on Π^+. For their classical relevant logic CR^*, Routley and Meyer introduce **Boolean negation** $\neg$ and the **Routley star** *. Let $\vee$, $\wedge$, $\neg$, $\sim$, $\to$, $\circ$, *, and $\mathbf{t}$ denote the operators on Π^+ defined by

$$A \vee B = A + B, \tag{3.14}$$

$$A \wedge B = A \cdot B, \tag{3.15}$$

$$\neg A = \overline{A}, \tag{3.16}$$

$$\sim A = \overline{A^\smile}, \tag{3.17}$$

$$A \to B = \overline{A^\smile ; \overline{B}}, \tag{3.18}$$

$$A \circ B = B ; A, \tag{3.19}$$

$$A^* = A^\smile, \tag{3.20}$$

$$\mathbf{t} = 1'. \tag{3.21}$$

When parentheses are omitted, unary operators should be evaluated first, followed by $;$, $\circ$, $\cdot$, $\wedge$, $\dagger$, $+$, $\vee$, and then $\to$, in that order. Repeated binary operators of equal precedence should be evaluated from left to right. Definitions (3.17) and (3.18) produce the standard connection between $\rightarrow$ and $\circ$,

$$\sim(A \rightarrow \sim B) = \overline{\left(\overline{A^{\smile} ; \overline{\overline{B^{\smile}}}}\right)^{\smile}} \simeq^{\times} B ; A = A \circ B .$$

Besides the laws of double negation and the commutativity of converse and complementation, this computation involves the axiom (R_9). The definitions also match the interpretations in Table 3.1. To show this, we compute G for (3.14), (3.15), (3.17), and (3.18). In three cases we convert to simpler equivalent formulas. The results agree with Table 3.1. Let $x = \mathsf{v}_0$, $y = \mathsf{v}_1$, and $z = \mathsf{v}_2$. Assume A, B are propositional variables, $1' \neq A, B \in \Pi$, so that, for example, $\mathsf{G}(xAy) = xAy$ and $\mathsf{G}(xBy) = xBy$. Then

$$
\begin{aligned}
\mathsf{G}(xA \vee By) &= \mathsf{G}(xA + By) = \mathsf{G}(xAy) \vee \mathsf{G}(xBy) = xAy \vee xBy, \\
\mathsf{G}(xA \wedge By) &= \mathsf{G}(xA \cdot By) = \mathsf{G}(x\overline{\overline{A} + \overline{B}}y) = \neg \mathsf{G}(x\overline{A} + \overline{B}y) \\
&= \neg(\mathsf{G}(x\overline{A}y) \vee \mathsf{G}(x\overline{B}y)) = \neg(\neg \mathsf{G}(xAy) \vee \neg \mathsf{G}(xBy)) \\
&= \neg(\neg xAy \vee \neg xBy) \equiv^{+} xAy \wedge xBy, \\
\mathsf{G}(x{\sim}Ay) &= \mathsf{G}(x\overline{A^{\smile}}y) = \neg \mathsf{G}(xA^{\smile}y) = \neg \mathsf{G}(yAx) = \neg(yAx), \\
\mathsf{G}(xA \rightarrow By) &= \mathsf{G}(x\overline{A^{\smile} ; \overline{B}}y) = \neg \mathsf{G}(xA^{\smile} ; \overline{B}y) = \neg \exists_z (\mathsf{G}(xA^{\smile}z) \wedge \mathsf{G}(z\overline{B}y)) \\
&= \neg\neg\forall_z \neg(\mathsf{G}(zAx) \wedge \neg \mathsf{G}(zBy)) = \neg\neg\forall_z \neg(zAx \wedge \neg zBy) \\
&= \neg\neg\forall_z \neg\neg\big(zAx \Rightarrow \neg\neg zBy\big) \equiv^{+} \forall_z \big(zAx \Rightarrow zBy\big).
\end{aligned}
$$

The **relevance logic operators** are $\vee$, $\wedge$, $\sim$, $\rightarrow$, $\circ$, and **t**. The **classical relevant logic operators** are $\vee$, $\wedge$, $\neg$, $\sim$, $\rightarrow$, $\circ$, *, and **t**. The classical relevant logic operators include $\vee$, $\neg$, *, **t**, and $\circ$. The first four of these coincide with Tarski's $+$, $^{-}$, $^{\smile}$, $1'$, and ; can be defined from $\circ$, so the closure of Π under the classical relevant logic operators is Π^{+}. Let Π^{r} be the closure of Π under just the relevance logic operators. Then Π^{r} is a proper subset of Π^{+} since $\neg$ and * cannot be defined from the relevance logic operators.

For the next definitions, recall that for any $A, B \in \Pi^{+}$, $A \leq B$ is the equation $A + B \mathrel{\mathring{=}} B \in \Sigma^{\times}$. Suppose $3 \leq n \leq \omega$ and $\Psi \subseteq \Sigma^{\times}$. The equations in Ψ are the **non-logical assumptions** of the logics CT_n^{Ψ} and T_n^{Ψ} defined next. We omit reference to Ψ when $\Psi = \emptyset$.

$$\mathsf{CT}_n^{\Psi} = \{A \colon A \in \Pi^{+},\ \Psi \vdash_n^{+} 1' \leq A\}, \tag{3.22}$$

$$\mathsf{T}_n^{\Psi} = \mathsf{CT}_n^{\Psi} \cap \Pi^{r}, \tag{3.23}$$

$$\mathsf{CT}_n = \{A \colon A \in \Pi^{+},\ \vdash_n^{+} 1' \leq A\}, \tag{3.24}$$

$$\mathsf{T}_n = \mathsf{CT}_n \cap \Pi^{r}. \tag{3.25}$$

T_n is **Tarski's basic n-variable relevance logic** with no non-logical assumptions. It uses only the relevance logic operators. Adding "C" to get CT_n indicates its classical counterpart, in which $\neg$ and * are admitted, thus allowing the full range of classical relevant logic operators. Next, we define some special sets of equations.

$$\Xi^d = \{A \leq A;A : A \in \Pi^+\}, \tag{3.26}$$

$$\Xi^c = \{A;B \mathrel{\overset{\circ}{=}} B;A : A, B \in \Pi^+\}, \tag{3.27}$$

$$\Xi^s = \{A \mathrel{\overset{\circ}{=}} A^{\smile} : A \in \Pi^+\}. \tag{3.28}$$

We refer to Ξ^d, Ξ^c, and Ξ^s as the **equations of density**, **commutativity**, and **symmetry**, respectively. The equations of density and commutativity are used to define TR, **Tarski's classical relevant logic**, by

$$\mathsf{TR} = \mathsf{CT}_4^{\Psi} \text{ where } \Psi = \Xi^d \cup \Xi^c. \tag{3.29}$$

3.11 Frames and the Relevance Logics CR* and KR

A **frame** is a quadruple $\mathfrak{K} = \langle K, R, ^*, \mathcal{I}\rangle$ consisting of a set K, a ternary relation $R \subseteq K^3$, a unary operation $^*\colon K \to K$, and a subset $\mathcal{I} \subseteq K$. The associated **complex algebra of** $\mathfrak{K}$ is $\mathfrak{Cm}(\mathfrak{K}) = \langle \wp(K), \cup, ^-, ;, ^{\smile}, \mathcal{I}\rangle$, where $\wp(K)$ is the set of subsets of K, and the operations $\cup$, $^-$, $;$, and $^{\smile}$ are defined on subsets $X, Y \subseteq K$ by

$$X \cup Y = \{x : x \in X \text{ or } x \in Y\}, \tag{3.30}$$

$$\overline{X} = K \setminus X = \{x : x \in K,\ x \notin X\}, \tag{3.31}$$

$$X;Y = \{z : \langle x, y, z\rangle \in R \text{ for some } x \in X, y \in Y\}, \tag{3.32}$$

$$X^{\smile} = \{z^* : z \in X\}. \tag{3.33}$$

A predicate $A \in \Pi^+$ is **valid in** $\mathfrak{K}$, and $\mathfrak{K}$ **validates** A, if the equation $1' \leq A$ is true in $\mathfrak{Cm}(\mathfrak{K})$. $\mathfrak{K}$ **invalidates** A if A is not valid in $\mathfrak{K}$. Any homomorphism from the predicate algebra into $\mathfrak{Cm}(\mathfrak{K})$ must send 1' to $\mathcal{I}$, so $A \in \Pi^+$ is valid in $\mathfrak{K}$ iff $\mathcal{I} \subseteq h(A)$ for every homomorphism $h : \mathfrak{P} \to \mathfrak{Cm}(\mathfrak{K})$. A homomorphism $h : \mathfrak{P} \to \mathfrak{Cm}(\mathfrak{K})$ **validates** A if $\mathcal{I} \subseteq h(A)$ and h **invalidates** A otherwise. For every set U, $\mathfrak{U}^2 = \langle U^2, R, ^*, \mathcal{I}\rangle$ is the **pair-frame on** U where

$$U^2 = \{\langle x, y\rangle : x, y \in U\}, \tag{3.34}$$

$$R = \{\langle\langle x, y\rangle, \langle y, z\rangle, \langle x, z\rangle\rangle : x, y, z \in U\}, \tag{3.35}$$

$$\langle x, y\rangle^* = \langle y, x\rangle \text{ for all } x, y \in U, \tag{3.36}$$

$$\mathcal{I} = \{\langle x, x\rangle : x \in U\}. \tag{3.37}$$

Note that $\mathfrak{Re}(U)$, the algebra of binary relations on U, is the complex algebra of the pair-frame on U, i.e., $\mathfrak{Re}(U) = \mathfrak{Cm}\left(\mathfrak{U}^2\right)$.

The following conditions on a frame $\mathfrak{K}$ are written in a first-order language with equality symbol $=$, ternary relation symbol R, unary function symbol *, and unary relation symbol $\mathcal{I}$, but we will also frequently use an atomic formula like $Rxyz$ as an abbreviation for $\langle x, y, z\rangle \in R$ when R is a ternary relation. Each condition should be read as holding for all $v, w, x, y, z \in K$.

$$Rxyz \Rightarrow Ryz^*x^*, \tag{3.38}$$

$$Rxyz \Rightarrow Rz^*xy^*, \tag{3.39}$$

$$Rxyz \Rightarrow Ry^*x^*z^*, \tag{3.40}$$

$$Rxyz \Rightarrow Rx^*zy, \tag{3.41}$$

$$Rxyz \Rightarrow Rzy^*x, \tag{3.42}$$

$$x = y \Leftrightarrow \exists_u(\mathcal{I}u \wedge Rxuy), \tag{3.43}$$

$$x^{**} = x, \tag{3.44}$$

$$\exists_x(Rvwx \wedge Rxyz) \Rightarrow \exists_u Rvuz, \tag{3.45}$$

$$\exists_x(Rvwx \wedge Rxyz) \Rightarrow \exists_u(Rvuz \wedge Rwyu), \tag{3.46}$$

$$Rxxx, \tag{3.47}$$

$$Rxyz \Leftrightarrow Ryxz, \tag{3.48}$$

$$x^* = x. \tag{3.49}$$

When (3.44) holds, the implication $\Rightarrow$ in (3.38)–(3.42) can be replaced by an equivalence $\Leftrightarrow$, since applying each of these frame conditions to $Rxyz$ two or three times produces $Rx^{**}y^{**}z^{**}$. See Lemma 3.4 (1) below.

Condition (3.46) holds in a frame $\mathfrak{K}$ iff the operation ; in the complex algebra $\mathfrak{Cm}(\mathfrak{K})$ satisfies the associative law (R_4), and (3.45) holds iff the semi-associative law (R_4') is true in $\mathfrak{Cm}(\mathfrak{K})$. Note that (3.46) implies (3.45), reflecting the fact that (R_4') is a special case of (R_4). See Theorem 3.11 below.

Conditions (3.38)–(3.46) hold in the pair-frame on any set. Table 3.7 illustrates some triples in the ternary relation R of the pair-frame on the set $U = \{1, 2, 3\}$ when $x = \langle 1, 2\rangle$, $y = \langle 2, 3\rangle$, and $z = \langle 1, 3\rangle$. Table 3.7 can be used to correlate each of the first five frame conditions with permutations of $\{1, 2, 3\}$. The triangle containing $Rxyz$ illustrates the hypothesis of each frame condition. The vertices of the triangle containing the conclusion of each condition are also labelled with 1, 2, and 3. By matching up the vertices of the conclusion with the vertices of the hypothesis one obtains a permutation of $\{1, 2, 3\}$. In cycle notation, the permutations match up with the conditions in this way: $(1, 2, 3)$ with (3.38), $(1, 3, 2)$ with (3.39), $(1, 3)$ with (3.40), $(1, 2)$ with (3.41), and $(2, 3)$ with (3.42). Successively applying the frame conditions is the same as composing the correlated permutations. The next lemma is an expression of the fact that the symmetric group on a 3-element set is generated by a permutation of order 3 (2 choices) together with a permutation of order 2 (3 choices), and it is also generated by any two permutations of order 2.

Table 3.7 Triples in the ternary relation of the pair-frame on $\{1, 2, 3\}$

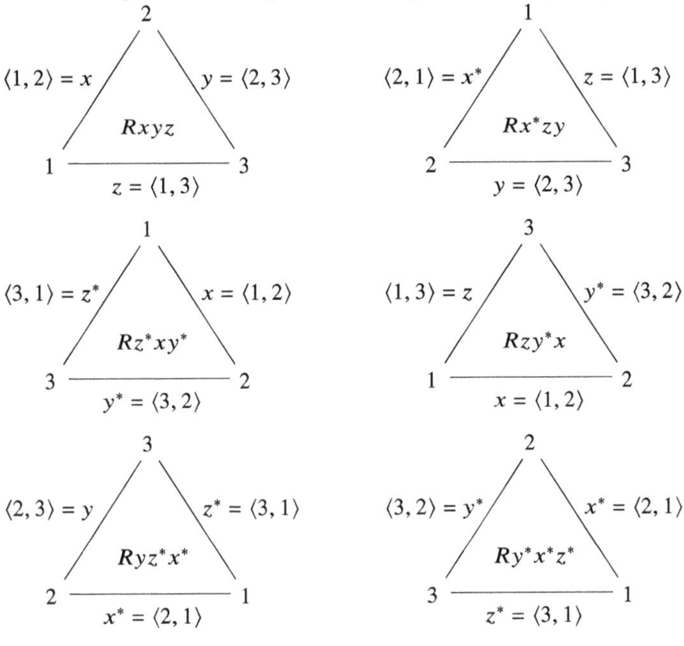

Lemma 3.3 *Suppose a frame $\mathfrak{K} = \langle K, R, {}^*, \mathcal{I}\rangle$ satisfies (3.44).*

1. *If $\mathfrak{K}$ satisfies either (3.38) or (3.39) and any one of* (3.40)*–(3.42), then it satisfies all five conditions (3.38)–(3.42).*
2. *If $\mathfrak{K}$ satisfies any two of* (3.40)*–(3.42) then it satisfies (3.38)–(3.42).*

The following lemma is useful in proving Theorems 3.11 and 3.12 below.

Lemma 3.4 *Assume $\mathfrak{K} = \langle K, R, {}^*, \mathcal{I}\rangle$ is a frame.*

1. *If (3.41) and (3.43) hold then (3.44) holds and $Rxyz \Leftrightarrow Rx^*yz$ for all $x, y, z \in K$.*
2. *If (3.41), (3.42), and (3.43) hold, then $v = v^*$ for all $v \in \mathcal{I}$.*
3. *Assume (3.41), (3.42), (3.43), and (3.45). If $x \in K$, $u, v \in \mathcal{I}$, $Rxux$, and $Rxvx$, then $u = v$.*

Proof (1) Since $x = x$, by (3.43) there must be some $u \in \mathcal{I}$ such that $Rxux$. Applying (3.41) twice yields $Rx^{**}ux$, so we obtain $x^{**} = x$ by (3.43). Since * is an involution, (3.41) implies that its converse also holds.

Proof (2) Assume $v \in \mathcal{I}$. By (3.43) there is some $u \in \mathcal{I}$ such that $Rvuv$. Then Rv^*vu by (3.41) and Rvu^*v by (3.42). From Rvu^*v we also have Rv^*vu^* by (3.41). From Rv^*vu and Rv^*vu^* we obtain $v^* = u$ and $v^* = u^*$ by (3.43) since $v \in \mathcal{I}$. From $v^* = u^*$ we get $v = u$ by part (1), so from $v^* = u$ we get $v^* = v$, as desired.

Proof (3) Assume $u, v \in \mathcal{I}$, $x \in K$, $Rxux$, and $Rxvx$. Then Rx^*xu and Rx^*xv by (3.41). From Rx^*xu we get Rux^*x^* by (3.42). Apply (3.45) to Rux^*x^* and Rx^*xv, obtaining some $y \in K$ such that $Ruyv$. Then Ru^*vy by (3.41) so $u^* = y$ by (3.43) since $v \in \mathcal{I}$. We therefore have Ruu^*v. But $u^* \in \mathcal{I}$ by part (2) since $u \in \mathcal{I}$ by assumption, hence $u = v$ by (3.43).

The next theorem asserts that conditions (3.47), (3.48), or (3.49) hold in a frame iff its complex algebra is dense, commutative, or symmetric, respectively

Theorem 3.10 *Let $\mathfrak{K} = \langle K, R, {}^*, \mathcal{I}\rangle$ be a frame.*

1. *$\mathfrak{Cm}(\mathfrak{K})$ is dense iff $\mathfrak{K}$ satisfies (3.47).*
2. *$\mathfrak{Cm}(\mathfrak{K})$ is commutative iff $\mathfrak{K}$ satisfies (3.48).*
3. *$\mathfrak{Cm}(\mathfrak{K})$ is symmetric iff $\mathfrak{K}$ satisfies (3.49).*

Proof For part (1), assume $\mathfrak{Cm}(\mathfrak{K})$ is dense and $x \in K$. Then $\{x\} \leq \{x\};\{x\}$ by density, hence $Rxxx$ and (3.47) holds. Assume (3.47) holds and $X \subseteq K$. For every $x \in X$, $Rxxx$ implies $\{x\} \subseteq \{x\};\{x\} \subseteq X;X$, hence $X \subseteq X;X$, which shows $\mathfrak{Cm}(\mathfrak{K})$ is dense. Part (3) is equally easy. For part (2), assume $\mathfrak{Cm}(\mathfrak{K})$ is commutative, $x, y, z \in K$, and $Rxyz$. Then $z \in \{x\};\{y\}$, but $\{x\};\{y\} = \{y\};\{x\}$ since $\mathfrak{Cm}(\mathfrak{K})$ is commutative, so $Ryxz$. This shows that (3.48) holds. For the converse, assume (3.48) holds and let $X, Y \subseteq K$ be elements of the complex algebra. By (3.48) and the definition of ; we have

$$\begin{aligned} X;Y &= \{z : z \in K, Rxyz \text{ for some } x \in X, y \in Y\} \\ &= \{z : z \in K, Ryxz \text{ for some } x \in X, y \in Y\} = Y;X, \end{aligned}$$

so $\mathfrak{Cm}(\mathfrak{K})$ is commutative.

The frames whose complex algebras are semi-associative relation algebras or relation algebras are characterized next. Because of Lemmas 3.3 and 3.4, Theorems 3.11 and 3.12 remain true if (3.41)–(3.43) are replaced by (3.38)–(3.44).

Theorem 3.11 (Maddux 1982, Thm 2.2) *Let $\mathfrak{K} = \langle K, R, {}^*, \mathcal{I}\rangle$ be a frame.*

1. *$\mathfrak{Cm}(\mathfrak{K}) \in \mathsf{NA}$ iff $\mathfrak{K}$ satisfies (3.41)–(3.43).*
2. *$\mathfrak{Cm}(\mathfrak{K}) \in \mathsf{SA}$ iff $\mathfrak{K}$ satisfies (3.41)–(3.43) and (3.45).*
3. *$\mathfrak{Cm}(\mathfrak{K}) \in \mathsf{RA}$ iff $\mathfrak{K}$ satisfies (3.41)–(3.43) and (3.46).*

The Jónsson-Tarski Representation Theorem (Jónsson and Tarski 1951, Thm 3.10) in combination with Theorems 3.10 and 3.11 produces a representation theorem for NA and the subvarieties that can be obtained by imposing semi-associativity, associativity, density, commutativity, or symmetry. For example, $\mathfrak{A}$ is a dense commutative relation algebra iff $\mathfrak{A}$ is isomorphic to a subalgebra of the complex algebra of a frame satisfying (3.41)–(3.43) and (3.46)–(3.47).

Theorem 3.12 (Maddux 1982, Thm 4.3) *Assume $\mathfrak{A} = \langle U, +, {}^-, ;, {}^\smile, 1'\rangle$ is an algebra satisfying axioms (R_1)–(R_3), (R_5), (R_7)–(R_{10}). Then there is a frame $\mathfrak{K} = \langle K, R, {}^*, \mathcal{I}\rangle$ such that the following statements hold.*

1. $\mathfrak{A} \cong \mathfrak{A}' \subseteq \mathfrak{Cm}(\mathfrak{K})$ *for some subalgebra* $\mathfrak{A}'$ *of the complex algebra* $\mathfrak{Cm}(\mathfrak{K})$.
2. $\mathfrak{Cm}(\mathfrak{K})$ *is an algebra satisfying axioms* (R_1)–(R_3), (R_5), (R_7)–(R_{10}).
3. $\mathfrak{Cm}(\mathfrak{K}) \in \mathsf{NA}$ *iff* $\mathfrak{A} \in \mathsf{NA}$.
4. $\mathfrak{A} \in \mathsf{NA}$ *iff* $\mathfrak{K}$ *satisfies (3.41)–(3.43).*
5. $\mathfrak{Cm}(\mathfrak{K}) \in \mathsf{SA}$ *iff* $\mathfrak{A} \in \mathsf{SA}$.
6. $\mathfrak{A} \in \mathsf{SA}$ *iff* $\mathfrak{K}$ *satisfies (3.41)–(3.43) and (3.45).*
7. $\mathfrak{Cm}(\mathfrak{K}) \in \mathsf{RA}$ *iff* $\mathfrak{A} \in \mathsf{RA}$.
8. $\mathfrak{A} \in \mathsf{RA}$ *iff* $\mathfrak{K}$ *satisfies (3.41)–(3.43) and (3.46).*
9. $\mathfrak{A}$ *is dense iff* $\mathfrak{K}$ *satisfies (3.47).*
10. $\mathfrak{A}$ *is commutative iff* $\mathfrak{K}$ *satisfies (3.48).*
11. $\mathfrak{A}$ *is symmetric iff* $\mathfrak{K}$ *satisfies (3.49).*

Proof This theorem was originally derived from Jónsson and Tarski (1951, Thms 2.15, 2.17, 2.18), but the desired frame $\mathfrak{K} = \langle K, R, {}^*, \mathcal{I}\rangle$ can be obtained directly from the algebra $\mathfrak{A}$. A subset $X \subseteq U$ is an **ultrafilter** of $\mathfrak{A}$ if $X \neq U$ and for all $x, y \in U$, if $x \in X$ then $x + y \in X$ and if $x, y \in X$ then $x \cdot y \in X$. Let K be the set of ultrafilters of $\mathfrak{A}$, and define $R \subseteq K^3$, ${}^*\colon K \to K$, and $\mathcal{I} \subseteq K$ by

$$\begin{aligned} R &= \{\langle X, Y, Z\rangle : X, Y, Z \in K, \{x ; y : x \in X, y \in X\} \subseteq Z\},\\ X^* &= \{x^{\smile} : x \in X\} \text{ for every } X \in K,\\ \mathcal{I} &= \{X : 1\text{'} \in X \in K\}. \end{aligned}$$

Define $\varepsilon\colon A \to \wp(K)$ by $\varepsilon(x) = \{X : x \in X \in K\}$ for every $x \in A$. Then ε is an isomorphic embedding of $\mathfrak{A}$ onto a subalgebra $\mathfrak{A}'$ of $\mathfrak{Cm}(\mathfrak{K})$, as required by part (1). This construction was first described in R. McKenzie's dissertation (McKenzie 1966, Thm 2.11).

The next theorem shows how frames arise from groups. Note that $|X|$ is the number of elements in X.

Theorem 3.13 *Suppose* $\mathfrak{G} = \langle K, ;, {}^*, e\rangle$ *is a group. Define a ternary relation* R *on* K *by* $R = \{\langle x, y, z\rangle : x ; y = z\}$. *Then* $\mathfrak{K} = \langle K, R, {}^*, \{e\}\rangle$ *is a frame satisfying (3.38)–(3.46).*

Proof We use the properties of groups, that ; is associative, and for all $x, y \in K$, $x ; e = x = e ; x$, $x^{**} = x$, $x ; x^* = e = x^* ; x$, and $(x ; y)^* = y^* ; x^*$. Frame conditions (3.38)–(3.42) all have the same assumption, $Rxyz$, and their conclusions are Ryz^*x^*, Rz^*xy^*, $Ry^*x^*z^*$, Rx^*zy, and Rzy^*x. According to the definition of R, we assume $z = x ; y$ and prove $y ; z^* = x^*$ for (3.38) by $y ; z^* = y ; (x ; y)^* = y ; (y^* ; x^*) = (y ; y^*) ; x^* = e ; x^* = x^*$, $z^* ; x = y^*$ for (3.39) by $z^* ; x = (x ; y)^* ; x = (y^* ; x^*) ; x = y^* ; (x^* ; x) = y^* ; e = y^*$, $y^* ; x^* = z^*$ for (3.40) by $y^* ; x^* = (x ; y)^* = z^*$, $x^* ; z = y$ for (3.41) by $x^* ; z = x^* ; (x ; y) = (x^* ; x) ; y = e ; y = y$, and $z ; y^* = x$ for (3.42) by $z ; y^* = (x ; y) ; y^* = x ; (y ; y^*) = x ; e = x$.

For (3.43) in one direction, let $x \in K$. We want $Rxux$ for some $u \in K$. Take $u = e$, and get $x ; e = x$, i.e., $Rxex$, as desired. For the other direction, assume

$Rxey$. This implies $x ; e = y$, but $x ; e = x$, so $x = y$. We have (3.44) since $x^{**} = x$. For (3.46), we assume $Rvwx$ and $Rxyz$ and wish to show $Rvuz$ and $Rwyu$. From our hypotheses we have $v ; w = x$ and $x ; y = z$. Let $u = w ; y$. Then $Rwyu$ and $z = x ; y = (v ; w) ; y = v ; (w ; y) = v ; u$, hence $Rvuz$, as desired. Note that (3.45) is a trivial consequence of (3.46).

The frame of a group satisfies (3.47) iff it has only one element. Groups (treated as frames) that satisfy (3.48) are usually called **Abelian groups**. The frame of a group satisfies (3.49) iff every element is its own inverse. Such groups are called **Boolean groups** (Bernstein 1939).

CR*-frames were introduced by Meyer and Routley (1974, p. 184). A frame $\mathfrak{K} = \langle K, R, {}^*, \mathcal{I} \rangle$ is a CR* **frame** if $\mathcal{I} = \{0\}$ and for all $a, b, c, d \in K$,

p1. $R0ab \Leftrightarrow a = b$,
p2. $\exists_x (Rabx \wedge Rxcd) \Leftrightarrow \exists_y (Racy \wedge Rybd)$,
p3. $Raaa$,
p4. $a^{**} = a$,
p5. $Rabc \Leftrightarrow Rac^*b^*$.

Meyer and Routley define the logic CR* as the sets of predicates that are valid in all CR*-frames (Meyer and Routley 1974, p. 187) and they characterize Anderson and Belnap's relevance logic R (Anderson and Belnap 1959, 1975; Anderson et al. 1992; Belnap 1960, 1967) as those predicates in Π^r that are in CR* when (3.17) is taken as the definition of $\sim$ (Meyer and Routley 1974, Translation Theorem, p. 190). Thus,

$$\mathsf{CR}^* = \{A : A \in \Pi^+, A \text{ is valid in every } \mathsf{CR}^*\text{-frame}\}, \tag{3.50}$$

$$\mathsf{R} = \mathsf{CR}^* \cap \Pi^r. \tag{3.51}$$

The logic KR can be obtained from R by adding (3.143) in Theorem 3.22; see Anderson et al. (1992, Sect. 65.1.2) (by Urquhart). A frame $\mathfrak{K}$ is a **KR-frame** if it is a CR*-frame satisfying (3.49) (Anderson et al. 1992, p. 350). Urquhart added, "A slight modification of the usual completeness proof for R shows that KR is complete with respect to the class of all KR model structures." Thus,

$$\mathsf{KR} = \{A : A \in \Pi^+, A \text{ is valid in every } \mathsf{KR}\text{-frame}\}. \tag{3.52}$$

Lemma 3.5 *If $\mathfrak{K} = \langle K, R, {}^*, \mathcal{I} \rangle$ is a frame such that* p1 *holds for some $0 \in K$ and* p2 *holds then (3.48) holds. Every* CR**-frame satisfies (3.48).*

Proof Assume $Rabc$. We have $R0aa$ by p1. Applying p2 to $R0aa$ and $Rabc$, we obtain some $y \in K$ such that $R0by$ and $Ryac$. Then $b = y$ by p1 so $Rbac$.

When (3.48) holds we can restate p5 in two ways. By switching the order of the first two entries in Rac^*b^* in p5 one gets (3.39). Switching the first two entries in $Rabc$ in p5 and interchanging a and b in the entire statement gives (3.38). In the presence of p1 and p2, the three postulates p5, (3.38), and (3.39) are equivalent because of Lemma 3.5.

Theorem 3.14 *Let $\mathfrak{K} = \langle K, R, {}^*, \{0\}\rangle$ be a frame.*

1. *$\mathfrak{K}$ is a* CR^**-frame satisfying* (3.41) *iff* $\mathfrak{Cm}\,(\mathfrak{K})$ *is a dense commutative relation algebra.*
2. *$\mathfrak{K}$ is a* KR*-frame iff* $\mathfrak{Cm}\,(\mathfrak{K})$ *is a dense symmetric relation algebra.*

Proof Assume $\mathfrak{Cm}\,(\mathfrak{K})$ is a dense commutative relation algebra. We will show $\mathfrak{K}$ is a CR^*-frame. By Theorem 3.11 (3), $\mathfrak{K}$ satisfies (3.41), (3.42), (3.43), and (3.46). Postulate p1 is what (3.43) reduces to when $\mathcal{I} = \{0\}$. Since $\mathfrak{Cm}\,(\mathfrak{K})$ is dense, $\mathfrak{K}$ satisfies (3.47) by Theorem 3.10(1) and (3.47) coincides with p3. Condition p4 follows from (3.43) and (3.41) by Lemma 3.4 (1). For p5, assume $Rabc$. By Theorem 3.10 (2), $\mathfrak{K}$ satisfies (3.48) since $\mathfrak{Cm}\,(\mathfrak{K})$ is commutative, so $Rbac$. Then Rb^*ca by (3.41), so Rac^*b^* by (3.42). Thus, p5 holds in one direction and the other direction follows from this by p4. Note that the implication in p2 from right to left is formally identical to the implication from left to right, so we need only prove the latter. Assume $Rabx$ and $Rxcd$. By (3.46), $Raud$ and $Rbcu$ for some $u \in K$, so $Rcbu$ since (3.48) holds. We get Ra^*du from $Raud$ by (3.41) and Rub^*c from $Rcbu$ by (3.42). By (3.46), there is some $y \in K$ such that Ra^*yc and Rdb^*y. By (3.41), (3.42), and p4, $Racy$ and $Rybd$.

Assume $\mathfrak{K}$ is a CR^*-frame satisfying (3.41). Then (3.48) holds by Lemma 3.5 so $\mathfrak{Cm}\,(\mathfrak{K})$ is commutative by Lemma 3.10 (2). $\mathfrak{Cm}\,(\mathfrak{K})$ is dense by p3 and Lemma 3.10 (1), (3.43) holds by p1 since $\mathcal{I} = \{0\}$, and (3.42) follows immediately from (3.41) and (3.48). To prove (3.46) assume $Rvwx$ and $Rxyz$. Then $Rwvx$ by (3.48), so by p2 applied to $Rwvx$ and $Rxyz$, there is some $u \in K$ such that $Rwyu$ and $Ruvz$, hence $Rvuz$ by (3.48). Since (3.41), (3.42), (3.43), and (3.46) hold, we conclude that $\mathfrak{Cm}\,(\mathfrak{K}) \in \mathsf{RA}$ by Theorem 3.11 (3).

Suppose $\mathfrak{K}$ is a KR-frame. Since $\mathfrak{K}$ is a CR^*-frame and satisfies (3.47) and (3.49) by definition, $\mathfrak{Cm}\,(\mathfrak{K})$ is dense and symmetric by Theorem 3.10 (1) (3). If $Rabc$ then Rac^*b^* by p5, hence Ra^*cb by (3.49). Thus, (3.41) holds and we conclude that $\mathfrak{Cm}\,(\mathfrak{A})$ is a relation algebra by part (1).

For the converse, assume $\mathfrak{Cm}\,(\mathfrak{K})$ is a dense symmetric relation algebra. Then $\mathfrak{K}$ satisfies (3.49) by Theorem 3.10(3). $\mathfrak{Cm}\,(\mathfrak{K})$ is commutative by Lemma 3.1, hence $\mathfrak{K}$ satisfies (3.48) by Theorem 3.10(2). By part (1), $\mathfrak{K}$ is a CR^*-frame. Because $\mathfrak{K}$ satisfies (3.49), $\mathfrak{K}$ is also a KR-frame.

3.12 The n-Variable Sequent Calculus

Assume $1 \leq n \in \omega$. An n**-sequent** is an ordered pair $\langle\Gamma, \Delta\rangle$, written $\Gamma \,|\, \Delta$, of sets $\Gamma, \Delta \subseteq \{xAy : x, y \in \Upsilon_n,\ A \in \Pi^+\}$. An n-sequent $\Gamma \,|\, \Delta$ is an **axiom** if $\Gamma \cap \Delta \neq \emptyset$ or $x\,1\text{'}x \in \Delta$ for some $x \in \Upsilon_n$.

Table 3.8 Axioms and rules of inference for the n-variable sequent calculus, where $1 \leq n \leq \omega$, $x, y, z \in \Upsilon_n$, $A, B \in \Pi^+$, and $\Gamma, \Delta \subseteq \{xAy : x, y \in \Upsilon_n,\ A \in \Pi^+\}$

Rule	Inference	Rule	Inference
Axiom	$\dfrac{}{xAy,\Gamma \mid xAy,\Delta}$	Cut	$\dfrac{\begin{array}{c}\Gamma \mid \Delta, xAy \\ xAy,\Gamma' \mid \Delta'\end{array}}{\Gamma,\Gamma' \mid \Delta,\Delta'}$
1'|	$\dfrac{xAy,\Gamma \mid \Delta}{xAz, z1'y,\Gamma \mid \Delta}$	|1'	$\dfrac{}{\Gamma \mid \Delta, x1'x}$
+|	$\dfrac{\begin{array}{c}xAy,\Gamma \mid \Delta \\ xBy,\Gamma' \mid \Delta'\end{array}}{xA+By,\Gamma,\Gamma' \mid \Delta,\Delta'}$	|+	$\dfrac{\Gamma \mid \Delta, xAy, xBy}{\Gamma \mid \Delta, xA+By}$
·|	$\dfrac{\Gamma, xAy, xBy \mid \Delta}{\Gamma, xA\cdot By \mid \Delta}$	|·	$\dfrac{\begin{array}{c}\Gamma \mid \Delta, xAy \\ \Gamma' \mid \Delta', xBy\end{array}}{\Gamma,\Gamma' \mid \Delta,\Delta', xA\cdot By}$
$^{-}$|	$\dfrac{\Gamma \mid \Delta, xAy}{x\overline{A}y,\Gamma \mid \Delta}$	|$^{-}$	$\dfrac{xAy,\Gamma \mid \Delta}{\Gamma \mid \Delta, x\overline{A}y}$
;|	$\dfrac{xAy, yBz,\Gamma \mid \Delta}{xA;Bz,\Gamma \mid \Delta, \text{no } y}$	|;	$\dfrac{\begin{array}{c}\Gamma \mid \Delta, xAy \\ \Gamma' \mid \Delta', yBz\end{array}}{\Gamma,\Gamma' \mid \Delta,\Delta', xA;Bz}$
$^{\smile}$|	$\dfrac{xAy,\Gamma \mid \Delta}{yA^{\smile}x,\Gamma \mid \Delta}$	|$^{\smile}$	$\dfrac{\Gamma \mid \Delta, xAy}{\Gamma \mid \Delta, yA^{\smile}x}$

Let Ψ be a set of n-sequents. A sequent is **n-provable from Ψ** (just **n-provable** when $\Psi = \emptyset$) if it is contained in every set of n-sequents that includes Ψ and the axioms and is closed under the rules of inference in Table 3.8.

In the rules in Table 3.8, Γ, Γ', Δ, and Δ' are sets of formulas in $\{xAy : x, y \in \Upsilon_n,\ A \in \Pi^+\}$, $A, B \in \Pi^+$ are predicates, and $x, y, z \in \Upsilon_n$. The notation "no y" in rule ;| means that $y \neq x, z$ and y does not occur in any formula in Γ or Δ.

The rules are taken from (Maddux 1983). The rules |· and ·| are derived from the rules for $^{-}$ and $+$ through the definition of $\cdot$ in (3.2). Braces and union symbols are frequently omitted from the notation for sequents in favor of commas. For example, we write Γ, xAy, xBy instead of $\Gamma \cup \{xAy\} \cup \{xBy\}$.

An n-**proof from** Ψ is a sequence of sequents in which every sequent is either in Ψ, or is an axiom, or follows from one or two previous sequents in the sequence by one of the rules of inference in Table 3.8. Whenever a rule is applied in an n-proof we include the numbers for the previous sequents used by the rule and the name of the rule. For every application of the rule $;\,|$ we also include notation of the form "no y" as a reminder that the eliminated variable y must not occur in the conclusion of $;\,|$.

If $\Gamma \,|\, \Delta$ is an n-sequent, then an n-**proof of** $\Gamma \,|\, \Delta$ **from** Ψ is an n-proof from Ψ in which $\Gamma \,|\, \Delta$ occurs. It is straightforward to prove that an n-sequent is n-provable from Ψ iff it has an n-proof from Ψ.

The sequent calculus is connected with classes of algebras defined in (Maddux 1983) and originally called MA_n, later renamed RA_n (Maddux 1989, Def 4).

Definition 3.1 Assume $\mathfrak{A} = \langle U, +, ^-, ;, ^\smile, 1' \rangle \in \mathsf{NA}$ is a non-associative relation algebra. For $3 \le n \le \omega$, an n-**dimensional relational basis** for $\mathfrak{A}$ is a set $B \subseteq \mathrm{At}(\mathfrak{A})^{n\times n}$ of $n \times n$ matrices of atoms of $\mathfrak{A}$ such that

1. for all $i, j, k < n$ and all $x, y, z \in B$, $x_{ii} \le 1'$, $x_{ij}{}^\smile = x_{ji}$, and $x_{ik} \le x_{ij} ; x_{jk}$,
2. for every $a \in \mathrm{At}(\mathfrak{A})$ there is some $x \in B$ such that $x_{01} = a$,
3. if $i, j < n$, $x \in B$, $a, b \in \mathrm{At}(\mathfrak{A})$, $x_{ij} \le a ; b$, and $i, j \ne k \langle n$, then there is some $y \in B$ such that $y_{ik} = a$, $y_{kj} = b$, and $x_{\ell m} = y_{\ell m}$ whenever $k \ne \ell, m < n$.

The algebra $\mathfrak{A}$ is an n-**dimensional relation algebra** if $\mathfrak{A}$ is a subalgebra of an atomic semi-associative relation algebra that has an n-dimensional relational basis. RA_n is the class of n-dimensional relation algebras.

Part (2) of the next theorem is proved by the equational axiomatizations of RA_n in (Maddux 2006, Thms 414, 419) and (Hirsch and Hodkinson 2002, Sect. 13.8).

Theorem 3.15 (Maddux 1983, Thms 2, 3, 6, 9, 10)

1. *If* $3 \le m \le n \le \omega$ *then* $\mathsf{RA}_m \supseteq \mathsf{RA}_n$.
2. *If* $3 \le n \le \omega$ *then* RA_n *is a variety.*
3. $\mathsf{RA}_3 = \mathsf{SA}$.
4. $\mathsf{RA}_4 = \mathsf{RA}$.
5. $\mathsf{RA}_\omega = \mathsf{RRA} = \bigcap_{3 \le n < \omega} \mathsf{RA}_n$.
6. *Assume* $3 \le n \le \omega$, $\mathcal{R} \subseteq (\Pi^+)^2$, *and* $A, B \in \Pi^+$. *Then the following conditions are equivalent.*

   a. $\mathsf{v}_0 A \mathsf{v}_1 \,|\, \mathsf{v}_0 B \mathsf{v}_1$ *is* n*-provable from* $\{\mathsf{v}_0 X \mathsf{v}_1 \,|\, \mathsf{v}_0 Y \mathsf{v}_1 : \langle X, Y \rangle \in \mathcal{R}\}$.
   b. *For every* $\mathfrak{A} \in \mathsf{RA}_n$ *and homomorphism* $h \colon \mathfrak{P} \to \mathfrak{A}$, *if* $h(X) \le h(Y)$ *whenever* $\langle X, Y \rangle \in \mathcal{R}$, *then* $h(A) \le h(B)$.

The following lemma is used extensively in the proof of Theorem 3.22 below.

Lemma 3.6 *Assume* $A, B \in \Pi^+$, $1 \le n \in \omega$, $i, j, k \le n$, *and* Ψ *is a set of* n*-sequents. In the derived rules below, any* n*-proof that contains the sequent above the horizontal line can be extended to an* n*-proof that contains the sequent below the line.*

$$\frac{\mathsf{v}_i A \mathsf{v}_j \,|\, \mathsf{v}_i B \mathsf{v}_j}{|\, \mathsf{v}_j A \to B \mathsf{v}_j} \textit{ if } i \neq j \tag{3.53}$$

$$\frac{|\, \mathsf{v}_j A \to B \mathsf{v}_j}{\mathsf{v}_i A \mathsf{v}_j \,|\, \mathsf{v}_i B \mathsf{v}_j} \tag{3.54}$$

$$\frac{|\, \mathsf{v}_i A \mathsf{v}_i}{\mathsf{v}_j 1\text{'} \mathsf{v}_k \,|\, \mathsf{v}_j A \mathsf{v}_k} \textit{ if } i \neq j, k \tag{3.55}$$

$$\frac{|\, \mathsf{v}_i A \mathsf{v}_i}{|\, \mathsf{v}_j A \mathsf{v}_j} \tag{3.56}$$

$$\frac{\mathsf{v}_i A \mathsf{v}_j \,|\, \mathsf{v}_i B \mathsf{v}_j}{\mathsf{v}_j A \mathsf{v}_i \,|\, \mathsf{v}_j B \mathsf{v}_i} \tag{3.57}$$

$$\frac{\mathsf{v}_i 1\text{'} \mathsf{v}_j \,|\, \mathsf{v}_i A \mathsf{v}_j}{|\, \mathsf{v}_i A \mathsf{v}_i} \textit{ if } i \neq j \tag{3.58}$$

$$\frac{\mathsf{v}_i A \mathsf{v}_j \,|\, \mathsf{v}_i B \mathsf{v}_j}{\mathsf{v}_i A \mathsf{v}_i \,|\, \mathsf{v}_i B \mathsf{v}_i} \tag{3.59}$$

Proof (3.53)

1.	$\mathsf{v}_i A \mathsf{v}_j \,|\, \mathsf{v}_i B \mathsf{v}_j$	in an n-proof from Ψ
2.	$\mathsf{v}_j A^\smile \mathsf{v}_i \,|\, \mathsf{v}_i B \mathsf{v}_j$	$1, {}^\smile|$
3.	$\mathsf{v}_j A^\smile \mathsf{v}_i, \mathsf{v}_i \overline{B} \mathsf{v}_j \,|$	$2, {}^-|$
4.	$\mathsf{v}_j A^\smile ; \overline{B} \mathsf{v}_j \,|$	$3, ;|$, no v_i, $i \neq j$
5.	$|\, \mathsf{v}_j \overline{A^\smile ; \overline{B}} \mathsf{v}_j$	$4, |^-$
6.	$|\, \mathsf{v}_j A \to B \mathsf{v}_j$	5, (3.18)

Proof (3.54)

1.	$\mid \mathsf{v}_j A \to B \mathsf{v}_j$	in an n-proof from Ψ
2.	$\mid \mathsf{v}_j \overline{A^{\smile} ; \overline{B}} \mathsf{v}_j$	1, (3.18)
3.	$\mathsf{v}_j A^{\smile} ; \overline{B} \mathsf{v}_j \mid \mathsf{v}_j A^{\smile} ; \overline{B} \mathsf{v}_j$	axiom
4.	$\mathsf{v}_j A^{\smile} ; \overline{B} \mathsf{v}_j , \mathsf{v}_j \overline{A^{\smile} ; \overline{B}} \mathsf{v}_j \mid$	3, $^{-}\mid$
5.	$\mathsf{v}_j A^{\smile} ; \overline{B} \mathsf{v}_j \mid$	2, 4, Cut
6.	$\mathsf{v}_i A \mathsf{v}_j \mid \mathsf{v}_i A \mathsf{v}_j$	axiom
7.	$\mathsf{v}_i A \mathsf{v}_j \mid \mathsf{v}_j A^{\smile} \mathsf{v}_i$	6, $\mid^{\smile}$
8.	$\mathsf{v}_i B \mathsf{v}_j \mid \mathsf{v}_i B \mathsf{v}_j$	axiom
9.	$\mid \mathsf{v}_i B \mathsf{v}_j , \mathsf{v}_i \overline{B} \mathsf{v}_j$	8, $\mid^{-}$
10.	$\mathsf{v}_i A \mathsf{v}_j \mid \mathsf{v}_i B \mathsf{v}_j , \mathsf{v}_j A^{\smile} ; \overline{B} \mathsf{v}_j$	7, 9, $\mid ;$
11.	$\mathsf{v}_i A \mathsf{v}_j \mid \mathsf{v}_i B \mathsf{v}_j$	5, 10, Cut

Proof (3.55)

1.	$\mid \mathsf{v}_i A \mathsf{v}_i$	in an n-proof from Ψ
2.	$\mid \mathsf{v}_i A^{\smile} \mathsf{v}_i$	1, $\mid^{\smile}$
3.	$\mathsf{v}_j A \mathsf{v}_k \mid \mathsf{v}_j A \mathsf{v}_k$	axiom
4.	$\mathsf{v}_j A \mathsf{v}_i , \mathsf{v}_i 1' \mathsf{v}_k \mid \mathsf{v}_j A \mathsf{v}_k$	3, $1' \mid$
5.	$\mathsf{v}_i A^{\smile} \mathsf{v}_j , \mathsf{v}_i 1' \mathsf{v}_k \mid \mathsf{v}_j A \mathsf{v}_k$	4, $^{\smile}\mid$
6.	$\mathsf{v}_i A^{\smile} \mathsf{v}_i , \mathsf{v}_i 1' \mathsf{v}_j , \mathsf{v}_i 1' \mathsf{v}_k \mid \mathsf{v}_j A \mathsf{v}_k$	5, $1' \mid$
7.	$\mathsf{v}_i 1' \mathsf{v}_j , \mathsf{v}_i 1' \mathsf{v}_k \mid \mathsf{v}_j A \mathsf{v}_k$	2, 6, Cut
8.	$\mathsf{v}_j 1'^{\smile} \mathsf{v}_i , \mathsf{v}_i 1' \mathsf{v}_k \mid \mathsf{v}_j A \mathsf{v}_k$	7, $^{\smile}\mid$
9.	$\mathsf{v}_j 1'^{\smile} ; 1' \mathsf{v}_k \mid \mathsf{v}_j A \mathsf{v}_k$	8, $; \mid$, no v_i, $i \neq j, k$
10.	$\mathsf{v}_j 1' \mathsf{v}_k \mid \mathsf{v}_j 1' \mathsf{v}_k$	axiom
11.	$\mid \mathsf{v}_j 1' \mathsf{v}_j$	$\mid 1'$
12.	$\mid \mathsf{v}_j 1'^{\smile} \mathsf{v}_j$	11, $\mid^{\smile}$
13.	$\mathsf{v}_j 1' \mathsf{v}_k \mid \mathsf{v}_j 1'^{\smile} ; 1' \mathsf{v}_k$	10, 12, $\mid ;$
14.	$\mathsf{v}_j 1' \mathsf{v}_k \mid \mathsf{v}_j A \mathsf{v}_k$	9, 13, Cut

Proof (3.56) This part is trivial if $i = j$ so assume $i \neq j$.

1.	$\mid \mathsf{v}_i A \mathsf{v}_i$	in an n-proof from Ψ
2.	$\mathsf{v}_j 1' \mathsf{v}_j \mid \mathsf{v}_j A \mathsf{v}_j$	1, (3.55), $i \neq j$
3.	$\mid \mathsf{v}_j 1' \mathsf{v}_j$	$\mid 1'$
4.	$\mid \mathsf{v}_j A \mathsf{v}_j$	2, 3, Cut

Proof (3.57) This part is trivial if $i = j$, so assume $i \neq j$.

1.	$\mathsf{v}_i A \mathsf{v}_j \mid \mathsf{v}_i B \mathsf{v}_j$	in an n-proof from Ψ
2.	$\mid \mathsf{v}_j A \to B \mathsf{v}_j$	2, (3.53), $i \neq j$
3.	$\mid \mathsf{v}_i A \to B \mathsf{v}_i$	2, (3.53)
4.	$\mathsf{v}_j A \mathsf{v}_i \mid \mathsf{v}_j B \mathsf{v}_i$	3, (3.54)

Proof (3.58)

1.	$\mathsf{v}_i 1' \mathsf{v}_j \mid \mathsf{v}_i A \mathsf{v}_j$	in an n-proof
2.	$\mathsf{v}_j 1' \mathsf{v}_i \mid \mathsf{v}_j 1' \mathsf{v}_i$	axiom
3.	$\mathsf{v}_i 1' \mathsf{v}_j, \mathsf{v}_j 1' \mathsf{v}_i \mid \mathsf{v}_i A ; 1' \mathsf{v}_i$	1, 2, $\mid ;$
4.	$\mathsf{v}_i 1' ; 1' \mathsf{v}_i \mid \mathsf{v}_i A ; 1' \mathsf{v}_i$	3, $; \mid$, no v_j, $i \neq j$
5.	$\mid \mathsf{v}_i 1' \mathsf{v}_i$	$\mid 1'$
6.	$\mid \mathsf{v}_i 1' ; 1' \mathsf{v}_i$	5, $\mid ;$
7.	$\mid \mathsf{v}_i A ; 1' \mathsf{v}_i$	4, 6, Cut
8.	$\mathsf{v}_i A \mathsf{v}_i \mid \mathsf{v}_i A \mathsf{v}_i$	axiom
9.	$\mathsf{v}_i A \mathsf{v}_j, \mathsf{v}_j 1' \mathsf{v}_i \mid \mathsf{v}_i A \mathsf{v}_i$	8, $1' \mid$
10.	$\mathsf{v}_i A ; 1' \mathsf{v}_i \mid \mathsf{v}_i A \mathsf{v}_i$	9, $; \mid$, no v_j, $i \neq j$
11.	$\mid \mathsf{v}_i A \mathsf{v}_i$	7, 10, Cut

Proof (3.59). This part is trivial if $i = j$, so assume $i \neq j$.

1.	$\mathsf{v}_i A \mathsf{v}_j \mid \mathsf{v}_i B \mathsf{v}_j$	in an n-proof
2.	$\mathsf{v}_i A \mathsf{v}_i, \mathsf{v}_i 1' \mathsf{v}_j \mid \mathsf{v}_i B \mathsf{v}_j$	1, $1' \mid$
3.	$\mathsf{v}_j 1' \mathsf{v}_i \mid \mathsf{v}_j 1' \mathsf{v}_i$	axiom
4.	$\mathsf{v}_i A \mathsf{v}_i, \mathsf{v}_i 1' \mathsf{v}_j, \mathsf{v}_j 1' \mathsf{v}_i \mid \mathsf{v}_i B ; 1' \mathsf{v}_i$	2, 3, $\mid ;$
5.	$\mathsf{v}_i A \mathsf{v}_i, \mathsf{v}_i 1' ; 1' \mathsf{v}_i \mid \mathsf{v}_i B ; 1' \mathsf{v}_i$	4, $; \mid$, no v_j, $i \neq j$
6.	$\mid \mathsf{v}_i 1' \mathsf{v}_i$	$\mid 1'$
7.	$\mid \mathsf{v}_i 1' ; 1' \mathsf{v}_i$	6, $\mid ;$
8.	$\mathsf{v}_i A \mathsf{v}_i \mid \mathsf{v}_i B ; 1' \mathsf{v}_i$	5, 7, Cut
9.	$\mathsf{v}_i B \mathsf{v}_i \mid \mathsf{v}_i B \mathsf{v}_i$	axiom
10.	$\mathsf{v}_i B \mathsf{v}_j, \mathsf{v}_j 1' \mathsf{v}_i \mid \mathsf{v}_i B \mathsf{v}_i$	9, $1' \mid$
11.	$\mathsf{v}_i B ; 1' \mathsf{v}_i \mid \mathsf{v}_i B \mathsf{v}_i$	10, $; \mid$, no v_j, $i \neq j$
12.	$\mathsf{v}_i A \mathsf{v}_i \mid \mathsf{v}_i B \mathsf{v}_i$	8, 11, Cut

Rules (3.53) and (3.54) do not involve 1' in their statements and do not use rules $\mid 1'$ and $1' \mid$ in their proofs. Rules (3.56), (3.57), and (3.59) do not involve 1' in their statements but rules $|1'$ and $1'|$ are used in their proofs. There are alternate proofs of (3.56) and (3.57) that avoid the use of rules $|1'$ and $1'|$. Observe that if the variables v_i and v_j are interchanged in an axiom or a rule in Table 3.8, the result is still an axiom or rule. It follows that if variables v_i and v_j are interchanged throughout an n-proof, the result is still an n-proof. Consequently, if $|\mathsf{v}_i A\mathsf{v}_i, \mathsf{v}_i A\mathsf{v}_j | \mathsf{v}_i B\mathsf{v}_j$, or $\mathsf{v}_i A\mathsf{v}_j | \mathsf{v}_j B\mathsf{v}_i$ is n-provable, so is $|\mathsf{v}_j A\mathsf{v}_j, \mathsf{v}_j A\mathsf{v}_i | \mathsf{v}_j B\mathsf{v}_i$, or $\mathsf{v}_j A\mathsf{v}_i | \mathsf{v}_i B\mathsf{v}_j$, respectively.

3.13 Characterizing $\mathbf{CT}_3^\Psi$, $\mathbf{CT}_4^\Psi$, TR, KR, and $\mathbf{CT}_\omega^\Psi$

Let End($\mathfrak{P}$) be the set of endomorphisms of the predicate algebra $\mathfrak{P}$, that is, homomorphisms from $\mathfrak{P}$ to itself. For every $\Psi \subseteq \Sigma^\times$, let

$$\Xi(\Psi) = \{g(P) \mathrel{\mathring{=}} g(Q) : P \mathrel{\mathring{=}} Q \in \Psi, P, Q \in \Pi^+, g \in \mathrm{End}(\mathfrak{P})\}.$$

A **Ψ-algebra** is an algebra similar to $\mathfrak{P}$ in which every equation in Ψ is true.

Lemma 3.7 *Let $A, B \in \Pi^+$ and $\Psi \subseteq \Sigma^\times$. Let V a class of algebras similar $\mathfrak{P}$ that is closed under subalgebras. Then the following statements are equivalent.*

1. *If $h\colon \mathfrak{P} \to \mathfrak{A}$ is a homomorphism and $\mathfrak{A} \in \mathsf{V}$ then $\mathfrak{A} \models_h \Xi(\Psi)$ implies $\mathfrak{A} \models_h A \mathrel{\mathring{=}} B$.*
2. *$A \mathrel{\mathring{=}} B$ is true in every Ψ-algebra in V.*

Proof Assume (2). To prove (1), assume $\mathfrak{A} \in \mathsf{V}$, $h\colon \mathfrak{P} \to \mathfrak{A}$ is a homomorphism, and $\mathfrak{A} \models_h \Xi(\Psi)$. We must show $h(A) = h(B)$. Let $\mathfrak{B}$ be the subalgebra of $\mathfrak{A}$ whose universe is $h(\Pi^+) = \{h(X) : X \in \Pi^+\}$. Note that $\mathfrak{B} \in \mathsf{V}$ by our assumption on V. We will show $\mathfrak{B}$ is a Ψ-algebra in V. Consider any homomorphism $f\colon \mathfrak{P} \to \mathfrak{B}$. Construct an endomorphism $g\colon \mathfrak{P} \to \mathfrak{P}$ as follows. For every propositional variable $1' \neq C \in \Pi$, since $f(C) \in h(\Pi^+)$ we may choose $D \in \Pi^+$ such that $h(D) = f(C)$. Set $g(C) = D$ so that $h(g(C)) = f(C)$. Make such a choice for every propositional variable. Since $\mathfrak{P}$ is absolutely freely generated by the propositional variables, these choices extend to the desired endomorphism $g \in \mathrm{End}(\mathfrak{P})$. Since $h(g(C)) = f(C)$ for every propositional variable, the properties of the homomorphisms f, g, h imply that $h(g(E)) = f(E)$ for every $E \in \Pi^+$. For every equation $C \mathrel{\mathring{=}} D \in \Psi$, $g(C) \mathrel{\mathring{=}} g(D) \in \Xi(\Psi)$, so our assumption $\mathfrak{A} \models_h \Xi(\Psi)$ implies $h(g(C)) = h(g(D))$, hence $f(C) = f(D)$. This completes the proof that $\mathfrak{B}$ is a Ψ-algebra in V. Since $A \mathrel{\mathring{=}} B$ is true in every Ψ-algebra in V and the homomorphism h maps $\mathfrak{P}$ into $\mathfrak{B} \in \mathsf{V}$, it follows that $h(A) = h(B)$.

For the converse we assume (1) and wish to show $A \mathrel{\mathring{=}} B$ is true in every Ψ-algebra in V. Assume $\mathfrak{A} \in \mathsf{V}$ is a Ψ-algebra. To show $A \mathrel{\mathring{=}} B$ is true in $\mathfrak{A}$, we consider an arbitrary homomorphism $h : \mathfrak{P} \to \mathfrak{A}$ and want to show $h(A) = h(B)$. Consider an equation in $\Xi(\Psi)$. It has the form $g(C) \mathrel{\mathring{=}} g(D)$ for some $g \in \mathrm{End}(\mathfrak{P})$ and some $C \mathrel{\mathring{=}} D \in \Psi$. The composition $h \circ g$ of h and g is also a homomorphism from $\mathfrak{P}$ to $\mathfrak{A}$, but $\mathfrak{A}$ is a Ψ-algebra in V, so $h(g(C)) = (h \circ g)(C) = (h \circ g)(D) = h(g(D))$, i.e., $\mathfrak{A} \models_h g(C) \mathrel{\dot{=}} g(D)$. This shows $\mathfrak{A} \models_h \Xi(\Psi)$, so by our hypothesis that (1) holds, we conclude that $\mathfrak{A} \models_h A \mathrel{\dot{=}} B$, i.e., $h(A) = h(B)$. Since h was an arbitrary homomorphism, we have shown that $A \mathrel{\dot{=}} B$ is true in $\mathfrak{A}$.

If $\Psi \subseteq \Sigma^\times$ is a set of equations and $1 \leq n \leq \omega$, then the set of **n-sequents corresponding to** Ψ consists of all sequents of the form $\mathsf{v}_i A \mathsf{v}_j, \Gamma \,|\, \mathsf{v}_i B \mathsf{v}_j, \Delta$ or $\mathsf{v}_i B \mathsf{v}_j, \Gamma \,|\, \mathsf{v}_i A \mathsf{v}_j, \Delta$, where $i, j \leq n$, $A, B \in \Pi^+$, $A \mathrel{\dot{=}} B \in \Psi$, and Γ and Δ are sets of n-sequents.

Theorem 3.16 *Characterizations of* CT_3^Ψ *and* CT_3.

1. *If* $\Psi \subseteq \Sigma^\times$, *then the following statements are equivalent for every* $A \in \Pi^+$.
   a. $A \in \mathsf{CT}_3^\Psi$.
   b. $\Psi \vdash_s^+ 1' \leq A$.
   c. $\Psi \vdash_s^\times 1' \leq A$.
   d. *If* $h\colon \mathfrak{P} \to \mathfrak{A}$ *is a homomorphism and* $\mathfrak{A} \in \mathsf{SA}$ *then* $\mathfrak{A} \models_h \Psi$ *implies* $\mathfrak{A} \models_h 1' \leq A$.
   e. *If* $\mathfrak{K}$ *is a frame satisfying (3.41), (3.42), (3.43), and (3.45),* $h\colon \mathfrak{P} \to \mathfrak{A}$ *is a homomorphism, and* $\mathfrak{Cm}(\mathfrak{K}) \models_h \Psi$, *then* $h(1') \subseteq h(A)$.
   f. $|\mathsf{v}_0 A \mathsf{v}_0$ *is 3-provable from the sequents corresponding to* Ψ.
   g. $\{\mathsf{G}(\varepsilon) : \varepsilon \in \Psi\} \vdash_s \forall_{\mathsf{v}_0} \mathsf{G}(\mathsf{v}_0 A \mathsf{v}_0)$.
2. *The following statements are equivalent for every* $A \in \Pi^+$.
   a. $A \in \mathsf{CT}_3$.
   b. $\vdash_s^+ 1' \leq A$.
   c. $\vdash_s^\times 1' \leq A$.
   d. $1' \leq A$ *is true in every semi-associative relation algebra.*
   e. $1' \leq A$ *is valid in every frame satisfying (3.41), (3.42), (3.43), and (3.45).*
   f. $|\mathsf{v}_0 A \mathsf{v}_0$ *is 3-provable.*
   g. $\vdash_s \forall_{\mathsf{v}_0} \mathsf{G}(\mathsf{v}_0 A \mathsf{v}_0)$.

Proof Parts (1a) and (1b) are equivalent by definition. Parts (1b) and (1c) are equivalent by Theorem 3.7(4). Parts (1c) and (1d) are equivalent by Theorem 3.3 (4).

To prove that parts (1d) and (1e) are equivalent, first assume (1d), that $\mathfrak{A} \models_h \Psi$ implies $\mathfrak{A} \models_h 1' \leq A$ whenever $\mathfrak{A} \in \mathsf{SA}$ and $h\colon \mathfrak{P} \to \mathfrak{A}$ is a homomorphism. Suppose the frame $\mathfrak{K}$ satisfies the four conditions and that $\mathfrak{Cm}(\mathfrak{K}) \models_h \Psi$ for some homomorphism $h\colon \mathfrak{P} \to \mathfrak{Cm}(\mathfrak{K})$. We (wish to show $h(1') \subseteq h(A)$. Since $\mathfrak{Cm}(\mathfrak{K}) \in \mathsf{SA}$ by Theorem 3.11(2), our hypothesis on A tells us that $\mathfrak{A} \models_h 1' \leq A$. By the definition of $\models_h$, we get $h(1') \subseteq h(A)$, as desired. Thus, (1e) holds.

For the converse, assume (1e), that $k(1') \subseteq k(A)$ whenever $\mathfrak{K}$ is a frame satisfying (3.41), (3.42), (3.43), and (3.45), and $k\colon \mathfrak{P} \to \mathfrak{Cm}(\mathfrak{K})$ is a homomorphism such that $\mathfrak{Cm}(\mathfrak{K}) \models_k \Psi$. Assume the hypotheses of (1d), that $\mathfrak{A} \in \mathsf{SA}$ and $\mathfrak{A} \models_h \Psi$ for some homomorphism $h\colon \mathfrak{P} \to \mathfrak{A}$. We wish to show $\mathfrak{A} \models_h 1' \leq A$. By Theorem 3.12(1) (5) (6) there is a frame $\mathfrak{K}$ such that $\mathfrak{Cm}(\mathfrak{K}) \in \mathsf{SA}$, $\mathfrak{K}$ satisfies the four conditions, and $\mathfrak{A}$ is isomorphic to a subalgebra of $\mathfrak{Cm}(\mathfrak{K})$. By composing h with the isomorphism from $\mathfrak{A}$ into $\mathfrak{Cm}(\mathfrak{K})$ we get a homomorphism $k\colon \mathfrak{P} \to \mathfrak{Cm}(\mathfrak{K})$. If $B \mathrel{\dot{=}} C \in \Psi$ then

$h(B) = h(C)$ since $\mathfrak{A} \models_h \Psi$, and this equality is preserved under the isomorphism from $\mathfrak{A}$ into $\mathfrak{Cm}(\mathfrak{K})$, hence $k(B) = k(C)$. This shows that $\mathfrak{Cm}(\mathfrak{K}) \models_k \Psi$. All the conditions are now met for concluding from our hypotheses on A that $k(1') \subseteq k(A)$. By applying the inverse of the isomorphism from $\mathfrak{A}$ into $\mathfrak{Cm}(\mathfrak{K})$ to both sides of this equation, we get back to $h(1') \subseteq h(A)$. By definition of $\models$, this means that $\mathfrak{A} \models_h 1' \leq A$, as desired.

Parts (1d) and (1f) are equivalent by Theorem 3.15 (3) (6) and Lemma 3.6. The equivalence of parts (1b) and (1f) follows from Theorem 3.7(2) when φ is $1' \leq A$, together with the observation that $\mathsf{G}(1' \leq A) \equiv_s^+ \forall_{\mathsf{v}_0}\mathsf{G}(\mathsf{v}_0 A \mathsf{v}_0)$ by Theorem 3.7(1). This completes the proof of part (1).

For part (2) it is enough to note that the statements (1a)–(1f) are equivalent to the corresponding statements (2a)–(2g) when $\Psi = \emptyset$. This is true by notational convention in all but two cases. Once we know that (1d) and (2d) are equivalent and that (1e) and (2e) are equivalent when $\Psi = \emptyset$, we get the equivalence of (2a)–(2g) from the equivalence of (1a)–(1f).

Part (1d) coincides with part (1) in Lemma 3.7 when $\mathsf{V} = \mathsf{SA}$ and $1' \leq A$ replaces $A \mathrel{\mathring{=}} B$. Applying Lemma 3.7 with $\psi = \emptyset$, we conclude that (1d) holds iff $1' \leq A$ is true in every $\emptyset$-algebra in SA. Every equation in $\emptyset$ is true in every algebra, so the latter statement simply says that $1' \leq A$ is true in every semi-associative relation algebra, that is, (2d) holds. Thus, (1d) and (2d) are equivalent when $\Psi = \emptyset$.

To see that (1e) and (2e) are equivalent when $\Psi = \emptyset$, note that since $\mathfrak{Cm}(\mathfrak{K}) \models_h \emptyset$ is vacuously true, (1e) asserts that $h(1') \subseteq h(A)$ for every homomorphism $h\colon \mathfrak{P} \to \mathfrak{Cm}(\mathfrak{K})$, i.e., $1' \leq A$ is true in $\mathfrak{Cm}(\mathfrak{K})$, i.e., A is valid in $\mathfrak{K}$, whenever $\mathfrak{K}$ is a frame satisfying (3.41), (3.42), (3.43), and (3.45). But that is exactly what (2e) says.

For the last two parts of the following theorem, recall that Ξ^d, Ξ^c, and Ξ^s are the equations of density (3.26), commutativity (3.27), and symmetry (3.28), respectively.

Theorem 3.17 *Characterizations of* CT_4^Ψ, CT_4, TR, *and* KR.

1. *If* $\Psi \subseteq \Sigma^\times$, *then the following statements are equivalent for every* $A \in \Pi^+$.
   a. $A \in \mathsf{CT}_4^\Psi$.
   b. $\Psi \vdash_4^+ 1' \leq A$.
   c. $\Psi \vdash^\times 1' \leq A$.
   d. *If* $\mathfrak{A} \in \mathsf{RA}$, $h\colon \mathfrak{P} \to \mathfrak{A}$ *is a homomorphism, and* $\mathfrak{A} \models_h \Psi$ *then* $\mathfrak{A} \models_h 1' \leq A$.
   e. *If* $\mathfrak{K}$ *is a frame satisfying (3.41), (3.42), (3.43), and (3.46),* $h\colon \mathfrak{P} \to \mathfrak{A}$ *is a homomorphism, and* $\mathfrak{Cm}(\mathfrak{K}) \models_h \Psi$, *then* $h(1') \subseteq h(A)$.
   f. *The sequent* $|\mathsf{v}_0 A \mathsf{v}_0$ *is 4-provable from the sequents corresponding to* Ψ.
   g. $\{\mathsf{G}(\varepsilon) : \varepsilon \in \Psi\} \vdash_4 \forall_{\mathsf{v}_0}\mathsf{G}(\mathsf{v}_0 A \mathsf{v}_0)$.
   h. $\Psi \vdash_3^+ 1' \leq A$.
   i. $\{\mathsf{G}(\varepsilon) : \varepsilon \in \Psi\} \vdash_3 \forall_{\mathsf{v}_0}\mathsf{G}(\mathsf{v}_0 A \mathsf{v}_0)$.

2. *The following statements are equivalent for every* $A \in \Pi^+$.
   a. $A \in \mathsf{CT}_4$.
   b. $\vdash_4^+ 1' \leq A$.

c. $\vdash^{\times} 1' \leq A$.
d. *$1' \leq A$ is true in every relation algebra.*
e. *$1' \leq A$ is valid in every frame satisfying (3.41), (3.42), (3.43), and (3.46).*
f. *The sequent $|\mathsf{v}_0 A \mathsf{v}_0$ is 4-provable.*
g. $\vdash_4 \forall_{\mathsf{v}_0} \mathsf{G}(\mathsf{v}_0 A \mathsf{v}_0)$.
h. $\vdash_3^+ 1' \leq A$.
i. $\vdash_3 \forall_{\mathsf{v}_0} \mathsf{G}(\mathsf{v}_0 A \mathsf{v}_0)$.

3. *The following statements are equivalent for every $A \in \Pi^+$.*

a. $A \in \mathsf{TR}$.
b. $\Xi^d \cup \Xi^c \vdash_4^+ 1' \leq A$.
c. $\Xi^d \cup \Xi^c \vdash^{\times} 1' \leq A$.
d. *$1' \leq A$ is true in every dense commutative relation algebra.*
e. *A is valid in every frame satisfying (3.41), (3.42), (3.43), (3.46), (3.47), and (3.48).*
f. *The sequent $|\mathsf{v}_0 A \mathsf{v}_0$ is 4-provable from the sequents corresponding to $\Xi^d \cup \Xi^c$.*
g. $\Xi^d \cup \Xi^c \vdash_4 \forall_{\mathsf{v}_0} \mathsf{G}(\mathsf{v}_0 A \mathsf{v}_0)$.
h. $\Xi^d \cup \Xi^c \vdash_3^+ 1' \leq A$.
i. $\Xi^d \cup \Xi^c \vdash_3 \forall_{\mathsf{v}_0} \mathsf{G}(\mathsf{v}_0 A \mathsf{v}_0)$.

4. *The following statements are equivalent for every $A \in \Pi^+$.*

a. $A \in \mathsf{KR}$.
b. $\Xi^d \cup \Xi^s \vdash_4^+ 1' \leq A$.
c. $\Xi^d \cup \Xi^s \vdash^{\times} 1' \leq A$.
d. *$1' \leq A$ is true in every dense symmetric relation algebra.*
e. *A is valid in every frame satisfying (3.41), (3.42), (3.43), (3.46), (3.47), and (3.49).*
f. *The sequent $|\mathsf{v}_0 A \mathsf{v}_0$ is 4-provable from the sequents corresponding to $\Xi^d \cup \Xi^s$.*
g. $\Xi^d \cup \Xi^s \vdash_4 \forall_{\mathsf{v}_0} \mathsf{G}(\mathsf{v}_0 A \mathsf{v}_0)$.
h. $\Xi^d \cup \Xi^s \vdash_3^+ 1' \leq A$.
i. $\Xi^d \cup \Xi^s \vdash_3 \forall_{\mathsf{v}_0} \mathsf{G}(\mathsf{v}_0 A \mathsf{v}_0)$.

Proof The proof of part (1) differs slightly from the proof of Theorem 3.16(1). Parts (1a) and (1b) are equivalent by definition, (1b) and (1c) are equivalent by Theorem 3.8(4), and (1c) is equivalent to (1d) by Theorem 3.3(3). The proof that parts (1d) and (1e) are equivalent is the same as the proof that parts (1d) and (1e) of Theorem 3.16 are equivalent, except that one uses Theorems 3.11(3), 3.12(7), and 3.12(8) in place of Theorems 3.11(2), 3.12(5) and 3.12(6), respectively. Parts (1d) and (1b) are equivalent by Theorem 3.15(4)(6) and Lemma 3.6. Part (1c) is equivalent to part (1b) by Theorem 3.8(2), equivalent to part (1c) by Theorem 3.6(5), and equivalent to part (1i) by Theorem 3.5(4) together with the observation that $1' \leq A \equiv_3^+ \forall_{\mathsf{v}_0} \mathsf{G}(\mathsf{v}_0 A \mathsf{v}_0)$ by Theorem 3.5 (3). Finally, (1g) and (1i) are equivalent by Theorem 3.8(3).

The proof of part (2) is the same as the proof of Theorem 3.16(2) except that RA and (3.46) replace SA and (3.45). Parts (3) and (4) follow mostly from part (1) by taking $\Psi = \Xi^d \cup \Xi^c$ and $\Psi = \Xi^d \cup \Xi^s$, respectively. In all three cases one uses various instances of Lemma 3.7, taking Ψ to be $\emptyset$, $\Xi^d \cup \Xi^c$, or $\Xi^d \cup \Xi^s$, and arguing as in the proof of Theorem 3.16(2).

There are some differences between the proofs of parts (1) and (2) and the proofs of parts (3) and (4). While the definition of TR still produces the equivalence of (3a) and (3b), the definition of KR yields the equivalence of (4a) and (4e). In the proof that (3d) and (3e) are equivalent, one needs to observe that $\mathfrak{Cm}(\mathfrak{K})$ is a dense commutative relation algebra iff $\mathfrak{K}$ satisfies (3.41), (3.42), (3.43), (3.46), (3.47), and (3.48). In the proof that (4d) and (4e) are equivalent, one must observe that $\mathfrak{Cm}(\mathfrak{K})$ is a dense symmetric relation algebra iff $\mathfrak{K}$ satisfies (3.41), (3.42), (3.43), (3.46), (3.47), and (3.49). These observations follow from Lemma 3.1 and Theorems 3.10, 3.11(3), and 3.14(2).

The inclusion of commutativity in the definitions of TR and KR allows some further characterizations as a consequence of Lemma 3.2.

Theorem 3.18 *Integral characterizations of* TR *and* KR.

1. *The following statements are equivalent for every $A \in \Pi^+$.*
   a. $A \in$ TR.
   b. *1' $\leq A$ is true in every integral dense commutative relation algebra.*
   c. *A is valid in every frame $\mathfrak{K} = \langle K, R, ^*, \mathcal{I}\rangle$ satisfying $|\mathcal{I}| = 1$, (3.41), (3.42), (3.43), (3.46), (3.47), and (3.48).*
2. *The following statements are equivalent for every $A \in \Pi^+$.*
   a. $A \in$ KR.
   b. *1' $\leq A$ is true in every integral dense symmetric relation algebra.*
   c. *A is valid in every frame $\mathfrak{K} = \langle K, R, ^*, \mathcal{I}\rangle$ satisfying $|\mathcal{I}| = 1$, (3.41), (3.42), (3.43), (3.46), (3.47), and (3.49).*

Theorems 3.16, 3.17, and 3.18 link the logics of 3-provability with semi-associative relation algebras and the logics of 4-provability with relation algebras. The logics of ω-provability are linked with representable relation algebras.

Theorem 3.19 *Characterizations of* $\mathsf{CT}^{\Psi}_{\omega}$ *and* CT_{ω}.

1. *If $\Psi \subseteq \Sigma^{\times}$, then the following statements are equivalent for every $A \in \Pi^+$.*
   a. $A \in \mathsf{CT}^{\Psi}_{\omega}$.
   b. $\Psi \vdash^+ 1' \leq A$.
   c. *If $h\colon \mathfrak{P} \to \mathfrak{A}$ is a homomorphism and $\mathfrak{A} \in$ RRA then $\mathfrak{A} \models_h \Psi$ implies $\mathfrak{A} \models_h$ 1' $\leq A$.*
   d. $|\mathsf{v}_0 A \mathsf{v}_0$ *is ω-provable from the sequents corresponding to Ψ.*
   e. $\{\mathsf{G}(\varepsilon) : \varepsilon \in \Psi\} \vdash \forall_{\mathsf{v}_0}\mathsf{G}(\mathsf{v}_0 A \mathsf{v}_0)$.

2. *The following statements are equivalent for every* $A \in \Pi^+$.

   a. $A \in \mathsf{CT}_\omega$.
   b. $\vdash^+ 1' \leq A$.
   c. $1' \leq A$ *is true in every representable relation algebra.*
   d. $|\mathsf{v}_0 A \mathsf{v}_0$ *is* ω*-provable.*
   e. $\vdash \forall_{\mathsf{v}_0} \mathsf{G}(\mathsf{v}_0 A \mathsf{v}_0)$.

Proof Parts (1a) and (1b) are equivalent by definition, (1b) and (1c) by Theorem 3.4, and (1b) and (5) by Theorem 3.1(3)(4) and $\mathsf{G}(1' \leq A) \equiv^+ \forall_{\mathsf{v}_0} \mathsf{G}(\mathsf{v}_0 A \mathsf{v}_0)$, and (1c) and (1d) by Theorem 3.15(5)(6) and Lemma 3.6.

Two observations close this section. For $5 \leq n < \omega$, membership in CT_n as defined in (3.24) cannot be characterized as n-provability in the sequent calculus. The situation is quite complicated and full of non-finite axiomatizability results; see Hirsch and Hodkinson (2002). However, the implication in one direction still holds.

Theorem 3.20 *If* $\Psi \subseteq \Sigma^\times$, $A \in \Pi^+$, $5 \leq n < \omega$, *and* $|\mathsf{v}_0 A \mathsf{v}_0$ *is* n*-provable in the sequent calculus from* Ψ, *then* $A \in \mathsf{CT}_n^\Psi$.

Several results in this section also apply to T_n^Ψ.

Theorem 3.21 *Theorems 3.16, 3.17, 3.19, and 3.20 continue to hold if A is assumed to be in* Π^r *instead of* Π^+, Ψ *is a set of equations between predicates in* Π^r, *and* CT_3^Ψ, CT_4^Ψ, CT_n^Ψ, *and* CT_ω^Ψ *are replaced by* T_3^Ψ, T_4^Ψ, T_n^Ψ, *and* T_ω^Ψ.

3.14 Theorems and Derived Rules of CT_3, CT_4, and TR

Theorem 3.16 provides several ways to show $A \in \mathsf{CT}_3$. One can prove $1' \leq A$ in $\mathcal{L}w^\times$ or $\mathcal{L}s_3^+$, or prove $1' \leq A$ is true in every semi-associative relation algebra, or prove $\forall_{\mathsf{v}_0} \mathsf{G}(\mathsf{v}_0 A \mathsf{v}_0)$ in $\mathcal{L}s_3$, or prove A is valid in every frame satisfying (3.41), (3.42), (3.43), and (3.46), or show that $|\mathsf{v}_0 A \mathsf{v}_0$ is 3-provable.

Theorem 3.22 below shows that CT_3 includes truth (3.60), laws of the excluded middle for Boolean negation (3.61) and De Morgan negation (3.62), self-implication (3.63), basic laws of disjunction and conjunction (3.65)–(3.74) and (3.91)–(3.92), laws of distributivity (3.75)–(3.76), laws of double Boolean negation (3.83)–(3.84) and double De Morgan negation (3.77)–(3.78), De Morgan laws for De Morgan negation (3.79)–(3.82) and for Boolean negation (3.85)–(3.88), explosion for Boolean negation (3.90), and some basic laws of fusion (3.96)–(3.99). Furthermore, CT_3 has many derived rules of inference, including adjunction (3.110), modus ponens (3.111), disjunctive syllogism (3.112), contraposition for De Morgan negation (3.115)–(3.116), a cut rule (3.117), the suffixing rule (3.122), the prefixing rule (3.123), and monotonicity for fusion (3.125). The prefixing axiom (3.127) is not in CT_3 but it is in CT_4, as shown by Theorem 3.23 in Sect. 3.15. The suffixing axiom (3.141) is not even ω-provable and the same applies to the axioms of permutation (3.140) and contraposition (3.139), as shown by Theorem 3.24 in Sect. 3.16.

Although explosion for Boolean negation (3.90) is in CT_3, explosion for De Morgan negation $A \wedge {\sim}A \to B$ and positive paradox $A \to (B \to A)$ are not even ω-provable, because as relations they need not contain the identity relation. For example, if $\langle y, x\rangle \in A$, $\langle x, y\rangle \notin A$, and $\langle y, x\rangle \notin B$, then $\langle x, x\rangle \notin (A \wedge {\sim}A) \to B$, and if $\langle y, x\rangle \in A$, $\langle z, y\rangle \in B$, and $\langle z, x\rangle \notin A$, then $\langle x, x\rangle \notin A \to (B \to A)$.

The terminology of n-proofs and n-provability extends from sequents to predicates. A predicate $A \in \Pi^+$ is **n-provable from** Ψ if the sequent $|\mathsf{v}_0 A \mathsf{v}_0$ is n-provable from Ψ and an **n-proof of** A **from** Ψ is an n-proof of the sequent $|\mathsf{v}_0 A \mathsf{v}_0$ from Ψ. Thus, A is n-provable from Ψ iff there is an n-proof of A from Ψ. By (3.53), (3.54), and (3.56), $A \to B$ is n-provable from Ψ iff there is an n-proof from Ψ of the sequent $\mathsf{v}_i A \mathsf{v}_j\,|\,\mathsf{v}_i B \mathsf{v}_j$ for distinct $i, j < n$. For this reason, we refer to an n-proof from Ψ of the sequent $\mathsf{v}_i A \mathsf{v}_j\,|\,\mathsf{v}_i B \mathsf{v}_j$ as an n-proof of $A \to B$ from Ψ. Accordingly, A (or $A \to B$) is **n-provable from density**, **n-provable from commutativity**, or **n-provable from symmetry** if the sequent $|\mathsf{v}_i A \mathsf{v}_i$ (or the sequent $\mathsf{v}_i A \mathsf{v}_j\,|\,\mathsf{v}_i B \mathsf{v}_j$) is n-provable from the equations of density Ξ^d (3.26), the equations of commutativity Ξ^c (3.27), or the equations of symmetry Ξ^s (3.28), respectively, for any distinct $i, j < n$.

Derived rules of inference are stated in the form $A \vdash B$ or $A, B \vdash C$. We say $A \vdash B$ is **n-provable** if every proof of $|\mathsf{v}_0 A \mathsf{v}_0$ can be extended by an n-proof to a proof of $|\mathsf{v}_0 B \mathsf{v}_0$. Similarly, we say $A, B \vdash C$ is **n-provable** if every proof of $|\mathsf{v}_0 A \mathsf{v}_0$ and every proof of $|\mathsf{v}_0 B \mathsf{v}_0$ can be concatenated and extended by an n-proof to obtain a proof of $|\mathsf{v}_0 C \mathsf{v}_0$. For example, if $|\mathsf{v}_0 A \mathsf{v}_0$ is m-provable and $A \vdash B$ is n-provable, then B is $\max(n, m)$-provable.

Theorem 3.22 *Assume* $A, B, C, D, E, F \in \Pi^+$.

1. *Predicates (3.60)–(3.62) are 1-provable.*
2. *Predicates (3.63)–(3.90) are 2-provable.*
3. *Predicates (3.91)–(3.109) are 3-provable.*
4. *Rules (3.110)–(3.113) are 1-provable.*
5. *Rules (3.114)–(3.121) are 2-provable.*
6. *Rules (3.122)–(3.126) are 3-provable.*
7. *Predicates (3.127)–(3.132) are 4-provable.*
8. *Predicates (3.133)–(3.135) are 3-provable from density.*
9. *Predicates (3.136)–(3.137) are 4-provable from density.*
10. *Predicates (3.138)–(3.139) are 3-provable from commutativity.*
11. *Predicates (3.140)–(3.141) are 4-provable from commutativity.*
12. *Predicate (3.142) is 4-provable from density and commutativity.*
13. *Predicate (3.143) is 2-provable from symmetry.*
14. *Predicate (3.144) is 3-provable from symmetry.*

15. The predicates and rules of CT_3, CT_4, *and* TR *include*

Logic	*Predicates*	*Rules*
CT_3	*(3.60)–(3.109)*	*(3.110)–(3.126)*
CT_4	*(3.60)–(3.109), (3.127)–(3.132)*	*(3.110)–(3.126)*
TR	*(3.60)–(3.109), (3.127)–(3.132), (3.138)–(3.142)*	*(3.110)–(3.126)*

The predicates and rules that have been provided with n-proofs are marked with asterisks.

1-provable predicates:

$$\mathbf{t} \quad * \tag{3.60}$$

$$A \vee \neg A \quad * \tag{3.61}$$

$$A \vee {\sim} A \quad * \tag{3.62}$$

2-provable predicates:

$$A \to A \quad * \tag{3.63}$$

$$((A \to A) \to B) \to B \quad * \tag{3.64}$$

$$A \vee A \to A \tag{3.65}$$

$$A \to A \vee B \tag{3.66}$$

$$B \to A \vee B \tag{3.67}$$

$$A \vee B \to B \vee A \tag{3.68}$$

$$A \to A \wedge A \tag{3.69}$$

$$A \wedge B \to A \tag{3.70}$$

$$A \wedge B \to B \tag{3.71}$$

$$A \wedge B \to B \wedge A \tag{3.72}$$

$$(A \vee B) \vee C \to A \vee (B \vee C) \tag{3.73}$$

$$(A \wedge B) \wedge C \to A \wedge (B \wedge C) \tag{3.74}$$

$$(A \vee B) \wedge C \to (A \wedge C) \vee (B \wedge C) \tag{3.75}$$

$$(A \wedge B) \vee C \to (A \vee C) \wedge (B \vee C) \tag{3.76}$$

$${\sim}{\sim} A \to A \tag{3.77}$$

$$A \to {\sim}{\sim} A \tag{3.78}$$

$${\sim}(A \vee B) \to {\sim} A \wedge {\sim} B \tag{3.79}$$

$${\sim}(A \wedge B) \to {\sim} A \vee {\sim} B \tag{3.80}$$

$${\sim} A \wedge {\sim} B \to {\sim}(A \vee B) \tag{3.81}$$

$${\sim} A \vee {\sim} B \to {\sim}(A \wedge B) \tag{3.82}$$

$$\neg\neg A \to A \tag{3.83}$$

$$A \to \neg\neg A \tag{3.84}$$

$$\neg(A \vee B) \to \neg A \wedge \neg B \tag{3.85}$$

$$\neg(A \wedge B) \to \neg A \vee \neg B \tag{3.86}$$

$$\neg A \wedge \neg B \to \neg(A \vee B) \tag{3.87}$$

$$\neg A \vee \neg B \to \neg(A \wedge B) \tag{3.88}$$

$$A \to B \vee \neg B \qquad * \tag{3.89}$$

$$\neg A \wedge A \to B \qquad * \tag{3.90}$$

3-provable predicates:

$$(A \to B) \wedge (A \to C) \to (A \to B \wedge C) \tag{3.91}$$

$$(A \to C) \wedge (B \to C) \to (A \vee B \to C) \tag{3.92}$$

$$(A \to B) \wedge (C \to D) \to (A \wedge C \to B \wedge D) \tag{3.93}$$

$$(A \to B) \wedge (C \to D) \to (A \vee C \to B \vee D) \qquad * \tag{3.94}$$

$$(A \to B) \vee (C \to D) \to (A \wedge C \to B \vee D) \tag{3.95}$$

$$A \circ B \to {\sim}(A \to {\sim}B) \tag{3.96}$$

$${\sim}(A \to {\sim}B) \to A \circ B \tag{3.97}$$

$$(A \to B) \circ A \to B \tag{3.98}$$

$$A \to (B \to A \circ B) \qquad * \tag{3.99}$$

$$A \to ((B \to {\sim}A) \to {\sim}B) \qquad * \tag{3.100}$$

$$A \circ B \wedge C \to A \circ (B \wedge {\sim}D) \vee (A \wedge C \circ D) \circ B \qquad * \tag{3.101}$$

$$A \circ B \wedge C \to (A \wedge {\sim}D) \circ B \vee A \circ (B \wedge D \circ C) \tag{3.102}$$

$$A \circ B \wedge C \to (A \wedge C \circ B^*) \circ (B \wedge A^* \circ C) \tag{3.103}$$

$$\mathbf{t} \circ A \to A \qquad * \tag{3.104}$$

$$A \to A \circ \mathbf{t}^* \qquad * \tag{3.105}$$

$$\mathbf{t} \to \mathbf{t}^* \qquad * \tag{3.106}$$

$$\mathbf{t}^* \to \mathbf{t} \qquad * \tag{3.107}$$

$$A \circ \mathbf{t} \to A \qquad * \tag{3.108}$$

$$\mathbf{t} \wedge {\sim}\mathbf{t} \to A \qquad * \tag{3.109}$$

1-provable rules:

$$A, B \vdash A \wedge B \qquad * \tag{3.110}$$

$$A \to B, A \vdash B \qquad * \tag{3.111}$$

$$A \vee B, {\sim}A \vdash B \qquad * \tag{3.112}$$

$$A \vdash A^* \qquad * \tag{3.113}$$

2-provable rules:

$$A \to B, B \to C \vdash A \to C \qquad * \qquad (3.114)$$

$$A \to B \vdash {\sim}B \to {\sim}A \qquad * \qquad (3.115)$$

$$A \to {\sim}B \vdash B \to {\sim}A \qquad * \qquad (3.116)$$

$$A \wedge B \to C, B \to C \vee A \vdash B \to C \qquad * \qquad (3.117)$$

$$A \vdash (A \to B) \to B \qquad * \qquad (3.118)$$

$$A \wedge B \to \neg C \vdash A \wedge C \to \neg B \qquad * \qquad (3.119)$$

$$B \wedge \neg C \to A \wedge \neg A \vdash B \to C \qquad (3.120)$$

$$B \wedge C^* \to A \wedge \neg A \vdash B \to {\sim}C \qquad (3.121)$$

3-provable rules:

$$A \to B \vdash (B \to C) \to (A \to C) \qquad * \qquad (3.122)$$

$$A \to B \vdash (C \to A) \to (C \to B) \qquad * \qquad (3.123)$$

$$A \to B, C \to D \vdash (B \to C) \to (A \to D) \qquad * \qquad (3.124)$$

$$A \to B, C \to D \vdash A \circ C \to B \circ D \qquad * \qquad (3.125)$$

$$A \to (B \to C) \vdash B \to ({\sim}C \to {\sim}A) \qquad * \qquad (3.126)$$

4-provable predicates:

$$(A \to B) \to ((C \to A) \to (C \to B)) \qquad * \qquad (3.127)$$

$$(A \to (B \to C)) \to (A \circ B \to C) \qquad * \qquad (3.128)$$

$$(A \circ B \to C) \to (A \to (B \to C)) \qquad * \qquad (3.129)$$

$$(A \to B) \to (A \circ C \to B \circ C) \qquad * \qquad (3.130)$$

$$(A \circ B) \circ C \to A \circ (B \circ C) \qquad * \qquad (3.131)$$

$$A \circ (B \circ C) \to (A \circ B) \circ C \qquad * \qquad (3.132)$$

3-provable from density:

$$(A \to {\sim}A) \to {\sim}A \qquad * \qquad (3.133)$$

$$A \wedge B \to A \circ B \qquad * \qquad (3.134)$$

$$(A \to B) \to {\sim}A \vee B \qquad * \qquad (3.135)$$

4-provable from density:

$$(A \to (A \to B)) \to (A \to B) \qquad * \qquad (3.136)$$

$$(A \to (B \to C)) \to (A \wedge B \to C) \qquad * \qquad (3.137)$$

3-provable from commutativity:

$$A \to ((A \to B) \to B) \qquad * \qquad (3.138)$$

$$(A \to {\sim}B) \to (B \to {\sim}A) \qquad * \qquad (3.139)$$

4-provable from commutativity:

$$(A \to (B \to C)) \to (B \to (A \to C)) \qquad * \qquad (3.140)$$

$$(A \to B) \to ((B \to C) \to (A \to C)) \qquad * \qquad (3.141)$$

4-provable from density and commutativity:

$$(A \to (B \to C)) \to ((A \to B) \to (A \to C)) \qquad * \qquad (3.142)$$

2-provable from symmetry:

$$A \wedge {\sim}A \to B \qquad * \qquad (3.143)$$

3-provable from symmetry:

$$A \circ B \to B \circ A \qquad * \qquad (3.144)$$

Proof (3.60) 1-proof of **t**.

1.	$\mid \mathsf{v}_0 1' \mathsf{v}_0$	$\mid 1'$
2.	$\mid \mathsf{v}_0 \mathbf{t} \mathsf{v}_0$	1, (3.21)

Proof (3.61) 1-proof of $A \vee \neg A$.

1.	$\mathsf{v}_0 A \mathsf{v}_0 \mid \mathsf{v}_0 A \mathsf{v}_0$	axiom
2.	$\mid \mathsf{v}_0 A \mathsf{v}_0, \mathsf{v}_0 \overline{A} \mathsf{v}_0$	1, $\mid^{-}$
3.	$\mid \mathsf{v}_0 A + \overline{A} \mathsf{v}_0$	2, $\mid +$
4.	$\mid \mathsf{v}_0 A \vee \neg A \mathsf{v}_0$	3, (3.14), (3.16)

Proof (3.62) 1-proof of $A \vee {\sim}A$.

1.	$\mathsf{v}_0 A \mathsf{v}_0 \mid \mathsf{v}_0 A \mathsf{v}_0$	axiom
2.	$\mathsf{v}_0 A^{\smile} \mathsf{v}_0 \mid \mathsf{v}_0 A \mathsf{v}_0$	1, $^{\smile}\mid$
3.	$\mid \mathsf{v}_0 A \mathsf{v}_0, \mathsf{v}_0 \overline{A^{\smile}} \mathsf{v}_0$	2, $\mid^{-}$
4.	$\mid \mathsf{v}_0 A + \overline{A^{\smile}} \mathsf{v}_0$	3, $\mid +$
5.	$\mid \mathsf{v}_0 A \vee {\sim} A \mathsf{v}_0$	4, (3.14), (3.17)

Proof (3.63, 3.64) 2-proof of $A \to A$ and $((A \to A) \to B) \to B$.

1.	$\mathsf{v}_0 A \mathsf{v}_1 \mid \mathsf{v}_0 A \mathsf{v}_1$	axiom
2.	$\mid \mathsf{v}_0 A \to A \mathsf{v}_0$	1, (3.53), (3.56)
3.	$\mid \mathsf{v}_0 (A \to A)^{\smile} \mathsf{v}_0$	2, $\mid^{\smile}$
4.	$\mathsf{v}_0 B \mathsf{v}_1 \mid \mathsf{v}_0 B \mathsf{v}_1$	axiom
5.	$\mid \mathsf{v}_0 \overline{B} \mathsf{v}_1, \mathsf{v}_0 B \mathsf{v}_1$	4, $\mid^{-}$
6.	$\mid \mathsf{v}_0 (A \to A)^{\smile} ; \overline{B} \mathsf{v}_1, \mathsf{v}_0 B \mathsf{v}_1$	3, 5, $\mid ;$
7.	$\mathsf{v}_0 \overline{(A \to A)^{\smile} ; \overline{B}} \mathsf{v}_1 \mid \mathsf{v}_0 B \mathsf{v}_1$	6, $^{-}\mid$
8.	$\mathsf{v}_0 (A \to A) \to B \mathsf{v}_1 \mid \mathsf{v}_0 B \mathsf{v}_1$	7, (3.18)

Proof (3.89) 2-proof of $A \to B \vee \neg B$.

1.	$\mathsf{v}_0 A \mathsf{v}_1, \mathsf{v}_0 B \mathsf{v}_1 \mid \mathsf{v}_0 B \mathsf{v}_1$	axiom
2.	$\mathsf{v}_0 A \mathsf{v}_1 \mid \mathsf{v}_0 B \mathsf{v}_1, \mathsf{v}_0 \overline{B} \mathsf{v}_1$	1, $\mid^{-}$
3.	$\mathsf{v}_0 A \mathsf{v}_1 \mid \mathsf{v}_0 B + \overline{B} \mathsf{v}_1$	2, $\mid +$
4.	$\mathsf{v}_0 A \mathsf{v}_1 \mid \mathsf{v}_0 B \vee \neg B \mathsf{v}_1$	3, (3.14), (3.16)

Proof (3.90) 2-proof of $A \wedge \neg A \to B$.

1.	$\mathsf{v}_0 A \mathsf{v}_1 \mid \mathsf{v}_0 A \mathsf{v}_1, \mathsf{v}_0 B \mathsf{v}_1$	axiom
2.	$\mathsf{v}_0 A \mathsf{v}_1, \mathsf{v}_0 \overline{A} \mathsf{v}_1 \mid \mathsf{v}_0 B \mathsf{v}_1$	1, $^{-}\mid$
3.	$\mathsf{v}_0 A \cdot \overline{A} \mathsf{v}_1 \mid \mathsf{v}_0 B \mathsf{v}_1$	2, $\cdot \mid$
4.	$\mathsf{v}_0 A \wedge \neg A \mathsf{v}_1 \mid \mathsf{v}_0 B \mathsf{v}_1$	3, (3.15), (3.16))

Proof (3.94) 3-proof of $(A \to B) \wedge (C \to D) \to (A \vee C \to B \vee D)$.

1.	$\mathsf{v}_2 A \mathsf{v}_0 \mid \mathsf{v}_2 A \mathsf{v}_0$	axiom
2.	$\mathsf{v}_2 A \mathsf{v}_0 \mid \mathsf{v}_0 A^{\smile} \mathsf{v}_2$	1, $\mid^{\smile}$
3.	$\mathsf{v}_2 B \mathsf{v}_1 \mid \mathsf{v}_2 B \mathsf{v}_1$	axiom
4.	$\mid \mathsf{v}_2 B \mathsf{v}_1, \mathsf{v}_2 \overline{B} \mathsf{v}_1$	3, $\mid^{-}$
5.	$\mathsf{v}_2 A \mathsf{v}_0 \mid \mathsf{v}_2 B \mathsf{v}_1, \mathsf{v}_0 A^{\smile} ; \overline{B} \mathsf{v}_1$	2, 4, $\mid ;$
6.	$\mathsf{v}_2 C \mathsf{v}_0 \mid \mathsf{v}_2 C \mathsf{v}_0$	axiom
7.	$\mathsf{v}_2 C \mathsf{v}_0 \mid \mathsf{v}_0 C^{\smile} \mathsf{v}_2$	6, $\mid^{\smile}$
8.	$\mathsf{v}_2 D \mathsf{v}_1 \mid \mathsf{v}_2 D \mathsf{v}_1$	axiom
9.	$\mid \mathsf{v}_2 D \mathsf{v}_1, \mathsf{v}_2 \overline{D} \mathsf{v}_1$	8, $\mid^{-}$
10.	$\mathsf{v}_2 C \mathsf{v}_0 \mid \mathsf{v}_2 D \mathsf{v}_1, \mathsf{v}_0 C^{\smile} ; \overline{D} \mathsf{v}_1$	7, 9, $\mid ;$
11.	$\mathsf{v}_2 A + C \mathsf{v}_0 \mid \mathsf{v}_2 B \mathsf{v}_1, \mathsf{v}_2 D \mathsf{v}_1, \mathsf{v}_0 A^{\smile} ; \overline{B} \mathsf{v}_1, \mathsf{v}_0 C^{\smile} ; \overline{D} \mathsf{v}_1$	5, 10, $+\mid$
12.	$\mathsf{v}_2 A + C \mathsf{v}_0 \mid \mathsf{v}_2 B + D \mathsf{v}_1, \mathsf{v}_0 A^{\smile} ; \overline{B} \mathsf{v}_1, \mathsf{v}_0 C^{\smile} ; \overline{D} \mathsf{v}_1$	11, $\mid +$
13.	$\mathsf{v}_0 (A + C)^{\smile} \mathsf{v}_2, \mathsf{v}_2 \overline{B + D} \mathsf{v}_1 \mid \mathsf{v}_0 A^{\smile} ; \overline{B} \mathsf{v}_1, \mathsf{v}_0 C^{\smile} ; \overline{D} \mathsf{v}_1$	12, $^{\smile}\mid$, $^{-}\mid$
14.	$\mathsf{v}_0 (A + C)^{\smile} ; \overline{B + D} \mathsf{v}_1 \mid \mathsf{v}_0 A^{\smile} ; \overline{B} \mathsf{v}_1, \mathsf{v}_0 C^{\smile} ; \overline{D} \mathsf{v}_1$	13, $;\ \mid$, no v_2
15.	$\mathsf{v}_0 \overline{A^{\smile} ; \overline{B}} \mathsf{v}_1, \mathsf{v}_0 \overline{C^{\smile} ; \overline{D}} \mathsf{v}_1, \mathsf{v}_0 (A + C)^{\smile} ; \overline{B + D} \mathsf{v}_1 \mid$	14, $^{-}\mid$
16.	$\mathsf{v}_0 \overline{A^{\smile} ; \overline{B}} \cdot \overline{C^{\smile} ; \overline{D}} \mathsf{v}_1 \mid \mathsf{v}_0 \overline{(A + C)^{\smile} ; \overline{B + D}} \mathsf{v}_1$	15, $\cdot \mid$, $\mid^{-}$
17.	$\mathsf{v}_0 (A \to B) \wedge (C \to D) \mathsf{v}_1 \mid \mathsf{v}_0 A \vee C \to B \vee D \mathsf{v}_1$	16, (3.14), (3.15),(3.18)

Proof (3.99) 3-proof of $A \to (B \to A \circ B)$.

1.	$\mathsf{v}_0 A \mathsf{v}_1 \mid \mathsf{v}_0 A \mathsf{v}_1$	axiom
2.	$\mathsf{v}_2 B \mathsf{v}_0 \mid \mathsf{v}_2 B \mathsf{v}_0$	axiom
3.	$\mathsf{v}_0 A \mathsf{v}_1, \mathsf{v}_2 B \mathsf{v}_0 \mid \mathsf{v}_2 B ; A \mathsf{v}_1$	1, 2, $\mid ;$
4.	$\mathsf{v}_0 A \mathsf{v}_1, \mathsf{v}_0 B^{\smile} \mathsf{v}_2, \mathsf{v}_2 \overline{B ; A} \mathsf{v}_1 \mid$	3, $^{\smile}\mid$, $^{-}\mid$
5.	$\mathsf{v}_0 A \mathsf{v}_1, \mathsf{v}_0 B^{\smile} ; \overline{B ; A} \mathsf{v}_1 \mid$	4, $;\mid$, no v_2
6.	$\mathsf{v}_0 A \mathsf{v}_1 \mid \mathsf{v}_0 \overline{B^{\smile} ; \overline{B ; A}} \mathsf{v}_1$	5, $\mid^{-}$
7.	$\mathsf{v}_0 A \mathsf{v}_1 \mid \mathsf{v}_0 B \to A \circ B \mathsf{v}_1$	6, (3.18), (3.19)

Proof (3.100) 3-proof of $A \to ((B \to {\sim}A) \to {\sim}B)$.

1.	$\mathsf{v}_0 A \mathsf{v}_1 \,\vert\, \mathsf{v}_0 A \mathsf{v}_1$	axiom
2.	$\mathsf{v}_1 B \mathsf{v}_2 \,\vert\, \mathsf{v}_1 B \mathsf{v}_2$	axiom
3.	$\mathsf{v}_0 A \mathsf{v}_1 \,\vert\, \mathsf{v}_1 \overline{\overline{\overline{A^{\smile}}}} \mathsf{v}_0$	1, $\vert^{\smile}$, $^{-}\vert$, $\vert^{-}$
4.	$\mathsf{v}_1 B \mathsf{v}_2 \,\vert\, \mathsf{v}_2 B^{\smile} \mathsf{v}_1$	2, $\vert^{\smile}$
5.	$\mathsf{v}_0 A \mathsf{v}_1, \mathsf{v}_1 B \mathsf{v}_2 \,\vert\, \mathsf{v}_2 B^{\smile} ; \overline{\overline{\overline{A^{\smile}}}} \mathsf{v}_0$	3, 4, $\vert ;$
6.	$\mathsf{v}_0 A \mathsf{v}_1, \mathsf{v}_2 \overline{B^{\smile} ; \overline{\overline{\overline{A^{\smile}}}}} \mathsf{v}_0, \mathsf{v}_1 B \mathsf{v}_2 \,\vert$	5, $^{-}\vert$
7.	$\mathsf{v}_0 A \mathsf{v}_1, \mathsf{v}_2 B \to {\sim} A \mathsf{v}_0, \mathsf{v}_1 B \mathsf{v}_2 \,\vert$	6, (3.17), (3.18)
8.	$\mathsf{v}_0 A \mathsf{v}_1, \mathsf{v}_0 B \to {\sim} A^{\smile} \mathsf{v}_2, \mathsf{v}_2 \overline{\overline{\overline{B^{\smile}}}} \mathsf{v}_1 \,\vert$	7, $^{\smile}\vert$, $\vert^{-}$, $^{-}\vert$
9.	$\mathsf{v}_0 A \mathsf{v}_1, \mathsf{v}_0 B \to {\sim} A^{\smile} ; \overline{\overline{\overline{B^{\smile}}}} \mathsf{v}_1 \,\vert$	8, $; \vert$, no v_2
10.	$\mathsf{v}_0 A \mathsf{v}_1 \,\vert\, \mathsf{v}_0 \overline{B \to {\sim} A^{\smile} ; \overline{\overline{\overline{B^{\smile}}}}} \mathsf{v}_1$	9, $\vert^{-}$
11.	$\mathsf{v}_0 A \mathsf{v}_1 \,\vert\, \mathsf{v}_0 (B \to {\sim} A) \to {\sim} B \mathsf{v}_1$	10, (3.17), (3.18)

Proof (3.101) 3-proof of $A \circ B \wedge C \to A \circ (B \wedge {\sim} D) \vee (A \wedge C \circ D) \circ B$.

1.	$\mathsf{v}_0 C \mathsf{v}_1 \,\vert\, \mathsf{v}_0 C \mathsf{v}_1$	axiom
2.	$\mathsf{v}_2 D \mathsf{v}_0 \,\vert\, \mathsf{v}_2 D \mathsf{v}_0$	axiom
3.	$\mathsf{v}_0 C \mathsf{v}_1, \mathsf{v}_2 D \mathsf{v}_0 \,\vert\, \mathsf{v}_2 D ; C \mathsf{v}_1$	1, 2, $\vert ;$
4.	$\mathsf{v}_2 A \mathsf{v}_1 \,\vert\, \mathsf{v}_2 A \mathsf{v}_1$	axiom
5.	$\mathsf{v}_2 A \mathsf{v}_1, \mathsf{v}_0 C \mathsf{v}_1, \mathsf{v}_2 D \mathsf{v}_0 \,\vert\, \mathsf{v}_2 (A \cdot C ; D) \mathsf{v}_1$	3, 4, $\vert \cdot$
6.	$\mathsf{v}_0 B \mathsf{v}_2 \,\vert\, \mathsf{v}_0 B \mathsf{v}_2$	axiom
7.	$\mathsf{v}_0 B \mathsf{v}_2, \mathsf{v}_2 A \mathsf{v}_1, \mathsf{v}_0 C \mathsf{v}_1, \mathsf{v}_2 D \mathsf{v}_0 \,\vert\, \mathsf{v}_0 B ; (A \cdot C ; D) \mathsf{v}_1$	5, 6, $\vert ;$
8.	$\mathsf{v}_0 B \mathsf{v}_2, \mathsf{v}_2 A \mathsf{v}_1, \mathsf{v}_0 C \mathsf{v}_1 \,\vert\, \mathsf{v}_0 \overline{D^{\smile}} \mathsf{v}_2, \mathsf{v}_0 B ; (A \cdot C ; D) \mathsf{v}_1$	7, $^{\smile}\vert$, $\vert^{-}$
9.	$\mathsf{v}_0 B \mathsf{v}_2, \mathsf{v}_2 A \mathsf{v}_1, \mathsf{v}_0 C \mathsf{v}_1 \,\vert\, \mathsf{v}_0 B \cdot \overline{D^{\smile}} \mathsf{v}_2, \mathsf{v}_0 B ; (A \cdot C ; D) \mathsf{v}_1$	6, 8, $\vert \cdot$
10.	$\mathsf{v}_0 B \mathsf{v}_2, \mathsf{v}_2 A \mathsf{v}_1, \mathsf{v}_0 C \mathsf{v}_1 \,\vert\, \mathsf{v}_0 (B \cdot \overline{D^{\smile}}) ; A \mathsf{v}_1, \mathsf{v}_0 B ; (A \cdot D ; C) \mathsf{v}_1$	4, 9, $\vert ;$
11.	$\mathsf{v}_0 B \mathsf{v}_2, \mathsf{v}_2 A \mathsf{v}_1, \mathsf{v}_0 C \mathsf{v}_1 \,\vert\, \mathsf{v}_0 (B \cdot \overline{D^{\smile}}) ; A + B ; (A \cdot D ; C) \mathsf{v}_1$	10, $\vert +$
12.	$\mathsf{v}_0 B ; A \mathsf{v}_1, \mathsf{v}_0 C \mathsf{v}_1 \,\vert\, \mathsf{v}_0 (B \cdot \overline{D^{\smile}}) ; A + B ; (A \cdot D ; C) \mathsf{v}_1$	11, $; \vert$, no v_2
13.	$\mathsf{v}_0 B ; A \cdot C \mathsf{v}_1 \,\vert\, \mathsf{v}_0 (B \cdot \overline{D^{\smile}}) ; A + B ; (A \cdot D ; C) \mathsf{v}_1$	12, $\cdot \vert$
14.	$\mathsf{v}_0 A \circ B \wedge C \mathsf{v}_1 \,\vert\, \mathsf{v}_0 A \circ (B \wedge {\sim} D) \vee (A \wedge C \circ D) \circ B \mathsf{v}_1$	13, (3.15), (3.17), (3.19)

Proof (3.104) 3-proof of $\mathbf{t} \circ A \to A$.

1.	$\mathsf{v}_0 A \mathsf{v}_1 \mid \mathsf{v}_0 A \mathsf{v}_1$	axiom
2.	$\mathsf{v}_0 A \mathsf{v}_2, \mathsf{v}_2 1' \mathsf{v}_1 \mid \mathsf{v}_0 A \mathsf{v}_1$	1, $1'\mid$
3.	$\mathsf{v}_0 A ; 1' \mathsf{v}_1 \mid \mathsf{v}_0 A \mathsf{v}_1$	2, $;\mid$, no v_2
4.	$\mathsf{v}_0 \mathbf{t} \circ A \mathsf{v}_1 \mid \mathsf{v}_0 A \mathsf{v}_1$	3, (3.19),(3.21)

Proof (3.105) 3-proof of $A \to A \circ \mathbf{t}^*$.

1.	$\mathsf{v}_0 A \mathsf{v}_1 \mid \mathsf{v}_0 A \mathsf{v}_1$	axiom
2.	$\mathsf{v}_0 A \mathsf{v}_1 \mid \mathsf{v}_1 A^{\smile} \mathsf{v}_0$	1, $\mid^{\smile}$
3.	$\mid \mathsf{v}_0 1' \mathsf{v}_0$	$\mid 1'$
4.	$\mathsf{v}_0 A \mathsf{v}_1 \mid \mathsf{v}_1 A^{\smile} ; 1' \mathsf{v}_0$	2, 3, $\mid ;$
5.	$\mathsf{v}_0 A \mathsf{v}_1 \mid \mathsf{v}_0 (A^{\smile} ; 1'^{\smile}) \mathsf{v}_1$	4, $\mid^{\smile}$
6.	$\mathsf{v}_1 A^{\smile} \mathsf{v}_2 \mid \mathsf{v}_1 A^{\smile} \mathsf{v}_2$	axiom
7.	$\mathsf{v}_1 A^{\smile} \mathsf{v}_2 \mid \mathsf{v}_2 A^{\smile\smile} \mathsf{v}_1$	6, $\mid^{\smile}$
8.	$\mathsf{v}_2 1' \mathsf{v}_0 \mid \mathsf{v}_2 1' \mathsf{v}_0$	axiom
9.	$\mathsf{v}_2 1' \mathsf{v}_0 \mid \mathsf{v}_0 1'^{\smile} \mathsf{v}_2$	8, $\mid^{\smile}$
10.	$\mathsf{v}_1 A^{\smile} \mathsf{v}_2, \mathsf{v}_2 1' \mathsf{v}_0 \mid \mathsf{v}_0 1'^{\smile} ; A^{\smile\smile} \mathsf{v}_1$	7, 9, $\mid ;$
11.	$\mathsf{v}_1 A^{\smile} ; 1' \mathsf{v}_0 \mid \mathsf{v}_0 1'^{\smile} ; A^{\smile\smile} \mathsf{v}_1$	10, $;\mid$, no v_2
12.	$\mathsf{v}_0 (A^{\smile} ; 1'^{\smile}) \mathsf{v}_1 \mid \mathsf{v}_0 1'^{\smile} ; A^{\smile\smile} \mathsf{v}_1$	11, $^{\smile}\mid$
13.	$\mathsf{v}_0 A \mathsf{v}_1 \mid \mathsf{v}_0 1'^{\smile} ; A^{\smile\smile} \mathsf{v}_1$	5, 12, Cut
14.	$\mathsf{v}_0 1'^{\smile} \mathsf{v}_2 \mid \mathsf{v}_0 1'^{\smile} \mathsf{v}_2$	axiom
15.	$\mathsf{v}_2 A \mathsf{v}_1 \mid \mathsf{v}_2 A \mathsf{v}_1$	axiom
16.	$\mathsf{v}_2 A^{\smile\smile} \mathsf{v}_1 \mid \mathsf{v}_2 A \mathsf{v}_1$	15, $^{\smile}\mid$
17.	$\mathsf{v}_0 1'^{\smile} \mathsf{v}_2, \mathsf{v}_2 A^{\smile\smile} \mathsf{v}_1 \mid \mathsf{v}_0 1'^{\smile} ; A \mathsf{v}_1$	14, 16, $\mid ;$
18.	$\mathsf{v}_0 1'^{\smile} ; A^{\smile\smile} \mathsf{v}_1 \mid \mathsf{v}_0 1'^{\smile} ; A \mathsf{v}_1$	17, $;\mid$, no v_2
19.	$\mathsf{v}_0 A \mathsf{v}_1 \mid \mathsf{v}_0 1'^{\smile} ; A \mathsf{v}_1$	13, 18, Cut
20.	$\mathsf{v}_0 A \mathsf{v}_1 \mid \mathsf{v}_0 A \circ \mathbf{t}^* \mathsf{v}_1$	19, (3.19), (3.20), (3.21)

Proof (3.106) 3-proof of $\mathbf{t} \to \mathbf{t}^*$.

1.	$\mathsf{v}_0 1' \mathsf{v}_1 \mid \mathsf{v}_0 1'^{\smile} ; 1' \mathsf{v}_1$	(3.105), $A = 1'$
2.	$\mathsf{v}_0 1'^{\smile} \mathsf{v}_1 \mid \mathsf{v}_0 1'^{\smile} \mathsf{v}_1$	axiom
3.	$\mathsf{v}_0 1'^{\smile} \mathsf{v}_2, \mathsf{v}_2 1' \mathsf{v}_1 \mid \mathsf{v}_0 1'^{\smile} \mathsf{v}_1$	2, $1'\mid$
4.	$\mathsf{v}_0 1'^{\smile} ; 1' \mathsf{v}_1 \mid \mathsf{v}_0 1'^{\smile} \mathsf{v}_1$	3, $;\mid$, no v_2
5.	$\mathsf{v}_0 1' \mathsf{v}_1 \mid \mathsf{v}_0 1'^{\smile} \mathsf{v}_1$	1, 4, Cut
6.	$\mathsf{v}_0 \mathbf{t} \mathsf{v}_1 \mid \mathsf{v}_0 \mathbf{t}^* \mathsf{v}_1$	5, (3.20), (3.21)

Proof (3.107) 3-proof of $\mathbf{t}^* \to \mathbf{t}$.

1.	$\mathsf{v}_1 1' \mathsf{v}_0 \mid \mathsf{v}_1 1'^{\smile} \mathsf{v}_0$	(3.106), (3.57)
2.	$\mathsf{v}_0 1'^{\smile} \mathsf{v}_1 \mid \mathsf{v}_0 1'^{\smile\smile} \mathsf{v}_1$	1, ${}^{\smile}\mid$, $\mid^{\smile}$
3.	$\mathsf{v}_0 1' \mathsf{v}_1 \mid \mathsf{v}_0 1' \mathsf{v}_1$	axiom
4.	$\mathsf{v}_0 1'^{\smile\smile} \mathsf{v}_1 \mid \mathsf{v}_0 1' \mathsf{v}_1$	3, ${}^{\smile}\mid$
5.	$\mathsf{v}_0 1'^{\smile} \mathsf{v}_1 \mid \mathsf{v}_0 1' \mathsf{v}_1$	2, 4, Cut
6.	$\mathsf{v}_0 \mathbf{t}^* \mathsf{v}_1 \mid \mathsf{v}_0 \mathbf{t} \mathsf{v}_1$	5, (3.20), (3.21)

Proof (3.108) 3-proof of $A \circ \mathbf{t} \to A$.

1.	$\mathsf{v}_0 A \mathsf{v}_1 \mid \mathsf{v}_0 A \mathsf{v}_1$	axiom
2.	$\mathsf{v}_1 A^{\smile} \mathsf{v}_0 \mid \mathsf{v}_0 A \mathsf{v}_1$	1, ${}^{\smile}\mid$
3.	$\mathsf{v}_1 A^{\smile} \mathsf{v}_2, \mathsf{v}_2 1' \mathsf{v}_0 \mid \mathsf{v}_0 A \mathsf{v}_1$	2, $1'\mid$
4.	$\mathsf{v}_2 1'^{\smile} \mathsf{v}_0 \mid \mathsf{v}_2 1' \mathsf{v}_0$	(3.107), (3.53), (3.54), (3.56)
5.	$\mathsf{v}_0 1' \mathsf{v}_2 \mid \mathsf{v}_0 1' \mathsf{v}_2$	axiom
6.	$\mathsf{v}_0 1' \mathsf{v}_2 \mid \mathsf{v}_2 1'^{\smile} \mathsf{v}_0$	5, $\mid^{\smile}$
7.	$\mathsf{v}_0 1' \mathsf{v}_2 \mid \mathsf{v}_2 1' \mathsf{v}_0$	4, 6, Cut
8.	$\mathsf{v}_1 A^{\smile} \mathsf{v}_2, \mathsf{v}_0 1' \mathsf{v}_2 \mid \mathsf{v}_0 A \mathsf{v}_1$	3, 7, Cut
9.	$\mathsf{v}_2 A \mathsf{v}_1 \mid \mathsf{v}_2 A \mathsf{v}_1$	axiom
10.	$\mathsf{v}_2 A \mathsf{v}_1 \mid \mathsf{v}_1 A^{\smile} \mathsf{v}_2$	9, $\mid^{\smile}$
11.	$\mathsf{v}_0 1' \mathsf{v}_2, \mathsf{v}_2 A \mathsf{v}_1 \mid \mathsf{v}_0 A \mathsf{v}_1$	8, 10, Cut
12.	$\mathsf{v}_0 1' ; A \mathsf{v}_1 \mid \mathsf{v}_0 A \mathsf{v}_1$	, $;\mid$, no v_2
13.	$\mathsf{v}_0 A \circ \mathbf{t} \mathsf{v}_1 \mid \mathsf{v}_0 A \mathsf{v}_1$	6, (3.18), (3.21)

Proof (3.109) 3-proof of $\mathbf{t} \wedge {\sim}\mathbf{t} \to A$.

1.	$\mathsf{v}_0 1\text{'} \mathsf{v}_1 \,\vert\, \mathsf{v}_0 1\text{'}^{\smile} ; 1\text{'} \mathsf{v}_1$	(3.105)
2.	$\mathsf{v}_0 1\text{'}^{\smile} \mathsf{v}_1 \,\vert\, \mathsf{v}_0 1\text{'}^{\smile} \mathsf{v}_1, \mathsf{v}_0 A \mathsf{v}_1$	axiom
3.	$\mathsf{v}_0 1\text{'}^{\smile} \mathsf{v}_2, \mathsf{v}_2 1\text{'} \mathsf{v}_1 \,\vert\, \mathsf{v}_0 1\text{'}^{\smile} \mathsf{v}_1, \mathsf{v}_0 A \mathsf{v}_1$	2, $1\text{'}\vert$
4.	$\mathsf{v}_0 1\text{'}^{\smile} ; 1\text{'} \mathsf{v}_1 \,\vert\, \mathsf{v}_0 1\text{'}^{\smile} \mathsf{v}_1, \mathsf{v}_0 A \mathsf{v}_1$	3, $;\vert$, no v_2
5.	$\mathsf{v}_0 1\text{'} \mathsf{v}_1 \,\vert\, \mathsf{v}_0 1\text{'}^{\smile} \mathsf{v}_1, \mathsf{v}_0 A \mathsf{v}_1$	1, 4, Cut
6.	$\mathsf{v}_0 1\text{'} \mathsf{v}_1, \mathsf{v}_0 \overline{1\text{'}^{\smile}} \mathsf{v}_1 \,\vert\, \mathsf{v}_0 A \mathsf{v}_1$	5, $^{-}\vert$
7.	$\mathsf{v}_0 1\text{'} \cdot \overline{1\text{'}^{\smile}} \mathsf{v}_1 \,\vert\, \mathsf{v}_0 A \mathsf{v}_1$	6, $\cdot\vert$
8.	$\mathsf{v}_0 \mathbf{t} \wedge {\sim}\mathbf{t} \mathsf{v}_1 \,\vert\, \mathsf{v}_0 A \mathsf{v}_1$	7, (3.15), (3.17), (3.21)

Proof (3.110) 1-proof of $A, B \vdash A \wedge B$. This is the adjunction rule.

1.	$\vert\, \mathsf{v}_0 A \mathsf{v}_0$	sequent in an n-proof
2.	$\vert\, \mathsf{v}_0 B \mathsf{v}_0$	sequent in an n-proof
3.	$\vert\, \mathsf{v}_0 A \cdot B \mathsf{v}_0$	1, 2, $\vert\cdot$
4.	$\vert\, \mathsf{v}_0 A \wedge B \mathsf{v}_0$	3, (3.15)

Proof (3.111) 1-proof of $A \to B, A \vdash B$. This rule is modus ponens. Note that (3.54) is applicable when $i = j$ and is therefore a 1-provable rule.

1.	$\vert\, \mathsf{v}_0 A \to B \mathsf{v}_0$	sequent in an n-proof
2.	$\vert\, \mathsf{v}_0 A \mathsf{v}_0$	sequent in an n-proof
3.	$\mathsf{v}_0 A \mathsf{v}_0 \,\vert\, \mathsf{v}_0 B \mathsf{v}_0$	1, (3.54)
4.	$\vert\, \mathsf{v}_0 B \mathsf{v}_0$	2, 3, Cut

Proof (3.112) 1-proof of $A \vee B, {\sim}A \vdash B$. This rule is disjunctive syllogism.

1.	$\vert\, \mathsf{v}_0 A \vee B \mathsf{v}_0$	sequent in an n-proof
2.	$\vert\, \mathsf{v}_0 {\sim} A \mathsf{v}_0$	sequent in an n-proof
3.	$\vert\, \mathsf{v}_0 A + B \mathsf{v}_0$	1, (3.14)
4.	$\vert\, \mathsf{v}_0 \overline{A^{\smile}} \mathsf{v}_0$	2, (3.17)
5.	$\mathsf{v}_0 A \mathsf{v}_0 \vert \mathsf{v}_0 A \mathsf{v}_0$	axiom
6.	$\mathsf{v}_0 A \mathsf{v}_0, \mathsf{v}_0 \overline{A^{\smile}} \mathsf{v}_0 \vert$	5, $\vert^{\smile}, \vert$
7.	$\mathsf{v}_0 A \mathsf{v}_0 \,\vert$	4, 6, Cut
8.	$\mathsf{v}_0 B \mathsf{v}_0 \,\vert\, \mathsf{v}_0 B \mathsf{v}_0$	axiom
9.	$\mathsf{v}_0 A + B \mathsf{v}_0 \,\vert\, \mathsf{v}_0 B \mathsf{v}_0$	7, 8, $+\vert$
10.	$\vert\, \mathsf{v}_0 B \mathsf{v}_0$	3, 9, Cut

Proof (3.113) 1-proof of $A \vdash A^*$.

1.	$\mid \mathsf{v}_0 A \mathsf{v}_0$	sequent in an n-proof
2.	$\mid \mathsf{v}_0 A^{\smile} \mathsf{v}_0$	1, $\mid^{\smile}$
3.	$\mid \mathsf{v}_0 A^{*} \mathsf{v}_0$	2, (3.20)

Proof (3.114) 2-proof of $A \to B, B \to C \vdash A \to C$. This is the transitivity rule. Note that (3.53) requires $i \neq j$, and is therefore a 2-provable rule.

1.	$\mid \mathsf{v}_0 A \to B \mathsf{v}_0$	sequent in an n-proof
2.	$\mid \mathsf{v}_0 B \to C \mathsf{v}_0$	sequent in an n-proof
3.	$\mathsf{v}_1 A \mathsf{v}_0 \mid \mathsf{v}_1 B \mathsf{v}_0$	1, (3.54)
4.	$\mathsf{v}_1 B \mathsf{v}_0 \mid \mathsf{v}_1 C \mathsf{v}_0$	2, (3.54)
5.	$\mathsf{v}_1 A \mathsf{v}_0 \mid \mathsf{v}_1 C \mathsf{v}_0$	3, 4, Cut
6.	$\mid \mathsf{v}_0 A \to C \mathsf{v}_0$	5, (3.53)

Proof (3.115) 2-proof of $A \to B \vdash {\sim}B \to {\sim}A$. This is one form of the rule of contraposition.

1.	$\mid \mathsf{v}_0 A \to B \mathsf{v}_0$	sequent in an n-proof
2.	$\mid \mathsf{v}_1 A \to B \mathsf{v}_1$	1, (3.56)
3.	$\mathsf{v}_0 A \mathsf{v}_1 \mid \mathsf{v}_0 B \mathsf{v}_1$	2, (3.54)
4.	$\mathsf{v}_1 \overline{B^{\smile}} \mathsf{v}_0 \mid \mathsf{v}_1 \overline{A^{\smile}} \mathsf{v}_0$	3, $^{\smile}\mid$, $\mid^{\smile}$, $^{-}\mid$, $\mid^{-}$
5.	$\mathsf{v}_1 {\sim}B \mathsf{v}_0 \mid \mathsf{v}_1 {\sim}A \mathsf{v}_0$	4, (3.17)
6.	$\mid \mathsf{v}_0 {\sim}B \to {\sim}A \mathsf{v}_0$	5, (3.53)

Proof (3.116) 2-proof of $A \to {\sim}B \vdash B \to {\sim}A$. This is another form of the rule of contraposition.

1.	$\mid \mathsf{v}_0 A \to {\sim}B \mathsf{v}_0$	sequent in an n-proof
2.	$\mathsf{v}_1 A \mathsf{v}_0 \mid \mathsf{v}_1 {\sim}B \mathsf{v}_0$	1, (3.54)
3.	$\mathsf{v}_1 A \mathsf{v}_0 \mid \mathsf{v}_1 \overline{B^{\smile}} \mathsf{v}_0$	2, (3.17)
4.	$\mathsf{v}_0 B \mathsf{v}_1 \mid \mathsf{v}_0 B \mathsf{v}_1$	axiom
5.	$\mathsf{v}_0 B \mathsf{v}_1, \mathsf{v}_1 \overline{B^{\smile}} \mathsf{v}_0 \mid$	4, $\mid^{\smile}$, $^{-}\mid$
6.	$\mathsf{v}_1 A \mathsf{v}_0, \mathsf{v}_0 B \mathsf{v}_1 \mid$	3, 5, Cut
7.	$\mathsf{v}_0 B \mathsf{v}_1 \mid \mathsf{v}_0 \overline{A^{\smile}} \mathsf{v}_1,$	6, $^{\smile}\mid$, $\mid^{-}$
8.	$\mathsf{v}_0 B \mathsf{v}_1 \mid \mathsf{v}_0 {\sim}A \mathsf{v}_1$	7, (3.17)
9.	$\mid \mathsf{v}_1 B \to {\sim}A \mathsf{v}_1$	8, (3.53)
10.	$\mid \mathsf{v}_0 B \to {\sim}A \mathsf{v}_0$	9, (3.56)

Proof (3.117) 2-proof of $A \wedge B \to C, B \to C \vee A \vdash B \to C$. This rule is a derived rule in Basic Logic, where it is called DR2 (Routley et al. 1982, p. 291) (derived rule number 2).

1.	$\mid \mathsf{v}_0 B \to C \vee A \mathsf{v}_0$	sequent in an n-proof
2.	$\mid \mathsf{v}_0 A \wedge B \to C \mathsf{v}_0$	sequent in an n-proof
3.	$\mid \mathsf{v}_0 B \to C + A \mathsf{v}_0$	1, (3.14)
4.	$\mid \mathsf{v}_0 A \cdot B \to C \mathsf{v}_0$	2, (3.15)
5.	$\mathsf{v}_1 B \mathsf{v}_0 \mid \mathsf{v}_1 C + A \mathsf{v}_0$	3, (3.54)
6.	$\mathsf{v}_1 A \cdot B \mathsf{v}_0 \mid \mathsf{v}_1 C \mathsf{v}_0$	4, (3.54)
7.	$\mathsf{v}_1 A \mathsf{v}_0 \mid \mathsf{v}_1 A \mathsf{v}_0$	axiom
8.	$\mathsf{v}_1 B \mathsf{v}_0 \mid \mathsf{v}_1 B \mathsf{v}_0$	axiom
9.	$\mathsf{v}_1 C \mathsf{v}_0 \mid \mathsf{v}_1 C \mathsf{v}_0$	axiom
10.	$\mathsf{v}_1 C + A \mathsf{v}_0 \mid \mathsf{v}_1 A \mathsf{v}_0, \mathsf{v}_1 C \mathsf{v}_0$	7, 9, $+\mid$
11.	$\mathsf{v}_1 B \mathsf{v}_0 \mid \mathsf{v}_1 A \mathsf{v}_0, \mathsf{v}_1 C \mathsf{v}_0$	5, 10, Cut
12.	$\mathsf{v}_1 A \mathsf{v}_0, \mathsf{v}_1 B \mathsf{v}_0 \mid \mathsf{v}_1 A \cdot B \mathsf{v}_0$	7, 8, $\mid\cdot$
13.	$\mathsf{v}_1 A \mathsf{v}_0, \mathsf{v}_1 B \mathsf{v}_0 \mid \mathsf{v}_1 C \mathsf{v}_0$	6, 12, Cut
14.	$\mathsf{v}_1 B \mathsf{v}_0 \mid \mathsf{v}_1 C \mathsf{v}_0$	11, 13, Cut
15.	$\mid \mathsf{v}_0 B \to C \mathsf{v}_0$	14, (3.53)

Proof (3.118) 2-proof of $A \vdash (A \to B) \to B$. This is the E-rule (Sylvan et al. 2003, p. 8), also called BR1 (Routley et al. 1982, p. 289) (basic rule number 1) and R5 (Sylvan et al. 2003, p. 193) (rule number 5).

1.	$\mid \mathsf{v}_0 A \mathsf{v}_0$	sequent in an n-proof
2.	$\mid \mathsf{v}_1 A \mathsf{v}_1$	1, (3.56)
3.	$\mid \mathsf{v}_1 A^{\smile} \mathsf{v}_1$	2, $\mid^{\smile}$
4.	$\mathsf{v}_1 B \mathsf{v}_0 \mid \mathsf{v}_1 B \mathsf{v}_0$	axiom
5.	$\mid \mathsf{v}_1 \overline{B} \mathsf{v}_0, \mathsf{v}_1 B \mathsf{v}_0$	4, $\mid^{-}$
6.	$\mid \mathsf{v}_1 A^{\smile} ; \overline{B} \mathsf{v}_0, \mathsf{v}_1 B \mathsf{v}_0$	3, 5, $\mid;$
7.	$\mathsf{v}_1 \overline{A^{\smile} ; \overline{B}} \mathsf{v}_0 \mid \mathsf{v}_1 B \mathsf{v}_0$	6, $^{-}\mid$
8.	$\mathsf{v}_1 A \to B \mathsf{v}_0 \mid \mathsf{v}_1 B \mathsf{v}_0$	7, (3.18)
9.	$\mid \mathsf{v}_0 (A \to B) \to B \mathsf{v}_0$	8, (3.53)

Proof (3.119) 2-proof of $A \wedge B \to \neg C \vdash A \wedge C \to \neg B$. This rule has been called antilogism, a term coined by Ladd-Franklin (1928).

1. $\mid \mathsf{v}_0 A \wedge B \to \neg C \mathsf{v}_0$ — sequent in an n-proof
2. $\mid \mathsf{v}_0 A \cdot B \to \overline{C} \mathsf{v}_0$ — 1, (3.15), (3.16)
3. $\mathsf{v}_1 A \cdot B \mathsf{v}_0 \mid \mathsf{v}_1 \overline{C} \mathsf{v}_0$ — 2, (3.54)
4. $\mathsf{v}_1 A \mathsf{v}_0 \mid \mathsf{v}_1 A \mathsf{v}_0$ — axiom
5. $\mathsf{v}_1 B \mathsf{v}_0 \mid \mathsf{v}_1 B \mathsf{v}_0$ — axiom
6. $\mathsf{v}_1 A \mathsf{v}_0, \mathsf{v}_1 B \mathsf{v}_0 \mid \mathsf{v}_1 A \cdot B \mathsf{v}_0$ — 4, 5, $\mid\cdot$
7. $\mathsf{v}_1 A \mathsf{v}_0, \mathsf{v}_1 B \mathsf{v}_0 \mid \mathsf{v}_1 \overline{C} \mathsf{v}_0$ — 3, 6, Cut
8. $\mathsf{v}_1 C \mathsf{v}_0 \mid \mathsf{v}_1 C \mathsf{v}_0$ — axiom
9. $\mathsf{v}_1 C \mathsf{v}_0, \mathsf{v}_1 \overline{C} \mathsf{v}_0 \mid$ — 8, $\overline{}\mid$
10. $\mathsf{v}_1 A \mathsf{v}_0, \mathsf{v}_1 B \mathsf{v}_0, \mathsf{v}_1 C \mathsf{v}_0 \mid$ — 7, 9, Cut
11. $\mathsf{v}_1 A \cdot C \mathsf{v}_0, \mathsf{v}_1 B \mathsf{v}_0 \mid$ — 10, $\cdot\mid$
12. $\mathsf{v}_1 A \cdot C \mathsf{v}_0 \mid \mathsf{v}_1 \overline{B} \mathsf{v}_0$ — 11, $\mid\overline{}$
13. $\mid \mathsf{v}_0 A \cdot C \to \overline{B} \mathsf{v}_0$ — 12, (3.53)
14. $\mid \mathsf{v}_0 A \wedge C \to \neg B \mathsf{v}_0$ — 13, (3.15), (3.16)

Proof (3.122) 3-proof of $A \to B \vdash (B \to C) \to (A \to C)$. This is the suffixing rule.

1. $\mid \mathsf{v}_0 A \to B \mathsf{v}_0$ — sequent in an n-proof
2. $\mathsf{v}_2 A \mathsf{v}_0 \mid \mathsf{v}_2 B \mathsf{v}_0$ — 1, (3.54)
3. $\mathsf{v}_0 A^{\smile} \mathsf{v}_2 \mid \mathsf{v}_0 B^{\smile} \mathsf{v}_2$ — 2, $^{\smile}\mid$, $\mid^{\smile}$
4. $\mathsf{v}_2 \overline{C} \mathsf{v}_1 \mid \mathsf{v}_2 \overline{C} \mathsf{v}_1$ — axiom
5. $\mathsf{v}_0 A^{\smile} \mathsf{v}_2, \mathsf{v}_2 \overline{C} \mathsf{v}_1 \mid \mathsf{v}_0 B^{\smile} ; \overline{C} \mathsf{v}_1$ — 3, 4, $\mid;$
6. $\mathsf{v}_0 A^{\smile} ; \overline{C} \mathsf{v}_1 \mid \mathsf{v}_0 B^{\smile} ; \overline{C} \mathsf{v}_1$ — 5, $;\mid$, no v_2
7. $\mathsf{v}_0 \overline{B^{\smile} ; \overline{C}} \mathsf{v}_1 \mid \mathsf{v}_0 \overline{A^{\smile} ; \overline{C}} \mathsf{v}_1$ — 6, $\overline{}\mid$, $\mid\overline{}$
8. $\mathsf{v}_0 B \to C \mathsf{v}_1 \mid \mathsf{v}_0 A \to C \mathsf{v}_1$ — 7, (3.18)
9. $\mid \mathsf{v}_0 (B \to C) \to (A \to C) \mathsf{v}_0$ — 8, (3.53), (3.56)

Proof (3.123) 3-proof of $A \to B \vdash (C \to A) \to (C \to B)$. This is the prefixing rule. The prefixing axiom $(A \to B) \to ((C \to A) \to (C \to B))$ (3.127) is 4-provable, so if $A \to B$ is 4-provable then $(C \to A) \to (C \to B)$ is also 4-provable by modus ponens (3.111). However, the prefixing rule only needs three variables.

1.	$\mid \mathsf{v}_0 A \to B \mathsf{v}_0$	sequent in an n-proof
2.	$\mathsf{v}_2 A \mathsf{v}_0 \mid \mathsf{v}_2 B \mathsf{v}_0$	1, (3.54)
3.	$\mathsf{v}_2 \overline{B} \mathsf{v}_0 \mid \mathsf{v}_2 \overline{A} \mathsf{v}_0$	2, $\overline{}\mid$, $\mid\overline{}$
4.	$\mathsf{v}_1 C^{\smile} \mathsf{v}_2 \mid \mathsf{v}_1 C^{\smile} \mathsf{v}_2$	axiom
5.	$\mathsf{v}_1 C^{\smile} \mathsf{v}_2, \mathsf{v}_2 \overline{B} \mathsf{v}_0 \mid \mathsf{v}_1 C^{\smile} ; \overline{A} \mathsf{v}_0$	3, 4, $\mid ;$
6.	$\mathsf{v}_1 C^{\smile} ; \overline{B} \mathsf{v}_0 \mid \mathsf{v}_1 C^{\smile} ; \overline{A} \mathsf{v}_0$	5, $; \mid$, no v_2
7.	$\mathsf{v}_1 \overline{C^{\smile} ; \overline{A}} \mathsf{v}_0 \mid \mathsf{v}_1 \overline{C^{\smile} ; \overline{B}} \mathsf{v}_0$	6, $\overline{}\mid$, $\mid\overline{}$
8.	$\mathsf{v}_1 C \to A \mathsf{v}_0 \mid \mathsf{v}_1 C \to B \mathsf{v}_0$	7, (3.18)
9.	$\mid \mathsf{v}_0 (C \to A) \to (C \to B) \mathsf{v}_0$	8, (3.53)

Proof (3.124) 3-provability of $A \to B, C \to D \vdash (B \to C) \to (A \to D)$. This is the affixing rule. Assume $A \to B$ and $C \to D$ are 3-provable. By the 3-provable prefixing rule (3.123), $(A \to C) \to (A \to D)$ is 3-provable. By the 3-provable suffixing rule (3.122), $(B \to C) \to (A \to C)$ is 3-provable. Hence, by the 2-provable transitivity rule (3.114), $(B \to C) \to (A \to D)$ is 3-provable.

Proof (3.125) 3-proof of $A \to B, C \to D \vdash A \circ C \to B \circ D$. This is the rule that fusion is monotonic. It preserves 4-provability because the monotonic fusion axiom (3.130) is 4-provable. However, the monotonic fusion rule is actually 3-provable.

1.	$\mid \mathsf{v}_0 C \to D \mathsf{v}_0$	sequent in an n-proof
2.	$\mid \mathsf{v}_0 A \to B \mathsf{v}_0$	sequent in an n-proof
3.	$\mathsf{v}_1 C \mathsf{v}_2 \mid \mathsf{v}_1 D \mathsf{v}_2$	1, (3.56), (3.54)
4.	$\mathsf{v}_2 A \mathsf{v}_0 \mid \mathsf{v}_2 B \mathsf{v}_0$	2, (3.54)
5.	$\mathsf{v}_1 C \mathsf{v}_2, \mathsf{v}_2 A \mathsf{v}_0 \mid \mathsf{v}_1 D ; B \mathsf{v}_0$	3, 4, $\mid ;$
6.	$\mathsf{v}_1 C ; A \mathsf{v}_0 \mid \mathsf{v}_1 D ; B \mathsf{v}_0$	5, $; \mid$, no v_2
7.	$\mid \mathsf{v}_0 C ; A \to D ; B \mathsf{v}_0$	6, (3.53)
8.	$\mid \mathsf{v}_0 A \circ C \to B \circ D \mathsf{v}_0$	7, (3.19)

Proof (3.126) 3-proof of $A \to (B \to C) \vdash B \to ({\sim}C \to {\sim}A)$. This is the cycling rule.

1.	$\mid \mathsf{v}_0 A \to (B \to C)\mathsf{v}_0$	sequent in an n-proof
2.	$\mathsf{v}_0 A \mathsf{v}_2 \mid \mathsf{v}_0 B \to C \mathsf{v}_2$	1, (3.56), (3.54)
3.	$\mathsf{v}_0 A \mathsf{v}_2 \mid \mathsf{v}_0 \overline{B^{\smile} ; \overline{C}} \mathsf{v}_2$	2, (3.18)
4.	$\mathsf{v}_1 B \mathsf{v}_0 \mid \mathsf{v}_1 B \mathsf{v}_0$	axiom
5.	$\mathsf{v}_1 B \mathsf{v}_0 \mid \mathsf{v}_0 B^{\smile} \mathsf{v}_1$	$4, \mid^{\smile}$
6.	$\mathsf{v}_1 C \mathsf{v}_2 \mid \mathsf{v}_1 C \mathsf{v}_2$	axiom
7.	$\mid \mathsf{v}_1 C \mathsf{v}_2, \mathsf{v}_1 \overline{C} \mathsf{v}_2$	$6, \mid^{-}$
8.	$\mathsf{v}_1 B \mathsf{v}_0 \mid \mathsf{v}_1 C \mathsf{v}_2, \mathsf{v}_0 B^{\smile} ; \overline{C} \mathsf{v}_2$	$5, 7, \mid ;$
9.	$\mathsf{v}_1 B \mathsf{v}_0, \mathsf{v}_0 \overline{B^{\smile} ; \overline{C}} \mathsf{v}_2 \mid \mathsf{v}_1 C \mathsf{v}_2$	$8, {}^{-}\mid$
10.	$\mathsf{v}_0 A \mathsf{v}_2, \mathsf{v}_1 B \mathsf{v}_0 \mid \mathsf{v}_1 C \mathsf{v}_2$	3, 9, Cut
11.	$\mathsf{v}_1 B \mathsf{v}_0, \mathsf{v}_1 \overline{C^{\smile}}^{\smile} \mathsf{v}_2, \mathsf{v}_2 \overline{\overline{A^{\smile}}} \mathsf{v}_0 \mid$	$10, {}^{\smile}\mid, \mid^{\smile}, {}^{-}\mid, \mid^{-}$
12.	$\mathsf{v}_1 B \mathsf{v}_0, \mathsf{v}_1 \overline{C^{\smile}}^{\smile} ; \overline{\overline{A^{\smile}}} \mathsf{v}_0 \mid$	11, $; \mid$, no v_2
13.	$\mathsf{v}_1 B \mathsf{v}_0 \mid \mathsf{v}_1 \overline{\overline{C^{\smile}}^{\smile} ; \overline{\overline{A^{\smile}}}} \mathsf{v}_0$	$12, \mid^{-}$
14.	$\mathsf{v}_1 B \mathsf{v}_0 \mid \mathsf{v}_1 {\sim}C \to {\sim}A \mathsf{v}_0$	13, (3.17), (3.18)
15.	$\mid \mathsf{v}_0 B \to ({\sim}C \to {\sim}A)\mathsf{v}_0$	14, (3.53)

Proof (3.127) 4-proof of $(A \to B) \to ((C \to A) \to (C \to B))$. This is the prefixing axiom.

1.	$\mathsf{v}_3 A \mathsf{v}_0 \mid \mathsf{v}_3 A \mathsf{v}_0$	axiom
2.	$\mid \mathsf{v}_3 \overline{A} \mathsf{v}_0, \mathsf{v}_0 A^{\smile} \mathsf{v}_3$	$1, \mid^{-}, \mid^{\smile}$
3.	$\mathsf{v}_3 \overline{B} \mathsf{v}_1 \mid \mathsf{v}_3 \overline{B} \mathsf{v}_1$	axiom
4.	$\mathsf{v}_3 \overline{B} \mathsf{v}_1 \mid \mathsf{v}_3 \overline{A} \mathsf{v}_0, \mathsf{v}_0 A^{\smile} ; \overline{B} \mathsf{v}_1$	$2, 3, \mid ;$
5.	$\mathsf{v}_2 C^{\smile} \mathsf{v}_3 \mid \mathsf{v}_2 C^{\smile} \mathsf{v}_3$	axiom
6.	$\mathsf{v}_2 C^{\smile} \mathsf{v}_3, \mathsf{v}_3 \overline{B} \mathsf{v}_1 \mid \mathsf{v}_2 C^{\smile} ; \overline{A} \mathsf{v}_0, \mathsf{v}_0 A^{\smile} ; \overline{B} \mathsf{v}_1$	$4, 5, \mid ;$
7.	$\mathsf{v}_2 C^{\smile} ; \overline{B} \mathsf{v}_1 \mid \mathsf{v}_2 C^{\smile} ; \overline{A} \mathsf{v}_0, \mathsf{v}_0 A^{\smile} ; \overline{B} \mathsf{v}_1$	6, $; \mid$, no v_3
8.	$\mathsf{v}_2 \overline{C^{\smile} ; \overline{A}} \mathsf{v}_0, \mathsf{v}_0 \overline{A^{\smile} ; \overline{B}} \mathsf{v}_1 \mid \mathsf{v}_2 \overline{C^{\smile} ; \overline{B}} \mathsf{v}_1$	$7, \mid^{-}, {}^{-}\mid$
9.	$\mathsf{v}_2 C \to A \mathsf{v}_0, \mathsf{v}_0 A \to B \mathsf{v}_1 \mid \mathsf{v}_2 C \to B \mathsf{v}_1$	8, (3.18)
10.	$\mathsf{v}_0 (C \to A)^{\smile} \mathsf{v}_2, \mathsf{v}_2 \overline{C \to B} \mathsf{v}_1, \mathsf{v}_0 A \to B \mathsf{v}_1 \mid$	$9, {}^{-}\mid, {}^{\smile}\mid$
11.	$\mathsf{v}_0 (C \to A)^{\smile} ; \overline{C \to B} \mathsf{v}_1, \mathsf{v}_0 A \to B \mathsf{v}_1 \mid$	10, $; \mid$, no v_2
12.	$\mathsf{v}_0 A \to B \mathsf{v}_1 \mid \mathsf{v}_0 \overline{(C \to A)^{\smile} ; \overline{C \to B}} \mathsf{v}_1$	$11, \mid^{-}$
13.	$\mathsf{v}_0 A \to B \mathsf{v}_1 \mid \mathsf{v}_0 (C \to A) \to (C \to B) \mathsf{v}_1$	12, (3.18)

Proof (3.128) 4-proof of $(A \to (B \to C)) \to (A \circ B \to C)$. This is the axiom of bunching hypotheses.

1.	$\mathsf{v}_3 A \mathsf{v}_0 \mid \mathsf{v}_3 A \mathsf{v}_0$	axiom
2.	$\mathsf{v}_2 B \mathsf{v}_3 \mid \mathsf{v}_2 B \mathsf{v}_3$	axiom
3.	$\mathsf{v}_2 C \mathsf{v}_1 \mid \mathsf{v}_2 C \mathsf{v}_1$	axiom
4.	$\mathsf{v}_2 B \mathsf{v}_3 \mid \mathsf{v}_3 B^{\smile} \mathsf{v}_2$	$2, \mid^{\smile}$
5.	$\mid \mathsf{v}_2 C \mathsf{v}_1, \mathsf{v}_2 \overline{C} \mathsf{v}_1$	$3, \mid^{-}$
6.	$\mathsf{v}_2 B \mathsf{v}_3 \mid \mathsf{v}_2 C \mathsf{v}_1, \mathsf{v}_3 B^{\smile} ; \overline{C} \mathsf{v}_1$	4, 5, ;
7.	$\mathsf{v}_2 B \mathsf{v}_3 \mid \mathsf{v}_2 C \mathsf{v}_1, \mathsf{v}_3 \overline{\overline{B^{\smile} ; \overline{C}}} \mathsf{v}_1$	$6, {}^{-}\mid, \mid^{-}$
8.	$\mathsf{v}_3 A \mathsf{v}_0 \mid \mathsf{v}_0 A^{\smile} \mathsf{v}_3$	$1, \mid^{\smile}$
9.	$\mathsf{v}_2 B \mathsf{v}_3, \mathsf{v}_3 A \mathsf{v}_0 \mid \mathsf{v}_2 C \mathsf{v}_1, \mathsf{v}_0 A^{\smile} ; \overline{\overline{B^{\smile} ; \overline{C}}} \mathsf{v}_1$	$7, 8, \mid ;$
10.	$\mathsf{v}_2 B ; A \mathsf{v}_0 \mid \mathsf{v}_2 C \mathsf{v}_1, \mathsf{v}_0 A^{\smile} ; \overline{\overline{B^{\smile} ; \overline{C}}} \mathsf{v}_1$	$9, ; \mid$, no v_3
11.	$\mathsf{v}_0 (B ; A)^{\smile} \mathsf{v}_2, \mathsf{v}_2 \overline{C} \mathsf{v}_1 \mid \mathsf{v}_0 A^{\smile} ; \overline{\overline{B^{\smile} ; \overline{C}}} \mathsf{v}_1$	$10, {}^{\smile}\mid, {}^{-}\mid$
12.	$\mathsf{v}_0 (B ; A)^{\smile} ; \overline{C} \mathsf{v}_1 \mid \mathsf{v}_0 A^{\smile} ; \overline{\overline{B^{\smile} ; \overline{C}}} \mathsf{v}_1$	$11, ; \mid$, no v_2
13.	$\mathsf{v}_0 \overline{A^{\smile} ; \overline{\overline{B^{\smile} ; \overline{C}}}} \mathsf{v}_1 \mid \mathsf{v}_0 \overline{(B ; A)^{\smile} ; \overline{C}} \mathsf{v}_1$	$12, \mid^{-}, {}^{-}\mid$
14.	$\mathsf{v}_0 A \to (B \to C) \mathsf{v}_1 \mid \mathsf{v}_0 A \circ B \to C \mathsf{v}_1$	13, (3.18), (3.19)

Proof (3.129) 4-proof of $(A \circ B \to C) \to (A \to (B \to C))$. This is the converse of bunching.

1.	$\mathsf{v}_2 A \mathsf{v}_0 \mid \mathsf{v}_2 A \mathsf{v}_0$	axiom
2.	$\mathsf{v}_3 B \mathsf{v}_2 \mid \mathsf{v}_3 B \mathsf{v}_2$	axiom
3.	$\mathsf{v}_3 \overline{C} \mathsf{v}_1 \mid \mathsf{v}_3 \overline{C} \mathsf{v}_1$	axiom
4.	$\mathsf{v}_2 A \mathsf{v}_0, \mathsf{v}_3 B \mathsf{v}_2 \mid \mathsf{v}_3 B ; A \mathsf{v}_0$	$1, 2, \mid ;$
5.	$\mathsf{v}_0 A^{\smile} \mathsf{v}_2, \mathsf{v}_2 B^{\smile} \mathsf{v}_3 \mid \mathsf{v}_0 (B ; A)^{\smile} \mathsf{v}_3$	$4, \mid^{\smile}, {}^{\smile}\mid$
6.	$\mathsf{v}_0 A^{\smile} \mathsf{v}_2, \mathsf{v}_2 B^{\smile} \mathsf{v}_3, \mathsf{v}_3 \overline{C} \mathsf{v}_1 \mid \mathsf{v}_0 (B ; A)^{\smile} ; \overline{C} \mathsf{v}_1$	$3, 5, \mid ;$
7.	$\mathsf{v}_0 A^{\smile} \mathsf{v}_2, \mathsf{v}_2 B^{\smile} ; \overline{C} \mathsf{v}_1 \mid \mathsf{v}_0 (B ; A)^{\smile} ; \overline{C} \mathsf{v}_1$	$6, ; \mid$, no v_3
8.	$\mathsf{v}_0 A^{\smile} \mathsf{v}_2, \mathsf{v}_2 \overline{\overline{B^{\smile} ; \overline{C}}} \mathsf{v}_1 \mid \mathsf{v}_0 (B ; A)^{\smile} ; \overline{C} \mathsf{v}_1$	$7, \mid^{-}, {}^{-}\mid$
9.	$\mathsf{v}_0 A^{\smile} ; \overline{\overline{B^{\smile} ; \overline{C}}} \mathsf{v}_1 \mid \mathsf{v}_0 (B ; A)^{\smile} ; \overline{C} \mathsf{v}_1$	$8, ; \mid$, no v_2
10.	$\mathsf{v}_0 \overline{(B ; A)^{\smile} ; \overline{C}} \mathsf{v}_1 \mid \mathsf{v}_0 \overline{A^{\smile} ; \overline{\overline{B^{\smile} ; \overline{C}}}} \mathsf{v}_1$	$9, \mid^{-}, {}^{-}\mid$
11.	$\mathsf{v}_0 A \circ B \to C \mathsf{v}_1 \mid \mathsf{v}_0 A \to (B \to C) \mathsf{v}_1$	10, (3.18), (3.19)

Proof (3.130) 4-proof of $(A \to B) \to (A \circ C \to B \circ C)$. This is monotonicity of fusion in the left argument.

1.	$\mathsf{v}_3 A \mathsf{v}_0 \mid \mathsf{v}_3 A \mathsf{v}_0$	axiom
2.	$\mathsf{v}_3 B \mathsf{v}_1 \mid \mathsf{v}_3 B \mathsf{v}_1$	axiom
3.	$\mathsf{v}_2 C \mathsf{v}_3 \mid \mathsf{v}_2 C \mathsf{v}_3$	axiom
4.	$\mid \mathsf{v}_3 \overline{B} \mathsf{v}_1, \mathsf{v}_3 B \mathsf{v}_1$	$2, \mid^{-}$
5.	$\mathsf{v}_2 C \mathsf{v}_3 \mid \mathsf{v}_3 \overline{B} \mathsf{v}_1, \mathsf{v}_2 C ; B \mathsf{v}_1$	$3, 4, \mid ;$
6.	$\mathsf{v}_3 A \mathsf{v}_0 \mid \mathsf{v}_0 A^{\smile} \mathsf{v}_3$	$1, \mid^{\smile}$
7.	$\mathsf{v}_2 C \mathsf{v}_3, \mathsf{v}_3 A \mathsf{v}_0 \mid \mathsf{v}_2 C ; B \mathsf{v}_1, \mathsf{v}_0 A^{\smile} ; \overline{B} \mathsf{v}_1$	$5, 6, \mid ;$
8.	$\mathsf{v}_2 C ; A \mathsf{v}_0 \mid \mathsf{v}_2 C ; B \mathsf{v}_1, \mathsf{v}_0 A^{\smile} ; \overline{B} \mathsf{v}_1$	$7, ; \mid$, no v_3
9.	$\mathsf{v}_0 (C ; A)^{\smile} \mathsf{v}_2, \mathsf{v}_2 \overline{C ; B} \mathsf{v}_1 \mid \mathsf{v}_0 A^{\smile} ; \overline{B} \mathsf{v}_1$	$8, {}^{-}\mid, {}^{\smile}\mid$
10.	$\mathsf{v}_0 (C ; A)^{\smile} ; \overline{C ; B} \mathsf{v}_1 \mid \mathsf{v}_0 A^{\smile} ; \overline{B} \mathsf{v}_1$	$9, ; \mid$, no v_2
11.	$\mathsf{v}_0 \overline{A^{\smile} ; \overline{B}} \mathsf{v}_1 \mid \mathsf{v}_0 \overline{(C ; A)^{\smile} ; \overline{C ; B}} \mathsf{v}_1$	$10, \mid^{-}, {}^{-}\mid$
12.	$\mathsf{v}_0 A \to B \mathsf{v}_1 \mid \mathsf{v}_0 A \circ C \to B \circ C \mathsf{v}_1$	11, (3.18), (3.19)

Proof (3.131) 4-proof of $(A \circ B) \circ C \to A \circ (B \circ C)$. Fusion is associative in one direction.

1.	$\mathsf{v}_2 B \mathsf{v}_3 \mid \mathsf{v}_2 B \mathsf{v}_3$	axiom
2.	$\mathsf{v}_0 C \mathsf{v}_2 \mid \mathsf{v}_0 C \mathsf{v}_2$	axiom
3.	$\mathsf{v}_0 C \mathsf{v}_2, \mathsf{v}_2 B \mathsf{v}_3 \mid \mathsf{v}_0 C ; B \mathsf{v}_3$	$1, 2, \mid ;$
4.	$\mathsf{v}_3 A \mathsf{v}_1 \mid \mathsf{v}_3 A \mathsf{v}_1$	axiom
5.	$\mathsf{v}_0 C \mathsf{v}_2, \mathsf{v}_2 B \mathsf{v}_3, \mathsf{v}_3 A \mathsf{v}_1 \mid \mathsf{v}_0 (C ; B) ; A \mathsf{v}_1$	$3, 4, \mid ;$
6.	$\mathsf{v}_0 C \mathsf{v}_2, \mathsf{v}_2 B ; A \mathsf{v}_1 \mid \mathsf{v}_0 (C ; B) ; A \mathsf{v}_1$	$5, ; \mid$, no v_3
7.	$\mathsf{v}_0 C ; (B ; A) \mathsf{v}_1 \mid \mathsf{v}_0 (C ; B) ; A \mathsf{v}_1$	$6, ; \mid$, no v_2
8.	$\mathsf{v}_0 (A \circ B) \circ C \mathsf{v}_1 \mid \mathsf{v}_0 A \circ (B \circ C) \mathsf{v}_1$	7, (3.19)

Proof (3.132) 4-proof of $A \circ (B \circ C) \to (A \circ B) \circ C$. Fusion is associative in the other direction.

1.	$\mathsf{v}_2 B \mathsf{v}_3 \,\vert\, \mathsf{v}_2 B \mathsf{v}_3$	axiom
2.	$\mathsf{v}_3 A \mathsf{v}_1 \,\vert\, \mathsf{v}_3 A \mathsf{v}_1$	axiom
3.	$\mathsf{v}_2 B \mathsf{v}_3, \mathsf{v}_3 A \mathsf{v}_1 \,\vert\, \mathsf{v}_2 B ; A \mathsf{v}_1$	1, 2, $\vert$;
4.	$\mathsf{v}_0 C \mathsf{v}_2 \,\vert\, \mathsf{v}_0 C \mathsf{v}_2$	axiom
5.	$\mathsf{v}_0 C \mathsf{v}_2, \mathsf{v}_2 B \mathsf{v}_3, \mathsf{v}_3 A \mathsf{v}_1 \,\vert\, \mathsf{v}_0 C ; (B ; A) \mathsf{v}_1$	3, 4, $\vert$;
6.	$\mathsf{v}_0 C ; B \mathsf{v}_3, \mathsf{v}_3 A \mathsf{v}_2 \,\vert\, \mathsf{v}_0 C ; (B ; A) \mathsf{v}_1$	5, ; $\vert$, no v_2
7.	$\mathsf{v}_0 (C ; B) ; A \mathsf{v}_1 \,\vert\, \mathsf{v}_0 C ; (B ; A) \mathsf{v}_1$	6, ; $\vert$, no v_3
8.	$\mathsf{v}_0 A \circ (B \circ C) \mathsf{v}_1 \,\vert\, \mathsf{v}_0 (A \circ B) \circ C \mathsf{v}_1$	7, (3.19)

Proof (3.133) 3-proof of $(A \to {\sim}A) \to {\sim}A$ from density. This is the *reductio ad absurdum* axiom.

1.	$\mathsf{v}_0 A^{\smile} \mathsf{v}_1 \,\vert\, \mathsf{v}_0 A^{\smile} ; A^{\smile} \mathsf{v}_1$	density
2.	$\mathsf{v}_0 A^{\smile} \mathsf{v}_2 \,\vert\, \mathsf{v}_0 A^{\smile} \mathsf{v}_2$	axiom
3.	$\mathsf{v}_2 A^{\smile} \mathsf{v}_1 \,\vert\, \mathsf{v}_2 A^{\smile} \mathsf{v}_1$	axiom
4.	$\mathsf{v}_2 A^{\smile} \mathsf{v}_1 \,\vert\, \mathsf{v}_2 \overline{\overline{A^{\smile}}} \mathsf{v}_1$	3, $\overline{}\vert$, $\vert\overline{}$
5.	$\mathsf{v}_0 A^{\smile} \mathsf{v}_2, \mathsf{v}_2 A^{\smile} \mathsf{v}_1 \,\vert\, \mathsf{v}_0 A^{\smile} ; \overline{\overline{A^{\smile}}} \mathsf{v}_1$	2, 4, $\vert$;
6.	$\mathsf{v}_0 A^{\smile} ; A^{\smile} \mathsf{v}_1 \,\vert\, \mathsf{v}_0 A^{\smile} ; \overline{\overline{A^{\smile}}} \mathsf{v}_1$	5, ; $\vert$, no v_2
7.	$\mathsf{v}_0 A^{\smile} \mathsf{v}_1 \,\vert\, \mathsf{v}_0 A^{\smile} ; \overline{\overline{A^{\smile}}} \mathsf{v}_1$	1, 6, Cut
8.	$\mathsf{v}_0 \overline{A^{\smile} ; \overline{\overline{A^{\smile}}}} \mathsf{v}_1 \,\vert\, \mathsf{v}_0 \overline{A^{\smile}} \mathsf{v}_1$	7, $\overline{}\vert$, $\vert\overline{}$
9.	$\mathsf{v}_0 A \to {\sim}A \mathsf{v}_1 \,\vert\, \mathsf{v}_0 {\sim}A \mathsf{v}_1$	8, (3.17), (3.18)

Proof (3.134) 3-proof of $A \wedge B \to A \circ B$ from density. The 3-proof of (3.134) is the first one in which we have a real need for a derived rule of the sequent calculus called **weakening**, indicated by a "W." Its form as a rule that could have been included in Table 3.8 is

$$\boxed{W} \quad \frac{\Gamma \,\vert\, \Delta}{\Gamma, \Gamma' \,\vert\, \Delta, \Delta'}$$

It is easily proved by induction on the lengths of n-proofs that this rule can be admitted without effect on the notion of n-provability. The idea is that any predicate that one wishes to add later via weakening can simply be added to the previous sequents. The form of the rules allows this; such additions take instances of rules to instances of rules. The base step of the induction is that weakenings of axioms are still axioms. This fact was used in the 2-proofs of (3.89) and (3.90) (sequent number 1 has a superfluous formula in it). It was used in the 3-proof of (3.109). The sequents corresponding to the equations of density, commutativity, and symmetry are also closed under weakenings; see the proof of (3.143). The proofs of (3.134), (3.135), and (3.137) would be unduly cluttered by actually using this device, so we use the weakening rule instead.

1. $\mathsf{v}_0 A\cdot B\mathsf{v}_1 \mid \mathsf{v}_0(A\cdot B);(A\cdot B)\mathsf{v}_1$ density
2. $\mathsf{v}_2 A\mathsf{v}_1 \mid \mathsf{v}_2 A\mathsf{v}_1$ axiom
3. $\mathsf{v}_0 B\mathsf{v}_2 \mid \mathsf{v}_0 B\mathsf{v}_2$ axiom
4. $\mathsf{v}_0 B\mathsf{v}_2, \mathsf{v}_2 A\mathsf{v}_1 \mid \mathsf{v}_0 B;A\mathsf{v}_1$ 2, 3, $\mid;$
5. $\mathsf{v}_0 A\mathsf{v}_2, \mathsf{v}_0 B\mathsf{v}_2, \mathsf{v}_2 A\mathsf{v}_1, \mathsf{v}_2 B\mathsf{v}_1 \mid \mathsf{v}_0 B;A\mathsf{v}_1$ 4, W
6. $\mathsf{v}_0 A\cdot B\mathsf{v}_2, \mathsf{v}_2 A\cdot B\mathsf{v}_1 \mid \mathsf{v}_0 B;A\mathsf{v}_1$ 5, $\cdot\mid$
7. $\mathsf{v}_0(A\cdot B);(A\cdot B)\mathsf{v}_1 \mid \mathsf{v}_0 B;A\mathsf{v}_1$ 6, $;\mid$, no v_2
8. $\mathsf{v}_0 A\cdot B\mathsf{v}_1 \mid \mathsf{v}_0 B;A\mathsf{v}_1$ 1, 7, Cut
9. $\mathsf{v}_0 A\wedge B\mathsf{v}_1 \mid \mathsf{v}_0 A\circ B\mathsf{v}_1$ 8, (3.15), (3.19)

Proof (3.135) 3-proof of $(A\to B)\to{\sim}A\vee B$ from density.

1. $\mathsf{v}_1 A\cdot\overline{B^\smile}\mathsf{v}_0 \mid \mathsf{v}_1(A\cdot\overline{B^\smile});(A\cdot\overline{B^\smile})\mathsf{v}_0$ density
2. $\mathsf{v}_2 B\mathsf{v}_1 \mid \mathsf{v}_2 B\mathsf{v}_1$ axiom
3. $\mathsf{v}_1\overline{B^\smile}\mathsf{v}_2 \mid \mathsf{v}_2\overline{B}\mathsf{v}_1$ 2, $\mid^\smile$, ${}^-\mid$, $\mid^-$
4. $\mathsf{v}_2 A\mathsf{v}_0 \mid \mathsf{v}_2 A\mathsf{v}_0$ axiom
5. $\mathsf{v}_2 A\mathsf{v}_0 \mid \mathsf{v}_0 A^\smile\mathsf{v}_2$ 4, $\mid^\smile$
6. $\mathsf{v}_1\overline{B^\smile}\mathsf{v}_2, \mathsf{v}_2 A\mathsf{v}_0 \mid \mathsf{v}_0 A^\smile;\overline{B}\mathsf{v}_1$ 3, 5, $\mid;$
7. $\mathsf{v}_1 A\mathsf{v}_2, \mathsf{v}_1\overline{B^\smile}\mathsf{v}_2, \mathsf{v}_2 A\mathsf{v}_0, \mathsf{v}_2\overline{B^\smile}\mathsf{v}_0 \mid \mathsf{v}_0 A^\smile;\overline{B}\mathsf{v}_1$ 6, W
8. $\mathsf{v}_1(A\cdot\overline{B^\smile})\mathsf{v}_2, \mathsf{v}_2(A\cdot\overline{B^\smile})\mathsf{v}_0 \mid \mathsf{v}_0 A^\smile;\overline{B}\mathsf{v}_1$ 7, $\cdot\mid$
9. $\mathsf{v}_1(A\cdot\overline{B^\smile});(A\cdot\overline{B^\smile})\mathsf{v}_0 \mid \mathsf{v}_0 A^\smile;\overline{B}\mathsf{v}_1$ 8, $;\mid$, no v_2
10. $\mathsf{v}_1 A\cdot\overline{B^\smile}\mathsf{v}_0 \mid \mathsf{v}_0 A^\smile;\overline{B}\mathsf{v}_1$ 1, 9, Cut
11. $\mathsf{v}_0\overline{A^\smile;\overline{B}}\mathsf{v}_1 \mid \mathsf{v}_1\overline{A\cdot\overline{\overline{B^\smile}}}\mathsf{v}_0$ 10, $\mid^-$, ${}^-\mid$
12. $\mathsf{v}_1 A\mathsf{v}_0 \mid \mathsf{v}_1 A\mathsf{v}_0$ axiom
13. $\mathsf{v}_0 B\mathsf{v}_1 \mid \mathsf{v}_0 B\mathsf{v}_1$ axiom
14. $\mid \mathsf{v}_1\overline{B^\smile}\mathsf{v}_0, \mathsf{v}_0 B\mathsf{v}_1$ 13, ${}^\smile\mid$
15. $\mathsf{v}_1 A\mathsf{v}_0 \mid \mathsf{v}_1 A\cdot\overline{B^\smile}\mathsf{v}_0, \mathsf{v}_0 B\mathsf{v}_1$ 12, 14, $\mid\cdot$
16. $\mathsf{v}_1\overline{A\cdot\overline{\overline{B^\smile}}}\mathsf{v}_0 \mid \mathsf{v}_0\overline{A^\smile}\mathsf{v}_1, \mathsf{v}_0 B\mathsf{v}_1$ 15, ${}^-\mid$, ${}^\smile\mid$, $\mid^-$
17. $\mathsf{v}_1\overline{A\cdot\overline{\overline{B^\smile}}}\mathsf{v}_0 \mid \mathsf{v}_0\overline{A^\smile}+B\mathsf{v}_1$ 16, $\mid+$
18. $\mathsf{v}_0\overline{A^\smile;\overline{B}}\mathsf{v}_1 \mid \mathsf{v}_0\overline{A^\smile}+B\mathsf{v}_1$ 11, 17, Cut
19. $\mathsf{v}_0 A\to B\mathsf{v}_1 \mid \mathsf{v}_0{\sim}A\vee B\mathsf{v}_1$ 18, (3.14), (3.17), (3.18)

Proof (3.136) 4-proof of $(A\to(A\to B))\to(A\to B)$ from density. This is the contraction axiom.

1.	$\mathsf{v}_0 A^\smile \mathsf{v}_2 \mid \mathsf{v}_0 A^\smile ; A^\smile \mathsf{v}_2$	density
2.	$\mathsf{v}_0 A^\smile \mathsf{v}_3 \mid \mathsf{v}_0 A^\smile \mathsf{v}_3$	axiom
3.	$\mathsf{v}_3 A^\smile \mathsf{v}_2 \mid \mathsf{v}_3 A^\smile \mathsf{v}_2$	axiom
4.	$\mathsf{v}_2 \overline{B} \mathsf{v}_1 \mid \mathsf{v}_2 \overline{B} \mathsf{v}_1$	axiom
5.	$\mathsf{v}_3 A^\smile \mathsf{v}_2, \mathsf{v}_2 \overline{B} \mathsf{v}_1 \mid \mathsf{v}_3 A^\smile ; \overline{B} \mathsf{v}_1$	3, 4, $\mid;$
6.	$\mathsf{v}_3 A^\smile \mathsf{v}_2, \mathsf{v}_2 \overline{B} \mathsf{v}_1 \mid \mathsf{v}_3 \overline{\overline{A^\smile ; \overline{B}}} \mathsf{v}_1$	5, $\overline{\ }\mid$, $\mid\overline{\ }$
7.	$\mathsf{v}_0 A^\smile \mathsf{v}_3, \mathsf{v}_3 A^\smile \mathsf{v}_2, \mathsf{v}_2 \overline{B} \mathsf{v}_1 \mid \mathsf{v}_0 A^\smile ; \overline{\overline{A^\smile ; \overline{B}}} \mathsf{v}_1$	2, 6, $\mid;$
8.	$\mathsf{v}_0 A^\smile ; A^\smile \mathsf{v}_2, \mathsf{v}_2 \overline{B} \mathsf{v}_1 \mid \mathsf{v}_0 A^\smile ; \overline{\overline{A^\smile ; \overline{B}}} \mathsf{v}_1$	7, $;\mid$, no v_3
9.	$\mathsf{v}_0 A^\smile \mathsf{v}_2, \mathsf{v}_2 \overline{B} \mathsf{v}_1 \mid \mathsf{v}_0 A^\smile ; \overline{\overline{A^\smile ; \overline{B}}} \mathsf{v}_1$	1, 8, Cut
10.	$\mathsf{v}_0 A^\smile ; \overline{B} \mathsf{v}_1 \mid \mathsf{v}_0 A^\smile ; \overline{\overline{A^\smile ; \overline{B}}} \mathsf{v}_1$	9, $;\mid$, no v_2
11.	$\mathsf{v}_0 \overline{A^\smile ; \overline{\overline{A^\smile ; \overline{B}}}} \mathsf{v}_1 \mid \mathsf{v}_0 \overline{A^\smile ; \overline{B}} \mathsf{v}_1$	10, $\overline{\ }\mid$, $\mid\overline{\ }$
12.	$\mathsf{v}_0 A \to (A \to B) \mathsf{v}_1 \mid \mathsf{v}_0 A \to B \mathsf{v}_1$	11, (3.18)

Proof (3.137) 4-proof of $(A \to (B \to C)) \to (A \wedge B \to C)$ from density.

1.	$\mathsf{v}_2 A \cdot B \mathsf{v}_0 \mid \mathsf{v}_2 (A \cdot B) ; (A \cdot B) \mathsf{v}_0$	density
2.	$\mathsf{v}_3 A \mathsf{v}_0 \mid \mathsf{v}_3 A \mathsf{v}_0$	axiom
3.	$\mathsf{v}_3 A \mathsf{v}_0 \mid \mathsf{v}_0 A^\smile \mathsf{v}_3$	2, $\mid^\smile$
4.	$\mathsf{v}_2 B \mathsf{v}_3 \mid \mathsf{v}_2 B \mathsf{v}_3$	axiom
5.	$\mathsf{v}_2 B \mathsf{v}_3 \mid \mathsf{v}_3 B^\smile \mathsf{v}_2$	4, $\mid^\smile$
6.	$\mathsf{v}_2 \overline{C} \mathsf{v}_1 \mid \mathsf{v}_2 \overline{C} \mathsf{v}_1$	axiom
7.	$\mathsf{v}_2 B \mathsf{v}_3, \mathsf{v}_2 \overline{C} \mathsf{v}_1 \mid \mathsf{v}_3 B^\smile ; \overline{C} \mathsf{v}_1$	5, 6, $\mid;$
8.	$\mathsf{v}_2 B \mathsf{v}_3, \mathsf{v}_2 \overline{C} \mathsf{v}_1 \mid \mathsf{v}_3 \overline{\overline{B^\smile ; \overline{C}}} \mathsf{v}_1$	7, $\overline{\ }\mid$, $\mid\overline{\ }$
9.	$\mathsf{v}_2 B \mathsf{v}_3, \mathsf{v}_3 A \mathsf{v}_0, \mathsf{v}_2 \overline{C} \mathsf{v}_1 \mid \mathsf{v}_0 A^\smile ; \overline{\overline{B^\smile ; \overline{C}}} \mathsf{v}_1$	3, 8, $\mid;$
10.	$\mathsf{v}_2 A \mathsf{v}_3, \mathsf{v}_2 B \mathsf{v}_3, \mathsf{v}_3 A \mathsf{v}_0, \mathsf{v}_3 B \mathsf{v}_0, \mathsf{v}_2 \overline{C} \mathsf{v}_1 \mid \mathsf{v}_0 A^\smile ; \overline{\overline{B^\smile ; \overline{C}}} \mathsf{v}_1$	9, W
11.	$\mathsf{v}_2 A \cdot B \mathsf{v}_3, \mathsf{v}_3 A \cdot B \mathsf{v}_0, \mathsf{v}_2 \overline{C} \mathsf{v}_1 \mid \mathsf{v}_0 A^\smile ; \overline{\overline{B^\smile ; \overline{C}}} \mathsf{v}_1$	10, $\cdot\mid$
12.	$\mathsf{v}_2 (A \cdot B) ; (A \cdot B) \mathsf{v}_0, \mathsf{v}_2 \overline{C} \mathsf{v}_1 \mid \mathsf{v}_0 A^\smile ; \overline{\overline{B^\smile ; \overline{C}}} \mathsf{v}_1$	12, $;\mid$, no v_3
13.	$\mathsf{v}_2 A \cdot B \mathsf{v}_0, \mathsf{v}_2 \overline{C} \mathsf{v}_1 \mid \mathsf{v}_0 A^\smile ; \overline{\overline{B^\smile ; \overline{C}}} \mathsf{v}_1$	1, 12, Cut
14.	$\mathsf{v}_0 (A \cdot B)^\smile \mathsf{v}_2, \mathsf{v}_2 \overline{C} \mathsf{v}_1 \mid \mathsf{v}_0 A^\smile ; \overline{\overline{B^\smile ; \overline{C}}} \mathsf{v}_1$	13, $^\smile\mid$
15.	$\mathsf{v}_0 (A \cdot B)^\smile ; \overline{C} \mathsf{v}_1 \mid \mathsf{v}_0 A^\smile ; \overline{\overline{B^\smile ; \overline{C}}} \mathsf{v}_1$	14, $;\mid$, no v_2

16.	$\mathsf{v}_0\overline{A^{\smile};\overline{\overline{B^{\smile};\overline{C}}}}\mathsf{v}_1 \mid \mathsf{v}_0\overline{(A\cdot B)^{\smile};\overline{C}}\mathsf{v}_1$	15, $^{-}\mid$, $\mid^{-}$
17.	$\mathsf{v}_0 A \to (B \to C)\mathsf{v}_1 \mid \mathsf{v}_0 A \wedge B \to C\mathsf{v}_1$	16, (3.15), (3.18)

Proof (3.138) 3-proof of $A \to ((A \to B) \to B)$ from commutativity. This is an axiomatic form of modus ponens.

1.	$\mathsf{v}_0 A\mathsf{v}_1 \mid \mathsf{v}_0 A\mathsf{v}_1$	axiom
2.	$\mathsf{v}_2 A \to B\mathsf{v}_0 \mid \mathsf{v}_2 A \to B\mathsf{v}_0$	axiom
3.	$\mathsf{v}_0 A\mathsf{v}_1, \mathsf{v}_2 A \to B\mathsf{v}_0 \mid \mathsf{v}_2(A \to B);A\mathsf{v}_1$	1, 2, $\mid;$
4.	$\mathsf{v}_2(A \to B);A\mathsf{v}_1 \mid \mathsf{v}_2 A;(A \to B)\mathsf{v}_1$	commutativity
5.	$\mathsf{v}_0 A\mathsf{v}_1, \mathsf{v}_2 A \to B\mathsf{v}_0 \mid \mathsf{v}_2 A;(A \to B)\mathsf{v}_1$	3, 4, Cut
6.	$\mathsf{v}_2 A\mathsf{v}_0 \mid \mathsf{v}_2 A\mathsf{v}_0$	axiom
7.	$\mathsf{v}_2 B\mathsf{v}_1 \mid \mathsf{v}_2 B\mathsf{v}_1$	axiom
8.	$\mathsf{v}_2 A\mathsf{v}_0 \mid \mathsf{v}_0 A^{\smile}\mathsf{v}_2$	6, $\mid^{\smile}$
9.	$\mid \mathsf{v}_2\overline{B}\mathsf{v}_1, \mathsf{v}_2 B\mathsf{v}_1$	7, $\mid^{-}$
10.	$\mathsf{v}_2 A\mathsf{v}_0 \mid \mathsf{v}_0 A^{\smile};\overline{B}\mathsf{v}_1, \mathsf{v}_2 B\mathsf{v}_1$	8, 9, $\mid;$
11.	$\mathsf{v}_2 A\mathsf{v}_0, \mathsf{v}_0\overline{A^{\smile};\overline{B}}\mathsf{v}_1 \mid \mathsf{v}_2 B\mathsf{v}_1$	10, $^{-}\mid$
12.	$\mathsf{v}_2 A\mathsf{v}_0, \mathsf{v}_0 A \to B\mathsf{v}_1 \mid \mathsf{v}_2 B\mathsf{v}_1$	11, (3.18)
13.	$\mathsf{v}_2 A;(A \to B)\mathsf{v}_1 \mid \mathsf{v}_2 B\mathsf{v}_1$	12, $;\mid$, no v_0
14.	$\mathsf{v}_0 A\mathsf{v}_1, \mathsf{v}_2 A \to B\mathsf{v}_0 \mid \mathsf{v}_2 B\mathsf{v}_1$	5, 13, Cut
15.	$\mathsf{v}_0 A\mathsf{v}_1, \mathsf{v}_2 A \to B\mathsf{v}_0, \mathsf{v}_2\overline{B}\mathsf{v}_1 \mid$	14, $^{-}\mid$
16.	$\mathsf{v}_0 A\mathsf{v}_1, \mathsf{v}_0(A \to B)^{\smile}\mathsf{v}_2, \mathsf{v}_2\overline{B}\mathsf{v}_1 \mid$	15, $^{\smile}\mid$
17.	$\mathsf{v}_0 A\mathsf{v}_1, \mathsf{v}_0(A \to B)^{\smile};\overline{B}\mathsf{v}_1 \mid$	16, $;\mid$, no v_2
18.	$\mathsf{v}_0 A\mathsf{v}_1 \mid \mathsf{v}_0\overline{(A \to B)^{\smile};\overline{B}}\mathsf{v}_1$	17, $\mid^{-}$
19.	$\mathsf{v}_0 A\mathsf{v}_1 \mid \mathsf{v}_0(A \to B) \to B\mathsf{v}_1$	18, (3.18)

Proof (3.139) 3-proof of $(A \to {\sim}B) \to (B \to {\sim}A)$ from commutativity. This is a contraposition axiom.

1.	$v_1 B ; A v_0 \mid v_1 A ; B v_0$	commutativity
2.	$v_1 A v_2 \mid v_1 A v_2$	axiom
3.	$v_1 A v_2 \mid v_2 \overline{\overline{A^\smile}} v_1$	2, $\mid^\smile$, $^-\!\mid$, $\mid^-$
4.	$v_2 B v_0 \mid v_2 B v_0$	axiom
5.	$v_2 B v_0 \mid v_0 B^\smile v_2$	4, $\mid^\smile$
6.	$v_1 A v_2, v_2 B v_0 \mid v_0 B^\smile ; \overline{\overline{A^\smile}} v_1$	3, 5, $\mid;$
7.	$v_1 A v_2, v_2 B v_0, v_0 \overline{B^\smile ; \overline{\overline{A^\smile}}} v_1 \mid$	6, $^-\!\mid$
8.	$v_1 A v_2, v_2 B v_0, v_0 B \to {\sim} A v_1 \mid$	7, (3.17), (3.18)
9.	$v_1 A ; B v_0, v_0 B \to {\sim} A v_1 \mid$	8, $;\mid$, no v_2
10.	$v_1 B ; A v_0, v_0 B \to {\sim} A v_1 \mid$	1, 9, Cut
11.	$v_1 B v_2 \mid v_1 B v_2$	axiom
12.	$v_2 \overline{\overline{B^\smile}} v_1 \mid v_1 B v_2$	11, $^\smile\!\mid$, $\mid^-$, $^-\!\mid$
13.	$v_2 A v_0 \mid v_2 A v_0$	axiom
14.	$v_0 A^\smile v_2 \mid v_2 A v_0$	13, $^\smile\!\mid$
15.	$v_0 A^\smile v_2, v_2 \overline{\overline{B^\smile}} v_1 \mid v_1 B ; A v_0$	12, 14, $\mid;$
16.	$v_0 A^\smile ; \overline{\overline{B^\smile}} v_1 \mid v_1 B ; A v_0$	15, $;\mid$, no v_2
17.	$\mid v_1 B ; A v_0, v_0 \overline{A^\smile ; \overline{\overline{B^\smile}}} v_1$	16, $\mid^-$
18.	$\mid v_1 B ; A v_0, v_0 A \to {\sim} B v_1$	17, (3.17), (3.18)
19.	$v_0 B \to {\sim} A v_1 \mid v_0 A \to {\sim} B v_1$	10, 18, Cut

Proof (3.140) 4-proof of $(A \to (B \to C)) \to (B \to (A \to C))$ from commutativity. This is the permutation axiom. Let $D = A \to (B \to C)$. Then $D = \overline{A^\smile ; \overline{\overline{B^\smile ; \overline{C}}}}$ by (3.17) and (3.18).

1.	$v_3 A v_0 \mid v_3 A v_0$	axiom
2.	$v_3 A v_0 \mid v_0 A^\smile v_3$	1, $\mid^\smile$
3.	$v_2 B v_3 \mid v_2 B v_3$	axiom
4.	$v_2 B v_3 \mid v_3 B^\smile v_2$	3, $\mid^\smile$
5.	$v_2 \overline{C} v_1 \mid v_2 \overline{C} v_1$	axiom
6.	$v_2 B v_3, v_2 \overline{C} v_1 \mid v_3 B^\smile ; \overline{C} v_1$	4, 5, $\mid;$
7.	$v_2 B v_3, v_2 \overline{C} v_1 \mid v_3 \overline{\overline{B^\smile ; \overline{C}}} v_1$	6, $^-\!\mid$, $\mid^-$
8.	$v_2 B v_3, v_3 A v_0, v_2 \overline{C} v_1 \mid v_0 A^\smile ; \overline{\overline{B^\smile ; \overline{C}}} v_1$	2, 7, $\mid;$
9.	$v_2 B v_3, v_3 A v_0, v_2 \overline{C} v_1, v_0 D v_1 \mid$	8, $^-\!\mid$, def. D

10.	$\mathsf{v}_2 B ; A \mathsf{v}_0, \mathsf{v}_2 \overline{C} \mathsf{v}_1, \mathsf{v}_0 D \mathsf{v}_1 \mid$	9, ; $\mid$, no v_3
11.	$\mathsf{v}_2 A ; B \mathsf{v}_0 \mid \mathsf{v}_2 B ; A \mathsf{v}_0$	commutativity
12.	$\mathsf{v}_2 A ; B \mathsf{v}_0, \mathsf{v}_2 \overline{C} \mathsf{v}_1, \mathsf{v}_0 D \mathsf{v}_1 \mid$	10, 11, Cut
13.	$\mathsf{v}_2 A \mathsf{v}_3 \mid \mathsf{v}_2 A \mathsf{v}_3$	axiom
14.	$\mathsf{v}_3 A^{\smile} \mathsf{v}_2 \mid \mathsf{v}_2 A \mathsf{v}_3$	13, $^{\smile}\mid$
15.	$\mathsf{v}_3 B \mathsf{v}_0 \mid \mathsf{v}_3 B \mathsf{v}_0$	axiom
16.	$\mathsf{v}_3 B^{\smile} \mathsf{v}_0 \mid \mathsf{v}_3 B \mathsf{v}_0$	15, $^{\smile}\mid$
17.	$\mathsf{v}_0 B^{\smile} \mathsf{v}_3, \mathsf{v}_3 A^{\smile} \mathsf{v}_2 \mid \mathsf{v}_2 A ; B \mathsf{v}_0$	14, 16, $\mid$;
18.	$\mathsf{v}_0 B^{\smile} \mathsf{v}_3, \mathsf{v}_3 A^{\smile} \mathsf{v}_2, \mathsf{v}_2 \overline{C} \mathsf{v}_1, \mathsf{v}_0 D \mathsf{v}_1 \mid$	12, 17, Cut
19.	$\mathsf{v}_0 B^{\smile} \mathsf{v}_3, \mathsf{v}_3 A^{\smile} ; \overline{C} \mathsf{v}_1, \mathsf{v}_0 D \mathsf{v}_1 \mid$	18, ; $\mid$, no v_2
20.	$\mathsf{v}_0 B^{\smile} \mathsf{v}_3, \mathsf{v}_3 \overline{\overline{A^{\smile} ; \overline{C}}} \mathsf{v}_1, \mathsf{v}_0 D \mathsf{v}_1 \mid$	19, $\mid^{-}$, $^{-}\mid$
21.	$\mathsf{v}_0 B^{\smile} ; \overline{\overline{A^{\smile} ; \overline{C}}} \mathsf{v}_1, \mathsf{v}_0 D \mathsf{v}_1 \mid$	20, ; $\mid$, no v_3
22.	$\mathsf{v}_0 D \mathsf{v}_1 \mid \mathsf{v}_0 B \to (A \to C) \mathsf{v}_1$	21, $\mid^{-}$, (3.18)
23.	$\mathsf{v}_0 A \to (B \to C) \mathsf{v}_1 \mid \mathsf{v}_0 B \to (A \to C) \mathsf{v}_1$	22, def. D

Proof (3.141) 4-proof of $(A \to B) \to ((B \to C) \to (A \to C))$ from commutativity. The suffixing axiom (3.141) (4-provable from commutativity) can be derived from the permutation axiom (3.140) (4-provable from commutativity) using the 1-provable derived rule of modus ponens (3.111).

Proof (3.142) 4-proof of $(A \to (B \to C)) \to ((A \to B) \to (A \to C))$ from density and commutativity. Let

$$
\begin{aligned}
D &= (A \to B) \to ((B \to (A \to C)) \to (A \to (A \to C))), \\
E &= (A \to B) \to ((B \to (A \to C)) \to (A \to C)), \\
F &= (B \to (A \to C)) \to ((A \to B) \to (A \to C)), \\
G &= (A \to (B \to C)) \to ((A \to B) \to (A \to C)).
\end{aligned}
$$

Contraction $(A \to (A \to C)) \to (A \to C)$ is 4-provable from density by (3.136). Apply modus ponens (3.111) (a 1-provable rule) twice to instances of 4-provable prefixing (3.127) to conclude that $D \to E$ is 4-provable from density. D is an instance of suffixing (3.141), which is 4-provable from commutativity. By modus ponens (3.111), E is therefore 4-provable from density and commutativity. $E \to F$ is an instance of permutation (3.140), so by modus ponens (3.111), F is 4-provable from density and commutativity. Apply the 2-provable transitivity rule (3.114) to F and permutation $(A \to (B \to C)) \to (B \to (A \to C))$ (4-provable from commutativity) to conclude that G is 4-provable from density and commutativity, as desired. This proof uses more instances of density and commutativity than are required.

The following 4-proof from density and commutativity shows that $A \leq A ; A$ and $A ; B \mathrel{\mathring{=}} B ; A$ are sufficient. By the way, $A \leq A ; A$ and $A ; (A \rightarrow B) \mathrel{\mathring{=}} (A \rightarrow B) ; A$ are also sufficient by (Maddux 2010, Thm 5.1(62)).

1.	$\mathsf{v}_2 B \mathsf{v}_3 \,|\, \mathsf{v}_2 B \mathsf{v}_3$	axiom
2.	$\mathsf{v}_2 B \mathsf{v}_3 \,|\, \mathsf{v}_3 B^\smile \mathsf{v}_2$	$1, |^\smile$
3.	$\mathsf{v}_2 \overline{C} \mathsf{v}_1 \,|\, \mathsf{v}_2 \overline{C} \mathsf{v}_1$	axiom
4.	$\mathsf{v}_2 B \mathsf{v}_3, \mathsf{v}_2 \overline{C} \mathsf{v}_1 \,|\, \mathsf{v}_3 B^\smile ; \overline{C} \mathsf{v}_1$	$2, 3, |;$
5.	$\mathsf{v}_2 B \mathsf{v}_3, \mathsf{v}_2 \overline{C} \mathsf{v}_1 \,|\, \mathsf{v}_3 \overline{\overline{B^\smile ; \overline{C}}} \mathsf{v}_1$	$4, {}^-|, |^-$
6.	$\mathsf{v}_3 A \mathsf{v}_0 \,|\, \mathsf{v}_3 A \mathsf{v}_0$	axiom
7.	$\mathsf{v}_3 A \mathsf{v}_0 \,|\, \mathsf{v}_0 A^\smile \mathsf{v}_3$	$6, |^\smile$
8.	$\mathsf{v}_2 B \mathsf{v}_3, \mathsf{v}_3 A \mathsf{v}_0, \mathsf{v}_2 \overline{C} \mathsf{v}_1 \,|\, \mathsf{v}_0 A^\smile ; \overline{\overline{B^\smile ; \overline{C}}} \mathsf{v}_1$	$5, 7, |;$
9.	$\mathsf{v}_2 B ; A \mathsf{v}_0, \mathsf{v}_2 \overline{C} \mathsf{v}_1 \,|\, \mathsf{v}_0 A^\smile ; \overline{\overline{B^\smile ; \overline{C}}} \mathsf{v}_1$	$8, ;|$, no v_3
10.	$\mathsf{v}_2 A \mathsf{v}_3 \,|\, \mathsf{v}_2 A \mathsf{v}_3$	axiom
11.	$\mathsf{v}_3 B \mathsf{v}_0 \,|\, \mathsf{v}_3 B \mathsf{v}_0$	axiom
12.	$\mathsf{v}_2 A \mathsf{v}_3, \mathsf{v}_3 B \mathsf{v}_0 \,|\, \mathsf{v}_2 A ; B \mathsf{v}_0$	$10, 11, |;$
13.	$\mathsf{v}_2 A ; B \mathsf{v}_0 \,|\, \mathsf{v}_2 B ; A \mathsf{v}_0$	commutativity
14.	$\mathsf{v}_2 A \mathsf{v}_3, \mathsf{v}_3 B \mathsf{v}_0 \,|\, \mathsf{v}_2 B ; A \mathsf{v}_0$	12, 13, Cut
15.	$\mathsf{v}_2 A \mathsf{v}_3, \mathsf{v}_3 B \mathsf{v}_0, \mathsf{v}_2 \overline{C} \mathsf{v}_1 \,|\, \mathsf{v}_0 A^\smile ; \overline{\overline{B^\smile ; \overline{C}}} \mathsf{v}_1$	9, 14, Cut
16.	$\mathsf{v}_3 B \mathsf{v}_0, \mathsf{v}_3 A^\smile \mathsf{v}_2, \mathsf{v}_2 \overline{C} \mathsf{v}_1 \,|\, \mathsf{v}_0 A^\smile ; \overline{\overline{B^\smile ; \overline{C}}} \mathsf{v}_1$	$15, {}^\smile|$
17.	$\mathsf{v}_3 B \mathsf{v}_0, \mathsf{v}_3 A^\smile ; \overline{C} \mathsf{v}_1 \,|\, \mathsf{v}_0 A^\smile ; \overline{\overline{B^\smile ; \overline{C}}} \mathsf{v}_1$	$16, ;|$, no v_2
18.	$\mathsf{v}_3 A \mathsf{v}_2 \,|\, \mathsf{v}_3 A \mathsf{v}_2$	axiom
19.	$\mathsf{v}_3 B \mathsf{v}_0 \,|\, \mathsf{v}_3 B \mathsf{v}_0$	axiom
20.	$\mathsf{v}_3 A \mathsf{v}_2 \,|\, \mathsf{v}_2 A^\smile \mathsf{v}_3$	$18, |^\smile$
21.	$|\, \mathsf{v}_3 B \mathsf{v}_0, \mathsf{v}_3 \overline{B} \mathsf{v}_0$	$19, |^-$
22.	$\mathsf{v}_3 A \mathsf{v}_2 \,|\, \mathsf{v}_3 B \mathsf{v}_0, \mathsf{v}_2 A^\smile ; \overline{B} \mathsf{v}_0$	$20, 21, |;$
23.	$\mathsf{v}_3 A \mathsf{v}_2, \mathsf{v}_2 \overline{A^\smile ; \overline{B}} \mathsf{v}_0 \,|\, \mathsf{v}_3 B \mathsf{v}_0$	$22, {}^-|$
24.	$\mathsf{v}_0 \overline{A^\smile ; \overline{B}}^\smile \mathsf{v}_2, \mathsf{v}_2 A^\smile \mathsf{v}_3 \,|\, \mathsf{v}_3 B \mathsf{v}_0$	$23, {}^\smile|$
25.	$\mathsf{v}_0 \overline{A^\smile ; \overline{B}}^\smile \mathsf{v}_2, \mathsf{v}_2 A^\smile \mathsf{v}_3, \mathsf{v}_3 A^\smile ; \overline{C} \mathsf{v}_1 \,|\, \mathsf{v}_0 A^\smile ; \overline{\overline{B^\smile ; \overline{C}}} \mathsf{v}_1$	17, 24, Cut
26.	$\mathsf{v}_0 \overline{A^\smile ; \overline{B}}^\smile \mathsf{v}_2, \mathsf{v}_2 A^\smile ; (A^\smile ; \overline{C}) \mathsf{v}_1 \,|\, \mathsf{v}_0 A^\smile ; \overline{\overline{B^\smile ; \overline{C}}} \mathsf{v}_1$	$25, ;|$, no v_3
27.	$\mathsf{v}_0 A \mathsf{v}_3 \,|\, \mathsf{v}_0 A \mathsf{v}_3$	axiom
28.	$\mathsf{v}_0 A \mathsf{v}_3 \,|\, \mathsf{v}_3 A^\smile \mathsf{v}_0$	$27, |^\smile$

29. $v_0\overline{C}v_1 \mid v_0\overline{C}v_1$ — axiom
30. $v_0Av_3, v_0\overline{C}v_1 \mid v_3A^\smile;\overline{C}v_1$ — 28, 29, $|;$
31. $v_3Av_2 \mid v_3Av_2$ — axiom
32. $v_3Av_2 \mid v_2A^\smile v_3$ — 31, $|^\smile$
33. $v_0Av_3, v_3Av_2, v_0\overline{C}v_1 \mid v_2A^\smile;(A^\smile;\overline{C})v_1$ — 30, 32, $|;$
34. $v_0A;Av_2, v_0\overline{C}v_1 \mid v_2A^\smile;(A^\smile;\overline{C})v_1$ — 33, $;|$, no v_3
35. $v_0Av_2 \mid v_0A;Av_2$ — density
36. $v_0Av_2, v_0\overline{C}v_1 \mid v_2A^\smile;(A^\smile;\overline{C})v_1$ — 34, 35, Cut
37. $v_2A^\smile v_0, v_0\overline{C}v_1 \mid v_2A^\smile;(A^\smile;\overline{C})v_1$ — 36, $^\smile|$
38. $v_2A^\smile;\overline{C}v_1 \mid v_2A^\smile;(A^\smile;\overline{C})v_1$ — 37, $;|$, no v_0
39. $v_0\overline{A^\smile;\overline{B}}^\smile v_2, v_2A^\smile;\overline{C}v_1 \mid v_0A^\smile;\overline{\overline{B^\smile;\overline{C}}}v_1$ — 26, 38, Cut
40. $v_0\overline{A^\smile;\overline{B}}^\smile v_2, v_2\overline{\overline{A^\smile;\overline{C}}}v_1 \mid v_0A^\smile;\overline{\overline{B^\smile;\overline{C}}}v_1$ — 39, $|^-, {}^-|$
41. $v_0\overline{A^\smile;\overline{B}}^\smile;\overline{\overline{A^\smile;\overline{C}}}v_1 \mid v_0A^\smile;\overline{\overline{B^\smile;\overline{C}}}v_1$ — 40, $;|$, no v_2
42. $v_0\overline{A^\smile;\overline{\overline{B^\smile;\overline{C}}}}v_1 \mid v_0\overline{\overline{A^\smile;\overline{B}}^\smile;\overline{\overline{A^\smile;\overline{C}}}}v_1$ — 41, ${}^-|, |^-$
43. $v_0A \to (B \to C)v_1 \mid v_0(A \to B) \to (A \to C)v_1$ — 42, (3.18)

Proof (3.143) 2-proof of $A \wedge {\sim}A \to B$ from symmetry.

1. $v_0Av_1 \mid v_0A^\smile v_1, v_0Bv_1$ — symmetry
2. $v_0A \cdot \overline{A^\smile}v_1 \mid v_0Bv_1$ — 1, ${}^-|$
3. $v_0A \wedge {\sim}Av_1 \mid v_0Bv_1$ — 2, (3.15), (3.17)

Proof (3.144) 3-proof of $A \circ B \to B \circ A$ from symmetry.

1.	$\mathsf{v}_2 A \mathsf{v}_1 \,|\, \mathsf{v}_2 A \mathsf{v}_1$	axiom
2.	$\mathsf{v}_2 A \mathsf{v}_1 \,|\, \mathsf{v}_1 A^{\smile} \mathsf{v}_2$	1, $|^{\smile}$
3.	$\mathsf{v}_1 A^{\smile} \mathsf{v}_2 \,|\, \mathsf{v}_1 A \mathsf{v}_2$	symmetry
4.	$\mathsf{v}_2 A \mathsf{v}_1 \,|\, \mathsf{v}_1 A \mathsf{v}_2$	2, 3, Cut
5.	$\mathsf{v}_0 B \mathsf{v}_2 \,|\, \mathsf{v}_0 B \mathsf{v}_2$	axiom
6.	$\mathsf{v}_0 B \mathsf{v}_2 \,|\, \mathsf{v}_2 B^{\smile} \mathsf{v}_0$	5, $|^{\smile}$
7.	$\mathsf{v}_2 B^{\smile} \mathsf{v}_0 \,|\, \mathsf{v}_2 B \mathsf{v}_0$	symmetry
8.	$\mathsf{v}_0 B \mathsf{v}_2 \,|\, \mathsf{v}_2 B \mathsf{v}_0$	6, 7, Cut
9.	$\mathsf{v}_0 B \mathsf{v}_2, \mathsf{v}_2 A \mathsf{v}_1 \,|\, \mathsf{v}_1 A ; B \mathsf{v}_0$	4, 8, $|;$
10.	$\mathsf{v}_0 B \mathsf{v}_2, \mathsf{v}_2 A \mathsf{v}_1 \,|\, \mathsf{v}_0 (A ; B)^{\smile} \mathsf{v}_1$	9, $|^{\smile}$
11.	$\mathsf{v}_0 B ; A \mathsf{v}_1 \,|\, \mathsf{v}_0 (A ; B)^{\smile} \mathsf{v}_1$	10, $;|$, no v_2
12.	$\mathsf{v}_0 (A ; B)^{\smile} \mathsf{v}_1 \,|\, \mathsf{v}_0 A ; B \mathsf{v}_1$	symmetry
13.	$\mathsf{v}_0 B ; A \mathsf{v}_1 \,|\, \mathsf{v}_0 A ; B \mathsf{v}_1$	11, 12, Cut
14.	$\mathsf{v}_0 A \circ B \mathsf{v}_1 \,|\, \mathsf{v}_0 B \circ A \mathsf{v}_1$	13, (3.19)

3.15 4-Provable Predicates in R and TR that are Not 3-Provable

This section presents predicates that are 4-provable but not 3-provable, even in the presence of the non-logical assumptions of density, commutativity, and symmetry.

Theorem 3.23 *Predicates (3.127)–(3.132), (3.136), (3.137), (3.140)–(3.142) are in* R *and* TR *but not* CT_3*. They are not 3-provable from density and symmetry.*

Proof To show the predicates (3.127)–(3.132), (3.136), (3.137), (3.140)–(3.142) belong to R it suffices to check that they are valid in every CR^*-frame. This is well known and will not be done here. To show they cannot be proved with four variables, we use a dense symmetric (hence commutative) semi-associative relation algebra that is not associative. Let $\mathfrak{K}_1 = \langle K, R, {}^*, \{0\}\rangle$, where $K = \{0, a, b, c\}$, $R \subseteq K^3$, $\langle x, y, z\rangle \in R$ iff $z \in \{x\};\{y\}$, and $x^* = x$ for all $x, y, z \in K$, and ; is defined in Table 3.9. Then $\mathfrak{Cm}(\mathfrak{K}_1) \cong \mathfrak{E}_4(\{1, 3\}) \in \mathsf{SA}$ by (Maddux 1982, Thm 2.5(4)(a)). Obviously $\mathfrak{K}_1$ satisfies (3.49) so it also satisfies (3.48) (see Lemma 3.1). It is also easy to check directly from the table that $\mathfrak{K}_1$ satisfies (3.47) and (3.48). Therefore, $\mathfrak{Cm}(\mathfrak{K}_1)$ is dense, commutative, and symmetric by Theorem 3.10. For each $i \in \{1, 2, 3, 4, 5, 6\}$ suppose $h_i : \mathfrak{P} \to \mathfrak{Cm}(\mathfrak{K}_1)$ is a homomorphism such that

Table 3.9 $\mathfrak{K}_1$ is a CR*-frame whose complex algebra is a dense symmetric semi-associative relation algebra that is not associative

$\mathfrak{K}_1 =$

;	$\{0\}$	$\{a\}$	$\{b\}$	$\{c\}$
$\{0\}$	$\{0\}$	$\{a\}$	$\{b\}$	$\{c\}$
$\{a\}$	$\{a\}$	$\{0, a\}$	$\{c\}$	$\{b\}$
$\{b\}$	$\{b\}$	$\{c\}$	$\{0, b\}$	$\{a\}$
$\{c\}$	$\{c\}$	$\{b\}$	$\{a\}$	$\{0, c\}$

$$
\begin{aligned}
h_1(A) &= \{a\}, & h_1(B) &= \{b\}, & h_1(C) &= \{0, b\},\\
h_2(A) &= \{a\}, & h_2(B) &= \{a, b\}, & h_2(C) &= \{0, a, b\},\\
h_3(A) &= \{a\}, & h_3(B) &= \{b\}, & h_3(C) &= \{a, b\},\\
h_4(A) &= \{a\}, & h_4(B) &= \{b\}, & h_4(C) &= \{a, c\},\\
h_5(A) &= \{a\}, & h_5(B) &= \{b\}, & h_5(C) &= \{b\},\\
h_6(A) &= \{a\}, & h_6(B) &= \{b, c\}, & h_6(C) &= \{b\}.
\end{aligned}
$$

The predicates are invalidated by the homomorphisms according to the following table. The symbol $\times$ means that the predicate named in the top row is invalidated by the homomorphism listed in the leftmost column. The symbol $\circ$ indicates that the predicate is validated by the homomorphism. These homomorphisms were calculated with GAP (2014).

	(3.127)	(3.128)	(3.129)	(3.130)	(3.131)	(3.132)	(3.140)	(3.141)	(3.136)	(3.137)	(3.142)
h_1	$\circ$	$\times$	$\times$	$\times$	$\circ$	$\circ$	$\times$	$\times$	$\circ$	$\times$	$\times$
h_2	$\circ$	$\times$	$\times$	$\circ$	$\circ$	$\circ$	$\times$	$\circ$	$\times$	$\times$	$\times$
h_3	$\circ$	$\circ$	$\times$	$\circ$	$\circ$	$\times$	$\times$	$\circ$	$\circ$	$\times$	$\times$
h_4	$\circ$	$\circ$	$\circ$	$\times$	$\times$	$\times$	$\circ$	$\times$	$\circ$	$\times$	$\times$
h_5	$\times$	$\circ$	$\times$	$\times$	$\circ$	$\times$	$\circ$	$\circ$	$\circ$	$\times$	$\circ$
h_6	$\times$	$\circ$	$\circ$	$\times$	$\times$	$\times$	$\circ$	$\circ$	$\circ$	$\circ$	$\circ$

3.16 Predicates in R and TR that are Not ω-Provable From Density

The predicates that rely on commutativity cannot be proved without that assumption, even in the presence of density and infinitely many variables.

Theorem 3.24 *Predicates (3.138)–(3.141) are in* R *and* TR *but not in* CT_ω. *They are not ω-provable from density. They are invalid in a* CR**-frame whose complex algebra is a non-commutative dense representable relation algebra.*

Table 3.10 $\mathfrak{Cm}(\mathfrak{K}_2)$ is a non-commutative dense representable relation algebra

$\mathfrak{K}_2 =$

;	$\{0\}$	$\{a\}$	$\{b\}$	$\{b^*\}$
$\{0\}$	$\{0\}$	$\{a\}$	$\{b\}$	$\{b^*\}$
$\{a\}$	$\{a\}$	$\{0, a, b, b^*\}$	$\{a, b\}$	$\{a\}$
$\{b\}$	$\{b\}$	$\{a\}$	$\{b\}$	$\{0, a, b, b^*\}$
$\{b^*\}$	$\{b^*\}$	$\{a, b^*\}$	$\{0, b, b^*\}$	$\{b^*\}$

Table 3.11 Assignments of propositional variables to $\mathfrak{Cm}(\mathfrak{K}_2)$ that invalidate (3.138)–(3.141)

(3.138)		(3.139)		(3.140)			(3.141)		
A	B	A	B	A	B	C	A	B	C
$\{a\}$	$\{a\}$	$\{a\}$	$\{b\}$	$\{a\}$	$\{b\}$	$\{a\}$	$\{0\}$	$\{a\}$	$\{a\}$
$\{b\}$	$\{a\}$	$\{b\}$	$\{b^*\}$	$\{b^*\}$	$\{a\}$	$\{a\}$	$\{0\}$	$\{b\}$	$\{a\}$
		$\{b^*\}$	$\{a\}$				$\{b\}$	$\{a\}$	$\{a\}$
							$\{b\}$	$\{b\}$	$\{a\}$

Proof Let $\mathfrak{K}_2 = \langle K, R, ^*, \{0\}\rangle$ be the frame determined by $K = \{0, a, b, b^*\}$, $0^* = 0$, $a^* = a$, $(b^*)^* = b$, $R \subseteq K^3$, and $\langle x, y, z\rangle \in R$ iff $z \in \{x\};\{y\}$, where ; is specified in Table 3.10. The complex algebra of $\mathfrak{K}_2$ is relation algebra number 13_{37}; see (Maddux 1978, p. 437). $\mathfrak{K}_2$ satisfies the conditions (3.41), (3.42), (3.43), and (3.46), hence $\mathfrak{Cm}(\mathfrak{K}_2) \in \mathsf{RA}$ by Theorem 3.11(3). $\mathfrak{K}_2$ is also satisfies (3.47) and $\mathfrak{Cm}(\mathfrak{K}_2)$ is a dense relation algebra (see Sect. 3.6). On the other hand, $\mathfrak{Cm}(\mathfrak{K}_2)$ is not commutative and (3.48) fails in $\mathfrak{K}_2$. The predicates (3.138)–(3.141) are invalidated in many ways, but in rather few ways if the invalidating homomorphisms are required to map the predicates to the empty set and the propositional variables to singleton subsets of K. All such homomorphisms have been found by using GAP (2014) and are listed in Table 3.11 for predicates (3.138)–(3.141). If a homomorphism sends A, B, and C to the corresponding subsets listed in some line in the column of a given predicate, then it invalidates that predicate by mapping it to a set not containing 0.

To show that (3.138)–(3.141) are not in CT_ω it suffices, by taking Ψ to be the equations of density in Theorem 3.19(1), to show that $\mathfrak{Cm}(\mathfrak{K}_2)$ is isomorphic with a dense proper relation algebra. A **finite sequence** is a function f with domain $\mathrm{dom}(f) = \{1, \cdots, n\}$ for some finite non-zero $n \in \omega$. Let $\mathbb{Q}$ be the set of rational numbers. Let U be the set of finite sequences of rational numbers. Define a binary relation $B \subseteq U \times U$ for $f, g \in U$ by fBg (we say f is below g or f comes before g) iff for some finite $n > 0$, $\mathrm{dom}(f) = \{1, \cdots, n\} \subseteq \mathrm{dom}(g)$, $f_i = g_i$ for all $i < n$, and $f_n < g_n$. Let

$$\begin{aligned}
\sigma(0) &= \{\langle x, x\rangle : x \in U\},\\
\sigma(b^*) &= B,\\
\sigma(b) &= B^{\smile},\\
\sigma(a) &= (U \times U) \setminus (\sigma(0) \cup B \cup B^{\smile}),\\
\rho(X) &= \bigcup_{x \in X} \sigma(x) \text{ for all } X \subseteq K.
\end{aligned}$$

It can be checked that ρ is an isomorphism of $\mathfrak{Cm}(\mathfrak{K}_2)$ with a dense proper relation algebra. By Theorem 3.19(1), every predicate in CT_ω is valid in $\mathfrak{K}_2$. Since (3.138)–(3.141) are not valid in $\mathfrak{K}_2$, they are not in CT_ω.

It has been confirmed with GAP (2014) that self-distribution (3.142) is valid in $\mathfrak{K}_2$ but is not valid in the 5-element frame whose complex algebra is the dense non-commutative relation algebra 29_{83}; see (Maddux 1978, p. 448) for its multiplication table. If 29_{83} is representable (which seems extremely likely) then (3.142) can be added to the list of predicates that rely on commutativity and are not ω-provable from density alone.

3.17 Predicates in R and TR that are Not ω-Provable From Commutativity

The predicates that are 3-provable or 4-provable from density require that assumption and are not even ω-provable from commutativity.

Theorem 3.25 *Predicates (3.133)–(3.137) and (3.142) are in* R *and* TR *but not* CT_ω. *They are not ω-provable from commutativity. They are invalid in the* CR^**-frame of the 2-element group, whose complex algebra is a commutative representable relation algebra.*

Proof The presence of (3.133)–(3.137) and (3.142) in R is mentioned in many sources. Let $\mathfrak{K}_3$ be the CR^*-frame of the 2-element group, shown in Table 3.12. Note that $\mathfrak{K}_3$ does not satisfy (dense). For each $i \in \{1, 2, 3, 4, 5, 6\}$ suppose $h_i : \mathfrak{P} \to \mathfrak{Cm}(\mathfrak{K}_3)$ is a homomorphism such that

$$\begin{aligned}
h_1(A) = h_2(A) = h_3(A) = h_4(A) = h_5(A) = h_6(A) &= \{a\},\\
h_2(B) = h_5(B) &= \{a\},\\
h_3(B) = h_4(B) = h_6(B) &= \{0\},\\
h_5(C) = h_6(C) &= \{0\}.
\end{aligned}$$

Then $g(D) = \{a\}$ whenever $g = h_i$ for some $i \in \{1, 2, 3, 4, 5, 6\}$ and D is any one of (3.133), (3.134), (3.135), (3.136), (3.137), or (3.142). Since $0 \notin g(D)$ this shows the six predicates are invalid in the group frame. The complex algebra of any group is in

Table 3.12 $\mathfrak{K}_3$ is the frame of the 2-element group

$\mathfrak{K}_3 =$

;	{0}	{a}
{0}	{a}	{0}
{a}	{0}	{a}

RRA, an observation first made by J. C. C. McKinsey; see (Jónsson and Tarski 1952, Thm 5.10), (Maddux 2006, Thm 233). In this case the representation is quite simple. Map 0 to the identity relation on a 2-element set, and map a to the transposition that interchanges the two elements.

3.18 3-Provable Predicates Not in R or CR*

The frames characteristic for classical relevant logic CR^* need not satisfy any of the frame properties (3.40)–(3.42). This leads to the problem, solved in this section, of determining a predicate in the vocabulary of R that corresponds to these properties.

Theorem 3.26 *1. Predicates (3.101)–(3.102) are in* T_3 *but not* R. *Predicate (3.103) is in* CT_3 *but not* CR^*.
2. (3.101) is valid in a frame $\mathfrak{K}$ *satisfying (3.43) iff* $\mathfrak{K}$ *satisfies (3.41).*

Proof Predicates (3.101)–(3.103) are 3-provable by Theorem 3.22. The use of * in (3.103) puts it in CT_3.

It is straightforward to check that $\mathfrak{K}_4 = \langle K, R, {}^*, \{0\}\rangle$ in Table 3.13 is a CR^*-frame. Therefore, everything in CR^* is valid in $\mathfrak{K}_4$. The frame conditions (3.40), (3.41), and (3.42) all fail in $\mathfrak{K}_4$ because $\langle a, a, a^*\rangle \in R$ but $\langle a^*, a^*, a\rangle \notin R$. Axiom ($\mathrm{R}_9$) fails because $(\{a\};\{a\})^\smile = \{a, a^*\}^\smile = \{a, a^*\} \neq \{a^*\} = \{a^*\};\{a^*\} = \{a\}^\smile;\{a\}^\smile$. Thus, $\mathfrak{Cm}(\mathfrak{K}_4) \notin \mathsf{SA}$. If $h\colon \mathfrak{P} \to \mathfrak{Cm}(\mathfrak{K}_4)$ is homomorphism such that $h(A) = h(B) = \{a\}$ and $h(C) = h(D) = \{a^*\}$, then h sends (3.101) to the empty set, hence (3.101) is not valid in $\mathfrak{K}_4$ and is not in CR^*. The other predicates can be handled similarly.

Table 3.13 A CR^*-frame whose complex algebra is not a semi-associative relation algebra

$\mathfrak{K}_4 =$

;	{0}	{a}	{a*}
{0}	{0}	{a}	{a*}
{a}	{a}	{a, a*}	{0, a, a*}
{a*}	{a*}	{0, a, a*}	{a*}

x	x^*
0	0
a	a^*
a^*	a

For part (2), assume $\mathfrak{K}$ is a frame satisfying (3.43). We first prove that if $\mathfrak{K}$ also satisfies (3.41) then (3.101) is valid in $\mathfrak{K}$. Note that (3.101) is a implication $F \to G$ where $F = B \circ A \wedge C$ and $G = B \circ (A \wedge {\sim}D) \vee (B \wedge C \circ D) \circ A$. By (3.14), (3.15), (3.17), and (3.19),

$$F = A ; B \cdot C,$$
$$G = (A \cdot \overline{D^{\smile}}) ; B + A ; (B \cdot D ; C).$$

By definition, $F \to G$ is valid in $\mathfrak{K}_4$ if the equation $1' \leq F \to G$ is true in $\mathfrak{Cm}\,(\mathfrak{K}_4)$, i.e., $h(1') \subseteq h(F \to G)$ for any homomorphism $h : \mathfrak{P} \to \mathfrak{Cm}\,(\mathfrak{K}_4)$. Any such homomorphism must send 1' to $\{0\}$, so $F \to G$ is valid in $\mathfrak{K}_4$ iff $0 \in h(F \to G)$ for every homomorphism $h : \mathfrak{P} \to \mathfrak{Cm}\,(\mathfrak{K})$. By (3.18), (3.31), and the homomorphism properties of h, $0 \in h(F \to G)$ iff

$$0 \in h\left(\overline{F^{\smile} ; \overline{G}}\right) = K \setminus \left(h(F)^{\smile} ; (K \setminus h(G))\right).$$

Writing this out according to (3.32), we get

for all $x, y \in K$, if $x \in h(F)^{\smile}$ and $y \notin h(G)$ then $\langle x, y, 0\rangle \notin R$.

By (3.33) this is equivalent to

for all $z, y \in K$, if $z \in h(F)$ and $\langle z^*, y, 0\rangle \in R$ then $y \in h(G)$.

By Lemma 3.4 (1), $\langle z^*, y, 0\rangle \in R$ iff $\langle z, 0, y\rangle \in R$, which is equivalent by (3.43) to $z = y$. Therefore, $0 \in h(F \to G)$ iff

$$\text{for all } z \in K, \text{ if } z \in h(F) \text{ then } z \in h(G), \text{ i.e., } h(F) \subseteq h(G). \tag{3.145}$$

We will prove (3.145). Assume $z \in h(F)$. Compute

$$h(F) = h(A ; B \cdot C) = h(A) ; h(B) \cap h(C),$$

so $z \in h(C)$ and $\langle x, y, z\rangle \in R$ for some $x \in h(A)$ and $y \in h(B)$ by (3.32). There are two cases. First assume $x \in h(\overline{D^{\smile}})$. Then

$$x \in h(A) \cap h(\overline{D^{\smile}}) = h(A \cdot \overline{D^{\smile}}).$$

From this, $\langle x, y, z\rangle \in R$, and $y \in h(B)$ we get $z \in h((A \cdot \overline{D^{\smile}}) ; B)$ by (3.32). By (3.30), $h(G)$ is the union of this last set with $G(A ; (B \cdot D ; C))$, so $z \in h(G)$, as desired.

For the second case, assume $x \notin h(\overline{D^{\smile}})$. Then $x \in K \setminus h(D)^{\smile}$ by (3.31), hence $x \in h(D)^{\smile}$. By (3.33), $x = w^*$ for some $w \in h(D)$. Then $\langle w^*, y, z\rangle \in R$ since $\langle x, y, z\rangle \in R$, hence $\langle w, z, y\rangle \in R$ by (3.43), (3.41), and Lemma 3.4 (1). Therefore,

$y \in h(D;C)$ by $w \in h(D)$, $z \in h(C)$, and (3.32). From this and $y \in h(B)$ we get $y \in h(B \cdot D;C)$, hence by $x \in h(A)$ and $\langle x, y, z\rangle \in R$, we have $z \in h(A;(B \cdot D;C))$. But $h(G)$ is the union of this set with $h((A \cdot \overline{D^{\smile}}); B)$, so $z \in h(G)$, as desired.

Assume we have a frame $\mathfrak{K} = \langle K, R, {}^*, \mathcal{I}\rangle$ satisfying (3.43) in which (3.41) fails because there are $x, y, z \in K$ such that $\langle x, y, z\rangle \in R$ and $\langle x^*, z, y\rangle \notin R$. We will show $1' \leq F \to G$ is not true in $\mathfrak{Cm}(\mathfrak{K})$. Suppose $h\colon \mathfrak{P} \to \mathfrak{Cm}(\mathfrak{K})$ a homomorphism such that

$$h(A) = \{x\}, \qquad h(B) = \{y\}, \qquad h(C) = \{z\}, \qquad h(D) = \{x^*\}.$$

Since $\langle x, y, z\rangle \in R$ we have $h(C) = \{z\} \subseteq \{x\};\{y\} = h(A);h(B) = h(\overline{A;B})$, so $h(F) = h(A;B \cdot C) = h(A;B) \cap h(C) = h(C) = \{z\}$. We also have $h(A \cdot \overline{D^{\smile}}) = h(A) \cap h(\overline{D^{\smile}}) = \{x\} \cap (K \setminus \{x\}) = \emptyset$, hence, independently of the value for B,

$$h((A \cdot \overline{D^{\smile}}); B) = h(A \cdot \overline{D^{\smile}}); h(B) = \emptyset; h(B) = \emptyset. \tag{3.146}$$

Furthermore, $y \notin \{x^*\};\{z\}$ since $\langle x^*, z, y\rangle \notin R$, hence $h(D;C) = h(D);h(C) = \{x^*\};\{z\} \subseteq K \setminus \{y\}$. Consequently, $h(B \cdot D;C) = h(B) \cap h(D;C) \subseteq \{y\} \cap (K \setminus \{y\}) = \emptyset$. We therefore have

$$h(A;(B \cdot D;C)) = h(A);h(B \cdot D;C) = \{x\};\emptyset = \emptyset. \tag{3.147}$$

From (3.146) and (3.147) we have $h(G) = h((A \cdot \overline{D^{\smile}}); B) \cup h(A;(B \cdot D;C)) = \emptyset$. The inclusion $\{z\} = h(F) \subseteq h(G) = \emptyset$ is false, contradicting (3.145), so $1' \leq F \to G$ is not true in $\mathfrak{Cm}(\mathfrak{K})$ and $F \to G$ is not valid in $\mathfrak{K}$. The contrapositive of what we have just proved is that if $F \to G$ is valid in a frame $\mathfrak{K}$ satisfying (3.43) then $\mathfrak{K}$ satisfies (3.41).

3.19 Counterexample to a Theorem of Kowalski

According to Kowalski (2013, Thm 8.1), R is "complete with respect to square-increasing [dense], commutative, integral relation algebras." On the contrary, R does *not* contain all the formulas true in this class of algebras. By Theorem 3.26, (3.101) is not a theorem of R but it is in T_3 and is therefore true in all semi-associative relation algebras, including all dense commutative integral relation algebras. Thus, (3.101) is a counterexample to (Kowalski 2013, Thm 8.1), which was obtained as an immediate consequence of (Kowalski 2013, Thm 7.1), that every normal De Morgan monoid is embeddable in a dense commutative integral relation algebra. However, the complex algebra of $\mathfrak{K}_4$ is a counterexample because $\mathfrak{K}_4$ fails to satisfy (3.41) and it would have to satisfy (3.41) if $\mathfrak{Cm}(\mathfrak{K}_4)$ were embeddable in a relation algebra. The difficulty seems to arise in the proof of (Kowalski 2013, Lemma 5.4(1)).

3.20 5-Provable Predicates Not in R or CT_4

This section presents two examples (3.148) and (3.149) of predicates that are 5-provable but not 4-provable. By creating infinitely many such predicates, Mikulás (2009) proved that T_ω is not finitely axiomatizable.

The original form of (3.148) was a logically valid sentence C2, due to Lyndon (1950, p. 712), that could not be derived from Tarski's axioms for the calculus of relations. Lyndon's sentence C2 was recast as the equation (L) (Maddux 2006, p. 30) by Chin and Tarski (1951, p. 354). The equation (L) expresses Desargues Theorem when the propositional variables denote points in a projective geometry and $A;B$ is the set of points on the line passing through points A and B. Predicate (3.148) was obtained from the equation (L) by reformulating it as a predicate that uses only the relevance logic operators $\wedge$, $\sim$, $\rightarrow$, and $\circ$ in the form ;. Predicate (3.148) is a consequence of (L) and is not necessarily equivalent to (L). Predicate (3.149) arises in the same way from another sentence C3 that is also due to Lyndon (1950, p. 712).

Theorem 3.27 (Maddux 2010, Thms 8.1, 8.2) *Predicates (3.148) and (3.149) are in* T_5 *but not in* CT_4.

$$
\begin{aligned}
&A;B \wedge C;D \wedge E;F \rightarrow \\
&\big((A \wedge {\sim}A);B \wedge C;D \wedge E;F\big) \vee \big(A;B \wedge C;(D \wedge {\sim}D) \wedge E;F\big) \\
&\vee \big(A;B \wedge C;D \wedge (E \wedge {\sim}E);F\big) \vee \big(A;B \wedge C;D \wedge E;(F \wedge {\sim}F)\big) \\
&\vee A;\big(A;C \wedge B;D \wedge (A;E \wedge B;F);(E;C \wedge F;D)\big);D
\end{aligned}
\tag{3.148}
$$

$$
\begin{aligned}
&A \wedge (B \wedge C;D);(E \wedge F;G) \rightarrow \\
&\big(A \wedge (B \wedge (C \wedge {\sim}C);D);(E \wedge F;G)\big) \\
&\vee \big(A \wedge (B \wedge C;D);(E \wedge F;(G \wedge {\sim}G))\big) \\
&\vee C;\big((C;A \wedge D;E);G \wedge D;F \wedge C;(A;G \wedge B;F)\big);G
\end{aligned}
\tag{3.149}
$$

Proof (3.148) and (3.149) both have the form $H \rightarrow J$. By Maddux (2010, Thm 8.1), if A, B, C, D, E, F, G are binary relations on U and the operations in H and J are interpreted according to Table 3.1 then $H \rightarrow J$ is a binary relation that contains the identity relation on U, or, equivalently, $H \subseteq J$. In both cases a straightforward set-theoretical proof of this fact refers to five elements of U. Two elements are assumed to be in the relation H and there are three more elements corresponding to the occurrences of ; in the predicate H. The proof consists of deducing facts expressed by J about the five elements from the assumption that the five elements are related to each other in ways described by H. The set-theoretical proofs can be written up as 5-proofs in the sequent calculus that are more elaborate but very similar to the 3-proof of (3.101) or the 4-proofs of (3.131) and (3.132). It follows that (3.148) and (3.149) are in T_5 by Theorem 3.21.

Table 3.14 The smallest KR-frame that invalidates (3.148) and (3.149)

$\mathfrak{K}_5 =$

;	{1'}	$\{a\}$	$\{b\}$	$\{c\}$
{1'}	{1'}	$\{a\}$	$\{b\}$	$\{c\}$
$\{a\}$	$\{a\}$	$\{1', a, c\}$	$\{b, c\}$	$\{a, b\}$
$\{b\}$	$\{b\}$	$\{b, c\}$	$\{1', a, b\}$	$\{a, c\}$
$\{c\}$	$\{c\}$	$\{a, b\}$	$\{a, c\}$	$\{1', b, c\}$

To show (3.148) and (3.149) are not in CT_4 let $\mathfrak{K}_5 = \langle K, R, ^*, \{1'\}\rangle$ where $K = \{0, a, b, c\}$, $x^* = x$ for every $x \in K$, and R is determined in Table 3.14. This table appears as (Maddux 2010, Table 6). Recall that $\langle x, y, z\rangle \in R$ iff $z \in \{x\};\{y\}$ where ; is defined by (3.32). Then $\mathfrak{K}_5$ is a KR-frame in which both (3.148) and (3.149) fail if $h : \mathfrak{P} \to \mathfrak{Cm}(\mathfrak{K}_5)$ is a homomorphism such that $\{a\} = h(A) = h(B) = h(E) = h(G)$, $\{c\} = h(C) = h(F)$, and $\{b\} = h(D)$. Such homomorphisms exist if we pick $1' \neq A, B, C, D, E, F, G \in \Pi$ such that $C, D, F \notin \{A, B, E, G\}$, and $D \notin \{C, F\}$.

The proof shows there are instances of (3.148) and (3.149) with only three distinct propositional variables that fail to be in CT_4. To get them, let $A = B = E = G \neq C = F \neq D$.

There are 14 KR-frames with four elements. $\mathfrak{K}_5$ is the unique 4-element KR-frame that invalidates both (3.148) and (3.149). It has 28 triples in its ternary relation R. Its complex algebra is the relation algebra 42_{65} in (Maddux 2006). There are exactly two other 4-element KR-frames that invalidate (3.148). However, (3.149) is valid in both of them. Their complex algebras are the relation algebras 36_{65} and 50_{65}. Among the 390 KR-frames with five elements, the number of them that invalidate both (3.148) and (3.149) is 58. The smallest two, both with 41 triples, are 118_{3013} and 200_{3013}.

3.21 Counting Characteristic TR-Frames and KR-Frames

Alasdair Urquhart wrote (Anderson et al. 1992, p. 349),

> The list of small models was enormously extended by a computer search using some remarkable programs written by Slaney, Meyer, Pritchard, Abraham, and Thistlewaite ... These programs churned out huge quantities of R matrices and model structures of all shapes and sizes. Clearly, there are lots and lots of R model structures out there! But what are they like? Can we classify them in some intelligent fashion? Are there general constructions that produce interesting examples? The answer to the first two questions is still obscure, though clearer than it was. The answer to the last question is an emphatic "yes!".

He continued (Anderson et al. 1992, p. 350),

> The first indication that KR is indeed nontrivial came from the computer, which churned out reams of interesting KR matrices. In retrospect, this is hardly surprising, because we now know that KR models can be manufactured *ad lib* from projective geometries.

These thoughts were on his mind when I first met Alasdair Urquhart, at a mathematical meeting in the late 1970s. At that time, I had constructed many finite symmetric integral relation algebras by hand and I described to him how easy it was to create them. From a chance encounter with Routley and Meyer (1973) while looking up the proof by Henkin (1973) of Tarski's theorem that the proof of the associative law (R_4) requires four variables, I knew that finite relation algebras share several properties with relevant model structures. I mentioned this and the ease with which they could be constructed to Alasdair. He said that explained why so many relevant model structures were being generated by his students, colleagues, and computers, as documented by his remarks quoted above. My own subsequent computer investigations showed that there are 14 KR-frames with four elements and 390 KR-frames with five elements. (The number of "R matrices and model structures" is much larger.) These large numbers illustrate a pattern that continues as the number of elements increases.

Alasdair pioneered and exploited the connection between projective geometries and KR-frames. This fruitful and crucial connection does provide a general construction that produces interesting examples, but does not account for the large number of KR-frames. One can create KR-frames by a random process, as will be shown in this section. Alasdair's astute judgement on his first two questions, "But what are they like? Can we classify them in some intelligent fashion?", that the answer is "still obscure" is confirmed here by Theorem 3.28 (9). The number of KR-frames on n elements grows like 2 raised to the power of a cubic polynomial in n. KR-frames are roughly as numerous as ternary relations and more numerous than graphs, whose number rises only like 2 raised to the power of a quadratic polynomial. In fact, a KR-frame can be constructed from an arbitrary graph. By contrast, projective planes are known only for those orders that are powers of primes.

By Theorem 3.11(1) and Lemmas 3.3 and 3.4 (1), the complex algebra of a frame is in NA iff it satisfies (3.38)–(3.44). Therefore, any frame satisfying those seven conditions is called an **NA-frame**. An NA-frame is **associative** if it satisfies (3.46), **dense** if it satisfies (3.47), **commutative** if it satisfies (3.48), and **symmetric** if it satisfies (3.49). A **TR-frame** is an associative dense commutative NA-frame. By Theorem 3.17(3) and Lemmas 3.3 and 3.4 (1), the TR-frames are characteristic for TR. By Theorem 3.14 (2), Theorem 3.11(3), Theorem 3.10(1)(3), Lemma 3.4 (1), and Lemma 3.3 (2), a frame is a KR-frame iff it is an associative dense symmetric NA-frame. By the definition of KR, the KR-frames are characteristic for KR.

An asymptotic formula for the number of isomorphism types of TR-frames and KR-frames can be computed by adapting the proof of Maddux (1985, Thm 12). The first step is to count the number of commutative NA-frames on a given finite set K with fixed involution $^*\colon K \to K$ and $\mathcal{I} = \{0\}$. After that one observes that a randomly chosen dense commutative NA-frame has a probability approaching 1 (as the number of elements increases) of being associative and having very few automorphisms. In fact, for any fixed dimension $3 \leq d \in \omega$, the probability that every d-provable predicate is valid in a random dense commutative NA-frame also approaches 1. In the following theorem, a function $N(n, s)$ is said to be **asymptotic**

to another function $M(n,s)$ if for every real number $r > 0$ there is some $n_0 \in \omega$ such that if $n_0 < n \in \omega$ and $1 \le s \le n$ then $|1 - N(n,s)/M(n,s)| < r$.

Theorem 3.28 *Assume $n = |K| \in \omega$, $0 \in K$, $\mathcal{I} = \{0\}$, ${}^*\colon K \to K$, $0^* = 0$, $x^{**} = x$ for all $x \in K$, and $s = |\{x : x^* = x \in K\}|$. For every ternary relation $R \subseteq K^3$ let $\mathfrak{K}(R) = \langle K, R, {}^*, \mathcal{I}\rangle$. Let*

$$\begin{aligned} F(n,s) &= \tfrac{1}{6}(s-1)s(s+1) + \tfrac{1}{12}(n-s)(n-s+1)(n-s+2) \\ &\quad + \tfrac{1}{4}(s-1)(n-s)(n+2), \\ G(n,s) &= \tfrac{1}{6}(s-1)(s-2)(s+3) + \tfrac{1}{12}(n-s)(n-s-1)(n-s+4) \\ &\quad + \tfrac{1}{4}(s-1)(n-s)(n+2), \\ P(n,s) &= (s-1)!\left(\tfrac{1}{2}(n-s)\right)!2^{\frac{1}{2}(n-s)}. \end{aligned}$$

1. *The numbers of relations $R \subseteq K^3$ for which $\mathfrak{K}(R)$ is a commutative, symmetric, dense commutative, or dense symmetric* NA*-frame are*

$$\begin{aligned} 2^{F(n,s)} &= |\{R : \mathfrak{K}(R) \text{ is a commutative NA-frame}\}|, \\ 2^{\frac{1}{6}(n-1)n(n+1)} &= |\{R : \mathfrak{K}(R) \text{ is a symmetric NA-frame}\}|, \\ 2^{G(n,s)} &= |\{R : \mathfrak{K}(R) \text{ is a dense commutative NA-frame}\}|, \\ 2^{\frac{1}{6}(n-1)(n-2)(n+3)} &= |\{R : \mathfrak{K}(R) \text{ is a dense symmetric NA-frame}\}|. \end{aligned}$$

2. *$P(n,s)$ is the number of automorphisms of $\langle K, {}^*, \{0\}\rangle$.*
3. *The number of isomorphism types of commutative* NA*-frames with n elements and $s \le n-2$ symmetric elements is asymptotic to*

$$\frac{1}{P(n,s)-1} \cdot 2^{F(n,s)}.$$

4. *The number of isomorphism types of symmetric* NA*-frames with n elements is asymptotic to*

$$\frac{1}{(n-1)!} \cdot 2^{\frac{1}{6}(n-1)n(n+1)}.$$

5. *The number of isomorphism types of dense commutative* NA*-frames with n elements and $s \le n-2$ symmetric elements is asymptotic to*

$$\frac{1}{P(n,s)-1} \cdot 2^{G(n,s)}.$$

6. *The number of isomorphism types of dense symmetric* NA*-frames with n elements is asymptotic to*

$$\frac{1}{(n-1)!} \cdot 2^{\frac{1}{6}(n-1)(n-2)(n+3)}.$$

7. *For every dimension* $3 \le d < \omega$, *the probability approaches* 1 *as* $n \to \infty$ *that the complex algebra of a randomly chosen commutative, dense commutative, symmetric, or dense symmetric* NA*-frame with* n *elements is in* RA_d.
8. *The number of isomorphism types of* TR*-frames with* n *elements and* $s \le n-2$ *symmetric elements is asymptotic to*

$$\frac{1}{P(n,s)-1} \cdot 2^{G(n,s)}.$$

9. *The number of isomorphism types of* KR*-frames with* n *elements is asymptotic to*

$$\frac{1}{(n-1)!} \cdot 2^{\frac{1}{6}(n-1)(n-2)(n+3)}.$$

10. *For every dimension* $3 \le d < \omega$, *the number of isomorphism types of* TR*-frames with* n *elements and* $s \le n-2$ *symmetric elements whose complex algebras are in* RA_d *and in which every* d*-provable predicate is valid is asymptotic to*

$$\frac{1}{P(n,s)-1} \cdot 2^{G(n,s)}.$$

11. *For every dimension* $3 \le d < \omega$, *the number of isomorphism types of* KR*-frames with* n *elements whose complex algebras are in* RA_d *and in which every* d*-provable predicate is valid is asymptotic to*

$$\frac{1}{(n-1)!} \cdot 2^{\frac{1}{6}(n-1)(n-2)(n+3)}.$$

Proof (1) We will count the number of commutative NA-frames $\mathfrak{K}(R) = \langle K, R, {}^*, \mathcal{I}\rangle$ on a given finite set K with fixed involution ${}^*\colon K \to K$ and $\mathcal{I} = \{0\} \subseteq K$. Consider an arbitrary $R \subseteq K^3$. If $\mathfrak{K}(R)$ satisfies (3.43) then $R_0 \subseteq R$ where

$$R_0 = \bigcup_{x \in K} \{\langle 0, x, x\rangle, \langle x, x^*, 0\rangle, \langle x^*, 0, x^*\rangle, \\ \langle x, 0, x\rangle, \langle x^*, x, 0\rangle, \langle 0, x^*, x^*\rangle\} \cup \{\langle 0, 0, 0\rangle\}.$$

R cannot contain any other triples with 0 in them, lest (3.43) be falsified. To get a commutative NA-frame we must consider only those relations $R \subseteq K^3$ that include R_0 and have no other triples in them that contain 0.

For all $x, y, z \in K$ let $C(x, y, z)$ be the smallest set of triples in K^3 containing $\langle x, y, z\rangle$ such that $\mathfrak{K}(C(x, y, z) \cup R_0)$ is a commutative NA-frame. Such a set is called a **cycle**. The isomorphism types of cycles are listed Table 3.15. For each isomorphism type of cycle, the number of triples in it and the number of cycles of that type are listed in Table 3.16.

If $\mathfrak{K}(R \cup R_0)$ is a commutative NA-frame then R must be the union of cycles. Therefore, to create such a relation one must choose to include or exclude each cycle. The number of choices available for each isomorphism type of cycle occurs in the rightmost column of Table 3.16. For example, there are $s-1$ cycles of the form

Table 3.15 Cycles $C(\text{-},\text{-},\text{-}) \subseteq K^3$ with $p, q, r, a, b, c \in K$, $p^* \neq p$, $q^* \neq q$, $r^* \neq r$, $a^* = a$, $b^* = b$, $c^* = c$, such that $\mathfrak{K}(C(\text{-},\text{-},\text{-}) \cup R_0)$ is a commutative NA-frame

$$C(a,a,a) = \{\langle a,a,a\rangle\}$$
$$C(a,b,b) = \{\langle a,b,b\rangle, \langle b,a,b\rangle, \langle b,b,a\rangle\}$$
$$C(a,b,c) = \{\langle a,b,c\rangle, \langle a,c,b\rangle, \langle b,a,c\rangle, \langle b,c,a\rangle, \langle c,a,b\rangle, \langle c,b,a\rangle\}$$
$$C(p,a,p) = \{\langle a,p,p\rangle, \langle a,p^*,p^*\rangle, \langle p,a,p\rangle, \langle p,p^*,a\rangle, \langle p^*,a,p^*\rangle, \langle p^*,p,a\rangle\}$$
$$C(p,p,a) = \{\langle a,p,p^*\rangle, \langle a,p^*,p\rangle, \langle p,a,p^*\rangle, \langle p,p,a\rangle, \langle p^*,a,p\rangle, \langle p^*,p^*,a\rangle\}$$
$$C(p,a,a) = \{\langle a,a,p\rangle, \langle a,a,p^*\rangle, \langle a,p,a\rangle, \langle a,p^*,a\rangle, \langle p,a,a\rangle, \langle p^*,a,a\rangle\}$$
$$C(p,a,b) = \{\langle a,b,p\rangle, \langle a,b,p^*\rangle, \langle a,p,b\rangle, \langle a,p^*,b\rangle, \langle b,a,p\rangle, \langle b,a,p^*\rangle, \langle b,p,a\rangle, \langle b,p^*,a\rangle, \langle p,a,b\rangle, \langle p,b,a\rangle, \langle p^*,a,b\rangle, \langle p^*,b,a\rangle\}$$
$$C(p,q,a) = \{\langle a,p,q^*\rangle, \langle a,p^*,q\rangle, \langle a,q,p^*\rangle, \langle a,q^*,p\rangle, \langle p,a,q^*\rangle, \langle p,q,a\rangle, \langle p^*,a,q\rangle, \langle p^*,q^*,a\rangle, \langle q,a,p^*\rangle, \langle q,p,a\rangle, \langle q^*,a,p\rangle, \langle q^*,p^*,a\rangle\}$$
$$C(p,p,p) = \{\langle p,p,p\rangle, \langle p,p^*,p\rangle, \langle p,p^*,p^*\rangle, \langle p^*,p,p\rangle, \langle p^*,p,p^*\rangle, \langle p^*,p^*,p^*\rangle\}$$
$$C(p,p,p^*) = \{\langle p,p,p^*\rangle, \langle p^*,p^*,p\rangle\}$$
$$C(p,p,q) = \{\langle p,p,q\rangle, \langle p,q^*,p^*\rangle, \langle p^*,p^*,q^*\rangle, \langle p^*,q,p\rangle, \langle q,p^*,p\rangle, \langle q^*,p,p^*\rangle\}$$
$$C(p,q,p) = \{\langle p,p^*,q\rangle, \langle p,p^*,q^*\rangle, \langle p,q,p\rangle, \langle p,q^*,p\rangle, \langle p^*,p,q\rangle, \langle p^*,p,q^*\rangle, \langle p^*,q,p^*\rangle, \langle p^*,q^*,p^*\rangle, \langle q,p,p\rangle, \langle q,p^*,p^*\rangle, \langle q^*,p,p\rangle, \langle q^*,p^*,p^*\rangle\}$$
$$C(p,q,r) = \{\langle p,q,r\rangle, \langle p,r^*,q^*\rangle, \langle p^*,q^*,r^*\rangle, \langle p^*,r,q\rangle, \langle q,p,r\rangle, \langle q,r^*,p^*\rangle, \langle q^*,p^*,r^*\rangle, \langle q^*,r,p\rangle, \langle r,p^*,q\rangle, \langle r,q^*,p\rangle$$

$C(a,a,a)$ with $a^* = a$ from which to choose. Since 0 is a symmetric element there are only $s-1$ other symmetric elements that create cycles of the form $\{\langle a,a,a\rangle\}$ as available choices.

To obtain a commutative NA-frame, the total number of available choices is obtained by adding all the numbers in the rightmost column that are labelled "yes" in the column headed with "Comm?", i.e., do you want a commutative NA-frame? If so, add the number of choices in the rightmost column. If you want the frame to also be dense, then add or do not add the number in the rightmost column according to the entry in the column headed "and dense?". Two of those entries are "no" since all triples in the cycles $C(a,a,a)$ and $C(x,x,x)$ will necessarily be included in the desired R and are therefore not available as choices for inclusion or exclusion from R.

$F(n,s)$ is the sum of all the numbers in the last column of Table 3.16. This accounts for the first equation in part (1). $G(n,s)$ is the sum of all the numbers in the last column of Table 3.16 that occur in rows whose entry under "and dense?" is "yes". This accounts for the third equation in part (1). Note that $s = n$ holds iff the resulting frame is symmetric. The second and fourth equations are therefore obtained by setting $s = n$ in the first and third equations. This completes the proof of part (1).

Table 3.16 The sizes and numbers of cycles $C(\text{-},\text{-},\text{-})$ on an n-element set with s symmetric elements, where $p^* \neq p, q^* \neq q, r^* \neq r, a^* = a, b^* = b, c^* = c$.
Types 4–13 disappear in the symmetric case $s = n$ and type 3 predominates.
Types 1–8 disappear in the non-symmetric case $s = 1$ and type 13 predominates

No.	Relation type	Size	Comm?	and dense?	The number of relations to choose from
1	$C(a,a,a)$	1	yes	no	$s-1$
2	$C(a,b,b)$	3	yes	yes	$(s-1)(s-2)$
3	$C(a,b,c)$	6	yes	yes	$(s-1)(s-2)(s-3)/6$
4	$C(p,a,p)$	6	yes	yes	$(s-1)(n-s)/2$
5	$C(p,p,a)$	6	yes	yes	$(s-1)(n-s)/2$
6	$C(p,a,a)$	6	yes	yes	$(s-1)(n-s)/2$
7	$C(p,a,b)$	12	yes	yes	$(s-1)(s-2)(n-s)/4$
8	$C(p,q,a)$	12	yes	yes	$(s-1)(n-s)(n-s-2)/4$
9	$C(p,p,p)$	6	yes	no	$(n-s)/2$
10	$C(p,p,p^*)$	2	yes	yes	$(n-s)/2$
11	$C(p,p,q)$	6	yes	yes	$(n-s)(n-s-2)/2$
12	$C(p,q,p)$	12	yes	yes	$(n-s)(n-s-2)/4$
13	$C(p,q,r)$	12	yes	yes	$(n-s)(n-s-2)(n-s-4)/12$

Proof (2) Let $\mathsf{Perm}(K)$ be the set of permutations of K. Let ι be the identity permutation on K, i.e., $\iota(x) = x$ for every $x \in K$. Let Aut be the set of automorphisms of $\langle K, {}^*, \mathcal{I}\rangle$.

$$\mathsf{Aut} = \{\sigma : \sigma \in \mathsf{Perm}(K), \sigma(0) = 0, \sigma(x)^* = \sigma(x^*) \text{ for every } x \in K\}.$$

To see that $|\mathsf{Aut}| = P(n,s)$, note first that the symmetric elements distinct from 0 can be arbitrarily permuted, which accounts for the term $(s-1)!$. The pairs of the form $\langle x, x^*\rangle$ can also be arbitrarily permuted, accounting for the term $\left(\frac{1}{2}(n-s)\right)!$. When sending the pair $\langle x, x^*\rangle$ to the pair $\langle y, y^*\rangle$, an automorphism can either send x to y and x^* to y^*, or the other way around. For each pair there are two ways to send it to its target pair, so the total number of ways of doing this is $2^{\frac{1}{2}(n-s)}$. $P(n,s)$ is the product of these three terms.

Proof (3–6) Partition Aut into three sets, Aut_1, Aut_2, and $\{\iota, {}^*\}$ by setting

$$\mathsf{Aut}_1 = \{\sigma : \sigma \in \mathsf{Aut}, x \neq \sigma(x) \neq x^* \text{ for some } x \in K\},$$
$$\mathsf{Aut}_2 = \mathsf{Aut} \setminus \left(\mathsf{Aut}_1 \cup \{\iota, {}^*\}\right).$$

The automorphisms in $\{\iota, {}^*\}$ are called **trivial** and the ones in $\mathsf{Aut}_1 \cup \mathsf{Aut}_2$ are called **non-trivial**. Recall that $n = |K|$ and $s = |\{x : x^* = x \in K\}|$. Then $1 \leq s$ because

$0^* = 0 \in K$, and $\mathfrak{K}(R \cup R_0)$ is symmetric iff $s = n$. If $s = n$ then $^* = \iota$, $\mathsf{Aut} = \mathsf{Perm}(K)$, $\mathsf{Aut}_1 = \mathsf{Aut} \setminus \{^*\}$, $\mathsf{Aut}_2 = \emptyset$, $|\mathsf{Aut}_1| = P(n, s) - 1$, and $|\mathsf{Aut}_2| = 0$.

Suppose $s < n$ and $\sigma \in \mathsf{Aut}_2$. Since $\sigma \notin \mathsf{Aut}_1$, there is no $x \in K$ such that $x \neq \sigma(x) \neq x^*$. Therefore, if $\sigma(x) \neq x$ then $\sigma(x) = x^*$ and $\sigma(x^*) = (\sigma(x))^* = x^{**} = x$. This shows the pair $\{x, x^*\}$ is switched by σ whenever σ moves x, hence every $\sigma \in \mathsf{Aut}_2$ is obtained by composing at least one (since $\sigma \neq \iota$) but not all (since $\sigma \neq {}^*$) transpositions of the form (x, x^*) in cycle notation. The number of pairs $\{x, x^*\}$ is $\frac{1}{2}(n - s)$, so $|\mathsf{Aut}_1| = P(n, s) - 2^{\frac{1}{2}(n-s)}$ and $|\mathsf{Aut}_2| = 2^{\frac{1}{2}(n-s)} - 2$. Let

$$W = \{R : \mathfrak{K}(R \cup R_0) \text{ is a commutative } \mathsf{NA} - \text{frame}\}.$$

For any property φ, let $\Pr[\varphi]$ be the probability that $R \in W$ has property φ,

$$\Pr[\varphi] = 2^{-F(n,s)} |\{R : R \in W, R \text{ has property } \varphi\}|.$$

We will compute the probability that $\mathfrak{K}(R \cup R_0) = \langle K, R \cup R_0, {}^*, \mathcal{I} \rangle$ has a non-trivial automorphism. For every ternary relation $R \subseteq K^3$ let $\mathsf{Aut}_{1,2}(R)$ be the set of non-trivial automorphisms of $\mathfrak{K}(R \cup R_0)$,

$$\mathsf{Aut}_{1,2}(R) = \{\sigma : \sigma \in \mathsf{Aut}_1 \cup \mathsf{Aut}_2, R = \{\langle \sigma(x), \sigma(y), \sigma(z) \rangle : \langle x, y, z \rangle \in R\}\}.$$

For every non-trivial $\sigma \in \mathsf{Aut}_1 \cup \mathsf{Aut}_2$ we will compute an upper bound on the number of ternary relations $R \in W$ such that $\sigma \in \mathsf{Aut}_{1,2}(R)$ and multiply by $|\mathsf{Aut}_1|$ or $|\mathsf{Aut}_2|$ to obtain an upper bound on the number of relations in W that have a non-trivial automorphism. With this upper bound we can show the probability of having a non-trivial automorphism approaches 0 as $n \to \infty$. Suppose $\sigma \in \mathsf{Aut}_1$. Then for some $x \in K$, $0 \neq x \neq \sigma(x) \neq x^*$. Let

$$X = \{0, x, x^*, \sigma(x), \sigma(x^*)\}.$$

Since $|K \setminus X| = 3$ if $x = x^*$ and $|K \setminus X| = 5$ if $x \neq x^*$, the number ways to choose one or two elements from $K \setminus X$ is at least $(n - 5) + \frac{1}{2}(n - 5)(n - 6) = \frac{1}{2}(n - 4)(n - 5)$. For each such choice $y, z \in K \setminus X$ we also have

$$\sigma(x) \notin Y = \{x, x^*, y, y^*, z, z^*\}. \tag{3.150}$$

To see this, note that we have assumed $\sigma(x) \neq x, x^*$. Both $\sigma(x) = y$ and $\sigma(x) = z$ violate the choice that $y, z \notin X$. Since σ is an automorphism, $\sigma(x) = y^*$ implies $\sigma(x^*) = \sigma(x)^* = y^{**} = y$, contradicting $y \notin X$, and similarly $\sigma(x) = z^*$ contradicts $z \notin X$. Suppose $C(x, y, z)$ is fixed by σ, that is, $C(x, y, z) = C(\sigma(x), \sigma(y), \sigma(z))$. Then σ must map Y onto itself because $C(x, y, z) \subseteq Y^3$ and every element of Y is in some triple in $C(x, y, z)$, as can be easily seen in Table 3.15. But this contradicts (3.150). Therefore, $C(x, y, z)$ is moved by σ. This proves that

$$\text{if } \sigma \in \mathsf{Aut}_1 \text{ then at least } \tfrac{1}{2}(n - 4)(n - 5) \text{ cycles are moved by } \sigma. \tag{3.151}$$

Suppose $\sigma \in \mathsf{Aut}_2$. This is possible only if $s \leq n-2$. The number of cycles moved by σ is

$$f(n,s,m) = \tfrac{1}{4}m(n-s-m)(n+s-2), \text{ where } m = |\{x : \sigma(x) \neq x\}|.$$

This computation was checked with GAP (2014). Note that m is always an even number, since moved elements come in pairs of the form $\{x, x^*\}$. Also, $2 \leq m < n-s$ because σ moves something but differs from *. Under these constraints the smallest non-zero value occurs when $m=2$ or $m=n-s-2$ and is $f(n,s,2) = f(n,s,n-s-2) = \frac{1}{2}((n-2)^2 - s^2)$. This proves that

$$\text{if } \sigma \in \mathsf{Aut}_2 \text{ then at least } \tfrac{1}{2}((n-2)^2 - s^2) \text{ cycles are moved by } \sigma. \tag{3.152}$$

For every $\sigma \in \mathsf{Aut}_1 \cup \mathsf{Aut}_2$, let M_σ be the number of cycles moved by σ. To make a relation $R \in W$ with $\sigma \in \mathsf{Aut}_{1,2}(R)$, one can freely choose to include or exclude each unmoved cycle in R. The cycles in each orbit under σ must be either all included in R or all excluded from R. If the moved cycles form a single orbit then one can only include or exclude the entire orbit, which has size M_σ. No cycles in that orbit are available as choices to include or exclude, so the number of available choices is $F(n,s) - M_\sigma + 1$. The number of orbits can vary from one orbit of size M_σ up to $M_\sigma/2$ orbits of size 2. The number of unmoved cycles is $F(n,s) - M_\sigma$. They are all free to be included in R or excluded. The moved elements offer somewhere between one orbit and $M_\sigma/2$ orbits as choices for inclusion. The number of choices ranges from at least $F(n,s) - M_\sigma + 1$ up to at most $F(n,s) - M_\sigma + M_\sigma/2 = F(n,s) - M_\sigma/2$. From (3.151) and (3.152) we know $((n-2)^2 - s^2)/4 \leq M_\sigma/2$ if $\sigma \in \mathsf{Aut}_2$ and $(n-4)(n-5)/4 \leq M_\sigma/2$ if $\sigma \in \mathsf{Aut}_1$, so the number of choices is at most $F(n,s) - ((n-2)^2 - s^2)/4$ if $\sigma \in \mathsf{Aut}_2$ and $F(n,s) - (n-4)(n-5)/4$ if $\sigma \in \mathsf{Aut}_1$. Thus,

$$|\{R : \sigma \in \mathsf{Aut}_{1,2}(R)\}| < \begin{cases} 2^{F(n,s)-\frac{1}{4}(n-4)(n-5)} & \text{if } \sigma \in \mathsf{Aut}_1, \\ 2^{F(n,s)-\frac{1}{2}((n-2)^2-s^2)} & \text{if } \sigma \in \mathsf{Aut}_2. \end{cases}$$

Since $|\mathsf{Aut}_1| < P(n,s)$ and $|\mathsf{Aut}_2| < 2^{\frac{1}{2}(n-s)}$, there are fewer than

$$P(n,s)2^{F(n,s)-\frac{1}{4}(n-4)(n-5)}$$

relations $R \in W$ such that $\mathsf{Aut}_{1,2}(R)$ is not empty, and if $s \leq n-2$ there are fewer than

$$2^{\frac{1}{2}(n-s)}2^{F(n,s)-\frac{1}{2}((n-2)^2-s^2)}$$

relations $R \in W$ such that $\mathsf{Aut}_{1,2}(R)$ is not empty. When $s \leq n-2$, our overestimate of the fraction of W that has a non-trivial automorphism in $\mathsf{Aut}_{1,2}(R)$ is obtained by adding these two numbers and dividing by $F(n,s)$.

$$E(n, s) = \frac{(s-1)!\left(\frac{1}{2}(n-s)\right)!2^{\frac{1}{2}(n-s)}}{2^{\frac{1}{4}(n-4)(n-5)}} + \frac{2^{\frac{1}{2}(n-s)}}{2^{\frac{1}{2}((n-2)^2-s^2)}}.$$

If $s = n$ then $E(n, s)$ is just the first term of this sum. In either case, a straightforward analysis of the growth rates for the numerators and denominators shows that $\lim_{n\to\infty} E(n, s) = 0$. Since $E(n, s)$ is the probability that $\mathfrak{K}(R \cup R_0)$ has a non-trivial automorphism, the probability that $\mathfrak{K}(R \cup R_0)$ has no non-trivial automorphisms approaches 1 as $n \to \infty$. Thus, a randomly selected $R \in W$ will almost certainly show up in $P(n, s) - 1$ ways in W if $s \leq n - 2$ and $P(n, s)$ ways in W if $s = n$. To estimate the number of isomorphism types of commutative NA-frames $\mathfrak{K}(R \cup R_0)$ we must therefore divide the total number $2^{F(n,s)}$ of such frames by $P(n, s) - 1$ if $s \leq n - 2$ and $P(n, n)$ if $s = n$. This gives us the approximations in parts (3) and (4). Parts (5) and (6) are proved in the same way, using $G(n, s)$ instead of $F(n, s)$ and redefining W as the set of $R \subseteq K^3$ such that $\mathfrak{K}(R \cup R_0)$ is a dense commutative NA-frame. The reasoning applies to both definitions of W.

Proof (7–11) For part (7), we will prove for any fixed dimension $3 \leq d < \omega$ that the probability approaches 1 as $n \to \infty$ that a randomly chosen $R \in W$ has a commutative NA-frame $\mathfrak{K}(R \cup R_0)$ whose complex algebra is in RA_d (because it almost certainly has a much stronger property) and hence, by Theorem 3.15 (6), every d-provable predicate is valid in $\mathfrak{K}(R \cup R_0)$. Part (7) together with parts (5) and (6) imply parts (10) and (11). Parts (8) and (9) follow from parts (10) and (11) by Theorem 3.15 (5). Indeed, associativity is equivalent to the validity of the 4-provable predicates (3.131) and (3.132) and is obtained with near certainty by taking $d = 4$.

For every integer $t \geq 1$, define the **diamond property D**(t) of relations $R \in W$ by

$$\mathbf{D}(t): \text{ if } 0 \neq x_1, \ldots, x_t, y_1, \ldots, y_t \in K \text{ then for some } 0 \neq z \in K, \quad (3.153)$$
$$\bigcup_{1\leq i\leq t} C(x_i, y_i, z) \subseteq R.$$

We will show for a fixed $1 \leq t \in \omega$ that $\mathbf{D}(t)$ almost certainly holds as $n \to \infty$. Consider one instance of $\mathbf{D}(t)$, say $0 \neq x_1, \ldots, x_t, y_1, \ldots, y_t \in K$. We want to show there is likely to be some z that works, where

$$z \textbf{ works} \text{ iff } \bigcup_{1\leq i\leq t} C(x_i, y_i, z) \subseteq R.$$

Note that z works only if all the cycles $C(x_i, y_i, z)$ are contained in R. Each cycle is contained with probability 1/2. If the cycles are disjoint their inclusions in R are independent events and the probability that z works is exactly 2^{-t}, but otherwise it is more, so z works with probability at least 2^{-t} and the probability that z does not work is at most $1 - 2^{-t}$. This is almost certain if t is large, but there are more and more z's as n increases. One of them is bound to work and only one is needed. We will calculate the probability that none of them work and see that it goes to zero. Let

$$K^{-} = K \setminus \Big(\{0\} \cup \bigcup_{1 \le i \le t} \{x_i, x_i^*, y_i, y_i^*\}\Big).$$

Then K^- is closed under * so we can partition K^- into pairs $\{z, z^*\}$ with $z \neq z^*$ and singletons $\{z\}$ for those z such that $z = z^*$. Let z_i, $1 \le i \le q$, be a selection of one element from each pair or singleton. There is at least one and at most $4t$ elements in $\bigcup_{1 \le i \le t}\{x_i, x_i^*, y_i, y_i^*\}$, so $n - 4t - 1 \le |K^-| \le n - 1$. The largest partition of K^- occurs when all its elements are symmetric and the smallest when all its elements are non-symmetric. Together these constraints put bounds on q, namely $(n - 4t - 1)/2 \le q \le n - 1$.

To see that distinct z_i's create independent events, suppose $1 \le j < k \le q$. The three sets $\{z_j, z_j^*\}$, $\{z_k, z_k^*\}$, and $\bigcup_{1 \le i \le t}\{x_i, x_i^*, y_i, y_i^*\}$ are disjoint, so $\bigcup_{1 \le i \le t} C(x_i, y_i, z_j)$ and $\bigcup_{1 \le i \le t} C(x_i, y_i, z_k)$ are also disjoint because every triple in the former set contains z_j or z_j^* but no triple in the latter set does so. Thus, the events that z_j works and z_k works are independent, as are their complements. The probability that z_j doesn't work for every $j \le q$ is the product of the probabilities that each z_j doesn't work. The probability that z_j doesn't work is at most $1 - 2^{-t}$, so the probability that none of them works is at most the product of q copies of $1 - 2^{-t}$, one for each z. The number q of z's goes to infinity with n since $(n - 4t - 1)/2 \le q$, so $(1 - 2^{-t})^q$ approaches 0 because $1 - 2^{-t} < 1$.

This means that any instance of $\mathbf{D}(t)$ will eventually hold, but we want to know they all hold. There are $(n - 1)^{2t}$ instances of $\mathbf{D}(t)$. The probability that some instance fails is no more than the sum over all instances of the probability that each instance fails. Each instance has a probability of failing that is at most $(1 - 2^{-t})^q$ with q depending on the instance. Since $1 - 2^{-t} < 1$, bigger exponents make a smaller product and these probabilities are largest when q is smallest. We have $(n - 4t - 1)/2 \le q$ for every instance, so $(1 - 2^{-t})^{(n-4t-1)/2}$ is an upper bound on the probability that any particular instance of $\mathbf{D}$(t) fails because no z works. The sum of these probabilities is therefore bounded above by the product of $(n - 1)^{2t}$, the number of instances, times the upper bound $(1 - 2^{-t})^{(n-4t-1)/2}$. Thus, the probability that $\mathbf{D}(t)$ fails is at most $(n - 1)^{2t}(1 - 2^{-t})^{(n-4t-1)/2}$. This bound goes to zero as $n \to \infty$ because it is a polynomial in n multiplied by a constant smaller than 1 raised to a power that is another polynomial in n. The probability of the complementary event, that $\mathbf{D}(t)$ holds, therefore approaches 1 as $n \to \infty$.

Suppose $3 \le d < \omega$. We have seen that the commutative NA-frame $\mathfrak{K}(R \cup R_0)$ of a randomly chosen $R \in W$ almost certainly satisfies $\mathbf{D}(d - 2)$. Even if it does not, its complex algebra $\mathfrak{Cm}\,(\mathfrak{K}(R \cup R_0))$ is atomic and is in NA. We will show, assuming $\mathbf{D}(d - 2)$ holds, that $\mathfrak{Cm}\,(\mathfrak{K}(R \cup R_0))$ is in RA_d because it has a d-dimensional relational basis. The atoms of the complex algebra are singletons of elements in K, so our d-dimensional relational basis B will consist of $d \times d$ matrices of singletons of elements of K. Let $R' = R \cup R_0$. Let B consist of those $x \in \{\{a\} : a \in K\}^{d\times d}$ such that for all $i, j, k < d$ and all $a, b, c \in K$, $x_{ii} = \{0\}$ and if $x_{ij} = \{a\}$, $x_{jk} = \{b\}$, and $x_{ik} = \{c\}$ then $\langle a, b, c\rangle \in R'$.

Let $x \in B$. To show B is a d-dimensional basis, we must verify conditions (1), (2), and (3) in Definition 3.1. By the definition of complex algebra, 1' $= \{0\}$, but we have $x_{ii} = \{0\}$ by the definition of B, so $x_{ii} \subseteq 1'$. Given $i, j\langle d$, suppose $x_{ij} = \{a\}$, $x_{ji} = \{b\}$, and $x_{ii} = \{c\}$. By the definition of B, we have $c = 0$ and $\langle a, b, c\rangle \in R'$, so $\langle a, b, 0\rangle \in R'$. Then $\langle stara, 0, b\rangle \in R'$ by (3.41), so $a^* = b$ by (3.43). But $\{a\}^{\smile} = \{a^*\}$ by (3.33), so $(x_{ij})^{\smile} = \{a\}^{\smile} = \{a^*\} = \{b\} = x_{ji}$, as desired. Given $i, j, k < d$, suppose $x_{ij} = \{a\}$, $x_{jk} = \{b\}$, and $x_{ik} = \{c\}$. By the definition of B, $\langle a, b, c\rangle \in R'$, so by (3.32), $\{a\};\{b\} \supseteq \{c\}$, i.e., $x_{ik} \subseteq x_{ij}\,;x_{jk}$, as desired. This completes the proof of part (1) in Definition 3.1.

For part (2), consider $a \in K$. We want $x \in B$ with $x_{01} = \{a\}$. It is enough to define x by $\{a\} = x_{01} = x_{02}$, $\{a^*\} = x_{10} = x_{20}$, and $\{0\} = x_{00} = x_{11} = x_{22} = x_{12} = x_{21}$, i.e.,

$$x = \begin{bmatrix} \{0\} & \{a\} & \{a\} \\ \{a^*\} & \{0\} & \{0\} \\ \{a^*\} & \{0\} & \{0\} \end{bmatrix}.$$

For part (3), assume $i, j < n, x \in B, a, b \in K, x_{ij} \subseteq \{a\};\{b\}$, and $i, j \neq k < n$. We need $y \in B$ such that $y_{ik} = \{a\}$, $y_{kj} = \{b\}$, and $x_{\ell m} = y_{\ell m}$ whenever $k \neq \ell, m < n$. If $a = 0$ then the required y is obtained directly from x by setting

$$y_{\ell m} = \begin{cases} x_{\ell m} \text{ if } k \neq \ell, m, \\ x_{im} \text{ if } k = \ell \neq m, \\ x_{\ell i} \text{ if } k = m \neq \ell, \\ x_{ii} \text{ if } k = m = \ell. \end{cases}$$

The key frame property that shows $y \in B$ is (3.43). A similar definition can be used for y when $b = 0$. We may therefore assume $0 \neq a, b$.

Let $c \in K^{d\times d}$ be the matrix of elements of K whose singletons are the entries in the matrix x, so that $\{c_{ij}\} = x_{ij}$ for all $i, j < d$. Without loss of generality we may assume that $k = d - 1$ and $0 = i \leq j \leq 1$. We want $y \in B$ such that $y_{0k} = \{a\}$, $y_{kj} = \{b\}$, and $x_{\ell m} = y_{\ell m}$ whenever $\ell, m < k = d - 1$. Define y on all arguments differing from k so that y agrees with x by setting $y_{\ell m} = x_{\ell m}$ for all $\ell, m < k = d - 1$. We must also set $y_{kk} = \{0\}$, $y_{0k} = \{a\}$, $y_{k0} = \{a^*\}$, $y_{kj} = \{b\}$, and $y_{jk} = \{b^*\}$. What remains is to choose $y_{\ell k}$ and $y_{k\ell}$ whenever $j, \ell < k = d - 1$ in such a way that $y \in B$. Note that $y_{\ell k} = \{c_{\ell k}\}$ and $y_{k\ell} = \{c_{k\ell}\}$ whenever $j, \ell < k = d - 1$. We extend c by setting $c_{kk} = 0$, $c_{0k} = a$, $c_{k0} = a^*$, $c_{kj} = b$, and $c_{jk} = b^*$. We will choose $c_{\ell k}, c_{k\ell} \in K$ whenever $j, \ell < k$, and set $y_{\ell k} = \{c_{\ell k}\}$ and $y_{k\ell} = \{c_{k\ell}\}$. Note that $\mathsf{D}(d - 2)$ implies $\mathsf{D}(t)$ whenever $1 \leq t \leq d - 2$.

Suppose $j = 0$. Then $x_{0j} = x_{00} = \{0\} \subseteq \{a\};\{b\}$, hence $\langle a, b, 0\rangle \in R'$, so $\langle a^*, 0, b\rangle \in R'$ by (3.41), $a^* = b$ by (3.43), and finally $a = b^*$ by (3.44). Apply $\mathsf{D}(1)$ to $u_1 = c_{10}$ and $v_1 = a = b^*$ to get $0 \neq w \in K$ such that $C(u_1, v_1, w) = C(c_{10}, a, w) \subseteq R'$. Set $c_{1k} = w$, $c_{k1} = w^*$, $y_{1k} = \{w\}$, and $y_{k1} = \{w^*\}$. The proof for the rest of the case $j = 0$ proceeds in the same way as the case in which $j = 1$, except that we do not know (or need) $a = b^*$.

Assume $j = 1$. We are done if $d = 3$ so assume $d > 3$ and $k = d - 1 > 2$. Apply $\mathsf{D}(2)$ to $u_1 = c_{20}$, $u_2 = c_{21}$, $v_1 = a$, and $v_2 = b^*$ to get $0 \neq w \in K$ such that

$C(u_1, v_1, w) = C(c_{20}, a, w) = C(c_{20}, c_{0k}, w) \subseteq R'$ and $C(u_2, v_2, w) = C(c_{21}, b^*, w) = C(c_{21}, c_{1k}, w) \subseteq R'$. Set $c_{2k} = w$, $c_{k2} = w^*$, $y_{2k} = \{w\}$, and $y_{k2} = \{w^*\}$. We are done if $d = 4$ and $k = 3$ so assume $d > 4$. Apply $\mathbf{D}(3)$ to $u_1 = c_{30}$, $u_2 = c_{31}$, $u_3 = c_{32}$, $v_1 = a = c_{0k}$, $v_2 = b^* = c_{1k}$, and $v_3 = c_{2k}$ to get c_{3k} such that $C(c_{30}, a, c_{3k}) = C(c_{30}, c_{0k}, c_{3k}) \subseteq R'$, $C(c_{31}, b^*, c_{3k}) = C(c_{31}, c_{1k}, c_{3k}) \subseteq R'$, and $C(c_{32}, c_{2k}, c_{3k}) = C(c_{32}, c_{2k}, c_{3k}) \subseteq R'$, and set $y_{3k} = \{c_{3k}\}$. Continue in this way until $\mathbf{D}(d-2)$ has been used. The conditions compiled in this process show that $y \in B$.

This completes the proof that if $\mathfrak{K}(R \cup R_0)$ has the diamond property $\mathbf{D}(d-2)$ then $\mathfrak{Cm}\,(\mathfrak{K}(R \cup R_0))$ has a d-dimensional relational basis, hence $\mathfrak{Cm}\,(\mathfrak{K}(R \cup R_0)) \in \mathsf{RA}_d$ and every d-provable predicate is valid in $\mathfrak{K}(R \cup R_0)$. We have shown that the commutative NA-frame corresponding to a randomly chosen relation $R \in W$ almost certainly has these properties. As was observed earlier, this is enough to conclude from parts (3)–(6) that parts (8)–(11) are also true and completes the proof of Theorem 3.28.

If an equation is true in every representable relation algebra then by Theorem 3.15 (1) (5) there is a smallest $n \in \omega$ such that it is true in every algebra in RA_n. Let Ξ be a finite set of equations true in RRA. Any finite subset of ω has a largest element, so there is some n such that every equation in Ξ is true in every algebra in RA_n. By Theorem 3.28 (7) a randomly chosen commutative NA-frame has a complex algebra that is almost certainly in RA_n. Therefore, a randomly chosen commutative NA-frame almost certainly validates every n-provable predicate and its complex algebra almost certainly satisfies every equation in Ξ. For example, any randomly chosen large KR-frame almost certainly validates (3.148), (3.149), and every other 5-provable predicate.

There are 594 TR-frames with five elements. In each of them, (3.148) is valid whenever (3.149) is valid. Predicate (3.149) is invalid in 286 of them and (3.148) is invalid in just 73. There are 390 KR-frames among those 594 TR-frames, and (3.148) and (3.149) are invalid in 58 of them. The fractions of TR-frames and KR-frames in which (3.148) and (3.149) are invalid shrink to zero as n increases. A randomly selected large TR-frame or KR-frame almost certainly validates (3.148) and (3.149) and the numbers of such frames both grow like c^{n^3} for some constant $c > 1$.

3.22 Questions

The results in this paper leave open or suggest a few technical questions and raise some others of a more general nature. The technical questions come first.

1. Can an axiomatization of TR be obtained by adding (3.101), (3.102), or (3.103) to an axiomatization of CR^*?
2. Which subsets of (3.60)–(3.132) axiomatize T_3, T_4, CT_3, CT_4, and CR^*?
3. Which of the derived rules of CT_3 listed in Theorem 3.22 are either derivable, admissible, or included by definition in R or CR^*? For example, the first two

1-provable rules are included in R by definition, but the third one turned out to be admissible. What about all the others?
4. Are there any deductive rules of CT_4 that require four variables?
5. Is almost every finite relation algebra representable?
6. Is relation algebra 29_{83} representable? See the end of Sect. 3.16.

Here are some questions about logic, philosophy, and history.

7. "Will the real negation please stand up?" (Anderson et al. 1992, p. 174). "Which is the *real* negation?" (Anderson et al. 1992, p. 492). Does Table 3.1 reveal the real negation? The results here suggest that Boolean negation is real and De Morgan negation is relevant negation.
8. Do the predicates (3.101), (3.102), and (3.103) have logical significance? Would any of them be proposed as an axiom for a relevance logic? Why were they never previously considered?
9. What are the philosophical implications of the fact for any fixed $n \in \omega$ a randomly selected TR-frame or KR-frame will almost certainly validate every n-provable predicate?
10. The algebra $\mathfrak{Re}(U)$ of binary relations on a set U is the prototypical example of a relation algebra. Since $\mathfrak{Re}(U)$ is the complex algebra of the pair-frame on U, could the pair-frame on U serve as a prototypical example of a frame for basic relevance logic?
11. Why do relevance logic and relation algebra overlap despite arising independently through the pursuit of completely different goals?
12. Why were Schröder's studies, Tarski's axiomatization, and relevance logic confined to the 4-variable fragment of the calculus of relations?

Acknowledgements Special thanks are due to the incredibly able Ian Hodkinson, whose insightful comments, pointing out various errors, omissions, and deficiencies, led to an extensive reorganization and expansion of the presentation.

References

Anderson, A. R., & Belnap, N. D, Jr. (1959). Modalities in Ackermann's "rigorous implication". *J. Symb. Logic*, *24*, 107–111.
Anderson, A. R., & Belnap, N. D, Jr. (1975). *Entailment. The Logic of Relevance and Necessity* (Vol. I). Princeton, N. J.-London: Princeton University Press.
Anderson, A. R., Belnap, N. D, Jr., & Dunn, J. M. (1992). *Entailment. The Logic of Relevance and Necessity* (Vol. II). Princeton, N. J.-London: Princeton University Press.
Belnap, N. D, Jr. (1960). Entailment and relevance. *J. Symbolic Logic*, *25*, 144–146.
Belnap, N. D, Jr. (1967). Intensional models for first degree formulas. *J. Symbolic Logic*, *32*, 1–22.
Bernstein, B. A. (1939). Sets of postulates for Boolean groups. *Annals of Mathematics (2)*, *40*(2), 420–422.
Chin, L. H., & Tarski, A. (1951). Distributive and modular laws in the arithmetic of relation algebras. *University of California Publications in Mathematics (N.S.)*, *1*, 341–384.

Dunn, J. M. (2001). A representation of relation algebras using Routley-Meyer frames. *Logic, Meaning and Computation* (Vol. 305, pp. 77–108). Synthese Library. Dordrecht: Kluwer Academic Publisher.

GAP. (2014). *GAP – Groups, Algorithms, and Programming, Version 4.7.6*. The GAP Group.

Givant, S. (2006). The calculus of relations as a foundation for mathematics. *Journal of Automated Reasoning*, *37*(4), 277–322. 2007.

Givant, S. (2017). *Introduction to Relation Algebras-Relation Algebras* (Vol. 1). Cham: Springer.

Henkin, L. (1949). The completeness of the first-order functional calculus. *J. Symbolic Logic*, *14*, 159–166.

Henkin, L. (1973). Internal semantics and algebraic logic. *Truth, Syntax and Modality (Proc. Conf. Alternative Semantics, Temple Univ., Philadelphia, Pa., 1970)* (Vol. 68, pp. 111–127). Studies in Logic and the Foundations of Mathematics. Amsterdam: North-Holland.

Henkin, L. (1996). The discovery of my completeness proofs. *Bull. Symbolic Logic*, *2*(2), 127–158.

Hirsch, R., & Hodkinson, I. (2002). *Relation Algebras by Games* (Vol. 147). Studies in Logic and the Foundations of Mathematics. Amsterdam: North-Holland Publishing Co. With a foreword by Wilfrid Hodges.

Jónsson, B., & Tarski, A. (1951). Boolean algebras with operators. *I. Amer. J. Math.*, *73*, 891–939.

Jónsson, B., & Tarski, A. (1952). Boolean algebras with operators. *II. Amer. J. Math.*, *74*, 127–162.

Kowalski, T. (2013). Relation algebras and **R**. In *Proceedings of the 12th Asian Logic Conference* (pp. 231–250). Hackensack, NJ: World Scientific Publishing.

Ladd-Franklin, C. F. (1928). The antilogism. *Mind*, *37*(148), 532–534.

Löwenheim, L. (1915). Über Möglichkeiten im Relativkalkül. *Math. Ann.*, *76*(4), 447–470.

Lyndon, R. C. (1950). The representation of relational algebras. *Ann. of Math.*, *2*(51), 707–729.

Maddux, R. D. (1978). *Topics in Relation Algebras*. ProQuest LLC, Ann Arbor, MI. Thesis (Ph.D.)–University of California, Berkeley.

Maddux, R. D. (1982). Some varieties containing relation algebras. *Trans. Amer. Math. Soc.*, *272*(2), 501–526.

Maddux, R. D. (1983). A sequent calculus for relation algebras. *Ann. Pure Appl. Logic*, *25*(1), 73–101.

Maddux, R. D. (1985). Finite integral relation algebras. In *Universal Algebra and Lattice Theory (Charleston, S.C., 1984)* (Vol. 1149, pp. 175–197). Lecture Notes in Math.. Berlin: Springer.

Maddux, R. D. (1989). Nonfinite axiomatizability results for cylindric and relation algebras. *J. Symbolic Logic*, *54*(3), 951–974.

Maddux, R. D. (1990). Necessary subalgebras of simple nonintegral semiassociative relation algebras. Algebra Universalis, 27(4), 544–558.

Maddux, R. D. (1991). Pair-dense relation algebras. *Trans. Amer. Math. Soc.*, *328*(1), 83–131.

Maddux, R. D. (1999). Relation algebras of formulas. In *Logic at Work* (Vol. 24, pp. 613–636). Studies in Fuzziness and Soft Computing. Heidelberg: Physica.

Maddux, R. D. (2006). *Relation Algebras* (Vol. 150). Studies in Logic and the Foundations of Mathematics. Amsterdam: Elsevier B. V.

Maddux, R. D. (2010). Relevance logic and the calculus of relations. *Rev. Symb. Log.*, *3*(1), 41–70.

McKenzie, R. (1970). Representations of integral relation algebras. *Michigan Math. J.*, *17*, 279–287.

McKenzie, R. N. W. (1966). *The Representation of Relation Algebras*. ProQuest LLC, Ann Arbor, MI. Thesis (Ph.D.)–University of Colorado at Boulder.

Meyer, R. K. and Routley, R. (1973). Classical relevant logics. I. Studia Logica, 32:51–68.

Meyer, R. K. and Routley, R. (1974). Classical relevant logics. II. Studia Logica, 33:183–194.

Mikulás, S. (2009). Algebras of relations and relevance logic. *J. Logic Comput.*, *19*(2), 305–321.

Quine, W. V. O. (1940). *Mathematical Logic*. New York: W. W. Norton & Co., Inc.

Quine, W. V. O. (1951). *Mathematical Logic*. Cambridge, Mass: Harvard University Press. Revised ed.

Quine, W. V. O. (1962). *Mathematical Logic*. Revised edition. Harper Torchbooks: The Science Library. New York: Harper & Row Publishers. xii+346 pp.

Quine, W. V. O. (1981). *Mathematical Logic*. Cambridge, Mass: Harvard University Press. Revised edition.
Routley, R., & Meyer, R. K. (1973). The semantics of entailment. I. In *Truth, Syntax and Modality (Proc. Conf. Alternative Semantics, Temple Univ., Philadelphia, Pa., 1970)* (Vol. 68, pp. 199–243). Studies in Logic and the Foundations of Mathematics. Amsterdam: North-Holland.
Routley, R., Plumwood, V., Meyer, R. K., & Brady, R. T. (1982). *Relevant Logics and their Rivals*. Atascadero, CA: Part I. Ridgeview Publishing Co.
Schröder, F. W. K. E. (1966). *Vorlesungen über die Algebra der Logik (exakte Logik). Band III. Algebra und Logik der Relative*. Anhang: Abriss der Algebra der Logik von Eugen Müller. New York: Chelsea Publishing Co. First published by B. G. Teubner, Leipzig, 1895.
Suguitani, L., Viana, J. P., & D'Ottaviano, I. M. L. (Eds.). (2016). *AlfredTarski: Lectures at Unicamp in 1975*. Editora Da Unicamp, Campinas. Centro de Lógica, Epistemologia e História da Ciência (Col: CLE v. 76).
Sylvan, R., Meyer, R., Plumwood, V., & Brady, R. (2003). *Relevant Logics and their Rivals. Vol. II* (Vol. 59). Western Philosophy Series. Aldershot: Ashgate Publishing Limited. A continuation of the work of Richard Sylvan, Robert Meyer, Val Plumwood and Ross Brady. Edited by Brady.
Tarski, A. (1953). A formalization of set theory without variables. *J. Symbolic Logic*, *18*, 189.
Tarski, A. (1965). A simplified formalization of predicate logic with identity. *Arch. Math. Logik Grundlagenforsch*, *7*, 61–79.
Tarski, A., & Givant, S. (1987). *A Formalization of Set Theory without Variables* (Vol. 41). American Mathematical Society Colloquium Publications. Providence, RI: American Mathematical Society.
Urquhart, A. (1984). The undecidability of entailment and relevant implication. *J. Symbolic Logic*, *49*(4), 1059–1073.
Urquhart, A. (1993). Failure of interpolation in relevant logics. *J. Philos. Logic*, *22*(5), 449–479.
Urquhart, A. (1999). Beth's definability theorem in relevant logics. In *Logic at Work* (Vol. 24 , pp. 229–234). Studies in Fuzziness and Soft Computing. Heidelberg: Physica.
Urquhart, A. (2017). The geometry of relevant implication. *IFCoLog J. Log. Appl.*, *4*(3), 591–604.
Urquhart, A. (2019). Relevant implication and ordered geometry. *Australas. J. Log.*, *16*(8), 342–354.

Chapter 4
Algorithmic Correspondence for Relevance Logics I. The Algorithm PEARL

Willem Conradie and Valentin Goranko

Second Reader
P. Jipsen
Chapman University
Dedicated to Alasdair Urquhart, on the occasion of his 75th birthday.

Abstract We apply and extend the theory and methods of algorithmic correspondence theory for modal logics, developed over the past 20 years, to the language $\mathcal{L}_R$ of relevance logics with respect to their standard Routley–Meyer relational semantics. We develop the non-deterministic algorithmic procedure PEARL for computing first-order equivalents of formulae of the language $\mathcal{L}_R$, in terms of that semantics. PEARL is an adaptation of the previously developed algorithmic procedures SQEMA (for normal modal logics) and ALBA (for distributive and non-distributive modal logics). We then identify a large syntactically defined class of *inductive formulae* in $\mathcal{L}_R$, analogous to previously defined such classes in the classical, distributive and non-distributive modal logic settings, and show that PEARL succeeds for every inductive formula and correctly computes a first-order definable condition which is equivalent to it with respect to frame validity. We also provide a detailed comparison with two earlier works, each extending the class of Sahlqvist formulae to relevance logics, and show that both are subsumed by simple subclasses of inductive formulae.

Keywords Relevance logic · Routley–Meyer relational semantics · Algorithmic correspondence · Inductive formulae

W. Conradie
School of Mathematics, University of the Witwatersrand, Witwatersrand, South Africa
e-mail: willem.conradie@wits.ac.za

V. Goranko (✉)
Department of Philosophy, Stockholm University, Stockholm, Sweden
e-mail: valentin.goranko@philosophy.su.se

School of Mathematics (Visiting professorship), University of the Witwatersrand, Witwatersrand, South Africa

I. Düntsch and E. Mares (eds.), *Alasdair Urquhart on Nonclassical and Algebraic Logic and Complexity of Proofs*, Outstanding Contributions to Logic 22,
https://doi.org/10.1007/978-3-030-71430-7_4

4.1 Introduction

This paper brings together two important areas of active development in non-classical logics, viz. *relevance logics* and *algorithmic correspondence theory*. Since it is intended mainly for readers competent in relevance logics, but not necessarily so much in correspondence theory, we focus in this introduction on the latter topic by first providing a brief overview.

Overview of Algorithmic Correspondence Theory

One of the classical results in modal logic since the invention of the possible worlds semantics was *Sahlqvist's theorem*[1] (Sahlqvist 1975), which makes the following two important claims for all formulae from a certain syntactically defined class (subsequently called *Sahlqvist formulae*), including the modal principles appearing in axioms of the most important systems of normal modal logics:

(i) *FO correspondence*: all Sahlqvist formulae define conditions on Kripke frames that are also definable in the corresponding first-order language (FO), and

(ii) *Completeness via canonicity*: all normal modal logics axiomatized with Sahlqvist formulae are complete with respect to the class of Kripke frames that they define, because these logics are *canonical*, i.e. valid in their respective canonical frames.

That result sets the stage for the emergence and development of the so-called *correspondence theory in modal logic*, cf. Benthem (2001). Over the past 20 years that theory has been expanded significantly in at least three directions:

- The class of formulae covered by Sahlqvist's theorem was extended considerably to the class of so-called *inductive formulae*[2] introduced first in Goranko and Vakarelov (2001), Goranko and Vakarelov (2002), and further extended and refined in Conradie et al. (2005) and in Goranko and Vakarelov (2006), where all inductive formulae in the basic multi-modal languages and in some important extensions were proved both first-order definable and canonical, thus extending Sahlqvist's theorem.
- The method for eliminating propositional variables from modal formulae and computing their first-order equivalents was substantially extended in and made algorithmic in a series of papers developing *algorithmic correspondence theory* implemented by the algorithmic procedure SQEMA (Conradie et al. 2006a), which not only provably succeeds in computing the first-order equivalents of all inductive (and, in particular, all Sahlqvist) formulae, but also automatically proves their canonicity, just by virtue of succeeding on them. That enabled an algorithmic, and easily automatizable, approach to proving completeness of numerous old and new modal logics studied in the literature.

[1] Essentially the same result was also proved independently by van Benthem in his PhD thesis Benthem (1976).

[2] The name refers to an inductive procedure for computing the *minimal valuations* of the occurring propositional variables that are to be computed in the right order and substituted in the formula to obtain the first-order frame condition defined by it.

The algorithm SQEMA was further extended to polyadic and hybrid modal languages in Conradie et al. (2006b) and strengthened further in Conradie et al. (2009), Conradie and Goranko (2008), Conradie et al. (2010).

- Both the traditional and the algorithmic correspondence theory were subsequently developed further and extended significantly. First, SQEMA was generalized to the algorithm ALBA introduced in Conradie and Palmigiano (2012) to cover the inductive formulae for distributive modal logic, which, respectively, generalize the Sahlqvist formulae of Gehrke et al. (2005). This was extended to a wide range of logics including, e.g. the intuitionistic modal mu-calculus (Conradie et al. 2015) and non-normal modal logics (Palmigiano et al. 2017), and ultimately to any logic algebraically captured by classes of normal (possibly non-distributive) lattice expansions (Conradie and Palmigiano 2019). As is evident from this list, this line of research has a strong algebraic flavour, and the reason for this is that, even while they pertain to relational semantics, the underlying mechanisms on which Sahlqvist-style results turn are ultimately order-theoretic. The line of research which develops this insight, and to which the cited papers belong, has been dubbed 'unified correspondence' (Conradie et al. 2014).

Still, the scope and popularity of correspondence theory has remained mostly confined to modal logics in a broader sense, and to some extent to intuitionistic logics, whereas its use and impact in relevance logics have remained rather limited and largely unexplored, with just a couple of works, mainly Seki (2003), Badia (2018), defining and exploring Sahlqvist-type formulae for relevance logics. Also, in the context of relevant algebras and their topological dual-spaces, in Urquhart (1996) Urquhart presents a correspondence result for algebraic inequalities built with fusion as the only operation. In Urquhart (1996), he also notes that '*Correspondence theory in the case of modal and intuitionistic logic has been extensively studied, but the analogous theory for the case of relevant logics is surprisingly neglected.*' This is indeed rather surprising, given that much work has been done in relevance logics (starting with the original paper Routley and Meyer 1973 introducing the Routley–Meyer semantics, cf. also the classic book Routley et al. 1982, as well as Dunn and Restall 2002) to identify the first-order conditions defined by numerous axioms of various systems of relevance logics, as well as proving their completeness with respect to several types of semantics, including Urquhart's semilattice semantics (Urquhart 1972) and the Routley–Meyer relational semantics. Because of the more complex semantics, this kind of calculation can be significantly more involved than for modal logics with their standard Kripke semantics.

Notably, it turns out that the idea of inductive formulae is much more relevant (no pun intended) to relevance logics than to modal logics. This is because almost all important modal logic principles that are first-order definable and canonical fall in the smaller, but much better known, class of Sahlqvist formulae, whereas this is not the case for the important axioms of relevance logics. Almost all of these axioms turn out to be inductive, but only some of them are of Sahlqvist type, in terms of the natural analogue of Sahlqvist formulae for relevance logics. Briefly, this is because of the natural nesting of relevance implications, as well as fusions, in such axioms.

Thus, we argue that algorithmic correspondence theory is very naturally applicable and potentially quite useful for relevance logics. That was the main motivation of carrying out the present work.

Contributions of This Paper

This work extends and adapts the theory and methods of algorithmic correspondence to relevance logics in the context of the Routley–Meyer relational semantics. We develop here a non-deterministic algorithmic procedure PEARL (acronym for Propositional variables Elimination Algorithm for Relevance Logic) for computing first-order equivalents in terms of frame validity of formulae of the language $\mathcal{L}_R$ for relevance logics. PEARL is an adaptation of the above-mentioned procedures SQEMA (Conradie et al. 2006a) (for normal modal logics) and ALBA (Conradie and Palmigiano 2012, 2019) (for distributive and non-distributive modal logics). We define a large syntactically defined class of *inductive relevance formulae* in $\mathcal{L}_R$ and show that PEARL succeeds for all such formulae and correctly computes their first-order equivalents with respect to frame validity. We also provide a detailed comparison with the two closest earlier works on the topic, viz. Seki (2003), Badia (2018), each extending the class of Sahlqvist formulae to relevance logics. We show that both are subsumed by a simple subclass of our inductive formulae.

We regard this work as initial exploration of algorithmic correspondence for relevance logics. There is much more to be done. Some extensions and continuations of the present work, like adding modal operators, are fairly routine. Others, such as proving canonicity of all formulae on which PEARL succeeds (in particular, all inductive formulae) are more involved, but seem feasible. These we intend to do in a follow-up part II of this work. Another non-trivial task is the development of algorithmic correspondence for Urquhart's semilattice semantics for relevance logics (Urquhart 1972).

The Structure of the Paper

In the preliminary Sect. 4.2, we summarize the basics of the Routley–Meyer relational semantics for relevance logics. In Sect. 4.3, we present the algorithmic procedure PEARL for computing first-order correspondents of formulae of relevance logic and prove its soundness with respect to the first-order equivalents that it computes for the frame conditions defined by the input formulae. Then, in Sect. 4.4 we define the class of inductive formulae for relevance logics, prove that PEARL succeeds on all inductive formulae, and compare with the classes of Sahlqvist formulae previously defined in the literature, viz. in Badia (2018), Seki (2003), showing that they are all subsumed by subclasses of inductive formulae. We end with brief concluding remarks and directions for further work in Sect. 4.5. At the end of the paper, we have added two appendices: Appendix 1 with some proofs, and auxiliary Appendix 2 with some axioms for relevance logics that we have copied there from Routley et al. (1982) as a source of important examples of inductive formulae, for reference, and for the reader's convenience.

4.2 Preliminaries

We assume basic familiarity with the syntax and relational semantics of modal and relevance logics, general references for which are, e.g. Blackburn et al. (2001) (for modal logics) and Routley et al. (1982), Dunn and Restall (2002) (for relevance logics), from where we quote some of the definitions below and give a few additional definitions, not explicitly mentioned there.

4.2.1 Syntax and Routley–Meyer Relational Semantics for Relevance Logics

Hereafter, we consider the language of propositional relevance logics $\mathcal{L}_R$ over a fixed set of propositional variables VAR containing the classical connectives $\wedge$, $\vee$, plus the relevant connectives **fusion** $\circ$, **(relevant) negation** $\sim$, **(relevant) implication** $\rightarrow$, and the special constant **(relevant) truth t**. The formulae of $\mathcal{L}_R$ are defined as expected:

$$A = p \mid \mathbf{t} \mid {\sim}A \mid (A \wedge A) \mid (A \vee A) \mid (A \circ A) \mid (A \rightarrow A)$$

where $p \in \mathsf{VAR}$.

A **relevance frame** is a tuple $\mathcal{F} = \langle W, O, R, ^* \rangle$, where

- W is a non-empty set of states (possible worlds);
- $O \subseteq W$ is the subset of **normal** states;
- $R \subseteq W^3$ is a **relevant accessibility relation**;
- $^* : W \rightarrow W$ is a function, called the **Routley star**, used to provide semantics for $\sim$.

The following binary relation $\preceq$ is defined in every relevance frame:

$$u \preceq v \text{ iff } \exists o(o \in O \wedge Rouv).$$

A **Routley–Meyer frame** (for short, **RM-frame**) is a relevance frame satisfying the following conditions for all $u, v, w, x, y, z \in W$:

1. $x \preceq x$.
2. If $x \preceq y$ and $Ryuv$ then $Rxuv$.
3. If $x \preceq y$ and $Ruyv$ then $Ruxv$.
4. If $x \preceq y$ and $Ruvx$ then $Ruvy$.
5. If $x \preceq y$ then $y^* \preceq x^*$.
6. O is upward closed w.r.t. $\preceq$, i.e. if $o \in O$ and $o \preceq o'$ then $o' \in O$.

These properties ensure that $\preceq$ is reflexive and transitive, hence a preorder, and that the semantics of the logical connectives has the monotonicity properties stated further. For the sake of comparing the definitions and results related to Sahlqvist formulae, here we have adopted the definition of Routley–Meyer frame as in Badia (2018). Note that in the original paper Routley and Meyer (1973) introducing the Routley–Meyer semantics, and in many subsequent sources, O is assumed to be an upwards closed set generated by a single element 0. Also, Routley and Meyer (1973) and others assume that the Routley star * is an involution, i.e. $x^{**} = x$. We will not make either of these assumptions here. However, $\preceq$ can be assumed to be a partial order (as originally assumed in Routley and Meyer 1973) w.l.o.g., since adding antisymmetry does not change the notion of frame validity.

A **Routley–Meyer model** (for short, **RM-model**) is a tuple $\mathcal{M} = \langle W, O, R, ^*, V\rangle$, where $\langle W, O, R, ^*\rangle$ is a Routley–Meyer frame and $V : \mathsf{VAR} \to \wp W$ is a mapping, called a **relevant valuation**, assigning to every atomic proposition $p \in \mathsf{VAR}$ a set $V(p)$ of states *upward closed* w.r.t. $\preceq$.

Truth of a formula A in a RM-model $\mathcal{M} = \langle W, O, R, ^*, V\rangle$ at a state $u \in W$, denoted $\mathcal{M}, u \Vdash A$, is defined as follows:

- $\mathcal{M}, u \Vdash p$ iff $u \in V(p)$;
- $\mathcal{M}, u \Vdash \mathbf{t}$ iff there is $o \in O$ such that $o \preceq u$; equivalently, iff $u \in O$;
- $\mathcal{M}, u \Vdash {\sim} A$ iff $\mathcal{M}, u^* \nVdash A$;
- $\mathcal{M}, u \Vdash A \wedge B$ iff $\mathcal{M}, u \Vdash A$ and $\mathcal{M}, u \Vdash B$;
- $\mathcal{M}, u \Vdash A \vee B$ iff $\mathcal{M}, u \Vdash A$ or $\mathcal{M}, u \Vdash B$;
- $\mathcal{M}, u \Vdash A \to B$ iff for every v, w such that $Ruvw$, if $\mathcal{M}, v \Vdash A$ then $\mathcal{M}, w \Vdash B$;
- $\mathcal{M}, u \Vdash A \circ B$ iff there exist v, w such that $Rvwu$, $\mathcal{M}, v \Vdash A$ and $\mathcal{M}, w \Vdash B$.

For every RM-model $\mathcal{M}$ and formula A we define the **extension of A in $\mathcal{M}$** as

$$[\![A]\!]_{\mathcal{M}} := \{u \in \mathcal{M} \mid \mathcal{M}, u \Vdash A\}.$$

A formula A is declared:

- **true in an RM-model $\mathcal{M}$**, denoted by $\mathcal{M} \Vdash A$, if $O \subseteq [\![A]\!]_{\mathcal{M}}$, i.e. $\mathcal{M}, o \Vdash A$ for every $o \in O$.
- **valid in an RM-frame $\mathcal{F}$**, denoted by $\mathcal{F} \Vdash A$, iff it is true in every RM-model over that frame.
- **RM-valid**, denoted by $\Vdash A$, iff it is true in every RM-model.

Remark 4.1 In Urquhart (1972), Urquhart proposed the well-known semilattice (or, operational) semantics for the relevant connectives. The states in these models can be thought of as pieces of information that can support assertions and can be combined. This combination of pieces of information imposes a natural (join) semilattice structure. In particular, a piece of information α supports an implication $\phi \to \psi$ (notation $\alpha \Vdash \phi \to \psi$) iff whenever we combine the α with any piece of information β which supports ϕ ($\beta \Vdash \phi$) the combination will support ψ ($\alpha \cdot \beta \Vdash \psi$).

An important property of this semantics is *Monotonicity*: for every RM-model $\mathcal{M}$ and formula A, the set $[\![A]\!]_{\mathcal{M}}$ is $\preceq$-*upward closed.*

A formula A of $\mathcal{L}_R$ not containing variables will be called a **constant formula**. Clearly, the truth of a constant formula in a RM-model does not depend on the valuation, i.e. the extension $[\![A]\!]_{\mathcal{M}}$ is the same for every RM-model $\mathcal{M}$ based on a RM-frame $\mathcal{F}$, so we will identify it with validity in $\mathcal{F}$ and denote it by $[\![A]\!]_{\mathcal{F}}$. Then, $\mathcal{F}, u \Vdash A$ iff $O \subseteq [\![A]\!]_{\mathcal{F}}$.

Formulae A and B from $\mathcal{L}_R$ are **semantically equivalent**, hereafter denoted $A \equiv B$, iff they are true at the same states in every RM-model; **RM-model-equivalent**, if they are true at the same RM-models; **RM-frame-equivalent**, if they are valid in the same RM-frames. Hereafter, 'equivalent formulae of $\mathcal{L}_R$' will mean 'semantically equivalent formulae', unless otherwise specified.

Clearly, Routley–Meyer frames are first-order structures for the first-order language with unary predicate symbol O, unary function symbol $*$, ternary relation symbol R, and individual variables $x_1, x_2, x_3, \ldots$, informally denoted x, x', x'', etc. We will call this language FO_R. Moreover, the semantics of relevance logic can be transparently expressed in FO_R and every relevance formula is then equivalently translated into a formula in FO_R by the following **standard translation** $ST : \mathcal{L}_R \to \mathrm{FO}_R$, parametric in a first-order individual variable:

$$
\begin{aligned}
ST_x(p) &= P(x)\\
ST_x(\mathbf{t}) &= O(x)\\
ST_x({\sim} A) &= \exists x'(x' = x^* \wedge \neg ST_{x'}(A))\\
ST_x(A \wedge B) &= ST_x(A) \wedge ST_x(B)\\
ST_x(A \vee B) &= ST_x(A) \vee ST_x(B)\\
ST_x(A \circ B) &= \exists x'x''(Rx'x''x \wedge ST_{x'}(A) \wedge ST_{x''}(B))\\
ST_x(A \to B) &= \forall x'x''(Rxx'x'' \wedge ST_{x'}(A) \to ST_{x''}(B))
\end{aligned}
$$

where x' and x'' are fresh individual variables.

It is routine to check that for every Routley–Meyer model $\mathcal{M}$, state w in $\mathcal{M}$ and $\mathcal{L}_R$-formula A, it holds that $\mathcal{M}, w \Vdash A$ iff $\mathcal{M} \models ST_x(A)[x := w]$, where $[x := w]$ indicates that the free variable x in $ST_x(A)$ is interpreted as w.

Positive and negative occurrences of logical connectives and propositional variables in a formula A of $\mathcal{L}_R$ are defined inductively on the structure of formulae, by technically treating propositional variables both as formulae and as unary (identity) connectives, as follows:

- When $A = \mathbf{t}$, the constant $\mathbf{t}$ occurs positively in the formula A.
- When $A = p$, the variable p occurs positive in the formula A.
- When $A = {\sim} B$, all positive (resp. negative) occurrences of connectives (incl. variables) in the subformula B are negative (resp. positive) occurrences of these connectives in A, and the occurrence of $\sim$ as a main connective is positive in A.
- When $A = B \bullet C$, where $\bullet \in \{\wedge, \vee, \circ\}$, all positive (resp. negative) occurrences of connectives (incl. variables) in the subformulae B and C are also positive (resp. negative) occurrences of these connectives in A. Besides, the occurrence of $\bullet$ as a main connective is positive in A.

- When $A = B \to C$, all positive (resp. negative) occurrences of connectives (incl. variables) in the subformula C are also positive (resp. negative) occurrences of these connectives in A, whereas all positive (resp. negative) occurrences of connectives (incl. variables) in the subformula B are negative (resp. positive) occurrences of these connectives in A. Besides, the occurrence of $\to$ as a main connective is positive in A.

We say that a **formula** $A \in \mathcal{L}_R$ **is positive (resp., negative) in a propositional variable** p iff all occurrences of p in A are positive (resp., negative).

A routine inductive argument over the structure of formulae shows that, if a formula A in $\mathcal{L}_R$ is positive in p, then its induced semantic operation $A_p^V(X)$ from $\mathcal{P}^\uparrow(W)$ into $\mathcal{P}^\uparrow(W)$, is monotone (i.e. order preserving), whereas it is antitone (i.e. order-reversing) if A is negative in p. Further, we will simply say that $A(p)$ is monotone (antitone) in p if $A_p^V(X)$ is monotone (antitone).

4.2.2 Complex Algebras of RM-Frames

The **complex algebra** of a Routley–Meyer frame $\mathcal{F} = \langle W, R, *, O\rangle$ is the structure $\mathcal{F}^+ = \langle \mathcal{P}^\uparrow(W), \cap, \cup, \to, \circ, \sim, O\rangle$ where $\mathcal{P}^\uparrow(W)$ is the set of all upwards closed subsets (hereafter called **up-sets**) of W, $\cap$ and $\cup$ are set-theoretic intersection and union, and for all $Y, Z \in \mathcal{P}^\uparrow(W)$ the following hold:

$Y \to Z = \{x \in W \mid \forall yz \in W, \text{ if } Rxyz \text{ and } y \in Y, \text{ then } z \in Z\}$,
$Y \circ Z = \{x \in W \mid \exists y, z \in W, Ryzx \text{ and } y \in Y \text{ and } z \in Z\}$,
$\sim Y = \{x \in W \mid x^* \notin Y\}$.

Note that for any RM-model $\mathcal{M} = \langle W, O, R, ^*, V\rangle$ based upon the RM-frame $\mathcal{F}$, and all $A, B \in \mathcal{L}_R$, the family $\mathcal{M}^+ = \{V(A) \mid A \in \mathcal{L}_R\}$ is a subalgebra of $\mathcal{F}^+$ and the following hold:

- $[\![\mathbf{t}]\!]_\mathcal{M} = O$,
- $[\![\sim A]\!]_\mathcal{M} = \sim[\![A]\!]_\mathcal{M}$,
- $[\![A \wedge B]\!]_\mathcal{M} = [\![A]\!]_\mathcal{M} \cap [\![B]\!]_\mathcal{M}$,
- $[\![A \vee B]\!]_\mathcal{M} = [\![A]\!]_\mathcal{M} \cup [\![B]\!]_\mathcal{M}$,
- $[\![A \to B]\!]_\mathcal{M} = [\![A]\!]_\mathcal{M} \to [\![B]\!]_\mathcal{M}$,
- $[\![A \circ B]\!]_\mathcal{M} = [\![A]\!]_\mathcal{M} \circ [\![B]\!]_\mathcal{M}$.

The following proposition states properties of the complex operations $\to$, $\circ$ and $\sim$ that are easy to verify and will be used further.

Proposition 4.1 *For every complex algebra $\mathcal{F}^+ = \langle \mathcal{P}^\uparrow(W), \cap, \cup, \to, \circ, \sim, O\rangle$ of an RM-frame, $X \in \mathcal{P}^\uparrow(W)$ and family $\{Y_i \mid i \in I\} \subseteq \mathcal{P}^\uparrow(W)$, the following hold.*

1. $X \to \bigcap_{i\in I} Y_i = \bigcap_{i\in I}(X \to Y_i)$,
2. $\bigcup_{i\in I} Y_i \to X = \bigcap_{i\in I}(Y_i \to X)$,
3. $X \circ \bigcup_{i\in I} Y_i = \bigcup_{i\in I}(X \circ Y_i)$,
4. $\bigcup_{i\in I} Y_i \circ X = \bigcup_{i\in I}(Y_i \circ X)$,

5. $\sim \bigcup_{i \in I} Y_i = \bigcap_{i \in I} (\sim Y_i)$,
6. $\sim \bigcap_{i \in I} Y_i = \bigcup_{i \in I} (\sim Y_i)$.

In fact, every complex algebra $\mathcal{F}^+ = \langle \mathcal{P}^{\uparrow}(W), \cap, \cup, \to, \circ, \sim, O \rangle$ is a complete and perfect distributive right-residuated magma with a constant O and a unary DeMorgan operation $\sim$(see, e.g. Jipsen and Kinyon 2019). These algebras are called 'relevance algebras' by Urquhart (1996), although he also includes lattice bounds and requires O to be a left identity of $\circ$.

Two families of elements of $\mathcal{P}^{\uparrow}(W)$ will be of particular interest to us, namely the set $J(\mathcal{F}^+) = \{\uparrow x \mid x \in W\}$ of all **principal up-sets** $\uparrow x = \{y \in W \mid y \succeq x\}$, and the set $M(\mathcal{F}^+) = \{(\downarrow x)^c \mid x \in W\}$ of set-theoretic compliments of principal downwards closed subsets (hereafter called **co-downsets**). The families $J(\mathcal{F}^+)$ and $M(\mathcal{F}^+)$ consist, respectively, of exactly the join- and meet-irreducible elements of the lattice $\langle \mathcal{P}^{\uparrow}(W), \cap, \cup \rangle$ (see, e.g. Davey and Priestley 2002). They have some easy to prove but important properties, summarized in the next proposition, which will be used further.

Proposition 4.2 *For every pre-ordered set $(W, \prec)$ and $X \in \mathcal{P}^{\uparrow}(W)$, the following hold.*

1. *For any up-set $X \in \mathcal{P}^{\uparrow}(W)$:*
   a. *X can be written as the union of elements of $J(\mathcal{F}^+)$, viz. $X = \bigcup\{\uparrow x \mid x \in X\}$. Thus, $\mathcal{P}^{\uparrow}(W)$ is $\cup$-**generated by** $J(\mathcal{F}^+)$.*
   b. *X can be written as the intersection of elements of $M(\mathcal{F}^+)$, viz. $X = \bigcap\{(\downarrow x)^c \mid x \notin X\}$. Thus, $\mathcal{P}^{\uparrow}(W)$ is $\cap$-**generated by** $M(\mathcal{F}^+)$.*
2. *For any $x \in W$ and family $\{X_i \mid i \in I\} \subseteq \mathcal{P}^{\uparrow}(W)$:*
   a. *$\uparrow x \subseteq \bigcup_{i \in I} X_i$ iff $\uparrow x \subseteq X_{i_0}$ for some $i_0 \in I$. Thus, every element of $J(\mathcal{F}^+)$ is **completely** $\cup$-**prime**.*
   b. *$(\downarrow x)^c \supseteq \bigcap_{i \in I} X_i$ iff $(\downarrow x)^c \supseteq X_{i_0}$ for some $i_0 \in I$. Thus, every element of $M(\mathcal{F}^+)$ is **completely** $\cap$-**prime**.*

4.3 PEARL: A Calculus for Computing First-Order Correspondents of Formulae of Relevance Logic

In this section we present a calculus of rewrite rules, in the style of the algorithms SQEMA (Conradie et al. 2006a) and ALBA (Conradie and Palmigiano 2012, 2019), which is sound and complete for deriving first-order frame correspondents for a large class of formulae of $\mathcal{L}_R$, viz. the class of *inductive (relevance) formulae* defined in Sect. 4.4. As we will show later, the class of inductive formulae substantially extends the classes of Sahlqvist formulae in $\mathcal{L}_R$ defined in Seki (2003), Badia (2018) and almost all axioms of important systems of relevance logic listed in Appendix 2 (copied there from Routley et al. 1982) are inductive formulae, while many of them are not Sahlqvist.

4.3.1 The Extended Language $\mathcal{L}_R^+$

Here, we extend the language $\mathcal{L}_R$ to $\mathcal{L}_R^+$ by adding connectives which are related, as residuals or adjoints, to some of the connectives in $\mathcal{L}_R$. In particular, we add the left adjoint $\sim^\flat$ and the right adjoint $\sim^\sharp$ of $\sim$, the Heyting implication $\Rightarrow$ (as right residual of $\wedge$), the co-Heyting implication $\prec$ as the left residual of $\vee$ and the operation $\hookrightarrow$ as the residual of $\circ$ in the second coordinate and of $\rightarrow$ in the first coordinate. $\mathcal{L}_R^+$ will be the working language of the algorithm PEARL. For its purpose we also include in $\mathcal{L}_R^+$ two (countably infinite) sets, $\mathsf{NOM} = \{\mathbf{j}_0, \mathbf{j}_1, \mathbf{j}_2, \ldots\}$ and $\mathsf{CNOM} = \{\mathbf{m}_0, \mathbf{m}_1, \mathbf{m}_2, \ldots\}$, of special variables, respectively, called **nominals** and **co-nominals**. Informally, we will denote nominals by $\mathbf{i}, \mathbf{j}, \mathbf{k}$, possibly with indices, while co-nominals will be denoted by $\mathbf{m}, \mathbf{n}$, possibly with indices. To distinguish visually from $\mathcal{L}_R$, the formulae of the extended language $\mathcal{L}_R^+$ will be denoted by lowercase Greek letters, typically α, β, γ, ϕ, ψ, ξ, etc. and are defined by the following grammar:

$$\phi = p \mid \mathbf{i} \mid \mathbf{m} \mid \top \mid \bot \mid \mathbf{t} \mid \sim\phi \mid (\phi \wedge \phi) \mid (\phi \vee \phi) \mid (\phi \circ \phi)$$
$$\mid (\phi \rightarrow \phi) \mid \sim^\flat\phi \mid \sim^\sharp\phi \mid (\phi \prec \phi) \mid (\phi \Rightarrow \phi) \mid (\phi \hookrightarrow \phi)$$

where $p \in \mathsf{VAR}$, $\mathbf{i} \in \mathsf{NOM}$ and $\mathbf{m} \in \mathsf{CNOM}$. We denote $\mathsf{ATOMS} := \mathsf{VAR} \cup \mathsf{NOM} \cup \mathsf{CNOM}$. The elements of ATOMS will be called **atoms**.

The additional connectives of $\mathcal{L}_R^+$ are interpreted in the same Routley–Meyer models as $\mathcal{L}_R$, except that the notion of valuation need to be adjusted so that instead of $V : \mathsf{VAR} \rightarrow \wp W$, we have $V : \mathsf{ATOMS} \rightarrow \wp W$ and V maps nominals to principal up-sets and co-nominals to compliments of principal downsets, i.e. for all $\mathbf{i} \in \mathsf{NOM}$ and all $\mathbf{m} \in \mathsf{CNOM}$ we have $V(\mathbf{i}) = {\uparrow}w$ for some $w \in W$ and $V(\mathbf{m}) = ({\downarrow}v)^c$ for some $v \in W$. The semantics of the additional connectives of $\mathcal{L}_R^+$ are given as follows:

- $\mathcal{M}, w \Vdash \mathbf{i}$ iff $w \in V(\mathbf{i})$,
- $\mathcal{M}, w \Vdash \mathbf{m}$ iff $w \in V(\mathbf{m})$,
- $\mathcal{M}, w \Vdash \top$,
- $\mathcal{M}, w \nVdash \bot$,
- $\mathcal{M}, w \Vdash \sim^\flat\phi$ iff there is a v such that $v^* = w$ and $\mathcal{M}, v \nVdash \phi$,
- $\mathcal{M}, w \Vdash \sim^\sharp\phi$ iff for all v such that $v^* = w$, it is the case that $\mathcal{M}, v \nVdash \phi$,
- $\mathcal{M}, w \Vdash \phi \prec \psi$ iff there exists v such that $v \preceq w$, $\mathcal{M}, v \Vdash \phi$ and $\mathcal{M}, v \nVdash \psi$,
- $\mathcal{M}, w \Vdash \phi \Rightarrow \psi$ iff for all $v \succeq w$, if $\mathcal{M}, v \Vdash \phi$ then $\mathcal{M}, v \Vdash \psi$,
- $\mathcal{M}, w \Vdash \phi \hookrightarrow \psi$ iff for all $v, u \in W$, if $Rvwu$ and $\mathcal{M}, v \Vdash \phi$ then $\mathcal{M}, u \Vdash \psi$.

Under the assumption that * is an involution, i.e. that $w^{**} = w$ for all $w \in W$, the clauses for $\sim^\flat$ and $\sim^\sharp$ become

- $\mathcal{M}, w \Vdash \sim^\flat\phi$ iff $\mathcal{M}, w^* \nVdash \phi$ iff $\mathcal{M}, w \Vdash \sim\phi$ and
- $\mathcal{M}, w \Vdash \sim^\sharp\phi$ iff $\mathcal{M}, w^* \nVdash \phi$ iff $\mathcal{M}, w \Vdash \sim\phi$.

The standard translation ST can be extended to the language $\mathcal{L}_R^+$. For that purpose we will add sets of individual variables $\{y_0, y_1, y_2, \ldots\}$ and $\{z_0, z_1, z_2, \ldots\}$ to be used for the translations of nominals and co-nominals, respectively. We extend the translation with the following clauses:

$$
\begin{aligned}
ST_x(\mathbf{j}_i) &= x \succeq y_i \\
ST_x(\mathbf{m}_i) &= \neg(x \preceq z_i) \\
ST_x(\top) &= x = x \\
ST_x(\bot) &= \neg(x = x) \\
ST_x({\sim}^\flat \phi) &= \exists x'((x')^* = x \wedge \neg ST_{x'}(\phi)) \\
ST_x({\sim}^\sharp \phi) &= \forall x'((x')^* = x \rightarrow \neg ST_{x'}(\phi)) \\
ST_x(\phi \prec \psi) &= \exists x'(x' \preceq x \wedge ST_{x'}(\phi) \wedge \neg ST_{x'}(\psi)) \\
ST_x(\phi \Rightarrow \psi) &= \forall x'(x' \succeq x \wedge ST_{x'}(\phi) \rightarrow ST_{x'}(\psi)) \\
ST_x(\phi \hookrightarrow \psi) &= \forall x' \forall x''(Rx'xx'' \wedge ST_{x'}(\phi) \rightarrow ST_{x''}(\psi))
\end{aligned}
$$

where x', x'' are fresh individual variables, and $x \preceq x'$ is shorthand for $\exists x''(O(x'') \wedge R(x''xx'))$.

The definition of **positive and negative occurrences of logical connectives, propositional variables, nominals and co-nominals** is extended to $\mathcal{L}_R^+$-formulae ϕ in the expected way. In particular:

- When $\phi = \mathbf{i}$ ($\phi = \mathbf{m}$ / $\top$ / $\bot$), the nominal $\mathbf{i}$ (co-nominal $\mathbf{m}$ / constant $\top$ / constant $\bot$) occurs positively in the formula ϕ.
- When $\phi = {\sim}^\flat \psi$ or $\phi = {\sim}^\sharp \psi$, all positive (resp. negative) occurrences of connectives (incl. variables, constants, nominals and co-nominals) in the subformula ψ are negative (resp. positive) occurrences of these connectives in ϕ. Besides, the occurrence of ${\sim}^\flat$ or ${\sim}^\sharp$ as the main connective is positive in ϕ.
- When $\phi = \psi \prec \chi$, all positive (resp. negative) occurrences of connectives (incl. variables, constants, nominals and co-nominals) in the subformula ψ are also positive (resp. negative) occurrences of these connectives in ϕ, whereas all positive (resp. negative) occurrences of connectives (incl. variables) in the subformula χ are negative (resp. positive) occurrences of these connectives in ϕ. Besides, the occurrence of $\prec$ as a main connective is positive in ϕ.
- When $\phi = \psi \Rightarrow \chi$, all positive (resp. negative) occurrences of connectives (incl. variables, constants, nominals and co-nominals) in the subformula χ are also positive (resp. negative) occurrences of these connectives in ϕ, whereas all positive (resp. negative) occurrences of connectives (incl. variables) in the subformula ψ are negative (resp. positive) occurrences of these connectives in ϕ. Besides, the occurrence of $\Rightarrow$ as a main connective is positive in ϕ.
- The clause for $\phi = \psi \hookrightarrow \chi$ is verbatim the same as for $\phi = \psi \Rightarrow \chi$, but replacing $\Rightarrow$ with $\hookrightarrow$.

Extending the complex algebraic operations for the additional connectives of $\mathcal{L}_R^+$ and identifying their salient properties is quite straightforward.

Some terminology: given an atom $a \in$ ATOMS, two RM-valuations, V and V', in a RM-frame $\mathcal{F}$ are called a**-variants** (notation $V \approx_a V'$), if $V(b) = V'(b)$ for all $b \in \text{ATOMS} \setminus \{a\}$.

The following observations are routine to verify from the semantic definitions:

Proposition 4.3 *For every RM-model $\mathcal{M}$ and formulae $\phi, \psi, \chi \in \mathcal{L}_R^+$, the following equivalences hold:*

1. $[\![\sim\phi]\!]_\mathcal{M} \subseteq [\![\psi]\!]_\mathcal{M}$ *iff* $[\![\sim^\flat\psi]\!]_\mathcal{M} \subseteq [\![\phi]\!]_\mathcal{M}$,
2. $[\![\phi]\!]_\mathcal{M} \subseteq [\![\sim\psi]\!]_\mathcal{M}$ *iff* $[\![\psi]\!]_\mathcal{M} \subseteq [\![\sim^\sharp\phi]\!]_\mathcal{M}$,
3. $[\![\phi]\!]_\mathcal{M} \subseteq [\![\psi \vee \chi]\!]_\mathcal{M}$ *iff* $[\![\phi \prec \psi]\!]_\mathcal{M} \subseteq [\![\chi]\!]_\mathcal{M}$,
4. $[\![\phi \wedge \psi]\!]_\mathcal{M} \subseteq [\![\chi]\!]_\mathcal{M}$ *iff* $[\![\phi]\!]_\mathcal{M} \subseteq [\![\psi \Rightarrow \chi]\!]_\mathcal{M}$,
5. $[\![\phi \circ \psi]\!]_\mathcal{M} \subseteq [\![\chi]\!]_\mathcal{M}$ *iff* $[\![\phi]\!]_\mathcal{M} \subseteq [\![\psi \rightarrow \chi]\!]_\mathcal{M}$,
6. $[\![\phi \circ \psi]\!]_\mathcal{M} \subseteq [\![\chi]\!]_\mathcal{M}$ *iff* $[\![\psi]\!]_\mathcal{M} \subseteq [\![\phi \hookrightarrow \chi]\!]_\mathcal{M}$,
7. $[\![\phi]\!]_\mathcal{M} \subseteq [\![\psi \rightarrow \chi]\!]_\mathcal{M}$ *iff* $[\![\psi]\!]_\mathcal{M} \subseteq [\![\phi \hookrightarrow \chi]\!]_\mathcal{M}$.

These equivalences say that the interpretations of the respective connectives in the complex algebra are each others' (co-)residuals or adjoints.

For the purpose of the algorithm PEARL, we will combine formulae of $\mathcal{L}_R^+$ into set-theoretic versions of sequents of formulae, as follows: an **inclusion** is an expression of the form $\phi \subseteq \psi$ for $\phi, \psi \in \mathcal{L}_R^+$, while a **quasi-inclusion** is an expression $\phi_1 \subseteq \psi_1, \ldots, \phi_n \subseteq \psi_n \vdash \phi \subseteq \psi$ where $\phi_1, \ldots, \phi_n, \psi_1, \ldots, \psi_n, \phi, \psi \in \mathcal{L}_R^+$. The semantics of these expressions is as expected: an inclusion $\phi \subseteq \psi$ is true in a Routley–Meyer model $\mathcal{M}$, denoted

$$\mathcal{M} \Vdash \phi \subseteq \psi,$$

iff $[\![\phi]\!]_\mathcal{M} \subseteq [\![\psi]\!]_\mathcal{M}$, while a quasi-inclusion $\phi_1 \subseteq \psi_1, \ldots, \phi_n \subseteq \psi_n \vdash \phi \subseteq \psi$ is true in $\mathcal{M}$, denoted

$$\mathcal{M} \Vdash \phi_1 \subseteq \psi_1, \ldots, \phi_n \subseteq \psi_n \vdash \phi \subseteq \psi$$

iff $[\![\phi_i]\!]_\mathcal{M} \nsubseteq [\![\psi_i]\!]_\mathcal{M}$ for some $1 \leq i \leq n$ or $[\![\phi]\!]_\mathcal{M} \subseteq [\![\psi]\!]_\mathcal{M}$. Now, the notions of validity of inclusions and quasi-inclusions in a RM-frame is defined in the expected way, as validity in all RM-models on that frame.

The formulae in $\mathcal{L}_R^+$ will be treated as a special type of inclusions, viz. a formula ϕ will be identified with the inclusion $\mathbf{t} \subseteq \phi$. Clearly, this identification complies with the semantics of formulae and inclusions.

4.3.2 *The Rules of* PEARL

Here, we introduce the rewrite rules of our calculus.[3] Most of these rules will be invertible, indicated by a double line.

[3] The rules introduced in this section can be seen as specializations of the rules of the general-purpose algorithm ALBA (Conradie and Palmigiano 2019) to the language and semantics of relevance logic. However, the fact that the complex algebras of Routley–Meyer frames are distributive lattice expan-

Every rule applies in the context of a set on inclusions, which are either free-standing (before the First approximation rule is applied), or are in the antecedent of a quasi-inclusion (after the application of the First approximation rule).

In order to claim soundness of each rule (to be shown in Sect. 4.3.4), we have to specify where and how it is applicable, with the following possible options:

1. to one or two individual inclusions, taken as its premises, but only in the antecedent of a quasi-inclusion.
2. to one or two individual inclusions, taken as its premises, in any context.
3. (only for the Ackermann-rules) to the set of all inclusions in the antecedent of a quasi-inclusion.

Unless otherwise indicated, case 2 above will be assumed by default.

Monotone Variable Elimination Rules

$$\frac{\alpha(p) \subseteq \beta(p)}{\alpha(\bot) \subseteq \beta(\bot)}\,(\bot) \qquad \frac{\gamma(p) \subseteq \chi(p)}{\gamma(\top) \subseteq \chi(\top)}\,(\top)$$

These two rules apply to inclusions and come with the following side conditions:

- for $(\bot)$: that α is negative in p and β is positive in p.
- for $(\top)$: that γ is positive in p and χ is negative in p.

First Approximation Rule

$$\frac{\phi \subseteq \psi}{\mathbf{j} \subseteq \phi,\ \psi \subseteq \mathbf{m} \vdash \mathbf{j} \subseteq \mathbf{m}}\,(\mathrm{FA})$$

where $\mathbf{j}$ is a nominal and $\mathbf{m}$ is a co-nominal not occurring in ϕ or ψ.[4] These are implicitly universally quantified over in the quasi-inclusion. This rule applies to inclusions, possibly in the context of a list of other inclusions, which it turns into quasi-inclusions.

It *does not* apply to inclusions within the antecedents of quasi-inclusions.

Approximation Rules

$$\frac{\chi \to \phi \subseteq \mathbf{m}}{\mathbf{j} \to \phi \subseteq \mathbf{m},\ \ \mathbf{j} \subseteq \chi}\,(\to\text{Appr-Left}) \qquad \frac{\chi \to \phi \subseteq \mathbf{m}}{\chi \to \mathbf{n} \subseteq \mathbf{m},\ \ \phi \subseteq \mathbf{n}}\,(\to\text{Appr-Right})$$

sions allows us to present these rules in a simpler style closer to that of Conradie and Palmigiano (2012) and, to some extent, Conradie et al. (2006a).

[4] This requirement is only needed for the inverse rule, but we impose it on both, to preserve the equivalence.

$$\frac{\mathbf{i} \subseteq \chi \circ \phi}{\mathbf{i} \subseteq \mathbf{j} \circ \phi,\ \ \mathbf{j} \subseteq \chi} (\circ\text{Appr-Left}) \qquad \frac{\mathbf{i} \subseteq \chi \circ \phi}{\mathbf{i} \subseteq \chi \circ \mathbf{k},\ \ \mathbf{k} \subseteq \phi} (\circ\text{Appr-Right})$$

$$\frac{\sim\phi \subseteq \mathbf{m}}{\phi \subseteq \mathbf{n},\ \ \sim\mathbf{n} \subseteq \mathbf{m}} (\sim\text{Appr-Left}) \qquad \frac{\mathbf{i} \subseteq \sim\phi}{\mathbf{j} \subseteq \phi,\ \mathbf{i} \subseteq \sim\mathbf{j}} (\sim\text{Appr-Right})$$

These rules apply to inclusions in the antecedents of quasi-inclusions, and have the requirement that the nominals and co-nominals introduced by them need to be *fresh*, i.e. do not occur in the derivation thus far. Thus, they are introduced as witnesses of existentially quantified inclusions.

Residuation Rules

$$\frac{\phi \subseteq \chi \vee \psi}{\phi \prec \chi \subseteq \psi} (\vee\text{Res}) \qquad \frac{\chi \wedge \psi \subseteq \phi}{\chi \subseteq \psi \Rightarrow \phi} (\wedge\text{Res}) \qquad \frac{\phi \subseteq \chi \rightarrow \psi}{\phi \circ \chi \subseteq \psi} (\rightarrow\text{Res})$$

$$\frac{\phi \circ \psi \subseteq \chi}{\psi \subseteq \phi \hookrightarrow \chi} (\circ\text{Res}) \qquad \frac{\phi \subseteq \psi \rightarrow \chi}{\psi \subseteq \phi \hookrightarrow \chi} (\rightarrow_1\text{Res})$$

Adjunction Rules

$$\frac{\phi \vee \chi \subseteq \psi}{\phi \subseteq \psi \quad \chi \subseteq \psi} (\vee\text{Adj}) \qquad \frac{\psi \subseteq \phi \wedge \chi}{\psi \subseteq \phi \quad \psi \subseteq \chi} (\wedge\text{Adj})$$

$$\frac{\sim\phi \subseteq \psi}{\sim^\flat \psi \subseteq \phi} (\sim\text{Left-Adj}) \qquad \frac{\phi \subseteq \sim\psi}{\psi \subseteq \sim^\sharp \phi} (\sim\text{Right-Adj})$$

Not to clutter the procedure with extra rules, we allow commuting the arguments of $\wedge$ and $\vee$ whenever needed before applying the rules above. Recall that, in case $*$ is assumed to be an involution, $\sim^\flat$ and $\sim^\sharp$ both coincide with $\sim$.

Ackermann-Rules

The Ackermann-rules given apply to the set of **all** inclusions in the antecedent of a quasi-inclusion, but only involve the variable (call it p) that is being eliminated.

The following conditions apply to the rules below:

- p does not occur in α,
- $\beta_1(p), \ldots, \beta_m(p)$ are positive in p, and
- $\gamma_1(p), \ldots, \gamma_m(p)$ are negative in p.
- p does not occur in any other inclusion in the antecedent of the quasi-inclusion to which the rule is applied.

Right Ackermann-rule:

$$\frac{\alpha \subseteq p,\ \ \beta_1(p) \subseteq \gamma_1(p), \ldots, \beta_m(p) \subseteq \gamma_m(p)}{\beta_1(\alpha) \subseteq \gamma_1(\alpha), \ldots, \beta_m(\alpha) \subseteq \gamma_m(\alpha)} \ (RAR)$$

Left Ackermann-rule:

$$\frac{p \subseteq \alpha,\ \ \gamma_1(p) \subseteq \beta_1(p), \ldots, \gamma_m(p) \subseteq \beta_m(p)}{\gamma_1(\alpha) \subseteq \beta_1(\alpha), \ldots, \gamma_m(\alpha) \subseteq \beta_m(\alpha)} \ (LAR)$$

Note that the rules ($\bot$) and ($\top$) are, in fact, special cases of the Ackermann-rules (RAR) and (LAR), respectively.

Simplification Rules

In the rules below Γ is a possibly empty list of inclusions.

$$\frac{\Gamma,\ \mathbf{i} \subseteq \phi\ \vdash\ \mathbf{i} \subseteq \psi}{\Gamma\ \vdash\ \phi \subseteq \psi} \text{ (Simpl-Left)} \qquad \frac{\Gamma,\ \psi \subseteq \mathbf{m}\ \vdash\ \phi \subseteq \mathbf{m}}{\Gamma\ \vdash\ \phi \subseteq \psi} \text{ (Simpl-Right)}$$

In the rule (Simpl-Left) the nominal **i** must not occur in ϕ, or ψ, or any inclusion in Γ. Likewise, in the rule (Simpl-Right) the co-nominal **m** must not occur in ϕ, or ψ, or any inclusion in Γ. These rules are usually applied in the post-processing, to eliminate nominals and co-nominals introduced by the approximation rules.

4.3.3 *Description of* PEARL

PEARL is a non-deterministic algorithmic procedure, the purpose of which is to eliminate propositional variables from inclusions, while maintaining frame validity. It consists of 3 main phases, which we will describe further and will illustrate with a running example.

We will illustrate the phases of the algorithm PEARL with the formula

$$\psi = (p \to q) \land (q \to r) \to (p \to r).$$

I. Pre-processing.
The algorithm starts with an input formula $\psi \in \mathcal{L}_R$, represented as the **initial inclusion t** $\subseteq \psi$. In our running example, the initial inclusion is

$$\mathbf{t} \subseteq (p \to q) \land (q \to r) \to (p \to r).$$

Remark More generally, the algorithm can start with any input inclusion $\phi \subseteq \psi$ in $\mathcal{L}_R^+$, with no difference in what follows; in particular, it can be applied likewise to formulae in $\mathcal{L}_R^+$.

The pre-processing considers the inclusion $\phi \subseteq \psi$ and applies the following transformations:

1. **Distribution rules**. Apply the following equivalences to surface positive (negative) occurrences of $\vee$ and negative (positive) occurrences of $\wedge$ in the left-hand (right-hand) side of the inclusion.

   $(\phi \vee \psi) \to \theta \equiv (\phi \to \theta) \wedge (\psi \to \theta)$,
   $\phi \to (\psi \wedge \theta) \equiv (\phi \to \psi) \wedge (\psi \to \theta)$,
   $(\phi \vee \psi) \circ \theta \equiv (\phi \circ \theta) \vee (\psi \circ \theta)$,
   $\phi \circ (\psi \vee \theta) \equiv (\phi \circ \psi) \vee (\phi \circ \theta)$,
   $(\phi \vee \psi) \wedge \theta \equiv (\phi \wedge \theta) \vee (\psi \wedge \theta)$,
   $\theta \wedge (\phi \vee \psi) \equiv (\theta \wedge \phi) \vee (\theta \wedge \psi)$,
   $(\phi \wedge \psi) \vee \theta \equiv (\phi \vee \theta) \wedge (\psi \vee \theta)$,
   $\theta \vee (\phi \wedge \psi) \equiv (\theta \vee \phi) \wedge (\theta \vee \psi)$,
   $\sim(\phi \vee \psi) \equiv \sim\phi \wedge \sim\psi$,
   $\sim(\phi \wedge \psi) \equiv \sim\phi \vee \sim\psi$.
2. Apply the ($\vee$Adj) and ($\wedge$Adj) rules to split inclusions into two, where possible.
3. Apply the **monotone variable elimination rules** ($\top$) and ($\bot$) wherever applicable.
   Thus, the pre-processing so far may split the original inclusion into a number of inclusions, on each of which the first two phases of the algorithm proceed separately.
4. Apply the **First approximation rule** (FA) to each inclusion. As a result, each inclusion is converted into a quasi-inclusion consisting of an implication with two inclusions in the antecedent, and one inclusion in the consequent. The inclusion in the consequent contains no propositional variables and thus all steps after this point are aimed only at eliminating propositional variables from the two inclusions in the antecedent.

The purpose of that pre-processing is to get the inclusions in the right shape so that the other rules can be applied.

The applicable pre-processing in our running example consists only of the last step, applying (FA), to produce

$$\mathbf{i} \subseteq \mathbf{t},\ (p \to q) \wedge (q \to r) \to (p \to r) \subseteq \mathbf{m} \quad \vdash \quad \mathbf{i} \subseteq \mathbf{m}.$$

Remark If O is generated by a singleton, then $\mathbf{t}$ itself is a nominal, semantically speaking, which can be used instead of $\mathbf{i}$ in this step.

II. Main (elimination) phase.
In this phase, the resulting system of quasi-inclusions is transformed by alternating the following two sub-phases:

1. **Applying the residuation, adjunction and approximation rules**. These rules prepare the antecedents of quasi-inclusions for the application of the Ackermann-rules. The residuation and adjunction rules are straightforward applications of properties of the operations on complex algebras. The approximation rules are a little more intricate. They are based on the $\cup$-primeness of nominals and $\cap$-primeness of co-nominals and the proof of their soundness in Sect. 4.3.4 will use the properties listed in Propositions 4.1 and 4.2.

   Note that the alternation of the two sub-phases is only needed because the polarity (left or right) of applications of the Ackermann-rules to the variables to be eliminated determines how the residuation, adjunction and approximation rules should be applied. If the right polarity is known or guessed in advance, there is no need to alternate; otherwise, the second sub-phase may fail because a wrong polarity was chosen, and then backtracking may be needed to change the preparation for applying the Ackermann-rules with different polarity. When dealing with inductive inclusions defined in Sect. 4.4.3, a *strategy* dictating the polarity in which these rules are to be applied is determined by the way in which the formula is analysed syntactically when judged to be inductive, as it will be described there. Thus, there is no need for alternation of the sub-phases when applied to such formulae.
2. **Applying the Ackermann-rules** (RAR) and (LAR) to the quasi-inclusions to eliminate propositional variables. After each application, some of the other rules may become applicable again, before the next application of the Ackermann-rules is enabled.

Eventually, the algorithm either succeeds to eliminate all variables or it reaches a stage where there are still variables but no further applications of the Ackermann-rules can be enabled. Then the algorithm fails.

Here is the elimination phase for our running example.

1. The quasi-inclusion produced in the pre-processing:

$$\mathbf{i} \subseteq \mathbf{t},\ (p \to q) \wedge (q \to r) \to (p \to r) \subseteq \mathbf{m} \ \vdash\ \mathbf{i} \subseteq \mathbf{m}.$$

2. Apply ($\to$Appr-Left), and then ($\to$Appr-Right) to produce

$$\mathbf{i} \subseteq \mathbf{t},\ \mathbf{j} \subseteq (p \to q) \wedge (q \to r),\ p \to r \subseteq \mathbf{n},\ \mathbf{j} \to \mathbf{n} \subseteq \mathbf{m} \ \vdash\ \mathbf{i} \subseteq \mathbf{m}.$$

3. Apply the adjunction rule ($\wedge$Adj) to the 2nd inclusion above, to obtain

$$\mathbf{i} \subseteq \mathbf{t},\ \mathbf{j} \subseteq p \to q,\ \mathbf{j} \subseteq q \to r,\ p \to r \subseteq \mathbf{n},\ \mathbf{j} \to \mathbf{n} \subseteq \mathbf{m} \ \vdash\ \mathbf{i} \subseteq \mathbf{m}.$$

4. Apply the approximation rule ($\rightarrow$Appr-Left) to $p \rightarrow r \subseteq \mathbf{n}$ to produce

$$\mathbf{i} \subseteq \mathbf{t},\ \mathbf{j} \subseteq p \rightarrow q,\ \mathbf{j} \subseteq q \rightarrow r,\ \mathbf{k} \subseteq p,\ \mathbf{k} \rightarrow r \subseteq \mathbf{n},\ \mathbf{j} \rightarrow \mathbf{n} \subseteq \mathbf{m}\ \vdash\ \mathbf{i} \subseteq \mathbf{m}.$$

5. Apply the Ackermann-rule with respect to p to the inclusions 2 and 4 that contain it, to obtain

$$\mathbf{i} \subseteq \mathbf{t},\ \mathbf{j} \subseteq \mathbf{k} \rightarrow q,\ \mathbf{j} \subseteq q \rightarrow r,\ \mathbf{k} \rightarrow r \subseteq \mathbf{n},\ \mathbf{j} \rightarrow \mathbf{n} \subseteq \mathbf{m}\ \vdash\ \mathbf{i} \subseteq \mathbf{m}.$$

6. Apply $\circ$-residuation to $\mathbf{j} \subseteq \mathbf{k} \rightarrow q$ to obtain

$$\mathbf{i} \subseteq \mathbf{t},\ \mathbf{j} \circ \mathbf{k} \subseteq q,\ \mathbf{j} \subseteq q \rightarrow r,\ \mathbf{k} \rightarrow r \subseteq \mathbf{n},\ \mathbf{j} \rightarrow \mathbf{n} \subseteq \mathbf{m}\ \vdash\ \mathbf{i} \subseteq \mathbf{m}.$$

7. Apply again the Ackermann-rule, now to eliminate q:

$$\mathbf{i} \subseteq \mathbf{t},\ \ \mathbf{j} \subseteq (\mathbf{j} \circ \mathbf{k}) \rightarrow r,\ \mathbf{k} \rightarrow r \subseteq \mathbf{n},\ \mathbf{j} \rightarrow \mathbf{n} \subseteq \mathbf{m}\ \vdash\ \mathbf{i} \subseteq \mathbf{m}.$$

8. Applying $\circ$-residuation to $\mathbf{j} \subseteq (\mathbf{j} \circ \mathbf{k}) \rightarrow r$ produces

$$\mathbf{i} \subseteq \mathbf{t},\ \ \mathbf{j} \circ (\mathbf{j} \circ \mathbf{k}) \subseteq r,\ \mathbf{k} \rightarrow r \subseteq \mathbf{n},\ \mathbf{j} \rightarrow \mathbf{n} \subseteq \mathbf{m}\ \vdash\ \mathbf{i} \subseteq \mathbf{m}.$$

9. Now, r can be eliminated by on last application of the Ackermann-rule, to produce the pure quasi-inclusion

$$\mathbf{i} \subseteq \mathbf{t},\ \mathbf{k} \rightarrow \mathbf{j} \circ (\mathbf{j} \circ \mathbf{k}) \subseteq \mathbf{n},\ \mathbf{j} \rightarrow \mathbf{n} \subseteq \mathbf{m}\ \vdash\ \mathbf{i} \subseteq \mathbf{m}.$$

Since all propositional variables have been successfully eliminated, this is the end of the elimination phase.

III. Post-processing.

This phase applies if/when the algorithm succeeds to eliminate all variables, thus ending with **pure quasi-inclusions**, containing only nominals and co-nominals, but no variables.

The purpose of the post-processing is to produce the first-order condition equivalent to the input formula. Each pure quasi-inclusion produced in the elimination phase is post-processed separately to produce a corresponding FO condition, and all these are then taken conjunctively to produce the corresponding FO condition of the input formula. So, we focus on the case of a single pure quasi-inclusion.

Computing a first-order equivalent of any pure quasi-inclusion can be done by straightforward application of the standard translation, but the result would usually be unnecessarily long and complicated. To avoid that, our post-processing of a pure quasi-inclusion starts with several optional simplification steps, involving the two simplification rules, as well as applications of residuation, adjunction and approximation rules and their inverses, wherever applicable, in order to reduce the number of nominals and co-nominals introduced in the elimination phase. Ideally, these simplification steps should end with a single pure inclusion.

The simplification sub-phase is illustrated in the running example as follows:

1. Applying the simplification rule (Simpl-Left) to the pure quasi-inclusion produced in the elimination phase:

$$\mathbf{k} \to \mathbf{j} \circ (\mathbf{j} \circ \mathbf{k}) \subseteq \mathbf{n},\ \mathbf{j} \to \mathbf{n} \subseteq \mathbf{m}\ \vdash\ \mathbf{t} \subseteq \mathbf{m}.$$

   This step would be redundant if the FA rules is accordingly modified when applied to formulae.
2. Applying the simplification rule (Simpl-Right) to the result produces

$$\mathbf{k} \to \mathbf{j} \circ (\mathbf{j} \circ \mathbf{k}) \subseteq \mathbf{n}\ \vdash\ \mathbf{t} \subseteq \mathbf{j} \to \mathbf{n}.$$

3. Then applying residuation ($\to$Res) on the right produces

$$\mathbf{k} \to \mathbf{j} \circ (\mathbf{j} \circ \mathbf{k}) \subseteq \mathbf{n}\ \vdash\ \mathbf{t} \circ \mathbf{j} \subseteq \mathbf{n}.$$

4. Again applying the simplification rule (Simpl-Right) produces

$$\mathbf{t} \circ \mathbf{j} \subseteq \mathbf{k} \to \mathbf{j} \circ (\mathbf{j} \circ \mathbf{k}).$$

5. After another residuation (to reduce the nesting depth on the right) we obtain

$$(\mathbf{t} \circ \mathbf{j}) \circ \mathbf{k} \subseteq \mathbf{j} \circ (\mathbf{j} \circ \mathbf{k}).$$

The next step is to compute the first-order equivalent. First, let us rewrite the resulting pure inclusion to replace the metavariables with concrete nominals and co-nominals, always picking the first ones available in the respective lists of designated variables.[5] Here is the result of rewriting our example:

$$(\mathbf{t} \circ \mathbf{j}_1) \circ \mathbf{j}_2 \subseteq \mathbf{j}_1 \circ (\mathbf{j}_1 \circ \mathbf{j}_2).$$

Recall that all nominals and co-nominals, as well as the current state of evaluation, are implicitly universally quantified. So, now we re-instate the universal quantifiers over all of them in the first step of the standard translation which associates with each nominal $\mathbf{j}_i$ the designated variable x_i denoting the element generating the up-set $\uparrow x_i$ where $\mathbf{j}_i$ is true. Likewise, the standard translation associates with each co-nominal $\mathbf{m}_i$ the designated individual variable y_i denoting the generator of the co-downset where $\mathbf{m}$ is true, i.e. $[\![\mathbf{m}]\!] = (\downarrow y_i)^c$. Lastly, recall that the constant $\mathbf{t}$ is translated into the set O, so we also universally quantify over its elements. In the running example, that step produces

[5] In practice, each rule should be applied to such concrete nominals and co-nominals, but we have used metavariables to avoid the extra technical bookkeeping.

$$\forall w \forall o \in O \forall x_i \forall x_2 \big(ST_w((\mathbf{t} \circ \mathbf{j}_1) \circ \mathbf{j}_2 \subseteq \mathbf{j}_1 \circ (\mathbf{j}_1 \circ \mathbf{j}_2)) \big).$$

The standard translation will now produce a simpler first-order equivalent, but it still does not take into account the monotonicity or anti-monotonicity of the valuations of nominals and co-nominals, as well as those of the relation R and the function *. Thus, further simplifications are possible, in fact desirable, at this stage. For these one can use a simple 'post-processing simplification guide', partly completed in Table 4.1. In that table:

- $[\![\psi]\!]$ is the extension of the pure formula ψ in the given frame, expressed as a FO formula.
- for any sets $X, Y \subseteq W$ and $w \in W$, the expression $RXYw$ is a shorthand for $\exists x \in X \exists y \in Y\, Rxyw$, respectively, simplified when X or Y is a singleton.

This table can be used to simplify on-the-fly the computation of the first-order equivalent, and can be applied separately to the left- and right-hand sides of the inclusion. Our example is computed as follows, using a more intuitive shorthand language (note that the quantification $\forall o \in O$ is now redundant while we still work with the constant $\mathbf{t}$):

$$\forall w \forall x_1 \forall x_2 \big(w \Vdash (\mathbf{t} \circ \mathbf{j}_1) \circ \mathbf{j}_2 \Rightarrow w \Vdash \mathbf{j}_1 \circ (\mathbf{j}_1 \circ \mathbf{j}_2) \big)$$

$$\forall w \forall x_1 \forall x_2 \big(R[\![\mathbf{t} \circ \mathbf{j}_1]\!] x_2 w \Rightarrow R x_1 [\![(\mathbf{j}_1 \circ \mathbf{j}_2)]\!] w \big)$$

$$\forall w \forall x_1 \forall x_2 \big(\exists u (u \in [\![\mathbf{t} \circ \mathbf{j}_1]\!] \;\&\; R u x_2 w) \Rightarrow \exists u (u \in [\![(\mathbf{j}_1 \circ \mathbf{j}_2)]\!] \;\&\; R x_1 u w) \big)$$

$$\forall w \forall x_i \forall x_2 \big(\exists u (x_1 \preceq u \;\&\; R u x_2 w) \Rightarrow \exists u (R x_1 x_2 u \;\&\; R x_1 u w) \big).$$

Table 4.1 Post-processing simplification table

Truth of simple pure formulae	Corresponding FO conditions
$w \Vdash {\sim}\mathbf{i}$	$x_{\mathbf{i}} \npreceq w^*$
$w \Vdash {\sim}^{\flat}\mathbf{i}$	$\exists u(u^* = w \;\&\; x_{\mathbf{i}} \npreceq u)$
$w \Vdash {\sim}^{\sharp}\mathbf{i}$	$\forall u(u^* = w \Rightarrow x_{\mathbf{i}} \npreceq u)$
$w \Vdash \mathbf{i} \circ \mathbf{j}$	$R x_{\mathbf{i}} x_{\mathbf{j}} w$
$w \Vdash \mathbf{t} \circ \mathbf{j}$	$x_{\mathbf{j}} \preceq w$
$w \Vdash \mathbf{i} \circ \mathbf{t}$	$R x_{\mathbf{i}} O w$
$w \Vdash \mathbf{i} \circ \psi$	$R x_{\mathbf{i}} [\![\psi]\!] w$
$w \Vdash \phi \circ \mathbf{j}$	$R [\![\phi]\!] x_{\mathbf{j}} w$
$w \Vdash \phi \circ \psi$	$R [\![\phi]\!] [\![\psi]\!] w$
$w \Vdash \mathbf{i} \to \mathbf{j}$	$\forall z(R w x_{\mathbf{i}} z \Rightarrow x_{\mathbf{j}} \preceq z)$
$w \Vdash \mathbf{i} \to \psi$	$\forall z(R w x_{\mathbf{i}} z \Rightarrow z \in [\![\psi]\!])$
$w \Vdash \phi \to \mathbf{j}$	$\forall z(R w [\![\phi]\!] z \Rightarrow x_{\mathbf{j}} \preceq z)$
$w \Vdash \phi \to \psi$	$\forall z(R w [\![\phi]\!] z \Rightarrow z \in [\![\psi]\!])$

From the above, using the anti-monotonicity of R over the first argument and the definition of R^2, we obtain

$$\forall w \forall x_i \forall x_2 \big(R x_1 x_2 w \Rightarrow R^2 x_1 (x_1 x_2) w\big)$$

which is the semantic condition for our input formula (known as axiom B2) known from Routley et al. (1982).

Another example is worked out in detail in Sect. 4.4.3. More examples are sketched in less detail in Appendix 2.

4.3.4 *Soundness of the Rules and Correctness of* PEARL

Here, we prove that the procedure PEARL is *correct*, in the sense of preserving validity in any given RM-frame both ways—from the initial inclusion to the final pure quasi-inclusion and vice versa. Here is the formal claim:

Theorem 4.1 (Correctness) *If* PEARL *transforms an initial inclusion Γ_0 into several quasi-inclusions $\Delta_1, \ldots, \Delta_k$ then for every RM-frame $\mathcal{F}$, the following holds: $\mathcal{F} \vdash \Gamma_0$ iff $\mathcal{F} \vdash \Delta_i$ for each $i = 1, \ldots, k$.*

We will prove the claim by showing that every rule is *sound* in the sense of preserving the validity of the current inclusion or quasi-inclusion to which it is applied in both directions. For most of the rules the argument is quite simple and an even stronger claim can be proved, viz. that the rule preserves validity both ways between premises and the conclusions *in every RM-model*. For some of the rules, however, viz the approximation rules and the Ackermann-rules, the argument must be done globally, for the entire quasi-inclusion. We proceed with the cases according to the various types of rules described in Sect. 4.3.2.

Pre-processing distribution rules. The soundness of these rules follow immediately from Proposition 4.1.

Monotone variable elimination rules. These rules are sound for preservation of frame validity, i.e.: under the assumptions on α, β, γ and δ, in any Routley–Meyer frame $\mathcal{T}$, it holds that $\mathcal{T} \Vdash \alpha(p) \subseteq \beta(p)$ iff $\mathcal{T} \Vdash \alpha(\perp) \subseteq \beta(\perp)$ and $\mathcal{T} \Vdash \gamma(p) \subseteq \chi(p)$ iff $\mathcal{T} \Vdash \gamma(\top) \subseteq \chi(\top)$. Indeed, the preservation from top to bottom in both rules is immediate, by substitution of p with $\perp$, resp. $\top$. The preservation from bottom to top for $(\perp)$ is by the chain of inclusions $\alpha(p) \subseteq \alpha(\perp) \subseteq \beta(\perp) \subseteq \beta(p)$, using the antitonicity of α and the monotonicity of β. Likewise for the rule $(\top)$.

Splitting rules for $\wedge$ and $\vee$. These rules are trivially sound.

First approximation rule. Soundness from top to bottom follows by the transitivity of set inclusion, i.e. the fact that for any $X, Y, Z, U \in \mathcal{P}^{\uparrow}(W)$, if $X \subseteq Y$, $Z \subseteq X$ and $Y \subseteq U$, then $Z \subseteq U$. Soundness from bottom to top follows from Proposition 4.2.

Approximation rules. For each of these, consider an arbitrary RM-frame $\mathcal{F} = \langle W, O, R, ^* \rangle$ and show preservation of validity of the entire quasi-inclusion in $\mathcal{F}$

from the premise to the conclusion and vice versa, recalling that all nominals and co-nominals in Γ are universally quantified.

($\rightarrow$Appr-Left) For any RM-model $\mathcal{M}$ over $\mathcal{F}$ the following chain of equivalences holds, by Propositions 4.1 and 4.2:

$\mathcal{M} \Vdash \chi \rightarrow \phi \subseteq \mathbf{m}$ iff

$[\![\chi \rightarrow \phi]\!]_{\mathcal{M}} \subseteq [\![\mathbf{m}]\!]_{\mathcal{M}}$ iff

$[\![\chi]\!]_{\mathcal{M}} \rightarrow [\![\phi]\!]_{\mathcal{M}} \subseteq [\![\mathbf{m}]\!]_{\mathcal{M}}$ iff

$\left(\bigcup\{\uparrow x \mid x \in [\![\chi]\!]_{\mathcal{M}}\}\right) \rightarrow [\![\phi]\!]_{\mathcal{M}} \subseteq [\![\mathbf{m}]\!]_{\mathcal{M}}$ iff

$\bigcap\left(\{\uparrow x \rightarrow [\![\phi]\!]_{\mathcal{M}} \mid x \in [\![\chi]\!]_{\mathcal{M}}\}\right) \subseteq [\![\mathbf{m}]\!]_{\mathcal{M}}$ iff

$\uparrow x_0 \rightarrow [\![\phi]\!]_{\mathcal{M}} \subseteq [\![\mathbf{m}]\!]_{\mathcal{M}}$ for some $x_0 \in W$, such that $x_0 \in [\![\chi]\!]_{\mathcal{M}}$ iff

$[\![\mathbf{j}]\!]_{\mathcal{M}'} \subseteq [\![\chi]\!]_{\mathcal{M}'}$ and $[\![\mathbf{j} \rightarrow \phi]\!]_{\mathcal{M}'} \subseteq [\![\mathbf{m}]\!]_{\mathcal{M}'}$,

where the model $\mathcal{M}'$ is an $\mathbf{j}$-variant of $\mathcal{M}$ such that $[\![\mathbf{j}]\!]_{\mathcal{M}'} = \uparrow x_0$.

Now, the soundness claim in both directions follows immediately, because the implicit existential quantification over $\mathbf{j}$ in the antecedent of the quasi-inclusion in the last step above converts into universal quantification over $\mathbf{j}$ (i.e. over all $\mathbf{j}$-variants of the starting model $\mathcal{M}$) in the entire quasi-inclusion. Thus, the quasi-inclusion with the premise inclusion in its antecedent is valid in all RM-models over $\mathcal{F}$ iff the quasi-inclusion resulting from the application of the rule to that premise in the antecedent is valid in all RM-models over $\mathcal{F}$.

($\rightarrow$Appr-Right) The argument is similar, by showing the following chain of equivalences, again using Propositions 4.1 and 4.2:

$\mathcal{M} \Vdash \chi \rightarrow \phi \subseteq \mathbf{m}$ iff

$[\![\chi \rightarrow \phi]\!]_{\mathcal{M}} \subseteq [\![\mathbf{m}]\!]_{\mathcal{M}}$ iff

$[\![\chi]\!]_{\mathcal{M}} \rightarrow [\![\phi]\!]_{\mathcal{M}} \subseteq [\![\mathbf{m}]\!]_{\mathcal{M}}$ iff

$[\![\chi]\!]_{\mathcal{M}} \rightarrow \left(\bigcap\{(\downarrow x)^c \mid x \notin [\![\phi]\!]_{\mathcal{M}}\}\right) \subseteq [\![\mathbf{m}]\!]_{\mathcal{M}}$ iff

$\bigcap\left(\{[\![\chi]\!]_{\mathcal{M}} \rightarrow (\downarrow x)^c \mid [\![\phi]\!]_{\mathcal{M}} \subseteq (\downarrow x)^c\}\right) \subseteq [\![\mathbf{m}]\!]_{\mathcal{M}}$ iff

$[\![\chi]\!]_{\mathcal{M}} \rightarrow (\downarrow x_0)^c \subseteq [\![\mathbf{m}]\!]_{\mathcal{M}}$ for some $x_0 \in W$, such that $[\![\phi]\!]_{\mathcal{M}} \subseteq (\downarrow x_0)^c$ iff

$[\![\phi]\!]_{\mathcal{M}'} \subseteq [\![\mathbf{n}]\!]_{\mathcal{M}'}$ and $[\![\chi \to \mathbf{n}]\!]_{\mathcal{M}'} \subseteq [\![\mathbf{m}]\!]_{\mathcal{M}'}$,

where the model $\mathcal{M}'$ is an $\mathbf{n}$-variant of $\mathcal{M}$ such that $[\![\mathbf{n}]\!]_{\mathcal{M}'} = (\downarrow x_0)^c$.

Now the argument for soundness in both directions is the same as above.
($\circ$Appr-Left) The argument is analogous to that for ($\to$Appr-Left).
($\circ$Appr-Right) The argument is analogous to that for ($\to$Appr-Right).

Residuation rules. The soundness in both directions of the residuation rules follows immediately from Proposition 4.3.

Adjunction rules. The soundness of ($\vee$Adj) and ($\wedge$Adj) in both directions is straightforward. The soundness of (~Left-Adj) and (~Right-Adj) in both directions follows from Proposition 4.3.

Ackermann-rules. The soundness of the Ackermann-rules is based on a $\mathcal{L}_R$-version of the so-called *Ackermann lemma*, proved by Ackermann in Ackermann (1935) in the context of second-order logic.

We now give two versions of Ackermann's lemma, essentially stating the soundness of the two respective Ackermann-rules.

Lemma 4.1 (Right Ackermann Lemma) *Let $\alpha \in \mathcal{L}_R^+$ with $p \notin \mathsf{PROP}(\alpha)$, let $\beta_1(p), \ldots, \beta_m(p) \in \mathcal{L}_R^+$ be positive in p, and let $\gamma_1(p), \ldots, \gamma_m(p) \in \mathcal{L}_R^+$ be negative in p. Take any valuation V on an RM-frame $\mathcal{F}$. Then,*

$$\mathcal{F}, V \Vdash \beta_j(\alpha/p) \subseteq \gamma_j(\alpha/p) \text{ for all } 1 \leq j \leq m$$

iff there exists some $V' \approx_p V$ such that

$$\mathcal{F}, V' \Vdash \alpha \subseteq p \text{ and } \mathcal{F}, V' \Vdash \beta_j(p) \subseteq \gamma_j(p), \text{ for all } 1 \leq j \leq m.$$

Proof For the implication from top to bottom, let $V'(p) = V(\alpha)$. Since α does not contain p, we have $V'(\alpha) = V(\alpha) = V'(p)$. Moreover, by assumption we get, for each $1 \leq j \leq m$:

$$V'(\beta_j(p)) = V(\beta_j(\alpha/p)) \subseteq V(\gamma_j(\alpha/p)) = V'(\gamma_j(p)).$$

For the implication from bottom to top, we make use of the fact that each β_j is monotone (being positive) in p, while each γ_j is antitone (being negative) in p.

We have $V(\alpha) = V'(\alpha) \subseteq V'(p)$, hence,

$$V(\beta_j(\alpha/p)) \subseteq V'(\beta_j(p)) \subseteq V'(\gamma_j(p)) \subseteq V(\gamma_j(\alpha/p)).$$

The proof of the following version of the lemma is completely analogous.

Lemma 4.2 (Left Ackermann Lemma) *Let $\alpha \in \mathcal{L}_R^+$ with $p \notin \mathsf{PROP}(\alpha)$, let $\beta_1(p), \ldots, \beta_m(p) \in \mathcal{L}_R^+$ be positive in p, and let $\gamma_1(p), \ldots, \gamma_m(p) \in \mathcal{L}_R^+$ be negative in p. Take any valuation V on an RM-frame $\mathcal{F}$. Then,*

$$\mathcal{F}, V \Vdash \gamma_j(\alpha/p) \subseteq \beta_j(\alpha/p) \textit{ for all } 1 \leq j \leq m$$

iff there exists some $V' \approx_p V$ *such that*

$$\mathcal{F}, V' \Vdash p \subseteq \alpha \textit{ and } \mathcal{F}, V' \Vdash \gamma_j(p) \subseteq \beta_j(p), \textit{ for all } 1 \leq j \leq m.$$

Now, the soundness claim in both directions follows immediately, just like in the cases of the approximation rules, because the implicit existential quantification over the valuation V' is in the antecedent of the quasi-inclusion, so it converts into universal quantification over all p-variants of the valuation V in the entire quasi-inclusion. Thus, the quasi-inclusion before the application of the (Left or Right) Ackermann-rule is valid in all RM-models over $\mathcal{F}$ iff the quasi-inclusion resulting from the application of the Ackermann-rule to the antecedent of that quasi-inclusion is valid in all RM-models over $\mathcal{F}$.

Simplification rules. The soundness of these rules follows from Proposition 4.2. This completes the proof of correctness of PEARL.

Corollary 4.1 (Correspondence) *If* PEARL *succeeds in transforming an initial inclusion* $\mathbf{t} \subseteq \phi$ *into the system of pure quasi-inclusions* $\Delta_1, \ldots, \Delta_k$ *with respective FO equivalents* $FO(\Delta_1), \ldots, FO(\Delta_k)$, *then for every RM-frame* $\mathcal{F}$, *the following holds:*

$$\mathcal{F} \Vdash \phi \textit{ iff } \mathcal{F} \Vdash FO(\Delta_1) \wedge \cdots \wedge FO(\Delta_k).$$

Almost all axioms used to define important systems of relevance logics studied in the literature are first-order definable and their first-order equivalents can be computed by PEARL. Actually, almost all of them fall in the syntactic class of *inductive formulae* defined in the next section. In particular, that is the case for all axioms copied from Routley et al. (1982) and listed in Appendix 2. A few of them are worked out there, while the rest we leave to the reader to verify.

4.3.5 *An Example of Failure of* PEARL

The only exception of a first-order definable, but non-inductive axiom, mentioned in the literature that we are currently know is the following:

$$\zeta = ((p \to p) \to q) \to q$$

which is claimed in Routley et al. (1982) to define the following frame condition:

$$\text{(C)} \quad \forall u \exists o \in O\ Ruou.$$

The algorithm PEARL, as presented here, fails on this formula. However, we also claim that the FO condition above is not equivalent to it. Indeed, it is easy to check that the formula is valid in every RM-frame satisfying that condition. However, the following simple RM-frame is a counter-example for the other direction, stating the necessity of that condition. Consider $\mathcal{F} = \langle W, O, R, ^* \rangle$, where

$$W = \{0, 1\};\ O = \{0\};\ R = \{(0, 0, 0), (0, 1, 1), (1, 1, 1)\};\ 0^* = 1, 1^* = 0.$$

Note that the relation $\preceq$ in $\mathcal{F}$ is the identity, so checking that this is an RM-frame is easy. We now claim that:

1. ζ is valid in every RM-model $\mathcal{M}$ over $\mathcal{F}$.
   Indeed, to check $\mathcal{M}, 0 \Vdash ((p \to p) \to q) \to q$ it suffices to note that $p \to p$ is true everywhere for any valuation, because $Ruvw$ implies $v = w$.
2. However, the frame condition (C) fails in $\mathcal{F}$ because $R101$ does not hold.

   We claim that the following frame condition for ζ is the correct one:

$$\text{(C')} \quad \forall u \exists v (\forall z (Rvuz \Rightarrow u \preceq z)\ \&\ Ruvu).$$

It is easy to see that (C) implies (C'). The converse, however, does not always hold, as seen from the frame above, which satisfies (C') but not (C).

A suitable extension of PEARL that does succeed on the formula ζ and computes the frame condition above is currently under construction.

4.4 Inductive Formulae for Relevance Logics

In this section, we first define the notions of Sahlqvist inclusions and formulae in the language $\mathcal{L}_R$ and then extend these to the more general class of inductive inclusions and formulae[6] (Sects. 4.4.1 and 4.4.3). We illustrate these definitions with a number of examples and then show that PEARL successfully computes first-order correspondents for all members of these classes (Sect. 4.4.4). We conclude the section by comparing our definitions to the two other proposals for Sahlqvist formulae in relevance logic in the literature, in Badia (2018), Seki (2003) (Sect. 4.4.5).

For the purposes of proving that PEARL successfully computes first-order correspondents of all Sahlqvist and inductive formulae, it is convenient to define these classes of formulae in terms of their *signed generation trees* (Sect. 4.4.2) and the existence of suitable partitions of certain branches in the latter, following, e.g. Conradie and Palmigiano (2019). However, by exploiting the special features of the syntax of $\mathcal{L}_R$, it is possible to give much simpler definitions which are more convenient for practically identifying inductive and Sahlqvist $\mathcal{L}_R$ formulae, but less suitable for

[6] These definitions are specializations of the general-purpose definition given in Conradie and Palmigiano (2019) for modal logics algebraically captured by classes of normal lattice expansions, which are based purely on the order-theoretic properties of the interpretations of the connectives. We refer readers who are interested in this level of generality and in the algebraic and order-theoretic analysis of the Sahlqvist an inductive classes to that paper.

generalization or for use in proofs. We start off by giving these simpler definitions in Sect. 4.4.1. To readers who only want to know how to identify Sahlqvist and inductive $\mathcal{L}_R$ formulae, but are not interested in the technicalities of proving the success of PEARL on these classes, we recommend reading only Sect. 4.4.1 and skipping the rest of this section.

4.4.1 Sahlqvist and Inductive Formulae: Practical Definitions

When referring to a positive (resp., negative) occurrence of a connectives, say $\wedge$, in a formula, we will often simply write 'an occurrence of $+\wedge$' (resp., 'an occurrence of $-\wedge$') instead of 'a positive occurrence of $\wedge$' (resp., 'a negative occurrence of $\wedge$'), and similarly for the other connectives and also for the propositional variables.

Definition 4.1 An $\mathcal{L}_R$-formula is **Sahlqvist** if, for each propositional variable p, at least one of the following holds:

- no occurrence of $+p$ is in the scope of any occurrence of $-\rightarrow$ or of $+\circ$, or
- no occurrence of $-p$ is in the scope of any occurrence of $-\rightarrow$ or of $+\circ$.

Example 4.1 Consider the formula (K) $(p \rightarrow (q \rightarrow p))$, also discussed in Examples 4.4 and 4.5. Since there are no occurrences of either $-\rightarrow$ or $+\circ$, the formula is Sahlqvist.

Example 4.2 Consider the formula $(p \rightarrow {\sim} p) \rightarrow {\sim} p$, also discussed in Examples 4.4 and 4.6. The first two occurrences of p are positive and in the scope of a $-\rightarrow$, so the first clause of Definition 4.1 does not apply. However, the last occurrence of p is the only negative one, and this is not in the scope of any $-\rightarrow$ or $+\circ$, so the second clause of Definition 4.1 is satisfied and the formula is Sahlqvist.

Notice that, if we were to delete the first negation in the formula we would obtain the formula $(p \rightarrow p) \rightarrow {\sim} p$ where we have both a positive and a negative occurrence of p in the scope of $-\rightarrow$, so the obtained formula would not be Sahlqvist.

Definition 4.2 A **polarity-type** over a set of propositional variables S is a map $\epsilon : S \rightarrow \{+, -\}$. For every polarity-type ϵ, we denote its **opposite polarity-type** by ϵ^{∂}, that is, $\epsilon^{\partial}(p) = +$ iff $\epsilon(p) = -$, for every $p \in S$. We will sometime talk of a **polarity-type over a formula** ϕ when we mean a polarity-type over the set $var(\phi)$ of propositional variables occurring in ϕ.

A positive (negative) occurrence of variable p in a formula ϕ **agrees with a polarity-type** ϵ **over** ϕ if $\epsilon(p) = +$ ($\epsilon(p) = -$), otherwise it **disagrees with** ϵ.

Definition 4.3 An occurrence of a propositional variable in a formula is in **good scope** if it is not in the scope of any occurrence of $-\circ$ or $+\rightarrow$ which is the scope of an occurrence of $-\rightarrow$ or $+\circ$.

Definition 4.4 Given a strict partial ordering $<_{\Omega}$ over $var(\phi)$ and a polarity type ϵ over ϕ, we say that ϕ is (Ω, ϵ)**-inductive** if

1. every variable occurrence that agrees with ϵ is in good scope, and
2. if an occurrence of a variable p which agrees with ϵ is in the scope of an occurrence of $- \to$ $(+\circ)$, then all variables q occurring in the other argument of this $- \to$ $(+\circ)$ disagree with ϵ and $q <_\Omega p$.
3. if an occurrence of a variable p which agrees with ϵ is in the scope of an occurrence of $-\vee$ $(+\wedge)$ which is the scope of an occurrence of $- \to$ or $+\circ$, then all variables q occurring in the other argument of this $-\vee$ $(+\wedge)$ disagree with ϵ and $q <_\Omega p$.

The formula ϕ is said to be **inductive** if it is (Ω, ϵ)**-inductive** for some partial ordering $<_\Omega$ and polarity-type ϵ.

Example 4.3 Consider the axiom (WB) $(p \to q) \wedge (q \to r) \to (p \to r)$, also considered further in Example 4.7. For ease of reference, we indicate the polarity of all variable and connective occurrences:

$$(\overset{+}{p} \overset{-}{\to} \overset{-}{q}) \overset{-}{\wedge} (\overset{+}{q} \overset{-}{\to} \overset{-}{r}) \overset{+}{\to} (\overset{-}{p} \overset{+}{\to} \overset{+}{r})).$$

Note that this formula is not Sahlqvist, as there is both a $-q$ and a $+q$ occurring in the scope of $- \to$. However, it is (Ω, ϵ)-inductive where $p <_\Omega q$, $r <_\Omega q$, and $\epsilon(p) = -, \epsilon(q) = -, \epsilon(r) = +$. Notice that all variable occurrences (and therefore all variable occurrences that agree with ϵ) are in good scope. As there are no occurrences of $+\circ$, $-\vee$ or $+\wedge$ we only need to check the two occurrences of $- \to$ in order to verify that Definition 4.4 is satisfied. Indeed, in $(\overset{+}{p} \overset{-}{\to} \overset{-}{q})$, the occurrence $-q$ agrees with ϵ, while $+p$, which is the other argument of $- \to$, disagrees with ϵ and moreover $p <_\Omega p$; in $(\overset{+}{q} \overset{-}{\to} \overset{-}{r})$ neither $+q$ nor $-r$ agree with ϵ.

Keeping the same choice of dependency order $<_\Omega$, other choices of ϵ under which the formula would be (Ω, ϵ)-inductive are possible, too. For example, $\epsilon(p) = -$, $\epsilon(q) = +$, $\epsilon(r) = +$, or $\epsilon(p) = \epsilon(q) = \epsilon(r) = +$ would both work. However, any choice with $\epsilon(p) = +$ and $\epsilon(q) = -$ would not work, as this would violate the condition 3 of Definition 4.4 in the subformula $(\overset{+}{p} \overset{-}{\to} \overset{-}{q})$.

4.4.2 *Signed Generation Trees for $\mathcal{L}_R$ Formulae*

In this subsection, we restrict all definitions to $\mathcal{L}_R$ formulae and inclusions. They can all be easily extended to $\mathcal{L}_R^+$, as well as with modal operators, but here we restrict our language of interest to $\mathcal{L}_R$.

We now define the notion of a signed generation tree of a formula. As mentioned in the introduction to this section, for the purpose of the forthcoming proofs, the definition of Sahlqvist and inductive inclusions (and formulae) is most conveniently given in terms of these trees. This style of definition first appears in the definition of the Sahlqvist formulae for distributive modal logic (Gehrke et al. 2005).

Definition 4.5 (*Signed Generation Tree*) The **positive** (resp. **negative**) *generation tree* of any $\mathcal{L}_R$-formula ϕ is defined by labelling the root node of the generation tree of ϕ with the sign $+$ (resp. $-$), and then propagating the labelling on each remaining node as follows:

- For any node labelled with $\vee$, $\wedge$ or $\circ$, assign the same sign to its children.
- If a node is labelled with $\sim$, assign the opposite sign to its child.
- If a node is labelled with $\rightarrow$, assign the opposite sign to its left child and the same sign to its right child.

Nodes in signed generation trees are **positive** (resp. **negative**) if they are signed $+$ (resp. $-$).

Example 4.4 Figure 4.1 shows the negative generation tree for the formula $p \rightarrow (q \rightarrow p)$, know as the axiom (K), while Fig. 4.2 shows the negative generation tree of the formula $(p \rightarrow \sim p) \rightarrow \sim p$. The negative generation tree of the (WB) axiom $(p \rightarrow q) \wedge (q \rightarrow r) \rightarrow (p \rightarrow r)$ is displayed in Fig. 4.3.

Signed generation trees will be mostly used in the context of inclusions $\phi \subseteq \psi$. In this context, we will typically consider the positive generation tree $+\phi$ for the left-hand side and the negative one $-\psi$ for the right-hand side.[7]

For any formula $\phi(p_1, \ldots, p_n)$, any polarity-type ϵ over $\{p_1, \ldots p_n\}$, and any $1 \leq i \leq n$, an **ϵ-critical node** in a signed generation tree of ϕ is a leaf node labelled with $\epsilon(p_i)p_i$. An **ϵ-critical branch** in the tree is a branch with an ϵ-critical leaf node. As we will see later, the variable occurrences corresponding to ϵ-critical nodes will be of special importance in the algorithm.

Hereafter, we will use sg to denote $+$ or $-$. For every formula $\phi = \phi(p_1, \ldots p_n)$ and every polarity-type ϵ, we say that the signed formula $\mathsf{sg}\,\phi$ **agrees with** ϵ, and write $\epsilon(\mathsf{sg}\,\phi)$, if every leaf in the signed generation tree of $\mathsf{sg}\,\phi$ is ϵ-critical, i.e. signed in agreement with ϵ. Given an occurrence of a subformula ψ of a formula ϕ, we will write $\mathsf{sg}'\psi \prec \mathsf{sg}\,\phi$ to indicate that the sign of that occurrence of ψ in the signed generation tree $\mathsf{sg}\,\phi$ is sg'. We will also write $\epsilon(\psi) \prec \mathsf{sg}\,\phi$ (resp. $\epsilon^{\partial}(\psi) \prec \mathsf{sg}\,\phi$) to indicate that the signed subtree of the given occurrence of the subformula ψ of the signed tree $\mathsf{sg}\,\phi$, agrees with ϵ (resp. with ϵ^{∂}).

4.4.3 Sahlqvist and Inductive Formulae in $\mathcal{L}_R$

For what follows, we will need to distinguish three syntactic types of signed formulae, occurring as nodes in signed generation trees, depending on their sign and main connective, as indicated in Table 4.2. This terminology indicates how formulas with these main connectives occurring on the left and right of inclusions are

[7] The convention of considering the positive generation tree of the left-hand side and the negative generation tree of the right-hand side of an inclusion goes back to Gehrke et al. (2005). Although this convention might seem counter-intuitive at first glance, it is by now well established in this line of research, and we therefore maintain it to facilitate easier comparisons.

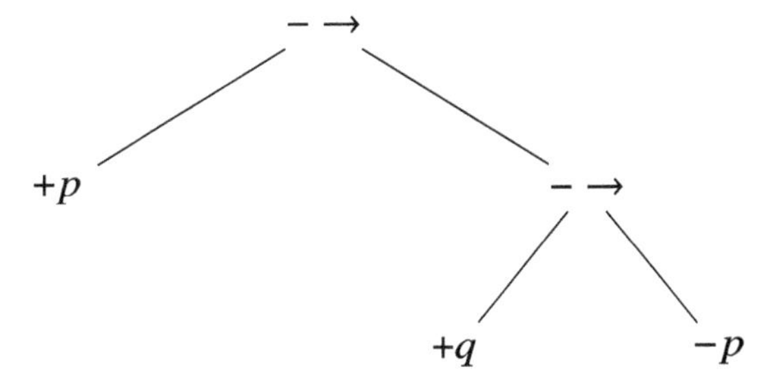

Fig. 4.1 The signed generation tree of $-[p \rightarrow (q \rightarrow p)]$. If $\epsilon(p) = -$ and $\epsilon(q) = +$, then the branches ending in $-p$ and $+q$ are ϵ-critical. All branches in the tree are excellent

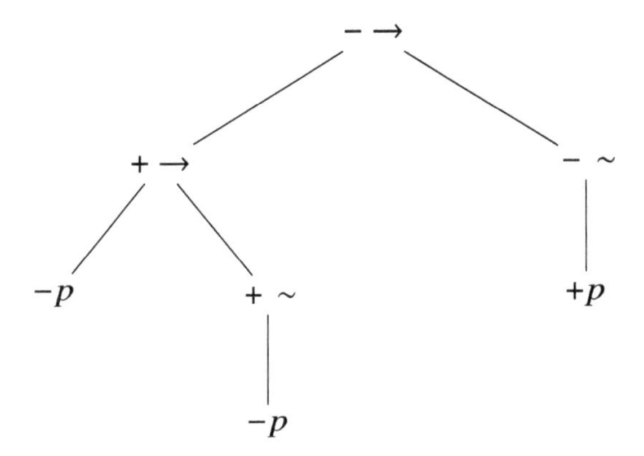

Fig. 4.2 The signed generation tree of $-[(p \rightarrow \sim p) \rightarrow \sim p]$. If $\epsilon(p) = +$, then the branch ending in $+p$ is ϵ-critical. The rightmost branch is excellent, while the leftmost and middle branches are good but not excellent

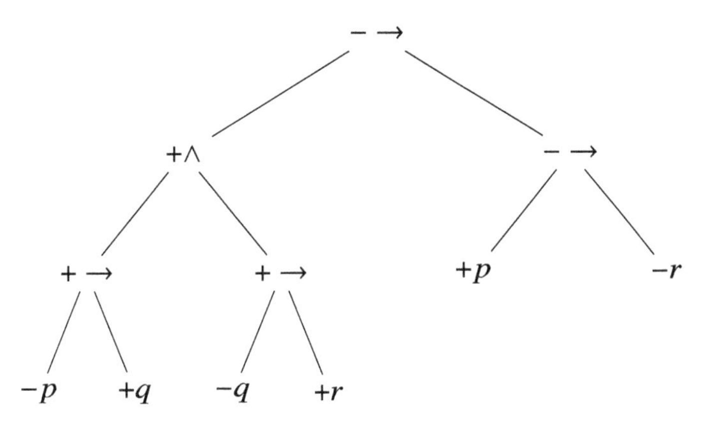

Fig. 4.3 The signed generation tree of $-[(p \rightarrow q) \wedge (q \rightarrow r) \rightarrow (p \rightarrow r)]$. If $\epsilon(p) = \epsilon(q) = \epsilon(r) = 1$, then the branches ending in $+p$, $+q$ and $+r$ are all ϵ-critical. The first four branches from the left are good, but not excellent, whereas the last two branches are excellent

Table 4.2 Types of nodes formulae in $\mathcal{L}_R$

Approximation nodes	Inner nodes	
	Adjunction nodes	Residuation nodes
$+\vee\ \wedge\circ\ \sim$	$+\wedge\ \sim$	$+\vee\ \rightarrow$
$-\wedge\ \vee \rightarrow \sim$	$-\vee\ \sim$	$-\wedge\ \circ$

dealt with by PEARL.[8] More precisely, a $+$-sign (resp., $-$-sign) indicates that the connective occurs on the right (left) side of an inclusion, at some point *after* first-approximation. Approximation nodes are dealt with by approximation rules (except $+\vee$ and $-\wedge$ that are dealt with by splitting with ($\vee$Adj) and ($\wedge$Adj) *before* first approximation on the left and right sides of an inclusion, respectively). Similarly, adjunction and residuation nodes (collectively called 'inner nodes') are dealt with by adjunction and approximation rules, respectively. Note that some nodes are listed as both approximation and inner nodes. This is because approximation rules require either the left-hand side of an inclusion to be a nominal or its right-hand side to be a co-nominal, whereas residuation and adjunction rules have no such requirements. In particular, after a residuation rule has been applied, this syntactic requirement may be lost and only residuation and adjunction may be available to surface a particular variable occurrence in order to prepare for the Ackerman rule. This makes the correct nesting order of connectives essential for the success of PEARL. For example, $+\circ$, corresponding to $\circ$ as the main connective on the right-hand side of an inclusion, can *only* be approximated, and so having it nested under $+\rightarrow$, which can only be residuated, may make it impossible to surface variable occurrences in their scope. The definitions of the Sahlqvist and inductive formulas that follow below therefore aim to stipulate sufficient conditions on the nesting order which will guarantee the success of the algorithm. In the case of the inductive formulae, the requirements on nesting imposed in the definition of the Sahlqvist formulae are relaxed, at the cost of imposing conditions on the co-occurrences (together in a subformula) of variables in the arguments of residuation nodes.

Remark 4.2 Here is an alternative, logic-based terminology and intuition for the classification of signed connectives and nodes, coming from modal logic which considers two main types of modal operators:

- **diamond-operators**, for which the semantic truth condition (for the positively signed), respectively, falsity condition (for the negatively signed), are given by existential quantification over accessible worlds. In the case of $\mathcal{L}_R$ formulae, these are the positively signed fusion operator $+\circ$, as well as the negatively signed implication $-\rightarrow$, because the falsity of the implication is given by an existentially quantified semantic condition.

[8] Most papers in the unified correspondence literature, e.g. Conradie and Palmigiano (2012), Conradie and Palmigiano (2019), classify nodes similarly, but with terminology referring not to the rules, but rather to the order-theoretic and algebraic properties of the interpretations of connectives upon which the soundness of these rules is based.

- **box-operators**, for which the semantic respective truth conditions are given by universal quantification over accessible worlds. In the case of $\mathcal{L}_R$ formulae, these are the positively signed implication $+\rightarrow$, as well as the negatively signed fusion operator $-\circ$, because the falsity of the fusion is given by a universally quantified semantic condition.

The signed connectives as described above can be called **proper diamonds**, respectively, **proper boxes**. Besides, the propositional connectives $\wedge$, $\vee$, as well as the negation $\sim$ can be treated as either diamonds or boxes. Added to the above, these define what one can call **(generalized) diamonds and boxes**, which, respectively, correspond to the Approximation nodes and the Inner nodes in the table below. Still, the signed connectives of *disjunctive type*, viz. $+\vee$ and $-\wedge$, are more naturally treated as boxes, as they typically distribute over conjunctions, whereas the signed connectives of *conjunctive type*, viz. $-\vee$ and $+\wedge$, are more naturally treated as diamonds, as they typically distribute over disjunctions. This is only intuition, to help the modal-logic-minded reader with remembering the types, but we emphasize that it is too coarse to serve as a viable alternative in the precise definition of the inductive formulae given below.

Definition 4.6 A branch in a signed generation tree $\mathsf{sg}\,\phi$ is called a **good branch** if it can be represented as a concatenation of two paths P_1 and P_2, any of which may possibly be of length 0, such that P_1 is a path starting from the leaf of the branch and consisting (apart from variable nodes) only of inner-nodes, and P_2 consists (apart from variable nodes) only of approximation-nodes. A good branch is **excellent** if it is a concatenation of paths P_1 and P_2 as above (each possibly empty), where only adjunction-nodes (and no residuation nodes) can occur on P_1.

Definition 4.7 Given an order type ϵ and a formula $\phi = \phi(p_1, \dots p_n)$ of $\mathcal{L}_R$, the signed generation tree $\mathsf{sg}\,\phi$ of ϕ is **ϵ-Sahlqvist** if every ϵ-critical branch is excellent. An inclusion $\phi \subseteq \psi$ is **ϵ-Sahlqvist** if both signed trees $+\phi$ and $-\psi$ are ϵ-Sahlqvist. An inclusion $\phi \subseteq \psi$ is **Sahlqvist** if it is ϵ-Sahlqvist for some ϵ. A formula ψ is **Sahlqvist** if the inclusion $\mathbf{t} \subseteq \psi$ is Sahlqvist.

Example 4.5 Consider the formula (K) mentioned in Examples 4.1 and 4.4 (Fig. 4.1), rewritten as the inclusion $\mathbf{t} \subseteq p \rightarrow (q \rightarrow p)$. The positive generation tree of $\mathbf{t}$ consists of the single node $+\mathbf{t}$ and so the only branch in this tree is trivially excellent (This observation applies to any inclusion with $\mathbf{t}$ on the left, so we will not repeat it further.). The negative generation tree of $p \rightarrow (q \rightarrow p)$ is given in Fig. 4.1, and each of the three branches in this tree is excellent, as they consist entirely of approximation-nodes. It follows that the inclusion $\mathbf{t} \subseteq p \rightarrow (q \rightarrow p)$ is ϵ-Sahlqvist for *any* polarity-type ϵ, hence the inclusion $\mathbf{t} \subseteq p \rightarrow (q \rightarrow p)$ and, consequently, the formula $p \rightarrow (q \rightarrow p)$, are both Sahlqvist.

Example 4.6 Consider the formula $(p \rightarrow \sim p) \rightarrow \sim p$, also discussed in Examples 4.4 and 4.2. It corresponds to the inclusion $\mathbf{t} \subseteq (p \rightarrow \sim p) \rightarrow \sim p$, so we only need focus on the negative generation tree of $(p \rightarrow \sim p) \rightarrow \sim p$, which is pictured in Fig. 4.2. The rightmost branch consists entirely of approximation-nodes, and is hence excellent, while the leftmost and middle branches are good, but not excellent, as both contain the proper residuation-node $(+ \rightarrow)$. Consequently, this formula is ϵ-Sahlqvist for any polarity-type ϵ with $\epsilon(p) = +$, but not for any polarity-type ϵ with $\epsilon(p) = -$.

Many other well-known axioms of relevance logic are, in fact, Sahlqvist, as stated in Propositions 4.6, 4.7 and 4.8 in Appendix 2.

We now introduce the more general classes of *inductive formulae and inclusions for* $\mathcal{L}_R$:

Definition 4.8 (*Inductive formulae and inclusions*) For any polarity-type ϵ and any strict partial order Ω on $p_1, \ldots p_n$, the signed generation tree $\mathsf{sg}\,\phi$ of a formula $\phi = \phi(p_1, \ldots p_n)$ is **(Ω, ϵ)-inductive** if

1. every ϵ-critical branch is good (cf. Definition 4.6);
2. for every residuation-node occurring in a critical branch ending in leaf $\epsilon(p_i)p_i$ and labelled with a subformula of the form $\gamma \odot \beta$ or $\beta \odot \gamma$ where $\odot$ is a signed residuation-connective, the following hold:
    a. the leaf $\epsilon(p_i)p_i$ of the critical branch occurs in the subtree corresponding to the subformula β,
    b. $\epsilon^\partial(\gamma) \prec \mathsf{sg}\,\phi$, i.e. the signed subtree corresponding to subformula γ has no critical leaves, and
    c. $p_k <_\Omega p_i$ for every p_k occurring in γ.

    Thus, clauses (b) and (c) above say that every leaf in the signed subtree corresponding to the subformula γ is non-critical (i.e. of the form $\epsilon^\partial(p_j)p_j$) with $p_j <_\Omega p_i$. (Note that we write $p_j <_\Omega p_i$ for $\Omega(p_i, p_i)$.)

We will refer to Ω as the **dependency order** on the variables.

An **inclusion** $\phi \subseteq \psi$ **is** (Ω, ϵ)**-inductive** if the signed generation trees $+\phi$ and $-\psi$ are both (Ω, ϵ)-inductive.[9]

An **inclusion** $\phi \subseteq \psi$ **is inductive** if it is (Ω, ϵ)-inductive for some Ω and ϵ.

A **formula** ψ **is inductive** if the inclusion $\mathbf{t} \subseteq \psi$ is inductive.

The intuition linking this definition to PEARL is that the polarity-type ϵ tells us for which version of the Ackermann-rule to prepare: if $\epsilon(p) = +$ (respectively, $\epsilon(p) = -$) we prepare for the right (respectively, left) Ackermann-rule by 'solving for' or 'displaying' positive (negative) occurrences of p. The acyclicity of the dependency order Ω and conditions (b) and (c) guarantee that this will be possible for all variables without including an occurrence of a variable to be solved for (according to ϵ) in the formula α of the Ackermann-rule and thereby possibly substituting it (through the application of the Ackermann-rule) into a scope from which it cannot be extracted in order to be displayed.

[9] For formulae, this is equivalent the notion of (Ω, ϵ)-inductiveness given in Definition 4.4, modulo considering the opposite polarity-type ϵ^∂.

Example 4.7 The negative generation tree of the axiom (WB) $(p \to q) \wedge (q \to r) \to (p \to r)$ was displayed in Fig. 4.3 (Recall that this formula was already considered in Example 4.3.). The two branches on the right of this tree consist only of approximation-nodes, and so they are excellent. The four branches on the left (ending in leaves $-p$, $+q$, $-q$ and $+r$, respectively) are good, all consisting of the signed variable, followed by the residuation-node $(+\to)$ and then above them the approximation-nodes $(+\wedge)$ and $(-\to)$. However, because of the presence of the residuation-node, these four branches are not excellent.

This formula is *not* ϵ-Sahlqvist for any polarity-type ϵ, since neither all $+q$-nodes nor all $-q$-nodes occur as the leaves of excellent branches, so it is impossible to choose an ϵ according to which all critical branches would be excellent.

However, it is (Ω, ϵ)-inductive where $p <_\Omega q, r <_\Omega q$, and $\epsilon(p) = +, \epsilon(q) = +$, $\epsilon(r) = -$. Under this choice of $<_\Omega$ and ϵ the critical branches ending in $+p$ and $-r$ are excellent, while the critical branch ending in $+q$ is good. Moreover, at the only residuation-node $(+\to)$ occurring on the latter branch, p plays the role of the subformula γ in Definition 4.8, while β is q. In the single-node subtree $-p$ corresponding to p, the only leaf (namely $-p$) is non-critical, and $p <_\Omega q$, so the requirements of the definition are satisfied.

Keeping the same choice of dependency order $<_\Omega$, other choices of ϵ under which the formula would be (Ω, ϵ)-inductive are possible, too. For example, $\epsilon(p) = +$, $\epsilon(q) = -$, $\epsilon(r) = -$ or $\epsilon(p) = \epsilon(q) = \epsilon(r) = -$ would both work. However, any choice with $\epsilon(p) = -$ and $\epsilon(q) = +$ would not work, as this would violate the conditions imposed on residuation nodes occurring on critical branches.

Several well-known axioms of relevance logic are not Sahlqvist, but are inductive. See some examples in Appendix 2.

Example 4.8 This is an example of a $\mathcal{L}_R$-formula which requires some pre-processing which splits the execution of the algorithm into two branches.

$$\psi = (\mathbf{t} \to \sim p) \to (((q \circ p) \to q) \wedge \sim(\mathbf{t} \to (\sim p \circ \sim q))).$$

The signed generation tree $-\psi$ is given on Fig. 4.4. The branches are numbered from 1 to 8 for ease of reference. Note that ϕ is not inductive because, for example, branches 3 and 8 are not good and have leaves $+q$ and $-q$, respectively, making it impossible to find a polarity-type for which all critical branches will be good. However, we will show that the pre-processing of PEARL splits this into two formulae, one of which is monotone, while the other is inductive. Indeed, PEARL takes

$$\mathbf{t} \subseteq (\mathbf{t} \to \sim p) \to (((q \circ p) \to q) \wedge \sim(\mathbf{t} \to (\sim p \circ \sim q)))$$

as input and then applies the equivalence $\phi \to (\psi \wedge \theta) \equiv (\phi \to \psi) \wedge (\psi \to \theta)$ after which the $\wedge$-split rule becomes applicable, yielding the inclusions

$$\mathbf{t} \subseteq (\mathbf{t} \to \sim p) \to ((q \circ p) \to q) \tag{4.1}$$

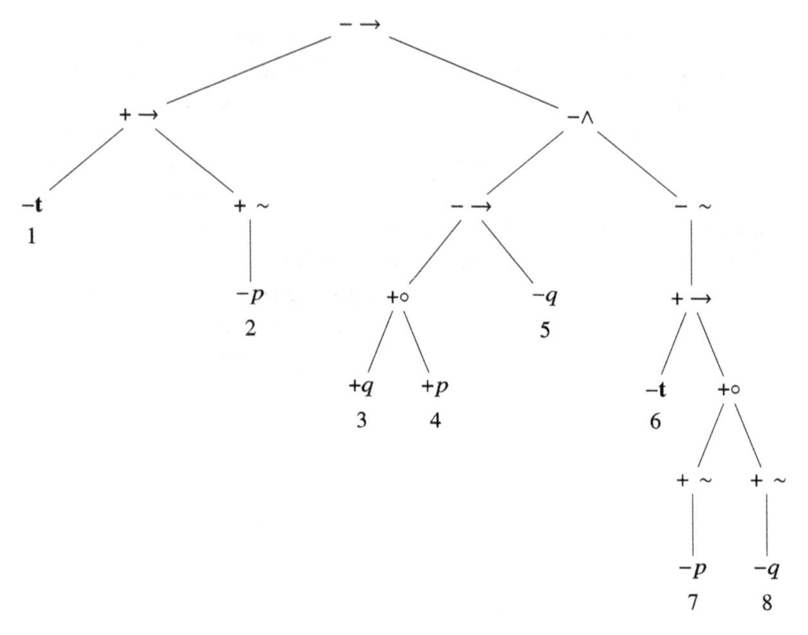

Fig. 4.4 The signed generation tree for Example 4.4

and

$$\mathbf{t} \subseteq (\mathbf{t} \rightarrow \ {\sim} p) \rightarrow \ {\sim}(\mathbf{t} \rightarrow ({\sim} p \circ {\sim} q)). \tag{4.2}$$

Now, the right-hand side of (4.2) is positive in both p and q while the left-hand side is vacuously negative in both these variables, so the monotone variable elimination rule ($\perp$) may be applied to transform (4.2) into

$$\mathbf{t} \subseteq (\mathbf{t} \rightarrow \ {\sim}\perp) \rightarrow \ {\sim}(\mathbf{t} \rightarrow ({\sim}\perp \circ {\sim}\perp)), \tag{4.3}$$

thereby eliminating all occurring propositional variables.

We note that (4.1) is (Ω, ϵ)-inductive for $\Omega = \varnothing$ and $\epsilon(p) = \epsilon(q) = -$. The algorithm now proceeds on (4.1) and applies the first-approximation rule to obtain

$$(\mathbf{t} \rightarrow \ {\sim} p) \rightarrow ((q \circ p) \rightarrow q) \subseteq \mathbf{m}_0 \vdash \mathbf{t} \subseteq \mathbf{m}_0.$$

Applications of the $\rightarrow$-approximation rule transforms this into

$$\mathbf{j}_1 \subseteq \mathbf{t} \rightarrow \ {\sim} p, \ \ (q \circ p) \rightarrow \mathbf{m}_2 \subseteq \mathbf{m}_1, \ \ q \subseteq \mathbf{m}_2, \ \ \mathbf{j}_1 \rightarrow \mathbf{m}_1 \subseteq \mathbf{m}_0 \vdash \mathbf{t} \subseteq \mathbf{m}_0.$$

Applying the left Ackermann-rule to eliminate q yields

$$\mathbf{j}_1 \subseteq \mathbf{t} \rightarrow \ {\sim} p, \ \ (\mathbf{m}_2 \circ p) \rightarrow \mathbf{m}_2 \subseteq \mathbf{m}_1, \ \ \mathbf{j}_1 \rightarrow \mathbf{m}_1 \subseteq \mathbf{m}_0 \vdash \mathbf{t} \subseteq \mathbf{m}_0.$$

We next solve for p by applying the $\rightarrow$-residuation and the $\sim$Right-Adj rule to produce

$$p \subseteq {\sim}^{\sharp}(\mathbf{j}_1 \circ \mathbf{t}),\ \ (\mathbf{m}_2 \circ p) \rightarrow \mathbf{m}_2 \subseteq \mathbf{m}_1,\ \ \mathbf{j}_1 \rightarrow \mathbf{m}_1 \subseteq \mathbf{m}_0 \vdash \mathbf{t} \subseteq \mathbf{m}_0,$$

to which the left Ackermann-rule can be applied with respect to p to produce the pure quasi-inclusion

$$(\mathbf{m}_2 \circ {\sim}^{\sharp}(\mathbf{j}_1 \circ \mathbf{t})) \rightarrow \mathbf{m}_2 \subseteq \mathbf{m}_1,\ \ \mathbf{j}_1 \rightarrow \mathbf{m}_1 \subseteq \mathbf{m}_0 \vdash \mathbf{t} \subseteq \mathbf{m}_0.$$

Applying a simplification rules turns this into

$$(\mathbf{m}_2 \circ {\sim}^{\sharp}(\mathbf{j}_1 \circ \mathbf{t})) \rightarrow \mathbf{m}_2 \subseteq \mathbf{m}_1 \vdash \mathbf{t} \subseteq \mathbf{j}_1 \rightarrow \mathbf{m}_1,$$

to the consequent of which we may apply the $\rightarrow$-residuation rule,

$$(\mathbf{m}_2 \circ {\sim}^{\sharp}(\mathbf{j}_1 \circ \mathbf{t})) \rightarrow \mathbf{m}_2 \subseteq \mathbf{m}_1 \vdash \mathbf{t} \circ \mathbf{j}_1 \subseteq \mathbf{m}_1,$$

from which another simplification rule application produces

$$\mathbf{t} \circ \mathbf{j}_1 \subseteq (\mathbf{m}_2 \circ {\sim}^{\sharp}(\mathbf{j}_1 \circ \mathbf{t})) \rightarrow \mathbf{m}_2. \tag{4.4}$$

Inclusions (4.3) and (4.4) can now be translated into first-order conditions, the conjunction of which is the first-order condition defined by ψ.

4.4.4 *Completeness of* PEARL *for All Inductive Formulae*

In this section, we prove that PEARL successfully computes first-order frame correspondents for all inductive inclusions and formulae. The proof essentially follows the outline of the proof of the analogous result given in Conradie and Palmigiano (2019) for the more general ALBA algorithm, but specialized to the setting of relevance logic. The proofs of all the lemmas can be found in Appendix 1.

The following definition captures the shape of an inductive inclusion resulting from pre-processing, i.e. one in which the splitting rules have been applied to eliminate all disjunctions on the left and conjunctions on the right that could have been surfaced through distribution.

Definition 4.9 An (Ω, ϵ)-inductive inclusion is **definite** if its critical branches contain no occurrences of $+\vee$ or $-\wedge$ as approximation nodes.

Lemma 4.3 *Let* $\{\phi_i \subseteq \psi_i \mid 1 \leq i \leq n\}$ *be the set of inclusions obtained by pre-processing an* (Ω, ϵ)*-inductive inclusion* $\phi \subseteq \psi$*. Then each* $\phi_i \subseteq \psi_i$ *is a definite* (Ω, ϵ)*-inductive inclusion.*

The following definition intends to capture the state of a quasi-inclusion after approximation rules have been applied exhaustively:

Definition 4.10 Call a quasi-inclusion $\Gamma \vdash \mathbf{i} \subseteq \mathbf{m}$ (Ω, ϵ)-*stripped* if for each $\xi \subseteq \chi \in \Gamma$ the following conditions hold:

1. one of $-\xi$ and $+\chi$ is pure, and the other is (Ω, ϵ)-inductive;
2. apart from the leaves, every ϵ-critical branch in $-\xi$ and $+\chi$ consists entirely of inner-nodes.

Lemma 4.4 *For any definite (Ω, ϵ)-inductive inclusion $\phi \subseteq \psi$ the system the quasi-inclusion $\mathbf{i} \subseteq \phi, \psi \subseteq \mathbf{m} \vdash \mathbf{i} \subseteq \mathbf{m}$ arising from first approximation can be transformed into an (Ω, ϵ)-stripped quasi-inclusion by the application of approximation rules.*

Definition 4.11 An (Ω, ϵ)-stripped quasi-inclusion $\Gamma \vdash \mathbf{i} \subseteq \mathbf{m}$ is *Ackermann-ready* with respect to a propositional variable p_i with $\epsilon_i = +$ (respectively, $\epsilon_i = -$) if every inclusion $\xi \subseteq \chi \in \Gamma$ is of one of the following forms:

1. $\xi \subseteq p$ where ξ is pure (i.e. not containing propositional variables); respectively, $p \subseteq \chi$, where χ is pure, or
2. $\xi \subseteq \chi$ where neither $-\xi$ nor $+\chi$ contains any $+p_i$ (respectively, $-p_i$) leaves.

Note that the right or left Ackermann-rule (depending on whether $\epsilon_i = +$ or $\epsilon_i = -$) is applicable to a system which is Ackermann-ready with respect to p_i. In fact, this would still have been the case had we weakened the requirement that ξ and χ must be pure, to simply require that they do not contain p_i.

Lemma 4.5 *If $\Gamma \vdash \mathbf{i} \subseteq \mathbf{m}$ is (Ω, ϵ)-stripped and p_i is Ω-minimal among propositional variables occurring in $\Gamma \vdash \mathbf{i} \subseteq \mathbf{m}$, then $\Gamma \vdash \mathbf{i} \subseteq \mathbf{m}$ can be transformed, through the application of residuation- and adjunction rules, into a system which is* Ackermann-ready *with respect to p_i.*

Lemma 4.6 *Applying the appropriate Ackermann-rule with respect to p_i to an (Ω, ϵ)-stripped quasi-inclusion which is Ackermann-ready with respect to p_i, again yields an (Ω, ϵ)-stripped quasi-inclusion.*

Theorem 4.2 PEARL *succeeds on all inductive inclusions.*

Proof Let $\phi \subseteq \psi$ be an (Ω, ϵ)-inductive inclusion. By Lemma 4.3, applying preprocessing yields a finite set of definite (Ω, ϵ)-inductive inclusions, each of which gives rise to a quasi-inclusion $\mathbf{i} \subseteq \phi', \psi' \subseteq \mathbf{m} \vdash \mathbf{i} \subseteq \mathbf{m}$. By Lemma 4.4, applications of the approximation rules convert this into an (Ω, ϵ)-stripped quasi-inclusion $\Gamma \vdash \mathbf{i} \subseteq \mathbf{m}$. By Lemma 4.2, $\Gamma \vdash \mathbf{i} \subseteq \mathbf{m}$ can be made Ackermann-ready with respect to any occurring Ω-minimal variable. After applying the appropriate Ackermann-rule to eliminate this variable, the resulting quasi-inclusion is again (Ω, ϵ)-stripped, now containing one propositional variable fewer. Now, Lemma 4.2 can be applied again. This process is iterated until all occurring propositional variables are eliminated and a pure system is obtained.

The next theorem is a direct corollary of Theorems 4.1 and 4.2.

Theorem 4.3 *Every inductive $\mathcal{L}_R$-inclusion, hence, every inductive $\mathcal{L}_R$–formula, has an effectively computable first-order frame correspondent on Routley–Meyer frames.*

4.4.5 Comparing with Other Sahlqvist Classes in the Literature

Here, we compare the inductive and Sahlqvist $\mathcal{L}_R$ formulae, as we have defined them, with two other proposals in the literature. We first consider the Sahlqvist relevance formulae introduced by Guillermo Badia in Badia (2018). It turns out that this class of formulae is incomparable with our Sahlqvist class, but forms a proper subclass of the inductive formulae. We next turn our attention to the Sahlqvist class for *modal* relevance logic introduced by Takahiro Seki in Seki (2003). Since $\mathcal{L}_R$ is a strict sublanguage of the language of modal relevance logic, we only consider the class of formulae obtained by restricting the latter definition to the language $\mathcal{L}_R$. We will see that this class lies properly within the intersection of our Sahlqvist class with that of Badia.

We first review Badia's definition of Sahlqvist relevance formulae Badia (2018) together with all the necessary notions needed in the build-up to this definition, and subsequently compare it to our definitions of Sahlqvist and inductive $\mathcal{L}_R$ formulae. Badia defines positive and negative relevance formulae in terms of their standard translations, viz. a relevance formula A is called positive (which we will refer to as **B-positive**) if $ST_x(A)$ is a formula built up from atomic formulae involving only unary predicates and first-order formulae where the only non-logical symbols are R and O, using the connectives $\exists$, $\forall$, $\wedge$ and $\vee$. Similarly, a relevance formula A is called negative (which we will refer to as **B-negative**) if $ST_x(A)$ is a formula built up from Boolean negations of atomic formulae involving only unary predicates and first-order formulae where the only non-logical symbols are R and O, using the connectives $\exists$, $\forall$, $\wedge$ and $\vee$.

It is easy to check that the B-positive $\mathcal{L}_R$ formulae are exactly those built up from propositional variables and constant formulae using $\wedge$, $\vee$ and $\circ$, while the B-negative $\mathcal{L}_R$ formulae are built up from constant formulae and negated propositional variables, using $\wedge$, $\vee$ and $\circ$. It is also easy to check that every B-positive (B-negative) formula positive (negative) in the sense of the present paper. However, reading the definitions of B-positive and B-negative formulae as purely *syntactic*, the converse would not hold. For example, ${\sim}(p \rightarrow {\sim}q)$ is positive but translates to $\neg\forall yz(Rx^*yz \wedge P(y) \rightarrow \neg Q(z^*))$ which is not B-positive, while $p \rightarrow {\sim}q$ is negative but translates to $\forall yz(Rxyz \wedge P(y) \rightarrow \neg Q(z*))$ which is not B-negative. We will therefore read the definition of B-positive and B-negative formulae *up to equivalence of the standard translations*. Under this assumption we can drive in negations and consider, for example, $\exists yz(Rx^*yz \wedge P(y) \wedge Q(z^*))$ instead of

$\neg\forall yz(Rx^*yz \wedge P(y) \rightarrow \neg Q(z^*))$ as the translation of $\sim(p \rightarrow \sim q)$, thus rendering it B-positive. In general, it should be clear that, under this assumption, the definitions of B-positive (B-negative) and positive (negative) $\mathcal{L}_R$ formulae coincide.

We will need the notions of *relevance Sahlqvist implications* (Badia 2018, Definition 3) and *dual relevance Sahlqvist implications* (Badia 2018, Definition 4). To avoid possible confusion with notions defined in the present paper, we will refer to the formulae in these classes as B-Sahlqvist implications and B-dual Sahlqvist implications, respectively.

Definition 4.12 (Badia 2018, Definitions 3 and 4) A formula $A \rightarrow B$ is called a **B-Sahlqvist implication** if B is positive while A is a formula built up from propositional variables, double negated atoms (i.e. formulae of the form $\sim\sim p$), negative formulae, the constant $\mathbf{t}$ and implications of the form $\mathbf{t} \rightarrow p$ (for any propositional variable p) using only the connectives $\wedge$, $\vee$ and $\circ$.

A formula $A \rightarrow B$ is called a **dual B-Sahlqvist implication** if B is negative while A is a formula built up from negated propositional atoms $\sim p$, triple negated atoms $\sim\sim\sim p$, positive formulae, the constant $\sim\mathbf{t}$, and implications of the form $p \rightarrow \mathbf{t}$ (for any propositional variable p) using only the connectives $\wedge$, $\vee$ and $\circ$.

Example 4.9 1. The formula $p \rightarrow (p \circ p)$ is a B-Sahlqvist implication. Note that it is ϵ-Sahlqvist for $\epsilon(p) = +$, but not for $\epsilon(p) = -$.
2. The formula $(p \rightarrow \mathbf{t}) \rightarrow (\sim p \circ \sim p)$ is a dual B-Sahlqvist implication. Note that it is $(\varnothing, \epsilon)$-inductive for $\epsilon(p) = -$, but not for $\epsilon(p) = +$.

The proofs of the following two lemmas can be found in Appendix 1.

Lemma 4.7 *Any B-Sahlqvist implication $A \rightarrow B$ is ϵ-Sahlqvist for $\epsilon(p) = +$ for all occurring propositional variables p.*

Lemma 4.8 *Any dual Sahlqvist B-implication $A \rightarrow B$ is (Ω, ϵ)-inductive for $\Omega = \varnothing$ and $\epsilon(p) = -$ for all occurring propositional variables p.*

Definition 4.13 (Badia 2018, Definitions 5) A B-Sahlqvist formula is any formula built up from (dual) B-Sahlqvist implications, propositional variables, and negated propositional variables using $\wedge$, the operations Θ on formulae (for any propositional variable free relevance formula θ) defined by $\Theta(B) = \theta \rightarrow B$, and applications of $\vee$ where the disjuncts share no propositional variables in common.

Example 4.10 Consider the conjunction of the B-Sahlqvist implication and dual B-Sahlqvist implication considered in Example 4.9, namely $(p \rightarrow (p \circ p)) \wedge ((p \rightarrow \mathbf{t}) \rightarrow (\sim p \circ \sim p))$. By Definition 4.13, this conjunction is a B-Sahlqvist formula. Note, however, that it is not inductive, as choosing $\epsilon(p) = +$ gives rise to a critical branch in the signed generation tree $-((p \rightarrow \mathbf{t}) \rightarrow (\sim p \circ \sim p))$ which is not good, while choosing $\epsilon(p) = -$ gives rise to a critical branch in the signed generation tree $-(p \rightarrow (p \circ p))$ which is not good, either.

To get us out of this impasse, it is sufficient to recall that the frame correspondent of a conjunction is the conjunction of the frame correspondents, and that we may therefore equivalently consider the formula $(p \rightarrow (p \circ p)) \wedge ((q \rightarrow \mathbf{t}) \rightarrow (\sim q \circ \sim q))$

in which the variable p has been renamed to q in the second conjunct. This formula necessarily defines the same first-order condition Routley–Meyer frames as the one we started with and, moreover, it is $(\varnothing, \epsilon)$-inductive for $\epsilon(p) = +$ and $\epsilon(q) = 1$.

Remark 4.3 We can use the equivalences $A \rightarrow (B \wedge C) \equiv (A \rightarrow B) \wedge (A \rightarrow C)$, $A \vee (B \wedge C) \equiv (A \vee B) \wedge (A \vee C)$ and $(B \wedge C) \vee A \equiv (B \vee A) \wedge (C \vee A)$ to equivalently reformulate Definition 4.13 as follows: a **pre-B-Sahlqvist formula** is any formula built up from (dual) B-Sahlqvist implications, propositional variables, and negated propositional variables using the operations on formulae Θ (for any propositional variable free relevance formula θ) defined by $\Theta(B) = \theta \rightarrow B$ and applications of $\vee$ where the disjuncts share no propositional variable in common. A **B-Sahlqvist formula** is any conjunction of pre-B-Sahlqvist formulae.

The following proposition shows that every B-Sahlqvist formula is an inductive formula with empty dependency order. For the sake of brevity, we will refer to inductive formulae with empty dependency order as $\varnothing$-inductive formulae.

Proposition 4.4 *Modulo renaming of variables, every B-Sahlqvist formula is an $\varnothing$-inductive formula.*

The proof is in Appendix 1.

Example 4.11 The (K) axiom $p \rightarrow (q \rightarrow p)$ was shown to be Sahlqvist in Example 4.5. However, note that it is not B-Sahlqvist: indeed, since the consequent $(p \rightarrow p)$ is neither positive nor negative, the formula as a whole is neither a B-Sahlqvist implication nor a B-Sahlqvist dual implication. The conclusion now follows by noting that, moreover, the main connective cannot be seen as one of the implications $\theta \rightarrow B$ allowed by Definition 4.13 where θ is a constant formula.

Example 4.12 The (WB) axiom $(p \rightarrow q) \wedge (q \rightarrow r) \rightarrow (p \rightarrow r)$ was shown to be inductive in Example 4.7. Considerations similarly to those employed in Example 4.11 can be used to see that it is not B-Sahlqvist.

In Seki (2003), Seki formulates and proves a Sahlqvist correspondence and completeness result for *modal* relevance logic. The language of modal relevance logic is obtained by adding the unary modalities $\Diamond$ and $\Box$ to $\mathcal{L}_R$. We will write $\mathcal{L}_{RM}$ to denote this language. The defined connectives $\blacklozenge$ and $\boxdot$ are abbreviations for $\sim\Box\sim$ and $\sim\Diamond\sim$, respectively. This language is interpreted on Routley–Meyer models enriched with two relations for interpreting the Diamond and Box.

In what follows we will project Seki's definition of Sahlqvist formulae for the language $\mathcal{L}_{RM}$, as well as all necessary notions leading up to that definition, onto $\mathcal{L}_R$. To prevent possible confusion with terminology, we will refer to Seki's class of Sahlqvist formulae as the S-Sahlqvist formulae and also refer to the classes of negative and positive formulae defined in Seki (2003) as S-positive and S-negative, respectively.

An $\mathcal{L}_{RM}$-formula is **S-positive** if it is built from propositional variables and **t** using $\wedge$, $\vee$, $\Box$, $\boxdot$, $\Diamond$ and $\blacklozenge$. An $\mathcal{L}_{RM}$-formula is **strongly S-positive** if it is of the form $\blacksquare^{m_1} p_1 \wedge \cdots \wedge \blacksquare^{m_k} p_k$ where $\blacksquare^n$ denotes a sequence of n of $\Box$s and $\boxdot$s. An $\mathcal{L}_{RM}$-formula is **S-negative** if it is equivalent to $\sim B$ for some S-positive formula B. An $\mathcal{L}_{RM}$-formula is **untied** if it can be constructed from S-negative formulae and strongly S-positive formulae using $\wedge$, $\Diamond$ and $\Diamond\!\!\!\cdot$.

An $\mathcal{L}_{RM}$-formula is **S-Sahlqvist** if it is equivalent to a conjunction of formulae of the form $\blacksquare^k(B \to C)$, where $k \geq 0$, B is untied and C is positive. Projecting the definition of $\mathcal{L}_{RM}$ S-Sahlqvist formulae onto $\mathcal{L}_R$, we see that an $\mathcal{L}_R$-formula is **S-Sahlqvist** iff it is equivalent to a conjunction of formulae of the form $B \to C$ where B is constructed from propositional variables using $\wedge$, $\vee$ and $\sim$only, while C is constructed from propositional variables using $\wedge$ and $\vee$ only.

Note that, modulo driving negations inwards, every S-Sahlqvist formula is a conjunction of B-Sahlqvist implications. By Lemma 4.7, the following proposition is therefore immediate:

Proposition 4.5 *Every S-Sahlqvist formula is a Sahlqvist formula.*

4.5 Concluding Remarks and Directions for Further Work

The present paper is part I of a bigger project. The forthcoming part II will explore some of the most interesting (or, at least most natural) continuations and extensions of this work, including, roughly in increasing order of projected difficulty:

- Adding modal operators and extending the methods and results to modal relevance logics, in particular covering the full class of Seki's Sahlqvist formulae. That should be fairly straightforward.
- Extending the language with all residual connectives, corresponding to permutations of the arguments in the semantic definitions. All results should apply likewise.
- Re-defining the classes of inductive formulae for plain and modal relevance logics in an alternative style, using a 'flat language' of relevance logic (following the idea of using polyadic modal languages in Goranko and Vakarelov 2001, Goranko and Vakarelov 2006) where the logical connectives can be composed into new polyadic connectives. That would enable defining the class of inductive formulae by means of their global structure when suitably rewritten in the richer language, rather than locally ('branch'-wise'), in terms of the patterns of occurrence of each variable, as in Sect. 4.4.3.
- Proving canonicity of all formulae on which PEARL succeeds (in particular, all inductive formulae) would require some non-trivial technical work but it is of predictable nature.
- Strengthening PEARL to cover some known cases of FO definable relevance formulae, such as the axiom ζ discussed in Sect. 4.3.5, and refining it to incorporate simplification steps taking into account the semantic monotonicity conditions.
- Developing algorithmic correspondence for other semantics for relevance logics, in particular for Urquhart's semilattice semantics. This research problem is yet to be explored.

Acknowledgements We are grateful to Peter Jipsen for carefully reading drafts of this paper and for his numerous valuable comments and suggestions which greatly aided its improvement. We also thank Alasdair Urquhart for some useful comments and corrections.

Appendix 1: Some Proofs

Proof (*Lemma* 4.3) Notice that the distribution during pre-processing only swaps the order of approximation nodes on (critical) paths, and hence does not affect the goodness of critical branches. Moreover, inner parts are entirely unaffected, and in particular the side conditions on residuation nodes of critical branches are maintained. Finally, notice that all occurrences of $+\vee$ or $-\wedge$ in the approximation parts of critical branches can be surfaced by applying the distribution laws forming part of pre-processing and then eliminated via splitting, thus producing definite inductive inclusions.

Proof (*Lemma* 4.4) By assumption, $+\phi$ and $-\psi$ are both definite (Ω, ϵ)-inductive. Hence, for any ϵ-critical branch in $+\phi$ or $-\psi$, we can apply approximation rules successively to the approximation nodes on that branch, until they have all been 'stripped off' and all remaining nodes on that branch are inner nodes. To conclude the proof it suffices to note that the inclusions in the conclusions of approximation rules are always pure on at least one side.

Proof (*Lemma* 4.5) If $\xi \subseteq \chi \in \Gamma$ and $-\xi$ and $+\chi$ contain no ϵ-critical p_i-nodes then this inclusion already satisfies condition 2 of Definition 4.11. So suppose that $-\xi$ and $+\chi$ contain some ϵ-critical p_i-node among them. This means $\xi \subseteq \chi$ is of the form $\alpha \subseteq \mathsf{Pure}$ with the ϵ-critical p_i-node in α and Pure pure, or of the form $\mathsf{Pure} \subseteq \delta$ with Pure pure and the ϵ-critical p_i-node in δ. We can now prove by simultaneous induction on α and δ that these inclusions can be transformed into the form specified by clause 1 of Definition 4.11.

The base cases are when $-\alpha = -p_i$ and $+\delta = +p_i$. Here the inclusions are in desired shape and no rules need be applied to them. We will only check here a few of the inductive cases. If $-\alpha = -(\alpha_1 \vee \alpha_2)$, then applying the $\vee$-adjunction rule we transform $\alpha_1 \vee \alpha_2 \subseteq \mathsf{Pure}$ into $\alpha_1 \subseteq \mathsf{Pure}$ and $\alpha_2 \subseteq \mathsf{Pure}$. The resulting system is clearly still (Ω, ϵ)-stripped, and we may apply the inductive hypothesis to $\alpha_1 \subseteq \mathsf{Pure}$ and $\alpha_2 \subseteq \mathsf{Pure}$.

If $-\alpha = -(\alpha_1 \circ \alpha_2)$, then, as per definition of inductive inclusions, exactly one of $-\alpha_1$ and $-\alpha_2$ contains an ϵ-critical node, and the other one is pure (here we use the Ω-minimality of p_i). Assume that the critical node is in $-\alpha_1$ and that α_2 is pure. Then, applying the $\circ$-residuation rule transforms $(\alpha_1 \circ \alpha_2) \subseteq \mathsf{Pure}$ into $\alpha_1 \subseteq \alpha_2 \to \mathsf{Pure}$, yielding an (Ω, ϵ)-stripped quasi-inclusion to which the inductive hypothesis is applicable. The cases for $-\alpha = -(\alpha_1 \wedge \alpha_2)$ and $-\alpha = - \sim\alpha_1$ can be are treated similarly with applications of the $\wedge$-residuation rule and the $\sim$-Left adjunction rule, respectively.

If $+\delta = + \sim\delta_1$, then $\mathsf{Pure} \subseteq \sim\delta_1$ becomes $\delta_1 \subseteq \sim^{\sharp}\mathsf{Pure}$ through the application of the $\sim$Right-adjunction rule, where $-\delta_1$ is (Ω, ϵ)-inductive and $\sim^{\sharp}\mathsf{Pure}$ is pure, hence resulting in an (Ω, ϵ)-stripped quasi-inclusion to which the inductive hypothesis is applicable. The cases for $+\delta = +(\delta_1 \wedge \delta_2)$, $+\delta = +(\delta_1 \vee \delta_2)$ and $+\delta = +(\delta_1 \rightarrow \delta_2)$ are left to the reader.

Proof (*Lemma* 4.6) Let $\Gamma \vdash \mathbf{i} \subseteq \mathbf{m}$ be an (Ω, ϵ)-stripped quasi-inclusion which is Ackermann-ready with respect to p_i. We only consider the case in which the right Ackermann-rule is applied, the case for the left Ackermann-rule being dual. This means that $\Gamma = \{\alpha_k \subseteq p \mid 1 \subseteq k \subseteq n\} \cup \{\beta_j(p_i) \subseteq \gamma_j(p_i) \mid 1 \leq j \leq m\}$ where the αs are pure and the $-\beta$s and $+\gamma$s contain no $+p_i$ nodes. We denote the pure formula $\bigvee_{k=1}^{n} \alpha_k$ by α. It is sufficient to show that for each $1 \leq j \leq m$, the trees $-\beta(\alpha/p_i)$ and $+\gamma(\alpha/p_i)$ satisfy the conditions of Definition 4.10. Conditions 2 follows immediately once we notice that, since α is pure and is being substituted everywhere for variable occurrences corresponding to non-critical nodes, $-\beta(\alpha/p_i)$ and $+\gamma(\alpha/p_i)$ have exactly the same ϵ-critical paths as $-\beta(p_i)$ and $+\gamma(p_i)$, respectively. Condition 1, namely that $-\beta(\alpha/p_i)$ and $+\gamma(\alpha/p_i)$ are (Ω, ϵ)-inductive, also follows using additionally the observation that all new paths that arose from the substitution are variable free.

Proof (*Lemma* 4.7) In the negative generation tree of $A \rightarrow B$, all leaves in the signed subtree of $-B$ are either $+\mathbf{t}$ or signed negative (since B is positive) and hence there are no ϵ-critical leaves in that subtree. Note that, in A, propositional variable occurrences in the negative subformulae give rise to negatively signed leaves (and hence ϵ-non-critical branches) in $+A$, while propositional variables and double negated propositional variables (outside the negative subformulae) only occur in the scope of $\wedge$, $\vee$ and $\circ$, are therefore give rise to positively signed leaves on excellent, ϵ-critical branches (consisting of approximation nodes only). Thus, since all ϵ-critical branches in the signed generation tree $-(A \rightarrow B)$ are excellent, $A \rightarrow B$ is ϵ-Sahlqvist.

Proof (*Lemma* 4.8) In the negative generation tree of $A \rightarrow B$, all leaves in the signed subtree of $-B$ are either $+\mathbf{t}$ or signed positive (since B is negative) and hence there are no ϵ-critical leaves in that subtree. Note that, in A, propositional variable occurrences in the positive subformulae give rise to positively signed leaves (and hence ϵ-non-critical branches) in $+A$, while negated and triple negated propositional variables (outside the positive subformulae) only occur in the scope of $\wedge$, $\vee$ and $\circ$, are therefore give rise to positively negatively leaves on excellent, ϵ-critical branches (consisting of approximation nodes only). Subformulae of the form $p \rightarrow \mathbf{t}$ also only occur in the scope of $\wedge$, $\vee$ and $\circ$, hence the $\rightarrow$ gives rise to the residuation-node $+\rightarrow$ through which a good critical branch with leaf $-p$ runs. Moreover, since the other argument of $\rightarrow$ is the constant $\mathbf{t}$, the conditions of Definition 4.8 are trivially met.

Proof (*Proposition* 4.4) Let A be a B-Sahlqvist formula. We first show that any pre-B-Sahlqvist formula is a $\varnothing$-inductive formula. We do this recursively on the construction of pre-B-Sahlqvist formulae. By Lemmas 4.7 and 4.8 B-Sahlqvist implications and dual B-Sahlqvist implications are $\varnothing$-inductive formulae. Trivially, propositional variables and negations of propositional variables are $\varnothing$-inductive formulae. Suppose that B and C are $(\varnothing, \epsilon_B)$ and $(\varnothing, \epsilon_C)$-inductive, respectively, and they have no propositional variables in common, then the generation tree of $-(B \vee C)$ simply joins those $-B$ and $-C$ with a new root $-\vee$. Clearly, any good branch in $-B$ or $-C$ is now part of a longer good branch in $-(B \vee C)$. Moreover, taking ϵ to be the union of ϵ_B and ϵ_B (since B and C have no variables in common this is a function), any residuation-node on a ϵ-critical branch is a residuation-node on an ϵ_B-critical branch in $-B$ or on an ϵ_C-critical branch in $-C$ and hence still satisfies the requirement of Definition 4.8. If follows that $B \vee C$ is $(\varnothing, \epsilon)$-inductive.

If θ is a constant formula, then the generation tree $-(\theta \to B)$ is the combination of the trees $+\theta$ and $-B$ with the approximation node $-\to$ as new root. Every good branch in the subtree $-B$ is now part of a longer good branch in $-(\theta \to B)$. Since θ contains no variables it follows that $\theta \to B$ is $(\varnothing, \epsilon_B)$-inductive.

We have thus established that every pre-B-Sahlqvist formula is a $\varnothing$-inductive.

Now, by definition, A is a conjunction of pre-B-Sahlqvist formulae $B_1, \ldots, B_n$. Assuming that the formulae $B_1, \ldots, B_n$ are pairwise variable-disjoint, it follows that $-A'(B_1, \ldots, B_n)$ is $(\varnothing, \epsilon)$-inductive when ϵ is simply the union of the ϵ_i. This assumption is justified by the fact that the frame correspondent of a conjunction is the conjunction of the frame correspondents (for relevance logic this is proved in Badia 2018, Lemma 9) and that we may therefore, up to frame-equivalence, rename the propositional variables in the B_i to insure that they are variable-disjoint.

Appendix 2: Some Axioms of Relevance Logics on Which PEARL Succeeds and Their First-Order Conditions

Most of the axioms below are copied from Routley et al. (1982). Hereafter, A, B, C are treated as variables. Following Routley et al. (1982) we use $\neg$, &, $\vee$, * for the FO connectives in the correspondence language, as well as $\leq$ instead of $\preceq$, as in the main text. For classical implication we use $\Rightarrow$ instead of $>$.

Axioms and Rules for the System B and Extensions

Axioms and Rules of the System B

A1. $A \to A$

A2. $A \wedge B \to A$

A3. $A \wedge B \to B$

A4. $A \to A \vee B$

A5. $B \to A \vee B$

A6. $(A \to B) \wedge (A \to C) \to (A \to B \wedge C)$

A7. $(A \to C) \wedge (B \to C) \to (A \vee B \to C)$

A8. $A \wedge (B \vee C) \to (A \wedge B) \vee (A \wedge C)$ A9. $\sim\sim A \to A$

Proposition 4.6 *All axioms A1–A9 are Sahlqvist formulae.*

Proof Routine. We illustrate the execution of PEARL on two of them, and leave the rest to the reader. We also sketch the post-processing for these, skipping some trivial steps.

A1. $A \to A$

PEARL begins with $\mathbf{t} \subseteq A \to A$,
transformed by (FA) to: $\mathbf{i} \subseteq \mathbf{t},\ A \to A \subseteq \mathbf{m} \ \Vdash\ \mathbf{i} \subseteq \mathbf{m}$,
followed by another approximation to: $\mathbf{i} \subseteq \mathbf{t},\ \mathbf{j} \subseteq A,\ \mathbf{j} \to A \subseteq \mathbf{m} \ \Vdash\ \mathbf{i} \subseteq \mathbf{m}$.
Now, applying the Ackermann-rule produces $\mathbf{i} \subseteq \mathbf{t},\ \mathbf{j} \to \mathbf{j} \subseteq \mathbf{m} \ \Vdash\ \mathbf{i} \subseteq \mathbf{m}$.
A post-processing reverse approximation eliminates $\mathbf{i}$: $\mathbf{j} \to \mathbf{j} \subseteq \mathbf{m} \ \Vdash\ \mathbf{t} \subseteq \mathbf{m}$.
(Hereafter, we will omit the introduction and elimination of that initially used nominal.) A post-processing reverse approximation eliminates $\mathbf{m}$: $\mathbf{t} \subseteq \mathbf{j} \to \mathbf{j}$.
Now, it is easy to see that the FO equivalent is a validity, as expected.

A9. $\sim\sim A \to A$

PEARL begins with $\mathbf{t} \subseteq \sim\sim A \to A$,
transformed by the reduced (FA) to: $\sim\sim A \to A \subseteq \mathbf{m} \ \Vdash\ \mathbf{t} \subseteq \mathbf{m}$,
then by approximation: $\mathbf{j} \subseteq \sim\sim A,\ \mathbf{j} \to A \subseteq \mathbf{m} \ \Vdash\ \mathbf{t} \subseteq \mathbf{m}$,
followed by a residuation step: $\sim A \subseteq \sim^{\sharp}\mathbf{j},\ \mathbf{j} \to A \subseteq \mathbf{m} \ \Vdash\ \mathbf{t} \subseteq \mathbf{m}$,
and then another residuation step: $\sim^{\flat} \sim^{\sharp}\mathbf{j} \subseteq A,\ \mathbf{j} \to A \subseteq \mathbf{m} \ \Vdash\ \mathbf{t} \subseteq \mathbf{m}$.
Now, applying the Ackermann-rule produces $\mathbf{j} \to \sim^{\flat} \sim^{\sharp}\mathbf{j} \subseteq \mathbf{m} \ \Vdash\ \mathbf{t} \subseteq \mathbf{m}$.
A post-processing reverse approximation eliminates $\mathbf{m}$: $\mathbf{t} \subseteq \mathbf{j} \to \sim^{\flat} \sim^{\sharp}\mathbf{j}$,
finally simplified to $\mathbf{t} \circ \mathbf{j} \subseteq \sim^{\flat} \sim^{\sharp}\mathbf{j}$.
The FO condition can now be easily computed:

$$\forall w \forall x_1 (x_{\mathbf{i}} \leq w \Rightarrow \exists u (u^{**} = w \ \&\ x_1 \leq u))$$

When * is an involution, this is clearly a validity.

Additional Schemata That Can Be Added to the System B

B1. $A \wedge (A \to B) \to B$
B2. $(A \to B) \wedge (B \to C) \to (A \to C)$
B3. $(A \to B) \to ((B \to C) \to (A \to C))$
B4. $(A \to B) \to ((C \to A) \to (C \to B))$
B5. $(A \to (A \to B) \to (A \to B)$ (or $(A \to (B \to C)) \to (A \wedge B \to C))$
B6. $A \to ((A \to B) \to B)$
B7. $(A \to (B \to C)) \to (B \to (A \to C))$
B8. $(A \to (B \to C)) \to ((A \to B) \to (A \to C))$
B9. $(A \to B) \to ((A \to (B \to C)) \to (A \to C))$

B10. $A \to (B \to B)$
B11. $B \to (A \to B)$
B12. $A \to (B \to (C \to A))$
B13. $A \to (B \to A \wedge B)$
B14. $(A \to B) \to ((A \to C) \to (A \to B \wedge C))$
B14'. $(A \to C) \to ((B \to C) \to (A \vee B \to C))$
B15. $(A \wedge B \to C) \to (A \to (B \to C))$
B16. $A \vee (A \to B)$
B17. $(A \to B) \vee (B \to A)$
B17'. $(A \to (A \wedge B)) \vee (B \to (A \wedge B))$
B18. $A \to (A \to A)$
B19. $A \vee B \to ((A \to B) \to B)$
B20. $(A \wedge B \to C) \to (A \to C) \vee (B \to C)$

Proposition 4.7 *All axioms B1-B20 are inductive formulae. Moreover, B1, B5–B7, and B10–B20 are Sahlqvist.*

Proof Routine. Sketches of the execution of PEARL on a selection of these axioms are given below, omitting some elimination and post-processing steps:

B1. $A \wedge (A \to B) \to B$

Elimination steps: $A \wedge (A \to B) \to B \subseteq \mathbf{m} \ \Vdash\ \mathbf{t} \subseteq \mathbf{m}$,
$\mathbf{j} \subseteq A \wedge (A \to B),\ \mathbf{j} \to B \subseteq \mathbf{m} \ \Vdash\ \mathbf{t} \subseteq \mathbf{m}$,
$\mathbf{j} \subseteq A,\ \mathbf{j} \subseteq A \to B,\ \mathbf{j} \to B \subseteq \mathbf{m} \ \Vdash\ \mathbf{t} \subseteq \mathbf{m}$,
Elimination of A:
$\mathbf{j} \subseteq \mathbf{j} \to B,\ \mathbf{j} \to B \subseteq \mathbf{m} \ \Vdash\ \mathbf{t} \subseteq \mathbf{m}$,
$\mathbf{j} \circ \mathbf{j} \subseteq B,\ \mathbf{j} \to B \subseteq \mathbf{m} \ \Vdash\ \mathbf{t} \subseteq \mathbf{m}$,
Elimination of B:
$\mathbf{j} \to \mathbf{j} \circ \mathbf{j} \subseteq \mathbf{m} \ \Vdash\ \mathbf{t} \subseteq \mathbf{m}$,
Simplification: $\mathbf{t} \subseteq \mathbf{j} \to (\mathbf{j} \circ \mathbf{j})$, equiv. $\mathbf{t} \circ \mathbf{j} \subseteq \mathbf{j} \circ \mathbf{j}$,
FO condition: $\forall w \forall x_{\mathbf{j}}(x_{\mathbf{j}} \leq w \Rightarrow Rx_{\mathbf{j}}x_{\mathbf{j}}w)$.
Equivalently, $\forall w Rwww$.

B3. $(A \to B) \to ((B \to C) \to (A \to C))$
Simplifying the pure quasi-inclusion after elimination:
$\mathbf{j}_2 \to \mathbf{n}_2 \subseteq \mathbf{n}_1,\ \mathbf{j}_3 \to (\mathbf{j}_2 \circ (\mathbf{j}_1 \circ \mathbf{j}_3)) \subseteq \mathbf{n}_2,\ \mathbf{j}_1 \to \mathbf{n}_1 \subseteq \mathbf{m} \ \vdash\ \mathbf{t} \subseteq \mathbf{m}$ iff
$\mathbf{j}_2 \to \mathbf{n}_2 \subseteq \mathbf{n}_1,\ \mathbf{j}_3 \to (\mathbf{j}_2 \circ (\mathbf{j}_1 \circ \mathbf{j}_3)) \subseteq \mathbf{n}_2 \ \vdash\ \mathbf{t} \subseteq \mathbf{j}_1 \to \mathbf{n}_1$ iff
$\mathbf{j}_2 \to \mathbf{n}_2 \subseteq \mathbf{n}_1,\ \mathbf{j}_3 \to (\mathbf{j}_2 \circ (\mathbf{j}_1 \circ \mathbf{j}_3)) \subseteq \mathbf{n}_2 \ \vdash\ \mathbf{t} \circ \mathbf{j}_1 \subseteq \mathbf{n}_1$ iff
$\mathbf{j}_3 \to (\mathbf{j}_2 \circ (\mathbf{j}_1 \circ \mathbf{j}_3)) \subseteq \mathbf{n}_2 \ \vdash\ \mathbf{t} \circ \mathbf{j}_1 \subseteq \mathbf{j}_2 \to \mathbf{n}_2$ iff
$\mathbf{j}_3 \to (\mathbf{j}_2 \circ (\mathbf{j}_1 \circ \mathbf{j}_3)) \subseteq \mathbf{n}_2 \ \vdash\ (\mathbf{t} \circ \mathbf{j}_1) \circ \mathbf{j}_2 \subseteq \mathbf{n}_2$ iff
$(\mathbf{t} \circ \mathbf{j}_1) \circ \mathbf{j}_2 \ \subseteq\ \mathbf{j}_3 \to (\mathbf{j}_2 \circ (\mathbf{j}_1 \circ \mathbf{j}_3))$.

B6. $A \to ((A \to B) \to B)$
Simplifying the pure quasi-inclusion after elimination:
$\mathbf{j}_2 \to (\mathbf{j}_2 \circ \mathbf{j}_1) \subseteq \mathbf{n}_1 \mathbf{j}_1 \to \mathbf{n}_1 \subseteq \mathbf{m} \ \vdash\ \mathbf{t} \subseteq \mathbf{m}$ iff
$\mathbf{j}_2 \to (\mathbf{j}_2 \circ \mathbf{j}_1) \subseteq \mathbf{n}_1 \ \vdash\ \mathbf{t} \subseteq \mathbf{j}_1 \to \mathbf{n}_1$ iff
$\mathbf{j}_2 \to (\mathbf{j}_2 \circ \mathbf{j}_1) \subseteq \mathbf{n}_1 \ \vdash\ \mathbf{t} \circ \mathbf{j}_1 \subseteq \mathbf{n}_1$ iff

$\mathbf{t} \circ \mathbf{j}_1 \subseteq \mathbf{j}_2 \to (\mathbf{j}_2 \circ \mathbf{j}_1)$.
Computing the FO condition:
$\forall w \forall o \in O \forall x_1 \forall x_2 (w \vdash \mathbf{t} \circ \mathbf{j}_1 \Rightarrow w \vdash \mathbf{j}_2 \to (\mathbf{j}_2 \circ \mathbf{j}_1))$ iff
$\forall w \forall x_1 \forall x_2 (x_1 \leq w \Rightarrow \forall z (Rwx_2z \Rightarrow z \Vdash (\mathbf{j}_2 \circ \mathbf{j}_1))$ iff
$\forall w \forall x_1 \forall x_2 (x_1 \leq w \Rightarrow \forall z (Rwx_2z \Rightarrow Rx_2x_1z))$ iff
$\forall x_1 \forall x_2 \forall z (Rx_1x_2z \Rightarrow Rx_2x_1z)$.

B7. $(A \to (B \to C)) \to (B \to (A \to C))$
Simplifying the pure quasi-inclusion after elimination:

$\mathbf{j}_2 \to \mathbf{n}_2 \subseteq \mathbf{n}_1,\ \mathbf{j}_3 \to ((\mathbf{j}_1 \circ \mathbf{j}_3) \circ \mathbf{j}_2) \subseteq \mathbf{n}_2,\ \mathbf{j}_1 \to \mathbf{n}_1 \subseteq \mathbf{m}\ \vdash\ \mathbf{t} \subseteq \mathbf{m}$ iff
...
$(\mathbf{t} \circ \mathbf{j}_1) \circ \mathbf{j}_2 \subseteq \mathbf{j}_3 \to ((\mathbf{j}_1 \circ \mathbf{j}_3) \circ \mathbf{j}_2)$.

Axioms for Negation

D1. $(A \wedge B \to C) \to (A \wedge \sim C \to \sim B)$
D2. $A \vee \sim A$
D3. $(A \to \sim A) \to \sim A$
D3'. $(A \to B) \to (\sim A \vee B)$
D4. $(A \to \sim B) \to (B \to \sim A)$
D5. $B \to (A \vee \sim A)$
D5'. $(A \wedge (\sim A \vee B)) \to B$
D6. $A \to (\sim A \to B)$
D6'. $\sim (A \to B) \to A$
D7. $\sim (A \to B) \to (B \to A)$
D8. $(A \to \sim (B \to C)) \to (\sim B \to \sim A)$

We state the following without proof.

Proposition 4.8 *All axioms D1-D7 are Sahlqvist, while D8 is properly inductive.*

An Example of a Sahlqvist Axiom with Fusion

The following formula is Sahlqvist, but neither B-Sahlqvist nor S-Sahlqvist.

AF2. (MR2 in Routley et al. 1982, p. 377) $(A \circ B \to C) \to (A \to (B \to C))$.
Pure quasi-inclusion after elimination:
$\mathbf{j}_1 \to \mathbf{n}_1 \subseteq \mathbf{m},\ \mathbf{j}_2 \to \mathbf{n}_2 \subseteq \mathbf{n}_1,\ \mathbf{j}_3 \to \mathbf{j}_1 \circ (\mathbf{j}_2 \circ \mathbf{j}_3) \subseteq \mathbf{n}_2\ \vdash\ \mathbf{t} \subseteq \mathbf{m}$.
Simplified pure inclusion: $((\mathbf{t} \circ \mathbf{j}_1) \circ \mathbf{j}_2) \circ \mathbf{j}_3 \subseteq \mathbf{j}_1 \circ (\mathbf{j}_2 \circ \mathbf{j}_3)$.

Semantic Conditions for Axiomatic Extensions of B

Following Routley et al. (1982), we use $\Rightarrow$ (instead of $>$), $\neg$, &, $\vee$, * for the FO connectives in the correspondence language, as well as the following abbreviations:

$$R^2abcd := \exists x(Rabx \,\&\, Rxcd), \quad R^2a(bc)d := \exists x(Rbcx \,\&\, Raxd),$$
$$R^3ab(cd)e := \exists x(R^2abxe \,\&\, Rcdx).$$

We present here the frame conditions for axiomatic extensions of the System B (with the axioms listed on the right), copied from Routley et al. (1982), where a, b, c, d, e are universally quantified variables. We invite the reader to verify whether the FO conditions computed by PEARL are, respectively, equivalent to those listed here.

- q1. $Raaa$ — B1
- q2. $Rabc \Rightarrow R^2a(ab)c$ — B2
- q3. $R^2abcd \Rightarrow R^2b(ac)d$ — B3
- q4. $R^2abcd \Rightarrow R^2a(bc)d$ — B4
- q5. $Rabc \Rightarrow R^2abbc$ — B5
- q6. $Rabc \Rightarrow Rbac$ — B6
- q7. $R^2abcd \Rightarrow R^2acbd$ — B7
- q8. $R^2abcd \Rightarrow R^3ac(bc)d$ — B8
- q9. $R^2abcd \Rightarrow R^3bc(ac)d$ — B9

- q10. $Rabc \Rightarrow b \leq c$ — B10
- q11. $Rabc \Rightarrow a \leq c$ (equivalently, given $s2$ below this reduces to $0 \leq a$) — B11
- q12. $R^2abcd \Rightarrow a \leq d$ — B12
- q13. $Rabc \Rightarrow a \leq c \,\&\, b \leq c$ — B13
- q14. $R^2abcd \Rightarrow Racd \,\&\, Rbcd$ — B14
- q15. $R^2abcd \Rightarrow \exists x(b \leq x \,\&\, c \leq x \,\&\, Raxd)$ — B15
- q16. $a \leq b \,\&\, 0x \Rightarrow a \leq x$ — B16
- q17. $a \leq b \vee b \leq a$ — B17
- q18. $Rabc \Rightarrow a \leq c \vee b \leq c$ — B18
- q19. $Rabc \Rightarrow (Rbac \,\&\, a \leq c)$ — B19
- q20. $(Rabc \,\&\, Rade) \Rightarrow \exists x(b \leq x \,\&\, d \leq x \,\&\, (Raxc \vee Raxe)$ — B20

Some conditions involving negation:

- s1. $Rabc \Rightarrow \exists x(b \leq x \,\&\, c^* \leq x \,\&\, Raxb^*$ — D1
- s2. $x^* \leq x$ for all x such that $0x$ (reduces to $0^* \leq 0$) — D2
- s3. Raa^*a — D3
- s4. $Rabc \Rightarrow Rac^*b^*$ — D4
- s5. $a^* \leq a$ — D5
- s6. $Rabc \Rightarrow a \leq b^*$ — D6
- s7. $(Rabc \,\&\, Ra^*de) \Rightarrow (d \leq c \vee b \leq e)$ — D7
- s8. $(Rabc \Rightarrow \exists y(Rac^*y \,\&\, (\forall d, e)(Ry^*de \Rightarrow d \leq b^*))$ — D8

References

Ackermann, W. (1935). Untersuchung über das Eliminationsproblem der mathematischen Logik. *Mathematische Annalen*, *110*, 390–413.

Badia, G. (2018). On Sahlqvist formulas in relevant logic. *Journal of Philosophical Logic*, *47*(4), 673–691.

Benthem, J. (1976). *Modal correspondence theory*. PhD thesis, Mathematisch Instituut & Instituut voor Grondslagenonderzoek, University of Amsterdam.

Benthem, J. (2001). Correspondence theory. In D. Gabbay, & F. Guenthner (Eds.), *Handbook of philosophical logic* (2nd ed., Vol. 3, pp. 325–408). Dordrecht: Springer - Science+Business Media, B.V.

Blackburn, P., de Rijke, M., & Venema, Y. (2001). *Modal logic*. Cambridge: Cambridge University Press.

Conradie, W., Fomatati, Y., Palmigiano, A., & Sourabh, S. (2015). Algorithmic correspondence for intuitionistic modal mu-calculus. *Theoretical Computer Science*, *564*, 30–62.

Conradie, W., Ghilardi, S., & Palmigiano, A. (2014). Unified correspondence. In A. Baltag, & S. Smets (Eds.), *Johan van Benthem on logic and information dynamics* (Vol. 5, pp. 933–975). Outstanding contributions to logic. Berlin: Springer International Publishing.

Conradie, W., & Goranko, V. (2008). Algorithmic correspondence and completeness in modal logic, IV. Semantic extensions of SQEMA. *Journal of Applied Non-Classical Logics*, *18*(2–3), 175–211.

Conradie, W., Goranko, V., & Vakarelov, D. (2005). Elementary canonical formulae: A survey on syntactic, algorithmic, and model-theoretic aspects. In R. Schmidt, I. Pratt-Hartmann, M. Reynolds, & H. Wansing (Eds.), *Advances in modal logic* (Vol. 5, pp. 17–51). London: Kings College.

Conradie, W., Goranko, V., & Vakarelov, D. (2006a). Algorithmic correspondence and completeness in modal logic, I. The core algorithm SQEMA. *Logical Methods in Computer Science*, *2*, (1:5).

Conradie, W., Goranko, V., & Vakarelov, D. (2006b). Algorithmic correspondence and completeness in modal logic, II. Polyadic and hybrid extensions of the algorithm SQEMA. *Journal of Logic and Computation*, *16*, 579–612.

Conradie, W., Goranko, V., & Vakarelov, D. (2009). Algorithmic correspondence and completeness in modal logic, III. Extensions of the algorithm SQEMA with substitutions. *Fundamenta Informaticae*, *92*(4), 307–343.

Conradie, W., Goranko, V., & Vakarelov, D. (2010). Algorithmic correspondence and completeness in modal logic, V. Recursive extensions of SQEMA. *Journal Applied Logic*, *8*(4), 319–333.

Conradie, W., & Palmigiano, A. (2012). Algorithmic correspondence and canonicity for distributive modal logic. *Annals of Pure and Applied Logic*, *163*(3), 338–376.

Conradie, W., & Palmigiano, A. (2019). Algorithmic correspondence and canonicity for non-distributive logics. *Annals of Pure and Applied Logic*, *170*(9), 923–974.

Davey, B., & Priestley, H. (2002). *Introduction to lattices and order*. Cambridge: Cambridge University Press.

Dunn, J., & Restall, G. (2002). Relevance logic. In D. Gabbay & F. Guenthner (Eds.), *Handbook of philosophical logic* (2nd ed., Vol. 6, pp. 1–128). Dordrecht: Springer - Science+Business Media, B.V.

Gehrke, M., Nagahashi, H., & Venema, Y. (2005). A Sahlqvist theorem for distributive modal logic. *Annals of Pure and Applied Logic*, *131*, 65–102.

Goranko, V., & Vakarelov, D. (2001). Sahlqvist formulae in hybrid polyadic modal languages. *Journal of Logic and Computation*, *11*(5), 737–754.

Goranko, V., & Vakarelov, D. (2002). Sahlqvist formulas unleashed in polyadic modal languages. In F. Wolter, H. Wansing, M. de Rijke, & M. Zakharyaschev (Eds.), *Advances in modal logic* (Vol. 3, pp. 221–240). Singapore: World Scientific.

Goranko, V., & Vakarelov, D. (2006). Elementary canonical formulae: Extending Sahlqvist's theorem. *Annals of Pure and Applied Logic*, *141*(1–2), 180–217.

Jipsen, P., & Kinyon, M. (2019). Nonassociative right hoops. *Algebra universalis*, *80*(4), 47.
Palmigiano, A., Sourabh, S., & Zhao, Z. (2017). Sahlqvist theory for impossible worlds. *Journal of Logic and Computation*, *27*(3), 775–816.
Routley, R., & Meyer, R. (1973). Semantics of entailment. In H. Leblanc (Ed.), *Truth, Syntax and Modality, Proceedings of the Temple University Conference on Alternative Semantics* (pp. 194–243). North Holland.
Routley, R., Meyer, R., Plumwood, V., & Brady, R. (1982). *Relevant logics and its rivals* (Vol. I). Ridgeview.
Sahlqvist, H. (1975). Correspondence and completeness in the first and second-order semantics for modal logic. In S. Kanger (Ed.), *Proceedings of the 3rd Scandinavian Logic Symposium, Uppsala 1973* (pp. 110–143). Amsterdam: Springer.
Seki, T. (2003). A Sahlqvist theorem for relevant modal logics. *Studia Logica*, *73*(3), 383–411.
Urquhart, A. (1972). Semantics for relevant logics. *Journal of Symbolic Logic*, *37*(1), 159–169.
Urquhart, A. (1996). Duality for algebras of relevant logics. *Studia Logica*, *56*(1/2), 263–276.

Chapter 5
Beth Definability in the Logic KR

Jacob Garber

Second Reader
Z. Gyenis
Jagiellonian University

Abstract The Beth Definability Property holds for an algebraizable logic if and only if every epimorphism in the corresponding category of algebras is surjective. Using this technique, Urquhart in 1999 showed that the Beth Definability Property fails for a wide class of relevant logics, including **T**, **E**, and **R**. However, the counterexample for those logics does not extend to the algebraic counterpart of the super relevant logic **KR**, the so-called Boolean monoids. Following a suggestion of Urquhart, we use modular lattices constructed by Freese to show that epimorphisms need not be surjective in a wide class of relation algebras. This class includes the Boolean monoids, and thus the Beth Definability Property fails for **KR**.

Keywords Beth definability · Relevant logic · Epimorphism · Boolean monoid · Relation algebra

5.1 Introduction

Relevant logics were first introduced to avoid the *paradoxes of material implication*, which are counterintuitive inferences that result from a mismatch between the intuitive meaning of implication and its formalization in classical logic. This work lead to the development of a wide swath of relevant logics, spanning from the basic logic **B** to the logic of relevant implication **R**, with many other relevant logics (such as ticket entailment **T** and relevant entailment **E**) filling out the middle. A comprehensive description of these logics and relevant logic in general can be found in any of Anderson and Belnap (1975), Anderson et al. (1992), Dunn and Restall (2002),

---

J. Garber (✉)
University of Alberta, Edmonton, Canada
e-mail: jgarber1@ualberta.ca

I. Düntsch and E. Mares (eds.), *Alasdair Urquhart on Nonclassical and Algebraic Logic and Complexity of Proofs*, Outstanding Contributions to Logic 22,
https://doi.org/10.1007/978-3-030-71430-7_5

Routley et al. (1982), Brady (2003). In this paper, we will focus our attention on the logic **KR**, which consists of adding the paradoxical axiom $(A \wedge \neg A) \rightarrow B$ to **R**. Despite being stronger than **R** and thus not a purely relevant logic, **KR** avoids various other paradoxes of implication and does not collapse to classical logic (see Kerr 2019). Following Dunn and Restall (2002), we call any such logic a *super relevant* logic. Several important properties of classical logic fail for the relevant logics, one of which is the *Beth Definability Property*.

Definition 5.1 Let L be a propositional logic and Σ a set of formulas from L containing a variable p. For a new variable q, let $\Sigma[p/q]$ denote the result of replacing all instances of p with q. We say Σ *implicitly defines* p if

$$\Sigma \cup \Sigma[p/q] \vdash p \leftrightarrow q.$$

Alternatively, we say Σ *explicitly defines* p if there is a formula A containing only the variables in Σ without p, such that

$$\Sigma \vdash p \leftrightarrow A.$$

A logic L is said to have the *Beth Definability Property* if for any set of formulas Σ and variable p, if Σ implicitly defines p, then Σ also explicitly defines p.

The well-known Beth Definability Theorem states that the Beth Definability Property holds for classical propositional logic. However, as shown by Urquhart (1999) in 1999 and extended by Blok and Hoogland (2006, Corollary 4.15) in 2006, the Beth Definability Property fails for all relevant logics between **B** and **R**. The techniques of those papers rely on the fact that Boolean negation is implicitly (but not explicitly) definable in those logics. This approach does not extend to **KR**, where Boolean negation is identified with relevant negation and is thus explicitly definable. However, Urquhart conjectured that the Beth Definability Property fails for **KR** as well, and outlined a possible method of attack using algebraic logic.

In algebraic logic, every algebraizable logic L has a corresponding category of algebras Alg L. For instance, the algebras of classical logic are the Boolean algebras, and those of intuitionistic logic are the Heyting algebras. Using this correspondence, it is often possible to translate properties of a logic into properties of its corresponding algebra.

Definition 5.2 For objects A, B in a category C, we say a map $f : A \rightarrow B$ is an *epimorphism* if for any other object C and maps $g, h : B \rightarrow C$,

$$\text{if } g \circ f = h \circ f, \text{ then } g = h.$$

The Beth Definability Property holds in an algebraizable logic L iff in the corresponding algebra Alg L all epimorphisms are surjective (the *ES* property). This correspondence was first proven by Németi in Henkin et al. (1985, Sect. 5.6), and further developed in Blok and Hoogland (2006). Intuitively, one can think of epimorphisms as being implicit definitions, and surjections as being explicit ones. For algebras $A \subseteq B$, we say A is an *epic subalgebra* of B if the inclusion map $i : A \to B$ is an epimorphism. Of course, this inclusion map is surjective iff $A = B$. To disprove the Beth Definability Property for a logic, it thus suffices to find a proper epic subalgebra in its corresponding category. To apply this to **KR**, we will analyze epimorphisms in the category of algebras for **KR**, the *Boolean monoids*. Boolean monoids are closely related to relation algebras, and can be equivalently defined as *dense symmetric relation algebras*.

We will tackle this problem using the approach described in Urquhart (2017, Problem 5.3). As discussed in that paper, there is a general correspondence between Boolean monoids and modular lattices. Every Boolean monoid contains a modular lattice, and given a modular lattice L, one can construct a corresponding Boolean monoid $\mathcal{A}(L)$ that contains an isomorphic copy of L as a sublattice. As shown in Freese (1979, Theorem 3.3), there exist modular lattices A and B such that A is a proper epic sublattice of B. Using the above correspondence, we extend this to the construction of a proper epic sub-Boolean monoid, which shows the Beth Definability Property fails for **KR**. This construction is rather general, and in fact shows that ES fails for a wide class of relation algebras that includes the Boolean monoids.

5.2 Boolean Monoids and Modular Lattices

As shown in Anderson and Belnap (1975, Sect. 28.2.3), the algebraic counterpart of the logic **R** with truth constant t (sometimes denoted $\mathbf{R}^{\mathbf{t}}$) is the variety of *De Morgan monoids*. These are De Morgan lattices with a commutative monoid operation. The addition of the axiom $(A \wedge \neg A) \to B$ to **R** corresponds to adding the axiom $a \wedge \neg a = 0$ to the algebra, which reduces the De Morgan lattice to a Boolean algebra. Such objects, which we call *Boolean monoids*, are the algebraic counterpart of **KR**. A particularly useful description of Boolean monoids is in terms of relation algebras. The following axiomatization of relation algebras is taken from Givant (2017b, Definition 2.1).

Definition 5.3 A *relation algebra* is an algebra $\langle A, \vee, \neg, \circ, ^{\smile}, t\rangle$ such that for all $a, b, c \in A$,

1. $a \vee b = b \vee a$
2. $a \vee (b \vee c) = (a \vee b) \vee c$
3. $\neg(\neg a \vee b) \vee \neg(\neg a \vee \neg b) = a$
4. $a \circ (b \circ c) = (a \circ b) \circ c$
5. $a \circ t = a$
6. $a^{\smile\smile} = a$
7. $(a \circ b)^{\smile} = b^{\smile} \circ a^{\smile}$
8. $(a \vee b) \circ c = (a \circ c) \vee (b \circ c)$
9. $(a \vee b)^{\smile} = a^{\smile} \vee b^{\smile}$
10. $\left(a^{\smile} \circ \neg(a \circ b)\right) \vee \neg b = \neg b$

With the standard definition of $a \wedge b = \neg(\neg a \vee \neg b)$, axioms 1–3 imply that $\langle A, \vee, \wedge, \neg\rangle$ is a Boolean algebra, axioms 4–7 imply that $\langle A, \circ, ^{\smile}, t\rangle$ is a monoid with involution, and axioms 8–10 relate the Boolean and monoid operations to each other.

For all $a, b \in A$, a relation algebra A is called

1. *abelian* if $a \circ b = b \circ a$,
2. *symmetric* if $a^{\smile} = a$,
3. *dense* if $a \leq a \circ a$.

In particular, a Boolean monoid can be equivalently defined as a *dense symmetric relation algebra.*

One important result from the theory of relation algebras is that every abelian relation algebra contains a special set of elements that form a modular lattice.

Definition 5.4 A lattice L is called *modular* if for all $x, y, z \in L$,

$$x \leq z \implies x \vee (y \wedge z) = (x \vee y) \wedge z.$$

This implication is equivalent to the following dual identities:

$$(x \wedge y) \vee (x \wedge z) = x \wedge (y \vee (x \wedge z)),$$
$$(x \vee y) \wedge (x \vee z) = x \vee (y \wedge (x \vee z)).$$

Definition 5.5 For a relation algebra A, an element $a \in A$ is called *reflexive* if $t \leq a$, *symmetric* if $a^{\smile} = a$, and *transitive* if $a \circ a \leq a$. An element with all three of these properties is a *reflexive equivalence* element. Define $\mathcal{L}(A)$ to be the set of all reflexive equivalence elements of A.

Theorem 5.1 *For an abelian relation algebra A, the set of reflexive equivalence elements $\mathcal{L}(A)$ is closed under fusion and meet, and forms a bounded modular lattice. Join is given by $a \circ b$, meet by $a \wedge b$, t is the bottom element, and 1 is the top.*

Proof See Givant (2017b, Corollary 5.17).

We can also in some sense reverse the above theorem, and use a modular lattice to construct a relation algebra.

Definition 5.6 A **KR** *frame* or *model structure* is a triple $F = \langle S, R, 0\rangle$ of a set S with ternary relation R and distinguished element 0, satisfying:

1. $R0ab$ iff $a = b$
2. $Raaa$
3. $Rabc$ implies $Rbac$ and $Racb$ (total symmetry)
4. $Rabc$ and $Rcde$ implies $\exists f \in S$ such that $Radf$ and $Rfbe$ (Pasch's Postulate)

The last condition has close ties to projective geometry and is explored in Urquhart (2017).

Definition 5.7 For a **KR** frame $F = \langle S, R, 0 \rangle$, the *complex algebra* of F is the algebra $\mathcal{A}(F) = \langle P(S), \cup, \cap, {}^c, \circ, t \rangle$, where

1. $\langle P(S), \cup, \cap, {}^c \rangle$ is the Boolean algebra on the power set of S.
2. $t = \{0\}$ is the monoid identity.
3. For $A, B \subseteq S$, fusion is defined as

$$A \circ B = \{ c \in S \mid Rabc \text{ for some } a \in A, b \in B \}.$$

We often write $\mathcal{A}(S)$ for the complex algebra when the ternary relation and distinguished element are understood from context.

Theorem 5.2 *For a* **KR** *frame* F, *the complex algebra* $\mathcal{A}(F)$ *is a Boolean monoid.*

Proof See Urquhart (2017, Sect. 2).

Definition 5.8 For a lattice L with least element 0, we define the following ternary relation on the elements of L:

$$Rabc \iff a \vee b = a \vee c = b \vee c.$$

Then with this relation, $\langle L, R, 0 \rangle$ is a **KR** frame iff L is modular.

Proof The first three properties of a **KR** frame follow immediately from the lattice structure of L. The last property, Pasch's Postulate, is equivalent to the modular law on L, which is shown in Urquhart (2017, Theorem 2.7).

Thus, for a modular lattice L with 0, the *lattice complex algebra* $\mathcal{A}(L)$ is a Boolean monoid. An alternative more direct proof of this can be found in Givant (2017b, Sect. 3.7). The definition of a **KR** frame is also an instance of the more general notion of a *relational structure*, where the construction of a complex algebra can be repeated in the context of Boolean algebras with operators. Givant (2017a, Chap. 19; 2014, Chap. 1) go into more detail.

For a lattice complex algebra $\mathcal{A}(L)$, a particularly simple description of its reflexive equivalence elements can be given in terms of the ideals of L.

Definition 5.9 For a lattice L, an *ideal* of L is a non-empty subset $J \subseteq L$ such that

1. If $a, b \in J$, then $a \vee b \in J$.
2. If $a \in J$ and $b \leq a$, then $b \in J$.

The set of all ideals of L is denoted Id L, which forms a lattice with respect to set inclusion. A special class of ideals are the *principal ideals*, which are of the form $(a] = \{ b \in L \mid b \leq a \}$. We will sometimes use the notation $(a]_L$ to emphasize that this is the principal ideal of a inside L.

Proposition 5.1 *For a lattice L, the principal ideal map*

$$\begin{aligned} I : L &\to \operatorname{Id} L \\ a &\mapsto (a] \end{aligned}$$

is an injective homomorphism of lattices.

Proof See Grätzer (2003, Corollary 4, p. 24).

Theorem 5.3 *For a modular lattice L with least element 0, the set of reflexive equivalence elements $\mathcal{L}(\mathcal{A}(L))$ and the set of ideals* Id *L are identical as lattices. That is, $\mathcal{L}(\mathcal{A}(L)) = \operatorname{Id} L$, and for all ideals $J, K \in \operatorname{Id} L$*

$$\begin{aligned} J \vee K &= J \circ K \\ J \wedge K &= J \cap K. \end{aligned}$$

Proof See Maddux (1981, p. 243).

5.3 Embeddings of Lattice Complex Algebras

The purpose of this section is to prove Theorem 5.5, which states that for all complete sublattices K of a modular lattice L, there is a corresponding complete embedding $\phi : \mathcal{A}(K) \to \mathcal{A}(L)$ of Boolean monoids. The construction of this map will rely on the following result from the theory of relation algebras.

Theorem 5.4 *Let A be a complete and atomic Boolean monoid, U the set of atoms of A, and B a complete Boolean monoid. Suppose $\phi : U \to B$ is a map with the following properties:*

1. *The elements $\phi(u)$ for $u \in U$ are non-zero, mutually disjoint, and have a join of 1 in B.*
2. $t = \bigvee\{\,\phi(u) \mid u \in U \text{ and } u \leq t\,\}$.
3. $\phi(u) \circ \phi(v) = \bigvee\{\,\phi(w) \mid w \in U \text{ and } w \leq u \circ v\,\}$ *for all* $u, v \in U$.

Then, ϕ extends in a unique way to a complete embedding $\phi : A \to B$ of Boolean monoids, given by

$$\phi(r) = \bigvee\{\,\phi(u) \mid u \in U \text{ and } u \leq r\,\},$$

where r is any element of A.

Proof This is a specialization of Givant (2017b, Theorem 7.13 and Corollary 7.14) to Boolean monoids.

We apply this to lattice complex algebras as follows. Recall that for a complete lattice L, a subset $K \subseteq L$ is a *complete sublattice* iff for all $S \subseteq K$, $\bigwedge S \in K$ and $\bigvee S \in K$, where these infima and suprema are calculated in L.

Theorem 5.5 *Let L be a complete modular lattice, $K \subseteq L$ a complete sublattice, and I_K and I_L their respective principal ideal maps. Then, there is a unique complete embedding of Boolean monoids $\phi : \mathcal{A}(K) \to \mathcal{A}(L)$ such that $\phi \circ I_K = I_L$.*

Proof We will first show uniqueness to determine what the map ϕ should be, and then use that definition to show it is a complete embedding.

Suppose that $\phi : \mathcal{A}(K) \to \mathcal{A}(L)$ is a complete embedding with $\phi \circ I_K = I_L$. Since ϕ is a complete homomorphism, it is determined by its values on the singleton subsets of K, which are the atoms of $\mathcal{A}(K)$. For all $a \in K$, we can write $(a]_K$ as the disjoint union

$$
\begin{aligned}
& (a]_K = \{a\} \cup \bigcup_{\substack{b<a \\ b \in K}} (b]_K \\
\Longrightarrow\ & \phi((a]_K) = \phi(\{a\}) \cup \bigcup_{\substack{b<a \\ b \in K}} \phi((b]_K) && \phi \text{ is a complete homomorphism} \\
\Longrightarrow\ & (a]_L = \phi(\{a\}) \cup \bigcup_{\substack{b<a \\ b \in K}} (b]_L && \phi \circ I_K = I_L \\
\Longrightarrow\ & \phi(\{a\}) = (a]_L \setminus \bigcup_{\substack{b<a \\ b \in K}} (b]_L && \phi \text{ preserves disjoint unions}
\end{aligned}
$$

Thus, ϕ is uniquely determined.

So then, let U be the set of singletons in $\mathcal{A}(K)$, and define the map $\phi : U \to \mathcal{A}(L)$ by

$$
\phi(\{a\}) = (a]_L \setminus \bigcup_{\substack{b<a \\ b \in K}} (b]_L.
$$

We will verify the three conditions of Theorem 5.4 to show that this can be extended to a complete embedding of Boolean monoids.

1. It suffices to show that the sets $\phi(\{a\})$ for $a \in K$ are non-empty, mutually disjoint, and cover L.
   - All sets are non-empty, since $a \in \phi(\{a\})$ for any $a \in K$.
   - Let $a, b \in K$ be distinct elements. Then $a \wedge b \leq a$, and $a \wedge b \leq b$. Since a and b are distinct, at least one of the previous inequalities must be strict, so without loss of generality suppose $a \wedge \{b\} < a$. Since K is a sublattice, $a \wedge b \in K$. Now suppose $x \in \phi(\{a\}) \cap \phi(\{b\})$. Then $x \leq a$ and $x \leq b$, so $x \in (a \wedge b]_L$. Since $a \wedge b < a$, this implies $x \notin \phi(\{a\})$, which is a contradiction. Thus, $\phi(\{a\})$ and $\phi(\{b\})$ are disjoint.
   - For an arbitrary $x \in L$, let

$$
a = \bigwedge F_x,
$$

where $F_x = \{ b \in K \mid x \leq b \}$. Since K is a complete sublattice, this infimum exists and is an element of K. By definition, x is a lower bound for F_x, so $x \leq a$, and thus $x \in (a]_L$. Since $a \in F_x$, we in fact have $a = \min F_x$. Furthermore, for any other $b \in K$ with $b < a$, it cannot be that $x \in (b]_L$, since then we would have $a \leq b$, which is impossible. Thus,

$$x \in (a]_L \setminus \bigcup_{\substack{b<a \\ b \in K}} (b]_L = \phi(\{a\}).$$

So the images of ϕ cover L.

2. The monoid identity $t = \{0\}$ is itself a singleton, and from the definition of ϕ we trivially have
$$\phi(t) = \phi(\{0\}) = (0]_L = \{0\} = t.$$

3. From left to right, let $a, b \in K$, and suppose that $z \in \phi(\{a\}) \circ \phi(\{b\})$. We wish to show $z \in \phi(\{c\})$, for some $c \in K$ with $\{c\} \subseteq \{a\} \circ \{b\}$.
By assumption $Rxyz$ for some $x \in \phi(\{a\})$ and $y \in \phi(\{b\})$. From the first condition, we know $a = \min F_x, b = \min F_y$, and $z \in \phi(\{c\})$, where $c = \min F_z$. Since $x \leq a$ and $y \leq b$, we have

$$\begin{aligned} & x \vee y \leq a \vee b & \\ \implies & x \vee z \leq a \vee b & \text{since } Rxyz \\ \implies & z \leq a \vee b & \\ \implies & c \leq a \vee b & \text{minimality of } c \\ \implies & a \vee c \leq a \vee b. & \end{aligned}$$

Symmetrically, we conclude $a \vee b \leq a \vee c$, and so $a \vee b = a \vee c$. A similar argument shows $a \vee c = b \vee c$. Thus $Rabc$, and so $c \in \{a\} \circ \{b\}$ as desired.
From right to left, let $a, b, c \in K$ and suppose $\{c\} \subseteq \{a\} \circ \{b\}$. We wish to show $\phi(\{c\}) \subseteq \phi(\{a\}) \circ \phi(\{b\})$. That is, for all $z \in \phi(\{c\})$, there exists $x \in \phi(\{a\})$ and $y \in \phi(\{b\})$ such that $Rxyz$. To do this, we use an approach similar to the one in Maddux (1981, p. 244). For a given z, let

$$\begin{aligned} x &= (b \vee z) \wedge a \\ y &= (a \vee z) \wedge b. \end{aligned}$$

We first show that $a = \min F_x$. From the definition of x we have $a \wedge b \leq x \leq a$, so $a \in F_x$. Now let $d \in F_x$ be any other element. Then $d \in K$ with $x \leq d$, so $x \leq a \wedge d$. Furthermore,

$$\begin{aligned} x \vee b &= ((b \vee z) \wedge a) \vee b && \text{definition of } x \\ &= (b \vee z) \wedge (a \vee b) && \text{modularity} \\ &= (b \vee z) \wedge (b \vee c) && \text{since } Rabc \\ &= b \vee z && \text{since } z \leq c. \end{aligned}$$

Since $z \leq b \vee z = x \vee b$, we then have

$$\begin{aligned} & z \leq (a \wedge d) \vee b && \text{since } x \leq a \wedge d \\ \Longrightarrow\ & c \leq (a \wedge d) \vee b && \text{minimality of } c \\ \Longrightarrow\ & b \vee c \leq (a \wedge d) \vee b && \\ \Longrightarrow\ & a \vee b \leq (a \wedge d) \vee b && \text{since } Rabc. \end{aligned}$$

Using absorption, this implies

$$\begin{aligned} a &\leq ((a \wedge d) \vee b) \wedge a && \\ &= (a \wedge d) \vee (a \wedge b) && \text{modularity} \\ &= a \wedge d && \text{since } a \wedge b \leq x \leq a \wedge d. \end{aligned}$$

Therefore $a \leq d$, so $a = \min F_x$ as wanted. Thus $x \in \phi(\{a\})$, and a symmetric argument shows that $y \in \phi(\{b\})$.
Now we show $Rxyz$. Using modularity,

$$\begin{aligned} x \vee z &= ((b \vee z) \wedge a) \vee z \\ &= (b \vee z) \wedge (a \vee z) \\ &= (a \vee z) \wedge (b \vee z) \\ &= ((a \vee z) \wedge b) \vee z \\ &= y \vee z. \end{aligned}$$

Since $x \leq a$ and $z \leq c$, we have $x \vee z \leq a \vee c = a \vee b$. Thus,

$$\begin{aligned} x \vee z &= (a \vee b) \wedge (x \vee z) && \\ &= (a \vee b) \wedge (a \vee z) \wedge (b \vee z) && \text{from above} \\ &= (a \vee (b \wedge (a \vee z))) \wedge (b \vee z) && \text{modularity} \\ &= (b \vee z) \wedge (a \vee ((a \vee z) \wedge b)). && \end{aligned}$$

Using that $(a \vee z) \wedge b \leq b \leq b \vee z$ and a final application of the modular law, we thus have

$$\begin{aligned} x \vee z &= ((b \vee z) \wedge a) \vee ((a \vee z) \wedge b) \\ &= x \vee y. \end{aligned}$$

Thus $Rxyz$, and the condition is shown.

Thus by Theorem 5.4, ϕ extends uniquely to a complete embedding $\phi : \mathcal{A}(K) \to \mathcal{A}(L)$ of Boolean monoids, where for all $S \subseteq K$,

$$\phi(S) = \bigcup_{a \in S} \phi(\{a\}).$$

We use this definition to show that $\phi \circ I_K = I_L$. For any $a \in K$, $(a]_K$ is a reflexive equivalence element of $\mathcal{A}(K)$ by Theorem 5.3. Since ϕ preserves equational properties, the image $\phi((a]_K)$ is also a reflexive equivalence element of $\mathcal{A}(L)$, and thus an ideal of L by the same theorem. From the definition of ϕ,

$$a \in \phi(\{a\}) \subseteq \phi((a]_K),$$

and so $(a]_L \subseteq \phi((a]_K)$ from the definition of an ideal.

On the other hand,

$$\phi((a]_K) = \bigcup_{\substack{b \leq a \\ b \in K}} \phi(\{b\}) \subseteq \bigcup_{\substack{b \leq a \\ b \in K}} (b]_L = (a]_L,$$

and so $\phi((a]_K) = (a]_L$.

5.4 An Epimorphism That Is Not Surjective

In this section, let LRA be the class of all subalgebras of lattice complex algebras, ARA the variety of abelian relation algebras, and R any class of relation algebras with $\mathsf{LRA} \subseteq \mathsf{R} \subseteq \mathsf{ARA}$. We will now use the following general construction and modular lattices constructed by Freese to show that ES fails for R.

Let L be a complete modular lattice, and $K \subseteq L$ a complete sublattice. The principal ideal map $I_L : L \to \mathrm{Id}\, L$ is an embedding of lattices, and from Theorem 5.3 we know $\mathrm{Id}\, L = \mathcal{L}(\mathcal{A}(L))$. Thus, let $K' = I_L(K)$ and $L' = I_L(L)$ be the isomorphic images of L and K contained in $\mathcal{L}(\mathcal{A}(L))$. In $\mathcal{A}(L)$, let U be the subalgebra generated by K', and V the subalgebra generated by L'.

Theorem 5.6 *In the above situation, if K is a proper epic sublattice of L, then U is a proper R-epic subalgebra of V.*

Proof Since $K' \subset L'$ we have $U \subseteq V$. Let W be any other algebra of R, and $f, g : V \to W$ two homomorphisms that agree on U. The image of a reflexive equivalence element is a reflexive equivalence element, so f and g restrict to maps

$$f|_{L'}, g|_{L'} : L' \to \mathcal{L}(W).$$

By Theorem 5.1, $\mathcal{L}(W)$ is a modular lattice under fusion and meet, and f and g preserve these operations, so these restrictions are homomorphisms of modular lattices. By assumption, f and g agree on U, and since $K' \subseteq U$ they must also agree on K'. But K' is an epic sublattice of L', so f and g must also agree on L'. Thus, $f|_{L'} = g|_{L'}$, and so $f = g$ since L' is the generating set of V. Thus, U is an R-epic subalgebra of V.

However, U is a proper subalgebra. Let $\phi : \mathcal{A}(K) \to \mathcal{A}(L)$ be the complete embedding of Theorem 5.5, and let $Z = \operatorname{im} \phi$. Since $I_K(K) \subseteq \mathcal{A}(K)$, we have

$$\phi(I_K(K)) = I_L(K) = K',$$

so $K' \subseteq Z$, which implies $U \subseteq Z$ since U is the smallest subalgebra that contains K'. Now for contradiction suppose that $U = V$. Then $L' \subseteq V = U \subseteq Z$. Thus for any $x \in L$, we have $(x]_L \in Z$, so there is some $S \subseteq K$ such that $\phi(S) = (x]_L$. In particular then, there is an $a \in S$ such that

$$x \in \phi(\{a\}) \subseteq (a]_L \implies x \leq a.$$

On the other hand,

$$a \in \phi(\{a\}) \subseteq \phi(S) = (x]_L \implies a \leq x.$$

Thus $x = a$, so $x \in K$. But the element $x \in L$ was arbitrary, so $L = K$, which is a contradiction.

Theorem 5.7 *ES fails for any class* R *of relation algebras with* $\mathsf{LRA} \subseteq \mathsf{R} \subseteq \mathsf{ARA}$.

Proof In Freese (1979, Theorem 3.3), Freese constructs modular lattices $A \subset B$ such that A is a proper epic sublattice of B. B has no infinite chains, so is complete by Davey and Priestley (2002, Theorem 2.41 (iii)). Likewise, A is a $\{0,1\}$-sublattice of B, and as a sublattice is complete by Theorems 2.40 and 2.41 (i) of the same. We can thus apply Theorem 5.6 to A and B, and the result follows.

Corollary 5.1 *ES fails for the varieties of abelian, symmetric, and dense symmetric relation algebras.*

Corollary 5.2 *The Beth Definability Property fails for* ***KR****.*

5.5 Conclusion

Using modular lattices constructed by Freese, we have shown that epimorphisms need not be surjective in a wide class of relation algebras. This class includes the Boolean monoids, which shows that the Beth Definability Property fails for the super relevant logic **KR**. This should be contrasted with the result of Bezhanishvili et al.

(2017, Theorem 8.5), which shows that the Beth Definability Property does hold for the super relevant logic **RM**. The super relevant logics thus exhibit more diversity than the relevant logics, where this property fails uniformly.

Acknowledgements The author would like to thank Katalin Bimbó for suggesting this problem for his undergraduate research project and supervising his work on it. Her advice and never-ending encouragement were instrumental in finding a solution. The author would also like to thank the second reader Zalán Gyenis for his very helpful comments when reviewing this paper, and the editors Ivo Düntsch and Edwin Mares for publishing it in this volume. Finally, the author dedicates this paper to the memory of Steven Givant, whose textbooks in relation algebra made this entire proof possible.

References

Anderson, A. R., & Belnap, N. D. (1975). *Entailment: The logic of relevance and necessity* (Vol. 1). Princeton: Princeton University Press.

Anderson, A. R., Belnap, N. D., & Dunn, J. M. (1992). *Entailment: The logic of relevance and necessity* (Vol. 2). Princeton: Princeton University Press.

Bezhanishvili, G., Moraschini, T., & Raftery, J. G. (2017). Epimorphisms in varieties of residuated structures. *Journal of Algebra*, *492*, 185–211.

Blok, W. J., & Hoogland, E. (2006). The Beth property in algebraic logic. *Studia Logica*, *83*, 49–90.

Brady, R. (Ed.). (2003). *Relevant logics and their rivals* (Vol. 2). Farnham: Ashgate.

Davey, B. A., & Priestley, H. A. (2002). *Introduction to lattices and order* (2nd ed.). Cambridge: Cambridge University Press.

Dunn, J. M., & Restall, G. (2002). Relevance logic. In *Handbook of philosophical logic* (Vol. 6, 2nd ed.). Dordrecht: Springer Netherlands.

Freese, R. (1979). The variety of modular lattices is not generated by its finite members. *Transactions of the American Mathematical Society*, *255*, 277–300.

Givant, S. (2014). *Duality theories for Boolean algebras with operators*. Springer monographs in mathematics. Berlin: Springer International Publishing.

Givant, S. (2017a). *Advanced topics in relation algebras*. Berlin: Springer International Publishing.

Givant, S. (2017b). *Introduction to relation algebras*. Berlin: Springer International Publishing.

Grätzer, G. (2003). *General lattice theory* (2nd ed.). Basel: Birkäuser.

Henkin, L., Monk, J. D., & Tarski, A. (1985). *Cylindric algebras* (Vol. 2). Amsterdam: North-Holland.

Kerr, A. D. (2019). A plea for KR. *Synthese*.

Maddux, R. (1981). Embedding modular lattices into relation algebras. *Algebra Universalis*, *12*, 242–246.

Routley, R., Meyer, R. K., Plumwood, V., & Brady, R. T. (1982). *Relevant logics and their rivals* (Vol. 1). Atascadero: Ridgeview Publishing Company.

Urquhart, A. (1999). Beth's definability theorem in relevant logics. In *Logic at work* (Vol. 24). Studies in fuzziness and soft computing. Heidelberg: Physica-Verlag.

Urquhart, A. (2017). The geometry of relevant implication. *IFCoLog Journal of Logics and Their Applications*, *4*(3), 591–604.

Chapter 6
Geometric Models for Relevant Logics

Greg Restall

Second Reader
P. Balbiani
Institut de Recherche en Informatique de Toulouse – IRIT

Abstract Alasdair Urquhart's work on models for relevant logics is distinctive in a number of different ways. One key theme, present in both his undecidability proof for the relevant logic R (Urquhart 1984) and his proof of the failure of interpolation in R (Urquhart 1993), is the use of techniques from geometry (Urquhart 2019). In this paper, inspired by Urquhart's work, I explore ways to generate natural models of R^+ from geometries, and different constraints that an accessibility relation in such a model might satisfy. I end by showing that a set of natural conditions on an accessibility relation, motivated by geometric considerations, is jointly unsatisfiable.

Keywords Relevant logic · Substructural · Frame · Model · Semantics · Geometry

6.1 Models for Relevant Logics

If a conditional is to be *relevant*—if $A \to B$ is to be true only when there is a genuine *connection* between the antecedent A and the consequent B—any 'worlds' semantics for that conditional must look rather unlike the well-known modal semantics for strict conditionals, counterfactuals and other non-classical conditional connectives. If I wish to evaluate the conditional $A \to B$ at some 'world' x, it will never suffice to find some class of worlds, related to x (whether that choice depends on A, or on B, or on anything else) and then check of those worlds where A is true, whether B is true

G. Restall (✉)
School of Historical and Philosophical Studies, The University of Melbourne, Melbourne, Australia
e-mail: restall@unimelb.edu.au
URL: http://consequently.org

I. Düntsch and E. Mares (eds.), *Alasdair Urquhart on Nonclassical and Algebraic Logic and Complexity of Proofs*, Outstanding Contributions to Logic 22,
https://doi.org/10.1007/978-3-030-71430-7_6

at those selected worlds, too. For then, the identity conditional $A \to A$ (in which the consequent is identical to the antecedent) is guaranteed to be true at absolutely any world whatsoever. You may not think that this is a problem, since the conditional $A \to A$, seems to satisfy the canons of relevance as well as any conditional does, and more than most,[1] but there is a problem for relevance nonetheless: For consider the status of the conditional $p \to (q \to q)$ at the world x, where p and q are atoms, chosen to have nothing in common at all. Select worlds, related to x, for checking in whatever manner your semantics dictates. If we want to know whether all of those worlds where p is true, also have $q \to q$ true, we know the answer already: It's a 'yes', since $q \to q$ is true *everywhere*. But there is no requirement that $q \to q$ have anything to do with p. To require this is to enforce as blatant a violation of relevance as we should ever expect to see.

So, a worlds semantics for a relevant conditional must differ from the standard worlds semantics for strict or counterfactual conditionals. One of Alasdair Urquhart's key insights was that the appropriate modification is not particularly difficult. We separate the world at which we check the conditional's *antecedent* from the worlds where we check its *consequent*. For what has been come to be called Urquhart's *semilattice* semantics, a conditional $A \to B$ holds at a world x if and only if for every world y at which A is true, the consequent B is true at the world $x \sqcup y$. What *is* this world $x \sqcup y$? It is found by *combining* x and y, with a semilattice (that is, commutative, associative and idempotent) operation $\sqcup$. This interpretation of a conditional makes some intuitive sense if we think of a conditional $A \to B$ as saying that we have the means to use whatever verifies A to produce something that verifies B. The x is the resource that we have, the y is the input (whatever verifies A) and the means of production is the *combination* (with $\sqcup$) of our initial resource x with the input y.

The points in this kind of model look rather less like *worlds* than the points of a Kripke model for a modal logic do, so Urquhart uses a different name for them. In his 1972 paper (Urquhart 1972), he calls them *pieces of information*. It is easy to see how pieces of information could plausibly be combined to give us new pieces of information, in a mode of combination that turns out to be associative, commutative and idempotent. Rather than follow Urquhart in calling our points 'pieces of information', I will use the more neutral 'point' in what follows, not only because the name is shorter, but also because our target models will be geometries, in which the points are, well, *points*.

When we utilise a semilattice combination operation on points to interpret our conditional, we have all the tools necessary to allow for identity conditionals $A \to A$ to fail at some of our points. All we need is a pair of points, x and y, where A holds at y but fails at $x \sqcup y$. If this occurs, we have a counterexample to $A \to A$ at x. If our information combination is not *cumulative* (if the result of combining y with x does not preserve all the information given by x alone) then the point x gives us the means

[1] Of course the identity conditional $A \to A$ might be seen to be problematic for other reasons, Martin and Meyer (1982), Martin (1978), but this to explore non-circular logics, in which true conditionals do not beg the question, would take us altogether too far away from our current topic.

to convert A-points into non-A-points, because the application of x to y transforms something that verifies A into something that no longer does so.[2]

Of course, the fact that identity conditionals $A \to A$ can fail *somewhere* does not mean that we want them to fail *everywhere*. Urquhart's semilattice semantics has a facility for this. A semilattice model has a *zero* point 0 where $0 \sqcup x = x$, for each point x. Then it is straightforward to verify that at the zero point, $A \to A$ is true. The logical truths of the logic $R_{\to}$ (the implicational fragment of R) are those formulas which hold at the zero point in each model, not those formulas that hold everywhere.

Before we continue, we would do well to introduce at least the *positive* logic R^+, since that is the focus of our discussion for the rest of this paper. There are many ways to define it. Here is a Hilbert-style axiomatisation. First, take as axioms, the following principles for implication:

$$A \to A \qquad (A \to B) \to ((C \to A) \to (C \to B))$$

$$(A \to (B \to C)) \to (B \to (A \to C)) \qquad (A \to (A \to B)) \to (A \to B)$$

together with the rule of *modus ponens*:

$$A \to B, A \ \Rightarrow \ B$$

This suffices for a Hilbert-style axiomatisation of the implicational fragment, $R_{\to}$.[3] To this we add sufficient axioms for $\wedge$ and $\vee$ to give them the usual lattice properties, according to which, conjunction behaves like a greatest lower bound and disjunction behaves like a least upper bound for the ordering expressed by the conditional.[4]

$$A \wedge B \to A \qquad A \wedge B \to B \qquad (C \to A) \wedge (C \to B) \to (C \to A \wedge B)$$

$$A \to A \vee B \qquad B \to A \vee B \qquad (A \to C) \wedge (B \to C) \to (A \vee B \to C)$$

and (since this does not come along by way of *modus ponens* since we cannot prove $A \to (B \to (A \wedge B)))$ we need to add one more rule, governing conjunction:

$$A, B \ \Rightarrow \ A \wedge B$$

[2] There are many different things we could say about how we might interpret 'combination', and much ink has been spilled on this very issue (Beall et al. 2012; Mares 2004; Restall 1994, 2000; Slaney 1990), both concerning this semilattice semantics and its generalisation, the ternary relational semantics of Routley and Meyer (1972, 1973). It is not my place in this short paper to address those issues. Instead, we will look at how geometries provide a rich playground for developing models for these logics.

[3] The axioms correspond to the I, B, C and W combinators. Noticeably absent is the K combinator, with the axiom $A \to (B \to A)$, which reeks of irrelevance. Adding this axiom strengthens the logic to the implicational fragment of intuitionistic logic.

[4] To cut down on parentheses, we use the convention that $\wedge$ and $\vee$ bind more tightly than the conditional. So, $A \wedge B \to A$ is a conditional with $A \wedge B$ as its antecedent, and so on.

But this is not *quite* enough for the logic R^+. Not all lattices are *distributive*, and the conditional of our logic is not strong enough to *force* the lattice to be distributive[5] so we need to add it as another axiom

$$A \wedge (B \vee C) \rightarrow (A \wedge B) \vee (A \wedge C)$$

With that, we have the logic R^+, our target for the rest of the paper.

⋆ ⋆ ⋆

Urquhart's elegant semilattice semantics for $R_{\rightarrow}$ faces difficulties when we extend it all the way to R^+. The problem lies specifically with *disjunction*. Once disjunction is interpreted in our models, we are faced with a dilemma. A natural thought is to take a disjunction to hold at a point if and only if at least one of the disjuncts holds at that point. But this sits uneasily with the modelling clause for a conditional. In semilattice models in which disjunction is interpreted in that way the formula

$$[(A \rightarrow (B \vee C)) \wedge (B \rightarrow D)] \rightarrow (A \rightarrow (D \vee C))$$

is valid, but it is not a theorem of R^+. The dilemma is this: either complicate the semantics for disjunction, so a disjunction can hold at a point when neither disjunct holds at that point,[6] or bite the bullet and build a semantics for a logic other than R. That is the basic choice for a semantics in which the conditional is interpreted with an operation like $\sqcup$ on points.

We can avoid the disjunction dilemma entirely if we generalise the semilattice semantics just a little. Instead of thinking that the application of a point x to a point y is *deterministic*, resulting in one and only one point $x \sqcup y$, we could take it to be *non-deterministic*. We have a ternary relation R, according to which $Rxyz$ if and only if z is a possible result of applying x to y. This is one way to understand Routley and Meyer's ternary relational semantics (Routley and Meyer 1972, 1973).[7] This semantics can be thought to be a generalisation of Urquhart's semilattice semantics (though they were developed independently), where we move from the specific case of a ternary relation defined by a binary operation—$x \sqcup y = z$—to the more general case of a ternary relation. In ternary relational models, we can evaluate disjunctions in using the straightforward evaluation clause without risk of overshooting the logic R^+. In general, a ternary relational model has this structure: A FRAME is a 4-tuple $\langle P, R, \sqsubseteq, N \rangle$, where P is a non-empty set of points, R is a ternary relation on P, $\sqsubseteq$ is a binary relation on P, and N is a subset of P, where the following conditions are satisfied.

- $\sqsubseteq$ is a partial order on P.

[5] If we added the weakening axiom K to give intuitionistic logic, the distribution axiom would come along for the ride.

[6] This is Lloyd Humberstone's approach, in his 'Operational Semantics for Positive R' (Humberstone 1988).

[7] Bimbó et al. (2018) give an account of the historical origins of the ternary semantics.

- R is $\sqsubseteq$-downward preserved in the first two positions, and $\sqsubseteq$-upward preserved in the third. That is, if $Rxyz$ and $x^- \sqsubseteq x$, $y^- \sqsubseteq y$ and $z \sqsubseteq z^+$ then $Rx^-y^-z^+$.
- $y \sqsubseteq z$ if and only if there is some x where Nx and $Rxyz$.

Here, N generalises the *zero* point 0 of Urquhart's models, and $\sqsubseteq$ is an inclusion relation on the information carried by points. In some frames, this relation is the *identity* relation, in which case the preservation condition on R is vacuously satisfied. In other frames, a non-vacuous inclusion relation plays an important role.

Then, given a frame we can evaluate formulas at points in the usual way. First, we fix truth for *atoms* at points, imposing the constraint that for any atom, p, if $x \sqsubseteq x^+$ then when $x \Vdash p$ we have $x^+ \Vdash p$ too. The evaluation relation $\Vdash$ generalises to the whole language with the following inductive truth conditions:

- $x \Vdash A \to B$ iff for each y, z where $Rxyz$, if $y \Vdash A$ then $z \Vdash B$,
- $x \Vdash A \wedge B$ iff $x \Vdash A$ and $x \Vdash B$,
- $x \Vdash A \vee B$ iff $x \Vdash A$ or $x \Vdash B$.

The frames provide the means to interpret two more logical concepts, $\circ$ and t,[8] as follows:

- $x \Vdash A \circ B$ iff for some y, z where $Ryzx$, $y \Vdash A$ and $z \Vdash B$,
- $x \Vdash t$ iff $x \in N$.

We say that A *entails* B in a model iff whenever $x \Vdash A$, then $x \Vdash B$ too, and that A and B are *equivalent* (on the model) if A entails B and vice versa. It is a good exercise to show that in any model of this kind, $A \circ B$ entails C if and only if A entails $B \to C$, and $t \circ A$ is equivalent to A.

In the general class of ternary relational models, we impose no constraints analogous to the requirement that $\sqcup$ be a semilattice operator. This means that ternary relational frames can provide models for logics much weaker than R^+. To model R^+, we can impose conditions on the ternary relation analogous to associativity, commutativity and idempotence for $\sqcup$.

- If $(\exists u)(Rxyu \wedge Ruzw)$ then $(\exists v)(Ryzv \wedge Rxvw)$,
- If $Rxyz$ then $Ryxz$,
- $Rxxx$.

It must be said that while the ternary relational semantics gains marks for generality, it loses some with regard to elegance and simplicity. The model theory is cumbersome, in that the semilattice operator $\sqcup$ (with identity 0) is replaced by a triple $R, \sqsubseteq, N$ of a three-place relation, a two-place relation and a set, with interconnecting conditions between all three items. Despite this increase in complexity, the gain in *generality* allows us to consider interesting models, and in particular, models which allow us to engage our *geometric* intuitions.

---

[8] To add these to our axiomatisation, it suffices to take $\circ$ to *residuate* the conditional, like this. We have $(A \circ B \to C) \to (A \to (B \to C))$ and its converse, $(A \to (B \to C)) \to (A \circ B \to C)$ (so $\circ$ acts like a kind of conjunction, and with $\wedge$ it binds more tightly than the conditional). For t, it suffices to set $t \to (A \to A)$ and t as our axioms.

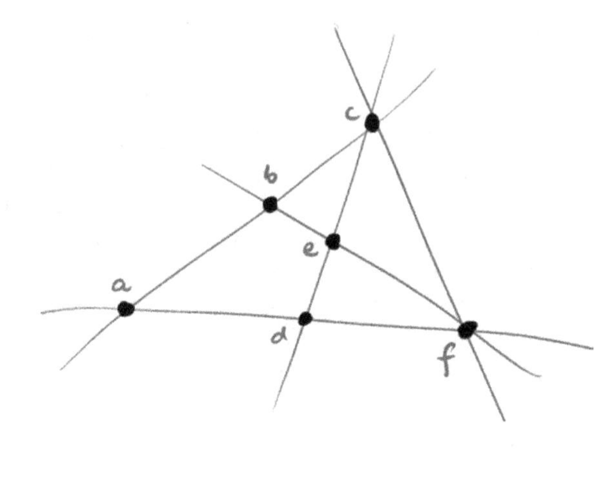

Fig. 6.1 Collinearity and betweenness

⋆⋆⋆

One reason that geometry is such a natural partner for ternary relational semantics is the ternary relation R at the heart of our models. Ternary relations are, naturally, harder to reason with than binary relations or sets. *Geometry* is a natural domain providing *genuine* ternary relations. One of the simplest naturally occurring ternary relations in a geometric setting is the *collinearity* relation: x, y and z are collinear ($Cxyz$) iff there is some line on which they all occur. Consider the diagram in Fig. 6.1. Here, $Cabc$ and $Ccde$ hold (for example) but none of $Cabd$, $Cabe$ or $Cabf$ hold. Furthermore, if $Cxyz$ holds, then so does $Cyxz$, $Cyzx$ and any other statement we can find by permuting the terms in the predication.

Another natural relation on some spaces (for example, each real Euclidean space $\mathbb{R}^n$) is the *betweenness* relation B where $Bxyz$ holds iff $Cxyz$ and y is *between* x and z on the line xz. If we think of the diagram above as depicting the real plane $\mathbb{R}^2$, we have $Babc$ but we do *not* have $Bbac$, since a does not occur between b and c.

Collinearity and betweenness are two examples of naturally occurring ternary relations, so we would hope that they might provide us a way to model relevant logics like R^+. However, as Urquhart noticed, the specifics of ternary relational models make this connection *close* but the fit is not exact. Consider first the order relation $\sqsubseteq$ in ternary relational models. We have whenever $x \sqsubseteq x'$ then if $Rxyz$ then $Rx'yz$, and in addition $\sqsubseteq$ is reflexive. If R is collinearity or betweenness in some affine or projective space, then any order $\sqsubseteq$ satisfying this condition is very heavily constrained. In any affine or projective geometry, given any point x, the set of pairs $\langle y, z\rangle$ such that x is collinear with y and z (or such that x is between y and z) *uniquely determines* the point x. There is no *other* point also satisfying those conditions. So, in spaces like these, the order relation must collapse into the identity relation.

This, then, puts pressure on the presence of normal points, those in the set N, in our frame if the relation R is anything like collinearity or betweenness for points. Whenever $x \in N$ we must have $Rxyz$ only when $y \sqsubseteq z$, which in our case means $y = z$. In a projective space or an affine space, lines have at least *three* points. There is no point x where for every y and z, x, y and z are collinear only when $y = z$. For any point x we can find *some* distinct points y and z where x, y and z are collinear, or indeed, where x is between y and z. So, no point in this space will count as *normal*.

So, although betweenness and collinearity provide elegant and straightforward ways to depict natural and mathematically rich ternary relations, these spaces do not provide ternary relational models for relevant logics like R^+, in and of themselves. They must undergo some corrective surgery, in order to satisfy the conditions for being a ternary relational model, where in this case, the required operation involves the addition of an extra point (or more) to provide us with some normal points of our model, and the expansion of the relations $\sqsubseteq$ and R to incorporate this new point (or points). This can be done in a general and uniform way, independently of the detail of the ternary relation involved in the underlying frame. To see how we can do this most easily, however, it will help to change our perspective on ternary relational frames just a little, and so we turn to this in our next section. It will turn out, too, that when we take this wider perspective, we will see how we could, in fact, do without those additions, and these geometric spaces will provide models for relevant logics all by themselves, though the logics that are so modelled prove somewhat different from R and its familiar cousins.

6.2 Collection Frames

In a ternary frame $\langle P, R, \sqsubseteq, N\rangle$, we have a non-empty set P of points, a ternary relation R, a binary relation $\sqsubseteq$, and a set (or a unary 'relation' or property) N. *Three, two, one.* The key insight involved in understanding collection frames is taking these three pieces of data, the ternary, binary and unary R, $\sqsubseteq$ and N as facets of one underlying *multi*-ary relation, which relates *collections* of points to points. A *collection* relation. The R of a ternary frame is given by the case of the collection involving *two* items, the $\sqsubseteq$ is the case of the collection involving *one* item, and N is given in the case where the collection is empty. The coherence conditions, connecting N, $\sqsubseteq$ and R then converge into a single condition governing the one underlying collection relation. Different logics can then be modelled not only by imposing different conditions on the collection relation, but also by different choices for what kinds of collections our relation relates. For example, it is one thing to think of our points as collected together in some kind of order (say, a list), so the question of whether $\langle a, b\rangle Rc$ or not may have a different answer to the question of whether $\langle b, a\rangle Rc$ or not. If we move from lists to *multisets* (which keep track of the multiplicity of membership, but not order) so $[a, b]$ is the same multiset as $[b, a]$ but it differs from $[a, a, b]$, then the fact that $[a, b]Rc$ holds if and only if $[b, a]Rc$ holds is not so much a special constraint holding of the collection relation R, but rather an inevitability, given that R here relates multisets.[9]

[9] In models for relevant logics, extending the ternary R to relate *triples* and *quadruples*, by setting $R(ab)cd$ to mean that there is some x where $Rabx$ and $Rxcd$ have become standard notation. For

But for this paper we will go further, and consider relations on *sets*. Here, not only does $\{a, b\}Rc$ hold iff $\{b, a\}Rc$, but $\{a, a\}Rc$ holds iff $\{a\}Rc$. The contraction rule holds in a very strong form in set frames. Multiplicities of membership are ignored completely. We are in the realm of models for R^+, and this will be a natural place to explore geometric models, because from this perspective, we can see our frames in a new light.

A reflexive *set* frame is a pair $\langle P, R\rangle$ consisting of a non-empty set P of points, and a relation R on $\mathcal{P}^{\text{fin}}(P) \times P$, relating finite subsets of P to elements of P. A possible intuitive understanding of XRy is that the points collected together in X can be represented by the individual point y.[10] For a reflexive set frame, relation R must satisfy the following two conditions:

1. $(\forall x \in P)(\{x\}Rx)$.
2. $(\forall X, Y \in \mathcal{P}^{\text{fin}}(P))(\forall z \in P)((X \cup Y)Rz \leftrightarrow (\exists y \in P)(YRy \wedge (X \cup \{y\})Rz)$.

Condition 1 we call REFLEXIVITY, for obvious reasons, and Condition 2 is COMPOSITIONALITY. From right to left it tells us that if Y is R-related to some point y, which then, when bundled together with X, is R-related to z, then R also vouchsafes the relation between X together with Y to z.[11] From left to right, it tells us that this process can be reversed. If a set is a union of X and Y then whatever it R-relates to (say z) can be found as the target of an R relation of X together with some R-representative of Y.

This may be unfamiliar to you if you have not considered set relations before, but it turns out that relations satisfying the compositionality condition are widespread, and rather natural. Consider these set relations on ω, the natural numbers:

MAXIMUM Here, XRy if and only if y is the largest member of X, and is 0 if X is empty. Call this value, $\max(X)$. R, so defined, is clearly reflexive, since $\max\{x\} = x$. For compositionality, it suffices to notice that for any sets X and Y (whether empty or not) $\max(X \cup Y) = \max(X \cup \{\max(Y)\})$.

SPECTRUM $\{x_1, \ldots, x_m\}Ry$ iff for some naturals $n_1, \ldots, n_m$, we have $y = 0 + n_1x_1 + \cdots + n_mx_m$, where again, $\{\}Ry$ if and only if $y = 0$. So, XRy iff y is some sum (of any multiplicity) members of X. So, $\{1\}Rx$ for *every* x, $\{2\}Rx$ for *even* x, and in general, $\{n\}$ is related to the multiples of n (including 0), while $\{2, 3\}$ is related to every number other than 1 (including 0). This relation is reflexive by design, and proving compositionality is a straightforward case of spelling out the definition. If $(X \cup Y)Rz$ then $z = 0 + \Sigma n_ix_i + \Sigma m_jy_j$ for some choices of $x_i \in X, y_j \in Y$ and naturals n_i and m_j. Choose $0 + \Sigma m_jy_j$ for the number related to Y and we have $(X \cup \{0 + \Sigma m_jy_j\})Rz$ straightforwardly. Conversely, if YRy (since $y = 0 + \Sigma m_jy_j$ for some appropriate choices of values for m_j and y_j) and

this reason, we continue to use 'R' to name this *collection* relation, since context will determine whether we mean a collection relation or a ternary relation, as appropriate.

[10] How 'represented' may be understood can, of course, vary from application to application, or model to model.

[11] This may be reminiscent to a *Cut* rule, taking us from and $\Gamma, A \succ B$ and $\Delta \succ A$ to $\Gamma, \Delta \succ B$.

$(X \cup \{y\})Rz$, that is, we have $z = 0 + \Sigma n_i x_i + n\Sigma m_j y_j$, then clearly $(X \cup Y)Rz$ too.

These two set relations are very different. The relation MAXIMUM is *functional*. For each X there is a unique y where XRy. It follows that the binary relation $\sqsubseteq$ induced by R on ω, given by setting $x \sqsubseteq y$ iff $\{x\}Ry$, is the identity relation. On the other hand, the SPECTRUM relation is anything but functional. The ordering $\sqsubseteq$ induced by this relation relates n to each number $n \times x$.

⋆ ⋆ ⋆

We have already introduced some of the notation that connects collection relations with the machinery underlying ternary frames. For any *set frame* $\langle P, R\rangle$ there is an underlying binary relation $\sqsubseteq$ on P, given by setting $x \sqsubseteq y$ iff $\{x\}Ry$. Since R is reflexive, so is $\sqsubseteq$. Since R is compositional, $\sqsubseteq$ is transitive. For if $\{x\}Ry$ and $\{y\}Rz$ then $(\{\,\} \cup \{y\})Rz$ and by compositionality, $(\{\,\} \cup \{x\})Rz$, and hence $\{x\}Rz$. So, we have a partial order on our frame, and the notation $\sqsubseteq$ for this binary relation is deserved.

What holds for the binary relation $\sqsubseteq$ on Routley–Meyer frames also holds for the ternary relation R, and the set N of normal points. We have the following fact:

Theorem 6.1 (From Set frames to Routley–Meyer frames) *If $\langle P, R\rangle$ is a set frame, then $\langle P, R, \sqsubseteq, N\rangle$, where we define*

- *$Rxyz$ iff $\{x, y\}Rz$,*
- *$x \sqsubseteq y$ iff $\{x\}Ry$,*
- *Nx iff $\{\,\}Rx$*

is a Routley–Meyer frame for R^+, satisfying the standard frame conditions connecting N, $\sqsubseteq$ and R, and the usual frame conditions on R of associativity, commutativity and idempotence.

Establishing this fact is an enjoyable matter of applying the reflexivity and compositionality conditions on R in specific cases. We have already proved that $\sqsubseteq$ is a partial order. To show that N is closed upward under $\sqsubseteq$ and that the induced ternary R is downward preserved in the first two positions and upward preserved in the third, we notice that these two facts:

If XRy and $y \sqsubseteq z$ then XRz
If $x \sqsubseteq y$ and $(X \cup \{y\})Rz$ then $(X \cup \{x\})Rz$

are all we need to show those conditions on N and on R, and they are both instances of the left-to-right parts of compositionality.

For the structural conditions on R for idempotence, symmetry and associativity, we reason as follows: We have $Rxxx$ since $\{x, x\}Rx$ is a restatement of $\{x\}Rx$, which is given by reflexivity. For symmetry, if $Rxyz$ then $\{x, y\}Rz$ and hence $Ryxz$. For associativity, if $Rxyu$ and $Ruzw$ then $\{x, y, z\}Rw$, and hence, there is some v where $\{y, z\}Rv$ and $\{x, v\}Rw$, i.e. $Ryzv$ and $Rxvw$.

With set frames in view, let's reconsider what it is for a model to interpret a *relevant* conditional. In Urquhart's semilattice semantics, we allowed application to fail to be cumulative: a formula A might be true at y but no longer hold true at $x \sqcup y$, and in that case, $A \to A$ would fail to hold at x. If we restate this in the context of set frames, it means that we allow $\{x, y\}Ry$ to fail, on occasions. By compositionality, this generalises. Sometimes we have XRy, but there is a subset X' of X, where $X'Ry$ fails. This is the fundamental requirement for a relation on a *relevant* set frame. The principle of *weakening* is satisfied if whenever XRy and $X' \subseteq X$ then $X'Ry$ too. In models for relevant logics, weakening can fail.

Set frames are a simple way to view the three distinct moving parts of a ternary frame (the normal worlds, the binary ordering relation and the ternary accessibility relation) as three distinct facets of one underlying *set* relation. They provide a natural way to understand frame models of relevant logics.[12]

⋆ ⋆ ⋆

With set frames in mind, let's return to the relations of collinearity and betweenness. How does revisiting these in the light of set frames change our perspective? We have already seen that when we consider these ternary relations on the set $\mathbb{R}^2$, we almost have a ternary frame, but there is nothing corresponding to the set of normal points that we would hope to find. This phenomenon occurs repeatedly with collection frames in general, and set frames in particular. Recall the two concrete examples of compositional set relations we have considered (MAX and SPECTRUM). Why did I not consider MIN, which might be thought to be a natural dual of MAX? Isn't it the case that for every X and Y, $\min(X \cup Y) = \min(X \cup \{\min(Y)\})$? Well, *yes*, but only when Y is non-empty. In the case where Y is empty, we are left with no value to choose that allows compositionality to be satisfied. If we stipulated that $\min\{\} = n$, then we would have for any m, $m = \min\{m\} = \min(\{m\} \cup \{\})$ which, by compositionality would be $\min(\{m, n\})$ which would differ from m when $m > n$. So, compositionality fails of *min*, because there is no value to choose to relate to the *empty* set, in just the same way that there is no point you can choose on $\mathbb{R}^2$ with collinearity (or betweenness) that could count as *normal* in the Routley–Meyer sense.

There is a straightforward response to this problem in each of these cases. We have already seen that different kinds of relations are found with different kinds of *collections* to be related. Multisets differ from sets, which differ from lists, or leaf-labelled trees. In the same way, *inhabited* (that is, non-empty) sets differ from *sets*, if only in one special case. It is a very small step from compositional *set* relations, on $\mathcal{P}^{\text{fin}}(P) \times P$, which we have seen, to compositional *inhabited-set* relations, on $\mathcal{P}^{\text{fin}*}(P) \times P$, where $\mathcal{P}^{\text{fin}*}(P)$ is the set of all *inhabited* finite subsets of the underlying point set P. We can say that a reflexive inhabited-set frame $\langle P, R\rangle$ is given by

[12] For more details concerning set frames, and their cousins, multiset frames, list frames and the like, the reader is referred to the paper 'Collection Frames' (Restall and Standefer 2020), written with Shawn Standefer.

a non-empty point set P and a relation R on $\mathcal{P}^{\text{fin}*}(P) \times P$ satisfying the same two conditions of reflexivity and compositionality as before:

1. $(\forall x \in P)(\{x\}Rx)$.
2. $(\forall X, Y \in \mathcal{P}^{\text{fin}*}(P))(\forall z \in P)((X \cup Y)Rz \leftrightarrow (\exists y \in P)(YRy \wedge (X \cup \{y\})Rz)$,

except that now we require that in each place that the relation R appears, the left hand sides ($\{x\}$, $X \cup Y$ and Y and $X \cup \{y\}$) are required to be inhabited. It is easy to see that this holds if and only if Y is inhabited, since $\{x\}$ and $X \cup \{y\}$ are inhabited in virtue of their form (You may wonder why we allow X to be empty. When X is empty, compositionality reads as follows: $(\forall Y \in \mathcal{P}^{\text{fin}*}(P))(\forall z \in P)(YRz \leftrightarrow (\exists y \in P)(YRy \wedge \{y\}Rz))$. In the case where R is reflexive, the left-to-right part is trivially true—choose z as a witness for the existential quantifier. For the right-to-left part, it is a natural generalisation of the transitivity of $\sqsubseteq$, which is a natural constraint, even when R relates only inhabited sets.)

On any frame $\langle P, R\rangle$, whether R is a set relation or merely an inhabited-set relation, we can define an evaluation relation $\Vdash$ between points and formulas in a natural way. We define $\Vdash$ on atoms, with the one proviso, that $\Vdash$ be *hereditary* along the order $\sqsubseteq$. If $x \Vdash p$ and $\{x\}Ry$, then $y \Vdash p$ too. With that, the relation is extended to complex formulas as follows:

- $x \Vdash A \wedge B$ iff $x \Vdash A$ and $x \Vdash B$,
- $x \Vdash A \vee B$ iff $x \Vdash A$ or $x \Vdash B$,
- $x \Vdash A \to B$ iff for each y where $y \Vdash A$, if $\{x, y\}Rz$ then $z \Vdash B$,
- $x \Vdash A \circ B$ iff there are y, z where $\{y, z\}Rx$, $y \Vdash A$ and $z \Vdash B$.

If the relation R is a set relation, defined on the empty set, then we can also give truth conditions for the Ackermann constant t, using the empty set.

- $x \Vdash t$ iff $\{\,\}Rx$.

A frame equipped with an evaluation relation $\Vdash$ is said to be a *model*. On these models we can define a straightforward notion of entailment. We say that A entails B according to some model when for any point x if $x \Vdash A$ then $x \Vdash B$.

Set frames and inhabited-set frames provide a natural setting for models for relevant logics with or without normal points. Once we turn our attention to *inhabited*-set frames, we have more natural examples of compositional relations:

MEMBERSHIP For any non-empty point set P, if we define XRy as holding if and only if $y \in X$, then R is a reflexive, compositional inhabited-set relation. It is straightforward that R is reflexive, and the compositionality condition is verified immediately, since for any inhabited set Y we can choose an appropriate member.

SUBSPACE On any affine or projective geometry with point set P, we can define a compositional inhabited-set relation R on its point set, by setting XRy iff y is in the smallest subspace of P containing each member of X. So, $\{x\}Ry$ iff $y = x$ (a point is a subspace of dimension 0), if $x \neq y$, then $\{x, y\}Rz$ iff z is on the line xy, if x, y and z are all distinct, $\{x, y, z\}Rw$ iff w is on the plane xyz, and so on. The collinearity relation generalises to higher dimensions in a natural way. The relation R, so defined, is compositional.

INSIDE If our space, like $\mathbb{R}^n$, comes equipped with a notion of *betweenness*, then the three-place relation of betweenness generalises appropriately. We have for any finite set X of points, XRy if and only if y is inside the shape inscribed by the points X (including its boundary). To show that this relation is compositional, we use an elementary geometric argument. If $(X \cup Y)Rz$ then z is in the shape bound by $X \cup Y$, then we can find something inside Y (say, y) where z is also in the shape bound by $X \cup \{y\}$ (For example, in the diagram in Fig. 6.1, the point e is inside the shape inscribed by $\{a, c, f\} = \{c\} \cup \{a, f\}$. We can find something inside the shape inscribed by $\{a, f\}$ (namely the point d) where $\{c, d\}Re$.). Conversely, if $(X \cup \{y\})Rz$ and YRy it is clear that $(X \cup Y)Rz$.

So, we have three compositional relations on the point set $\mathcal{P}^{\text{fin}*}(P)$, two of which have natural geometric meanings. Here, the fit is perfect. If we are willing to forego the existence of normal points, geometries provide very natural examples of inhabited-set frames, and these frames can be used to model relevant logics. Set frames are models the relevant logic R^+ (This fact is an immediate consequence of what we have proved. Ternary relational frames satisfying idempotence, commutativity and associativity are models for R^+ and any set frame generates a ternary relational frame satisfying these conditions.). Inhabited-set frames, on the other hand, are not always models for R^+. The absence of normal points makes a difference. To show this, it suffices to note that $((p \to p) \to q) \to q$ is a theorem of R^+, or equivalently, that $(p \to p) \to q$ entails q on all R^+ models (The reasoning is straightforward. If $x \Vdash (p \to p) \to q$ then since $(\{x\} \cup \{\,\})Rx$ there is some y where $\{\,\}Ry$ and $\{x, y\}Rx$. Since $\{\,\}Ry$, we have $y \Vdash p \to p$ since for every z where $z \Vdash p$, if $\{y, z\}Rw$, then since $\{\,\}Ry$, $\{z\}Rw$, and hence, $w \Vdash p$, too. So, since $y \Vdash p \to p$, $x \Vdash (p \to p) \to q$ and $\{x, y\}Rx$, we have $x \Vdash q$ as desired.).

In inhabited-set frames, this reasoning does not apply, since we cannot appeal to the existence of some y where $\{\,\}Ry$. Not only does *that* reasoning not apply, we can find a counterexample to the entailment. For the counterexample, consider the inhabited-set frame on the real line $\mathbb{R}^1$ where XRy when y is in the interval bounded by X. On this frame, $x \sqsubseteq y$ iff $\{x\}Ry$ iff $x = y$, so any set of points is a possible extension of an atomic formula, the heredity condition puts no constraints on where formulas can be true. So, let p be true at 0 and 1, but nowhere else. On this model, $p \to p$ is true *nowhere*. Take any x at all. We can find some y, where $y \Vdash p$ (either 0 or 1 will do) where there is some z between x and y where p fails. So, in this model, $p \to p$ is true nowhere, and it follows that $(p \to p) \to p$ is true at every point (vacuously). It follows from that the point 2 (for example) provides a counterexample to the entailment from $(p \to p) \to p$ (which holds at 2) to p (which fails there). So, the argument from $(p \to p) \to p$ to p has a counterexample in this model, and so, the argument from $(p \to p) \to q$ to q also fails there. The logic of inhabited-set frames is weaker than the logic of set frames. These frames are natural models of a substructural logic, but this logic is weaker than the relevant logic R in distinctive ways.

6.3 Adding a Normal Point

Of course, there are reasons why, for certain applications, we may *not* like to forego the existence of normal points. In this section, we will show that there are two distinct ways to add a single normal point to a non-empty set frame $\langle P, R\rangle$. The aim is to consider a new element $\infty \notin P$, and to define a new compositional relation between $\mathcal{P}^{\text{fin}}(P \cup \{\infty\})$ and $P \cup \{\infty\}$, extending the relation R, where in this new relation, the empty set $\{\}$ is related to the new element ∞, so this new element is a normal point in the extended frame.

If this newly defined relation is compositional, the result will be a R^+ frame, since all compositional set relations are R^+ frames. The key question to answer for any such extension R' is whether $XR'z$ holds or not, in cases where either $\infty \in X$ or $z = \infty$. In the case where $\infty \notin X$ and $z \neq \infty$, we will set R' to echo the verdict of R unchanged: $XR'z$ iff XRz.

We will see that there are two natural ways to extend R to $P \cup \{\infty\}$. We will call these R^+ and $R^\times$. Let's start with R^+. The choice we make for R^+ is straightforward:

$$XR^+z \quad \text{iff} \quad \text{either } (X\backslash\{\infty\})Rz \text{ or } z = \infty$$

That is, any set of points X is related to the new normal point ∞, and furthermore, the set of points X is related to one of the original points z iff the set X, with the point ∞ removed (if it was present at all), is related to that point z in the original frame. This is one choice for extending R to $\mathcal{P}^{\text{fin}}(P \cup \{\infty\}) \times P \cup \{\infty\}$. In this set frame, the added point ∞ is *above* each point in the ordering. We have $z \sqsubseteq^+ \infty$, since $\{z\}R^+\infty$, for every z.

That is one natural way to extend R on non-empty finite sets to a new relation R^+ on all finite sets. This extension is simple, but it has significant shortcomings. For one, since the added point ∞ extends *every* point in the frame, there is no way for our frames to have divergent pairs of points (Points x and y are divergent if there is no point z where $x \sqsubseteq z$ and $y \sqsubseteq z$). Since x and y converge in ∞, any apparently opposing positions x and y take on some claim is 'resolved' in ∞, which simply agrees with both x *and* y. This puts pressure on the interpretation of negation on our frames, and the idea that some points might be *worldlike*, in the sense of being comprehensive states of affairs. It is one thing for a logic to be paraconsistent (as the relevant logic R is), it is another to say that any and all circumstances can be subsumed into the one all-encompassing situation, ∞. The second issue with such an extension is that any compositional relation R^+ with such a point ∞ must satisfy the converse of the weakening condition: That is, we will have whenever XR^+y and $X \subseteq X'$ then $X'R^+y$ too. It is straightforward to verify that if XR^+y, then since $(X \cup \{\})R^+y$ we have $(X \cup \{\infty\})R^+y$ and since $X'R^+\infty$, we conclude by compositionality $(X \cup X')R^+y$. So, if $X \subseteq X'$ we have $X'R^+y$, too. So, if R^+, so defined is compositional, it satisfies the converse of the weakening condition. This means that R^+ is compositional only when the underlying relation R already satisfied the converse of the weakening condition. We have seen some compositional inhabited-set frames that satisfy this condition (MEMBERSHIP, SUBSPACE and INSIDE all do so) while others do not (MAX and MIN notably do not), so R^+ will work as a technique for converting an inhabited-set frame into a set frame for only some of our frames.

It is just as well, then, that there is another natural way to extend R to a compositional relation on all finite sets, with the addition of ∞, and this relation does not have the constraints exhibited by R^+. The relation $R^\times$ is defined differently.

$$XR^\times z \text{ iff } \begin{cases} z = \infty, & X = \{\} \text{ or } X = \{\infty\} \\ (X\backslash\{\infty\})Rz, & X \neq \{\} \text{ and } X \neq \{\infty\} \end{cases}$$

Here, the relation takes the opposite policy to R^+, which makes *everything* related to the new normal point. In this case, a set X is related to ∞ only in the two cases where it absolutely *has* to be so related. In the case $\{\}R^\times\infty$ (which was the design goal, that we have *some* point as the target of the empty set) and $\{\infty\}R^\times\infty$, which is demanded by the reflexivity of $R^\times$. In all other cases, we say that a set is *not* related to ∞, and we say that a set X is related to one of our original points z if and only if the set X *without* ∞, is related to z in the original frame.

With these two definitions, we have the following fact:

Theorem 6.2 *If R is a compositional inhabited-set relation on P, and $\infty \notin P$ then $R^\times$, defined above, is a compositional set relation on $P \cup \{\infty\}$. If R also satisfies the converse weakening condition, then R^+, also defined above, is also compositional. If R is reflexive, then so is R^+ and $R^\times$.*

The proof of this fact is a relatively straightforward set case analysis.

Proof That R^+ and $R^\times$ are reflexive follows immediately from the reflexivity of R.

For compositionality, we will consider $R^\times$ first. Let's suppose that $(X \cup Y)R^\times z$, in order to find some y where $YR^\times y$ and $(X \cup \{y\})R^\times z$. By definition $(X \cup Y)R^\times z$ holds if and only if $z = \infty$ (if $X \cup Y$ is either empty, or $\{\infty\}$) or $((X \cup Y)\backslash\{\infty\}))Rz$ (otherwise). Let's take these cases in turn. If $X \cup Y$ is either empty, or $\{\infty\}$ then clearly of X is either empty or $\{\infty\}$ and so is Y, so in this case, both $YR^\times\infty$ and $(X \cup \{\infty\})R^\times\infty$, as desired. So, now consider the second case: we have $((X \cup Y)\backslash\{\infty\}))Rz$ and $X \cup Y$ is neither empty nor $\{\infty\}$. We aim to find some y where $YR^\times y$ and $(X \cup \{y\})R^\times z$. If Y itself is empty or $\{\infty\}$, then we choose ∞ for y. We have, then, $YR^\times\infty$ and since $((X \cup Y)\backslash\{\infty\}))Rz$, we have $(X\backslash\{\infty\})Rz$, so we have $(X \cup \{\infty\})R^\times z$ as desired. On the other hand, if Y has some element other than ∞, since $((X \cup Y)\backslash\{\infty\}))Rz$, we have $((X\backslash\{\infty\}) \cup (Y\backslash\{\infty\}))Rz$, and since R is compositional, there is some y where $(Y\backslash\{\infty\})Ry$ and $((X\backslash\{\infty\}) \cup \{y\})Rz$, which gives us $YR^\times y$ and $(X \cup \{y\})R^\times z$ as desired.

Now for the second half of the compositionality condition for $R^\times$, suppose that there is some y where $YR^\times y$ and $(X \cup \{y\})R^\times z$. We aim to show that $(X \cup Y)R^\times z$. If $YR^\times y$ then either $y = \infty$ and Y contains at most ∞, or otherwise $(Y\backslash\{\infty\})Ry$. In the first case, $(X \cup \{y\})R^\times z$ tells us that $(X \cup \{\infty\})R^\times z$, which means either that

$(X\backslash\{\infty\})Rz$, or X also contains at most ∞ and then $z = \infty$. In the either of these cases, we have $(X \cup Y)R^{\times}z$, as desired. So, let's suppose $y \neq \infty$. In that case we have $(Y\backslash\{\infty\})Ry$, and then, since $(X \cup \{y\}))R^{\times}z$, we have $((X \cup \{y\})\backslash\{\infty\})Rz$, and by the compositionality of R, $((X \cup Y)\backslash\{\infty\})Rz$, which gives $(X \cup Y)R^{\times}z$, as desired.

Now consider R^{+}. Let's suppose that $(X \cup Y)R^{+}z$, in order to find some y where $YR^{+}y$ and $(X \cup \{y\})R^{+}z$. By definition, $(X \cup Y)R^{+}z$ holds iff $((X \cup Y)\backslash\{\infty\})Rz$ or $z = \infty$. In the second case, we then have $(X \cup \{\infty\})R^{+}\infty$ and $YR^{+}\infty$ and we are done. In the first case, since $((X \cup Y)\backslash\{\infty\})Rz$ we have $((X\backslash\{\infty\}) \cup (Y\backslash\{\infty\}))Rz$, and so, by compositionality, there is some $y \in P$ where $(Y\backslash\{\infty\})Ry$ and $((X\backslash\{\infty\}) \cup \{y\})Rz$. It follows immediately that $YR^{+}y$ and $(X \cup \{y\})R^{+}z$ as desired.

For the second half of the compositionality condition for R^{+}, suppose that there is some y where $YR^{+}y$ and $(X \cup \{y\})R^{+}z$. We aim to show that $(X \cup Y)R^{+}z$. If $y = \infty$ then $(X \cup \{y\})R^{+}z$ ensures that XRz, and then we can appeal to the converse weakening condition, to get $(X \cup Y)Rz$, as desired. If $y \neq \infty$, then we have $(Y\backslash\{\infty\})Ry$. From $(X \cup \{y\})R^{+}z$, we have either $z = \infty$ (in which case $(X \cup Y)R^{+}z$ immediately) or $z \neq \infty$ and $((X \cup \{y\})\backslash\{\infty\})Rz$. In that case, $((X\backslash\{\infty\}) \cup \{y\})Rz$, and by the compositionality of R, we have $((X\backslash\{\infty\}) \cup (Y\backslash\{\infty\}))Rz$, which gives us $(X \cup Y)R^{+}z$, as desired.

So, we have two different ways to 'upgrade' inhabited-set frames to set frames. One technique (R^{+}) applies only to frames satisfying the converse weakening condition, while the other ($R^{\times}$) is applicable more widely. In fact, $R^{\times}$ is *essentially* relevant, in that the move from R to $R^{\times}$ fails to preserve weakening, if it is present. Where z is one of the original points in our frame, we have, by reflexivity, $\{z\}R^{\times}z$ in our new frame. However, the definition of $R^{\times}$ rules out $\{\,\}R^{\times}z$, so weakening fails. The addition of a normal point by the move from R to $R^{\times}$ is one which introduces a modicum of relevance, whether it was there in the original model, or not.

6.4 *Functional* Geometric Set Frames

There is something appealing about the idea that a set relation be *functional*. This was the approach of the semilattice semantics, after all. At least at the level of pairs $\{x, y\}$ there is a unique point $x \sqcup y$ to which this pair is related. I will end this short paper exploring some of the constraints around functionality in the setting of geometric models, to end with another fact, showing that a collection of plausible of different constraints on set relations on geometries are jointly incompatible. We will start with some properties that a compositional inhabited-set relation might satisfy, starting with functionality:

- R is FUNCTIONAL iff whenever $\emptyset \neq X \subseteq \mathbb{R}$ and X is finite, there is some unique y where XRy. The inhabited-set relations MAX and MIN are functional. So is any *constant* relation given by setting XRy iff $y = c$ for a given constant c (It is easy to verify that such a relation is also compositional.).

- R is INCLUSIVE iff whenever $\emptyset \neq X \subseteq \mathbb{R}$ and X is finite, and XRy, then $y \in b(X)$, where $b(X)$ is the region *bound* by the set X. In the case of subsets of the real line, $b(X) = \{z : (\exists x_1 \in X)(\exists x_2 \in X)(x_1 \leq z \leq x_2)\}$, so $b(\{x\}) = \{x\}$, $b(\{x, y\}) = [x, y]$ if $x \leq y$, and in general, $b(\{x_1, \dots, x_n\}) = [\min(x_1, \dots, x_n), \max(x_1, \dots, x_n)]$. The relations MAX and MIN are inclusive, as is the *betweenness* relation, but the *collinearity* relation (which is the *universal* relation on $\mathbb{R}$, but is not trivial in $\mathbb{R}^n$ for $n > 2$) fails to be inclusive.
- R is PRESERVED UNDER TRANSLATION iff whenever $\emptyset \neq X \subseteq \mathbb{R}$, X is finite, and $x \in \mathbb{R}$, if XRy then $(X + x)R(y + x)$. The examples MIN, MAX, as well as all of our geometric examples, are preserved under translation (This notion generalises in a natural way on spaces like $\mathbb{R}^n$ where points and regions can be translated across space and preserved under rotations, and other length-preserving transformations in space.).
- R is REGIONAL iff whenever the finite sets X and Y bound the same region (that is $b(X) = b(Y)$) then for all z, XRz iff YRz. Again, our examples of MIN, MAX, collinearity and betweenness are all regional in this sense.

We will show that only two relations jointly satisfy this set of conditions on $\mathbb{R}$.

Theorem 6.3 *The only two compositional inhabited-set relations on* $\mathbb{R}$ *that are functional, inclusive, preserved under translation and regional are* min *and* max.

Proof Take a functional inclusive regional compositional inhabited-set relation R on $\mathbb{R}$, which is preserved under translation, but is not min or max. For simplicity, let's write $\{x, y\}Rz$ as $x * y = z$, since R is functional. We automatically have $x * y = y * x$ (since R is a *set* relation) and by inclusivity, if $x \leq y$ then $x \leq x * y \leq y$, so we have $x * x = x$, for each x.

Since R is not min, there are values $x < y$ where $x < x * y$, and since R is inclusive, we must have $x < x * y \leq y$. Similarly, since R is not max, we have some $u < v$ where $u * v < v$, and since R is inclusive, we must have $u \leq u * v < v$.

Now, since R is regional, we have $z * (x * y) = x * y$ for any $z \in [x, y]$, since for any regional relation R, $\{x, y, z\}Ra$ iff $\{x, y\}Ra$, since $b(\{x, y, z\}) = b(\{x, y\})$. This means that $x * y$ is an input for which the $*$ function acts *locally* like a *maximum*: for the values z in that non-empty interval $[x, x * y]$, we have $z * (x * y) = x * y = \max(z, x * y)$.

For the same reason, $w * (u * v) = u * v$ for each $w \in [u, v]$ (recall, $u < v$). This means that $u * v$ is an input for which the $*$ function acts *locally* like a *minimum*: for the values w in the non-empty interval $[u * v, v]$, we have $w * (u * v) = u * v = \min(w, u * v)$.

Given this, consider the real value $\epsilon = \min(x * y - x, v - u * v) > 0$. It is the length of the shorter of the two intervals $[x, x * y]$ and $[u * v, v]$. If we consider the interval $[x, x * y]$, and translate this across $\mathbb{R}$ so the right end has moved to ϵ. The interval $[0, \epsilon]$ is no longer than $[x, x * y]$ and since R is preserved under translation, we have $0 * \epsilon = \epsilon$ since for any $z \in [x, x * y]$, $z * (x * y) = x * y$. Similarly, we can translate the interval $[u * v, v]$ so the *left* end moves to 0. The interval $[0, \epsilon]$ is also no longer than $[u * v, v]$, and since R is preserved under translation, we have

$0 * \epsilon = 0$ since for any $z \in [u * v, v]$ we have $z * (u * v) = u * v$. Since $\epsilon \neq 0$ we have a contradiction from $0 * \epsilon = \epsilon$ and $0 * \epsilon = 0$. It follows that our relation R cannot fail to be both MIN and MAX.

There are meagre choices, then, on the menu of compositional, functional, inclusive, regional inhabited-set relations on $\mathbb{R}$ that are preserved under translation. There are *fewer* choices for such relations on $\mathbb{R}^n$ for $n \geq 2$. We will end with this fact:

Theorem 6.4 *There are* no *compositional, functional, inclusive, regional inhabited-set relations on $\mathbb{R}^n$ that are preserved under translation, when $n \geq 2$.*

Proof If R is such a relation on $\mathbb{R}^n$, then its restriction to any line in $\mathbb{R}^n$ must still be compositional, functional, inclusive, regional and preserved under translation on that subspace. It must, therefore, be either MIN or MAX on that line. However, a line can be translated onto its mirror reflection, in $\mathbb{R}^2$ (and any higher dimensional space) by rotation. The relations MIN and MAX are not preserved under this translation, rather, they are swapped. So, no relation on $\mathbb{R}^n$ ($n \geq 2$) jointly satisfies all these criteria.

Acknowledgements Thanks to Shawn Standefer for many helpful conversations on the material discussed here, including correcting a few of my stumbles along the way, to referees for this volume, for helpful suggestions, and to an (online) audience of the Melbourne Logic Group, including Graham Priest, Shay Logan, Yale Weiss and Tomasz Kowalski, for feedback on this material. Many thanks to Philippe Balbiani and Alasdair Urquhart for helpful comments on a draft of this chapter. This research was supported by the Australian Research Council, Discovery Grant DP150103801.

References

Beall, J., Brady, R., Dunn, J. M., Hazen, A. P., Mares, E., Meyer, R. K., et al. (2012). On the ternary relation and conditionality. *Journal of Philosophical Logic*, *41*(3), 595–612.

Bimbó, K., Dunn, J., & Ferencz, N. (2018). Two manuscripts, one by Routley, one by Meyer: The origins of the Routley-Meyer semantics for relevance logics. *The Australasian Journal of Logic*, *15*(2), 171–209.

Humberstone, I. L. (1988). Operational semantics for positive r. *Notre Dame Journal of Formal Logic*, *29*(1), 61–80.

Mares, E. D. (2004). *Relevant logic: A philosophical interpretation*. Cambridge: Cambridge University Press.

Martin, E. P. (1978). *The P-W problem*. PhD thesis, Australian National University.

Martin, E. P., & Meyer, R. K. (1982). Solution to the P-W problem. *The Journal of Symbolic Logic*, *47*, 869–886.

Restall, G. (1994). A useful substructural logic. *Bulletin of the Interest Group in Pure and Applied Logic*, *2*(2), 135–146.

Restall, G. (2000). *An introduction to substructural logics*. London: Routledge.

Restall, G., & Standefer, S. (2020). *Collection frames for substructural logics*. Paper in progress.

Routley, R., & Meyer, R. K. (1972). Semantics of entailment – II. *Journal of Philosophical Logic*, *1*(1), 53–73.

Routley, R., & Meyer, R. K. (1973). Semantics of entailment. In H. Leblanc (Ed.), *Truth, Syntax and Modality, Proceedings of the Temple University Conference on Alternative Semantics* (pp. 194–243). North Holland.

Slaney, J. K. (1990). A general logic. *Australasian Journal of Philosophy*, *68*, 74–88.
Urquhart, A. (1972). Semantics for relevant logics. *The Journal of Symbolic Logic*, *37*, 159–169.
Urquhart, A. (1984). The undecidability of entailment and relevant implication. *The Journal of Symbolic Logic*, *49*(4), 1059–1073.
Urquhart, A. (1993). Failure of interpolation in relevant logics. *Journal of Philosophical Logic*, *22*(5), 449–479.
Urquhart, A. (2019). Relevant implication and ordered geometry. *The Australasian Journal of Logic*, *16*(8), 342–354.

Chapter 7
Revisiting Semilattice Semantics

Shawn Standefer

Second Reader
I. Sedlár
The Czech Academy of Sciences

Abstract The operational semantics of Urquhart is a deep and important part of the development of relevant logics. In this paper, I present an overview of work on Urquhart's operational semantics. I then present the basics of collection frames. Finally, I show how one kind of collection frame, namely, functional set frames, is equivalent to Urquhart's semilattice semantics.

Keywords Relevant logics · Operational semantics · Semilattice semantics · Collection frames · Frame theory

7.1 Introduction

The operational semantics of Urquhart (1972a, b) is a deep and important part of the development of relevant logics.[1] Formally, the operational semantics provided one of the first intuitive model-theoretic interpretations for the implication of relevant logics.[2] Philosophically, the operational models have a natural interpretation in terms of combining information: The elements of the domain are pieces of information, and

[1] See Dunn and Restall (2002), Bimbó (2006) for more on the general area of relevant logics.
[2] The period when the operational semantics was developed was quite active for the area of models for relevant logics, with the publication of Maksimova (1969), Routley and Meyer (1972a, b, 1973), Fine (1974). See Bimbó and Dunn (2018), Bimbó et al. (2018) for more on some of the early contributions to the area, including discussion of an early manuscript by Routley, published as Ferenz (2018). Scott (1973, fn. 33), Chellas (1975, 143, fn. 17–18) note that Scott had developed a version of ternary relational models earlier but had not published it. I thank Lloyd Humberstone for the references of the preceding sentence.

S. Standefer (✉)
Institute of Philosophy, Slovak Academy of Sciences, Bratislava, Slovakia
e-mail: standefer@gmail.com

I. Düntsch and E. Mares (eds.), *Alasdair Urquhart on Nonclassical and Algebraic Logic and Complexity of Proofs*, Outstanding Contributions to Logic 22,
https://doi.org/10.1007/978-3-030-71430-7_7

a piece of information verifies an implication whenever combining it with any piece of information verifying the antecedent results in a piece of information verifying the consequent. This relation of verification extends naturally to conjunction and disjunction.[3] Completeness results are available as well.[4]

The operational frames come with a set of postulates, many of which can be dropped.[5] Dropping postulates, of course, result in different sets of validities. The full set of the standard postulates, which will be set out shortly, give the operational models the structure of a join semilattice. The models obeying the full set of postulates will be called *semilattice models*.

Operational models have been studied by others in other contexts. Došen (1988, 1989) studies general groupoid models of substructural logics and connects them with sequent systems. Buszkowski (1986) uses groupoid models to study Lambek calculus.

The goal of this paper is to set out another view on semilattice semantics. Restall and Standefer (2020) provide a new approach to frame semantics for relevant logics. Our approach uses a binary relation between collections of points and points, rather than the standard ternary relation among points.[6] For this paper, I will focus on the case when the collections of interest are *sets* of points. In this paper, I will show that functional set models coincide with semilattice models in the sense that from a semilattice frame one can define a functional set frame, and from a functional set frame, one can define a semilattice frame, and repeating the process gets you the original model. Further, I will show that the logic of functional set frames properly extends the logic of (possibly non-functional) set frames. Before getting to these results, I will provide some background on operational and semilattice models and their logic, highlighting some features that are perhaps underappreciated. I will then briefly present an overview of set frames. In the final section, I will present the results, which will, I hope, add to our understanding of the logic of the semilattice models.

---

[3] The extension to conjunction is arguably more natural than the extension to disjunction, a point raised by Humberstone (1988). Some information can reasonably verify a disjunction by exhaustively splitting into two portions, each of which verifies one of the disjuncts, as opposed to the standard clause used by Urquhart, namely, that a disjunction is verified by some information when one or the other disjunct is. I will briefly return to Humberstone's approach to disjunction in the next section.

[4] See Fine (1976b), Charlwood (1981). It should be noted that Urquhart (1972b) already had completeness results for the implicational logics.

[5] Urquhart (1972a, b) also considered extending the frames with modal elements, adding a set of possible worlds and a modal accessibility relation on them in order to interpret the implication of the logic E of entailment. The modal accessibility relation for E obeys the usual $\mathsf{S4}$ conditions, namely, reflexivity and transitivity. Urquhart raises some questions about different logics resulting from different conditions put on the modal accessibility relation. Fine (1976a) proves a completeness theorem for the $\mathsf{S5}$ analog of E. This idea is briefly discussed by Mares and Standefer (2017). As far as I know, there has been no exploration of the modal expansions of the semilattice semantics, or more general operational semantics, with a primitive modal operator, $\Box$, in addition to the non-modal implication of the underlying logic.

[6] For more on ternary relational frames, see Routley and Meyer (1972a, b, 1973), Routley et al. (1982), or Restall (2000), among others. For discussion of their philosophical significance, see Beall et al. (2012).

7.2 Semilattice Frames

In this section I will define semilattice frames, and the more general operational frames, and provide some comments on their logic. Once the basic formal apparatus has been presented, I will briefly survey the work that has been done in the area, in order to highlight some underappreciated aspects of the semilattice and operational frames.

Definition 7.1 (*Semilattice frame*) A *semilattice frame* is a triple $\langle P, \sqcup, 0\rangle$, where $0 \in P$ and $\sqcup : P \times P \mapsto P$ obeys the following conditions:

(S1) $0 \sqcup x = x$.
(S2) $x \sqcup y = y \sqcup x$.
(S3) $x \sqcup (y \sqcup z) = (x \sqcup y) \sqcup z$.
(S4) $x \sqcup x = x$.

More general operational frames can be had by dropping any of the latter three conditions. The class of operational frames dropping postulate (S4) is one I will come back to briefly.

Definition 7.2 (*Semilattice model*) A semilattice model is a pair of a semilattice frame $\langle P, \sqcup, 0\rangle$ together with a valuation $V : \mathsf{At} \mapsto \wp(P)$.

A verification relation $\Vdash$ is a binary relation between points and formulas defined inductively as follows:

- $x \Vdash p$ iff $x \in V(p)$.
- $x \Vdash B \wedge C$ iff $x \Vdash B$ and $x \Vdash C$.
- $x \Vdash B \vee C$ iff $x \Vdash B$ or $x \Vdash C$.
- $x \Vdash B \rightarrow C$ iff for all $y \in P$, if $y \Vdash B$, then $x \sqcup y \Vdash C$.

Definition 7.3 (*Holds, Validity*) A formula A *holds* in a semilattice model $\langle P, \sqcup, 0, V\rangle$ iff $0 \Vdash A$. A formula A is *valid* for semilattice frames iff A holds in all semilattice models. Write $\models_{\mathsf{SL}} A$ to mean that A is valid for semilattice frames.

When discussing the operational semantics, the natural point of comparison is with the logic R^+, which is the "positive fragment" of R in the vocabulary $\{\rightarrow, \wedge, \vee\}$ and its subvocabularies.[7] R^+ can be given a Hilbert-style axiomatization as follows:

(R1) $A \rightarrow A$
(R2) $A \wedge B \rightarrow A$, $A \wedge B \rightarrow B$
(R3) $(A \rightarrow B) \wedge (A \rightarrow C) \rightarrow (A \rightarrow B \wedge C)$
(R4) $A \rightarrow A \vee B$, $A \rightarrow B \vee A$
(R5) $(A \rightarrow C) \wedge (B \rightarrow C) \rightarrow (A \vee B \rightarrow C)$

[7] The term "positive fragment" is somewhat misleading, since this is naturally taken to include at least the fusion connective, $\circ$, and the Ackermann constant, t, as these are usually included, with negation, in standard forms of the full axiomatization of R. For this paper, I will use "positive fragment" for what is better called "the implication-conjunction-disjunction fragment".

(R6) $A \wedge (B \vee C) \rightarrow (A \wedge B) \vee (A \wedge C)$

(R7) $(A \rightarrow B) \rightarrow ((B \rightarrow C) \rightarrow (A \rightarrow C))$

(R8) $A \rightarrow ((A \rightarrow B) \rightarrow B)$

(R9) $(A \rightarrow (A \rightarrow B)) \rightarrow (A \rightarrow B)$

(R10) $A, A \rightarrow B \Rightarrow B$

(R11) $A, B \Rightarrow A \wedge B$

The logic $\mathsf{RW}^+$is obtained by dropping axiom (R9). The logics T^+ and TW^+, which will figure only briefly below, can be obtained by dropping axiom (R8) from R^+ and RW^+, respectively, and adding $(A \rightarrow B) \rightarrow ((C \rightarrow A) \rightarrow (C \rightarrow B))$.

Let us call the logic generated by the semilattice semantics UR. There are a few related logics that get discussed in the literature. One of those, URW, is the set of formulas valid in the class of operational frames obtained by dropping the postulate $x \sqcup x = x$ while retaining the others. The logic URW is most naturally compared with RW^+. Finally, the logics UT and UTW are obtained by adding a binary relation, $\preccurlyeq$, on points to the classes of operational frames for UR and URW, respectively, and modifying the verification clause for implication.[8] These four logics will be called the *operational logics*.

Let us use $\mathsf{L}_{\rightarrow}$ and $\mathsf{L}_{\rightarrow,\wedge}$ for the fragments of the logic L in the subscripted vocabularies. It turns out that the theorems of $\mathsf{UR}_{\rightarrow}$ coincide with those of $\mathsf{R}^+_{\rightarrow}$, and, similarly, $\mathsf{UR}_{\rightarrow,\wedge}$ coincides with $\mathsf{R}^+_{\rightarrow,\wedge}$.

With disjunction, a difference emerges. UR properly extends R^+. By way of example, both

$$(A \rightarrow B \vee C) \wedge (B \rightarrow D) \rightarrow (A \rightarrow D \vee C)$$

and

$$(A \rightarrow ((A \rightarrow A) \vee A)) \rightarrow (A \rightarrow A)$$

are theorems of UR that are not theorems of R^+.[9] Humberstone (1988) shows how to modify the semilattice frames along with the verification condition for disjunction to yield frames for which R^+ is sound and complete. Humberstone does this by adding a second operation, $+$, on points to the frames, which operation is used in the verification condition for disjunction, as well as a distinguished unit element for this operation. Humberstone's verification condition for disjunction is the following:

- $x \Vdash B \vee C$ iff there are $y, z \in P$ such that $x = y + z$, $y \Vdash B$ and $z \Vdash C$.[10]

[8] The modification is: $x \Vdash B \rightarrow C$ iff for all $y \in P$ such that $x \preccurlyeq y$, if $y \Vdash B$, then $x \sqcup y \Vdash C$. The logics UT and UTW will not feature much below, so further comment on them will be relegated to footnotes.

[9] It is worth noting that UT properly extends T^+, as shown by the same examples.

[10] This sort of condition for disjunction also occurs in work on dependence logic and inquisitive semantics. For the former, see Yang and Väänänen (2016). For the latter, see Ciardelli et al. (2019), as well as Ciardelli and Roelofsen (2011), Punčochář (2015, 2016, 2019), and Holliday (2021). Humberstone (2019) discusses the issues in a general setting.

There are, additionally, a few conditions on the modified frames, for which the interested reader should see the cited paper.[11]

UR properly extends R^+, and further, the extension is not captured by a simple axiom scheme in the way that R^+ extends T^+ by the addition of axiom scheme (R8). Rather, the following additional rule is used, where the notation $[A_1, \dots, A_k] \to B$ means $A_1 \to (\cdots (A_k \to B) \cdots)$: From

$$B \wedge ([A_1 \wedge q_1, \dots, A_n \wedge q_n] \to C) \to ([B_1 \wedge q_1, \dots, B_n \wedge q_n]) \to E)$$

and

$$B \wedge ([A_1 \wedge q_1, \dots, A_n \wedge q_n] \to D) \to ([B_1 \wedge q_1, \dots, B_n \wedge q_n] \to E),$$

to infer

$$B \wedge ([A_1, \dots, A_n] \to C \vee D) \to ([B_1, \dots, B_n] \to E),$$

where the q_i, $1 \leq i \leq n$, are distinct and occur only where displayed.

When viewed as a Hilbert-style axiom system, the charm of UR is, perhaps, not obvious. It adds to R^+ a complex rule, and one might wonder whether the additional theorems are really *that* appealing. The Hilbert-style axiomatization is, I think, not the logic's best side. Indeed, Dunn and Restall (2002, 69) remark, "We forbear taking cheap shots at such an ungainly rule, the true elegance of which is hidden in the details of the completeness proof that we shall not be looking into. Obviously Anderson and Belnap's R is to be preferred when the issue is simplicity of Hilbert-style axiomatisations." The models have a clear appeal, but there is more to say on a proof-theoretic front.

Charlwood (1978) presents a natural deduction system for UR.[12] The system uses subscripts, much like the Fitch systems of Anderson and Belnap (1975), Brady (1984). Charlwood shows that the natural deduction system for UR admits a normalization theorem. On the basis of that theorem, he shows that a second additional rule used by Fine (1976b) is in fact admissible and, in light of a proved equivalence with the Hilbert-style axiomatization with the above rule, unnecessary. This is instrumental in showing that the Hilbert-style axiomatization is complete for the semilattice semantics. As far as I know, similar completeness results for Hilbert-style axiomatizations for the other operational logics have yet to be obtained. While completeness for UR has been settled, Urquhart (2016) points out that another important meta-theoretic question remains open, namely, whether UR is decidable. Urquhart (1984) famously showed that R was undecidable, and a decidability result for semilattice logic would provide an important contrast.

The normalization theorem shows that the rules fit together in a natural way. Further evidence of the naturalness with which the rules fit together comes from

[11] The reader should also see the discussion of Humberstone (2011, 905ff.).

[12] This system and variants for URW, UT, and UTW are presented by Giambrone and Urquhart (1987, 437–438).

the fact that distribution, $A \wedge (B \vee C) \rightarrow (A \wedge B) \vee (A \wedge C)$, is derivable without a special distribution rule, which is not the case in the Anderson-Belnap-Brady-style indexed Fitch systems. Distribution is, rather, a consequence of the introduction and elimination rules for the connectives involved, with the normal proof having the same form that it would in intuitionistic natural deduction. Indeed, Urquhart (1989) notes this as a point in favor of the semilattice logic.

Sequent systems have received much attention in the study of the logic of operational semantics. A sequent system was already provided by Urquhart (1972b, 31), along with a completeness proof for **UR**. This system uses indexed formulas and multiple conclusions. Giambrone and Urquhart (1987) presents two subscripted sequent systems for **UR**, as well as modifications to obtain systems for the other operational logics. These systems are proved equivalent to each other. Kashima (2003) presents cut-free, multiple conclusion, labeled sequent systems for the operational logics.

While the semilattice semantics has gotten a lot of attention, the more general operational semantics should not be ignored. In particular the operational frames that drop postulate (S4), $x \sqcup x = x$, have a lot of appeal for logicians interested in non-contractive logics.[13] An alternative semantics, *disjoint semantics*, for the non-contractive logic **URW** was defined by Giambrone et al. (1987). Disjoint frames keep all the postulates of the semilattice frame and add the postulate

- $x \leq y \sqcup z$ iff there are u, w such that $u \sqcup w = x$, $u \leq y$, and $w \leq z$, where $x \leq y$ iff $x \sqcup y = y$.

Two points x and y are said to be *disjoint*, Jxy, iff for all $z \in P$, $z \leq x$ and $z \leq y$ only if $z = 0$. The verification clause for the implication is then modified to the following.

- $x \Vdash B \rightarrow C$ iff for all $y \in P$, if Jxy and $y \Vdash B$, then $x \sqcup y \Vdash C$

Disjoint semantics for **UTW** is obtained by adding in a binary relation and adapting the verification clause, much the same as operational semantics for **UT** is obtained from the semilattice semantics.

Meyer et al. (1988) shows that over the vocabulary $\{\rightarrow, \wedge\}$, the disjoint and contraction-free operational semantics are equivalent and that $\mathsf{RW}_{\rightarrow,\wedge}$ and $\mathsf{TW}_{\rightarrow,\wedge}$ are complete with respect to the appropriate classes of operational models. Kashima (2003) shows that disjoint and operational semantics are equivalent even when disjunction is in the language. It is, as far as I know, an open question whether RW^+ and TW^+ are complete with respect to the appropriate classes of frames. As remarked by Giambrone and Urquhart (1987, 439), the standard examples where the semilattice semantics goes beyond the ternary relational semantics, or equivalently the standard axiomatizations of the contraction-less relevant logics, turn out not to be valid in contraction-free operational semantics.

This brief survey of work on semilattice semantics will conclude with some recent work. A logic has the *variable sharing property* when all theorems of the form

[13] In the relevant logic tradition, one of the primary virtues of non-contractive logics is that they support a non-trivial naive set theory. For examples of work in this area, see Brady (1984), Brady (1989, 2006, 2014, 2017), Weber (2010a, 2010b, 2012, 2013), among others.

$A \to B$ are such that A and B share a propositional variable.[14] The logic R enjoys the variable sharing property, and variable sharing is usually taken as a necessary condition on being a relevant logic. Weiss (2019) shows that UR has the variable sharing property, as does as an extension with an involutive negation, and he does this via a semilattice structure using arithmetic operations, as opposed to the matrix methods often used, such as the 8-valued algebra used by Anderson and Belnap (1975, 252–254). Weiss (2020) shows how to conservatively extend the semilattice semantics with a constructive negation.

Given the importance of the operational semantics, it is worth comparing any new semantics for relevant logics to it. In the remainder of the paper, I will provide enough background on collection frames to illuminate the connections and divergences between collection frames and operational frames.

7.3 Set Frames

Let us turn to set frames. As a notational convention, where P is a non-empty set, $\mathcal{P}$ will be the set of all finite subsets of P.

Definition 7.4 (*Set frames*) A *set frame* is a pair $\langle P, R\rangle$, where P is a non-empty set of points and R is a binary relation on $\mathcal{P} \times P$ that obeys the conditions

Reflexivity $\forall x \in P, \{x\}Rx$, and
Transitivity $\forall X, Y \in \mathcal{P}\forall y \in P$, if $\exists z(XRz \wedge (\{z\} \cup Y)Ry)$, then $(X \cup Y)Ry$.
Evaluation $\forall X, Y \in \mathcal{P}\forall y \in P$, if $(X \cup Y)Ry$, then $\exists z(XRz \wedge (\{z\} \cup Y)Ry)$.

The conjunction of Transitivity and Evaluation will be called *Compositionality*, and set relations obeying Compositionality will be called *compositional*.

In general, we do not have to impose the first condition, Reflexivity, although dropping it will require generalizing the definition of validity. In this paper there will not be a need to discuss non-reflexive set frames, since the desired equivalence appears to require the condition, so all set frames will be reflexive. Non-empty members of $\mathcal{P}$ will be called *inhabited*.[15] One can consider set frames only on inhabited sets, but I will not do so here. The conditions Transitivity and Evaluation appear to be required for collection frames to work properly, unlike the previous conditions. Their contributions in the development of the framework are many, including the verification of heredity for conditionals, $A \to B$, and validating structural rules in proof systems. An example of their contribution in the present work can be found in

[14] For a general characterization of the variable sharing property, see Robles and Méndez (2011, 2012).

[15] Restall and Standefer (2020) consider many types of collections, not just sets, and there are empty versions of all of these. Especially in the general setting is useful to have a term for distinguishing the collection frames that exclude empty collections and those that include them. The term "inhabited" is used here for terminological continuity with the cited paper and Restall (2021).

the proof of Lemma 7.3, verifying that a frame has a certain property. The interested reader should consult Restall and Standefer (2020) for details.

Some comments on set frames are in order. First, the binary relations of set frames are defined over finite sets of points because the binary relations of more general collection frames are defined over finite collections. This enables a straightforward connection with more familiar frames, such as ternary relational frames and operational frames, where one is possible. There appears to be no barrier to defining the binary relations over infinite collections, and this generalization will be left to future work.

Next is a comment on the interpretation of the binary relation R. If we think of the points as being bodies of information, we can think of XRy as saying that the result of combining together all the information in X is contained in y. On this interpretation, Reflexivity is a sensible condition, as it is intuitive that the information obtained by combining all the information in $\{x\}$ is contained in x. After all, there is no other point that can supply information available. We can also use the informational interpretation to motivate the two parts of Compositionality. The two parts say that one can combine together the information in $X \cup Y$ in one go or break it into parts and combine together the information in X and combine that with the information in Y. These are especially natural conditions in the context of set frames, since most sets can be broken into parts in a variety of ways.

Finally, we will comment on the relation between set frames and the better known ternary relational frames for relevant logics. Every set frame, as defined above, induces a ternary relational frame for the logic R^+, but not every ternary relational frame for R^+ induces a set frame, which point will come up again later. Nonetheless, set frames are interesting for at least two reasons. First, they permit generalizations that are not obvious with ternary relational frames, namely, permitting non-reflexive and inhabited frames. Second, it is comparatively easy to verify whether a structure is a set frame, whereas it is somewhat more involved to verify that a structure is a ternary relational frame that verifies the frame conditions for R^+.

Compositional set relations are fairly common. For example, suppose that $P = \omega$ and XRy iff $y = \max(X)$, where $\max(\{\ \}) = 0$. This relation is compositional and reflexive. As another example, let $P = \omega^+$, the positive natural numbers, and XRy iff for some $x \in X$, x and y share a prime factor or $y = 1$, when $X \neq \{\ \}$, and $\{\ \}R1$. This relation is also compositional and reflexive. The interested reader should see Restall and Standefer (2020) or Restall (2021) for more. The first example is an example of a functional, compositional relation. I will put things more precisely in a definition, which will be important below.

Definition 7.5 (*Functionality*) A set frame $\langle P, R\rangle$ is *functional* iff both

- for all $X \in \mathcal{P}$ there is $x \in P$ such that XRx, and
- if XRy and XRz, then $y = z$.

Functional set frames are pleasantly common. Note that functional set frames obey a stronger form of Evaluation.[16]

[16] I thank Lloyd Humberstone for pointing this out.

Uniform Evaluation $\forall X \in \mathcal{P} \exists z \in P[XRz$ and $\forall Y \in \mathcal{P} \forall y \in P$, if $(X \cup Y)Ry$, then $(\{z\} \cup Y)Ry]$.

Uniform Evaluation differs from Evaluation in that the point to which X evaluates, z, is independent of the choice of Y.

Definition 7.6 A *set model* is a pair of a set frame $\langle P, R \rangle$ and a valuation V : $\mathsf{At} \mapsto \wp(P)$ satisfying the heredity property, if $x \in V(p)$ and $\{x\}Ry$, then $y \in V(p)$. Valuations with this property will be called *hereditary*. Such a model is said to be *built on* the set frame.

A *verification relation* $\Vdash$ is a binary relation between points and formulas defined inductively as follows:

- $x \Vdash p$ iff $x \in V(p)$.
- $x \Vdash B \wedge C$ iff $x \Vdash B$ and $x \Vdash C$.
- $x \Vdash B \vee C$ iff $x \Vdash B$ or $x \Vdash C$.
- $x \Vdash B \to C$ iff for all $y, z \in P$, if $\{x, y\}Rz$ and $y \Vdash B$, then $z \Vdash C$.

As one might expect, preservation of verification along R extends from atoms to all formulas.

Theorem 7.1 (Heredity) *If $x \Vdash A$ and $\{x\}Ry$, then $y \Vdash A$.*

Proof The proof is by induction on the construction of the formula. It is routine.

In the present setting, I will focus on valid formulas. This permits the use of the following definition for validity, which is a special case of the more general notion.[17]

Definition 7.7 (*Holds, valid*) A formula A holds on a set model iff for all $x \in P$ such that $\{\ \}Rx$, $x \Vdash A$.

A formula A is valid on a set frame iff A holds in all models built on that set frame.

A formula A is valid in a class of set frames iff A is valid on every set frame in that class.

If A is valid in the class of all set frames, we will write $\models_{\mathsf{Set}} A$. If A is valid in the class of all functional set frames, we will write $\models_{\mathsf{Fun}} A$.

With the definition of validity in hand, we can talk about the logic of set frames.

The logic R^+ is sound for the class of set frames, which is to say that if A is a theorem of R^+then $\models A$. The question of completeness, whether whenever we have $\models A$ we also have that A is a theorem of R^+, is still open at the time of writing. In the next section, I will show that UR is sound and complete for the class of functional set frames. It is this contrast, between the logic of set frames, which may be R^+ or may extend it, and the logic of functional set frames, which coincides with UR, that is the main reason for focusing on set frames.

[17] Restall and Standefer (2020) use a sequent presentation of R^+, and define validity for sequents. The present definition of validity is a special case of the definition they use.

An alternative that is not being pursued here is to use multiset frames, rather than set frames.[18] Multisets differ from sets in distinguishing the number of times an element is a member of that multiset, and multisets and sets are similar in not keeping track of the order.[19] The multisets $[a, a, b]$ and $[a, b, a]$ are identical, but they both differ from the multiset $[a, b]$, as the latter contains a only once and the former both contain it twice. Finite multisets are those that contain a finite number of elements a finite, non-zero number of times. Multiset frames and models are defined much as set frames and models, where the binary R relates finite multisets of points to points and the definition of verification trades sets for multisets. The technical details of the arguments to follow are slightly easier in the context of multiset frames, but the technical advance is in the context of set frames.[20] For that reason the focus is on set frames.

7.4 Another View on Semilattice Logic

With the necessary background in place, I can now turn to the task of connecting semilattice models and functional set models. There is a tight connection between them. Every semilattice frame induces a functional set frame, and each semilattice model induces a corresponding functional set model that agrees on all formulas. Similarly, every functional set frame induces a semilattice frame, and the models on those frames agree on all formulas. Broadening out to include non-functional set frames yields a counterexample to a theorem of UR.

Given a semilattice frame $\langle P, 0, \sqcup \rangle$, define $\bigsqcup : \mathcal{P} \mapsto P$ as follows:

$$\bigsqcup X = \begin{cases} 0 & X = \{\,\} \\ x & X = \{x\} \\ x_1 \sqcup (\cdots (x_{n-1} \sqcup x_n)) & X = \{x_1, \ldots, x_n\} \end{cases}$$

When $X = \{x, y\}$, I'll write $x \sqcup y$ for $\bigsqcup X$.

Lemma 7.1 *Let $\langle P, \sqcup, 0 \rangle$ be a semilattice frame. Then $\langle P, R \rangle$ is a functional set frame, where R is defined as follows:*

- *XRy iff $\bigsqcup X = y$.*

---

[18] Multiset frames that obey a contraction principle are similar to the definition of R-frame of Mares (2004, 210), using a ternary relation R^3 on points and defining relations of higher arity. Given the conditions on R^{n+1}, for $n \geq 2$, the first n arguments can be viewed as forming a multiset related to the final argument.

[19] See Blizard (1988) for an overview of multiset theory. Meyer and McRobbie (1982a, b) uses multisets in an illuminating study of relevant logics.

[20] Every ternary relational frame for the logic R^+ induces a reflexive multiset frame that obeys a contraction principle. As mentioned above, some ternary relational frames for R^+ can be shown not to induce a reflexive set frame.

Proof The relation R is well defined. If $X = Y$, then $\bigsqcup X = \bigsqcup Y$, so XRz iff YRz.

Reflexivity follows from the singleton case of the definition of $\bigsqcup$. It remains to check the two directions of compositionality, for which we show that $\bigsqcup(X \cup Y) = \bigsqcup X \sqcup \bigsqcup Y$, for all $X, Y \in \mathcal{P}$.

Suppose that $X = Y = \{\,\}$. Then $\bigsqcup(X \cup Y) = \bigsqcup\{\,\} = 0 = 0 \sqcup 0 = \bigsqcup X \sqcup \bigsqcup Y$.

Suppose that exactly one of X and Y is $\{\,\}$, say X. Then $\bigsqcup(X \cup Y) = \bigsqcup Y = 0 \sqcup \bigsqcup Y = \bigsqcup X \sqcup \bigsqcup Y$.

Suppose that $X = \{x_1, \ldots, x_n\}$ and $Y = \{y_1, \ldots, y_m\}$. Then, we have

$$\bigsqcup(X \cup Y) = \bigsqcup\{x_1, \ldots, x_n, y_1, \ldots, y_m\}.$$

In virtue of the semilattice frame conditions we have

$$\bigsqcup\{x_1, \ldots, x_n, y_1, \ldots, y_m\} = \bigsqcup\{x_1, \ldots, x_n\} \sqcup \bigsqcup\{y_1, \ldots, y_m\},$$

with (S2)–(S3) being used to separate out the x_i's from the y_j's, and (S4) being used to duplicate or collapse some elements in case $X \cap Y \neq \emptyset$. Finally, from definitions, we obtain

$$\bigsqcup\{x_1, \ldots, x_n\} \sqcup \bigsqcup\{y_1, \ldots, y_m\} = \bigsqcup X \sqcup \bigsqcup Y,$$

which suffices for the desired identity, $\bigsqcup(X \cup Y) = \bigsqcup X \sqcup \bigsqcup Y$.

For Transitivity, suppose that XRx and $(\{x\} \cup Y)Ry$. Then $\bigsqcup X = x$ and $x \sqcup \bigsqcup Y = y$. It follows that $\bigsqcup X \sqcup \bigsqcup Y = y$, so $\bigsqcup(X \cup Y) = y$, so $(X \cup Y)Ry$.

For Evaluation, suppose that $(X \cup Y)Ry$. Then $\bigsqcup X \sqcup \bigsqcup Y = y$. As $\bigsqcup X = z$, for some z, XRz and $z \sqcup \bigsqcup Y = y$, so $(\{z\} \cup Y)Ry$, as desired.

The functionality conditions are secured by the fact that $\bigsqcup$ is a function.

For a given semilattice frame, the *source frame*, say that the preceding construction *induces* the set frame defined, which will be called the *induced frame*. All semilattice frames induce functional set frames. What about the converse? Do all functional set frames induce semilattice frames? Yes, as will be shown. I will prove a lemma first.

Lemma 7.2 *Let $\langle P, R\rangle$ be a functional set frame. For the x such that $\{\,\}Rx$, $(\{x\} \cup X)Ry$ iff XRy.*

Proof The left to right direction follows from Transitivity and the assumption that $\{\,\}Rx$. The right to left direction follows from Evaluation and the fact that $X = X \cup \{\,\}$.

Lemma 7.3 *Let $\langle P, R\rangle$ be a functional set frame. Then $\langle P, \sqcup, 0\rangle$ is a semilattice frame, where 0 is the x such that $\{\,\}Rx$, and for $x, y \in P$, $x \sqcup y = z$ iff $\{x, y\}Rz$.*

Proof We need to show that 0 and $\sqcup$ are well defined and obey the appropriate conditions. First, the uniqueness of 0 follows from the functionality of R.

Next, we show that $\sqcup$ is well defined.

For all $x, y \in P$, there is a z such that $\{x, y\}Rz$, as R is functional. Suppose that $x \sqcup y = z$ and $x \sqcup y = z'$. Then $\{x, y\}Rz$ and $\{x, y\}Rz'$. As R is functional, this implies $z = z'$. We conclude $\sqcup$ is well defined.

Finally, we show that $\sqcup$ satisfies the conditions on semilattice frames.

Since $\{x\}Rx$ and $\{x\} = \{x\} \cup \{\ \}$, from the preceding lemma, $\{0, x\}Rx$, so $0 \sqcup x = x$.

Since $\{x, y\} = \{y, x\}$, $\{x, y\}Rz$ iff $\{y, x\}Rz$, so $x \sqcup y = z = y \sqcup x$.

As $\{x, x\} = \{x\}$ and $\{x\}Rx$, by definition, $\{x, x\}Rx$, so $x \sqcup x = x$.

Let $\{x, y, z\}Rw$. By Compositionality, for some v, $\{x, y\}Rv$ and $\{v, z\}Rw$, so $x \sqcup y = v$ and $v \sqcup z = w$. By Compositionality again, for some v', $\{y, z\}Rv'$ and $\{x, v'\}Rw$, so $y \sqcup z = v'$ and $x \sqcup v' = w$. So, $x \sqcup (y \sqcup z) = w = (x \sqcup y) \sqcup z$.

The preceding lemmas show that semilattice frames induce functional set frames and, conversely, functional set frames induce semilattice frames. The induced frames have a close connection with the source frames. I will prove two "round trip" theorems, showing that the constructions given above do not result in any changes when performed in succession.[21] They are, in a sense, inverses.

For the next results, it will be useful to define some notation. Given a source semilattice frame M, let M^σ be the induced functional set frame as defined in Lemma 7.1. Given a source functional set frame N, let N^λ be the induced semilattice frame as defined in Lemma 7.3.

Theorem 7.2 *Let $M = \langle P_M, \sqcup_M, 0_M \rangle$ be a semilattice frame. Then $M = M^{\sigma\lambda}$.*

Proof The constructions keep the set of points the same, so $P_M = P_{M^\sigma} = P_{M^{\sigma\lambda}}$.

By definition, $0_M = 0_{M^\sigma}$. As $\{\ \}R_{M^\sigma}0_{M^\sigma}$, $0_{M^{\sigma\lambda}} = 0_{M^\sigma}$, whence $0_M = 0_{M^{\sigma\lambda}}$.

Suppose $x \sqcup_M y = z$. This is the case iff $\{x, y\}R_{M^\sigma}z$, which is equivalent to $x \sqcup_{M^{\sigma\lambda}} y = z$. This suffices for the showing that $\sqcup_M = \sqcup_{M^{\sigma\lambda}}$.

Theorem 7.3 *Let $M = \langle P_M, R_M \rangle$ be a functional set frame. Then $M = M^{\lambda\sigma}$.*

Proof As in the proof of the previous theorem, the constructions do not change the sets of points, so $P_M = P_{M^\lambda} = P_{M^{\lambda\sigma}}$.

Let $X \in \mathcal{P}$ be arbitrary and suppose XR_My. There are three subcases depending on X.

Suppose $X = \{\ \}$. Then $y = 0_{M^\lambda}$, so $XR_{M^{\lambda\sigma}}y$.

Suppose $X = \{x\}$. Since M is functional, there is a y such that XR_My. Then $\{x\}R_Mx$ implies $x = y$. Thus, $\bigsqcup_{M^\lambda} X = y$, so $XR_{M^{\lambda\sigma}}y$. The converse is similar.

Suppose $X = \{x_1, \ldots, x_n\}$, for some $n \geq 2$. From repeated application of Evaluation, there are $z_1, \ldots, z_{n-1}$ such that $\{x_1, x_2\}R_Mz_1$, $\{z_1, x_3\}R_Mz_2$, ..., $\{z_{n-1}, x_n\}R_My$. By definition, $x_1 \sqcup_{M^\lambda} x_2 = z_1$, ..., and $z_{n-1} \sqcup_{M^\lambda} x_n = y$. It then follows that $\bigsqcup_{M^\lambda}\{x_1, \ldots, x_n\} = y$. Therefore $XR_{M^{\lambda\sigma}}y$, as desired. The converse is similar.

The final piece required for the connection between the logics of these two classes of frames is to show that the models built on a source frame and an induced frame agree on the evaluation of formulas. I will now prove that with two lemmas.

[21] I thank Lloyd Humberstone for the suggestion of proving these theorems.

Lemma 7.4 *Let* $M = \langle P, \sqcup, 0 \rangle$ *be a semilattice frame. If* $\Vdash_{\mathsf{SL}}$ *is a verification relation on* M*, then* $\Vdash_{\mathsf{Set}}$ *is a hereditary verification relation on* M^{σ}*, where* $x \Vdash_{\mathsf{Set}} p$ *iff* $x \Vdash_{\mathsf{SL}} p$*. Moreover, for all* $x \in P$ *and all formulas* A*,* $x \Vdash_{\mathsf{Set}} A$ *iff* $x \Vdash_{\mathsf{SL}} A$*.*

Proof The coherence of the definition is straightforward from the definition. Heredity then follows as $\{x\}Ry$ implies $x = y$ from the functionality of R.

The second part of the claim is proved by induction on formula structure. The base case holds by definition. The cases where A is of the form $B \wedge C$ or $B \vee C$ are immediate by the inductive hypothesis.

Suppose A is of the form $B \rightarrow C$. Then, $x \Vdash_{\mathsf{Set}} B \rightarrow C$ iff for all y, z, if $\{x, y\}Rz$ and $y \Vdash_{\mathsf{Set}} B$ then $z \Vdash_{\mathsf{Set}} C$. Let y be an arbitrary point such that $y \Vdash_{\mathsf{SL}} B$. As R is functional, there is a z such that $\{x, y\}Rz$. By the inductive hypothesis, $y \Vdash_{\mathsf{Set}} B$, so $z \Vdash_{\mathsf{Set}} C$. By the inductive hypothesis, $z \Vdash_{\mathsf{SL}} C$. As $\{x, y\}Rz$, $x \sqcup y = z$. Therefore, so $x \sqcup y \Vdash_{\mathsf{SL}} C$. Therefore, $x \Vdash_{\mathsf{SL}} B \rightarrow C$.

Suppose $x \Vdash_{\mathsf{SL}} B \rightarrow C$. Let y, z be arbitrary points such that $\{x, y\}Rz$ and suppose $y \Vdash_{\mathsf{Set}} B$. By the inductive hypothesis, $y \Vdash_{\mathsf{SL}} B$. Therefore, $x \sqcup y \Vdash_{\mathsf{SL}} C$. Since $\{x, y\}Rz$, $x \sqcup y = z$, so it follows that $z \Vdash_{\mathsf{SL}} C$. By the inductive hypothesis, $z \Vdash_{\mathsf{Set}} C$, which establishes that $x \Vdash_{\mathsf{Set}} B \rightarrow C$.

From the preceding lemma, we can see that the logic of functional set frames is contained in the logic of semilattice frames.

Theorem 7.4 *For all formulas* A*,* $\models_{\mathsf{Fun}} A$ *only if* $\models_{\mathsf{SL}} A$*.*

As there are no conditions on verification relations in semilattice frames, we can prove the following lemma.

Lemma 7.5 *Let* $\Vdash_{\mathsf{Set}}$ *be a verification relation on a functional set frame* $N = \langle P, R \rangle$ *and let* $\langle P, 0, \sqcup \rangle$ *be* N^{λ}*. Define a semilattice verification* $\Vdash_{\mathsf{SL}}$ *as* $x \Vdash_{\mathsf{SL}} p$ *iff* $x \Vdash_{\mathsf{Set}} p$*. The result is a semilattice model. Moreover, for every* $x \in P$ *and formula* A*,* $x \Vdash_{\mathsf{SL}} A$ *iff* $x \Vdash_{\mathsf{Set}} A$*.*

Proof The initial portion of the corollary is immediate from the preceding lemma. The moreover portion follows from a straightforward induction on formula complexity. We will present the $B \rightarrow C$ case, as it is the only non-trivial one.

Suppose $x \Vdash_{\mathsf{Set}} B \rightarrow C$. Then, for all y, z such that $\{x, y\}Rz$, if $y \Vdash_{\mathsf{Set}} B$, then $z \Vdash_{\mathsf{Set}} C$. Let y be arbitrary and suppose $y \Vdash_{\mathsf{SL}} B$. By the inductive hypothesis, $y \Vdash_{\mathsf{Set}} B$. Since R is functional, for some z, $\{x, y\}Rz$, so $z \Vdash_{\mathsf{Set}} C$. By the inductive hypothesis again, $z \Vdash_{\mathsf{SL}} C$. Since $\{x, y\}Rz$, $x \sqcup y = z$, so $x \sqcup y \Vdash_{\mathsf{SL}} C$, which suffices for $x \Vdash_{\mathsf{SL}} B \rightarrow C$.

Suppose $x \Vdash_{\mathsf{SL}} B \rightarrow C$. Then for all y, if $y \Vdash_{\mathsf{SL}} B$ then $x \sqcup y \Vdash_{\mathsf{SL}} C$. Let y, z be arbitrary points such that $\{x, y\}Rz$. Suppose $y \Vdash_{\mathsf{Set}} B$. By the inductive hypothesis, $y \Vdash_{\mathsf{SL}} B$, so $x \sqcup y \Vdash_{\mathsf{SL}} C$. Since $\{x, y\}Rz$, $x \sqcup y = z$, so $z \Vdash_{\mathsf{SL}} C$. By the inductive hypothesis $z \Vdash_{\mathsf{Set}} C$, which suffices to establish $x \Vdash_{\mathsf{Set}} B \rightarrow C$.

This corollary suffices for the following theorem.

Theorem 7.5 *For all formulas* A*,* $\models_{\mathsf{SL}} A$ *only if* $\models_{\mathsf{Fun}} A$*.*

Table 7.1 Counterexample

R		R		$\Vdash$	
$\{\}$	b	$\{a, b\}$	a	a	r
$\{a\}$	a	$\{a, c\}$	a, b, c	b	q
$\{b\}$	b	$\{b, c\}$	c	c	p, r
$\{c\}$	c	$\{a, b, c\}$	a, b, c		

There is, then, a match between the valid formulas of semilattice frames and those of functional set frames.

One more theorem remains to be proved, showing that the logic of functional set frames properly extends the logic of set frames.

Lemma 7.6 *There is a formula A such that A is valid in the class of functional set frames but not valid in the class of set frames.*

Proof For the formula, we take $(p \to (q \vee r)) \wedge (q \to r) \to (p \to r)$, which is valid in semilattice frames but is not a theorem of R^+, as noted by Urquhart (1972a, 163) who attributes it to Dunn and Meyer. A simple non-functional set frame counterexample to this in the class of all set frames can be found. For this counterexample, let $P = \{a, b, c\}$ and R defined as in Table 7.1. R so defined is a reflexive, compositional set relation. The valuation given in Table 7.1 is trivially hereditary. It suffices to refute $(p \to (q \vee r)) \wedge (q \to r) \to (p \to r)$ to find a point x at which the antecedent of the implication is true but the consequent is not, for which we will use a. Since $\{a, c\}Rb$, while $c \Vdash p$ and $b \nVdash r$, $a \nVdash p \to r$. It remains to verify that $a \Vdash p \to (q \vee r)$ and $a \Vdash q \to r$. For the former, note that $q \vee r$ is true at all points, although it is in virtue of the q disjunct at b and in virtue of the r disjunct at the other points. For the latter, the only point at which q is true is b, and the only point that $\{a, b\}$ bears R to is a, which has r true.

This lemma suffices for the desired theorem.

Theorem 7.6 *The set of formulas valid in the class of all set frames is a proper subset of the set of all formulas valid in the class of all functional set frames, which is* UR. *In symbols, there is a formula A such that $\models_{\mathsf{Fun}} A$ but $\not\models_{\mathsf{Set}} A$.*

Proof Immediate from the preceding lemma.

The results of this section situate set frames with respect to the well-known semilattice frames. Functional set models and semilattice models line up neatly. They generate the logic UR. Further, the procedure of inducing one frame type from the other takes you back to where you started after two steps. We can see points in a functional set frame as pieces of information, as suggested in the context of semilattice frames by Urquhart (1972a), and sets of points are collections of information. Combining these collections of information is done via set union, which pleasantly coincides with Urquhart's original notation.

Stepping back, we see that UR is not sound for the class of all set frames. R^+ is sound for the class of all set frames but it is currently unknown whether it is complete with respect to that class.[22] Finally, I will note that essentially the same arguments show the same fit between functional *multiset* frames and operational frames that drop postulate (S4), $x \sqcup x = x$, but retain the others.

Acknowledgements I would like to thank Greg Restall, Lloyd Humberstone, Igor Sedlár, and the audiences of the Melbourne Logic Seminar and the Logic Supergroup for feedback and discussion that greatly improved this work. This research was supported by the Australian Research Council, Discovery Grant DP150103801.

References

Anderson, A. R., & Belnap, N. D. (1975). *Entailment: The logic of relevance and necessity* (Vol. 1). Princeton: Princeton University Press.

Beall, J., Brady, R., Dunn, J. M., Hazen, A. P., Mares, E., Meyer, R. K., et al. (2012). On the ternary relation and conditionality. Journal of Philosophical Logic, 41(3), 595–612.

Bimbó, K. (2006). Relevance logics. In D. Jacquette (Ed.), *Philosophy of logic* (Vol. 5, pp. 723–789). Handbook of the philosophy of science. Amsterdam: Elsevier.

Bimbó, K., & Dunn, J. M. (2018). Larisa Maksimova's early contributions to relevance logic. In S. Odintsov (Ed.), *Larisa Maksimova on implication, interpolation, and definability* (Vol. 15, pp. 33–60). Outstanding contributions to logic. Berlin: Springer International Publishing.

Bimbó, K., Dunn, J. M., & Ferenz, N. (2018). Two manuscripts, one by Routley, one by Meyer: The origins of the Routley-Meyer semantics for relevance logics. Australasian Journal of Logic, 15(2), 171–209.

Blizard, W. D. (1988). Multiset theory. Notre Dame Journal of Formal Logic, 30(1), 36–66.

Brady, R. (2006). *Universal Logic*. Stanford: CSLI Publications.

Brady, R. T. (1984). Natural deduction systems for some quantified relevant logics. *Logique Et Analyse, 27*(8), 355–377.

Brady, R. T. (1989). The non-triviality of dialectical set theory. In G. Priest, R. Routley, & J. Norman (Eds.), *Paraconsistent logic: Essays on the inconsistent* (pp. 437–470). Munchen Philosophia Verlag.

Brady, R. T. (2014). The simple consistency of naive set theory using metavaluations. *Journal of Philosophical Logic, 43*(2–3), 261–281.

Brady, R. T. (2017). Metavaluations. Bulletin of Symbolic Logic, 23(3), 296–323.

Buszkowski, W. (1986). Completeness results for lambek syntactic calculus. *Mathematical Logic Quarterly, 32*(1–5), 13–28.

Charlwood, G. (1978). *Representations of semilattice relevance logics*. PhD thesis, University of Toronto.

Charlwood, G. (1981). An axiomatic version of positive semilattice relevance logic. Journal of Symbolic Logic, 46(2), 233–239.

Chellas, B. F. (1975). Basic conditional logic. Journal of Philosophical Logic, 4(2), 133–153.

Ciardelli, I., Groenendijk, J., & Roelofsen, F. (2019). *Inquisitive Semantics*. Oxford University Press.

Ciardelli, I., & Roelofsen, F. (2011). Inquisitive logic. Journal of Philosophical Logic, 40(1), 55–94.

Došen, K. (1988). Sequent-systems and groupoid models. *I. Studia Logica, 47*(4), 353–385.

[22] The stumbling block for proving completeness, briefly, is that the canonical frame for R^+ appears to be one of those ternary relational frames that does not induce a set frame.

Došen, K. (1989). Sequent-systems and groupoid models. *II. Studia Logica, 48*(1), 41–65.
Dunn, J. M., & Restall, G. (2002). Relevance logic. In D. M. Gabbay, & F. Guenthner (Eds.), *Handbook of philosophical logic* (2nd ed., Vol. 6, pp. 1–136). Amsterdam: Kluwer
Ferenz, N. (2018). Richard Routley, "Semantic analysis of entailment and relevant implication: I." *Australasian Journal of Logic, 15*(2), 210–279.
Fine, K. (1974). Models for entailment. Journal of Philosophical Logic, 3(4), 347–372.
Fine, K. (1976a). Completeness for the S5 analogue of E_I, (abstract). *Journal of Symbolic Logic, 41,* 559–560.
Fine, K. (1976b). Completeness for the semilattice semantics with disjunction and conjunction (abstract). Journal of Symbolic Logic, 41, 560.
Giambrone, S., Meyer, R. K., & Urquhart, A. (1987). A contractionless semilattice semantics. Journal of Symbolic Logic, 52(2), 526–529.
Giambrone, S., & Urquhart, A. (1987). Proof theories for semilattice logics. Mathematical Logic Quarterly, 33(5), 433–439.
Holliday, W. H. (forthcoming). Inquisitive intuitionistic logic. In N. Olivetti, & R. Verbrugge (Eds.), *Advances in modal logic* (Vol. 13). London: College Publications.
Humberstone, L. (1988). Operational semantics for positive R. Notre Dame Journal of Formal Logic, 29, 61–80.
Humberstone, L. (2011). *The Connectives*. MIT Press.
Humberstone, L. (2019). Supervenience, dependence, disjunction. Logic and Logical Philosophy, 28(1), 3–135.
Kashima, R. (2003). On semilattice relevant logics. Mathematical Logic Quarterly, 49(4), 401–414
Maksimova, L. L. (1969). Interpretatsiya sistem so strogoĭ implikatsieĭ. In *10th All-Union Algebraic Colloquium*(Abstracts) (p. 113). (An interpretation of systems with rigorous implication).
Mares, E. (2004). *Relevant logic: A philosophical interpretation*. Cambridge: Cambridge University Press.
Mares, E., & Standefer, S. (2017). The relevant logic E and some close neighbours: A reinterpretation. *IfCoLog Journal of Logics and Their Applications, 4*(3), 695–730.
Meyer, R. K., Martin, E. P., Giambrone, S., & Urquhart, A. (1988). Further results on proof theories for semilattice logics. Mathematical Logic Quarterly, 34(4), 301–304.
Meyer, R. K., & McRobbie, M. A. (1982a). Multisets and relevant implication I. Australasian Journal of Philosophy, 60(2), 107–139.
Meyer, R. K., & McRobbie, M. A. (1982b). Multisets and relevant implication II. *Australasian Journal of Philosophy, 60*(3), 265–281.
Punčochář, V. (2015). Weak negation in inquisitive semantics. Journal of Logic, Language and Information, 24(3), 323–355.
Punčochář, V. (2016). A generalization of inquisitive semantics. Journal of Philosophical Logic, 45(4), 399–428.
Punčochář, V. (2019). Substructural inquisitive logics. Review of Symbolic Logic, 12(2), 296–330.
Restall, G. (2000). *An introduction to substructural logics*. London: Routledge.
Restall, G. (2021). Geometric models for relevant logics. In I. Düntsch & E. Mares (Eds.), *Alasdair Urquhart on nonclassical and algebraic logic and complexity of proofs*. Outstanding contributions to logic. Cham: Springer Nature.
Restall, G., & Standefer, S. (2020). Collection frames for substructural logics. In preparation. Preprint available at https://consequently.org/writing/collection-frames/.
Robles, G., & Méndez, J. M. (2011). A class of simpler logical matrices for the variable-sharing property. Logic and Logical Philosophy, 20(3), 241–249.
Robles, G., & Méndez, J. M. (2012). A general characterization of the variable-sharing property by means of logical matrices. Notre Dame Journal of Formal Logic, 53(2), 223–244.
Routley, R., & Meyer, R. K. (1972a). The semantics of entailment-II. Journal of Philosophical Logic, 1(1), 53–73.
Routley, R., & Meyer, R. K. (1972b). The semantics of entailment-III. Journal of Philosophical Logic, 1(2), 192–208.

Routley, R., & Meyer, R. K. (1973). The semantics of entailment. In H. Leblanc (Ed.), *Truth, Syntax, and Modality: Proceedings of the Temple University Conference on Alternative Semantics* (pp. 199–243). Amsterdam: North-Holland Publishing Company.
Routley, R., Plumwood, V., Meyer, R. K., & Brady, R. T. (1982). *Relevant logics and their rivals* (Vol. 1). Atascadero: Ridgeview.
Scott, D. S. (1973). Background to formalization. In H. Leblanc (Ed.), *Truth, Syntax and Modality* (pp. 244–273). North-Holland Publishing Company.
Urquhart, A. (1972a). Semantics for relevant logics. The Journal of Symbolic Logic, 37, 159–169.
Urquhart, A. (1972b). *The semantics of entailment*. PhD thesis, University of Pittsburgh.
Urquhart, A. (1984). The undecidability of entailment and relevant implication. *Journal of Symbolic Logic, 49*(4), 1059–1073.
Urquhart, A. (1989). What is relevant implication? In J. Norman, & R. Sylvan (Ed.), *Directions in relevant logic* (pp. 167–174). Amsterdam: Kluwer.
Urquhart, A. (2016). Relevance logic: Problems open and closed. Australasian Journal of Logic, 13(1), 11–20.
Weber, Z. (2010a). Extensionality and restriction in naive set theory. Studia Logica, 94(1), 87–104.
Weber, Z. (2010b). Transfinite numbers in paraconsistent set theory. Review of Symbolic Logic, 3(1), 71–92.
Weber, Z. (2012). Transfinite cardinals in paraconsistent set theory. Review of Symbolic Logic, 5(2), 269–293.
Weber, Z. (2013). Notes on inconsistent set theory. In F. Berto, E. Mares, K. Tanaka, & F. Paoli (Ed.), *Paraconsistency: Logic and applications* (pp. 315–328). Berlin: Springer.
Weiss, Y. (2019). A note on the relevance of semilattice relevance logic. Australasian Journal of Logic, 16(6), 177–185.
Weiss, Y. (2020). A conservative negation extension of positive semilattice logic without the finite model property. *Studia Logica*, 1–12. Forthcoming.
Yang, F., & Väänänen, J. (2016). Propositional logics of dependence. Annals of Pure and Applied Logic, 167(7), 557–589.

Chapter 8
The Universal Theory Tool Building Toolkit Is Substructural

Shay Allen Logan

Second Reader
G. Payette
University of Calgary

Abstract Consider the set of inferences that are acceptable to use in all our theory building endeavors. Call this set of inferences the universal theory building toolkit, or just 'the toolkit' for short. It is clear that the toolkit is tightly connected to logic in a variety of ways. Beall, for example, has argued that logic just is the toolkit. This paper avoids making a stand on that issue and instead investigates reasons for thinking that, logic or not, the toolkit is substructural. It is presented as a dialogue for the simple reason that it summarizes a range of dialogues on this subject that the author has had with various folks over the past few years. The method I use to investigate the toolkit is inspired in both philosophical and technical details by Alasdair Urquhart's work on semantics for relevance logics from the early 1970s.

Keywords Entailment · Substructural logics · Theory building · Contraction · The curry paradox

In some of his recent work, Jc Beall has argued for a position that, reduced to a slogan, amounts to the claim that logic is the universal theory building toolkit. In Logan (2020), I argued that Beall was mistaken about what logic this should lead us to accept. While my argument there is best understood as a reaction to Beall's proposal, the intellectual genealogy of both the technical and philosophical work I

[1]See Urquhart (1972). It's also worthwhile to mention that this is more of an autobiographical genealogy of this work, rather than a full tracing-back of the exact origins of the ideas. If we were pursuing the latter, we would probably have to go further back in time than we have here.

S. Allen Logan (✉)
Department of Philosophy, Kansas State University, Manhattan, KS, USA
e-mail: salogan@ksu.edu; shay.a.logan@gmail.com

I. Düntsch and E. Mares (eds.), *Alasdair Urquhart on Nonclassical and Algebraic Logic and Complexity of Proofs*, Outstanding Contributions to Logic 22,
https://doi.org/10.1007/978-3-030-71430-7_8

pursued there are naturally traced back to Alasdair Urquhart's 1972 paper 'Semantics for Relevant Logics' (SRL).[1] In fact, we can be more specific and trace it back to the following two-sentence fragment of SRL:

> In argumentation, we may have to consider not only what information may be available, but also what the facts are. That is, we conclude that a piece of information X determines p against a certain background of facts.

In this brief passage, Urquhart drew attention to something very simple, but very important: argumentation does not occur in a vacuum; it occurs set against a background of facts.

But there's more! SRL not only provides the philosophical impetus for my work in Logan (2020), it also provides the core technical innovation I used there. *This* part can be traced back to a single sentence of SRL:

> It is clear that given any two pieces of information, X and Y, we may put them together to form a new piece of information, $X \cup Y$, containing all the information in X together with all the information in Y.

The crucial idea introduced here is the idea that semantic notions might be interpreted using a binary operation on indices. This can be contrasted with, e.g., Kripke frames, where one uses a binary *relation* on indices instead.

Already in 1972, Urquhart had worked out much of the semantic picture this led to. In 1974, Kit Fine gave semantic theories very similar to Urquhart's that covered an even more diverse family of logics.[2] There is much to be said about both what stays the same and what changes in the move from Urquhart's semantics to Fine's. But for the sake of the story I'm telling, it suffices to highlight just one key interpretive difference: where Urquhart understood the indices in his models to play the role of *pieces of information*, Fine took the indices in his models to play the role of *theories*.

Appealing to theories rather than pieces of information gave rise to an immediate problem: if we understand indices to be *theories*, how should we interpret the binary operation we've defined on them? Urquhart, thinking of his indices as pieces of information, interpreted the binary operation he defined as set union. This isn't an option on Fine's theory-based understanding of the semantics since in general the union of two theories is *not* a theory. The alternative Fine proposed was to read the binary operation as the operation of closing one theory under another.

Fine's proposal looks nice from a distance, but on closer inspection can be seen to suffer from a rather alarming problem. If we write '$cl_{t_1}(-)$' (where we read '$-$' as 'blank') for the operation Fine reads as closing under t_1, then it will in general *not* be the case that $t_2 \subseteq cl_{t_1}(t_2)$ and it will in general *not* be the case that $cl_{t_1}(cl_{t_1}(t_2)) = cl_{t_1}(t_2)$. Thus what Fine was calling 'closure' was not, in fact, closure in any traditional sense. This left it unclear what the right philosophical interpretation of the binary operation at play in Fine's semantics should be—assuming there to even be one! The solution to this problem was floating in the air already by the time Fine's paper appeared in print. But the (to my mind at least) best presentation of it did not appear for another two decades.

[2] See Fine (1974).

In 'A General Logic,' John Slaney revisited the Urquhartian interpretation of logic as involving bodies of information.[3] Like Fine and Urquhart, Slaney equipped these sites of evaluation with a binary operation. Unlike either, he interpreted this operation as what he called the *application* of one theory to another.[4] Slaney characterized this operation by saying that when we apply the information t to the information u, we use the resources of t to say what inferences are available to us, and we use the resources of u to determine what is available for the inferences to act on. But this idea makes just as much sense in theory-land as it does in information-land: applying t to u roughly means taking t as one's *theory of entailment* and taking u as the theory to which the theory of entailment is applied. This *strongly* resonates with the Urquhartian idea to only examine inference 'as it occurs set against a background'. It also gives us a philosophically palatable way of understanding the binary operation at play in Fine's semantics.

The theory I proposed in Logan (2020) is a version of the theory these observations naturally lead us to. In particular, the *deep fried semantics* I presented there used a binary operation on a set of indices that I interpreted as theories. The binary operation was interpreted as *application* à la Slaney's proposal. After a few simplifying assumptions, the deep fried approach landed us at the well-known substructural relevant logic RW.

Thus deep fried semantics, which plays an important role in the discussion below, enjoys at least a deep spiritual connection to Alasdair Urquhart's work—albeit a connection that is mediated by Fine and Slaney along the way. But it's also true that deep fried semantics is a direct reaction against Beall's aforementioned arguments. In broad strokes, what Beall argues is that logic is that which is universal with regard to our theory building endeavors. Formally, Beall works this out by supposing we have on hand a range of consequence relations and then seeing what they all agree on. The result, after some technical finagling, is that the universal theory building toolkit (and thus, for Beall, *logic*) is the weak, purely extensional relevant logic FDE.

I agree with Beall that we formally minded folk ought to be examining what we might call the universal theory building toolkit. Unlike Beall, I will (in this paper at least) remain silent about whether we should take this theory to be logic. After all, regardless of whether the universal theory building toolkit is logic, it's clearly a philosophically important thing to be studying. But the question of whether one should take a stand on the issue of universal theory building toolkit being logic is a relatively minor thing to disagree about, so I take it that, philosophically speaking, Beall and I are largely on the same page.

Technically, however, the matter is different. In particular, I disagree with Beall about the universal theory building toolkit being FDE. As I pointed out a moment ago, following the Urquhart-Fine-Slaney route will land us at a substructural logic. Beall identifies FDE as the universal theory building toolkit, so is committed to saying

[3] See Slaney (1990).

[4] As mentioned, this idea was in the air well before this piece, and Slaney acknowledges this in his paper. He explicitly cites Giambrone (1985), Meyer (1975), Meyer and Routley (1972), and Dunn's work on relevant implication in this regard.

the universal theory building toolkit is *not* substructural. This disagreement—the disagreement about structural rules—is the focus of the paper you're reading.

The source of this disagreement is a disagreement about the nature of the consequence relations. Beall says very little about them. I claim, and this is central to the work below, that consequence relations are themselves given to us by *theories of entailment*, and that these theories of entailment are theories that are as worthy of a spot at the table as any other theories might be. It's exactly in working our how to make room at the table for these theories that we find ourselves in need of the Urquhart-Fine-Slaney approach and, as a result, find ourselves embracing a substructural system. It also turns out, as I take time to remind readers at the end of this paper, that we have good, paradox-avoiding reasons to adopt such a system anyways.

A note on presentation: I've chosen to write the paper in the form of a dialogue. This is because it is essentially a compilation of dialogues I've had the pleasure to be a part of over the last few years of presenting this view—though, as it turns out, I respond much more nimbly in print than I do in person. Among the interlocutors worth personally thanking, Jc Beall deserves the most recognition, followed immediately by Eric Carter, Nathan Kellen, and Graham Leach-Krouse.

8.1 Some Troubling Inferences

Setting: It's early 2020, during the Covid-19 pandemic and global quarantine. An online meeting of the Logic Supergroup has just ended. One by one, faces disappear from the screen as people sign off. Soon only SL (a Substructural Logician) and TL (a Traditional Logician) remain on the screen. They're old friends, so a conversation naturally starts up.

TL: What an excellent talk! The speaker certainly had me convinced.
SL: I also enjoyed it, though I'm a bit less convinced. I detected a few troubling inferences in the middle of her talk that I'm not sure I can endorse.
TL: Really! I was paying close attention and didn't notice any such thing. What were the offending inferences?
SL: There were three inference forms that she seemed to be relying on that worried me. First, she seemed to allow herself to infer $\phi \to \psi$ from $\phi \to (\phi \to \psi)$. Second, she seemed to need to infer ψ from $(\phi \to \psi) \wedge \phi$. Finally, she seemed at several points to infer something of the form $(\phi_2 \to \phi_3) \to (\phi_1 \to \phi_3)$ from something of the form $\phi_1 \to \phi_2$. Or at least, this is how things seemed to me. I'm open to having misunderstood her.
TL: That's always a possibility, of course. But would it really be so bad if you were right? I don't see any issues at all with any of these inferences.
SL: Do you say that because you think she provided independent justification for these inferences being acceptable in the case at hand? Or is that you take these inferences to be generally acceptable in our theory building endeavors?

TL: The latter, I think, although I'm not sure what you mean by 'theory building endeavors'.
SL: I suppose I mean that you take these inferences to be part of the universal theory building toolkit.
TL: I'm still not sure I follow.
SL: Perhaps a toy example will help make things clear. Suppose we are studying creatures found in Eastern Australia. The members of one species we encounter—call it species X—have the following features:

(i) Members of species X have fur.
(ii) Members of species X are warm-blooded.
(iii) Members of species X lay eggs.

Suppose also that we're ill-equipped to the study we've set out to do and are, in fact, completely ignorant about Australia's fauna.
TL: Ill-equipped indeed!
SL: Right. But suppose we nonetheless set out to identify what phylogenetic class species X belongs to. If we do this using the information in an outdated biology textbook as our background theory, we may (in virtue of (iii)) come to the following conclusion:

(iv) Members of species X are not mammals.

If we use the information in a more up-to-date biology textbook as our background theory, we might instead come to the following conclusion:

(v) Members of species X *are* mammals—in particular, they are monotremes.

The lesson to learn here is this: what conclusions we can draw from a certain body of information we have on hand (e.g., from (i)–(iii)) will in general change when we change the background theory against which we draw said conclusions.[5]
TL: Ah, now I think I see where you're heading. You're pointing out that not *all* conclusions we can draw from a body of information exhibit this sort of background-dependence. For example, regardless of whether we use the outdated textbook or the up-to-date textbook, we will certainly be able to draw the following conclusions:

(vi) Members of species X have fur and members of species X are warm-blooded.
(vii) Members of species X have fur and members of species X lay eggs.

SL: Exactly this. It's worthwhile to state the contrast explicitly: (iv) and (v) followed from (i)–(iii) only in the presence of this or that background theory; (vi) and (vii) followed from (i)–(iii) simply in virtue of the fact that we were trying to build a *theory*, and not a mere *set of sentences*. Being a *theory* requires having a certain amount of regularity.

---

[5] Gillman Payette has pointed out to me that there are traditions where 'information' is used in such a way that one cannot have information that is inaccurate. Were that the way 'information' was being used here, it would seem that the outdated biology textbook wasn't supplying information. So be it; I hereby declare that I am using 'information' in a way that does not carry this connotation.

TL: Wait, hang on, I'm sorry. Shouldn't we be proceeding in the reverse order somehow? What I mean is that it seems odd to start out by thinking about theories and to work up from *there* to some sort of consequence relation.
SL: I don't think it's odd at all to proceed *from* theories *to* consequence relations and, indeed, that's just what I plan to do. The notion of a theory is, on the story I'm telling, the primitive notion in terms of which we will be defining other notions. This gives the story what I take to be a rather 'Brandom-ian' flavor, which I rather like.[6]

Regardless, recall that (as I mentioned earlier) I'm interested in what I called *the universal theory building toolkit*. This, as you might expect, is the collection of all inferences that are safe to use whenever we are building theories, no matter the subject matter of the theory. To put it another way, the universal theory building toolkit is the collection of inferences that all *actual* theories are invariant under.
TL: Invariant how?
SL: I imagine I'll have more to say about this later, so for the moment I'll just give a few hints. The basic idea is this: since theories exhibit certain amounts of regularity, it follows that there are inferences that, when applied to a theory, will keep you within that very theory, not matter what the theory is. Put in more mathematical terms, we can imagine inferences as functions from (sometimes sets of) sentences to sentences. A theory is *invariant* under a given inference when, no matter which of its sentences we put into the inference, the sentence we get out was already in the theory.
TL: Ok. This is a familiar-enough way of thinking about invariance. It's also clear enough that the set containing all those inferences that literally all the theories are invariant under is (a) a reasonable thing to call by the title 'universal theory building toolkit' and (b) an interesting thing worth studying.

All that said, I *still* don't see what's objectionable about the inferences in question. For example, no matter the context and no matter the subject matter of the theory we're building, it will *always* be acceptable to infer $\phi \rightarrow \psi$ from $\phi \rightarrow (\phi \rightarrow \psi)$, it seems to me. The other inferences seem similarly warranted. So I'm just not sure what's making you so grumpy.

8.2 Theory Building

SL: Let's first make sure we're on the same page when it comes to theory building. I recently read something that put it rather nicely; let's see if I can share my screen and show it to you.
A long, frustrating ordeal ensues. Some minutes later, a limited form of screen-sharing is achieved and ***SL*** *displays the following passage from* Beall (2018):

> [O]nce she has identified her target phenomenon (about which she aims to give the true and as-complete-as-possible theory), the task of the theorist is twofold:
>
> - gather the truths about the target phenomenon

[6] See Brandom (1998) or Brandom (2009).

- construct the right closure relation to 'complete' the true theory – to give as full or complete a true theory as the phenomenon allows.

TL: If I've understood the passage correctly, Beall thinks (and, I take it, you think) that theory building involves two intuitively different operations: the 'gather some truths' operation, and the 'close under a closure relation' operation. I'll admit I'd never taken the time to think hard about theory building, but it seems that Beall is correct to point out theory building does seem to depend on both of the operations he's identified.
SL: That's my reading as well. I also think that, when combined with a few uncontroversial theses, Beall's observation here is enough to give us good reason to claim that the inferences the speaker seemed to rely on can't be part of the universal theory building toolkit. So, unless she gave us some reason to think they're acceptable in the situation at hand (or I'm wrong about her relying on them), I think she's left her conclusion in some doubt.
TL: Hang on, I've lost track of the inferences you're troubled by. Would you kindly remind me what they were?
SL: Of course. Here, I'll write them on our shared screen.

Another long, frustrating ordeal ensues. Some minutes later, a limited form of screen-writing is achieved, though neither participant can write particularly elegantly on the shared platform. Still, ***SL*** *manages to write the following in mostly legible script:*

Rule 1 If $\phi \to (\phi \to \psi) \in th$, then $\phi \to \psi \in th$.
Rule 2 If $(\phi \to \psi) \wedge \phi \in th$, then $\psi \in th$.
Rule 3 If $\phi_1 \to \phi_2 \in th$, then $(\phi_2 \to \phi_3) \to (\phi_1 \to \phi_3) \in th$.

TL: I take it that 'th' here is an arbitrary theory.
SL: That's right.
TL: And, just to be sure I'm on board, what you're saying is that none of these rules should be in the universal theory building toolkit. So you deny that the kind of regularity demanded by theory-hood would require us to add ψ to all theories that contain $(\phi \to \psi) \wedge \phi$
SL: That's correct.

TL *makes an incredulous face.*

SL: I can see I have my work cut out for me when it comes to convincing you. I thought about it while we were settling our technical matters, and I think I only need the following four theses for my argument:

1. Background theories are theories like any other.
2. Within a context, the behavior of a given background theory is determined by features of the theory itself.
3. The only type of feature relevant to determining how a background theory behaves in a context is the sentences it is committed to.
4. The behavior of background theories under application is determined by a connective in the object language, rather than by something more mysterious.

TL: I see quite a bit I don't like here, my friend.
SL: Really? I'm surprised. Let's hear your objections to the theses first, then, before we turn to my argument, since it seems you're not likely to buy the argument if you don't buy these theses.

8.3 Application and Bunching

TL: While you were settling our technical problems and thinking about your theses, I took the time to skim the paper you mentioned above. The following passage is relevant to my first objection:
After a lag, the shared screen updates to show the following passage

> [T]heories are pictured as pairs (to highlight the closure relation):
>
> $$\langle T_1, \vdash_{T_1}\rangle, \langle T_2, \vdash_{T_2}\rangle, \langle T_3, \vdash_{T_3}\rangle, \ldots, \langle T_n, \vdash_{T_n}\rangle$$
>
> Logic shows up in each such theory-specific consequence relation $\vdash_{T_i}$; it is the relation under which all true theories, so understood, are closed; it is the relation on top of which all closure relations for our true theories are built.[7]

SL: Ah yes. I suppose I should have mentioned that what I've called 'the universal theory building toolkit' Beall has called 'the universal closure relation'.
TL: That's helpful, but it's not your *vocabulary* that's bothering me. What's bothering me is that your theses are all about background theories, and I see no mention of them here at all!
SL: Oh well that's also easy enough to clear up: what Beall calls 'closing under a closure relation' I call 'applying a background theory'. Does that make the connection clear enough?
TL: You've settled one of my worries only to raise another! Your first thesis says background theories are theories like any other. Now you're saying that background theories are closure relations. But this is clearly incorrect—a closure relation is a relation; a background theory is a theory. And (though I thought it went without saying) relations aren't theories and theories aren't relations.
SL: I see now I've been incautious in stating my position. Mea culpa. I don't want to say that background theories and consequence relations are the same thing; I want to say that the *operation* Beall has labeled 'closing under a closure relation' is an instance of the *operation* I call 'applying a background theory'.
TL: Ok good. At least you're not trying to convince me that relations are theories! That gives me some consolation. I'll admit, though, that it's not a lot, since it's not at all obvious to me that closing under a closure relation is an instance of whatever operation 'applying a background theory' is.

[7] Beall's restriction to *true* theories strikes me as problematic, but in ways that are orthogonal to this paper. Thus, my interlocutors will put the issue aside.

SL: It might help to think about what you do in your logic classroom. My guess is that you do something like the following: First, you tell your students that ϕ follows from Γ just if in no possibility is everything in Γ true while ϕ is untrue. Then you spell out in detail what the possibilities are and how to determine what's true in any one of them.
TL: That's exactly what I do. Given all this, students can then determine, for particular Γ and ϕ, whether ϕ follows from Γ—which is to say, whether ϕ is in the closure of Γ under the given consequence relation. No mention of theory application is required.
SL: On the contrary: there's an entirely clear case of theory application going on! When you teach your students what 'follows from' means and what the possibilities are and all that jazz, what you've done is teach your students a particular theory—namely, the classical theory of entailment. Closing a given set of sentences under the classical consequence relation is exactly the same thing as applying the classical theory of entailment, taken as a background theory, to determine what follows (classically) from the sentences at hand.
TL: Ah, I see. You take the theories of entailment themselves to be the theories that are being applied. I'll grant you this is *a* way of describing what goes on in my logic classroom. But this doesn't absolve you of wrongdoing, my friend. Theories of entailment will be theories that, perhaps among other things, contain sentences of the form 'this entails that'. These sentences are metalinguistic because 'entails' is not an object language expression. By saying that theories containing these sorts of expressions are theories like any others, you're mixing levels—that is, you're confusing metalanguage and object language.
SL: What can only be expressed in the metalanguage is determined by the choices we make about what to include in the object language. For example, it's not somehow determined in advance that conjunction has to be an object language device. We can build languages where conjunction is missing. Were we discussing such a language 'and' would be a metalinguistic, rather an object-linguistic notion.

So what's relevant isn't what *is in fact* in the object language and in the metalanguage, but what we *should* include in the object language and in the metalanguage. Given that we're talking about the universal theory building toolkit, and that one of the theory building operations (application of one theory to another) *crucially* interacts with entailments, including entailment in the object language is a no-brainer.
TL: I don't know if I'd call it a no-brainer, but perhaps what you're saying is this: if we suppress the entailment connective, then it seems we have no recourse except to include it in the metalanguage. But since what follows from a given family of sentences does in fact depend on what background theory we are using, we will then need multiple entailment relations in our metalanguage. This leaves open a natural question: how do we in fact come to employ these different entailment relations? It seems there are two answers we might give (a) the entailment relations are just there—they're floating in the ether somehow, or perhaps we consult an oracle each time we put them to work—or (b) they're actually spelled out in some concrete way. Option (a) is, for obvious reasons, unappealing. But in case (b), we have to recognize that what's being spelled out is a *theory* governing the behavior of the entailment

relation at hand—in other words, what's being spelled out is a *theory of entailment*. And this, you're claiming, is a genuine theory that operates 'in the background'.
SL: That's a much better way to put it.
TL: Thanks! I'm not sure I'm totally convinced, but I'll at least grant that you've made a plausible case for the thesis that background theories are theories and that background theories and the background 'role' more generally are things we need to keep track of. Now I'm curious to hear what this thesis does for you—what role does it play in your argument?
SL: The most important thing to note is this: if background theories are theories like any other, it follows that background theories may themselves be constructed by applying a 'back background' theory to a 'forebackground' theory. Forebackground and backbackground theories, in turn, may have further internal structure, and this could go back arbitrarily (though presumably only arbitrarily *finitely*) far.
TL: Can you give an example of a backbackground theory? I'm having trouble picturing such a thing.
SL: Easy enough! On a toy way of understanding it, physics might be thought of as some axioms—say the axioms of quantum field theory or some such—together with their mathematical consequences. So physics, on this toy picture, is what we get when we apply our mathematical theory to a set of axioms.
TL: So far so good, but this is only two theories.
SL: Right. But we might also think of the mathematics we're applying here as itself being composed of some axioms (say, the ZFC axioms) together with all their logical consequences. So physics on this picture is what we get when we apply logic to mathematical axioms, then apply the resulting theory to our physical axioms.
TL: Oh I see. So in this example 'logic' shows up as a backbackground theory. I can see where forebackground theories and all the rest occur now, thanks.
SL: Excellent. So, like I was saying, what the first thesis forces us to realize is that theories can be constructed in very complex ways.
TL: I expect you have a candidate family of structures in mind that can track these different ways theories can be built.
SL: Right again. This time we'll steal a definition from Stephen Read.[8] Define the term 'bunch' together with the auxiliary terms 'I-bunch' and 'E-bunch' by simultaneous recursion as follows:

- Any sentence is a(n atomic) I-bunch.
- Any set of I-bunches is an E-bunch.
- I-bunches and E-bunches (and nothing else) are bunches.
- If X and Y are bunches, then $(X; Y)$ is an I-bunch.

TL: It'll take a minute for me to fully grok this definition. In the meantime, I have another question: how should I read the 'I-bunching' and 'E-bunching' operations?
SL: The idea is that bunches name theories. The atomic bunch 'ϕ' names the smallest theory containing ϕ. The theories named by more complex bunches are fairly straightforward:

[8] See Read (1988).

- '$\{X_i\}_{i\in I}$' names the smallest theory that contains the theories named by each X_i;
- '$X; Y$' names the theory we get by applying the theory named by X to the theory named by Y.

TL: I think I get the picture. I only have one question left: suppose X and Y are bunches, and let $t_{X;Y}$ be the theory named by $X; Y$. Presumably, since $t_{X;Y}$ is a theory, it's just a set of sentences. And sets of sentences are perfectly good bunches. So $t_{X;Y}$ is also a bunch. But then $X; Y$ and $t_{X;Y}$ are different bunches that name the same theory. Why do we want or allow this?
SL: This is an excellent question. My answer will, in some sense, just pass the buck: facts about how a theory *can be* built are matters of concern even in classical logic.
TL: Can you show me this 'concern' you speak of in action?
SL: Easily. Think, for example, of the usual conditional introduction rule:

$$\frac{X \cup \{A\} : B}{X : A \supset B}$$

TL: Wait, don't tell me! I think I see where you're going. Using your theory-goggles, you'll read 'X' as 'the theory containing all the formulas in X' and you'll read '$X \cup \{A\}$ as 'the theory containing A as well as all the formulas in X'. Thus, the fact that we *can build* a theory that (contains? entails? I'm not sure what word you'll use here) B by adding A to X is what tells us that the theory generated by X contains/entails/whatever $A \supset B$.
SL: Correct! And 'contains' will suffice. I'll say a bit more about that while dealing with the second thesis—speaking of which, are you ready to move on to the second thesis?
TL: Indeed.

8.4 Closure

SL: As a reminder, my second thesis is that, within a context, the behavior of a background theory is determined by features of the theory itself. The point, I should make clear, is that once we know what our background theory is and we know the context in which it's operating, we know everything we need to know about what it's going to do. We don't, in other words, have further variance about the turnstile, as Beall allows.
TL: That's what I took you to be saying, but I don't see how it can be correct! Are you honestly telling me that I can't vary the turnstile, as Beall does? Surely that's wrong, and wrong for reasons Beall is very clear about—in different theory building endeavors, we just *do* use different consequence relations.
SL: Well first, a note about what 'context' means. I don't think you've misrepresented what I mean, but I can easily see how what you're saying might be misunderstood, so I thought we should be clear.

A given theory t can occur as a foreground theory, as a background theory, as a black background theory, etc. In general, given a bunch Γ, we can highlight (pick out, point at, whathaveyou) a given occurrence of the theory t within Γ. Supposing we write $\Gamma(t)$ for Γ with t highlighted in this way, then $\Gamma(-)$ (to be read 'Γ blank') is the context in which t occurs.

TL: Indeed, that's what I had in mind. I'm glad we're on the same page.

SL: Great! Then my answer to your question is that every instance of turnstile-variance examined by Beall is just an instance of context variance in this sense.

TL: Can you elaborate on that a bit?

SL: Of course. Recall that Beall writes '$\Gamma \vdash_{T_n} \phi$' to mean that ϕ follows from Γ according to the consequence relation at play in theory T_n. In brief, ϕ is a T_n-consequence of Γ.

TL: Correct. And he takes this to be synonymous with 'ϕ is in the T_n-closure of Γ.' Now, you've said that closing under a closure relation is the same thing as applying a theory of entailment. How does that play out here?

SL: It couldn't be easier! Let's let c_n be the theory of entailment underlying the closure relation $\vdash_{T_n}$. Then where Beall writes $\Gamma \vdash_{T_n} \phi$, I simply write $c_n; \Gamma \vdash \phi$. Here the 'naked' turnstile is simple containment—so $c_n; \Gamma \vdash \phi$ simply means that the theory we get by applying c_n to Γ contains the sentence ϕ. Given everything that I've said so far, you should now be able to tell that $\Gamma \vdash_{T_n} \phi$ and $c_n; \Gamma \vdash \phi$ do in fact mean the same thing.

But despite having the same meaning, I think that the latter way of saying things is better. In particular, when we write $\Gamma \vdash_{T_n} \phi$, the theory c_n that we're using to define that consequence relation $\vdash_{T_n}$ is being squirreled away somewhat dishonestly in a subscript. When we write $c_n; \Gamma \vdash \phi$, in contrast, both the theory c_n and the role it is playing are being proudly acknowledged.

TL: I suppose being explicit about what we're doing is praiseworthy, sure. I'm not sure I'd go so far as to call Beall's use of subscripts 'dishonest squirreling away' of background theories, but that's a matter I'll let you take up with Beall. On that note, I'm wondering if you can make sense of what Beall *does* do? If so, I'd like to see it as it might help me make sense of what purpose your machinery is serving.

SL: I think I can. From my perspective, Beall seems to be interested in the question 'what follows from t no matter what background theory we use?' Stated otherwise, Beall examines the consequences of $b; t$ as we allow b to vary among all possible options. By cataloguing the set of inferences that are valid in this way, we get a theory that is universal in a certain minimal sense. The issue I have is that the framework I've provided makes his theory look odd in at least three ways.

TL: One of them I can already see: now that we have your bunching apparatus on hand, it's not clear why we should restrict our attention to those bunch-sentence inferences where the bunch in question has the form $b; t$. How else does it look odd to you?

SL: The first thing that bothers me concerns vocabulary. As a result of his suppression of background theories, Beall's account is left with no reason to take seriously connectives that directly interact with the application operation. I'll have more to say about this below, though, so I'll rush on and tell you the final thing that bothers me

as well: not only does Beall restrict to bunches of the form $b; t$, he *also* requires that b be the type of theory that, when applied, results in a *closure relation.*
TL: Do you mean 'closure relation' in the Tarskian sense?
SL: Yes. The part I'm particularly grumpy about is that Beall has restricted the range of background theories he's examining to those that have the feature that for all t, we get both of the following:

- $b; t \supseteq t$;
- $b; (b; t) = b; t$.[9]

This is a further restriction on the sorts of theories Beall allows to play the 'b' role in '$b; t$'.

Altogether the point is that we should embrace a broader sort of universality than Beall has allowed. Our aim should be to describe what tools we can use when theory building *no matter the theory*. Our aim should not be to determine, as Beall did, what tools we can use to determine what follows from a given theory as we vary which closure-relation-generating background theory we apply to it.

8.5 Commitment and Containment

TL: I'll have to wait and see how the details pan out before I can evaluate all of that. I'm anxious to get to the good bits, though, so let's try to blast through these last two theses. First up is your claim that the only feature relevant to determining how a background theory behaves in a context is the sentences it is committed to.
SL: Looking at a toy example might help make clear what I have in mind here. So consider, e.g., all the information we had about the motions of the planets as of 1900 or so. Now consider what follows from these when we, first, use classical mechanics as our background theory and, second, use relativistic mechanics as our background theory. It's well known, and a matter of import in the history of science, that the results will differ. We needn't worry about the exact nature of the difference. What's important for us is that it's not *inexplicable* why, even applied to the same data, these theories give different predictions. It's because the theories have different *commitments*.
TL: Thanks! That was quite helpful. The claim you seem to be making looks to have two components:

- Theories that behave differently do so because they actually differ in some way, and
- Given the kinds of things theories are, the only way they can differ is in what sentences they're committed to.

The first component I'm happy to endorse, but in order for me to endorse the second component, I'll first need to know more about the notion of *commitment* at play here.

[9] There is a mild abuse of language here—strictly speaking we ought to say that, e.g., the theory named by $b; t$ is contained in the theory named by t, etc.

SL: Easy enough—it's the same thing as containment. Theories just are sets of sentences.[10] I hope you'll agree that on this reading of commitment, your second component is entirely unobjectionable.
TL: Ah yes, you'd mentioned this already. I suppose that, given this reading, the second component *is* unobjectionable. But I'm a bit confused here—I thought we adopted bunches because theories *couldn't* be described as mere sets.
SL: Not quite. We adopted bunches because there were more *theory building operations* than the binary '$\in$' relation of set theory was capable of tracking (easily). At any rate, since theories are sets of sentences, it's hard to imagine what an objection to the third thesis would look like. It seems the only forceful way to object would be by presenting a counterexample—two theories that differ in behavior without differing in commitments. But if the theories don't differ in commitments, I wouldn't know how to tell them apart in the first place!
TL: Agreed. Let's move on to the final thesis, that the background behavior of a theory is determined by a connective in the object language, rather than by something more mysterious. We've already agreed that all behavior—and thus, in particular, the background behavior—of theories is determined by their content. What this thesis *seems* to be trying to do is point us in the direction of the exact sort of content we need to look at if we are interested in understanding how a theory is going to behave when we use it in the background. Apart from that, I'll admit I'm not completely sure about what it means.
SL: The point I'm trying to make with this thesis is that there is a certain *type* of sentence that we need to look to when we are determining how a theory behaves when used in the background.
TL: Can you say more? I'm still a bit at sea here.
SL: I suppose what I'm picturing is this: let t_1 and t_2 be theories. Suppose ψ is in the theory we get from applying t_1 to t_2. Presumably this happens because t_2 contains some bit of information ϕ that t_1 states is sufficient for ψ. This connection *seems* well-modeled by a connective that we write '$\phi \to \psi$'. All I'm claiming in my fourth thesis is that this idea—the idea that some '$\to$-like' connective mediates the way a theory behaves when we use it as a background theory—is correct. Given the way it behaves it's natural to call this connective the *entailment* connective and to call sentences dominated by this connective *entailments*. Thus, my fourth thesis can equally (but more succinctly) be put by saying this: if we want to know how a given theory is going to behave when used as a background theory, all we have to look at are its entailments.
TL: I don't see anything objectionable here, but again, I'd like to test my understanding. So, given what we've said so far, the application of a theory t_1 to a theory t_2 ought to be the theory we get from the following set of sentences:

$$\{\psi : \phi \to \psi \in t_1 \text{ and } \phi \in t_2\}$$

[10] This is, I acknowledge, a very logician-y perspective to take on theories. There is, I think, room to take a more semantic view of theories—or at least, there will be room for such a thing once we have semantics in hand.

SL: That's exactly right! t_1's entailments (→-formulas) determine how it behaves when used, as a background theory. And the way they determine its behavior is exactly as you've shown—they simply tell us what follows from what's in a given theory.
TL: I take it then that, on your point of view, any system that lacks a connective that behaves in this way is only telling part of the story when it comes to universal theory building.
SL: That's correct! As I'm sure you recognize, this ties back to one of the criticisms I had of Beall's account from before: since the story Beall told completely ignored background theories, it naturally lacked any such connective.
TL: Got it. Now I'm interested to know what this fourth thesis does for you.
SL: The main thing it does is suggest what the semantic clause for the entailment connective ought to be.
TL: Wait, wait. Are you saying you can produce a full-fledged semantic theory based on these ideas? Talk about burying the lede! Let's see what this theory looks like; whatever objections I have left will certainly be easier to make once the actual theory is on the table.
SL: Very well. But first, a warning: what I'm going to give is *a* way of turning the ideas we've discussed so far into a semantics. The semantic clauses I'm about to show you are emphatically *not* here because I think they in fact capture the universal theory building toolkit in all its glory. The point is only to give a proof of concept; that is, to show that we can give a semantics that accords with the above intuitions, and that we could use such a thing to figure out what the universal theory building toolkit might be. If you think these are the wrong clauses, then you likely think they also lead us to the wrong universal theory building toolkit. That's fine. What matters is that you can see how to get from the basic picture involving theories and theory building to a semantics and from there to a universal theory building toolkit.
TL: Noted.

8.6 Semantics

SL: Ok great! So, first things first: let's say a **model** is a quadruple $\langle T, \ell, \sqsubseteq, \circ \rangle$ with

- T a set of indices that we will think of as playing the role of *theories*;
- $\ell \in T$ is an index that we will think of as playing the role of the *universal* theory;
- $\sqsubseteq$ a partial ordering of T that we will think of as telling us when one theory *extends* another;
- $\circ$ a binary operation on T that we will think of as playing the role of application of one theory to another.

Given the roles we want $\sqsubseteq$ and $\circ$ to play, we should impose a few conditions on their interactions. In particular, if $a \sqsubseteq b$, then b should be a 'bigger' (or rather, shouldn't be a smaller) theory than a—that is, everything in a is already in b. So if we apply

the same theory, say t to both a and b, then we should not end up with less in the b-case than we did in the a-case. So whenever $a \sqsubseteq b$ is true, we expect $t \circ a \sqsubseteq t \circ b$ to also be true. A similar line of reasoning leads us to expect that whenever $a \sqsubseteq b$ is true, $a \circ t \sqsubseteq b \circ t$ will also be true.
TL: So far so good. But you'll also want to require $\circ$ to have the following features, right?

Commutativity $a \circ b = b \circ a$.
Associativity $(a \circ b) \circ c = a \circ (b \circ c)$.
Idempotence $a \circ a = a$.

SL: I want nothing to do with any of these! I'll admit that they all *look* friendly enough. But think about what they say and you'll see they're far from innocent.
TL: Alright, let's see. We wanted $\circ$ to play the role of application. So $a \circ b$ is, intuitively, the theory generated by taking a as background theory and applying it to b taken as foreground theory. Thus in $a \circ b$, it's a that is giving us our theory of entailment. In an abuse of Beall's notation, we might think of $a \circ b$ as the theory we get by closing b under the closure relation $\vdash_a$. And now I see that you're correct—there's in general no good reason to suppose that closing b under $\vdash_a$ gets us the same theory as we get by closing a under $\vdash_b$.
SL: Excellent reasoning! Of course, I'm worried that '$\vdash_a$' and '$\vdash_b$' (to the extent that I endorse such things at all) won't even be closure relations in the traditional, Tarskian sense. But that's a different point altogether. In any case, Associativity and Idempotence turn out to be equally suspicious when subjected to a similar inspection.
TL: I feel like once we work out *all* the details these problems will go away, but I can certainly see how they're at least *prima facie* problems, so I'll go along with things for the moment. I suppose the next thing you'll tell me is how to get from models to truth?
SL: Not quite! There's another matter we need to address first: let's say that a is *b-invariant* when $b \circ a = a$. The thing we're interested in is the *universal* theory building toolkit—the tools we can use in all our theory building.

The reason I'm bringing this up is the following: we want ℓ to be a theory that *every* theory is invariant under. Thus ℓ will be universal among theories in the sense that applying ℓ always only gets us all and only the stuff we already have. So if we use the tools in ℓ in our theory building, we will for sure not be going astray—it will contain only those tools that all actual theories are invariant under. All told this suggests that we demand ℓ be a left-identity for $\circ$—that is, we require that $\ell \circ t = t$ for all t.
TL: I see! ℓ, then, is the index at which exactly those formulas that belong to the universal theory building toolkit turn out true, yes?
SL: Yes. And truth, as you probably expect, requires valuations.[11] To make things easy, we'll take valuations to record both which atomic sentences are true-in-a-theory

[11] It would be better to speak of commitment/anti-commitment or some other more theory-centric notion rather than truth. But it's so much easier to think and speak in truth terms that we will do that instead.

($\models_1$) and which atomic sentences are false-in-a-theory ($\models_0$).[12] Thus a *valuation* is a function from theories to functions from atoms to $\{\{1\}, \{0\}, \{0, 1\}, \emptyset\}$. Intuitively, valuations record which atoms a theory *says are true* and which atoms it *says are false*. The only condition we should require is that these functions be *monotonic*: if $t \sqsubseteq s$ is supposed to mean that s extends t, then if p is true (false) in t, it should be true (false) in s.

TL: I see where we're going now. Let's see if I've got it right: we can extend valuations to complex sentences using the following clauses:

($\neg_1$) $t \models_1 \neg\phi$ iff $t \models_0 \phi$;
($\neg_0$) $t \models_0 \neg\phi$ iff $t \models_1 \phi$;
($\wedge_1$) $t \models_1 \phi \wedge \psi$ iff $t \models_1 \phi$ and $t \models \psi$;
($\wedge_0$) $t \models_0 \phi \wedge \psi$ iff $t \models_0 \phi$ or $t \models_0 \psi$;
($\rightarrow_1$) $t \models_1 \phi \rightarrow \psi$ iff ...

Hang on wait a minute. I can't quite see how ($\rightarrow_1$) and ($\rightarrow_0$) should go.

SL: Indeed! This is where things get a bit tricky. Also tricky: we need not only clauses for entailments, but also clauses for both of the bunching operations before we're done!

TL: Of course! Let's see if I can figure out some of these. Hmmm...well, E-Bunches were supposed to be formed by just lumping some things together. So intuitively, an E-bunch should be true just when each of its members is true and false when at least one of its members is. So we ought to have these:

(**E**$_1$) $t \models_1 \{X_i\}_{i \in I}$ iff for all $i \in I$, $t \models_1 X_i$;
(**E**$_0$) $t \models_0 \{X_i\}_{i \in I}$ iff for some $i \in I$, $t \models_0 X_i$.

Altogether that leaves four clauses I still need some help with: ($\rightarrow_1$) and ($\rightarrow_0$) and (**I**$_1$) and (**I**$_0$)

SL: Two of these are, once you see them, not too bad. Here they are

($\rightarrow_1$) $t \models_1 \phi \rightarrow \psi$ iff for all u, (i) if $u \models_1 \phi$ then $t \circ u \models_1 \psi$ and (ii) if $u \models_0 \psi$, then $t \circ u \models_0 \phi$;
(**I**$_1$) $t \models_1 X; Y$ iff there are u and v so that $u \circ v \sqsubseteq t$ and $u \models_1 X$ and $v \models_1 Y$.

TL: Let's see if I can make sense of these. ($\rightarrow_1$) says that a theory makes an entailment true just when, every time we apply it to a theory that makes its antecedent true we end up with a theory that makes its consequent true and every time we apply it to a theory that makes its consequent false we end up with a theory that makes its antecedent false.

SL: Correct!

TL: All that seems fine. (**I**$_1$) is a bit trickier, but I think I can see what's going on there too. It's probably important to recognize that both '$\circ$' and ';' are analogues of application—the former being the semantic analogue of application and the latter

[12] This makes things 'easier' only in a very attenuated sense. In particular, by use of this device, we don't have to detour through an explanation of the so-called Routley-star, which would distract us from the primary purpose of this paper.

being the proof-theoretic analogue of application. So if $u \vDash_1 X$ and $v \vDash_1 Y$, then the application of u to v ($u \circ v$) should make true the application of X to Y ($X; Y$). Thus clearly $u \circ v \vDash_1 X; Y$. But ($\mathbf{I_1}$) says not only that $u \circ v$ will do this, but also that anything extending $u \circ v$ will, and that's puzzling me a bit.
SL: Here's a hint: a fairly straightforward induction shows the monotonicity we demanded for atomic formulas can be extended upwards.
TL: Oh excellent! So then yes, given monotonicity, all and only those theories t that extend something of the form $u \circ v$ where $u \vDash X$ and $v \vDash Y$ will make true $X; Y$. What fun! I think I'm on board.
SL: I'm impressed—you're picking things up quite quickly! The remaining two clauses are fiddly. There are explanations of them in Logan (2020), but there's really no great reason for us to dwell on them, so let's just let them pass in silence for now. I grant that doing this feels at least a little bit icky. But it helps to recall again that I'm not aiming to do much by presenting this semantics. All I'm trying to do is demonstrate that we *can* build a semantic theory that captures the intuitions we have so far. You may think the details here aren't quite right, but that's ok with me—we can meet up later to work out the details. What matters is that you see the framework can be made to work. Also don't lose track of what I'm trying to convince you of: nothing more than that the universal theory building toolkit is substructural. In particular, I'm *not* interested in arguing that this or that *particular* substructural theory is the universal theory building toolkit.
TL: You're right, you're right. ($\rightarrow_0$) and ($\mathbf{I_0}$) are small fries. What we're *supposed* to be talking about are the inferences you labeled Rule 1, Rule 2, and Rule 3. Before we deal with those, I suppose you'll need to tell me what validity amounts to, though I suspect I know what you'll say.
SL: It's not much of a surprise—a formula is valid in a model just when it's true at ℓ in that model. A formula is valid simpliciter just when it's valid in every model.

8.7 Application and Containment

TL: That's what I though. Let's return now to the matter of the Rules. I can't help but think that their acceptability is connected somehow to the worries I raised a moment ago about conditions on the application operation that you've used '$\circ$' to refer to in your semantics.
SL: Such rules are indeed intimately connected to the behavior of the $\circ$ operation, so I'm not surprised to hear you say that. Still, I'd like to hear you say a bit more about your worries before we go down that road.
TL: I think what's bothering me is something like this: in the semantics you gave above, you allow for $t \circ t$ to be something other than t. And, I take it from what you say that this isn't an accidental feature of the semantics, but an intentional feature.
SL: Correct! In fact, the semantics allows both for $t \circ t \not\sqsubseteq t$ and $t \not\sqsubseteq t \circ t$. Intuitively, this corresponds to the fact that there are theories that *grow* when applied

to themselves and to the fact that there are theories that *shrink* when applied to themselves.

TL: I'm trying to understand how this could possibly be the case, and I think I see what's bothering me. Let's say a theory is nonidempotent when it *grows* when applied to itself.

SL: Nonidempotent is exactly the sort of ugly word I'm happy to take in; go on.

TL: Well here's the thing: suppose th is a nonidempotent theory, and write th^2 for the result of applying th to itself. Since th is nonidempotent, $th^2 \not\subseteq th$. But th^2 is, given what we've said before, $\{\psi : \phi \rightarrow \psi \in th \text{ and } \phi \in th\}$. Thus, since $th^2 \not\subseteq th$, for some ϕ and ψ we will have all of the following:

- $\phi \rightarrow \psi \in th$, and
- $\phi \in th$, but
- $\psi \notin th$.

SL: That's correct!

TL: Well I think it goes without saying that any such 'theory' is degenerate and ought to be rejected—if you're not closed under modus ponens, you just not a theory at all!

SL: I completely disagree; there's nothing at all degenerate about such a th! Remember: entailments determine a theory's *background* behavior. We spelled this out in the semantics above by saying that th makes true $\phi \rightarrow \psi$ iff any time we *apply* th to a theory where thus-and-so happens, such-and-such occurs. The details aren't super important. What matters is that $\phi \rightarrow \psi$ and ϕ don't interact when minding their own business inside th. To get them to play together we have to *apply* th to th. So if you could build a theory th so that $\phi \rightarrow \psi \in th, \phi \in th$, but $\psi \notin th \circ th$, then *that* would be a problem. But what you've pointed out here as a bug is actually just a feature.

TL: I'm not convinced the bug isn't a bug. The problem—and now I'm finally going to say what I've wanted to say all along—is that you've misunderstood what it means to take th as a background theory. All it means to take th as a background theory is to restrict our attention to situations compatible with th. If you want to think about it model-theoretically, this means all we need to do is restrict our attention to models that verify every member of th. So when we talk about, e.g., 'the th-consequences of f' what we're interested in is really just the consequences of f that are compatible with th, and this just means we're interested in $th \cup f$. Ergo applying th to f is just forming the union of th and f.

SL: Well...

TL: Let me finish! You've already accepted that applying a theory committed to $\phi \rightarrow \psi$ to a theory committed to ϕ results in a theory committed to ψ. So if th is committed to $\phi \rightarrow \psi$ and to ϕ then the theory that results from applying th to th is committed to ψ. But I just pointed out that the theory that results from applying th to th is just the theory $th \cup th$, which is clearly just th. So th is, if committed to $\phi \rightarrow \psi$ and to ϕ, already committed to ψ.

SL: Your objection strikes right to the heart of the matter. As you probably expect, I reject your claim that application is set union. But this follows from what I've already said: we cannot accept any account of the universal theory building theory that forces us to accept Rules 1–3.

TL: Wait just a minute, I'm confused—I don't see the connection between my set-theoretic reading of application and these rules. I meant the objection to just be an objection to what I took to be a core part of your semantics.

SL: As I mentioned a moment ago, features of the application operation are tightly connected to these kinds of rules. So it shouldn't be too surprising that your proposal to read application as set union leads us to accept such rules. Nonetheless, let's have a look to see how this plays out.

TL: Great! Let's focus on the rule that says we can infer ψ from $(\phi \rightarrow \psi) \wedge \phi$.

SL: Alright then. Suppose we've adopted your set-theoretic reading of application, and suppose also that we have on hand a theory that contains $(\phi \rightarrow \psi) \wedge \phi$. In the language we introduced a bit ago, that means we're working with some t such that $t \vDash_1 (\phi \rightarrow \psi) \wedge \phi$. Then $t \vDash_1 \phi \rightarrow \psi$ and $t \vDash_1 \phi$. Thus $t \circ t \vDash_1 \psi$.

TL: Ah I see now: if '$\circ$' names set union, then $t \circ t$ is the same thing as t. So then since $t \circ t \vDash_1 \psi$, we also have that $t \vDash_1 \psi$. Very good! I take it that it's also the case that were I to adopt the set-theoretic interpretation, I would be forced to accept the other two rules? And the reasons are similar?

SL: Indeed. The details are slightly more complex, but the basic principle is the same.

TL: Ok, we've established a connection between my claim that application is set union and the rules you're challenging. Are you finally willing to tell me why you have a problem with these rules?

SL: I am. But I'm warning you: my reason is quite boring.

TL: I have braced myself for boredom. Please proceed.

8.8 Theory Building Again

SL: Very well. According to Rule 1 (for example), every theory that contains $\phi \rightarrow (\phi \rightarrow \psi)$ also contains $\phi \rightarrow \psi$. But we can build theories do contain $\phi \rightarrow (\phi \rightarrow \psi)$ but do not contain $\phi \rightarrow \psi$. So we have to reject Rule 1. Thus in particular, we have to reject that application is set union. For similar reasons, we also have to reject the other two rules.

TL: I'm afraid you've left me a rather easy route out: I only need to reject that you can build such theories. Of course, I accept that you can construct *sets of sentences* that contain $\phi \rightarrow (\phi \rightarrow \psi)$ without containing $\phi \rightarrow \psi$. That's easy. But you can also construct sets of sentences that contain ϕ and contain ψ without containing $\phi \wedge \psi$. Just as you'd agree that the goal of building a *theory* rules out the latter, you must also agree that it also rules out the former.

SL: I agree with you: anything that counts as a theory will, if it contains ϕ and contains ψ also contains $\phi \wedge \psi$. Why do you think the same thing is true of $\phi \rightarrow (\phi \rightarrow \psi)$ and $\phi \rightarrow \psi$?

TL: Well let's see. It's not something I've ever had to justify before, but I guess it's something like this: when we set out to build a theory, our aim is to build *as-complete-as-possible* theories.[13] But note that whenever $\phi \rightarrow (\phi \rightarrow \psi)$ is true $\phi \rightarrow \psi$ is also true. So there's no harm in adding $\phi \rightarrow \psi$ to our theory if we've already added $\phi \rightarrow (\phi \rightarrow \psi)$ to it, since we cannot find a counterexample to one that isn't a counterexample to the other. And since there's no harm in adding it, there *is* harm in keeping it out, since doing so keeps out theory from being as complete as possible. Ergo we should concern ourselves with idempotent theories only.[14]

SL: This is a sophisticated objection. My knee-jerk response was to say that in fact there *are* counterexamples to $\phi \rightarrow \psi$ that are not counterexamples to $\phi \rightarrow (\phi \rightarrow \psi)$, and to then just build one using the semantics we built above.

But I can see that this isn't likely to move you. You'd simply say that this shows the semantics is wrong—it was, after all designed with the express intent of capturing the behavior of theories under application. To appeal to it in order to justify my account of how theories behave under application is thus flatly circular. So I'll need a different argument.

8.9 Curry

There's a long moment of silence while **SL** *gathers his thoughts while staring just to the left of the camera. Just as it starts to get awkward, he looks at the camera and smiles.*

SL: Ok, I think I see where to go. I'll start on your territory, where Rules 1–3 are all allowed.

TL: How kind of you.

SL: We'll see. Here's a fun fact: since we're accepting Rule 2, all theories are idempotent.

TL: Interesting! Let's see if I can see why this would be. Hmmm…well, let's suppose ψ is in the application of t to itself. Then there will be a $\phi \in t$ so that $\phi \rightarrow \psi$ is also in t. But then since $\phi \rightarrow \psi$ and ϕ are both in t, so is $(\phi \rightarrow \psi) \wedge \phi$—you've already admitted this rule. So by Rule 2, ψ is in t. And there we have it! Everything in the application of t to t is in t, so t is idempotent.

SL: Excellently done! Now suppose our theory building concerns truth in languages with enough arithmetic around to do some diagonalization. We won't need to get too technical. All we need is that we can build a 'Curry sentence'; that is, some sentence c so that for some sentence U that we're not willing to accept, the following hold:

$$\text{C1}: c \rightarrow (T\ulcorner c \urcorner \rightarrow U) \qquad \text{C2}: (T\ulcorner c \urcorner \rightarrow U) \rightarrow c$$

[13] This is an important consideration in Beall (2017), for example.

[14] Eric Stei presented a forceful version of this line of argument in response to a talk I gave on related material at the North Carolina Philosophical Society in 2018.

Since C1 and C2 are true, we might add them to our theory. Maybe we also observe that T1 and T2 are true and add them to our theory:

$$\text{T1}: T\ulcorner c \urcorner \to c \qquad \text{T2}: c \to T\ulcorner c \urcorner$$

TL: Well now I'm suspicious, but go on.[15]
SL: Thanks. We'll have time to deal with your suspicions later. I claim that if we include C1, C2, T1, and T2 in our theory, and require that the theory be idempotent, then it will also contain U. But we said above that U was something we were unwilling to accept.
TL: I'd like to see this argument!
SL: It's fairly short, so it's worth being quasi-formal about it. Here, let me write it:

***SL** scrolls to a blank page in the shared screen and, slowly and laboriously, scrawls the following:*

(1) By assumption, $T\ulcorner c \urcorner \to c$ is in our theory.
(2) So by Rule 3 applied to (1), $(c \to (T\ulcorner c \urcorner \to U)) \to (T\ulcorner c \urcorner \to (T\ulcorner c \urcorner \to U))$ is in our theory.
(3) By assumption, $c \to (T\ulcorner c \urcorner \to U)$ is in our theory.
(4) Since our theory is idempotent, we conclude from (2) and (3) that $T\ulcorner c \urcorner \to (T\ulcorner c \urcorner \to U)$ is in our theory.
(5) Rule 1 applied to (4) tells us that $T\ulcorner c \urcorner \to U$ is in our theory.
(6) By assumption, $(T\ulcorner c \urcorner \to U) \to c$ is in our theory.
(7) Since our theory is idempotent, we conclude from (5) and (6) that c is in our theory.
(8) By assumption, $c \to T\ulcorner c \urcorner$ is in our theory.
(9) Thus, again appealing to the fact that our theory is idempotent, we conclude from (7) and (8) that $T\ulcorner c \urcorner$ is in our theory.
(10) Thus, by one final appeal to the theory being idempotent, we conclude from (5) and (9) that U is in our theory.[16]

8.10 Concluding Thoughts

TL: Slick! You've shown that if we accept rules 1–3 and there's anything at all that we're unwilling to accept, then we're in trouble if we try to go about building a theory that contains C1, C2, T1, and T2.
SL: Exactly. So, since we can build theories containing these sentences without getting into trouble, we can build theories that reject some of these rules.

---

[15] Thanks to Ben Caplan for raising the suspicion this eventually leads to.

[16] Andrew Tedder has helpfully pointed out that this is quite similar to the derivation in Dunn et al. (1979).

TL: I think you're being too quick here, friend. This gets back to my suspicion from before. The right conclusion to draw from this isn't that the universal theory building toolkit can't accept Rules 1–3, but that we have to avoid saying awful things like C1 and C2 (or at least, we have to avoid saying such things if we also want to say things like T1 and T2).
SL: Let me get this straight: your solution is to save the *universal* theory building theory by *not* thinking about certain theories. I don't understand how that can possibly be correct.
TL: Well when you put it like *that* it sounds pretty bad, I'll admit. Perhaps what I mean to say is this: adding truth predicates and diagonalization to the mix *always* causes problems. This seems to give strong prima facie evidence for their being the type of things best handled separately using very thick gloves, not barehandedly as you've done here.

Less metaphorically, what your argument shows is that these concepts are unapt for theory-work and should either be abandoned or replaced.[17]
SL: I have two replies to this. First, truth and related notions feature prominently in our semantic theories, in metaphysics, and in a host of other places. If the universal theory building toolkit trivializes all of theories containing such notions that are also complex enough to do some diagonalization, then it seems we just cannot do non-trivial work in semantics, metaphysics, and the like.

Thus if we think we *can* do, e.g., semantics or metaphysics, then its incumbent on us to think through what that says about the structure of the universal theory building toolkit.

Second, I will grant that *if* we demand all theories to obey Rules 1–3, *then* of course we cannot build theories containing C1, C2, T1, and T2. So there are two options here: drop our demand that all theories obey Rules 1–3 or drop theories containing C1, C2, T1, and T2. I suppose what I want to claim is that the former option is the more intelligible and appealing option.
TL: I think I see what you're saying, even if I disagree. Maybe we could say it like this: if we began naïvely in our search for the universal theory building toolkit, then we wouldn't have any reason to reject theories containing C1, C2, T1, and T2 or to demand that theories obey Rules 1–3. We find the latter option 'intuitive' largely because of our extensive training in traditional logic. But you've provided an alternative way of understanding theories and theory building that plays into the naïve intuitions and is otherwise quite compelling as well. It's thus on the traditionalist at this point to provide a defense of her position.[18] This is a nice line of argumentation, but it leaves one question open: what evidence can you actually marshal in defense of the view of theories you've provided? I think you've made clear that if we accept the broad outlines of your account—e.g., if we accept that there are two theory building

[17] See Scharp (2013) for an illuminating discussion of the latter option.

[18] I have never encountered a traditional logician who is this conciliatory to the substructural position. Perhaps the best way to read this part of the dialogue, then, is not as a record of some possible interaction between a traditional logician and a substructural logician, but rather as a model for how such interactions could go, were dialogue between these two sides to be more fruitful than it in fact tends to be.

operations; that they behave in basically the way you outline; that theories are sets of sentences; etc.—then we should accept that the universal theory building theory is substructural. What can you say in defense of *those* hypotheses, though?
SL: That's a very serious question. I tend to think of what I'm doing in a broadly Carnapian way—what I'm presenting here is a *theory* about what sorts of things theories are and about how theory building goes. I'm then appealing to pragmatic considerations to urge adopting my theory building theory over its rivals. The kinds of pragmatic considerations I'm aiming at include both the paradox-avoidance bits just highlighted as well as the fact that my theory allows us to go on talking about theories in many of the ways we customarily do. I don't expect we have time for me to say more about this now, but, generally speaking, that's the sort of answer I'd tend to give to your question.
TL: I suppose that will have to do for now. There's one other objection I'd like to raise: one might, it seems to me, accept that there are theories that contain C1, C2, T1, and T2, but also hold that messing about with such things isn't the domain of *logic*. Logic is concerned with more pristine matters where behavior like non-self-closure doesn't arise.
SL: It's true that Beall (e.g.,) has argued that the there is an important and philosophically central sense of the word 'logic' on which logic *just is* the universal theory building toolkit.[19] I personally don't see a need to take a stand on how the universal theory building toolkit and logic are related. This is because I take it to be clear that whether the universal theory building toolkit is identical to logic or not, the theory itself is of central philosophical importance. I've scrupulously avoided the claim that what I'm doing is logic. If you'd like, I'm happy to give you the word 'logic' to use for the game we play when we restrict our attention to idempotent theories. Logic, in that case, is not the *universal* theory building toolkit. It's a special-case theory building toolkit. So anytime you're out and about and sit down to build a theory, you need to check to see whether you can use logic. That's not *usually* how we expect logic to work, but maybe you can make a case for using the word this way. In any event, I'll be over here if you need me, exploring what we can use to build theories all the time.

At the end of this speech, the connection suddenly and irreparably breaks. ***SL*** *and* ***TL*** *are forced, as happens so often, to perform the usual goodbyes and see you soon and via email. But they return to their quarantined lives refreshed by this pleasant interaction.*

Acknowledgements I am grateful for the helpful feedback this paper received from Gillman Payette.

---

[19] See Beall (2017) and Beall (2018).

References

Beall, J. (2017). There is no logical negation: True, false, both, and neither. *Australasian Journal of Logic*, *14*(1):Article no. 1.

Beall, J. (2018). The simple argument for subclassical logic. *Philosophical Issues, 28*(1), 30–54.

Brandom, R. (1998). *Making it explicit: Reasoning, representing, and discursive commitment*. Harvard University Press.

Brandom, R. (2009). *Articulating reasons: An introduction to inferentialism*. Harvard University Press.

Dunn, J. M., Meyer, R. K., & Routley, R. (1979). Curry's paradox. *Analysis, 39*(3), 124–128.

Fine, K. (1974). Models for entailment. *Journal of Philosophical Logic, 3*(4), 347–372.

Giambrone, S. (1985). TW+ and RW+ are decidable. *Journal of Philosophical Logic, 14*(3), 235–254.

Logan, S. A. (2020). Deep fried logic. *Erkenntnis*, 1–30.

Meyer, R. K. (1975). The consistency of arithmetic. *Typescript*.

Meyer, R. K., & Routley, R. (1972). Algebraic analysis of entailment I. *Logique et analyse, 15*(59/60), 407–428.

Read, S. (1988). *Relevant Logic: A Philosophical Examination of Inference*. Blackwell.

Scharp, K. (2013). *Replacing truth*. OUP Oxford.

Slaney, J. (1990). A general logic. *Australasian Journal of Philosophy, 68*(1), 74–88.

Urquhart, A. (1972). Semantics for relevant logics. *The Journal of Symbolic Logic, 37*(1), 159–169.

Chapter 9
More on the Power of a Constant

Marcus Kracht

Second Reader
F. Wolter
University of Liverpool

Abstract In a recent paper, Godblatt and Kowalski show that if we add to monomodal logic just a single propositional constant then instead of two coatoms (also known as Post-complete logics), we suddenly have continuum many. In this note we shall provide an alternative proof of that fact by showing that the simulation results of Kracht and Wolter can be sharpened.

Keywords Modal logic · Post-complete logics · Simulation

9.1 Just One Constant

It is known that modal logic has only two Post-complete logics: the logic of the reflexive one-point frame, $\mathsf{K} \oplus p \leftrightarrow \Box p$, and the logic of the irreflexive one-point frame, $\mathsf{K} \oplus \Box\bot$. As soon as we have two modal operators, that situation changes. There are now continuously many Post-complete logics. In their recent paper, Goldblatt and Kowalski (2014) show that this is also the case for monomodal logics provided we add a single propositional constant.

In this note we show that the result can be proved by using a slightly revised version of the simulation employed in Kracht and Wolter (1999). Basically, the idea is that $\Diamond\varphi$ is read as x $(\mathsf{c} \to \varphi)$ and $\blacklozenge\varphi$ as x $(\neg\,\mathsf{c} \to \varphi)$, where c is the added constant. The original construction used x $\boxminus\bot$ in place of c. This has the disadvantage that the simulation increases the codimension of the logic by 1, since all logics are contained

M. Kracht (✉)
Fakultät Linguistik und Literaturwissenschaft, Universität Bielefeld, Postfach 10 01 31, 33501 Bielefeld, Germany
e-mail: marcus.kracht@uni-bielefeld.de

I. Düntsch and E. Mares (eds.), *Alasdair Urquhart on Nonclassical and Algebraic Logic and Complexity of Proofs*, Outstanding Contributions to Logic 22,
https://doi.org/10.1007/978-3-030-71430-7_9

in $\mathsf{K} \oplus \boxminus\bot$. If L has codimension n in $\mathrm{Ext}\,\mathsf{K}_2$, $L^\bullet$, its simulated counterpart, has codimension $n+1$ in $\mathrm{Ext}\,\mathsf{K}_1$. With the added constant we can improve this to n. The simulation now preserves codimension. Hence, if $\mathrm{Ext}\,\mathsf{K}_2$ has continuously many coatoms, so has $\mathrm{Ext}\,\mathsf{K}_1^1$, where the upper 1 indicates the presence of a propositional constant.

9.2 Notation and Preliminaries

We shall work with two languages. The first is bimodal logic. It uses a denumerable set of variables p_i, $i \in \omega$, the constant $\top$, the Boolean connectives $\neg$, and $\wedge$, from which all others are defined in the usual way, and the modal operators $\Diamond$ and $\blacklozenge$. The semantic structures are general bimodal frames and bimodal algebras (see Kracht 1999 for technical details).

The other language is monomodal logic with the modal operator x and an additional constant, c. The structures are quadruples $\mathfrak{F} = \langle W, C, R, U\rangle$ where W is a set (the set of worlds), R a binary relation on W $U \subseteq \wp(W)$ a field of sets closed under the operation

$$\mathrm{x}\, D := \{w : \textit{there is } v\textit{:}\ w\, R\, v : v \in D\} \tag{9.1}$$

And finally, $C \in U$. The only thing to note here is that C serves as the value for the constant c. Hence we have $\langle \mathfrak{F}, \beta, w\rangle \vDash \mathsf{c}$ iff $w \in \mathsf{c}$.

9.3 The Simulation

Assume that we have a bimodal frame $F := \langle W, R_\Box, R_\blacksquare\rangle$. Write w^0 for the pair $\langle w, 0\rangle$ and w^1 for the pair $\langle w, 1\rangle$. Define a monomodal frame $F^s := \langle W \times \{0, 1\}, W \times \{0\}, R^s\rangle$, where

$$\begin{aligned} R^s := \quad & \{\langle w^0, v^0\rangle : w\, R_\Box\, v\} \\ & \cup \{\langle w^1, v^1\rangle : w\, R_\blacksquare\, v\} \\ & \cup \{\langle w^0, w^1\rangle : w \in W\} \\ & \cup \{\langle w^1, w^0\rangle : w \in W\} \end{aligned} \tag{9.2}$$

If $F := \langle W, R_\Box, R_\blacksquare, U\rangle$ is a bimodal general frame, then $F^s := \langle W \times \{0, 1\}, W \times \{0\}, R^s, U^s\rangle$ is its simulation counterpart, where

$$U^s := \{D \times \{0\} \cup E \times \{1\} : D, E \in U\} \tag{9.3}$$

Proposition 9.1 *F^s is a general frame.*

Proof Closure under Boolean operations is clear. Also, the value of the constant is in U^s. So all we have to show is that if $H \in U^s$, then $\mathrm{x}\ H \in U^s$ as well. To this end, we dissect H into two sets, as $H = D \times \{0\} \cup E \times \{1\}$. Also, $D \times \{0\} \in U^s$, since it is $H \cap W \times \{0\}$. Hence, by construction, $D \in U$. Likewise, $E \in U$. Now

$$\mathrm{x}\ C = (\Diamond D) \times \{0\} \cup (\blacklozenge E) \times \{1\} \cup (E \times \{0\}) \cup (D \times \{0\}) \tag{9.4}$$

which is in U^s.

Proposition 9.2 *The logic of all simulation frames is*

$$\begin{aligned} \mathsf{K}_1^1 \oplus \{\, &\mathsf{c} \to (\mathrm{x}\ (\neg \mathsf{c} \wedge p) \leftrightarrow \boxminus(\neg \mathsf{c} \to p)), \\ &\neg \mathsf{c} \to (\mathrm{x}\ (\mathsf{c} \wedge p) \leftrightarrow \boxminus(\mathsf{c} \to p)), \\ &\mathsf{c} \wedge p \to \mathrm{x}\ (\neg \mathsf{c} \wedge \mathrm{x}\ (\mathsf{c} \wedge p)), \\ &\neg \mathsf{c} \wedge p \to \mathrm{x}\ (\mathsf{c} \wedge \mathrm{x}\ (\neg \mathsf{c} \wedge p))\} \end{aligned} \tag{9.5}$$

This logic is called Sim^1.

Proof Clearly, the postulates are valid on all simulation frames. Conversely, let $F = \langle W, C, R, U\rangle$ be a refined monomodal frame. Then put $W_s := W \cap C$. Now let $w \in W_s$. Since $w \vDash \mathsf{c}$, we also have $w \vDash \mathrm{x}\ (\neg \mathsf{c} \wedge p) \leftrightarrow \boxminus(\neg \mathsf{c} \to p)$. This translates into the following. There exists a unique w' such that $w\ R\ w'$ and $w' \notin C$. By the second axiom, there exists a unique w'' such that $w'\ R\ w''$ and $w'' \in C$. The third guarantees that $w'' = w$. Hence we have a bijection $h : W \to W$ such that $w\ R\ h(w)\ R\ w = h(h(w))$. Now the unsimulation frame F_s is defined as follows.

$$\begin{aligned} F_s &:= \langle W_s, R_\Box, R_\blacksquare, U_s\rangle \\ R_\Box &:= \{\langle w, v\rangle : w, v \in W_s, w\ R\ v\} \\ R_\blacksquare &:= \{\langle w, v\rangle : w, v \in W_s, h(w)\ R\ h(v)\} \\ U_s &:= \{D \in U : D \subseteq W_s\} \end{aligned} \tag{9.6}$$

It is immediately checked that U_s is closed under the Boolean operations. Moreover, if $D \in U_s$ then

$$\begin{aligned} \Diamond D &= \{w : \textit{there is } v\textit{: } w\ R\ v \in D\} \\ &= \mathrm{x}\ (W_s \cap D) \\ \blacklozenge D &= \{w : \textit{there is } v\textit{: } h(w)\ R\ h(v)\} \\ &= \mathrm{x}\ (-W_s \cap \mathrm{x}\ (-W_s \cap \mathrm{x}\ (W_s \cap D))) \end{aligned} \tag{9.7}$$

And so U_s is closed under the modal operators as well.

Now translate these algebras. If $\mathfrak{A} = \langle A, 1, -, \cap, \Box, \blacksquare\rangle$ is a bimodal algebra, put

$$\mathfrak{A}^s := \langle A \times A, 1 \times 1, 1 \times 0, -, \cap, \boxminus\rangle, \tag{9.8}$$

where

$$\mathrm{x}\ \langle a, b\rangle = \langle b \cup \Diamond a, a \cup \blacklozenge b\rangle \tag{9.9}$$

On the other hand, given a unimodal algebra $\mathfrak{B} = \langle B, 1, c, -, \cap, \boxminus \rangle$, put $B_s := \{b \cap c : b \in B\}$. And put

$$\begin{aligned} \Diamond b &:= c \cap (\mathrm{x}\, b) \\ \blacklozenge b &:= c \cap (\mathrm{x}\, (-c \cap \mathrm{x}\, (-c \cap \mathrm{x}\, b))) \end{aligned} \tag{9.10}$$

The next theorem is the actual core.

Theorem 9.1 *Simulation and unsimulation commute with* H, S *and* P. *Hence, the simulation and unsimulation image of a variety is a variety.*

Proof (S). Let $\mathfrak{A}$ be a bimodal algebra, $\mathfrak{C}$ a subalgebra. Then $\mathfrak{C}^s$ is a subalgebra of $\mathfrak{A}^s$. Also, if $\mathfrak{D}$ is a subalgebra of $\mathfrak{A}^s$, it has the form $\mathfrak{B}^s$. For take an element $\langle c, d\rangle \in D$. Since $\langle 1, 0\rangle \in D$, we have $\langle c, 0\rangle \in D$, and $\langle 0, d\rangle \in D$. Moreover, we have $\langle 0, c\rangle \in D$, since this is $\mathrm{x}\, \langle c, 0\rangle$. Likewise have $\langle d, 0\rangle \in D$. Hence, let $B := \{c : \langle c, 0\rangle \in D\}$. Then $D = B^s$, and $\mathfrak{B}^s = \mathfrak{D}$, as required.

(H). Let $\mathfrak{A}$ be a bimodal algebra, and Θ a congruence. Then $\Theta^s := \Theta \times \Theta = \{\langle \langle c, d\rangle, \langle c', d'\rangle\rangle : c \,\Theta\, c', d \,\Theta\, d'\}$. This is a congruence on $\mathfrak{A}^s$, and $\mathfrak{A}^s/\Theta^s \cong (\mathfrak{A}/\Theta)^s$. To see this, note that if $\langle c, d\rangle \Theta^s \langle c', d'\rangle$ then $\Diamond\langle c, d\rangle = \langle c \cup \Diamond d, d \cup \blacklozenge c\rangle \,\Theta^s\, \langle c' \cup \Diamond d', d' \cup \blacklozenge c'\rangle = \mathrm{x}\, \langle c', d'\rangle$. Conversely, let Ξ be a nontrivial congruence on $\mathfrak{A}^s$. Let $\langle c, d\rangle \,\Xi\, \langle c', d'\rangle$. Then it follows that $\langle 1, 0\rangle \cap \langle c, d\rangle \,\Xi\, \langle 1, 0\rangle \cap \langle c', d'\rangle$, that is, $\langle c, 0\rangle \,\Xi\, \langle c', 0\rangle$. And it follows that $\langle 0, c\rangle = \mathrm{x}\, \langle c, 0\rangle \,\Xi\, \mathrm{x}\, \langle c', 0\rangle = \langle 0, c'\rangle$. Likewise, $\langle 0, d\rangle \,\Xi\, \langle 0, d'\rangle$ as well as $\langle d, 0\rangle \,\Xi\, \langle d', 0\rangle$. So Ξ has the form $\Theta \times \Theta$ for some Θ.
(P). That $\left(\prod_{i\in I} \mathfrak{A}_i\right) \cong \prod_{i \in I} \mathfrak{A}_i^s$ is pretty straightforward. Also for unsimulations.

Corollary 9.1 *The simulation map is an isomorphism between* Ext K_2 *and* Ext K_1^1. □

A logic is *Post-complete* if it is consistent but has no proper consistent extension. Alternatively, L is Post-complete if it has codimension 1 in the lattice. Recall from Williamson (1998) that Ext Kt and a fortiori Ext K_2 has continuum many Post-complete elements. As a corollary we get

Corollary 9.2 Ext K_1^1 *has continuum many Post-complete elements.* □

Acknowledgements I thank Frank Wolter for carefully reviewing this chapter.

References

Goldblatt, R. I., & Kowalski, T. (2014). The power of a propositional constant. *Journal of Philosophical Logic*, *43*, 133–152.

Kracht, M. (1999). *Tools and techniques in modal logic*. Studies in logic (Vol. 142). Amsterdam: Elsevier.

Kracht, M., & Wolter, F. (1999). Normal monomodal logics can simulate all others. *Journal of Symbolic Logic*, *64*, 99–138.

Williamson, T. (1998). Continuum many maximal consistent normal bimodal logics with inverses. *Notre Dame Journal of Formal Logic*, *39*, 128–134.

Chapter 10
Strong Completeness of S4 for the Real Line

Philip Kremer

Second Reader
R. Goldblatt
Victoria University of Wellington

Abstract In the topological semantics for modal logic, S4 is well known to be complete for the rational line and for the real line: these are special cases of S4's completeness for any dense-in-itself metric space. The construction used to prove completeness can be slightly amended to show that S4 is not only complete but strongly complete, for the rational line. But no similarly easy amendment is available for the real line. In an earlier paper, we proved a general theorem: S4 is strongly complete for any dense-in-itself metric space. Strong completeness for the real line is a special case. In the current paper, we give a proof of strong completeness tailored to the special case of the real line: the current proof is simpler and more accessible than the proof of the more general result and involves slightly different techniques. We proceed in two steps: first, we show that S4 is strongly complete for the space of finite and infinite binary sequences, equipped with a natural topology; and then we show that there is an interior map from the real line onto this space.

Keywords Modal logic · Topological semantics · Strong completeness · Real line

10.1 Introduction

It is my honour to contribute to this festschrift for Alasdair Urquhart. The first logic course I ever took was a course on relevance logic with Alasdair in 1985: as a result, I saw a completeness proof for the relevance logic R before I ever saw one for classical

---

P. Kremer (✉)
Department of Philosophy, University of Toronto, Toronto, Canada
e-mail: philip.kremer@utoronto.ca

I. Düntsch and E. Mares (eds.), *Alasdair Urquhart on Nonclassical and Algebraic Logic and Complexity of Proofs*, Outstanding Contributions to Logic 22,
https://doi.org/10.1007/978-3-030-71430-7_10

logic. Ever since, I've maintained an interest in alternatives to, and extensions of, classical logic—especially in completeness results.

In the topological semantics for modal logic (McKinsey (1941); McKinsey and Tarski (1944); Rasiowa and Sikorski (1963)), S4 is well known to be complete for the class of all topological spaces, as well as for a number of particular topological spaces, notably the rational line, $\mathbb{Q}$, and the real line, $\mathbb{R}$. The results for $\mathbb{Q}$ and $\mathbb{R}$ are special cases of the fact that S4 is complete for any dense-in-itself metric space: see Rasiowa and Sikorski (1963), Theorem XI, 9.1, which is derived from McKinsey (1941); McKinsey and Tarski (1944). It is customary to strengthen completeness to *strong* completeness, i.e., the claim that any consistent set of formulas is satisfiable at some point in the space in question. As long as the language is countable, the construction used to prove completeness can be slightly amended to show that S4 is not only complete but strongly complete, for $\mathbb{Q}$ (see Kremer (2013)). But no similarly easy amendment is available for $\mathbb{R}$: until Kremer (2013), the questions of strong completeness for $\mathbb{R}$ was open.

In Kremer (2013), we prove that S4 is strongly complete for *any* dense-in-itself metric space—and therefore for $\mathbb{R}$. In the current paper, we give a proof of strong completeness tailored to the special case of $\mathbb{R}$. This proof is useful for at least two reasons. First, since the proof in the current paper is tailored to a special case, it is simpler and more accessible than the proof in Kremer (2013), avoiding many of the bells and whistles needed there for the more general claim. In particular, we can bypass all mentions of ultrafilters and of algebraic semantics. We believe that it usefully clarifies matters to work through a simplified proof of a special case, before considering a more general case. Second, in proving Lemma 10.5, below, we use a different technique than the proof of the same lemma, Lemma 6.1, in Kremer (2013). The proof technique used here might be generalized or adapted in ways that the proof technique in Kremer (2013) cannot, either for logics stronger than S4 or for logics that extend S4 with additional vocabulary such as the universal modality $\forall$ or the difference modality $\neq$: Goldblatt and Hodkinson (2019), for example, leaves a number of open questions in this area for which the techniques here might be useful.

Completeness for any given dense-in-itself metric space X is typically proved by showing that any finite rooted reflexive transitive Kripke frame is the image of an interior map from X. When $X = \mathbb{Q}$, strengthening completeness to *strong* completeness is accomplished by slightly amending the construction to show that any *countable* rooted reflexive transitive Kripke frame is the image of an interior map from $\mathbb{Q}$. But this strategy is not generalizable: because of the Baire Category Theorem, the countable rooted reflexive transitive Kripke frame $\langle \mathbb{N} \leq \rangle$, for example, is *not* the image of any interior map from $\mathbb{R}$ (I owe this observation to Guram Bezhanishvili, David Gabelaia, and Valentin Shehtman): see Kremer (2013), Sect. 3, for details.

To show that S4 is strongly complete for $\mathbb{R}$, we proceed in two steps. First, we show that S4 is strongly complete for the space $2^{\leq\omega}$ of finite and infinite binary sequences, equipped with a natural topology: see Sect. 10.3. We call $2^{\leq\omega}$ the *infinite binary tree with limits* and Lando Lando (2012) calls it the *complete binary tree*. Then we show that there is an interior map from $\mathbb{R}$ onto $2^{\leq\omega}$: see Sect. 10.4. Thus,

S4 is strongly complete for $\mathbb{R}$. In fact, we proceed by showing that there's an interior map from the open unit interval, $\mathcal{I} = (0, 1)$ onto $2^{\leq\omega}$: this suffices since there are many interior maps from $\mathbb{R}$ onto $\mathcal{I}$. We note that Lando (2012) already constructs an interior map from the closed unit interval $[0, 1]$ onto $2^{\leq\omega}$: see Lando (2012), Sect. 5.4. Our construction is quite similar but is simpler because Lando's project requires her to track not only topological properties of the map but also measure-theoretic properties.

Note: while this chapter was in production, Robert Goldblatt contacted me about the infinite binary tree with limits. While browsing casually through Freyd and Scedov (1990) published in 1990, he had chanced upon both a definition of this tree and a construction of an interior map from the closed interval $[-1/2, 1/2]$ onto this tree: it follows that there is an interior map from $\mathbb{R}$ onto it, see (Freyd and Scedov 1990, 1.749 and 1.74(10)).

10.2 Basics

We begin by fixing notation and terminology. We assume a propositional language with a countable set PV of propositional variables; standard Boolean connectives $\wedge$, $\vee$ and $\neg$; and one modal operator, $\Box$. A finite set of formulas is *consistent* iff either it is empty or the negation of the conjunction of the formulas in it is not a theorem of S4; and an infinite set of formulas is *consistent* iff every finite subset is consistent.

A *Kripke frame* is an ordered pair $\langle X, R\rangle$, where X is a nonempty set and $R \subseteq X \times X$. We somewhat imprecisely identify X with $\langle X, R\rangle$, letting context or fiat determine R. A Kripke frame X is reflexive [transitive] iff R is: for the rest of this paper, we assume that all Kripke frames are reflexive and transitive. A Kripke frame is *rooted* iff $(\exists r \in W)(\forall w \in W)(rRw)$. A subset O of X is *open* iff $(\forall x, y \in X)(x \in O \ \&\ xRy \Rightarrow y \in O)$. A subset C of X is *closed* iff $X - C$ is open. The *interior* of a set $S \subseteq X$ is the largest open subset of S: $Int(S) =_{\text{df}} \{x \in S : \forall y \in X, xRy \Rightarrow y \in S\}$. The *closure* of a set $S \subseteq X$ is the smallest closed superset of S: $Cl(S) =_{\text{df}} X - Int(X - S)$. A *topological space* is an ordered pair $\langle X, \tau\rangle$, where X is a nonempty set and $\tau \subseteq (X)$ is a topology on X. We somewhat imprecisely identify X with $\langle X, \tau\rangle$, letting context or fiat determine τ. Thus, for example, we identify $\mathbb{R}$ with $\langle \mathbb{R}, \tau_{\mathbb{R}}\rangle$, where $\tau_{\mathbb{R}}$ is the standard topology on $\mathbb{R}$. We assume the basics of point-set topology, in particular the notion of the interior and closure, $Int(S)$ and $Cl(S)$, of a subset S of a topological space.

A *Kripke model* [*topological model*] is an ordered pair $M = \langle X, V\rangle$, where X is a Kripke frame [topological space] and $V : PV \rightarrow (X)$. We use the term *model* to cover Kripke models and topological models. For any model $M = \langle X, V\rangle$, V is extended to all formulas as follows: $V(\neg A) = X - V(A)$; $V(A \wedge B) = V(A) \cap V(B)$; $V(A \vee B) = V(A) \cup V(B)$; and $V(\Box A) = Int(V(A))$. If Γ is a nonempty set of formulas, then $V(\Gamma) =_{\text{df}} \bigcap_{A\in\Gamma} V(A)$; if Γ is empty, then $V(\Gamma) =_{\text{df}} X$.

Suppose that Γ is a set of formulas. If X is a Kripke frame or topological space and $x \in X$, then we say that Γ is *satisfiable at x in X* iff there is some model $M = \langle X, V\rangle$

such that $x \in V(\Gamma)$; and we say that Γ is *satisfiable in* X iff Γ is satisfiable at some x in X. We say that S4 is *complete* for X iff every finite consistent set of formulas is satisfiable in X, and *strongly complete* for X iff every consistent set of formulas is satisfiable in X.

The following completeness theorem follows from Rasiowa and Sikorski (1963), Theorem XI, 9.1, (vii), which itself derived from McKinsey (1941); McKinsey and Tarski (1944):

Theorem 10.1 S4 *is complete for* $\mathbb{R}$.

Theorem 10.1 is well known: there are new and more accessible proofs in Aiello et al. (2003); Bezhanishvili and Gehrke (2005); Mints and Zhang (2005). The current paper's main result is a special case of the main theorem, Theorem 1.2, in Kremer (2013):

Theorem 10.2 S4 *is* strongly *complete for* $\mathbb{R}$.

Before we prove Theorem 10.2, we recall the standard notion of an *interior map*. A function from a topological space or Kripke frame to a topological space or Kripke frame is *continuous* iff the preimage of every open set is open; is *open* iff the image of every open set is open; and is an *interior map* iff it is continuous and open. Suppose that $M = \langle X, V\rangle$ and $M' = \langle X', V'\rangle$ are models, and that f is a surjective interior map from X onto X'. Then f is an *interior map from M onto M'* iff, for every $p \in PV$ and $x \in X$, $x \in V(p)$ iff $f(x) \in V'(p)$. The following lemma and corollary are standard:

Lemma 10.1 *If f is an interior map from $M = \langle X, V\rangle$ onto $M' = \langle X', V'\rangle$, then for every formula B and $x \in X$, $x \in V(B)$ iff $f(x) \in V'(B)$.*

Corollary 10.1 *Suppose that each of X and X' is a Kripke frame or topological space, and that there is an interior map from X onto X'. Then if Γ is satisfiable in X' then Γ is satisfiable in X.*

Given Corollary 10.1, we can divide the work of proving Theorem 10.2 into two parts. The first part is mainly logical: we show that S4 is strongly complete for the space $2^{\leq\omega}$ of finite and infinite binary sequences, equipped with a natural topology (Lemma 10.5). The second part is purely topological: we show that there's an interior map from $\mathbb{R}$ onto $2^{\leq\omega}$.

10.3 The Space $2^{\leq\omega}$

For each $n \geq 0$, let 2^n be the set of binary sequences (sequences of 0's and 1's) of length n. Let $2^{<\omega} =_{\text{df}} \bigcup_{n=0}^{\infty} 2^n$, i.e., $2^{<\omega}$ is the set of finite binary sequences. We write *length*(b) for the length of $b \in 2^{<\omega}$. Let 2^{ω} be the set of infinite binary sequences of order type ω. And let $2^{\leq\omega} =_{\text{df}} 2^{<\omega} \cup 2^{\omega}$. We use Λ for the empty binary sequence, i.e., the binary sequence of length 0. We use b, b', etc., to range over $2^{<\omega}$; $\mathbf{b}$, $\mathbf{b}'$, etc., to range over 2^ω; and $\boldsymbol{b}, \boldsymbol{b}'$, etc., to range over $2^{\leq\omega}$. If $b \in 2^{<\omega}$ and $\boldsymbol{b} \in 2^{\leq\omega}$, then we write $b^\frown\boldsymbol{b}$ for b concatenated with $\boldsymbol{b}$. We write $b0$ and $b1$ for $b^\frown\langle 0\rangle$ and $b^\frown\langle 1\rangle$. Given any $\mathbf{b} \in 2^\omega$ and any $n \in \mathbb{N}$, the finite binary sequence $\mathbf{b}|_n$ is the initial segment of length n of $\mathbf{b}$. Thus, $\mathbf{b}|_0 = \Lambda$. Given $b \in 2^{<\omega}$ and $\boldsymbol{b} \in 2^{\leq\omega}$, we say $b \leq \boldsymbol{b}$ iff b is an initial segment of $\boldsymbol{b}$ and $b < \boldsymbol{b}$ iff both $b \leq \boldsymbol{b}$ and $b \neq \boldsymbol{b}$. We also use '$\leq$' for $\leq$ restricted to $2^{<\omega}$.

We identify $2^{<\omega}$ with the *infinite binary tree*, i.e., the countably infinite rooted transitive reflexive Kripke frame $\langle 2^{<\omega}, \leq\rangle$. We can think of an infinite binary sequence $\mathbf{b} \in 2^\omega$ as the *limit* of the branch of finite sequences $\mathbf{b}|_0, \mathbf{b}|_1, \mathbf{b}|_2, \ldots$. Accordingly, we think of $2^{\leq\omega}$ as the infinite binary tree *with limits*.

For any $b \in 2^{<\omega}$, it will be useful to define two related sets: $\leq(b) =_{\mathrm{df}} \{b' \in 2^{<\omega} : b \leq b'\}$ and $\leq^*(b) =_{\mathrm{df}} \{\boldsymbol{b}' \in 2^{\leq\omega} : b \leq \boldsymbol{b}'\}$. We impose a natural topology on $2^{\leq\omega}$, by taking as a basis all the sets of the form $\leq^*(b)$, where $b \in 2^{<\omega}$. (Nick Bezhanishvili pointed out to me that this is the Scott topology on $2^{\leq\omega}$: See Vickers (1989), p. 95, for a definition of the Scott topology on any partially ordered set.) The main task of the current section is to prove that S4 is strongly complete for $2^{\leq\omega}$ – see Lemma 10.5.

The following result, due originally to Dov Gabbay and independently discovered by Johan van Benthem, is well known; for a proof see Goldblatt (1980), Theorem 1:

Lemma 10.2 *Any finite rooted reflexive transitive Kripke frame is the image of* $2^{<\omega}$ *under some interior map.*

Together with the fact that any finite consistent set Γ of formulas is satisfiable in some finite rooted reflexive transitive Kripke frame, Lemma 10.2 entails that S4 is complete for $2^{<\omega}$. Lemma 10.2 can be strengthened to

Lemma 10.3 *Any countable rooted reflexive transitive Kripke frame is the image of* $2^{<\omega}$ *under some interior map.*

Proof This is Lemma 3.3 in Kremer (2013). Unfortunately, the proof there is incorrect. Here, we reproduce, almost verbatim, the corrected proof in Kremer (2014): see p. 451, proof of (ii.b). Suppose that $\langle W, R\rangle$ is a countable Kripke frame with root r. We will, in effect, unravel $\langle W, R\rangle$ into $2^{<\omega}$. For each $w \in W$, let $R(w) = \{w' \in W : wRw'\}$ and let $succ_0(w), succ_1(w), succ_2(w), \ldots$ be an enumeration of $R(w)$ in which every member of $R(w)$ occurs infinitely often. We also need a function $zero : 2^{<\omega} \to \mathbb{N}$, defined as follows: $zero(\Lambda) = 0$; $zero(b0) = zero(b) + 1$; and $zero(b1) = 0$. Note that $zero(b)$ is simply the number of uninterrupted occurrences of 0 at the end of b: e.g., $zero(001101000) = 3$, $zero(100001) = 0$, and $zero(000100) = 2$. Now we define the surjective interior map $\varphi : 2^{<\omega} \to W$ recursively as follows: $\varphi(\Lambda) = r$; $\varphi(b0) = \varphi(b)$; and $\varphi(b1) = succ_{zero(b)}(\varphi(b))$.

We have to check that φ is a surjective interior map. **Surjectivity.** To see that φ is surjective, suppose that $w \in W$. Then $w = succ_n(r)$ for some $n \in \mathbb{N}$. Let 0^n be the sequence of n 0's. And note that $\varphi(0^n 1) = succ_{zero(0^n)}(r) = succ_n(r) = w$. **Continuity.** It suffices to show that the preimage of $R(w)$ is open for every $w \in W$: note that the preimage of $R(w)$ is $\bigcup_{\varphi(b)=w} \leq(b)$. **Openness.** It suffices to show that the image of $\leq(b)$ is open for every $b \in 2^{<\omega}$: note that the image of $\leq(b)$ is $R(\varphi(b))$.

Together with the fact that any consistent set Γ of formulas is satisfiable in some countable rooted reflexive transitive Kripke frame, Lemma 10.3 entails

Lemma 10.4 S4 *is strongly complete for* $2^{<\omega}$.

The remainder of this section uses Lemma 10.4 to prove

Lemma 10.5 S4 *is strongly complete for* $2^{\leq\omega}$.

Proof This is Lemma 6.1 in Kremer (2013). Here we give quite a different proof that bypasses the dependence, useful in Kremer (2013) but unnecessary here, on ultrafilters and on algebraic semantics. Let Γ be a consistent set of formulas. Given Lemma 10.4, Γ is satisfiable in $2^{<\omega}$. So there is a Kripke model $M = \langle 2^{<\omega}, V\rangle$ such that $V(\Gamma) \neq \emptyset$. We will define a $V^* : PV \to 2^{\leq\omega}$ and show that, in the topological model $M^* = \langle 2^{\leq\omega}, V^*\rangle$, we have $V^*(\Gamma) \neq \emptyset$.

First, we assign sets $\Delta_{\boldsymbol{b}}$ and $\Sigma_{\boldsymbol{b}}$ of formulas to each $\boldsymbol{b} \in 2^{\leq\omega}$. If $\boldsymbol{b} \in 2^{<\omega}$ then $\Delta_{\boldsymbol{b}} = \Sigma_{\boldsymbol{b}} =_{\text{df}} \{A : \boldsymbol{b} \in V(A)\}$. Note, if $\boldsymbol{b} \in 2^{<\omega}$, then $\Sigma_{\boldsymbol{b}}$ is consistent; $\Sigma_{\boldsymbol{b}}$ is also *complete* in the following sense: for every formula A, either $A \in \Sigma_{\boldsymbol{b}}$ or $\neg A \in \Sigma_{\boldsymbol{b}}$. If $\boldsymbol{b} \in 2^{\omega}$, then let $\Delta_{\boldsymbol{b}} =_{\text{df}} \bigcup_{n=0}^{\infty} \bigcap_{m=n}^{\infty} \Sigma_{\boldsymbol{b}|_m} = \{A : (\exists n \in \mathbb{N})(\forall m \geq n)(\boldsymbol{b}|_m \in V(A))\}$. Note that $\Delta_{\boldsymbol{b}}$ is consistent, so that we can let $\Sigma_{\boldsymbol{b}}$ be any complete consistent superset of $\Delta_{\boldsymbol{b}}$.

For $p \in PV$, define $V^*(p) = \{\boldsymbol{b} \in 2^{\leq\omega} : p \in \Sigma_{\boldsymbol{b}}\}$. Now we show that, for every formula A,

$$\text{for every } \boldsymbol{b} \in 2^{\leq\omega}, \boldsymbol{b} \in V^*(A) \text{ iff } A \in \Sigma_{\boldsymbol{b}}. \tag{10.1}$$

The proof is by induction on the construction of A. If $A \in PV$ then (10.1) follows from the definition of $V^*(A)$; and if A is of the form $\neg B$, $(B \wedge C)$ or $(B \vee C)$, then (10.1) follows from the fact that each $\Sigma_{\boldsymbol{b}}$ is consistent and complete. So suppose that A is of the form $\Box B$ and make the inductive hypothesis that

$$\text{for every } \boldsymbol{b} \in 2^{\leq\omega}, \boldsymbol{b} \in V^*(B) \text{ iff } B \in \Sigma_{\boldsymbol{b}}. \tag{10.2}$$

We want to show,

$$\text{for every } \boldsymbol{b} \in 2^{\leq\omega}, \boldsymbol{b} \in V^*(\Box B) \text{ iff } \Box B \in \Sigma_{\boldsymbol{b}}. \tag{10.3}$$

Proof of ($\Rightarrow$). Choose $\boldsymbol{b} \in 2^{\leq\omega}$ and assume that $\boldsymbol{b} \in V^*(\Box B)$. So there is some $b' \in 2^{<\omega}$ such that $\boldsymbol{b} \in \leq^*(b') \subseteq V^*(B)$. So, for every $b'' \in \leq(b') = 2^{<\omega} \cap \leq^*(b')$, we have $B \in \Sigma_{b''}$, by (10.2). So $\leq(b') \subseteq V(B)$, by the definition of the $\Sigma_{\boldsymbol{b}}$. So $\leq(b') \subseteq V(\Box B)$, by the definition of $V(\Box B)$. If $\boldsymbol{b} \in 2^{<\omega}$, then $\boldsymbol{b} \in \leq(b') \subseteq V(\Box B)$, so that $\Box B \in \Sigma_{\boldsymbol{b}}$, by the definition of $\Sigma_{\boldsymbol{b}}$. On the other hand, suppose that $\boldsymbol{b} \notin 2^{<\omega}$. So $\boldsymbol{b} \in 2^{\omega}$. Since $b' \leq \boldsymbol{b}$, we have $b' = \boldsymbol{b}|_n$, for some $n \in \mathbb{N}$. So $\boldsymbol{b}|_n \in \leq(b') \subseteq V(\Box B)$. So $\boldsymbol{b}|_m \in V(\Box B)$, for every $m \geq n$. So $\Box B \in \Sigma_{\boldsymbol{b}|_m}$, for every $m \geq n$. So $\Box B \in \Delta_{\boldsymbol{b}}$. So $\Box B \in \Sigma_{\boldsymbol{b}}$, as desired.

Proof of ($\Leftarrow$). Choose $\boldsymbol{b} \in 2^{\leq\omega}$ and assume that $\Box B \in \Sigma_{\boldsymbol{b}}$. We consider two cases: (i) $\boldsymbol{b} \in 2^{<\omega}$, and (ii) $\boldsymbol{b} \in 2^{\omega}$. In Case (i) $\boldsymbol{b} \in V(\Box B)$, by the definition of $\Sigma_{\boldsymbol{b}}$. So, for every $b' \in \leq(\boldsymbol{b})$, we have $b' \in V(B)$. So, for every $b' \in \leq(\boldsymbol{b})$, we have $B \in \Sigma_{b'}$. So, for every $\boldsymbol{b}' \in \leq^*(\boldsymbol{b})$, we have $B \in \Delta_{\boldsymbol{b}'}$. So, for every $\boldsymbol{b}' \in \leq^*(\boldsymbol{b})$, we have $B \in \Sigma_{\boldsymbol{b}'}$. So, by (10.2), for every $\boldsymbol{b}' \in \leq^*(\boldsymbol{b})$, we have $\boldsymbol{b}' \in V^*(B)$. So $\boldsymbol{b} \in V^*(\Box B)$, as desired.

In Case (ii), $\neg\Box B \notin \Delta_{\boldsymbol{b}}$. So there is some $m \in \mathbb{N}$ such that $\Box B \in \Sigma_{\boldsymbol{b}|_m}$. So $\boldsymbol{b}|_m \in V(\Box B)$. So, for every $b' \in 2^{<\omega}$, if $\boldsymbol{b}|_m \leq b'$ then $b' \in V(B)$. So, for every $b' \in 2^{<\omega}$, if $\boldsymbol{b}|_m \leq b'$ then $B \in \Sigma_{b'}$. But then, by the definition of $\Delta_{\mathbf{b}''}$ for $\mathbf{b}'' \in 2^{\omega}$, we have for every $\mathbf{b}'' \in 2^{\omega}$, if $\boldsymbol{b}|_m \leq \mathbf{b}''$ then $B \in \Delta_{\mathbf{b}''} \subseteq \Sigma_{\mathbf{b}''}$. So, for every $\boldsymbol{b}^* \in 2^{\leq\omega}$, if $\boldsymbol{b}|_m \leq \boldsymbol{b}^*$ then $B \in \Sigma_{\boldsymbol{b}^*}$. So, by (10.2), for every $\boldsymbol{b}^* \in \leq^*(\boldsymbol{b}|_m)$, $\boldsymbol{b}^* \in V^*(B)$. So $\boldsymbol{b} \in \leq^*(\boldsymbol{b}|_m) \subseteq V^*(\Box B)$, as desired.

Given (10.1), to see that Γ is satisfiable in $2^{\leq\omega}$, simply choose $b \in 2^{<\omega}$ with $b \in V(\Gamma)$. Note: $\Gamma \subseteq \Sigma_b$, so that $b \in V^*(\Gamma)$, by (10.1).

10.4 An Interior Map From $\mathcal{I} = (0, 1)$ onto $2^{\leq\omega}$

Our remaining work is purely topological: we want to prove

Lemma 10.6 *There is an interior map from $\mathbb{R}$ onto $2^{\leq\omega}$.*

Let $\mathcal{I} = (0, 1)$ be the open unit interval. As noted in the introductory remarks, it suffices to prove

Lemma 10.7 *There is an interior map from $\mathcal{I}$ onto $2^{\leq\omega}$.*

As noted above, Lando (2012) already constructs an interior map from the closed unit interval $[0, 1]$ onto $2^{\leq\omega}$. The following construction is similar but simpler, because Lando's project requires her to track not only topological properties of the map but also measure-theoretic properties. In particular, for measure-theoretic reasons, Lando uses 'thick' Cantor sets, where we only use Cantor sets. (These constructions were discovered independently, around the same time, around 2011.)

We prove Lemma 10.7 by partitioning $\mathcal{I}$ into nonempty pairwise disjoint sets $X_{\boldsymbol{b}}$, one for each $\boldsymbol{b} \in 2^{\leq\omega}$. We then define $\mathbf{F} : \mathcal{I} \to 2^{\leq\omega}$ as follows: $\mathbf{F}(x)$ = the unique $\boldsymbol{b} \in 2^{\leq\omega}$ such that $x \in X_{\boldsymbol{b}}$. The trick is to do this in such a way that $\mathbf{F}$ is a surjective interior map.

First, some preliminaries. For subsets of $\mathcal{I}$, we interpret interior, Int, and closure, Cl, as relativized to $\mathcal{I}$. Let $\mathcal{C}$ be the Cantor set without the endpoints 0 and 1. So $\mathcal{C}$ is the set of all real numbers that have a ternary expansion of the form $0.a_1a_2a_3 \ldots a_k \ldots$ where each a_k is either 0 or 2, and where not all the a_k's are 0 (so that $0 \notin \mathcal{C}$) and not all the a_k's are 2 (so that $1 \notin \mathcal{C}$): we will find it useful to represent real numbers as ternary expansions. Figure 10.1 pictorially represents $\mathcal{C}$, which is closed (in the space $\mathcal{I}$). $\mathcal{C}$ can be got from progressively deleting open intervals from $\mathcal{I} = (0, 1)$ as follows: delete the open interval $(0.1, 0.2)$, which is the middle third of $\mathcal{I}$, leaving $(0, 0.1] \cup [0.2, 1)$. Then delete the middle thirds of each of these: delete the open interval $(0.01, 0.02)$ from $(0, 0.1]$ and delete the open interval $(0.21, 0.22)$ from $[0.2, 1)$: this leaves $(0, 0.01] \cup [0.02, 0.1] \cup [0.2, 0.21] \cup [0.22, 1)$. More precisely, a *middle third* is any open interval of the form $(0.a_1a_2 \ldots a_n1, 0.a_1a_2 \ldots a_n2)$, where

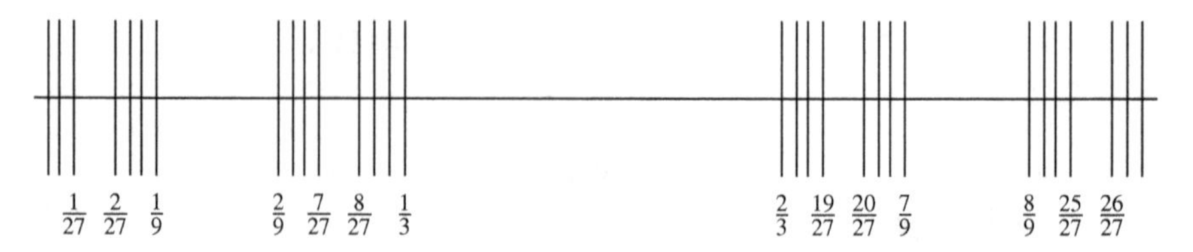

Fig. 10.1 The Cantor set, $\mathcal{C}$, without the endpoints 0 and 1

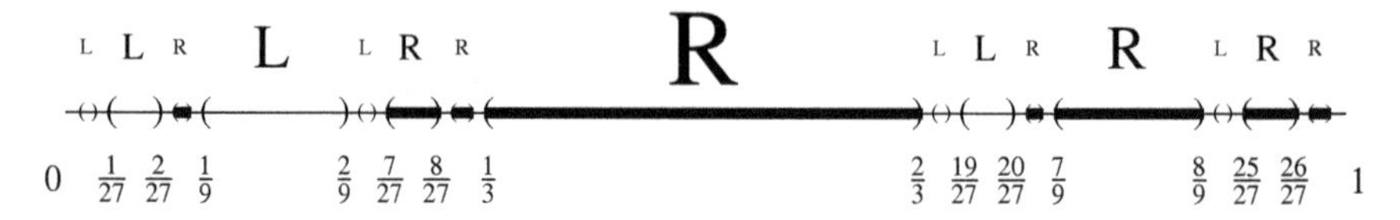

Fig. 10.2 Labeling deleted middle thirds with L and R. The labels appear above the labeled middle thirds: for clarity, we have written the labels of larger middle thirds in larger fonts. The set $\mathcal{R}$ is represented by thicker lines

$n \geq 0$ and where $a_k = 0$ or 2 for all $k \leq n$. It is well known that if we take what's left undeleted after we carry out this process of deleting middle thirds *ad infinitum*, then we get $\mathcal{C} = \mathcal{I} - \bigcup\{J : J \text{ is a middle third}\}$.

Label the deleted middle thirds with L and R, for *left* and *right*, as in Figure 10.2. And let $\mathcal{L}$ be the union of the middle thirds labeled L, and $\mathcal{R}$ be the union of the middle thirds labeled R.

Now suppose that $J \subseteq \mathcal{I}$ is an open interval. Let $f_J : \mathcal{I} \to J$ be the unique increasing linear function from $\mathcal{I}$ onto J. We define $\mathcal{L}(J)$, $\mathcal{R}(J)$, and $\mathcal{C}(J)$ as the images under f_J of $\mathcal{L}$, $\mathcal{R}$, and $\mathcal{C}$ respectively. Thus, $\mathcal{L}(J)$, $\mathcal{R}(J)$, and $\mathcal{C}(J)$ are copies of $\mathcal{L}$, $\mathcal{R}$, and $\mathcal{C}$, respectively. Finally, suppose that $O \subseteq \mathcal{I}$ is open. We say that an open interval $J \subseteq O$ is a *maximal open interval in* O iff, for any open interval $J' \subseteq O$, if $J \cap J' \neq \emptyset$ then $J' \subseteq J$. Note that O is the disjoint union of the maximal open intervals in O. We define

$$\mathcal{L}(O) = \bigcup_{J \text{ is a maximal open interval in } O} \mathcal{L}(J), \tag{10.4}$$

and similarly for $\mathcal{R}(O)$ and $\mathcal{C}(O)$. So $\mathcal{L}(O)$ is the union of copies of $\mathcal{L}$, and similarly for $\mathcal{R}(O)$ and $\mathcal{C}(O)$. Note the following:

Lemma 10.8 *1.* $\mathcal{L}(O)$, $\mathcal{R}(O)$, *and* $\mathcal{C}(O)$ *are pairwise disjoint;*
2. $\mathcal{L}(O)$ *and* $\mathcal{R}(O)$ *are open;*
3. $O = \mathcal{L}(O) \,\dot{\cup}\, \mathcal{R}(O) \,\dot{\cup}\, \mathcal{C}(O)$*; and*
4. $Cl(\mathcal{L}(O)) - \mathcal{L}(O)$
$= Cl(\mathcal{R}(O)) - \mathcal{R}(O)$
$= Cl(\mathcal{C}(O))$
$= Cl(O) - (\mathcal{L}(O) \cup \mathcal{R}(O))$.
5. *If* J *is a maximal open interval in* O*, then* $J \cap \mathcal{C}(O)$ *is nonempty.*

If $S \subseteq \mathcal{I}$ and there is some open interval $J \subseteq S$, then we define the *width* of S as follows: $width(S) = sup\{length(J) : J$ is an open interval and $J \subseteq S\}$. Note the following:

Lemma 10.9 *If O is an open subset of $\mathcal{I}$, then $width(\mathcal{R}(O)) = width(O)/3$ and $width(\mathcal{L}(O)) = width(O)/9$.*

Our next task is to define nonempty open $O_b \subseteq \mathcal{I}$ and other nonempty sets $X_b \subseteq \mathcal{I}$ for each $b \in 2^{<\omega}$, and also to define nonempty sets $X_{\mathbf{b}} \subseteq \mathcal{I}$ for each $\mathbf{b} \in 2^{\omega}$. Once this has been done, we will have a partition of $\mathcal{I}$ into sets $X_{\boldsymbol{b}}$ for each $\boldsymbol{b} \in 2^{\leq\omega}$. We will define $\mathbf{F} : 2^{\leq\omega} \to \mathcal{I}$ as follows: $\mathbf{F}(\boldsymbol{b})$ = the unique $x \in \mathbb{R}$ such that $x \in X_{\boldsymbol{b}}$. And we will show that $\mathbf{F}$ is a surjective interior map.

Define the O_b, for $b \in 2^{<\omega}$, recursively as follows:

$$O_{\Lambda} =_{\text{df}} \mathcal{I} \tag{10.5}$$

$$O_{b0} =_{\text{df}} \mathcal{L}(O_b) \tag{10.6}$$

$$O_{b1} =_{\text{df}} \mathcal{R}(O_b) \tag{10.7}$$

For $b \in 2^{<\omega}$, we define $X_b =_{\text{df}} \mathcal{C}(O_b)$: If $b = \Lambda$, then X_b is simply $\mathcal{C}$, the Cantor set without endpoints; and if b is some other finite binary sequence, then X_b is a union of infinitely many copies of $\mathcal{C}$. Note that each of $\mathcal{L}(O_{\boldsymbol{b}})$ and $\mathcal{R}(O_{\boldsymbol{b}})$ is open in $\mathcal{I}$; that each of $\mathcal{L}(O_{\boldsymbol{b}})$, $\mathcal{R}(O_{\boldsymbol{b}})$, and $\mathcal{C}(O_b)$ is nonempty; and that, $(\forall b \in 2^{<\omega})(O_b = O_{b0} \dot\cup O_{b1} \dot\cup X_b)$. Note the following facts about the O_b and the X_b:

Lemma 10.10 *1. X_b and O_b are nonempty, for each $b \in 2^{<\omega}$.*
2. O_b is open, for each $b \in 2^{<\omega}$.
3. If $b \leq b'$ then $X_{b'} \subseteq O_{b'} \subseteq O_b$.
4. If $b < b'$ then $X_b \cap X_{b'} = X_b \cap O_{b'} = \emptyset$.
5. If $b' \not\leq b \not\leq b'$ then $O_b \cap O_{b'} = \emptyset$.
6. If $b \not\leq b'$ then $O_b \cap X_{b'} = \emptyset$.
7. If $b \neq b'$ then $X_b \cap X_{b'} = \emptyset$.
8. $width(O_b) \leq 1/3^{length(b)}$.

Lemma 10.11 $(\forall b, b' \in 2^{<\omega})(b \leq b' \Rightarrow Cl(X_b) \subseteq Cl(X_{b'}))$.

Proof The fact that $(\forall b \in 2^{\omega})(Cl(X_b) \subseteq Cl(X_{b0}))$ follows immediately from the following, for any $b \in 2^{\omega}$:

1. $O_{b0} = O_{b00} \dot\cup O_{b01} \dot\cup X_{b0}$ (Lemma 10.8, item 3),
2. $Cl(X_{b0}) = Cl(O_{b0}) - (O_{b00} \cup O_{b01})$ (Lemma 10.8, item 4), and
3. $Cl(X_b) = Cl(O_{b0}) - O_{b0}$ (Lemma 10.8, item 4).

Similarly $(\forall b \in 2^{\omega})(Cl(X_b) \subseteq Cl(X_{b1}))$. This suffices for the lemma.

For $\mathbf{b} \in 2^{\omega}$, define $X_{\mathbf{b}} =_{\text{df}} \bigcap_{n \in \mathbb{N}} O_{\mathbf{b}|_n}$.

Lemma 10.12 $\mathcal{I} = \dot\bigcup_{\boldsymbol{b} \in 2^{\leq\omega}} X_{\boldsymbol{b}}$.

Proof The X_b are pairwise disjoint, by Lemma 10.10. To see that $\mathcal{I} = \bigcup_{b \in 2^{\leq\omega}} X_b$, suppose that $x \in \mathcal{I}$, but suppose that $x \notin X_b$, for any $b \in 2^{<\omega}$. It suffices to find a $\mathbf{b} \in 2^\omega$ such that $x \in X_\mathbf{b}$: we will inductively define $b_n \in 2^{<\omega}$, each of length n, so that $b_0 \leq b_1 \leq \ldots \leq b_n \leq b_{n+1} \leq \ldots$, and so that $x \in O_{b_n}$ for each n. Let $b_0 = \Lambda$, the empty sequence. Assume that $x \in O_{b_n}$. Then $x \in O_{b_n 0} \dot{\cup} O_{b_n 1} \dot{\cup} X_{b_n}$. But $x \notin X_{b_n}$. So x is a member of exactly one of $O_{b_n 0}$ and $O_{b_n 1}$. Let b_{n+1} be whichever of $b_n 0$ and $b_n 1$ is such that $x \in O_{b_{n+1}}$. Note that each b_n has length n, that $b_0 \leq b_1 \leq \ldots \leq b_n \leq b_{n+1} \leq \ldots$ and that $x \in O_{b_n}$ for each n. Let $\mathbf{b}$ be the unique member of 2^ω such that $\mathbf{b}|_n = b_n$. Then note that $x \in \bigcap_n O_{\mathbf{b}|_n} = X_\mathbf{b}$, as desired.

Given Lemma 10.12, every $x \in \mathcal{I}$ is in exactly one of the X_b. Let $\mathbf{F}(x) =_{\text{df}}$ the unique $b \in 2^{\leq\omega}$ such that $x \in X_b$. Our final task is to show that $\mathbf{F}$ is a surjective interior map. This follows from Lemma 10.13 ($\mathbf{F}$ is continuous) and Corollary 10.3 ($\mathbf{F}$ is an open surjection), below. Notation: for $S \subseteq 2^{\leq\omega}$, we use *Preimg*$(S)$ for the preimage of S under $\mathbf{F}$; and for $S \subseteq \mathcal{I}$, we use *Img*(S) for the image of S under $\mathbf{F}$.

Lemma 10.13 $\mathbf{F}$ *is continuous.*

Proof Recall that the the sets of the form $\leq^*(b)$, where $b \in 2^{<\omega}$, form a basis for our topology on $2^{\leq\omega}$. Also recall that O_b is open in $\mathcal{I}$, by Lemma 10.10, item 2. So it suffices to show that *Preimg*$(\leq^*(b)) = O_b$, for any $b \in 2^{<\omega}$.

Choose $b \in 2^{<\omega}$. We show, in turn, that (1) $O_b \subseteq$ *Preimg*$(\leq^*(b))$ and that (2) *Preimg*$(\leq^*(b)) \subseteq O_b$.

For (1), choose $x \in O_b$. To show that $x \in$ *Preimg*$(\leq^*(b))$, it suffices to show that $b \leq \mathbf{F}(x)$. Suppose, for a reductio, that $b \nleq \mathbf{F}(x)$. We consider two cases. (Case 1) $\mathbf{F}(x) \in 2^{<\omega}$. Then $O_b \cap X_{\mathbf{F}(x)} = \emptyset$, since $b \nleq \mathbf{F}(x)$ and by Lemma 10.10, item 6. But then, since $x \in X_{\mathbf{F}(x)}$, we have $x \notin O_b$, a contradiction. (Case 2) $\mathbf{F}(x) \in 2^\omega$. Let $n = |b|$. Then $b \neq \mathbf{F}(x)|_n$, since $b \nleq \mathbf{F}(x)$. So $b \nleq \mathbf{F}(x)|_n \nleq b$, since b and $\mathbf{F}(x)|_n$ are of the same length. So $O_b \cap O_{\mathbf{F}(x)|_n} = \emptyset$, Lemma 10.10, item 5. But $x \in X_{\mathbf{F}(x)} = \bigcap_{k \in \mathbb{N}} O_{\mathbf{F}(x)|_k} \subseteq O_{\mathbf{F}(x)|_n}$. So $x \notin O_b$, a contradiction.

For (2), choose $x \in$ *Preimg*$(\leq^*(b))$. Then $\mathbf{F}(x) \in \leq^*(b)$, so that $b \leq \mathbf{F}(x)$. Recall that $x \in X_{\mathbf{F}(x)}$: so, to show that $x \in O_b$, it suffices to show that $X_{\mathbf{F}(x)} \subseteq O_b$. If $\mathbf{F}(x) \in 2^{<\omega}$ then $X_{\mathbf{F}(x)} \subseteq O_b$, by Lemma 10.10, item 3. Suppose, on the other hand, that $\mathbf{F}(x) \in 2^\omega$. Since $b \leq \mathbf{F}(x)$, we get $b = \mathbf{F}(x)|_k$ for some $k \in \mathbb{N}$. Thus, $X_{\mathbf{F}(x)} = \bigcap_{n \in \mathbb{N}} O_{\mathbf{F}(x)|_n} \subseteq O_{\mathbf{F}(x)|_k} = O_b$.

Lemma 10.14 *Suppose that $J \subseteq \mathcal{I}$ is an open interval, $b \in Img(J) \cap 2^{<\omega}$, $b' \in 2^{<\omega}$ and $b \leq b'$. Then $b' \in Img(J)$.*

Proof Choose $x \in J$ with $\mathbf{F}(x) = b$. Then $x \in X_b$. So $x \in Cl(X_{b'})$, by Lemma 10.11. So there is some $y \in X_{b'} \cap J$. So $b' \in Img(J)$, since $\mathbf{F}(y) = b'$.

Lemma 10.15 *Suppose that $J \subseteq \mathcal{I}$ is an open interval, $b \in Img(J) \cap 2^{<\omega}$, $\mathbf{b}' \in 2^\omega$ and $b \leq \mathbf{b}'$. Then $\mathbf{b}' \in Img(J)$.*

Proof Let $n = length(b)$, so that $b = \mathbf{b}'|_n$. We will now inductively choose open intervals $J_0, J_1, ... \subseteq J \cap O_b$ and points $x_0 \in J_0, x_1 \in J_1, \ldots$ so that $F(x_k) = \mathbf{b}'|_{n+k}$, for each $k \geq 0$.

First, choose $x_0 \in J$ such that $\mathbf{F}(x_0) = b = \mathbf{b}'|_n$. Since $x_0 \in J \cap O_b$, we can choose an open interval J_0 so that $x_0 \in J_0$ and $Cl(J_0) \subseteq J \cap O_b$. Suppose that we have chosen an open interval J_k and a point $x_k \in J_k$ with $F(x_k) = \mathbf{b}'|_{n+k}$. Then $\mathbf{b}'|_{n+k} \in Img(J_k)$. So $\mathbf{b}'|_{n+k+1} \in Img(J_k)$, by Lemma 10.14. So there is an $x_{k+1} \in Img(J_k)$ with $\mathbf{F}(x_{k+1}) = \mathbf{b}'|_{n+k+1}$. Note that $x_{k+1} \in X_{\mathbf{b}'|_{n+k+1}} \subseteq O_{\mathbf{b}'|_{n+k+1}}$. So $x_{k+1} \in J_k \cap O_{\mathbf{b}'|_{n+k+1}}$. So we can choose an open interval J_{k+1} with $x_{k+1} \in J_{k+1}$ and $Cl(J_{k+1}) \subseteq J_k \cap O_{\mathbf{b}'|_{n+k+1}}$.

Note: $Cl(J_{k+1}) \subseteq J_k$ for each $k \geq 0$. So $\langle Cl(J_k)\rangle_k$ is a decreasing sequence of closed intervals. So $\bigcap_k Cl(J_k)$ is nonempty. Also, $\bigcap_k Cl(J_k) \subseteq J$ and $\bigcap_k Cl(J_k) \subseteq \bigcap_k O_{\mathbf{b}'|_{n+k}}$. So there is a point $x \in \bigcap_k Cl(J_k) \subseteq J \cap X_{\mathbf{b}'}$. So $\mathbf{F}(x) = \mathbf{b}'$ and $x \in J$. So $\mathbf{b}' \in Img(J)$.

Lemma 10.16 *Suppose that $J \subseteq \mathcal{I}$ is an open interval and $\mathbf{b} \in Img(J) \cap 2^\omega$. Then there is a $b' \in Img(J) \cap 2^{<\omega}$ with $b' \leq \mathbf{b}$.*

Proof Suppose that $J \subseteq \mathcal{I}$ is an open interval and $\mathbf{b} \in Img(J) \cap 2^\omega$. Choose $x \in J$ with $\mathbf{F}(x) = \mathbf{b}$, and choose a positive real number d so that $(x - d, x + d) \subseteq J$. Choose $n \in \mathbb{N}$ with $1/3^n < d$ and let $b' = \mathbf{b}|_n \in 2^{<\omega}$. Note that $x \in O_{b'} \cap (x - d, x + d)$; also, $width(O_{b'}) < d$, by Lemma 10.10, item 8. Let J' be any maximal open interval in $O_{b'}$ with $x \in J'$, and note two things about J': (1) J' has length $\leq width(O_{b'}) < d$, since J' is an open interval and $J' \subseteq O_{b'}$; and (2) $J' \cap \mathcal{C}(O_{b'})$ is nonempty, by Lemma 10.8, item 5. By (2), there is an $x' \in J' \cap X_{b'}$, and by (1) $J' \subseteq (x - d, x + d)$. So $x' \in J$ and $\mathbf{F}(x') = b'$. So $b' \in Img(J)$.

Corollary 10.2 *$Img(J)$ is open in $2^{\leq\omega}$, for every interval $J \subseteq \mathcal{I}$.*

Proof It suffices to show that $Img(J) = \bigcup_{b \in Img(J)\cap 2^{<\omega}} \leq^*(b)$. So consider any interval $J \subseteq \mathcal{I}$. We will show, in turn, that (1) $\bigcup_{b \in Img(J)\cap 2^{<\omega}} \leq^*(b) \subseteq Img(J)$ and that (2) $Img(J) \subseteq \bigcup_{b \in Img(J)\cap 2^{<\omega}} \leq^*(b)$.

For (1), by Lemmas 10.14 and 10.15, if $b \in Img(J) \cap 2^{<\omega}$ then $\leq^*(b) \subseteq Img(J)$. So $\bigcup_{b \in Img(J)\cap 2^{<\omega}} \leq^*(b) \subseteq Img(J)$.

For (2), note that if $\boldsymbol{b} \in Img(J)$, then there is some $b' \leq \boldsymbol{b}$ such that $b' \in Img(J) \cap 2^{<\omega}$: this follows from Lemma 10.16 if $\boldsymbol{b} \in 2^\omega$; and it is trivial if $\boldsymbol{b} \in 2^{<\omega}$, since we can just let $b' = \boldsymbol{b}$. Thus, if $\boldsymbol{b} \in Img(J)$ then there exists $b' \in 2^{<\omega}$ with $\boldsymbol{b} \in \leq^*(b') \subseteq Img(J)$. Thus, $Img(J) \subseteq \bigcup_{b \in Img(J)\cap 2^{<\omega}} \leq^*(b)$.

Corollary 10.3 **F** *is an open surjection.*

Proof The openness of **F** follows immediately from Corollary 10.2. It remains to show that **F** is surjective – equivalently, that $\boldsymbol{b} \in Img(\mathcal{I})$, for every $\boldsymbol{b} \in 2^{\leq\omega}$. By Corollary 10.2, $Img(\mathcal{I})$ is open and therefore upwardly closed under $\leq$. Also, $\Lambda \in Img(\mathcal{I})$, since $\Lambda = \mathbf{F}(x)$ for any $x \in X_\Lambda = \mathcal{C}(\mathcal{I}) = \mathcal{C}$. So, $\boldsymbol{b} \in Img(\mathcal{I})$, for every $\boldsymbol{b} \in 2^{\leq\omega}$, since $\Lambda \leq \boldsymbol{b}$ and $Img(\mathcal{I})$ is upwardly closed under $\leq$.

Acknowledgements Thanks to the audience at the Ninth International Tbilisi Symposium on Language, Logic and Computation (2011) in Kutaisi, Georgia, for listening to me present the more general paper, Kremer (2013). Special thanks to each of David Gabelaia, Nick Bezhanishvili, Roman

Kontchakov, and Mamuka Jibladze, for indulging me by letting me explain the proof in detail in the case considered by this paper, $\mathbb{R}$. Also, a big thanks to Robert Goldblatt, for carefully reading a draft of this paper and for very useful comments.

References

Aiello, M., van Benthem, J., & Bezhanishvili, G. (2003). Reasoning about space: The modal way. *Journal of Logic and Computation*, *13*(6), 889–920.

Bezhanishvili, G., & Gehrke, M. (2005). Completeness of S4 with respect to the real line: revisited. *Annals of Pure and Applied Logic*, *131*(1–3), 287–301.

Freyd, P. J., & Scedov, A. (1990). *Categories, allegories*. Amsterdam: North-Holland.

Goldblatt, R. (1980). Diodorean modality in Minkowski spacetime. *Studia Logica*, *39*(2–3), 219–236.

Goldblatt, R., & Hodkinson, I. (2019). Strong completeness of modal logics over 0-dimensional metric spaces. *Review of Symbolic Logic* (pp. 1–22). Published online, 24 October 2019.

Kremer, P. (2013). Strong completeness of S4 for any dense-in-itself metric space. *Review of Symbolic Logic*, *6*(3), 545–570.

Kremer, P. (2014). Quantified modal logic on the rational line. *Review of Symbolic Logic*, *7*(3), 439–454.

Lando, T. (2012). Completeness of *S*4 for the Lebesgue measure algebra. *Journal of Philosophical Logic*, *41*(2), 287–316.

McKinsey, J. C. C. (1941). A solution of the decision problem for the Lewis systems S2 and S4, with an application to topology. *Journal of Symbolic Logic*, *6*, 117–134.

McKinsey, J. C. C., & Tarski, A. (1944). The algebra of topology. *Annals of Mathematics*, *2*(45), 141–191.

Mints, G., & Zhang, T. (2005). A proof of topological completeness for *S*4 in (0, 1). *Annals of Pure and Applied Logic*, *133*(1–3), 231–245.

Rasiowa, H., & Sikorski, R. (1963). *The mathematics of metamathematics*. Monografie Matematyczne, Tom 41. Państwowe Wydawnictwo Naukowe, Warsaw.

Tarski, A. (1938). Der Aussagenkalkül und die Topologie. *Fundamenta Mathematicae*, *31*, 103–134.

van Benthem, J., Bezhanishvili, G., ten Cate, B., & Sarenac, D. (2006). Multimodal logics of products of topologies. *Studia Logica*, *84*(3), 369–392.

Vickers, S. (1989). *Topology via logic*. Cambridge tracts in theoretical computer science (Vol. 5). Cambridge: Cambridge University Press.

Chapter 11
Modal Logics of Some Hereditarily Irresolvable Spaces

Robert Goldblatt

Second Reader
G. Bezhanishvili
New Mexico State University

Abstract A topological space is *hereditarily k-irresolvable* if none of its subspaces can be partitioned into k dense subsets. We use this notion to provide a topological semantics for a sequence of modal logics whose n-th member $K4\mathbb{C}_n$ is characterised by validity in transitive Kripke frames of circumference at most n. We show that under the interpretation of the modality $\Diamond$ as the derived set (of limit points) operation, $K4\mathbb{C}_n$ is characterised by validity in all spaces that are hereditarily $n+1$-irresolvable and have the T_D separation property. We also identify the extensions of $K4\mathbb{C}_n$ that result when the class of spaces involved is restricted to those that are crowded, or densely discrete, or openly irresolvable, the latter meaning that every non-empty open subspace is 2-irresolvable. Finally, we give a topological semantics for K4M, where M is the McKinsey axiom.

Keywords Modal logic · Kripke frame · Circumference · Topological semantics · Derived set · Resolvable space · Hereditarily irresolvable · Openly irresolvable · Dense · Crowded · Scattered space · Alexandrov topology

2020 Mathematics Subject Classification 03B45 · 54F99

R. Goldblatt (✉)
School of Mathematics and Statistics, Victoria University of Wellington, Kelburn, New Zealand
e-mail: rob.goldblatt@msor.vuw.ac.nz

I. Düntsch and E. Mares (eds.), *Alasdair Urquhart on Nonclassical and Algebraic Logic and Complexity of Proofs*, Outstanding Contributions to Logic 22,
https://doi.org/10.1007/978-3-030-71430-7_11

11.1 Introduction

One theme of this article is the use of *geometric* ideas in the semantic analysis of logical systems. Another is the use of *relational* semantics in the style of Kripke. Both themes feature prominently in the many-faceted research portfolio of Alasdair Urquhart. In the area of relevant logic, he discovered how to construct certain relational models of relevant implication out of projective geometries, and related these to modular geometric lattices (Urquhart 1983, see also Urquhart 2017). This led to his striking demonstration in Urquhart (1984) that the main systems of relevant implication are undecidable, and then to a proof in Urquhart (1993) that these systems fail to satisfy the Craig interpolation theorem. In the area of Kripke semantics for modal logics, in addition to an early book with Rescher on temporal logics (Rescher and Urquhart 1971), his contributions have included the construction in Urquhart (1981) of a normal logic that is recursively axiomatisable and has the finite model property but is undecidable; and the proof in Urquhart (2015) that there is a formula of quantified S5 which, unlike the situation of propositional S5, is not equivalent to any formula of modal degree one. His article Urquhart (1978) on topological representation of lattices has been very influential. It assigned to any bounded lattice a dual topological space with a double ordering and showed that the original lattice can be embedded into a lattice of certain 'stable' subsets of its dual space. This generalised the topological representations of Stone (1936) for Boolean algebras and Priestley (1970) for distributive lattices. It stimulated the development of further duality theories for lattices, such as those of Hartung (1992), Hartung (1993), Allwein and Hartonas (1993), Ploščica (1995) and Hartonas and Dunn (1997). The structure of Urquhart's dual spaces was exploited by Allwein and Dunn (1993) to develop a Kripke semantics for linear logic, and by Dzik et al. (2006) to do likewise for non-distributive logics with various negation operations. Urquhart's article also played an important role in the development of the important notion of a *canonical extension*. This was first introduced by Jónsson and Tarski (1951) for Boolean algebras with operators and is closely related to the notion of canonical model of a modal logic. After several decades of evolution, Gehrke and Harding (2001) gave an axiomatic definition of a canonical extension of any bounded lattice-based algebra, showing that it is unique up to isomorphism. In proving that such an extension exists, they observed that it can be constructed as the embedding of the original algebra into the lattice of stable subsets of its Urquhart dual space. Craig and Haviar (2014) have clarified the relationship between Urquhart's construction and other manifestations of canonical extensions. This area has now undergone substantial development and application, by Gehrke and others, of theories of canonical extensions and duality for lattice expansions. There are surveys of this progress in Gehrke (2018) and Goldblatt (2018). In Goldblatt (2020) a new notion of bounded morphism between polarity structures has been introduced that provides a duality with homomorphisms between lattices with operators. The seed that Urquhart planted has amply borne fruit.

The geometrical ideas of the present article come from topology. Our aim is to provide a topological semantics for a sequence of modal logics that were originally defined by properties of their binary-relational Kripke models. The nth member of this sequence is called $K4\mathbb{C}_n$, and is the smallest normal extension of the logic K4 that includes an axiom scheme $\mathbb{C}_n$ which will be described below. It was shown in Goldblatt (2021) that the theorems of $K4\mathbb{C}_n$ are characterised by validity in all finite transitive Kripke frames that have circumference at most n, meaning that any cycle in the frame has length at most n, or equivalently that any non-degenerate cluster has at most n elements. For $n \geq 1$, adding the reflexivity axiom $\varphi \to \Diamond \varphi$ to $K4\mathbb{C}_n$ gives the logic $S4\mathbb{C}_n$ which is characterised by validity in all finite reflexive transitive Kripke frames that have circumference at most n. It was also shown that $S4\mathbb{C}_n$ has a topological semantics in which any formula of the form $\Diamond \varphi$ is interpreted as the topological closure of the interpretation of φ. Under this interpretation, $S4\mathbb{C}_n$ is characterised by validity in all topological spaces that are *hereditarily* $n+1$-*irresolvable*. Here, a space is called k-*resolvable* if it can be partitioned into k dense subsets, and is hereditarily k-irresolvable if none of its non-empty subspaces is k-resolvable. k could be any cardinal, but we will deal only with finite k.

The interpretation of $\Diamond$ as closure, now known as C-semantics, appears to have first been considered by Tang (1938). McKinsey (1941) showed that the formulas that are C-valid in all topological spaces are precisely the theorems of S4. McKinsey and Tarski (1944, 1948) then undertook a much deeper analysis which showed that S4 is characterised by C-validity in any given space that has a certain 'dissectability' property that is possessed by every finite-dimensional Euclidean space, and more generally by any metric space that is crowded, or dense-in-itself, i.e. has no isolated points. They also suggested studying an alternative topological interpretation, now called d-semantics, in which $\Diamond \varphi$ is taken to be the derived set, i.e. the set of limit points, of the interpretation of φ. This d-semantics does not validate the transitivity axiom $\Diamond \Diamond \varphi \to \Diamond \varphi$. That is d-valid in a given space iff the space satisfies the T_D property that every derived set is closed.

The logic $K4\mathbb{C}_0$ is in fact the Gödel-Löb provability logic, which is known to have a topological characterisation by d-validity in all spaces that are *scattered*, meaning that every non-empty subspace has an isolated point. The logic $K4\mathbb{C}_1$ was shown by Gabelaia (2004) to be characterised by d-validity in hereditarily 2-irresolvable spaces. Such spaces satisfy the T_D property, which however may not hold in hereditarily k-irresolvable spaces for $k > 2$. The principal new result proven here is that for $n > 1$, $K4\mathbb{C}_n$ is characterised by d-validity in all spaces that are both hereditarily $n+1$-irresolvable and T_D. Our proof adapts the argument of Gabelaia (2004) and also makes use of certain spaces that are n-resolvable but not $n+1$-resolvable. These are provided by a construction of El'kin (1969b).

Sections 11.2 and 11.3 review the theory of relational and topological semantics for modal logic that we will be using. Section 11.4 is the heart of the article: it discusses hereditary irresolvability and gives the constructions that lead to our characterisation of $K4\mathbb{C}_n$, Sect. 11.5 gives characterisations of some extensions of $K4\mathbb{C}_n$ that correspond to further topological constraints.

11.2 Frames and Logics

We begin with a review of relation semantics for propositional modal logic. A standard reference is Blackburn et al. (2001). Formulas $\varphi, \psi, \ldots$ are constructed from some denumerable set of propositional variables by the Boolean connectives $\top$, $\bot$, $\neg$, $\wedge$, $\vee$, $\to$ and the unary modalities $\Diamond$ and $\Box$. We write $\Box^* \varphi$ as an abbreviation of the formula $\varphi \wedge \Box \varphi$, and $\Diamond^* \varphi$ for $\varphi \vee \Diamond \varphi$.

A *frame* $\mathcal{F} = (W, R)$ consists of a binary relation R on a set W. Each $x \in W$ has the set $R(x) = \{y \in W : xRy\}$ of *R-successors*. A *model* $\mathcal{M} = (W, R, V)$ on a frame has a valuation function V assigning to each variable p a subset $V(p)$ of W. The model then assigns to each formula φ a subset $\mathcal{M}(\varphi)$ of W, thought of as the set of points at which φ is true. These truth sets are defined by induction on the formation of φ, putting $\mathcal{M}(p) = V(p)$ for each variable p, interpreting each Boolean connective by the corresponding Boolean operation on subsets of W and putting

$$\mathcal{M}(\Diamond \varphi) = R^{-1}\mathcal{M}(\varphi) = \{x \in W : R(x) \cap \mathcal{M}(\varphi) \neq \emptyset\},$$
$$\mathcal{M}(\Box \varphi) = \{x \in W : R(x) \subseteq \mathcal{M}(\varphi)\}.$$

A formula φ is *true in model* $\mathcal{M}$, written $\mathcal{M} \models \varphi$, if $\mathcal{M}(\varphi) = W$; and is *valid in frame* $\mathcal{F}$, written $\mathcal{F} \models \varphi$, if it is true in all models on $\mathcal{F}$.

A *normal logic* is any set L of formulas that includes all tautologies and all instances of the scheme K: $\Box(\varphi \to \psi) \to (\Box \varphi \to \Box \psi)$, and whose rules include modus ponens and $\Box$-generalisation (from φ infer $\Box \varphi$). The members of a logic L may be referred to as the *L-theorems*. The set $L_{\mathcal{C}}$ of all formulas valid in (all members of) some given class $\mathcal{C}$ of frames is a normal logic. A logic is *characterised by validity in* $\mathcal{C}$, or is *sound and complete for validity in* $\mathcal{C}$, if it is equal to $L_{\mathcal{C}}$. The smallest normal logic, known as K, is characterised by validity in all frames.

A logic is *transitive* if it contains all instances of the scheme 4: $\Diamond \Diamond \varphi \to \Diamond \varphi$, which is valid in precisely those frames that are transitive, i.e. their relation R is transitive. The smallest transitive normal logic, known as K4, is characterised by validity in all transitive frames.

In a transitive frame $\mathcal{F} = (W, R)$, a *cluster* is a subset C of W that is an equivalence class under the equivalence relation $\{(x, y) : x = y \text{ or } xRyRx\}$. The R-cluster containing x is $C_x = \{x\} \cup \{y : xRyRx\}$. If x is irreflexive, i.e. not xRx, then $C_x = \{x\}$, and C_x is called a *degenerate* cluster. Thus, if R is an irreflexive relation, then all clusters are degenerate. If xRx, then C_x is *non-degenerate*. If C is a non-degenerate cluster then it contains no irreflexive points and the relation R is universal on C and maximally so. A *simple* cluster is non-degenerate with one element, i.e. a singleton $C_x = \{x\}$ with xRx. If R is *antisymmetric*, i.e. $xRyRx$ implies $x = y$ in general, then every cluster is a singleton, so is either simple or degenerate.

The relation R lifts to a well-defined relation on the set of clusters by putting $C_x R C_y$ iff xRy. This relation is transitive and antisymmetric on the set of clusters. A cluster C_x is *final* if it is maximal in this ordering, i.e. there is no cluster $C \neq C_x$ with $C_x RC$. This is equivalent to requiring that xRy implies yRx.

We take the *circumference* of a frame $\mathfrak{F}$ to be the supremum of the set of all lengths of cycles in $\mathfrak{F}$, where a cycle of length $n \geq 1$ is a finite sequence $x_1, \ldots, x_n$ of distinct points such that $x_1 R \cdots R x_n R x_1$. The circumference is 0 iff there are no cycles. In a transitive frame, the points of any cycle are R-related to each other and are reflexive, and all belong to the same non-degenerate cluster. Conversely, any finite non-empty subset of a non-degenerate cluster can be arranged (arbitrarily) into a cycle. Thus, if a finite transitive frame has circumference $n \geq 1$, then n is equal to the size of a largest non-degenerate cluster. The circumference is 0 iff the frame is irreflexive, i.e. has only degenerate clusters. A finite transitive frame has circumference *at most n* iff each of its non-degenerate clusters has at most n members.

Now given formulas $\varphi_0, \ldots, \varphi_n$, define the formula $\mathbb{P}_n(\varphi_0, \ldots, \varphi_n)$ to be

$$\Diamond(\varphi_1 \wedge \Diamond(\varphi_2 \wedge \cdots \wedge \Diamond(\varphi_n \wedge \Diamond \varphi_0)) \cdots)$$

provided that $n \geqslant 1$. For the case $n = 0$, put $\mathbb{P}_0(\varphi_0) = \Diamond \varphi_0$. Let $\mathbb{D}_n(\varphi_0, \ldots, \varphi_n)$ be $\bigwedge_{i<j\leqslant n} \neg(\varphi_i \wedge \varphi_j)$ (for $n = 0$ this is the empty conjunction $\top$). Define $\mathbb{C}_n$ to be the scheme

$$\Box^* \mathbb{D}_n(\varphi_0, \ldots, \varphi_n) \to (\Diamond \varphi_0 \to \Diamond(\varphi_0 \wedge \neg \mathbb{P}_n(\varphi_0, \ldots, \varphi_n)).$$

Let K4$\mathbb{C}_n$ be the smallest transitive normal logic that includes the scheme $\mathbb{C}_n$. In Goldblatt (2021), Theorems 1 and 4, the following were shown.

Theorem 11.1 *For all* $n \geq 0$,

(1) *A transitive frame validates* $\mathbb{C}_n$ *iff it has circumference at most n and no strictly ascending chains (i.e. no sequence* $\{x_m : m < \omega\}$ *with* $x_m R x_{m+1}$ *but not* $x_{m+1} R x_m$ *for all m).*
(2) *A finite transitive frame validates* $\mathbb{C}_n$ *iff it has circumference at most n.*
(3) *A formula is a theorem of K4*$\mathbb{C}_n$ *iff it is valid in all finite transitive frames that have circumference at most n.* □

The cases $n = 0, 1$ were described in detail in Goldblatt (2021). K4$\mathbb{C}_0$ is equal to the Gödel-Löb modal logic of provability, the smallest normal logic to contain the Löb axiom $\Box(\Box \varphi \to \varphi) \to \Box \varphi$. It is characterised by validity in all finite transitive frames that are irreflexive, i.e. all clusters are degenerate. This was first shown by Segerberg (1971), and there have been a number of other proofs since (see Boolos 1993, Chap. 5). K4$\mathbb{C}_1$ is equal to the logic K4Grz$_\Box$, where Grz$_\Box$ is the axiom

$$\Box(\Box(p \to \Box p) \to p) \to \Box p.$$

This logic was shown by Gabelaia (2004) to be characterised by validity in all finite frames whose clusters are all singletons (which may individually be arbitrarily degenerate or simple).

Parts (1) and (3) of Theorem 11.1 imply that K4$\mathbb{C}_n$ is a *subframe logic* in the sense of Fine (1985) that it is complete for validity in its class of validating Kripke frames and this class is closed under subframes.

Part (2) of Theorem 11.1 states that validity of $\mathbb{C}_n$ exactly expresses the property of having circumference at most n over the class of *finite* transitive frames. But for transitive frames in general, there is no such formula or even set of formulas:

Theorem 11.2 *There does not exist any set of modal formulas whose conjoint validity in any transitive frame exactly captures the property of having circumference at most $n \geq 0$.*

Proof Let $\mathfrak{F}_\omega$ be the transitive irreflexive frame $(\omega, <)$, and for each $1 \leq m < \omega$ let $\mathfrak{F}_m$ be the frame consisting of a single non-degenerate m-element cluster on the set $\{0, \ldots, m-1\}$. Let $f_m : \mathfrak{F}_\omega \to \mathfrak{F}_m$ be the map $f_m(x) = x \bmod m$. Then f_m is a surjective bounded morphism, a type of map that preserves validity of modal formulas in frames (Blackburn et al. 2001, Theorem 3.14). Hence, any formula valid in $\mathfrak{F}_\omega$ will be valid in $\mathfrak{F}_m$ for every $m \geq 1$.

Now suppose, for the sake of contradiction, that there does exist a set Φ_n of formulas that is valid in a transitive frame iff that frame has a circumference at most n. Then $\mathfrak{F}_\omega \models \Phi_n$, since $\mathfrak{F}_\omega$ has circumference 0, and therefore has circumference at most n, what ever $n \geq 0$ is chosen. But then by the previous paragraph, $\mathfrak{F}_{n+1} \models \Phi_n$, which contradicts the definition of Φ_n because $\mathfrak{F}_{n+1}$ has circumference $n + 1$. □

11.3 Topological Semantics

We first review some of the topological ideas that will be used to interpret modal formulas. These can be found in many textbooks, such as Willard (1970), Engelking (1977).

If X is a topological space, we do not introduce a symbol for the topology of X, but simply refer to various subsets as being open or closed *in* X. If $x \in X$, then an *open neighbourhood of* x is any open set that contains x. A subset of X *intersects* another subset if the two subsets have non-empty intersection. If $Y \subseteq X$, then x is a *closure point* of Y if every open neighbourhood of x intersects Y. The set of all closure points of Y is the *closure* of Y, denoted $\mathsf{Cl}_X\, Y$. It is the smallest closed superset of Y in X. It has $Y \subseteq \mathsf{Cl}_X\, Y = \mathsf{Cl}_X\, \mathsf{Cl}_X\, Y$. A useful fact is that if O is open in X, then $O \cap \mathsf{Cl}_X\, Y \subseteq \mathsf{Cl}_X(O \cap Y)$. Y is called *dense in* X when $\mathsf{Cl}_X\, Y = X$. This means that every non-empty open set intersects Y.

We write $\mathsf{Int}_X\, Y$ for the interior of Y in X, the largest open subset of Y. Thus, $x \in \mathsf{Int}_X\, Y$ iff some open neighbourhood of x is included in Y.

Any subset S of X becomes a *subspace* of X under the topology whose open sets are all sets of the form $S \cap O$ with O open in X. The closure operator Cl_S of the subspace S satisfies $\mathsf{Cl}_S\, Y = S \cap \mathsf{Cl}_X\, Y$. Hence, a set $Y \subseteq S$ is dense in S, i.e. $\mathsf{Cl}_S\, Y = S$, iff $S \subseteq \mathsf{Cl}_X\, Y$.

A *punctured neighbourhood of* x is any set of the form $O - \{x\}$ with O an open neighbourhood of x. If $Y \subseteq X$, then x is a *limit point* of Y in X if every punctured neighbourhood of x intersects Y. The set of all limit points of Y is the *derived set* of Y, which we denote $\mathsf{De}_X\, Y$. (Other notations are Y', dY and $\langle d \rangle Y$.) In general,

$\mathsf{Cl}_X\, Y = Y \cup \mathsf{De}_X\, Y$. Another useful fact is that if O is open in X, then $O \cap \mathsf{De}_X\, Y \subseteq \mathsf{De}_X(O \cap Y)$. If S is a subspace of X, then any $Y \subseteq S$ has $\mathsf{De}_S\, Y = S \cap \mathsf{De}_X\, Y$.

A T_D *space* is one in which the derived set $\mathsf{De}_X\{x\}$ of any singleton is closed, which is equivalent to requiring that $\mathsf{De}_X\, \mathsf{De}_X\{x\} \subseteq \mathsf{De}_X\{x\}$. This in turn is equivalent to the requirement that any derived set $\mathsf{De}_X\, Y$ is closed, i.e. that $\mathsf{De}_X\, \mathsf{De}_X\, Y \subseteq \mathsf{De}_X\, Y$ for all $Y \subseteq X$ (Aull and Thron 1962, Theorem 5.1). The T_D property is strictly weaker than the T_1 separation property that $\mathsf{De}_X\{x\} = \emptyset$ in general, which is itself equivalent to the requirement that any singleton is closed, and to the requirement that any two distinct points each have an open neighbourhood that excludes the other point. The simplest example of a non-T_D space is a two-element space $X = \{x, y\}$ with the *indiscrete* (or *trivial*) topology in which only X and $\emptyset$ are open. It has $\mathsf{De}_X\{x\} = \{y\}$ and $\mathsf{De}_X\{y\} = \{x\}$. A useful known fact is

Lemma 11.1 *A space X is T_D iff it has $x \notin \mathsf{Cl}_X\, \mathsf{De}_X\{x\}$ for all $x \in X$.*

Proof Since $x \notin \mathsf{De}_X\{x\}$ in general, if X is T_D then $x \notin \mathsf{Cl}_X\, \mathsf{De}_X\{x\}$ as $\mathsf{Cl}_X\, \mathsf{De}_X\{x\} = \mathsf{De}_X\{x\}$. Conversely, since $\mathsf{Cl}_X\{x\}$ is a closed superset of $\mathsf{De}_X\{x\}$, it includes $\mathsf{Cl}_X\, \mathsf{De}_X\{x\}$, so we have $\mathsf{Cl}_X\, \mathsf{De}_X\{x\} \subseteq \mathsf{Cl}_X\{x\} = \mathsf{De}_X\{x\} \cup \{x\}$. Thus, if x is not in $\mathsf{Cl}_X\, \mathsf{De}_X\{x\}$, then $\mathsf{Cl}_X\, \mathsf{De}_X\{x\} \subseteq \mathsf{De}_X\{x\}$, hence, $\mathsf{De}_X\{x\}$ is closed. □

We now review the topological semantics for modal logics (Bezhanishvili et al. 2005). A *topological model* $\mathcal{M} = (X, V)$ on a space X has a valuation V assigning a subset of X to each propositional variable. A truth set $\mathcal{M}_d(\varphi)$ is then defined by induction on the formation of φ by letting $\mathcal{M}_d(p) = V(p)$, interpreting the Boolean connectives by the corresponding Boolean set operations and putting $\mathcal{M}_d(\Diamond\, \varphi) = \mathsf{De}_X(\mathcal{M}_d(\varphi))$, the set of limit points of $\mathcal{M}_d(\varphi)$. Then $\mathcal{M}_d(\Box\, \varphi)$ is determined by the requirement that it be equal to $\mathcal{M}_d(\neg\Diamond\neg\varphi)$. This gives

- $x \in \mathcal{M}_d(\Diamond\, \varphi)$ iff every punctured neighbourhood of x intersects $\mathcal{M}_d(\varphi)$;
- $x \in \mathcal{M}_d(\Box\, \varphi)$ iff there is a punctured neighbourhood of x included in $\mathcal{M}_d(\varphi)$.

A formula φ is *d-true in* $\mathcal{M}$, written $\mathcal{M} \models_d \varphi$, if $\mathcal{M}_d(\varphi) = X$; and is *d-valid in space* X, written $X \models_d \varphi$, if it is d-true in all models on X. The set $\{\varphi : X \models_d \varphi\}$ of all formulas that are d-valid in X is a normal logic, called the *d-logic of* X. It need not be a transitive logic, because the scheme 4 is d-valid in X iff X is T_D (Esakia 2001, 2004).

A point x is *isolated* in a space X if $\{x\}$ is open in X. If X has no isolated points, it is *crowded* (also called *dense-in-itself*). This means that every point is a limit point of X, i.e. $\mathsf{De}_X\, X = X$. Thus, X is crowded iff $X \models_d \Diamond\top$. A subset S is called *crowded in* X if it is crowded as a subspace, i.e. $\mathsf{De}_S\, S = S$, which is equivalent to requiring that $S \subseteq \mathsf{De}_X\, S$, and hence that $\mathsf{Cl}_X\, S = \mathsf{De}_X\, S$, since $\mathsf{Cl}_X\, S = S \cup \mathsf{De}_X\, S$. The following is standard.

Lemma 11.2 *Let O and S be subsets of X with O open. If S is dense in X then $O \cap S$ is dense in O, and if S is crowded in X then $O \cap S$ is crowded in O.*

Proof If $\mathsf{Cl}_X\, S = X$, then $O = O \cap \mathsf{Cl}_X\, S \subseteq \mathsf{Cl}_X(O \cap S)$, showing that $O \cap S$ is dense in O. If S is crowded, then $O \cap S \subseteq O \cap \mathsf{De}_X\, S \subseteq \mathsf{De}_X(O \cap S)$, so $O \cap S$ is crowded. □

We now explain a relationship between d-validity and frame validity. Let $\mathcal{F} = (W, R)$ be a transitive frame. There is an associated *Alexandrov topology* on W in which the open subsets O are those that are *up-sets* under R, i.e. if $w \in O$ and wRv then $v \in O$. Call the resulting topological space W_R. Let $R^* = R \cup \{(w, w) : w \in W\}$ be the reflexive closure of R. Then R^* is a *quasi-order*, i.e. is reflexive and transitive, and the topology of W_R has as a basis the sets $R^*(w)$ for all $w \in W$, where $R^*(w) = \{v : wR^*v\} = \{w\} \cup R(w)$..

A *d-morphism* from a space X to $\mathcal{F}$ is a function $f : X \to W$ that has the following properties:

(i) f is a continuous and open function from X to the space W_R.
(ii) If $w \in W$ is R-reflexive, then the preimage $f^{-1}\{w\}$ is crowded in X.
(iii) If w is R-irreflexive, then $f^{-1}\{w\}$ is a discrete subspace of X, i.e. each point of $f^{-1}\{w\}$ is isolated in $f^{-1}\{w\}$, or equivalently $f^{-1}\{w\} \cap \mathsf{De}_X\, f^{-1}\{w\} = \emptyset$.

In (i), f is continuous when the f-preimage of any open subset of W_R is open in X, while f is open when the f-image of any open subset of X is open in W_R. The importance of this kind of morphism is that a surjective d-morphism preserves d-validity as frame validity, in the following sense.

Theorem 11.3 (Bezhanishvili et al. 2005, Corollary 2.9) *If there exists a d-morphism from X onto $\mathcal{F}$, then for any formula φ, $X \models_d \varphi$ implies $\mathcal{F} \models \varphi$.* □

The interpretation of $\Diamond$ by De is sometimes called *d-semantics* (Bezhanishvili et al. 2005). It has $\mathcal{M}_d(\Diamond^* \varphi) = \mathsf{Cl}_X(\mathcal{M}_d(\varphi))$ and $\mathcal{M}_d(\Box^* \varphi) = \mathsf{Int}_X(\mathcal{M}_d(\varphi))$, because $\mathsf{Cl}_X\, Y = Y \cup \mathsf{De}_X\, Y$. By contrast, *C-semantics* interprets $\Diamond$ as Cl, defining truth sets $\mathcal{M}_C(\varphi)$ inductively in a topological model $\mathcal{M}$ by putting $\mathcal{M}_C(\Diamond \varphi) = \mathsf{Cl}_X(\mathcal{M}_C(\varphi))$ and $\mathcal{M}_C(\Box \varphi) = \mathsf{Int}_X(\mathcal{M}_C(\varphi))$. A formula φ is *C-valid in X*, written $X \models_C \varphi$, iff $\mathcal{M}_C(\varphi) = X$ for all models $\mathcal{M}$ on X. In C-semantics there is no distinction in interpretation between $\Diamond$ and $\Diamond^*$, or between $\Box$ and $\Box^*$.

A space is *scattered* if each of its non-empty subspaces has an isolated point, i.e. no non-empty subset is crowded. This condition d-validates the Löb axiom, and hence, d-validates the logic $K4\mathbb{C}_0$, since it is equal to the Gödel-Löb logic. In fact, $K4\mathbb{C}_0$ is characterised by d-validity in all scattered spaces, a result due to Esakia (1981). It can be readily explained via relational semantics. If φ is a non-theorem of $K4\mathbb{C}_0$, then φ fails to be valid in some frame (W, R) with W finite and R irreflexive and transitive. In such a frame, $R^{-1}Y$ is the derived set of Y in the Alexandrov space W_R, for any $Y \subseteq W$. This implies that the relational semantics on (W, R) agrees with the d-semantics on W_R, in the sense that a formula is true at a point w in a relational model (W, R, V) iff it is d-true at w in the topological model (W_R, V). Hence, φ is not d-valid in the space W_R. But W_R is scattered, since for any non-empty $Y \subseteq W$, there is an R-maximal element $w \in Y$, i.e. wRv implies $v \notin Y$, hence, $R^*(w)$ is an open neighbourhood of w in W_R that contains no member of Y other than w, making w isolated in Y.

From now on, we focus on the logics $K4\mathbb{C}_n$ with $n \geq 1$.

11.4 Hereditary Irresolvability

A *partition* of a space X is, as usual, a collection of non-empty subsets of X, called the *cells*, that are pairwise disjoint and whose union is X. It is a k*-partition*, where k is a positive integer, if it has exactly k cells. A *dense partition* is one for which each cell is dense in X. A *crowded partition* is one whose cells are crowded.

For $k \geq 2$, a space is called k*-resolvable* if it has k pairwise disjoint non-empty dense subsets. Since any superset of a dense set is dense, k-resolvability is equivalent to X having a dense k-partition. X is k*-irresolvable* if it is not k-resolvable. It is *hereditarily* k*-irresolvable*, which may be abbreviated to k-HI, if every non-empty subspace of X is k-irresolvable. Note that if $k \leq n$, then an n-resolvable space is also k-resolvable, since we can amalgamate cells of a dense partition to form new dense partitions with fewer cells. Hence, if X is k-HI, then it is also n-HI.

k-resolvability was defined by Ceder (1964) with the extra requirement that each cell of a dense k-partition should intersect each non-empty open set in at least k points. That requirement was dropped by later authors, including El'kin (1969a) and Eckertson (1997), with the latter defining the k-HI notion.

The prefix k- is usually omitted when $k = 2$. Thus, a space is *resolvable* if it has a disjoint pair of non-empty dense subsets, and is *hereditarily irresolvable*, or HI, if it has no non-empty subspace that is resolvable.

It is known that any HI space is T_D. For convenience, we repeat an explanation of this from Goldblatt (2021). In general $\mathsf{Cl}_X\{x\} = \{x\} \cup \mathsf{De}_X\{x\}$ with $x \notin \mathsf{De}_X\{x\}$ and $\{x\}$ dense in $\mathsf{Cl}_X\{x\}$, while $\mathsf{Cl}_X\, \mathsf{De}_X\{x\} \subseteq \mathsf{Cl}_X\{x\}$. But if X is HI, then $\mathsf{Cl}_X\{x\}$ is irresolvable, so $\mathsf{De}_X\{x\}$ cannot be dense in $\mathsf{Cl}_X\{x\}$, hence $\mathsf{Cl}_X\, \mathsf{De}_X\{x\}$ can only be $\mathsf{De}_X\{x\}$, i.e. $\mathsf{De}_X\{x\}$ is closed.

On the other hand, a k-HI space need not be T_D when $k > 2$. For instance, we saw that a two-element indiscrete space is not T_D, but since it has no 3-partition it is k-HI for every $k > 2$.

It is also known that every scattered space is HI. For, in a scattered space, any non-empty subspace Y has an isolated point which will belong to one cell of any 2-partition of Y and prevent the other cell from being dense, hence prevent Y from being resolvable. It follows that every scattered space is T_D. This is a topological manifestation of the celebrated proof-theoretic fact that the transitivity axiom 4 is derivable from Löb's axiom over K (see Boolos 1993, p. 11).

In Goldblatt (2021), the following results were proved for all $n \geq 1$, where $\mathrm{S4}\mathbb{C}_n$ is the smallest normal extension of $\mathrm{K4}\mathbb{C}_n$ that includes the scheme $\Box\varphi \to \varphi$, or equivalently $\varphi \to \Diamond\varphi$.

1. A space X has $X \models_d \mathbb{C}_n$ iff $X \models_C \mathbb{C}_n$ iff X is hereditarily $n+1$-irresolvable.
2. If (W, R) is a finite quasi-order, then it has circumference at most n iff the space W_R is hereditarily $n+1$-irresolvable.
3. $\mathrm{S4}\mathbb{C}_n$ is characterised by C-validity in all hereditarily $n+1$-irresolvable spaces, i.e. a formula is a theorem of $\mathrm{S4}\mathbb{C}_n$ iff it is C-valid in all hereditarily $n+1$-irresolvable spaces. Moreover, $\mathrm{S4}\mathbb{C}_n$ is characterised by C-validity in all *finite* hereditarily $n+1$-irresolvable spaces.

4. $K4\mathbb{C}_n$ is not characterised by d-validity in any class of finite spaces.

The reason for the last result is that every finite space that d-validates K4 is scattered and so d-validates the Löb axiom, which is not a theorem of $K4\mathbb{C}_n$ when $n \geq 1$.

Using the first result listed above, we can infer that $K4\mathbb{C}_n$ is sound for d-validity in all hereditarily $n+1$-irresolvable T_D spaces. The main result of this paper is that, conversely, $K4\mathbb{C}_n$ is complete for d-validity in all hereditarily $n+1$-irreducible T_D spaces (albeit not for d-validity in all the finite ones). To indicate how this will be proved, note that, by Theorem 11.1, we have that $K4\mathbb{C}_n$ is complete for d-validity in all finite frames that validate $K4\mathbb{C}_n$. So to show that $K4\mathbb{C}_n$ is complete for d-validity in some class of spaces, it suffices by Theorem 11.3 to show that every finite $K4\mathbb{C}_n$-frame is a d-morphic image of some space in that class. For $n = 1$, this was done by Gabelaia (2004) (see also Bezhanishvili et al. 2010), proving that $K4\mathbb{C}_1$ (in the form $K4Grz_{\Box}$) is the d-logic of HI-spaces by showing that any finite $K4\mathbb{C}_1$-frame is a d-morphic image of an HI space. A finite $K4\mathbb{C}_1$-frame has only singleton clusters, and each non-degenerate one was replaced by an HI space to construct the desired HI preimage. We will now generalise this construction to make it work for all $n > 1$ as well.

Suppose that $\mathcal{F} = (W, R)$ is finite and transitive. Let $\mathcal{C}$ be the set of R-clusters of $\mathcal{F}$. We define a collection $\{X_C : C \in \mathcal{C}\}$ of spaces, with each X_C having a partition $\{X_w : w \in C\}$ indexed by C. If $C = \{w\}$ is a degenerate cluster, put $X_C = X_w = \{w\}$ as a one-point space. For C non-degenerate, with $C = \{w_1, \dots, w_k\}$ for some positive integer k, take X_C to be a copy of a space that has a crowded dense k-partition. Label the cells of that partition $X_{w_1}, \dots, X_{w_k}$. We take X_C to be disjoint from $X_{C'}$ whenever $C \neq C'$ (replacing spaces by homeomorphic copies where necessary to achieve this). That completes the definition of the X_C's and the X_w's.

Now let $X_{\mathcal{F}} = \bigcup\{X_C : C \in \mathcal{C}\}$, and define a surjective map $f : X_{\mathcal{F}} \to W$ by putting $f(x) = w$ iff $x \in X_w$. This entails that $f^{-1}\{w\} = X_w$ and $f^{-1}C = X_C$.

For $C, C' \in \mathcal{C}$, write $CR^{\uparrow}C'$ if CRC' but not $C'RC$, i.e. C' is a strict R-successor of C. Define a subset $O \subseteq X_{\mathcal{F}}$ to be *open* iff for all $C \in \mathcal{C}$, $O \cap X_C$ is open in X_C and

$$\text{if } O \cap X_C \neq \emptyset, \text{ then for all } C' \text{ such that } CR^{\uparrow}C',\ X_{C'} \subseteq O.$$

It is readily checked that these open sets form a topology on $X_{\mathcal{F}}$. If B is an open subset of X_C, then $O_B = B \cup \bigcup\{X_{C'} : CR^{\uparrow}C'\}$ is an open subset of $X_{\mathcal{F}}$ (this uses transitivity of $R^{\uparrow}$), with $O_B \cap X_C = B$. It follows that X_C is a subspace of $X_{\mathcal{F}}$, i.e. the original topology of X_C is identical to the subspace topology on the underlying set of X_C inherited from the topology of $X_{\mathcal{F}}$.

Lemma 11.3 *f is a d-morphism from $X_{\mathcal{F}}$ onto $\mathcal{F}$.*

Proof To show f is continuous it is enough to show that the preimage $f^{-1}R^*\{w\}$ of any basic open subset of W_{R^*} is open in $X_{\mathcal{F}}$. If C is the R-cluster of w, then $R^*\{w\} = C \cup \bigcup\{C' : CR^{\uparrow}C'\}$, so

$$f^{-1}R^*\{w\} = f^{-1}C \cup \bigcup\{f^{-1}C' : CR^{\uparrow}C'\} = X_C \cup \bigcup\{X_{C'} : CR^{\uparrow}C'\},$$

which is indeed open in $X_{\mathcal{F}}$.

To show that f is an open map, we must show that if O is an open subset of $X_{\mathcal{F}}$, then $f(O)$ is open in W_{R^*}, i.e. is an R-up-set. So, suppose $w \in f(O)$ and wRv. We want $v \in f(O)$. Let C be the cluster of w. We have $w = f(x)$ for some $x \in O \cap X_C$. If $v \in C$, then C is non-degenerate and X_v is dense in X_C, so as $O \cap X_C$ is open in X_C, there is some $y \in X_v \cap O$. Then $v = f(y) \in f(O)$. If however $v \notin C$, then the cluster C' of v has $CR^{\uparrow}C'$, hence $X_{C'} \subseteq O$. Taking any $y \in X_v \subseteq X_{C'}$ gives $v = f(y) \in f(O)$ again. That completes the proof that $f(O)$ is an R-up-set.

If $w \in C$ is reflexive, then $f^{-1}\{w\} = X_w$ is crowded in X_C, i.e. $f^{-1}\{w\} \subseteq \mathsf{De}_{X_C} f^{-1}\{w\}$. But $\mathsf{De}_{X_C} f^{-1}\{w\} \subseteq \mathsf{De}_{X_{\mathcal{F}}} f^{-1}\{w\}$, since X_C is a subspace of $X_{\mathcal{F}}$, so $f^{-1}\{w\} = X_w$ is crowded in $X_{\mathcal{F}}$.

Finally, if w is irreflexive, then $f^{-1}\{w\} = \{w\}$ is discrete in $X_{\mathcal{F}}$. □

Lemma 11.4 *If X_C is T_D for all $C \in \mathcal{C}$, then $X_{\mathcal{F}}$ is T_D.*

Proof By Lemma 11.1, a space X is T_D iff it has $x \notin \mathsf{Cl}_X \mathsf{De}_X\{x\}$ in general. If $x \in X_{\mathcal{F}}$, then $x \in X_C$ for some C. If X_C is T_D, then there is an open neighbourhood O of x in X_C that is disjoint from $\mathsf{De}_{X_C}\{x\}$. As $\mathsf{De}_{X_C}\{x\} = X_C \cap \mathsf{De}_{X_{\mathcal{F}}}\{x\}$, O is disjoint from $\mathsf{De}_{X_{\mathcal{F}}}\{x\}$. Let $O' = \bigcup\{X_{C'} : CR^{\uparrow}C'\}$. Then O' is $X_{\mathcal{F}}$-open with $x \notin O'$, so no point of O' is a limit point of $\{x\}$ in $X_{\mathcal{F}}$. Hence, $O \cup O'$ is an $X_{\mathcal{F}}$-open neighbourhood of x that is disjoint from $\mathsf{De}_{X_{\mathcal{F}}}\{x\}$, showing that $x \notin \mathsf{Cl}_{X_{\mathcal{F}}} \mathsf{De}_{X_{\mathcal{F}}}\{x\}$. □

Note that this result need not hold with T_1 in place of T_D. $X_{\mathcal{F}}$ need not be T_1 even when every X_C is. For if $CR^{\uparrow}C'$ with $x \in X_C$ and $y \in X_{C'}$, then every open neighbourhood of x in $X_{\mathcal{F}}$ contains y, so $x \in \mathsf{Cl}_{X_{\mathcal{F}}}\{y\} - \{y\}$, showing that $\{y\}$ is not closed.

Lemma 11.5 *If X_C is n-HI for all $C \in \mathcal{C}$, then $X_{\mathcal{F}}$ is n-HI.*

Proof If $X_{\mathcal{F}}$ is not n-HI, then it has some non-empty subspace Y that has n pairwise disjoint subsets $S_1, \ldots, S_n$ that are each dense in Y, i.e. $Y \subseteq \mathsf{Cl}_{X_{\mathcal{F}}} S_i$. Since $\mathcal{C}$ is finite and $R^{\uparrow}$ is antisymmetric, there must be a $C \in \mathcal{C}$ such that X_C intersects Y and C is $R^{\uparrow}$-maximal with this property. Thus, $X_C \cap Y \neq \emptyset$ but if $CR^{\uparrow}C'$ then $X_{C'} \cap Y = \emptyset$. Then putting $O = X_C \cup \bigcup\{X_{C'} : CR^{\uparrow}C'\}$ gives $O \cap Y = X_C \cap Y \neq \emptyset$.

Now O is $X_{\mathcal{F}}$-open, so $O \cap Y$ is a non-empty Y-open set, hence, it intersects the sets S_i as they are dense in Y. Thus, the sets $\{O \cap S_i : 1 \leq i \leq n\}$ are pairwise disjoint and non-empty. They are also dense in $X_C \cap Y$, as

$$X_C \cap Y = O \cap Y \subseteq O \cap \mathsf{Cl}_{X_{\mathcal{F}}} S_i \subseteq \mathsf{Cl}_{X_{\mathcal{F}}}(O \cap S_i),$$

with the last inclusion holding because O is $X_{\mathcal{F}}$-open. This shows that $X_C \cap Y$ is an n-resolvable subspace of X_C, proving that X_C is not n-HI. □

To prove that $\mathrm{K4}\mathbb{C}_n$ is characterised by d-validity in $n+1$-HI T_D spaces, we want to show that such spaces provide d-morphic preimages of all finite transitive frames of circumference at most n. To achieve this, the work so far indicates that we need to replace non-degenerate clusters by $n+1$-HI T_D spaces that have crowded dense k-partitions for various $k \leq n$. So we need to show such spaces exist.

The literature contains several constructions of n-resolvable spaces that are not $n+1$-resolvable. For instance, van Douwen (1993) constructs ones that are crowded, countable and regular.[1] The most convenient construction for our purposes is given by El'kin (1969b). To describe it, first define E to be a space, based on the set ω of natural numbers, for which the set of open sets is $\mathcal{U} \cup \{\emptyset\}$ where $\mathcal{U}$ is some non-principal ultrafilter on ω. This makes E a *door* space: every subset is either open or closed. It is crowded, as no singleton belongs to $\mathcal{U}$, and is T_1 as every co-singleton $\omega - \{x\}$ does belong to $\mathcal{U}$. E has the special property that *the intersection of any two non-empty E-open sets is non-empty* (infinite actually). This implies that any non-empty open set is dense in E.

The closure properties of an ultrafilter also ensure that E is HI. For if a non-empty subspace Y of E has a 2-partition, then either Y is open and so at least one cell of the partition is open, which prevents the other cell from being dense in Y; or else Y is closed and so both cells are closed and hence neither is dense.

Now view $\omega \times \{1, \dots, n\}$ as the union of its disjoint subsets $\omega \times \{i\}$ for $1 \leq i \leq n$. Let X_n be the space based on $\omega \times \{1, \dots, n\}$ whose non-empty open sets are all the sets of the form $\bigcup_{i \leq n}(O_i \times \{i\})$ where each O_i is a *non-empty* open subset of E. This definition does satisfy the axioms of a topology because of the special property of E noted above. Put $S_i = \omega \times \{i\}$. Then $\{S_i : 1 \leq i \leq n\}$ is an n-partition of X_n that is dense because every non-empty X_n-open set intersects every S_i, so the cells are all dense in X_n. Hence, X_n is n-resolvable.

The intersection of any non-empty X_n-open set with S_i is of the form $O_i \times \{i\}$ with O_i open in E. It follows that S_i as a subspace of X_n is a homeomorphic copy of E, so inherits the topological properties of E, including being a door space that is HI and having all its non-empty open subsets be dense. It also follows that the non-empty open sets of X_n are all the sets of the form $\bigcup_{i \leq n} O'_i$ where each O'_i is a non-empty open subset of S_i.

S_i inherits from E the property that its non-empty open sets are infinite. This implies that each S_i is crowded in X_n, as is X_n itself.

X_n is also T_1, since for any point $(x, i) \in X_n$ the set

$$X_n - \{(x, i)\} = S_1 \cup \cdots \cup [(\omega - \{x\}) \times \{i\}] \cup \cdots \cup S_n$$

is open in X_n, so $\{(x, i)\}$ is closed.

X_n is not $n+1$-resolvable. This is implied by several results in the literature, including that of El'kin (1969a), Proposition 1, which states that a space is $n+1$-irresolvable if it has a dense n-partition with each cell having the property that each of its crowded subspaces is irresolvable. Illanes (1996), Lemma 2 proves $n+1$-irresolvability of any space that has an n-partition whose cells are *openly irresolvable* (OI), meaning that every non-empty *open* subspace is irresolvable. The most general

[1] van Douwen's terminology is different. He calls a space *n-irresolvable* if it has a dense partition of size n, but none larger.

result of this type would appear to be that of Eckertson (1997), Lemma 3.2(a), proving $n+1$-irresolvability of any space that is merely the union of n subspaces that are each openly irresolvable. But it is instructive and more direct here to give a proof for X_n that uses its particular structure.

Lemma 11.6 *If A is a dense subset of X_n, then there exists an $i \leq n$ such that $A \cap S_i$ is non-empty and open in S_i.*

Proof Let A be dense. Suppose that the conclusion does not hold. Then for each $i \leq n$, if $A \cap S_i$ is non-empty then it is not open in S_i, so is not equal to S_i. Hence, its complement $S_i - (A \cap S_i)$ is non-empty, and open in S_i as S_i is a door space. If $A \cap S_i = \emptyset$, then $S_i - (A \cap S_i)$ is again non-empty and open in S_i. Therefore the union $\bigcup_{i \leq n}[S_i - (A \cap S_i)]$ is, by definition, a non-empty open subset of X_n. But this union is $X_n - A$, so that contradicts the fact that A is dense. □

Now if X_n were $n+1$-resolvable, it would have $n+1$ subsets $A_1, \ldots, A_{n+1}$ that are pairwise disjoint and dense. Then by the lemma just proved, for each $j \leq n+1$ there would be some $i \leq n$ such that $A_j \cap S_i$ is non-empty and open in S_i, hence is dense in S_i as explained above. Hence, by the pigeonhole principle there must be *distinct* $j, k \leq n+1$ such that there is some $i \leq n$ with both subsets $A_j \cap S_i$ and $A_k \cap S_i$ dense in S_i. But these subsets are disjoint, so that contradicts the irresolvability of S_i. Therefore X_n cannot be $n+1$-resolvable.

Theorem 11.4 *For any $n \geq 1$ there exists a non-empty crowded hereditarily $n+1$-irresolvable T_1 space Y_n that has a crowded dense n-partition.*

Proof For any $k > 1$, every k-irresolvable space has a non-empty open subspace that is k-HI, constructed as the complement of the union of all k-resolvable subspaces (Eckertson 1997, Proposition 2.1). So we apply this with $k = n+1$ to the $n+1$-irresolvable space X_n just described to conclude that X_n has a non-empty open subspace Y_n that is $n+1$-HI. Y_n inherits the T_1 condition from X_n and, since Y_n is open, it inherits the crowded condition from X_n, and it intersects each of the dense sets S_i. Also each intersection $S_i' = Y_n \cap S_i$ is crowded and dense in Y_n, as S_i is crowded and dense in X_n and Y_n is open. Thus, $\{S_i' : 1 \leq i \leq n\}$ is a crowded dense n-partition of Y_n. □

Theorem 11.5 *For any $n \geq 1$, every finite* $\mathrm{K4}\mathbb{C}_n$ *frame is a d-morphic image of an hereditarily $n+1$-irresolvable T_D space.*

Proof Let $\mathfrak{F}$ be a finite $\mathrm{K4}\mathbb{C}_n$ frame. $\mathfrak{F}$ is transitive with circumference at most n. We carry out the construction of the space $X_\mathfrak{F}$ as above.

For each non-degenerate cluster C of $\mathfrak{F}$, if C has $k \geq 1$ elements, we take X_C to be a copy of the T_1 space Y_k of Theorem 11.4, and let $\{X_w : w \in C\}$ to be the crowded dense k-partition of Y_k provided by that theorem. Now $k \leq n$ and Y_k is $k+1$-HI, so it is $n+1$-HI. Also any singleton subspace is $n+1$-HI, so we see that every subspace X_C of $X_\mathfrak{F}$ is $n+1$-HI. Hence, $X_\mathfrak{F}$ is $n+1$-HI by Lemma 11.5.

Every subspace X_C of $X_\mathfrak{F}$, including the singleton ones, is T_1, hence is T_D. So $X_\mathfrak{F}$ is T_D by Lemma 11.4.

The d-morphism from $X_\mathfrak{F}$ onto $\mathfrak{F}$ is provided by Lemma 11.3. □

In the case $n = 1$ of this construction, all clusters of $\mathfrak{F}$ are singletons, and $X_{\mathfrak{F}}$ is obtained by replacing each non-degenerate cluster with a copy of the El'kin space E. This is exactly the construction of Gabelaia (2004) and Bezhanishvili et al. (2010).

Theorem 11.6 *For any $n \geq 1$, the logic* K4$\mathbb{C}_n$ *is characterised by d-validity in all spaces that are hereditarily $n+1$-irresolvable and T_D.*

Proof If a space X is $n+1$-HI and T_D, then the d-logic of X includes schemes 4 and $\mathbb{C}_n$, so it includes K4$\mathbb{C}_n$ as the smallest normal logic to include these schemes. Hence, every theorem of K4$\mathbb{C}_n$ is d-valid in X.

For the converse direction, if a formula φ is not a theorem of K4$\mathbb{C}_n$, then by Theorem 11.1(3) there is a finite frame $\mathfrak{F}$ that validates K4$\mathbb{C}_n$ but does not validate φ. By Theorem 11.5 $\mathfrak{F}$ is a d-morphic image of some $n+1$-HI T_D space X. Since $\mathfrak{F} \not\models \varphi$, Theorem 11.3 then gives $X \not\models_d \varphi$. Thus, it is not the case that φ is d-valid in all $n+1$-HI and T_D spaces. □

As already noted, the case $n = 1$ of this result was given in Gabelaia (2004), and in that case the T_D condition is redundant, as hereditarily irresolvable spaces are always T_D.

11.5 Some Extensions of K4$\mathbb{C}_n$

The D-axiom $\Diamond\top$ is d-valid in a space X iff $X = \mathsf{De}_X X$, i.e. iff X is crowded. In general a space of the type $X_{\mathfrak{F}}$ need not be crowded, for if C is a degenerate final cluster of $\mathfrak{F}$, then X_C is an open singleton containing an isolated point of $X_{\mathfrak{F}}$. But we have shown in Goldblatt (2021), Sect. 7 that K4D$\mathbb{C}_n$ is characterised by validity in all finite transitive frames that have circumference at most n and all final clusters *non-degenerate*. If $\mathfrak{F}$ is such a frame, and C' is any final cluster of $\mathfrak{F}$, then $X_{C'}$ is a crowded space of the type given by Theorem 11.4. Now any open neighbourhood of a point x in $X_{\mathfrak{F}}$ includes an open set of the form $O_B = B \cup \bigcup\{X_{C'} : CR^{\uparrow}C'\}$, where B is an open neighbourhood of x in some subspace X_C. If C is final, then $O_B = B$ and X_C is crowded, so O_B contains points other than x. If C is not final then there is a final C' with $CR^{\uparrow}C'$, so O_B includes $X_{C'}$, which consists of points distinct from x. Thus, x is not isolated, showing that $X_{\mathfrak{F}}$ is crowded. This leads us to conclude

Theorem 11.7 *For $n \geq 1$, K4D$\mathbb{C}_n$ is characterised by d-validity in all crowded T_D spaces that are hereditarily $n+1$-irresolvable.* □

At the opposite extreme are logics containing the constant formula

$$\mathrm{E}: \quad \Box\bot \vee \Diamond\Box\bot.$$

This is d-valid in a space iff it is *densely discrete*,[2] meaning that the set of isolated points is dense in the space. The set of isolated points is $X - \mathsf{De}_X\, X$, so X is densely discrete iff $\mathsf{Cl}_X(X - \mathsf{De}_X\, X) = X$, i.e.

$$(X - \mathsf{De}_X\, X) \cup \mathsf{De}_X(X - \mathsf{De}_X\, X) = X.$$

This equation expresses the d-validity of $\neg \Diamond \top \vee \Diamond \neg \Diamond \top$, which is equivalent to E (see Gabelaia 2004, proof of Theorem 4.28).

K4E$\mathbb{C}_n$ was shown in Goldblatt (2021), Sect. 7 to be characterised by validity in all finite transitive frames that have circumference at most n and all final clusters *degenerate*. If $\mathcal{F}$ is such a frame, and C' is any final cluster of $\mathcal{F}$, then $X_{C'}$ is an open singleton, as noted above. If a point x of $X_{\mathcal{F}}$ belongs to X_C where C is not final in $\mathcal{F}$, then there is a final C' with $C R^{\uparrow} C'$, so any open neighbourhood of x will include $X_{C'}$ and hence contain an isolated point. This shows that the isolated points are dense in $X_{\mathcal{F}}$, and leads to

Theorem 11.8 *for $n \geq 1$, K4E$\mathbb{C}_n$ is characterised by d-validity in all densely discrete T_D spaces that are hereditarily $n + 1$-irresolvable.* □

In Theorem 11.5, we could have replaced every non-degenerate cluster C by a copy of the same space Y_n since its crowded dense n-partition can be converted into a crowded dense k-partition for any $k < n$ by amalgamating cells. But allowing X_C to vary with the size of C gives more flexibility in defining spaces. This is well illustrated in the case of logics that include the well-studied McKinsey axiom M, often stated as $\Box \Diamond \varphi \to \Diamond \Box \varphi$. We use the equivalent forms $\Diamond(\Box \varphi \vee \Box \neg\varphi)$ and $\Diamond \Box \varphi \vee \Diamond \Box \neg\varphi$.

It follows from results of Bezhanishvili et al. (2003), Proposition 2.1 that in C-semantics, M defines the class of openly irresolvable (OI) spaces (recall that these are the spaces in which every non-empty *open* subspace is irresolvable). Equivalently, in d-semantics, the scheme $\Diamond^*(\Box^* \varphi \vee \Box^* \neg\varphi)$ defines the class of OI spaces. We give a direct proof of this.

Lemma 11.7 *A space X is openly irresolvable iff $X \models_d \Diamond^*(\Box^* \varphi \vee \Box^* \neg\varphi)$ for all φ.*

Proof Suppose $X \not\models_d \Diamond^*(\Box^* \varphi \vee \Box^* \neg\varphi)$ for some φ. Then there is a model $\mathcal{M}$ on X and a point of X that is not a closure point of $\mathcal{M}_d((\Box^* \varphi \vee \Box^* \neg\varphi)$, and so has an open neighbourhood U disjoint from this d-truth set. Let $A = \mathcal{M}_d(\varphi)$. Then U is disjoint from $\mathsf{Int}_X\, A$ and from $\mathsf{Int}_X(X - A)$, hence U is included in $\mathsf{Cl}_X\, A$ and in $\mathsf{Cl}_X(X - A)$. Thus, $U \cap A$ and $U \cap (X - A)$ are dense subsets of the non-empty open U, showing that U is resolvable, and so X is not OI.

Conversely, assume X is not OI, so has a non-empty open subset U which has a subset A such that A and $U - A$ are dense in U. Hence, U is included in $\mathsf{Cl}_X\, A$ and in $\mathsf{Cl}_X(U - A) \subseteq \mathsf{Cl}_X(X - A)$, so is disjoint from $\mathsf{Int}_X\, A$ and from $\mathsf{Int}_X(X - A)$.

[2] This property was previously called *weakly scattered*. The change of terminology was made, with explanation, in Bezhanishvili et al. (2020), Definition 2.1.

Take a model $\mathcal{M}$ on X with $A = \mathcal{M}_d(p)$ for some variable p. Then U is disjoint from $\mathcal{M}_d((\Box^* p \vee \Box^* \neg p)$, so $\Diamond^*(\Box^* p \vee \Box^* \neg p)$ is d-false in $\mathcal{M}$ at any member of U, hence is not d-valid in X. □

We now explore criteria for the d-validity of M itself.

Theorem 11.9 *If X is crowded and OI, then $X \models_d$ M.*

Proof Let X be crowded and OI. Take any model $\mathcal{M}$ on X, any formula φ, and let $S = \mathcal{M}_d(\Box^* \varphi \vee \Box^* \neg\varphi)$. Then S is open, as the union of two interiors, so as X is crowded, it follows that S is crowded, hence $\mathsf{Cl}_X\, S = \mathsf{De}_X\, S$. Using this, and the fact that $\mathcal{M}_d(\Box^* \psi) \subseteq \mathcal{M}_d(\Box\, \psi)$ for any ψ, we deduce that

$$\mathcal{M}_d(\Diamond^*(\Box^* \varphi \vee \Box^* \neg\varphi)) = \mathcal{M}_d(\Diamond(\Box^* \varphi \vee \Box^* \neg\varphi)) \subseteq \mathcal{M}_d(\Diamond(\Box\varphi \vee \Box\neg\varphi)).$$

As X is OI, $\Diamond^*(\Box^* \varphi \vee \Box^* \neg\varphi)$ is d-true in $\mathcal{M}$ by Lemma 11.7. Therefore by the above inclusion, $\Diamond(\Box\varphi \vee \Box\neg\varphi)$ is d-true in $\mathcal{M}$.

This shows that scheme M is d-valid in X. □

The converse of this result does not hold. For instance, a two-element indiscrete space $X = \{x, y\}$ is resolvable, as $\{x\}$ and $\{y\}$ are dense, so X is not OI, but it d-validates M. The latter is so because the operation De_X interchanges $\{x\}$ and $\{y\}$ and leaves X and $\emptyset$ fixed, from which it follows that $\Diamond\varphi \wedge \Diamond\neg\varphi$ is d-false, hence $\Box\varphi \vee \Box\neg\varphi$ is d-true, at both points in any model on X. So X d-validates M.

What does hold is that in d-semantics, M defines the class of crowded openly irresolvable spaces *within the class of T_D spaces.*

Lemma 11.8 *Let X be crowded and T_D.*

1. *Any open neighbourhood O of a point x in X includes an open neighbourhood O' of x such that $O' - \{x\}$ is non-empty and open.*
2. *If O is an open set in X, then $O \subseteq \mathsf{Cl}_X\, S$ implies $O \subseteq \mathsf{De}_X\, S$, for any $S \subseteq X$.*
3. $\mathsf{Int}_X\, \mathsf{Cl}_X\, S = \mathsf{Int}_X\, \mathsf{De}_X\, S$ *for any $S \subseteq X$.*
4. *For any model $\mathcal{M}$ on X and formula φ,*

$$\mathcal{M}_d(\Diamond^*(\Box^* \varphi \vee \Box^* \neg\varphi)) = \mathcal{M}_d(\Diamond(\Box\varphi \vee \Box\neg\varphi)).$$

Proof 1. Let $x \in O$ with O open. Put $O' = O - \mathsf{De}_X\{x\}$. Since $x \notin \mathsf{De}_X\{x\}$, the set O' contains x, and is open because $\mathsf{De}_X\{x\}$ is closed by the T_D condition. Then $O' - \{x\}$ is non-empty, since x is not isolated as X is crowded. Also $O' - \{x\} = O - (\mathsf{De}_X\{x\} \cup \{x\}) = O - \mathsf{Cl}_X\{x\}$ which is open.

2. Let $O \subseteq \mathsf{Cl}_X\, S$ and $x \in O$. If U is any open neighbourhood of x, then so is $O \cap U$, hence by part 1 there is a non-empty open set $O_1 \subseteq O \cap U$ with $x \notin O_1$. Now $O \cap S$ is dense in O, as $O = O \cap \mathsf{Cl}_X\, S \subseteq \mathsf{Cl}_X(O \cap S)$. Therefore as O_1 is open in O, it intersects $O \cap S$. As $O_1 \subseteq U - \{x\}$, we get that $U - \{x\}$ intersects S. This proves that x is a limit point of S, as required.

3. Putting $O = \mathsf{Int}_X \mathsf{Cl}_X S$ in 2, we get that $\mathsf{Int}_X \mathsf{Cl}_X S$ is an open subset of $\mathsf{De}_X S$, hence is a subset of $\mathsf{Int}_X \mathsf{De}_X S$. Conversely, $\mathsf{Int}_X \mathsf{De}_X S \subseteq \mathsf{Int}_X \mathsf{Cl}_X S$ as $\mathsf{De}_X S \subseteq \mathsf{Cl}_X S$.

4. It was shown in the proof of Theorem 11.9, just using the fact that X is crowded, that the left truth set is included in the right one. For the reverse inclusion, working in the model $\mathcal{M}$, suppose $\Diamond(\Box\varphi \vee \Box\neg\varphi)$ is true at some point x. Then so is $\Diamond\Box\varphi \vee \Diamond\Box\neg\varphi$, hence so is one of $\Diamond\Box\varphi$ and $\Diamond\Box\neg\varphi$. If $\Diamond\Box\varphi$ is true at x, then so is $\Diamond^*\Box\varphi$, hence $\Box^*\Diamond\neg\varphi$ is false at x. By part 3, $\mathcal{M}_d(\Box^*\Diamond\neg\varphi) = \mathcal{M}_d(\Box^*\Diamond^*\neg\varphi)$, so then $\Box^*\Diamond^*\neg\varphi$ is false at x, hence $\Diamond^*\Box^*\varphi$ is true at x.

Similarly, if $\Diamond\Box\neg\varphi$ is true at x, then so is $\Diamond^*\Box^*\neg\varphi$. Since $\Diamond\Box\varphi \vee \Diamond\Box\neg\varphi$ is true at x, so then is $\Diamond^*\Box^*\varphi \vee \Diamond^*\Box^*\neg\varphi$, hence so is $\Diamond^*(\Box^*\varphi \vee \Box^*\neg\varphi)$, as required to prove the inclusion from right to left. □

Theorem 11.10 *If X is a T_D space and $X \models_d$ M, then X is crowded and openly irresolvable.*

Proof Let X be T_D space and $X \models_d$ M. Then X d-validates $\Diamond(\Box\top \vee \Box\neg\top)$. But this formula d-defines $\mathsf{De}_X X$ in any model on X, so $\mathsf{De}_X X = X$, i.e. X is crowded.

We now have that X is crowded and T_D, and any formula of the form $\Diamond(\Box\varphi \vee \Box\neg\varphi)$ is d-valid on X, i.e. d-true in all models on X. But then by Lemma 11.8.4, $\Diamond^*(\Box^*\varphi \vee \Box^*\neg\varphi)$ is d-valid in X. Hence, X is OI by Lemma 11.7. □

Theorems 11.9 and 11.10 combine to give

Corollary 11.1 *If X is T_D, then $X \models_d$ M iff X is crowded and openly irresolvable. Hence if X is T_D and crowded, then $X \models_d$ M iff X is openly irresolvable iff $X \models_C$ M.* □

K4M$\mathbb{C}_n$ was shown in Goldblatt (2021), Sect. 7 to be characterised by validity in all finite transitive frames that have circumference at most n and all final clusters simple. If $\mathfrak{F}$ is such a frame, $X_{\mathfrak{F}}$ is $n+1$-HI and T_D, as shown in the proof of Theorem 11.5. All final clusters of $\mathfrak{F}$ are non-degenerate, which is enough to ensure that $X_{\mathfrak{F}}$ is a crowded space, as explained in our discussion of K4D$\mathbb{C}_n$.

Lemma 11.9 *$X_{\mathfrak{F}}$ is openly irresolvable.*

Proof Let O be any non-empty open subset of $X_{\mathfrak{F}}$. Then $O \cap X_C \neq \emptyset$ for some cluster C of $\mathfrak{F}$. Put $B = O \cap X_C$. Then $O_B = B \cup \bigcup\{X_{C'} : CR^{\uparrow}C'\}$ is a non-empty subset of O that is open in $X_{\mathfrak{F}}$. If C is final, then since it is a non-degenerate singleton, X_C is a copy of the El'kin space E, and also $O_B = B \subseteq X_C$. If however C is not final, there is a final C' with $CR^{\uparrow}C'$. Then $X_{C'} \subseteq O_B$ and $X_{C'}$ is open in $X_{\mathfrak{F}}$.

So, in any case, we see that O has a non-empty open subset O' (either O_B or $X_{C'}$) that is included in a subspace X' (either X_C or $X_{C'}$) that is a copy of E and hence is HI. Hence, O' is irresolvable. Now if O had a pair of disjoint dense subsets, then these subsets would intersect the open O' in a pair of disjoint dense subsets of O', contradicting irresolvability of O'. Therefore O is irresolvable as required. □

Altogether we have now shown that $X_{\mathcal{F}}$ is T_D, crowded, OI, and $n+1$-HI, which implies that it d-validates K4M$\mathbb{C}_n$. We conclude

Theorem 11.11 *For $n \geq 1$, K4M$\mathbb{C}_n$ is characterised by d-validity in all T_D spaces that are crowded, openly irresolvable, and hereditarily $n+1$-irresolvable.* □

When $n = 1$, this can be simplified, since an HI space is always OI and T_D. Thus, the logic K4M$\mathbb{C}_1$ is characterised by d-validity in the class of all crowded HI spaces, as was shown by Gabelaia (2004), Theorem 4.26 with K4M$\mathbb{C}_1$ in the form K4MGrz$_\Box$. But the class of crowded HI spaces characterises K4D$\mathbb{C}_1$, as shown above. Therefore K4M$\mathbb{C}_1$ is identical to the ostensibly weaker K4D$\mathbb{C}_1$. This can also be seen quite simply from our relational completeness result for K4D$\mathbb{C}_1$. In a finite K4D$\mathbb{C}_1$-frame, any final cluster is a singleton by the validity of $\mathbb{C}_1$ and is non-degenerate by validity of D, so all final clusters are simple, making the frame validate M. Thus, M is a theorem of K4D$\mathbb{C}_1$.

We can also deal with the logic K4M, which is characterised by finite transitive frames in which all final clusters are simple (Chagrov and Zakharyaschev 1997, Sect. 5.3). The space $X_{\mathcal{F}}$ can be constructed without assuming that $\mathcal{F}$ has any bound on its circumference. If $\mathcal{F}$ validates K4M, then $X_{\mathcal{F}}$ will be T_D, crowded and OI, so will d-validate K4M. From this, we can conclude

Theorem 11.12 *K4M is characterised by d-validity in all T_D spaces that are crowded and openly irresolvable.* □

Acknowledgements I thank Guram Bezhanishvili for his very helpful comments and suggestions about the content and presentation of this chapter.

References

Allwein, G., & Dunn, J. M. (1993). Kripke models for linear logic. *The Journal of Symbolic Logic, 58*(2), 514–545.

Allwein, G., & Hartonas, C. (1993). Duality for bounded lattices. Indiana University Logic Group, Preprint Series, IULG–93–25.

Aull, C. E., & Thron, W. J. (1962). Separation axioms between T_0 and T_1. *Indagationes Mathematicae (Proceedings), 65*, 26–37.

Bezhanishvili, G., Bezhanishvili, N., Lucero-Bryan, J., & van Mill, J. (2020). Tree-like constructions in topology and modal logic. *Archive for Mathematical Logic*. https://doi.org/10.1007/s00153-020-00743-6.

Bezhanishvili, G., Esakia, L., & Gabelaia, D. (2005). Some results on modal axiomatization and definability for topological spaces. *Studia Logica, 81*, 325–355.

Bezhanishvili, G., Esakia, L., & Gabelaia, D. (2010). K4.Grz and hereditarily irresolvable spaces. In Feferman, S., & Sieg, W. (eds.), *Proofs, Categories and Computations. Essays in Honor of Grigori Mints* (pp. 61–69). College Publications.

Bezhanishvili, G., Mines, R., & Morandi, P. J. (2003). Scattered, Hausdorff-reducible, and hereditarily irresolvable spaces. *Topology and its Applications, 132*, 291–306.

Blackburn, P., de Rijke, M., & Venema, Y. (2001). *Modal Logic*. Cambridge University Press.

Boolos, G. (1993). *The Logic of Provability*. Cambridge University Press.

Ceder, J. G. (1964). On maximally resolvable spaces. *Fundamenta Mathematicae, 55*, 87–93.

Chagrov, A., & Zakharyaschev, M. (1997). *Modal Logic*. Oxford University Press.

Craig, A., & Haviar, M. (2014). Reconciliation of approaches to the construction of canonical extensions of bounded lattices. *Mathematica Slovaca*, *64*(6), 1335–1356.

Dzik, W., Orłowska, E., & van Alten, C. J. (2006). Relational representation theorems for lattices with negations: A survey. In de Swart, H. C. M., Orłowska, E., Schmidt, G., & Roubens, M. (eds.) *Theory and Applications of Relational Structures as Knowledge Instruments II, Lecture Notes in Computer Science*, (pp. 245–266, vol. 4342). Springer.

Eckertson, F. W. (1997). Resolvable, not maximally resolvable spaces. *Topology and its Applications*, *79*, 1–11.

El'kin, A. G. (1969a). Resolvable spaces which are not maximally resolvable. *Moscow University Mathematics Bulletin*, *24*, 116–118.

El'kin, A. G. (1969b). Ultrafilters and undecomposable spaces. *Moscow University Mathematics Bulletin*, *24*, 37–40.

Engelking, R. (1977). *General Topology*. Warsaw: PWN — Polish Scientific Publishers.

Esakia, L. (2004). Intuitionistic logic and modality via topology. *Annals of Pure and Applied Logic*, *127*, 155–170.

Esakia, L. L. (1981). Diagonal constructions, Löbs formula and Cantor's scattered spaces. In *Studies in Logic and Semantics* (pp. 128–143). Tbilisi: Metsniereba. In Russian.

Esakia, L. L. (2001). Weak transitivity–a restitution. *Logical Investigations*, *8*, 244–255. In Russian.

Fine, K. (1985). Logics containing K4. Part II. *The Journal of Symbolic Logic*, *50*(3), 619–651.

Gabelaia, D. (2004). *Topological, Algebraic and Spatio-Temporal Semantics for Multi-Dimensional Modal Logics*. Ph.D. thesis, King's College London.

Gehrke, M. (2018). Canonical extensions: an algebraic approach to Stone duality. *Algebra Universalis*, 79(Article 63). https://doi.org/10.1007/s00012-018-0544-6.

Gehrke, M., & Harding, J. (2001). Bounded lattice expansions. *Journal of Algebra*, *239*, 345–371.

Goldblatt, R. (2018). Canonical extensions and ultraproducts of polarities. *Algebra Universalis*, 79(Article 80). https://doi.org/10.1007/s00012-018-0562-4.

Goldblatt, R. (2021). Modal logics that bound the circumference of transitive frames. In J. Madarász & G. Székely *Hajnal Andréka and István Németi on unity of science: from computing to relativity theory through algebraic logic* (233–265) Springer. https://doi.org/10.1007/978-3-030-64187-0_10.

Goldblatt, R. (2020). Morphisms and duality for polarities and lattices with operators. *Journal of Applied Logics – IfCoLog Journal of Logics and their Applications*, *7*(6):1019. Open access at https://www.collegepublications.co.uk/ifcolog/?00042.

Hartonas, C., & Dunn, J. M. (1997). Stone duality for lattices. *Algebra Universalis*, *37*, 391–401.

Hartung, G. (1992). A topological representation of lattices. *Algebra Universalis*, *29*, 273–299.

Hartung, G. (1993). An extended duality for lattices. In K. Denecke & H.-J. Vogel (Eds.), *General Algebra and Applications* (pp. 126–142). Berlin: Heldermann-Verlag.

Illanes, A. (1996). Finite and ω-resolvability. *Proceedings of the American Mathematical Society*, *124*(4), 1243–1246.

Jónsson, B., & Tarski, A. (1951). Boolean algebras with operators, part I. *American Journal of Mathematics*, *73*, 891–939.

McKinsey, J. C. C. (1941). A solution of the decision problem for the Lewis systems S2 and S4 with an application to topology. *The Journal of Symbolic Logic*, *6*, 117–134.

McKinsey, J. C. C., & Tarski, A. (1944). The algebra of topology. *Annals of Mathematics*, *45*, 141–191.

McKinsey, J. C. C., & Tarski, A. (1948). Some theorems about the sentential calculi of Lewis and Heyting. *The Journal of Symbolic Logic*, *13*, 1–15.

Ploščica, M. (1995). A natural representation of bounded lattices. *Tatra Mountains Mathematical Publications*, *5*, 75–88.

Priestley, H. A. (1970). Representations of distributive lattices by means of ordered Stone spaces. *Bulletin of the London Mathematical Society*, *2*, 186–190.

Rescher, N., & Urquhart, A. (1971). *Temporal logic*. Springer.

Segerberg, K. (1971). *An Essay in Classical Modal Logic, Filosofiska Studier* (vol. 13). Uppsala Universitet.

Stone, M. H. (1936). The theory of representations for Boolean algebras. *Transactions of the American Mathematical Society, 40*, 37–111.

Tang, T.-C. (1938). Algebraic postulates and a geometric interpretation for the Lewis calculus of strict implication. *Bulletin of the American Mathematical Society, 44*, 737–744.

Urquhart, A. (1978). A topological representation theory for lattices. *Algebra Universalis, 8*, 45–58.

Urquhart, A. (1981). Decidability and the finite model property. *Journal of Philosophical Logic, 10*(3), 367–370.

Urquhart, A. (1983). Relevant implication and projective geometry. *Logique et Analyse, 103–104*, 345–357.

Urquhart, A. (1984). The undecidability of entailment and relevant implication. *The Journal of Symbolic Logic, 49*(4), 1059–1073.

Urquhart, A. (1993). Failure of interpolation in relevant logics. *Journal of Philosophical Logic, 22*(5), 449–479.

Urquhart, A. (2015). First degree formulas in quantified S5. *The Australasian Journal of Logic, 12*(5), 204–210. https://ojs.victoria.ac.nz/ajl/article/view/470.

Urquhart, A. (2017). The geometry of relevant implication. *IFCoLog Journal of Logics and their Applications, 4*(3), 591–604. http://collegepublications.co.uk/ifcolog/?00012.

van Douwen, E. K. (1993). Applications of maximal topologies. *Topology and its Applications, 51*, 125–139.

Willard, S. (1970). *General Topology*. Addison-Wesley. Dover Publications Edition 2004.

Chapter 12
St. Alasdair on Lattices Everywhere

Katalin Bimbó and J. Michael Dunn

Second Reader
I. Düntsch
Fujian Normal University & Brock University

Abstract Urquhart works in several areas of logic where he has proved important results. Our paper outlines his topological lattice representation and attempts to relate it to other lattice representations. We show that there are different ways to generalize Priestley's representation of distributive lattices—Urquhart's being one of them, which tries to keep prime filters (or their generalizations) in the representation. Along the way, we also mention how semi-lattices and lattices figured into Urquhart's work.

Keywords Galois connection · Lattice · Relational semantics · Semilattice · Topological frame

2020 Mathematics Subject Classification: 03G10 · 03B47

K. Bimbó (✉)
Department of Philosophy, University of Alberta, 2–40 Assiniboia Hall, Edmonton, AB T6G 2E7, Canada
e-mail: bimbo@ualberta.ca
URL: http://www.ualberta.ca/~bimbo

J. M. Dunn
Department of Philosophy and Luddy School of Informatics, Computing and Engineering, Indiana University, 901 East 10th Street, Bloomington, IN 47408–3912, USA
e-mail: dunn@indiana.edu

I. Düntsch and E. Mares (eds.), *Alasdair Urquhart on Nonclassical and Algebraic Logic and Complexity of Proofs*, Outstanding Contributions to Logic 22,
https://doi.org/10.1007/978-3-030-71430-7_12

12.1 Introduction

Alasdair Urquhart has the title of "St. Alasdair" in the Logicians Liberation League.[1] We have never known exactly why. Is it because he works miracles, or because he is very nice? We think both. We have each known Urquhart for many years, and indeed one of us (JMD) has known him for over 50 years. JMD once thought of Alasdair as a kind of younger brother, arriving in the Anderson–Belnap family of relevantists just shortly after his two older brothers, Bob Meyer and JMD left the nest. Indeed, JMD might have overlapped with Alasdair in an alternative possible world. The last chapter of JMD's dissertation (Dunn 1966) was intended to show the decidability of the two relevant logics **E** and **R**. Fortunately, his dissertation director Nuel Belnap insisted that the dissertation was complete as it was, and JMD did not need to spend another year on it. We say fortunately because as almost every reader knows, Urquhart (1984) showed the undecidability of these logics.

Urquhart has been very helpful to both of us (JMD and KB) in various ways, including through reading our work and giving us helpful suggestions and criticisms (almost all of which we have agreed with).

Urquhart has made considerable contributions to logic and the philosophy of logic, including non-classical logics (particularly, relevance logic), lattice theory, foundations of mathematics, history of logic, theory of computation, and computational complexity theory. In this chapter, we focus on just one of these, his *topological representation of lattices*. However, we will mention some other contexts where lattices, modular lattices, and semilattices appear in Urquhart's work—seemingly, everywhere. Lattices had been given a number of different representations since Birkhoff and Frink (1948) (using sets of subsets of the elements), but Urquhart's was the first one using topological structures.

In Sect. 12.2, we illustrate the idea of emulating abstract algebras by concrete objects, namely, groups by permutations. Then, we turn to lattices in Sect. 12.3, where we present Urquhart's lattice representation, and we briefly compare it to Priestley's representation. Sect. 12.4 explores the confluence of two trains of thought, one coming from a representation of orthocomplemented lattices and the other originating in Galois theory. This leads to another generalization of Priestley's representation. In Sect. 12.5, we explain the importance of the topologies on frames. Among the various lattice representations, Urquhart's is the first one to provide all components for duality. In the concluding Sect. 12.6, we quickly point out the importance of all the lattice representations for the model theory of substructural logics.

---

[1] The "LLL" was founded by Robert K. Meyer in 1969. The LLL's manifesto, group pictures of some of the members, and more may be found at the URL (as of June 2020): aal.ltumathstats.com/curios/logicians-liberation-league.

12.2 Representations of Abstract Algebras

As any schoolchild knows from personal experience, algebra is abstract, though they may not know this word. There is the term "abstract algebra" to cover algebras that do not just give you laws for manipulating numbers, but to give you laws for various structures that abstract out properties of various structures beyond numbers. A good example is a *monoid*, and we shall quickly examine representations of monoids to give a kind of introduction and paradigm for representations of algebras. We start by defining a *semigroup* as a set S together with an associative binary operation $\cdot$ on S. "Associative," of course, means that $x \cdot (y \cdot z) = (x \cdot y) \cdot z$. Additional axioms are the usual ones for identity, $x = x$ (reflexivity), if $x = y$, then $y = x$ (symmetry), and if $x = y$ and $y = z$, then $x = z$ (transitivity). And, of course, we must not forget the substitution of identicals, which in this case can be stated as if $x' = x$ then $x' \cdot y = x \cdot y$, and if $y' = y$ then $x \cdot y' = x \cdot y$. A monoid is a semigroup with an *identity element* e satisfying $e \cdot x = x = x \cdot e$ (identity). A *group* is a monoid with a unary *inverse* operation $^{-1}$ satisfying $x \cdot x^{-1} = e = x^{-1} \cdot x$.

Suppose you are creating a "butterfly zoo."[2] The ideal would not just to have a couple of butterflies, but in fact, to have examples of every species in Lepidoptera. This may be unrealistic for butterflies, but it is obtainable for groups. This may seem surprising because the number of different kinds of groups is obviously infinite. However, it turns out that we can construct groups that are representatives of every kind of group in the sense that every group is isomorphic to one of these representatives. This gives what is called the "Cayley Representation Theorem" for groups.

A standard example of a group is a *permutation group*, i.e., a collection of 1–1 functions from a set onto itself that is closed under composition. But this is much more than just an example. Any group is isomorphic to a subgroup of a permutation group of some set. Cayley showed that every group is isomorphic to a subgroup of the collection of 1–1 functions on some set closed under composition, where e is an identity function, and $^{-1}$ is the converse forming operation, i.e., if $f(x) = y$, then $f(y) = x$.[3] Important and beautiful as this theorem is, we can give the idea of a "representation" with its simpler monoid version. A *transformation monoid* is just like a permutation group except the functions are not required to be 1–1, nor are they required to be onto.

Given a monoid $\langle S, \cdot \rangle$, each element $a \in S$ determines a function f_a that maps each element x onto the element $a \cdot x$. Consider the set $F = \{ f_a : a \in S \}$ of all such functions. Note that F is clearly closed under composition since $f_a(f_b(x)) = a \cdot (b \cdot x) = (a \cdot b) \cdot x = f_{a \cdot b}(x)$. Thus, we can map S onto F in a way that carries each $a \in S$ to f_a, namely, $h(a) = f_a$. Moreover, h is 1–1. Now suppose $a \neq b$ but $f_a = f_b$. Then $a \cdot e = b \cdot e$, and so $a = b$ contrary to our assumption. So, h is an isomorphism of the monoid to a submonoid of a transformation monoid.

---

[2] We will call it *Lepidopterary* to attract scientifically minded visitors. :-)

[3] In other words, every group is a subgroup of the automorphism group on some set; "automorphism" is the taxonomical name for a permutation in the scheme of morphisms.

We can easily expand the above to link groups to permutation groups. We only need to show that f_a is 1–1 and onto. But we will not expand on this here. Another expansion would be to consider all the subgroups of a group, which form a lattice. Indeed, Whitman (1946) showed that every lattice can be viewed as a lattice of subgroups of some group. We will not expand on this either.

On the other hand, we can use transformation monoids to provide a kind of semantics for a very simple logic. Define $\mathcal{A} \vDash_a \mathcal{B}$ iff $\forall x \in S$ (if $x \vDash \mathcal{A}$ then $a \cdot x \vDash \mathcal{B}$). Informally, sentence $\mathcal{A}$ has sentence $\mathcal{B}$ as a consequence according to state a iff every state x where $\mathcal{A}$ holds is such that when viewing a as a function, a transforms x into a state $a(x)$, where $\mathcal{B}$ holds. Where e is the identity element, note that $\mathcal{A} \vDash_e \mathcal{B}$ iff, $\forall a \in S$ (if $e \vDash \mathcal{A}$ then $a \cdot e = a \vDash \mathcal{B}$). We can define $\mathcal{A} \vDash_S \mathcal{B}$ iff $\mathcal{A} \vDash_e \mathcal{B}$.

We have not yet infused our sentences with any logical structure such as connectives. Nonetheless, it is obvious that $\mathcal{A} \vDash_S \mathcal{A}$. It is also clear that $\vDash_S$ is transitive, that is, if $\mathcal{A} \vDash_S \mathcal{B}$ and $\mathcal{B} \vDash_S \mathcal{C}$ then $\mathcal{A} \vDash_S \mathcal{C}$. Now, we may consider adding a pair of naturally emerging connectives, which resemble the familiar conjunction and conditional. $\mathcal{A} \circ \mathcal{B} = \{a \cdot b \colon a \in \mathcal{A} \text{ and } b \in \mathcal{B}\}$, $\mathcal{A} \to \mathcal{B} = \{x : \forall a \text{ (if } x \in \mathcal{A} \text{ then } x \cdot a \in \mathcal{B})\}$. It is worth pointing out that the definition of $\mathcal{A} \to \mathcal{B}$ parallels the valuation clause for implications in Urquhart's (1972b), where he gave a semantics for relevant implication. The latter is called *semilattice semantics*, because the operation $\cdot$ is not functional composition, rather, it is set union, which has the additional properties of commutativity and idempotence. This semantics fits precisely the implicational fragment of **R**, as he showed in Urquhart (1972a).

Having found the first use of a semilattice in Urquhart's work—while illustrating the idea of a representation—we go on to lattices.

12.3 Representations of Lattices

A slightly different approach to the semantics of a logic than what we have already mentioned may be sketched as follows. We start with sentences. Sets of sentences describe a situation, and in turn, sets of situations characterize propositions. However, sentences are often too delicate, and they make too many distinctions. If a logic cannot distinguish between $\mathcal{A}$ and $\mathcal{B}$ with respect to their role in reasoning, then there is no need to distinguish $\mathcal{A}$ and $\mathcal{B}$ in their interpretations. Then it is just as good (or better) to start with the Lindenbaum algebra of a logic as with all the sentences.

Definition 12.3.1 A *lattice logic* ($\mathfrak{Lat}$) has two binary connectives $\wedge$ (conjunction) and $\vee$ (disjunction) with a denumerable set of sentence letters. The formulas (wff's) are defined as usual; they are abbreviated by $\mathcal{A}, \mathcal{B}, \mathcal{C}, \ldots$. The consequence relation ($\vdash$) satisfies the following axioms and rules. (The two-way turnstile $\dashv\vdash$ indicates that $\vdash$ holds in both directions.)

(1) $\mathcal{A} \vdash \mathcal{A}$, $\quad \mathcal{A} \vdash \mathcal{B}$ and $\mathcal{B} \vdash \mathcal{C}$ imply $\mathcal{A} \vdash \mathcal{C}$;
(2) $\mathcal{A} \wedge \mathcal{B} \vdash \mathcal{A}$, $\quad \mathcal{A} \wedge \mathcal{B} \vdash \mathcal{B}$, $\quad \mathcal{A} \vdash \mathcal{A} \vee \mathcal{B}$, $\quad \mathcal{A} \vdash \mathcal{B} \vee \mathcal{A}$;
(3) $(\mathcal{A} \wedge (\mathcal{B} \wedge \mathcal{C})) \dashv\vdash ((\mathcal{A} \wedge \mathcal{B}) \wedge \mathcal{C})$, $\quad (\mathcal{A} \vee (\mathcal{B} \vee \mathcal{C})) \dashv\vdash ((\mathcal{A} \vee \mathcal{B}) \vee \mathcal{C})$;

(4) $\mathcal{A} \vdash \mathcal{C}$ and $\mathcal{B} \vdash \mathcal{C}$ imply $\mathcal{A} \vee \mathcal{B} \vdash \mathcal{C}$, $\quad \mathcal{A} \vdash \mathcal{B}$ and $\mathcal{A} \vdash \mathcal{C}$ imply $\mathcal{A} \vdash \mathcal{B} \wedge \mathcal{C}$.

A *lattice logic with limits* ($\mathfrak{LatL}$) additionally includes two zero-ary connectives $\boldsymbol{T}$ (triviality or constant truth) and $\boldsymbol{F}$ (absurdity or constant falsity). The next axioms hold for $\boldsymbol{T}$ and $\boldsymbol{F}$.

(5) $\mathcal{A} \vdash \boldsymbol{T}$; $\quad \boldsymbol{F} \vdash \mathcal{A}$.

Remark 12.3.1 Often, it is useful to think of a lattice as two semilattices glued together. Indeed, if we would exclude $\wedge$ or $\vee$ from the vocabulary, then the leftovers would be semilattice logics. The addition of $\boldsymbol{T}$ and $\boldsymbol{F}$ is technically motivated, and it is, by and large, harmless. $\boldsymbol{T}$ is triviality, or in a more favorable tone of voice, $\boldsymbol{T}$ is a formula implied by all formulas, thus, in a sense a theorem. $\boldsymbol{T}$ and $\boldsymbol{F}$ are the limits of what a user of $\mathfrak{LatL}$ can say—to use a Wittgensteinian metaphor.

If a logic extends $\mathfrak{Lat}$ (or $\mathfrak{LatL}$), then we can define an equivalence relation, which we denote by $\equiv$, on the set of wff's by $\mathcal{A} \equiv \mathcal{B} := \mathcal{A} \dashv\vdash \mathcal{B}$. Then the Lindenbaum algebra of the logic contains a *lattice*, in which the elements are $[\mathcal{A}]$, where $[\mathcal{A}] := \{\mathcal{B} \colon \mathcal{A} \dashv\vdash \mathcal{B}\}$. The algebra of $\mathfrak{LatL}$ is *bounded*, which is advantageous if we want to obtain the algebra from a topology.[4]

Definition 12.3.2 A *lattice* $\mathbf{L} = \langle A; \wedge, \vee\rangle$ is an algebra where $\wedge$ and $\vee$ are binary operations on the set A, and the following equations hold.

(1) $a \wedge a = a, \quad a \wedge b = b \wedge a, \quad a \wedge (b \wedge c) = (a \wedge b) \wedge c$;
(2) $a \vee a = a, \quad a \vee b = b \vee a, \quad a \vee (b \vee c) = (a \vee b) \vee c$;
(3) $a \wedge (b \vee a) = a, \quad a \vee (b \wedge a) = a$.

A *bounded lattice* $\mathbf{L} = \langle A; \wedge, \vee, \top, \bot\rangle$ is a lattice with two distinguished elements of A satisfying the equations in (4).

(4) $\bot \vee a = a, \quad \top \wedge a = a$.

Notation 12.3.2 We have assumed some commonly used notational conventions. For instance, $a, b, c, \ldots$ are elements of A, and an equation holds in a structure when its universal closure does. Other symbols for the least and greatest elements of an algebraic structure that supports an order relation are 1 and 0. We do not introduce a special label for bounded lattices—even though not all lattices are bounded—because we almost always mean bounded lattices.

Lattices, with or without bounds, have a rich theory. For our purposes, it is interesting to carve out two equational subclasses of lattices. The lattices in which (m) holds are *modular*, and those in which (d) holds are *distributive*.

(m) $a \wedge (b \vee (a \wedge c)) = (a \wedge b) \vee (a \wedge c)$;
(d) $a \wedge (b \vee c) = (a \wedge b) \vee (a \wedge c)$.

[4] Sometimes the Lindenbaum algebra is called Lindenbaum–Tarski algebra.

The equation (d) implies (m) in the context of a lattice, but not the other way around. From the point of view of the semantics of substructural logics, a dividing line that is useful to draw is between lattices in which (d) holds, and the rest of lattices. Every particular lattice is distributive or not, and in fact, the Lindenbaum algebras of many logics include a lattice that is not distributive. However, we will not rely on a lattice not being distributive; rather, we will not assume that it is distributive. Sometimes, we may call a lattice for which we have not stipulated distributivity a *general lattice* for emphasis.

12.3.1 Urquhart's Lattice Representation

We recall Urquhart's lattice representation from Urquhart (1978). Urquhart saw his own lattice representation as a generalization of Priestley's representation of distributive lattices that she published in 1970. (See Priestley 1970, 1972.)

Definition 12.3.3 A *doubly ordered space* $\mathfrak{F} = \langle U; \sqsubseteq_1, \sqsubseteq_2 \rangle$ satisfies the conditions listed in (f1)–(f3). (The u's range over U.)

(f1) $U \neq \emptyset, \quad \sqsubseteq_1 \subseteq U \times U, \quad \sqsubseteq_2 \subseteq U \times U$;
(f2) for $n \in \{1, 2\}$: $\; u \sqsubseteq_n u, \quad u_1 \sqsubseteq_n u_2$ and $u_2 \sqsubseteq_n u_3$ imply $u_1 \sqsubseteq_n u_3$;
(f3) $u_1 \sqsubseteq_1 u_2$ and $u_1 \sqsubseteq_2 u_2$ imply $u_1 = u_2$.

Remark 12.3.3 A distinctive feature of Priestley's representation, especially, in comparison to Stone's in Stone (1937), is that the space from which a distributive lattice is defined is partially ordered. Then, Urquhart goes a step further by adding another order relation. We may also note that the omission of anti-symmetry from both relations is not essential, because both $\sqsubseteq_1$ and $\sqsubseteq_2$ are (weak) partial orders in the doubly ordered space of a lattice. It is also useful to note that the complements of $\sqsubseteq_1$ and $\sqsubseteq_2$ are *irreflexive*.

Definition 12.3.4 The *left image* of V (a subset of U) in a doubly ordered space is defined by (fl); similarly, the *right image* of V is given by (fr).

(fl) $l(V) = \{ u \in U : \forall v \, (u \sqsubseteq_1 v \Rightarrow v \notin V) \}$;
(fr) $r(V) = \{ u \in U : \forall v \, (u \sqsubseteq_2 v \Rightarrow v \notin V) \}$.

A subset of a doubly ordered space V is *stable* when $lr(V) = V$. The set of all stable subsets of U is denoted by $\mathcal{P}(U)^\dagger$.

Remark 12.3.4 Subsets with the property $rl(V) = V$ would do just as well as stable sets. Universal instantiation in the defining conditions in (fl) and (fr) yields that $V \cap l(V) = \emptyset$ and $V \cap r(V) = \emptyset$.

Proposition 12.3.5 *If* $\mathfrak{F} = \langle U; \sqsubseteq_1, \sqsubseteq_2 \rangle$ *is a doubly ordered space, then the set of stable subsets of* U *is a* lattice *with meet and join defined as* $\cap$ *and* $\Cup$*, where the latter is* (f4) $V_1 \Cup V_2 = l(r(V_1) \cap r(V_2))$.

Proof First, we note that l's type is $l\colon \mathcal{P}(U) \longrightarrow \mathscr{C}_1$ and r's type is $r\colon \mathcal{P}(U) \longrightarrow \mathscr{C}_2$.[5] To see this, let us assume that $V \subseteq U$, $u_1 \in lV$ and $u_1 \sqsubseteq_1 u_2$ but $u_2 \notin lV$. From the latter, it follows that there is a u_3 such that $u_2 \sqsubseteq_1 u_3$ while $u_3 \in V$. However, this contradicts $u_1 \in lV$ via $u_1 \sqsubseteq_1 u_2$ and $u_2 \sqsubseteq_1 u_3$, which imply $u_1 \sqsubseteq_1 u_3$. The two orders are alike, hence showing r's type is alike too.

l and r form a Galois connection between $\mathscr{C}_2$ and $\mathscr{C}_1$, that is, if $V \in \mathscr{C}_1$ and $W \in \mathscr{C}_2$, then $V \subseteq lW$ iff $W \subseteq rV$. We show that the "only-if" conditional holds. Let us assume that $V \subseteq lW$ and $u_1 \in W$. Toward a contradiction, let $u_1 \notin rV$ be stipulated. Then, for some u_2, $u_1 \sqsubseteq_2 u_2$ and $u_2 \in V$. But then both $u_2 \in W$ and $u_2 \in lW$, which is impossible. The defining property of a Galois connection is symmetric in l and r; hence, the proof is complete. □

Remark 12.3.5 The proposition and its proof is not a literal quote from Urquhart (1978), but it is essentially in Urquhart's paper. Our notation for the doubly ordered space intends to suggest that such a space may be viewed as a frame and the logic $\mathfrak{LatL}$ can be interpreted by mapping sentences of this logic into stable subsets of the space.

It is pleasing that a doubly ordered frame carries a lattice, indeed, a *complete* one. However, from an algebraic point of view, it is more interesting to know if every lattice can be viewed as a set of certain subsets of a doubly ordered frame. Since lattices come in all sizes and shapes, the usual strategy to establish that an *isomorphic set representation* exists is to define a doubly ordered frame from an arbitrary lattice and then to show that there is a suitable isomorphism.

Definition 12.3.6 A *filter* F in a lattice $\mathbf{L}$ is subset of the carrier set (i.e., $F \subseteq A$) with properties (1)–(2).

(1) $a, b \in F$ implies $a \wedge b \in F$;
(2) $a \in F$ and $a \wedge b = a$ imply $b \in F$.

A filter is *proper* when $F \neq A$; a filter is *non-empty* when $F \neq \emptyset$.

Ideals, *proper* ideals and *non-empty* ideals are duals of respective filters. In particular, (3) and (4) define ideals.

(3) $a, b \in I$ implies $a \vee b \in I$;
(4) $b \in I$ and $a \vee b = b$ imply $a \in I$.

Remark 12.3.6 Filters (in the algebra of a logic) correspond to theories (in the logic). They are sublattices, therefore, the set of filters is closed under intersection. The intersection of a pair of filters includes the joins of the elements in those filters, that is, intersection can represent join. However, the union of a pair of filters does

[5] We use the notation $\mathscr{C}$ as in Bimbó and Dunn (2008), that is, $C \in \mathscr{C}$ iff C is a cone (or an upset, or an increasing set—to use other terms). Then, $\mathscr{C}_1$ and $\mathscr{C}_2$ are the sets of cones with respect to $\sqsubseteq_1$ and $\sqsubseteq_2$, respectively. We may omit parentheses—for readability—from $r(V)$ and $l(V)$.

not need to be a filter. The intersection of a pair of cones of filters is a cone of filters, and it can represent meet. Turning a lattice around, we can see that intersection on ideals can stand for meet and intersection on cones of ideals can represent join. To improve on these matches, special filters and ideals may be used.

Definition 12.3.7 Let I be an ideal on the lattice $\mathbf{L}$.

(1) I is *principal* when there is a $b \in I$ such that $a \in I$ iff $a \vee b = b$.
(2) I is *prime* when $a \wedge b \in I$ implies $a \in I$ or $b \in I$.
(3) I is *meet-irreducible* when for no I_1, I_2 distinct from I, $I_1 \cap I_2 = I$.

Principal, prime and *join-irreducible* filters are defined dually.

Stone (1936) used cones of prime ideals to represent Boolean algebras. The context of a semantic interpretation for a logic motivates the equivalent view of a representation by certain sets of ultrafilters. ("Ultrafilter" is an alternative name for a maximal filter, and in a Boolean algebra all of these are prime.) Unfortunately, in a lattice that is not distributive, there are too few prime filters to anchor an isomorphic representation. Meet-irreducible ideals and join-irreducible filters generalize their prime counterparts, and as Birkhoff and Frink (1948) proved, cones of join-irreducible filters provide an isomorphic representation of a lattice with intersection standing in for meet. They called such a representation a *meet-representation*, perhaps, because they did not define an operation to represent joins.

Remark 12.3.7 Prime filters are really special. Upward closed sets of prime ideals provide a meet- and a join-representation at the same time in a distributive lattice (including a Boolean algebra, where the upward closure trivializes). Furthermore, the complement of a prime filter is a prime ideal in any lattice. Prime filters are *relatively maximal*, that is, they are maximal with respect to not containing a particular element of a distributive lattice. Join-irreducible filters are similarly relatively maximal in lattices. However, cones of join-irreducible filters do not provide a join-representation (in modular but non-distributive lattices), nor is it true that the complement of a join-irreducible filter is a meet-irreducible ideal. (The complement of any filter is a prime co-cone in any lattice—see Bimbó and Dunn 2008, Chap. 4.) The ingenuity of Urquhart's representation relies on the observation that the complement of a join-irreducible filter contains at least one meet-irreducible ideal such that the filter and the ideal are relatively maximal with respect to each other, even though they may not exhaust the carrier set of the lattice.

Definition 12.3.8 The pair $\langle F, I\rangle$ is a *maximal disjoint filter–ideal pair* (MDFIP, for short) when (1)–(2) are satisfied.

(1) F is a non-empty, proper filter, and I is a non-empty, proper ideal;
(2) for any F', $F \subsetneq F'$ implies $F' \cap I \neq \emptyset$, and for any I', $I \subsetneq I'$ implies $F \cap I' \neq \emptyset$.

Proposition 12.3.9 *If* $\langle F, I\rangle$ *is a* MDFIP, *then* F *is a* join-irreducible filter *and* I *is a* meet-irreducible ideal.

The proof of this proposition can be pieced together from Birkhoff and Frink (1948); see also Urquhart (1978, Lemma 3). To put it quickly, if F were the intersection of two *other* filters, then F either would not be maximal or it would have a common element with I. It is also true that if a filter F and an ideal I are disjoint, then they can be extended into a MDFIP $\langle F', I' \rangle$ such that $F \subseteq F'$ and $I \subseteq I'$. (In general, this is a non-trivial claim that is usually proved using Zorn's lemma. We discuss this on Sect. 12.4.)

Definition 12.3.10 If $\mathbf{L}$ is a lattice, then the *doubly ordered space of* $\mathbf{L}$ is $\mathfrak{F}_{\mathbf{L}} = \langle \mathbb{U}, \subseteq_1, \subseteq_2 \rangle$, where (1) and (2) describe the components.

(1) $\mathbb{U}$ is the set of MDFIP's on $\mathbf{L}$;
(2) $\langle F_1, I_1 \rangle \subseteq_1 \langle F_2, I_2 \rangle$ iff $F_1 \subseteq F_2$, and $\langle F_1, I_1 \rangle \subseteq_2 \langle F_2, I_2 \rangle$ iff $I_1 \subseteq I_2$.

Remark 12.3.8 Obviously, $\subseteq$ is a partial order, hence, $\subseteq_1$ and $\subseteq_2$ are pre-orders. If both $\langle F_1, I_1 \rangle \subseteq_1 \langle F_2, I_2 \rangle$ and $\langle F_1, I_1 \rangle \subseteq_2 \langle F_2, I_2 \rangle$ hold, then both the filter and the ideal in the first pair is, respectively, a subset of the filter and the ideal in the second pair. But the pairs are maximally disjoint, hence, they are the same.

Despite all the duality between operations and sets of elements in a lattice, a lattice does not need to be self-dual. Further, the MDFIP's have to be linked to elements of a lattice, which suggests that we have to choose between filters and ideals.

Proposition 12.3.11 *A lattice* $\mathbf{L}$ *is* isomorphic *to a subset of stable sets on* $\mathbb{U}$.

Proof (*Sketch*) By favoring filters, an $a \in A$ is mapped by h into elements of $\mathbb{U}$ as $h(a) = \{ \langle F, I \rangle \in \mathbb{U} \colon a \in F \}$. It can be shown that $lrh(a) = h(a)$, that is, $h(a)$ is a stable set. Then, $h(a \wedge b) = h(a) \cap h(b)$ is immediate. And also, $h(a \vee b) = l(rh(a) \cap rh(b))$. So far, h is a lattice homomorphism. The fact that h is injective follows from separation; if $a \nleq b$, then there is a $\langle F, I \rangle \in \mathbb{U}$ such that $a \in F$ but $b \notin F$. □

12.3.2 Priestley's Representation Generalized

Priestley's representation of distributive lattices was motivated by a certain dissatisfaction she had with Stone's representation in Stone (1937), in particular, with features of the topological characterization of the prime ideal space.[6] The set of prime filters in a Boolean algebra forms an anti-chain, but in other distributive lattices it is easy to find prime filters (or prime ideals) that are distinct, yet one is a subset of the other. Priestley's invention is the addition of an order relation to a topology, which concretely will be realized by the subset relation.

---

[6] A Stone space for a Boolean algebra is a compact totally disconnected topology. But for a distributive lattice, Stone gave a *more complicated* characterization. Namely, the topology is T_0 with a basis comprising relatively bicompact sets with a further property linking intersections of basic sets with a closed set.

Definition 12.3.12 An *ordered space* is $\mathfrak{F} = \langle U, \leq \rangle$, where $\leq$ is a partial order on U.

Proposition 12.3.13 *The set of cones on* $\mathfrak{F}$ *is a* distributive lattice *with* $\cap$ *and* $\cup$ *as the lattice operations.*

The proof of this claim is practically obvious, hence, we do not even sketch it.

Definition 12.3.14 If $\mathbf{L}$ is a distributive lattice, then the *ordered space of* $\mathbf{L}$ is $\mathfrak{F}_{\mathbf{L}} = \langle \mathbb{U}, \subseteq \rangle$, where $\mathbb{U}$ is the set of prime filters, which is ordered by set inclusion.

Proposition 12.3.15 *A distributive lattice* $\mathbf{L}$ *is* isomorphic *to a subset of the set of cones on* $\mathbb{U}$.

Proof (*Sketch*) First of all, $h(a)$ for $a \in A$, is $\{ F \in \mathbb{U} \colon a \in F \}$. Cones of filters provide a meet-representation via intersection; since all the elements in the cones are prime filters, the union of such cones is a join-representation. The injectivity of h follows by an old and well-known result of Birkhoff, stating that distinct elements of a distributive lattice can be separated by a prime filter. □

Remark 12.3.9 To see Urquhart's representation as a generalization of Priestley's, we may compare the spaces first. Priestley's partial order could be weakened to a pre-order, or $\sqsubseteq_1$ and $\sqsubseteq_2$ could be strengthened to partial orders. To handle the one vs two orders, we could set $\sqsubseteq_1$ to be $\leq$ and $\sqsubseteq_2$ to be $\leq^{-1}$ (the inverse of the $\leq$ relation). Moving into the other direction, $\sqsubseteq_2$ could be simply omitted.

We have already pointed out that the complement of a prime filter is a prime ideal, that is, the MDFIP's are uniquely determined by either element of the pair in a distributive lattice. This means that there is a 1–1 map between prime filters and MDFIP's in a distributive lattice. It may be useful to glance at $rh(a)$. $F_1 \sqsubseteq_2 F_2$ iff $F_2 \subseteq F_1$, and the latter, iff $F_2 \sqsubseteq_1 F_1$. Thus, $F' \in rh(a)$ iff for all F, $F \subseteq F'$ implies $F \notin h(a)$.

Example 12.3.16 Let us consider some easy lattices. $\mathbb{Z}$ is a distributive lattice with min and max (as binary operations). Every principal filter is prime, and $h(n) = \{ [m) \colon m \leq n \}$, where $[m) = \{ n \in \mathbb{Z} \colon m \leq n \}$. $rh(n) = \{ [i) \colon n < i \}$, in other words, $rh(n)$ is the complement of $h(n)$ in the set of prime filters on $\mathbb{Z}$. If we take $\mathbb{Q}$ in place of $\mathbb{Z}$, then the definition of $rh(n)$ looks as before; a difference between those sets in the filter spaces of $\mathbb{Z}$ and $\mathbb{Q}$ is that the latter principal cocone is not generated by a principal cone. Finally, if we take **4** (the four-element Boolean algebra) with a and b the labels for the non-extremal elements, then $h(a) = \{ [a) \}$, and $rh(a) = h(b)$, that is, $\{ [b) \}$. In each case, $rh(a) = \mathbb{U} - h(a)$, and by a similar argument, $l(rh(a) \cap rh(b)) = \mathbb{U} - ((\mathbb{U} - h(a)) \cap (\mathbb{U} - h(b))) = h(a) \cup h(b)$.

12.4 One or Two Binary Relations

The work of De Morgan, Boole, and Frege led to a logic that was new in its time—in the nineteenth century. However, challenges to two-valued logic popped up soon after its first formulation in linear notation. In the early twentieth century, practicing logicians found the two-valued conditional too weak, which inspired the invention of strict implication and modal logic by C. I. Lewis. A serious challenge from physics produced the first example of a logic that questions the distributivity of $\wedge$ and $\vee$.[7]

Birkhoff and von Neumann (1936) explain certain differences between the views of reality derived from classical mechanics and those derived from quantum mechanics. Aspects of quantum mechanics that often attract attention are properties of its phase-spaces, namely, their principal incompleteness in description and in computable dependence. In other words, many observations are mutually exclusive and predictions of the position and momentum at the same time are necessarily imprecise. Birkhoff and von Neumann focus on the differences between reasoning in classical mechanics and quantum mechanics.

The classical view of a phase-space allows one to consider arbitrary subsets as experimental propositions, that is, propositions stating position and momentum of a certain kind. This classical view is *classical* in the sense of classical (Newtonian) mechanics and classical in the sense of 2-valued (Boolean) logic. Propositions are subsets of a phase-space and the operations on them correspond to intersection, union, and complementation. Birkhoff and von Neumann argue that, in contrast, experimental propositions in quantum mechanics correspond to closed linear subspaces of Hilbert space. And the operations on these propositions are intersection, linear sum, and orthogonal complementation.

The algebraic characterization of the experimental propositions in quantum mechanics leads to a bounded modular lattice with orthocomplementation. Birkhoff and von Neumann (*ibid.*, Sect. 10) pinpoint the failure of distributivity as the central difference between the calculi of classical and quantum propositions.[8]

Interlude: Modular lattices and KR frames. Modularity is a tricky property. Every distributive lattice is modular, and modularity is readily definable by a side condition on a prototypical equation expressing distributivity. We repeat (d) from above, which is to be compared with (m′).

(d) $a \wedge (b \vee c) = (a \wedge b) \vee (a \wedge c)$
(m′) $a \wedge (b \vee c) = (a \wedge b) \vee (a \wedge c)$, provided that $c \leq a$.

Dedekind was the first to characterize non-modular lattices as lattices that have a sublattice isomorphic to a five-element lattice (which is often labeled as $\mathbf{N}_5$). Despite this elegant algebraic description, modularity has not been characterized in terms of sequent calculus rules. The proof of the sequent $\mathcal{A} \wedge (\mathcal{B} \vee (\mathcal{A} \wedge \mathcal{C})) \Vdash (\mathcal{A} \wedge \mathcal{B}) \vee (\mathcal{A} \wedge \mathcal{C})$ (assuming more or less standard rules for $\wedge$ and $\vee$) appears to require the same structural rules on the left-hand side as $\mathcal{A} \wedge (\mathcal{B} \vee \mathcal{C}) \Vdash (\mathcal{A} \wedge \mathcal{B}) \vee (\mathcal{A} \wedge \mathcal{C})$, that is, distributivity does.

Urquhart used modular lattices in a crucial way in Urquhart (1984) to prove the undecidability of some of the major relevance logics such as **T** (ticket entailment),

[7] At least, it is one of the earliest and best-known examples of a non-distributive logic.

[8] Orthocomplemented modular lattices should not be confused with orthomodular lattices. The set of lattices in the latter category is a proper subset of those in the former.

E (entailment) and **R** (relevant implication).[9] He also constructed a representation of modular lattices. We give an overview of the representation in a nutshell.

Definition 12.4.1 If **L** is a modular lattice with least element $\bot$, then its **KR**-*frame* is $\mathfrak{F}_{\mathbf{L}} = \langle A; R, \bot\rangle$, where $R \subseteq A^3$ such that

(1) $R(a, b, c)$ iff $a \vee b = c \vee b$ and $a \vee b = a \vee c$.

KR is a crypto-relevance logic (cf. Routley et al. 1982, Chap. 5, Sect. 5) or perhaps, a corrupted one (cf. Anderson et al. 1992, Sects. 54 and 65). To put it quickly, **KR** adds $(\mathcal{A} \wedge {\sim}\mathcal{A}) \to \mathcal{B}$ to **R**, and its relevant character is hidden in its positive fragment, which is corrupted by negation. That is, from the point of view of relevance logic **KR** degrades **R** because **KR** lacks the variable sharing property. **KR** may be given a Meyer–Routley style semantics with a ternary accessibility relation. In the previous definition, we predicted that $\mathfrak{F}_{\mathbf{L}}$ has suitable properties to be called a **KR**-frame. (We do not prove here that it does.)

Definition 12.4.2 If $\mathfrak{F} = \langle U; R, \bot\rangle$ is a **KR**-frame, then its *modular lattice* with least element is defined as $\mathscr{L}(\mathfrak{F}) = \langle \mathcal{P}(U)^{\text{lin}}; \cap, \circ, \emptyset\rangle$, where (1)–(3) hold.

(1) $V \in \mathcal{P}(U)^{\text{lin}}$ iff $V \in \mathcal{P}(U)$ and $\forall u, v, w\, ((R(u, v, w) \wedge u, v \in V) \Rightarrow w \in V)$;
(2) if $V, W \subseteq U$, then $V \circ W = \{u : \exists v, w\, (v \in V$ and $w \in W$ and $R(v, w, u))\}$;
(3) $\cap$ is intersection, and $\emptyset$ is the empty set.

The superscript ^{lin} abbreviates "linear." It is true for all subsets of U that $V \subseteq V \circ V$, but the other inclusion does not hold, in general; on $\mathcal{P}(U)^{\text{lin}}$, $\circ$ is idempotent.

Proposition 12.4.3 *If* **L** *is a modular lattice with least element, then* **L** *is* isomorphic *to a sublattice of* $\mathscr{L}(\mathscr{F}(\mathbf{L}))$.

The claim is proved in Urquhart (2017). We note that this representation uses *subsets* of the carrier set rather than *sets of subsets*. To this extent, it does not fit the paradigm that we sketched at the beginning of Sect. 12.3, and it is not a derivative of the lattice representation in Sect. 12.3.1. The isomorphism establishing $\mathbf{L} \cong \mathscr{L}(\mathscr{F}(\mathbf{L}))$ maps $a \in A$ into $(a]$ (the principal ideal generated by a). Another way to look at the essence of this representation is to say that the ideal space of a modular lattice can be delineated precisely as the space of linear subsets with respect to $\circ$.[10]

Let us return to the logic of quantum mechanics. At the time of the writing of Birkhoff and von Neumann (1936), the equivalence of fields of sets and Boolean algebras was already known. Indeed, it is mentioned inter alia (on p. 831) about the logic of classical mechanics. Orthocomplemented distributive lattices are Boolean algebras, thus, the non-distributive modular lattices should be bunched together with

[9] We cannot go into the details here, but we mention Anderson et al. (1992, Sect. 65) too.

[10] Another representation of modular lattices was obtained by Jónsson (1953). He proved that every lattice that can be represented with join being $R_1; R_2; R_1$ (where R_1 and R_2 are two equivalence relations on a set) is modular.

non-modular lattices from the point of view of their representation by sets. The algebraic equations stipulated by Birkhoff and von Neumann give an *ortholattice* if modularity is omitted. These lattices are of interest in themselves, but they also played a fascinating role in the discovery of lattice representations on relational frames—including Urquhart's. Birkhoff and von Neumann (1936) mentions various models of modular lattices with orthocomplementation, including projective geometries and skew fields. Thus, it should not be surprising that Birkhoff abstracted out the idea of a *polarity* by 1940 or so.

Definition 12.4.4 A *polarity* is a triple $\langle X, Y, R\rangle$, where X and Y are sets and $R \subseteq X \times Y$. For $V \subseteq X$ and $W \subseteq Y$, their respective *polars* are defined by (1) and (2).

(1) $r(V) = \{\, y \in Y : \forall x\,(x \in V \Rightarrow R(x, y))\,\}$
(2) $l(W) = \{\, x \in X : \forall y\,(y \in W \Rightarrow R(x, y))\,\}$

Birkhoff (1967, V.7) also proved that the composition of r and l are closure operations (lr on subsets of X, rl on subsets of Y.) Furthermore, $lr[\mathcal{P}(X)]$ is a complete lattice that is dually isomorphic to $rl[\mathcal{P}(Y)]$. Birkhoff observed that if R is symmetric and irreflexive on a set (i.e., $X = Y$), then the set of closed subsets is an ortholattice. One of his examples is Cartesian n-space with R being $\perp$, the orthogonality relation.

Remark 12.4.1 Birkhoff and von Neumann in their paper of (1936) were concerned with logic, though they invoked many algebraic and geometric ideas too. It appears that Birkhoff went on to pursue the development of lattice theory (on which he published a book in 1940), whereas von Neumann, after developing what he called continuous geometry, focused on more practical areas such as computing and physics, and even game theory.

The abstraction of polarities is very useful, but in logical terms, it is the "easy direction" in giving a semantics for a logic. That is, it shows that a set X with an appropriate binary relation could serve as a frame for orthologic. Goldblatt (1974, 1975) showed that an *isomorphic representation* of ortholattices can be obtained along the lines of orthoframes. We, in effect, already defined an *orthoframe* above as a set with an irreflexive symmetric relation on it.

Definition 12.4.5 A lattice $\mathbf{L} = \langle A; \wedge, \vee, ', \perp, \top\rangle$ is an *ortholattice* when $\mathbf{L}$ is a lattice in which the following quasi-equations hold.

(1) $a'' = a, \quad a \wedge a' = \perp, \quad a \vee a' = \top, \quad a \wedge b = a$ implies $a' \vee b' = a'$.

An ortholattice is quite like a Boolean algebra—except that it does not need to be distributive. Every Boolean algebra is an ortholattice, but not vice versa.

Definition 12.4.6 If $\mathbf{L}$ is an ortholattice, then its *orthoframe* is $\mathfrak{F} = \langle \mathbb{X}, \perp\rangle$, where (1) and (2) specify the components.

(1) $\mathbb{X}$ is the set of proper filters on A;

(2) $F_1 \perp F_2$ iff $\exists a \in A$ such that $a' \in F_1$ and $a \in F_2$.

Ortholattices are interesting in themselves. However, we wish to emphasize their generalization to polarities. We labeled the two functions as r and l in (1) and (2) in Definition 12.4.4 to point at certain similarities with Urquhart's functions r and l.[11] (fl) and (fr) have a similar form as the definitions in (1) and (2), except that there are two relations $\not\leftrightharpoons_1$ and $\not\leftrightharpoons_2$. Both relations are irreflexive, but not much else seems to be true of them. The functions r and l form a Galois pair in both cases.[12]

We recall some results about Galois pairs to show how the ideas about complementation, orthogonality, and negation led to the lattice representation by Hartonas and Dunn (1993, 1997).

Definition 12.4.7 Let $\mathbf{U} = \langle U; \leq_1 \rangle$ and $\mathbf{W} = \langle W; \leq_2 \rangle$ be two posets and let f be a function form U to W, and let g be a function from W to U. Then the pair $\langle f, g \rangle$ is a *Galois connection* between $\mathbf{U}$ and $\mathbf{W}$ iff $\forall x \in U \, \forall y \in W,\ x \leq_1 g(y)$ iff $y \leq_2 f(x)$.

Remark 12.4.2 It is easy to show that—equivalently—we can require $x_1 \leq_1 x_2$ implies $f(x_2) \leq_2 f(x_1)$, $y_1 \leq_2 y_2$ implies $g(y_2) \leq_1 g(y_1)$, $x \leq_1 g(f(x))$ and $y \leq_2 f(g(y))$.

If we view the members of U as propositions, $x_1 \leq_1 x_2$ may be interpreted as "x_1 implies x_2," and similarly, with W and $y_1 \leq_2 y_2$. You may be puzzled as to why we have two possibly disjoint sets of propositions, but try to set that aside. It is natural to view f and g as *negations*. Writing them as $\backsim$ and $\sim$, we have that x implies $\backsim y$ iff y implies $\sim x$. If you squint a bit (so as not to be able to distinguish the two different negations), this looks like a familiar principle of contraposition. And the last two inequations are double negation introductions: $x \leq_1 \backsim\sim x$ and $y \leq_2 \sim\backsim y$.

There is a straightforward way to construct a Galois connection between all subsets of a set X (the powerset of X) and all subsets of a set Y, where $\leq_1$ is set inclusion (the subset relation) restricted to $\mathcal{P}(X)$ and $\leq_2 = \subseteq \restriction \mathcal{P}(Y)$. First, let us think of the members of X and Y as information states, and so their subsets can be viewed as "U.C.L.A. propositions." We could pick any relation R between the two sets. However, for reasons that pertain to seeing a Galois connection as involving a pair of negations, it is common to use the symbol $\perp$—like in an orthoframe. $\perp$ may be thought of as *orthogonality* or *perp* (for perpendicularity), or more generally as a kind of *incompatibility*, which may go one way but not the other.

Definition 12.4.8 Let X and Y be connected with $\perp$. For any $V \subseteq X$, $V^\perp = \{ y \in Y : \forall x \in V \ x \perp y \}$. Dually, for any $W \subseteq Y$, ${}^\perp W = \{ x \in X : \forall y \in W \ x \perp y \}$.

[11] "Right" and "left" are, obviously, at hand, in particular, they are used in the theory of fields.

[12] Such functions, in an abstract setting, i.e., outside of Galois theory, have been studied by Everett (1944) and Ore (1944, 1962). Since the power set (or a set of special subsets of a set) has a natural ordering on it, namely, the subset relation, it is immediate that Galois connections on a collection of subsets induce a lattice (cf. Birkhoff 1967, V.8).

Remark 12.4.3 $V^{\perp}$ can be thought of as a kind of negation of V, i.e., the set of states $y \in Y$ such that every state x that verifies V is incompatible with y. And symmetrically, with ${}^{\perp}W$. It is worth noting that when $\perp$ is a symmetric relation, that is, $x \perp y$ implies $y \perp x$ and $X = Y$—like in the orthoframe of an ortholattice in Definition 12.4.6—then $V^{\perp} = {}^{\perp}V$. Conversely, $\perp$ is symmetric if it comes from orthonegation.

Theorem 12.4.9 *$V \subseteq {}^{\perp}W$ iff $W \subseteq V^{\perp}$; that is, $\langle \cdot^{\perp}, {}^{\perp}\cdot \rangle$ are a* Galois connection *between $\mathcal{P}(X)$ and $\mathcal{P}(Y)$.*

Proof ($\Rightarrow$) Suppose that $V \subseteq {}^{\perp}W$ and that $y \in W$, to show that $y \in V^{\perp}$, which means $\forall x \in V\, x \perp y$. So let us suppose $x \in V$; then $x \in {}^{\perp}W$. Since $y \in W$, $x \perp y$.
($\Leftarrow$) It is proved similarly. □

Thus $\perp$ allows us to define a "concrete" Galois connection between all the subsets of a set X and all the subsets of a set Y. But there are other "concrete" Galois connections that hold just between some subsets of a set X and Y—as we saw in Proposition 12.3.5. Not only does perp allow us to define a Galois connection between subsets of X and subsets of Y, but this is a fully general way to obtain Galois connections.

Theorem 12.4.10 *Every Galois connection is generated—up to isomorphism—as in Definition 12.4.8, for some sets X, Y and $\perp$.*

Proof Let us assume that $\mathbf{U}$ and $\mathbf{W}$ are Galois connected with $\langle f, g \rangle$ as in Definition 12.4.7. We consider the sets of cones on U and W, which we denote by $\mathscr{C}_1$ and $\mathscr{C}_2$. We define a perp relation between two cones ($C \in \mathscr{C}_1$ and $D \in \mathscr{C}_2$) so that $C \perp D$ iff $\exists x \in C$ such that $f(x) \in D$. Note that we can define the dual perp relation $\perp'$ so that $D \perp' C$ iff $\exists y \in D$ such that $g(y) \in C$. It is easy to see that $\perp'$ is the *converse* of $\perp$. Thus, if $C \perp D$ then $\exists x \in C\ f(x) \in D$. We invoke $x \leq_1 g(f(x))$ to obtain $g(f(x)) \in C$. So $\exists y \in D$, namely, $y = f(x)$, such that $g(y) \in C$; therefore, $D \perp' C$. The other direction is proven "dually."

We now define the embedding $h(x) = \{C \in \mathscr{C}_1 \colon x \in C\}$ and $h(y) = \{D \in \mathscr{C}_2 \colon y \in D\}$. The trick then is to show that $h(f(x)) = h(x)^{\perp}$, i.e., $f(x) \in D$ iff $\forall C$ (if $x \in C$ then $C \perp D$). The left-to-right direction is immediate.

For right to left, we contrapose. Thus, assume that $f(x) \notin D$. We need to show that it is not the case that $\forall C$ (if $x \in C$ then $C \perp D$), i.e., we need to find some cone C so that $x \in C$, and yet not $C \perp D$. Let C be the principal cone $[x) = \{x' \colon x \leq_1 x'\}$. Then $x \in C$, and yet it is not the case that $C \perp D$, for otherwise, $\exists x' \in C$ and $f(x') \in D$. That is, $x \leq_1 x'$ and $f(x') \in D$, but $f(x') \leq_2 f(x)$ and $f(x') \in D$; hence, $f(x) \in D$. Contradiction!

That $h(g(y)) = {}^{\perp}h(y)$ may be shown similarly. □

This theorem is a *representation of Galois connections* between partially ordered sets, which can readily be extended to a Galois connection between two *semilattices*. A meet semilattice is a partially ordered set, where every pair of elements a and b has a greatest lower bound $a \wedge b$. A join semilattice is defined dually, requiring that every pair of elements a and b has a least upper bound $a \vee b$. This can be done quite elegantly by requiring the set X to be a meet semilattice $\langle S; \wedge \rangle$ and the set Y to be a join semilattice $\langle S'; \vee \rangle$. But a more casual way is just to let the right-hand set also be a meet semilattice (with the order inverted). Then all one needs to do is to extend the definition of a cone to a filter F. An advantage of this track is that one can avoid the apparatus of "dual filters" altogether. This naturally leads to a representation of lattices once one realizes that a lattice is just two semilattices "glued together," one up and the other down.

Remark 12.4.4 In the case of Urquhart's representation, the functions r and l have to be restricted to upward closed subsets (with respect to one or the other order relation) on the frame. That is, the Galois connected posets are $\langle \mathscr{C}_1(U), \subseteq \rangle$ and $\langle \mathscr{C}_2(U), \subseteq \rangle$. Then, taking cones on each set, we can find $\perp$ using Theorem 12.4.10. For example, in the doubly ordered space of a lattice, if $C \in \mathscr{C}(\mathscr{C}_1(\mathbb{U}))$ such that $h(a) \in C$, then $C \perp D$ holds when $D \in \mathscr{C}(\mathscr{C}_2(\mathbb{U}))$ and $rh(a) \in D$. Of course, not all C's and D's are of this form, but $h(a)$ and $rh(a)$ are the lr and rl stable sets in $\mathscr{C}_1(\mathbb{U})$ and $\mathscr{C}_2(\mathbb{U})$, respectively. For any $\langle F, I \rangle \in h(a)$, $a \in F$ and for any $\langle F, I \rangle \in rh(a)$, $a \in I$.

In comparison with ortholattices, the two orders $\sqsubseteq_1$ and $\sqsubseteq_2$ seem to be complicated. But the ortholattices have an additional component—the orthocomplement—that is used in the definition of $\perp$ in the isomorphic representation. Now, we have seen that from a Galois connection on a lattice (chopped into two semilattices), one can find a polarity. Concretely, Urquhart's representation contains a Galois connection between two posets, which are constructed from the two orders. So, we might wonder how these observations may be put to use to construct a lattice representation. Indeed, this can be accomplished in more than one way. Next, we outline a lattice representation due to Hartonas and Dunn (1993, 1997).

Definition 12.4.11 A *frame* is $\mathfrak{F} = \langle X, Y, \perp \rangle$, where $X, Y \neq \emptyset$ and $\perp \subseteq X \times Y$.

This definition simply takes a polarity for a frame. (Continuing the idea of the previous theorem, we use the $\perp$ notation.)

Definition 12.4.12 If $\mathbf{L}$ is a lattice, then its *lattice frame* is $\mathfrak{F}_{\mathbf{L}} = \langle \mathbb{F}, \mathbb{I}, \between \rangle$, where the components are specified in (1)–(3).

(1) $\mathbb{F}$ is the set of filters on A;
(2) $\mathbb{I}$ is the set of ideals on A;
(3) $F \between I$ iff for some $a \in A$, $a \in F$ and $a \in I$.

Remark 12.4.5 The definitions of the $\perp$ relation in Definition 12.4.6 and in the proof of Theorem 12.4.10 and the definition of $\between$ above are remarkably similar. A lattice does not need to be complemented, and so there is no unary operation that could be applied to a. However, a lattice with its natural order relation $\leq$ and its converse $\leq^{-1}$ can be seen to have a Galois connection $\langle \mathrm{Id}, \mathrm{Id} \rangle$, where $\mathrm{Id}(x) = x$. Of course, $F_1 \cap F_2 \neq \emptyset$ would give the total relation on non-empty filters. But $\leq^{-1}$ turns filters into ideals and vice versa. It seems to us that using the above frame is also in the spirit of Birkhoff and Frink (1948), since cones of filters give a meet-representation and cones of ideals give a join-representation—both with intersection. The role of $\between$ is to create a dual isomorphism between the complete lattices of closed subsets of the set of filters and closed subsets of the set of ideals.

The remaining piece is to ensure that the elements of a lattice can be mapped into appropriate collections of filters (or ideals). Of course, if we take as a hint the usual definition of h that uses $\in$, then we get that $h(a) = [[a))$. That is, each lattice element is mapped into the principal cone of filters generated by the principal filter generated by the lattice element.

Remark 12.4.6 Triples $\langle G, M, I\rangle$, which have the structure of a polarity or of a frame in the sense of Definition 12.5.1 have been termed *contexts* by Wille (1982). A subset of G is an *extension* of a concept, whereas a subset of M is its *intension* (in Church's terminology). A hierarchy of concepts (in the Aristotelian sense) pertaining to a context can be constructed based on the observation that "The more specific a concept is, the fewer exemplars it has." Thus, a pair of an extension and intension fits into a lattice of concepts given a context.[13] Although Wille *does construct* a lattice from a formal context, his interest lies with concepts and their relationships (cf. Wille 1985). His representation of a lattice diverges markedly from that of Hartonas and Dunn (1997) and Hartonas and Dunn (1993), because his context of a lattice $\mathbf{L} = \langle L; \wedge, \vee\rangle$ is $\langle L, L, \leq\rangle$. Hartung (1992) uses topological contexts to give an isomorphic representation for formal concept lattices.

We may quickly compare (or even contrast) Urquhart's and Hartonas and Dunn's representations. The most striking difference is the disjointness of the MDFIP's and the overlap that defines $\perp$. In the spirit of reverse mathematics (or of concerns about uses of equivalents of the axiom of choice), we should point out that Hartonas and Dunn do not rely on prime filters, prime ideals, join-irreducible filters or meet-irreducible ideals. To prove that such objects exist, or that a disjoint pair of a filter and an ideal can be extended into a MDFIP (Urquhart 1978, Lemma 3), one apparently has to appeal to the axiom of choice (or to Zorn's lemma, etc.). The two representations appear not to be equivalent in the sense that Hartonas and Dunn's resides in ZF, but Urquhart's seems to require ZFC. On the side of similarities, we may point out that in both representations the ideals play second fiddle to filters (of a certain kind).

Three more lattice representations. The differences may inspire us to consider variations on either representation that would make them more similar.

1. For example, it may seem that the "dual" of maximally disjoint filter–ideal pairs should be *minimally overlapping filter–ideal pairs* (MOFIP's, for short). Since there are no "subatomic particles" in a lattice, minimal overlap implies that there is exactly one element that is common to the filter and the ideal in the pair. Furthermore, the shared element must be the least element of the filter and the greatest element of the ideal. Then, we are talking about filter–ideal pairs, in which both the filter and the

[13] Wille's notions of (formal) context and (formal) concept, which he designed for computer science applications should not be confused with philosophical investigations of concepts following Wittgenstein or with the use of the term "concept" in cognitive science.

ideal are principal, and they are generated by the same element. Principal filters have some pleasant properties. For example, a meet-representation of the lattice by sets of filters that contain a particular element preserves arbitrary meets iff all the filters are complete filters (which principal filters are).[14] However, not all lattices are complete, and in a lattice that is not complete, there are non-principal filters. Perhaps, the best-known example of a non-complete distributive lattice is $\mathbb{Q}$, the set of rationals with min and max for meet and join. If we set aside the problems caused by non-complete lattices, then the minimally overlapping filter–ideal pairs provide a relatively simple representation.[15] If $h(a) = \{ \langle F, I \rangle : a \in F \}$ is the embedding of the lattice into its MOFIP space, then the set of the first projection is generated by $[a)$. Using the same relation as in the Hartonas–Dunn representation, $rh(a) = \{ \langle F, I \rangle : a \in I \}$. This obviously suffices to model $\vee$.

2. Allwein and Dunn (1993, p. 522) point out that a representation may be had without insisting upon maximality in MDFIP's. Indeed, this improves the representation in the sense of duality theory, which we briefly touch upon in the next section. This representation has been worked out in Allwein and Hartonas (1993), where the focus is on duality theory, and in Gehrke and Harding (2001), with an emphasis on canonical extensions. Dually, a representation can be constructed from arbitrary overlapping filter–ideal pairs. This representation is worked out in Bimbó and Dunn (2008, Chap. 9), where we called the frames *centered spaces*. We could argue that this is the most balanced representation in the sense that there is no need to choose between a filter and an ideal as the preferred object. A pair of a filter and ideal both of which contain an element a, carve out a sublattice of a lattice (by their common elements). So a is mapped into the sublattices that contain a with two "tentacles," so to speak, which are the rest of the filter and that of the ideal.

3. We started this section by considering the number of binary relations that each representation stipulates. Priestley added order to the space, and the general lattice representations all added further elements—so far. The preference for filters suggests that we may consider another generalization of the Priestley space. $\mathfrak{F} = \langle U; \leq \rangle$ is an *inclusion space* when $U \neq \emptyset$ and $\leq$ is a partial order on U. Of course, $\cup$ cannot stand for $\vee$, if $\wedge$ is $\cap$. However, r and l can be defined as usual in a polarity, and $\uplus$ can represent $\vee$. For an isomorphic copy of a lattice **L**, we simply take the filter space of **L** with set inclusion as the partial order. This representation does not coincide with Priestley's on a distributive lattice if the filters are restricted to prime filters, because the concrete $\uplus$ which is a closure of $\cup$ relies on filters that are not join-irreducible. On the other hand, this representation coincides with Priestley's in the definition of the topology on $\langle \mathbb{U}, \subseteq \rangle$. Namely, the subbasis is defined as $\{ h(a), -h(a) : a \in A \}$, where $h(a) = \{ F \in \mathbb{U} : a \in F \}$. For more details, see Bimbó and Dunn (2008, Chap. 9).

Remark 12.4.7 For certain purposes, it is satisfactory to have a lattice that emerges from a frame. Indeed, in the area of modal logic, some researchers prefer to weaken a topological frame to a general frame. (Retaining a set of propositions confers

[14] See Birkhoff and Frink (1948, Sect. 11).

[15] Bimbó (1999, 2001) used such a representation of lattices as a component of a semantics.

benefits without the burden of imposing additional conditions on a Stone space with a binary relation.) We mention a couple of recent papers, Orłowska and Rewitzky (2005), Hartonas (2019), and Düntsch and Orłowska (2019), which advocate for lattice representations without topologies within the "discrete duality" program.

12.5 Topological Structures

The previous sections showed how to obtain a lattice from a relational structure, moreover, how to define a relational structure from any lattice, so that an isomorphic copy of the lattice can be found in the set algebra on its relational structure. However, in general, it would be unreasonable to expect that the isomorphism is *surjective*. For instance, every element of a lattice generates a filter, but an infinite lattice may have non-principal filters too. And the power set of the set of filters has a *strictly greater cardinality*, which may be inherited by a subset of the power set, which perhaps, allows only for cones.

Stone's representation of Boolean algebras by sets in Stone (1936) is acclaimed, because he found an elegant way to characterize the *image of a Boolean algebra* under the intended isomorphism. A Stone space is simply a compact totally disconnected topology, in which a Boolean algebra emerges as the set of clopen sets. That is, the elements of the Boolean algebra are those open sets in the topology that are also closed (i.e., their complements are open). Sets of prime filters of the form $h(a)$ constitute a basis for the Stone space of a Boolean algebra. Indeed, $h[A]$ is the set of clopen sets of the topology in which the basis comprises the sets $h(a)$, for $a \in A$.

Priestley's representation added an order, hence, her space is a compact, totally order disconnected topology. A distributive lattice arises as the clopen cones of the topology, that is, increasing sets that are both open and closed. Given a distributive lattice, sets of the form $h(a)$ together with their complements form a subbasis of a topology, which is the Priestley space of a distributive lattice.

Urquhart's representation requires additions in its topological component, because it has two order relations on a set.

Definition 12.5.1 A *doubly ordered topological space* is $\mathfrak{F}$ as in Definition 12.3.3 with a compact topology on U that satisfies (1)–(5).

(1) $\mathbb{C} \subseteq \mathcal{P}(U)$ such that, if $X \in \mathbb{C}$ then both $-X$ and $-rX$ are open;
(2) $x \not\sqsubseteq_1 y$ implies that for some $X \in \mathbb{C}$ such that $X \in \mathcal{P}(U)^\dagger$ both $x \in X$ and $y \notin X$;
(3) $x \not\sqsubseteq_2 y$ implies that for some $X \in \mathbb{C}$ such that $X \in \mathcal{P}(U)^\dagger$ both $x \in rX$ and $y \notin rX$;
(4) if $X, Y \in \mathbb{C}$, then both $-r(X \cap Y)$ and $-l(rX \cap rY)$ are open;
(5) the set $\{ -X : X \in \mathbb{C} \wedge X \in \mathcal{P}(U)^\dagger \} \cup \{ -rX : X \in \mathbb{C} \wedge X \in \mathcal{P}(U)^\dagger \}$ is a subbasis for the topology.

Given a doubly ordered topological space $\mathfrak{F}$, $\{ X : X \in \mathbb{C} \wedge X \in \mathcal{P}(U)^\dagger \}$ is a lattice.

Definition 12.5.2 If **L** is a lattice, then the *subbasis of the topology* of its doubly ordered topological space is $\{ h(a) \colon a \in A \} \cup \{ -rh(a) \colon a \in A \}$.

The definition of the subbasis is quite simple and it closely resembles the definition of the subbasis in the Priestley representation.

Dualities. Having outlined examples of topological frames for lattices (and ortholattices), now we mention another role that topologies can play in a lattice representation. Topologies may lead to dualities between a class of algebras and a class of relational structures. We fix both the class of algebras (as lattices) and the class of relational structures (e.g., doubly ordered topological spaces). Then, given any $\mathfrak{F}$, we can construct a lattice, that is, $\mathscr{L}(\mathfrak{F})$ is a lattice, where $\mathscr{L}$ is a map from frames to lattices. This step gives a semantics for our $\mathfrak{LatL}$ with *soundness* guaranteed. In order to get an isomorphic copy of a lattice **L**, we define its frame $\mathscr{F}(\mathbf{L})$ and show that $\mathbf{L} \cong \mathscr{L}(\mathscr{F}(\mathbf{L}))$. Here $\mathscr{F}$ is a map from lattices to frames. This step gives *completeness* for $\mathfrak{LatL}$ with respect to the class of relational structures.

Isomorphisms are special homomorphisms, and the latter are maps that are natural companions of algebras. If $\mathfrak{F}$ is a relational structure, then the counterpart notion is relational isomorphism between a pair of frames. If, in addition, $\mathfrak{F}$ is equipped with a topology, then relational isomorphisms should be homeomorphisms (in both direction). In other words, we would like to have that $\mathfrak{F} \rightleftharpoons \mathscr{F}(\mathscr{L}(\mathfrak{F}))$. Once we have both correspondences, we have *object duality*, because we can match frames and lattices to each other.

However, we can go a step or two further. Homomorphisms between algebras are of interest in themselves. They are the maps between algebras the properties of which tell us a lot about the particular algebras in question. We can define maps between frames too so that the map turns a frame of a certain kind into a frame of the same kind. If the homomorphisms (on the side of algebras) and the frame morphisms (on the side of frames) compose and certain maps can function as identities for composition, then we may talk about a duality between categories of lattices and frames (cf. Awodey 2010). Pulling back from full categorical duality (i.e., functorial duality), we could simply consider full duality, that is, a 1–1 correspondence between frame morphisms and homomorphisms on top of object duality.

For the sake of comparison, we outline two representations that we already mentioned—now outfitted with topologies.

Definition 12.5.3 An *ordered topological orthospace* is $\mathfrak{F} = \langle X; \leq, \perp, \mathscr{O} \rangle$, where $\langle X, \leq \rangle$ with $\mathscr{O}$ is a poset with a compact topology, and $\perp \subseteq X^2$ is an orthogonality relation (irreflexive and symmetric). Also, (1)–(4) hold. ($\mathfrak{CL}$ is the set of clopens and $\mathfrak{CL}^\dagger$ denotes the set of clopen stable sets.)

(1) $x \nleq y$ implies $\exists O \in \mathfrak{CL}^\dagger (x \in O \wedge y \notin O)$;
(2) $x \perp y$ and $x \leq z$ imply $z \perp y$;
(3) $O \in \mathfrak{CL}^\dagger$ implies $O^\perp \in \mathfrak{CL}$;
(4) $x \not\perp y$ implies $\exists O \in \mathfrak{CL}^\dagger (x \in O \wedge y \in O^\perp)$.

A *frame morphism* f is a continuous function with properties (5)–(6).

(5) $fx \perp fy$ implies $x \perp y$;
(6) $\neg z \perp fy$ implies $\exists x\,(\neg x \perp y \wedge z \leq fx)$.

The above definition (from Bimbó 2007), which enriches an orthoframe with not only a topology but also with an order relation, allows us to prove *full duality* (i.e., duality for both objects and maps) between ortholattices and orthoscapes.

We argued that inclusion spaces provide an alternative generalization of Priestley spaces.

Definition 12.5.4 A *topological inclusion space* is $\mathfrak{F} = \langle X; \leq, \mathscr{O} \rangle$, where $\langle X; \leq \rangle$ is a poset, and $\mathscr{O}$ is a compact topology with (1)–(2) true.

(1) $x \not\leq y$ implies $\exists O \in \mathfrak{C}^{\dagger}(x \in O \wedge y \notin O)$;
(2) $U, V \in \mathfrak{C}^{\dagger}$ implies $U \Cup V \in \mathfrak{C}$, that is, $l(rU \cap rV)$ is clopen.

A *frame morphism* f is a continuous order preserving map satisfying (3).

(3) $O \in \mathfrak{C}^{\dagger}$ implies $f^{-1}[O] \in \mathcal{P}(X)^{\dagger}$, that is, the inverse image of a clopen stable set is stable.

The enriched inclusion spaces support *full duality*, including a duality between homomorphisms and frame morphisms (see Bimbó and Dunn 2008, Chap. 9).

Remark 12.5.1 We may note that both of these representations rely on mere filters, and do not require the use of Zorn's lemma or some other equivalent of the axiom of choice to prove maximality of any kind. Also, both frames include an order, which is the only relation in an inclusion space, and it makes orthospaces smoother. Allwein and Hartonas (1993) discuss the question of full duality in Urquhart's representation, and in its relaxed version—where maximality is omitted.

12.6 Conclusions

The logic of quantum logic isolated by Birkhoff and von Neumann may be the first but surely not the last example of a logic that does not stipulate distributivity for $\wedge$ and $\vee$. Such logics were developed for technical reasons (like lattice-**R**, in Meyer (1966)), for the sake of simplicity (like full Lambek calculus, in Ono (2003)) and to capture resource-minded reasoning (like linear logic, in Girard (1987)). Non-distributive logics arise naturally as "substructural logics," since Gentzen's structural rules of permutation, thinning, and contraction are essential for a proof of distribution.[16] The various lattice representations differ on how easily they can be extended to semantics for logics. We only mention Allwein and Dunn (1993), Bimbó and Dunn (2008) and Düntsch et al. (2004) as examples that provide semantics for a wide range of logics starting from Urquhart's lattice representation.

[16] "Substructural logics" is often used as an honorific to include relevance logics such as **T**, **E** and **R** too, even though these logics have a distributive lattice reduct in their Lindenbaum algebra.

We close by mentioning once more Alasdair Urquhart's Sainthood. He was the first to give a *topological representation for lattices*. Thus, we think that he should become the "Patron Saint of Lattices" for doing this, and for all of the miracles he has performed with all kinds of (semi)lattices.

Acknowledgements We are grateful to the editors of the volume for inviting us to contribute a paper, and for providing a helpful report from the second reader. Our work on this paper was supported by the *Insight Grant* entitled "From the Routley–Meyer semantics to gaggle theory and beyond: The evolution and use of relational semantics for substructural and other intensional logics" (#435-2019-0331) awarded by the *Social Sciences and Humanities Research Council of Canada*.

References

Allwein, G., & Dunn, J. M. (1993). Kripke models for linear logic. *Journal of Symbolic Logic*, *58*(2), 514–545.

Allwein, G., & Hartonas, C. (1993). *Duality for bounded lattices. IU Logic Group Preprint Series IULG-93-25*. Bloomington, IN: Indiana University.

Anderson, A. R., Belnap, N. D., & Dunn, J. M. (1992). *Entailment: The logic of relevance and necessity* (Vol. II). Princeton, NJ: Princeton University Press.

Awodey, S. (2010). *Category theory*. Oxford logic guides (Vol. 52, 2nd Ed.). Oxford, UK: Oxford University Press.

Bimbó, K. (1999). *Substructural logics, combinatory logic and λ-calculus*. PhD thesis, Indiana University, Bloomington, Ann Arbor (UMI).

Bimbó, K. (2001). Semantics for structurally free logics $LC+$. *Logic Journal of the IGPL*, *9*, 525–539.

Bimbó, K. (2007). Functorial duality for ortholattices and De Morgan lattices. *Logica Universalis*, *1*(2), 311–333.

Bimbó, K., & Dunn, J. M. (2002). Four-valued logic. *Notre Dame Journal of Formal Logic*, *42*, 171–192.

Bimbó, K. & Dunn, J. M. (2008). *Generalized galois logics: Relational semantics of nonclassical logical calculi*. CSLI Lecture notes (Vol. 188). Stanford, CA: CSLI Publications.

Birkhoff, G. (1967). *Lattice theory* (Vol. 25, 3rd Ed). Providence, RI: AMS Colloquium Publications. American Mathematical Society.

Birkhoff, G., & Frink, O. (1948). Representations of lattices by sets. *Transactions of the American Mathematical Society*, *64*, 299–316.

Birkhoff, G., & von Neumann, J. (1936). The logic of quantum mechanics. *Annals of Methematics*, *37*(4), 823–843.

Craig, A., & Haviar, M. (2014). Reconciliation of approaches to the construction of canonical extensions of bounded lattices. *Mathematica Slovaca*, *64*(6), 1335–1356.

Dunn, J. M. (1966). *The algebra of intensional logics*. PhD thesis, University of Pittsburgh, Ann Arbor (UMI). (Published as v. 2 in the *Logic PhDs* series by College Publications, London (UK), 2019.).

Dunn, J. M. (1991). Gaggle theory: An abstraction of Galois connections and residuation, with applications to negation, implication, and various logical operators. In J. van Eijck (Ed.), *Logics in AI: European workshop JELIA '90, number 478 in Lecture Notes in Computer Science* (pp. 31–51). Berlin: Springer.

Dunn, J. M. (1993). Star and perp: Two treatments of negation. *Philosophical Perspectives*, *7*, 331–357. (Language and Logic, 1993, J. E. Tomberlin (ed.)).

Dunn, J. M. & Hardegree, G. M. (2001). *Algebraic methods in philosophical logic*. Oxford logic guides (Vol. 41). Oxford, UK: Oxford University Press.

Dunn, J. M., & Zhou, C. (2005). Negation in the context of gaggle theory. *Studia Logica*, *80*(2–3), 235–264. Special issue: Negation in constructive logic, edited by H. Wansing, S. Odintsov, and Y. Shramko.

Düntsch, I., & Orłowska, E. (2019). A discrete representation of lattice frames. In Blackburn, P., Lorini, E., & Guo, M., (Eds.), *Logic, rationality and interaction. LORI 2019*. Lecture notes in computer science (Vol. 11813, pp. 86–97). Springer.

Düntsch, I., Orłowska, E., Radzikowska, A. M., & Vakarelov, D. (2004). Relational representation theorems for some lattice-based structures. *Journal of Relational Methods in Computer Science*, *1*(1), 132–160.

Everett, C. J. (1944). Closure operators and Galois theory in lattices. *Transactions of the American Mathematical Society*, *55*, 514–525.

Gehrke, M., & Harding, J. (2001). Bounded lattice expansions. *Journal of Algebra*, *238*, 345–371.

Gierz, G., Hofmann, K. H., Keimel, K., Lawson, J. D., Mislove, M. W., & Scott, D. S. (2003). *Continuous lattices and domains*. Encyclopedia of mathematics and its applications (Vol. 93). Cambridge, UK: Cambridge University Press.

Girard, J.-Y. (1987). Linear logic. *Theoretical Computer Science*, *50*, 1–102.

Goldblatt, R. (1974). Semantic analysis of orthologic. *Journal of Philosophical Logic*, *3*, 19–35.

Goldblatt, R. (1975). The Stone space of an ortholattice. *Bulletin of the London Mathematical Society*, *7*, 45–48.

Harding, J. (1998). Canonical completions of lattices and ortholattices. *Tatra Mountains Mathematical Publications*, *15*, 85–96.

Hartonas, C. (2019). Discrete duality for lattices with modal operators. *Journal of Logic and Computation*, *29*(1), 71–89.

Hartonas, C., & Dunn, J. M. (1993). Duality theorems for partial orders, semilattices, Galois connections and lattices. IU Logic Group Preprint Series IULG-93-26. Bloomington, IN: Indiana University.

Hartonas, C., & Dunn, J. M. (1997). Stone duality for lattices. *Algebra Universalis*, *37*, 391–401.

Hartung, G. (1992). A topological representation of lattices. *Algebra Universalis*, *29*, 273–299.

Hutchinson, G. (1973). The representation of lattices by modules. *Bulletin of the American Mathematical Society*, *79*(1), 172–176.

Jónsson, B. (1953). On the representation of lattices. *Mathematica Scandinavica*, *1*, 193–206.

Meyer, R. K. (1966). *Topics in modal and many-valued logic*. PhD thesis, University of Pittsburgh, Ann Arbor (UMI).

Ono, H. (2003). Substructural logics and residuated lattices – an introduction. In V. F. Hendricks & J. Malinowski (Eds.), *50 years of studia logica, number 21 in trends in logic* (pp. 193–228). Amsterdam: Kluwer.

Ore, O. (1944). Galois connexions. *Transactions of the American Mathematical Society*, *55*, 493–513.

Ore, O. (1962). *Theory of graphs*. American Mathematical Society Colloquium Publications (Vol. 38). Providence, RI: American Mathematical Society.

Orłowska, E., & Rewitzky, I. (2005). Duality via truth: Semantic frameworks for lattice-based logics. *Logic Journal of the IGPL*, *13*, 467–490.

Priestley, H. A. (1970). Representation of distributive lattices by means of ordered Stone spaces. *Bulletin of the London Mathematical Society*, *2*, 186–190.

Priestley, H. A. (1972). Ordered topological spaces and the representation of distributive lattices. *Proceedings of the London Mathematical Society*, *24*(3), 507–530.

Routley, R., Meyer, R. K., Plumwood, V., & Brady, R. T. (1982). *Relevant logics and their rivals* (Vol. 1). Atascadero, CA: Ridgeview Publishing Company.

Stone, M. H. (1936). The theory of representations for Boolean algebras. *Transactions of the American Mathematical Society*, *40*(1), 37–111.

Stone, M. H. (1937–38). Topological representations of distributive lattices and Brouwerian logics. *Časopis pro pěstování matematiky a fysiky, Čast matematická*, 67:1–25.

Urquhart, A. (1972a). Completeness of weak implication. *Theoria*, *37*, 274–282.

Urquhart, A. (1972b). Semantics for relevant logic. *Journal of Symbolic Logic*, *37*, 159–169.
Urquhart, A. (1978). A topological representation theorem for lattices. *Algebra Universalis*, *8*, 45–58.
Urquhart, A. (1984). The undecidability of entailment and relevant implication. *Journal of Symbolic Logic*, *49*, 1059–1073.
Urquhart, A. (2017). The geometry of relevant implication. In K. Bimbó, & J. M. Dunn (Eds.), *Proceedings of the third workshop, May 16–17, 2016, Edmonton, Canada*. The IFCoLog journal of logics and their applications (Vol. 4, pp. 591–604). London, UK: College Publications.
Whitman, P. M. (1946). Lattices, equivalence relations and subgroups. *Bulletin of the American Mathematical Society*, *52*, 507–522.
Wille, R. (1982). Restructuring lattice theory: An approach based on hierarchies of concepts. In I. Rival (Ed.), *Ordered Sets* (pp. 445–470). Dordrecht: D. Reidel.
Wille, R. (1985). Tensorial decomposition of concept lattices. *Order*, *2*, 81–95.

Chapter 13
Application of Urquhart's Representation of Lattices to Some Non–classical Logics

Ivo Düntsch and Ewa Orłowska

Second Reader
I. Rewitzky
University of Stellenbosch

Abstract Based on Alasdair Urquhart's representation of not necessarily distributive bounded lattices we exhibit several discrete dualities in the spirit of the "duality via truth" concept by Orłowska and Rewitzky. We also exhibit a discrete duality for Urquhart's relevant algebras and their frames.

Keywords Discrete duality · Urquhart's lattice representation · Commutator algebras · Relevant algebras

13.1 Introduction

Since the origin of Kanger and Kripke-style semantics for formal languages of logics, it has been evident that studying both Hilbert-style algebraic semantics and relational semantics are meaningful and important methodological endeavours. Algebras and relational systems (usually referred to as frames) which provide semantics of a logic are abstract structures defined by axioms that characterize properties of operators in algebras, and properties of relations in frames. A discrete duality is a relationship

I. Düntsch (✉)
College of Mathematics and Computer Science, Fujian Normal University, Fuzhou, Fujian, China
e-mail: D.ivo@fjnu.edu.cn; duentsch@brocku.ca

Department of Computer Science, Brock University, St. Catharines, Ontario, Canada

E. Orłowska
National Institute of Telecommunications, Szachowa 1, 04–894 Warsaw, Poland
e-mail: E.Orlowska@il-pib.pl

I. Düntsch and E. Mares (eds.), *Alasdair Urquhart on Nonclassical and Algebraic Logic and Complexity of Proofs*, Outstanding Contributions to Logic 22,
https://doi.org/10.1007/978-3-030-71430-7_13

between a class Alg of algebras and a class Frm of frames that is established in two steps:

- Given a class Alg of algebras, from each algebra L in Alg, a concrete frame, its *canonical frame*, is defined explicitly over certain subsets of the given algebra, and it is shown to belong to the class Frm. Similarly, given a class Frm, from each frame X in Frm a concrete algebra, its *complex algebra*, is defined explicitly as a closure system with respect to some closure operator on the given frame, and it is shown to be in the class Alg.
- Two representation theorems are proved:

Theorem 1 Every algebra $L \in$ Alg is embeddable into the complex algebra of its canonical frame.

Theorem 2 Every frame $X \in$ Frm is embeddable into the canonical frame of its complex algebra.

Observe that such procedure allows us to separate the algebraic and relational methodologies.

An essential part of the proofs of these theorems consists in showing that embeddings preserve the operations in the algebras and the relations in the frames, respectively. Since a topology is not needed in any logic's semantic structure, for every $L \in$ Alg and for every $X \in$ Frm the representation structures, the complex algebra of the canonical frame and the canonical frame of the complex algebra, respectively, declared in the representation theorems do not involve any topology. Alternatively, one may look at the situation as using the discrete topology, since all subsets of the frame are considered clopen. This motivates the name of this duality. Indeed, the algebraic representation for Boolean algebras with operators by Jónsson and Tarski (1951) was constructed to be more general than the topological context of Stone's representation for Boolean algebras.

The problem of proving a discrete duality usually starts when one of the classes Alg or Frm is known, and the other is not. Given a class Alg, finding the appropriate class Frm is supported by correspondence theory (van Benthem 1984) and tools such as the system SQEMA (Georgiev 2006). However, these tools have some limitations, one of them being that for not necessarily distributive lattices they usually do not offer a solution. Experience has shown that knowing a class Frm and looking for Alg is usually much more difficult, and examples for these are rather scarce (Düntsch and Gediga 2018), (Düntsch and Orłowska 2008, 2011). The existence of representation theorems in either case is not self evident. For example, a necessary condition for an algebra representation to exist is that Alg is closed under completions, which rules out, for example, some algebras of basic logic, see Kowalski and Litak (2008), Düntsch et al. (2016). Orłowska and Rewitzky (2007) discuss how discrete dualities contribute to proving completeness of logics and to correspondence theory, among others. A comprehensive source for discrete dualities is the compendium by Orłowska et al. (2015).

The paper is structured as follows: Following an introductory section, the discrete duality for lattices and lattice frames is presented in Sect. 13.3. It is based on the topology–free version of Alasdair Urquhart's representation theorem for lattices (AU) (Urquhart 1978), and the representation theorem for lattice frames presented in Düntsch and Orłowska (2019). This section is followed by two examples of discrete dualities based on these concepts. In Sect. 13.4 a representation theorem for a known class of commutator algebras is proved. Here, the algebras are bounded lattices endowed with a commutator operator. We present an appropriate class of commutator frames and the representation theorem for them in a weak version. Namely, there is only a one way preservation of a property of one of the relations in the frames. A counterexample is presented showing that a full frame representation is not provable. In Sect. 13.5 we present a discrete duality for DeMorgan lattices, and in Sect. 13.6 one for ortholattices. Finally, in Sect. 13.7, a discrete duality is proved for relevant algebras and relevant frames presented and investigated by Urquhart (1996, 2019).

13.2 Notation and First Definitions

If no confusion can arise, we shall refer to structures just by their universe. A *frame* is a tuple $\langle X, R_0, \dots, R_n\rangle$ where X is a set, and the R_i are relations on X. Let $\langle X, R\rangle$ and $\langle Y, S\rangle$ be frames such that R, S are n–ary relations. A mapping $f : X \to Y$ is called an *embedding* if f is injective, and for all $x_0, \dots, x_{n-1} \in X$, $R(x_0, \dots, x_{n-1})$ if and only if $S(f(x_0), \dots, f(x_{n-1}))$. An injective f is called a *weak embedding*, if $R(x_0, \dots, x_{n-1})$ implies $S(f(x_0), \dots, f(x_{n-1}))$. If $R \subseteq X^2$ and $x \in X$ we set $R(x) := \{y \in X : x \; R \; y\}$. $R^{\smile}$ is the *converse* of R, i.e. $R^{\smile} := \{\langle y, x\rangle : x \; R \; y\}$.

A closure operator on a partially ordered set $\langle P, \leq\rangle$ is a mapping $f : P \to P$ which satisfies for all $a, b \in P$,

1. $a \leq f(a)$, (Extensive)
2. $a \leq b$ implies $f(a) \leq f(b)$, (Isotone)
3. $f(f(a)) = f(a))$. (Idempotent)

Let $\langle L, \vee, \wedge, 0, 1\rangle$ be a bounded lattice. A function $f : L \to L$ is called a

1. *Possibility operator*, if for all $a, b \in L$
   a. $f(0) = 0$, (Normality)
   b. $f(a \vee b) = f(a) \vee f(b)$, (Additivity)
2. *Necessity operator*, if for all $a, b \in L$
   a. $f(1) = 1$, (Dual normality)
   b. $f(a \wedge b) = f(a) \wedge f(b)$, (Multiplicativity)
3. *Sufficiency operator* if for all $a, b \in L$
   a. $f(0) = 1$, (Co–Normality)
   b. For all $a, b \in L$, $f(a \vee b) = f(a) \wedge g(b)$. (Co–Additivity)

If L is a Boolean algebra, the dual of a possibility operator f is the function f^∂ defined by $f^\partial(a) := -f(-a)$; clearly, f^∂ is a necessity operator. The main examples in our context are those obtained from frames: Let $R \subseteq X^2$, $Y \subseteq X$, and define the following operators on 2^X:

$$\langle R\rangle(Y) := \{x : R(x) \cap Y \neq \emptyset\}, \qquad \text{Possibility,} \tag{13.1}$$

$$[R](Y) := \{x : R(x) \subseteq Y\}, \qquad \text{Necessity,} \tag{13.2}$$

$$[[R]](Y) := \{x : Y \subseteq R(x)\}, \qquad \text{Sufficiency.} \tag{13.3}$$

If $\circ$ is a binary operation on B, we extend $\circ$ over subsets of B by $A \circ_c A' := \{a \circ a' : a \in A, a' \in A'\}$ for all $A, A' \subseteq B$.

13.3 Discrete Duality for Lattices and Lattice Frames

Early in his career, Urquhart (1978) obtained the first representation for not necessarily distributive bounded lattices. Subsequently, Hartung (1992) and Hartonas and Dunn (1993) presented representations, respectively based on formal concept lattices and two sorted universes. It turned out that all of the representations lead to the canonical extension of a bounded lattice. Craig and Haviar (2014).

Suppose that $\langle L, \vee, \wedge, 0, 1\rangle$ is a bounded lattice. A *maximal pair* is a pair $\langle F, I\rangle$, where F is a filter of L, I is an ideal of L, $F \cap I = \emptyset$, and F, respectively I, is maximally disjoint to I, respectively, to F, i.e. if G is a filter of L with $F \subsetneq G$, then $G \cap I \neq \emptyset$, and if J is an ideal of L with $I \subsetneq J$, then $F \cap J \neq \emptyset$. Let X_L be the set of all maximal pairs. To facilitate notation, if $x \in X_L$ with $x = \langle F, I\rangle$ we let $x_1 = F$ and $x_2 = I$.

Lemma 13.1 (Urquhart (1978, Lemma 3)) *If F is a filter of L, I an ideal of L, and $F \cap L = \emptyset$, there is a maximal pair x such that $F \subseteq x_1$ and $I \subseteq x_2$.*

We define two relations $\leq_1, \leq_2$ on X_L by $x \leq_i y$ if and only if $x_i \subseteq y_i$. Clearly, $\leq_1$ and $\leq_2$ are quasiorders on X_L. The structure $\langle X_L, \leq_1, \leq_2\rangle$ is called the *canonical frame of L*.

Conversely, a *doubly ordered set* is a structure $\langle X, \leq_1, \leq_2\rangle$ where X is a nonempty set and $\leq_1, \leq_2$ are quasiorders on X which satisfy for all $x, y \in X$

$$x \leq_1 y \text{ and } x \leq_2 y \text{ implies } x = y. \tag{13.4}$$

If $x \in X$ and $i \in \{1, 2\}$, we let $\uparrow_i x := \{y \in X : x \leq_i y\}$. For $Y \subseteq X$ set

$$l(Y) := \{x : \uparrow_1 x \cap Y = \emptyset\} = [\leq_1](-Y), \tag{13.5}$$

$$r(Y) := \{x : \uparrow_2 x \cap Y = \emptyset\} = [\leq_2](-Y). \tag{13.6}$$

It is easy to see that if Y is $\leq_1$–increasing and Z is $\leq_2$–increasing, then

$$Y \subseteq l(Z) \text{ if and only if } Z \subseteq r(Y). \tag{13.7}$$

$Y \subseteq X$ is called *stable*, if $Y = l(r(Y))$. It is not hard to see that Y is stable if and only if $Y = [\leq_1]\langle\leq_2\rangle(Y)$. The collection of stable sets is denoted by L_X. The operations and constants on L_X are defined as follows: For $Y, Z \in L_X$ let

$$Y \vee_X Z := [\leq_1]\langle\leq_2\rangle(Y \cup Z), \tag{13.8}$$

$$Y \wedge_X Z := Y \cap Z, \tag{13.9}$$

$$0_X := \emptyset, \tag{13.10}$$

$$1_X := X. \tag{13.11}$$

We call this structure the *complex algebra of* X.

Theorem 13.1 (Urquhart 1978) *Let L be a bounded lattice and X be a doubly ordered set.*

1. *The structure* $\langle L_X, \vee_X, \wedge_X, \emptyset, X\rangle$ *is a complete bounded lattice.*
2. *The mapping* $h : L \to L_{X_L}$ *defined by*

$$h(a) := \{x \in X_L : a \in x_1\}$$

is a lattice embedding preserving 0 *and* 1.

Urquhart's representation uses topological arguments, and thus, it is not a first order construct. Recently, a first order discrete duality for bounded lattices was presented by Düntsch and Orłowska (2019); independently, Hartonas (2019) exhibited a discrete duality based on two sorted frames.

A *lattice frame* is a doubly ordered frame $\langle X, \leq_1, \leq_2\rangle$ which satisfies the following conditions:

LF_1. Each element of X is below a $\leq_1$ maximal one and a $\leq_2$ maximal one,
LF_2. $x \not\leq_1 y \Rightarrow (\exists z)[y \leq_1 z \text{ and } (\forall w)(x \leq_1 w \Rightarrow z \not\leq_2 w)]$,
LF_3. $x \not\leq_2 y \Rightarrow (\exists z)[y \leq_2 z \text{ and } (\forall w)(x \leq_2 w \Rightarrow z \not\leq_1 w)]$.

Theorem 13.2 (Düntsch and Orłowska 2019, Theorem 5) *Let X be a lattice frame, and* $k_1, k_2 : X \to 2^{L_X}$ *be defined by*

$$k_1(x) := \{Y \in L_X : x \in Y\}, \tag{13.12}$$

$$k_2(x) := \{Y \in L_X : x \in r(Y)\}. \tag{13.13}$$

Set $k(x) := \langle k_1(x), k_2(x)\rangle$. *Then, k is a frame embedding of X into* X_{L_X}.

For later use we note the following technical result:

Lemma 13.2 (Düntsch and Orłowska (2019, Lemma 4)) *Let* $Y \in L_X$ *and* $x \in X$.

1. *Y is* $\leq_1$*–increasing.*

2. *$\uparrow_1 x$ is the smallest stable set containing x.*
3. *$k_1(x)$ is the principal filter of L_X generated by $\uparrow_1 x$, i.e.*

$$Y \in k_1(x) \iff x \in Y. \tag{13.14}$$

4. *$k_2(x)$ is the principal ideal of L_X generated by $[\leq_1](-\uparrow_2 x)$, i.e.*

$$Y \in k_2(x) \iff Y \subseteq [\leq_1](-\uparrow_2 x). \tag{13.15}$$

13.4 Representation Theorems for Commutator Structures

Our first example of the AU representation of general bounded lattices is a representation theorem for a class of commutator algebras:

> "Commutator theory is a part of universal algebra. It is rooted in the theories of groups and rings. From the general algebraic perspective the commutator was first investigated in the seventies by J. Smith for Mal'cev varieties (Mal'cev varieties are characterized by the condition that all congruences on their algebras permute)" (Czelakowski 2015, p. 1).

A prominent example of a commutator algebra is the lattice of all normal subgroups of a group with appropriately defined operations $\otimes$ and $\rightarrow$, see e.g. Freese and McKenzie (1987).

We do not have a discrete duality here in the strict sense, because in the representation theorem for the class of commutator frames which we proposed there is only a weak embedding as mentioned in the introduction. Parts of some proofs for similar situations have appeared in various places such as Orłowska et al. (2015) and Düntsch et al. (2003), and we include them here for completeness.

A *commutator algebra*[1] is a structure $\langle L, \vee, \wedge, \otimes, \rightarrow, 0, 1\rangle$ such that the reduct $\langle L, \vee, \wedge, 0, 1\rangle$ is a bounded lattice with natural ordering $\leq$, and $\otimes$ is a binary operation, called the *commutator*, such that for all $a, b, c \in L$,

LC_1. $a \otimes b \leq a \wedge b$,
LC_2. $a \otimes (b \vee c) = (a \otimes b) \vee (a \otimes c)$,
LC_3. $a \otimes c \leq b$ implies $c \leq a \rightarrow b$,
LC_4. $a \otimes b = b \otimes a$.

In particular, a commutator algebra is a residuated lattice. In the rest of the section, L will denote a commutator algebra, if not stated otherwise.

Some easy to prove properties of commutator algebras are as follows:

Lemma 13.3 *For all $a, b, c \in L$,*

1. $a \otimes 0 = 0$.
2. $a \otimes 1 \leq 1$.

[1] Not to be confused with the concept of the same name used in Physics.

3. $a \leq b$ *implies* $a \otimes c \leq b \otimes c$.
4. $a \to a = 1$.

A *commutator frame* is a structure $\langle X, \leq_1, \leq_2, R, S, Q \rangle$ where $\langle X, \leq_1, \leq_2 \rangle$ is a lattice frame, and Q, R, S are ternary relations on X which satisfy the following conditions:

Monotonicity:

FC_1. $R(x, y, z)$ and $x' \leq_1 x$ and $y' \leq_1 y$ and $z \leq_1 z' \Rightarrow R(x', y', z')$,
FC_2. $S(x, y, z)$ and $x \leq_2 x'$ and $y' \leq_1 y$ and $z' \leq_2 z \Rightarrow S(x', y', z')$,
FC_3. $Q(x, y, z)$ and $x' \leq_1 x$ and $y \leq_2 y'$ and $z' \leq_2 z \Rightarrow Q(x', y', z')$,
FC_4. $S(x, y, z) \Rightarrow z \leq_2 x$,
FC_5. $Q(x, y, z) \Rightarrow z \leq_2 y$,
FC_6. $R(x, y, z) \Rightarrow R(y, x, z)$.

Stability:

FC_7. $R(x, y, z) \Rightarrow \exists u \in X$ ($x \leq_1 u$ and $S(u, y, z)$),
FC_8. $R(x, y, z) \Rightarrow \exists u \in X$ ($y \leq_1 u$ and $Q(x, u, z)$),
FC_9. $S(x, y, z) \Rightarrow \exists u \in X$ ($z \leq_2 u$ and $R(x, y, u)$),
FC_{10}. $Q(x, y, z) \Rightarrow \exists u \in X$ ($z \leq_2 u$ and $R(x, y, u)$).

In Orłowska et al. (2015, Sect. 15.2) a representation result for residuated lattices is presented, and we use several frame conditions given there; we can also use the definition of the operators on the complex algebra of a commutator frame from Orłowska et al. (2015). We define an auxiliary binary operation $\circ_X$ on 2^X by

$$Y \circ_X Z := \{z \in X : (\forall x, y \in X)[Q(x, y, z) \text{ and } x \in Y \text{ implies } y \in r(Z)]\}. \tag{13.16}$$

The next observation will be helpful in the sequel:

Lemma 13.4 (Düntsch et al. (2003, Lemma 16)) *If* $Y, Z \in L_X$, *then* $rl(Y \circ_X Z) = Y \circ_X Z$.

The complex algebra of X is the structure $\langle L_X, \otimes_X, \to_X, \emptyset, X \rangle$ where L_X is the complex algebra of the lattice frame reduct of X, and

$$Y \otimes_X Z := l(Y \circ_X Z), \tag{13.17}$$
$$Y \to_X Z := \{x : (\forall y, z)[R(y, x, z) \text{ and } y \in Y \text{ implies } z \in Z]\}. \tag{13.18}$$

The easy proof of the following technical result is left to the reader:

Lemma 13.5 $z \in Y \otimes_X Z \iff (\forall t)[z \leq_1 t \Rightarrow (\exists u_t, v_t)(Q(u_t, v_t, t)$ *and* $u_t \in Y$ *and* $\uparrow_2 v_t \cap Z \neq \emptyset)]$.

We now proceed to prove the representation theorems.

Theorem 13.3 *The canonical frame of a commutator algebra is a commutator frame.*

Proof Let $x, y, z \in X_L$, and define

$$R_\otimes(x, y, z) \overset{\text{df}}{\Longleftrightarrow} (\forall a, b \in L)[a \in x_1 \text{ and } b \in y_1 \Rightarrow a \otimes b \in z_1], \tag{13.19}$$

$$S_\otimes(x, y, z) \overset{\text{df}}{\Longleftrightarrow} (\forall a, b \in L)[a \otimes b \in z_2 \text{ and } b \in y_1 \Rightarrow a \in x_2], \tag{13.20}$$

$$Q_\otimes(x, y, z) \overset{\text{df}}{\Longleftrightarrow} (\forall a, b \in L)[a \otimes b \in z_2 \text{ and } a \in x_1 \Rightarrow b \in y_2]. \tag{13.21}$$

It was noted in Orłowska et al. (2015, Proposition 15.2.9) that

$$R_\otimes(x, y, z) \Longleftrightarrow x_1 \otimes_c y_1 \subseteq z_1, \tag{13.22}$$

$$Q_\otimes(x, y, z) \Longleftrightarrow x_1 \otimes_c (-y_2) \subseteq -z_2, \tag{13.23}$$

$$S_\otimes(x, y, z) \Longleftrightarrow -x_2 \otimes_c y_1 \subseteq -z_2. \tag{13.24}$$

Here, $\otimes_c$ is the complex extension of $\otimes$, i.e. for $M, M' \subseteq L$, $M \otimes_c M' = \{a \otimes b : a \in M, b \in M'\}$.

The preservation of the monotonicity axioms is straightforward to show and can be found in the literature (Orłowska et al. 2015, Sect. 15.2). By way of example, we shall only prove the stability axioms FC_8 and FC_{10}, as the remaining ones are similar. Suppose that $x, y, z \in X_L$.

FC_8: Let $R(x, y, z)$; then, $x_1 \otimes_c y_1 \subseteq z_1$ by (13.22). Let $J := \{a \in L : (\exists a')[a' \in x_1 \text{ and } a \otimes a' \in z_2]\}$. Then J is an ideal of L: Suppose that $a, b \in J$, so that $a', b' \in x_1$ and $a \otimes a', b \otimes b' \in z_2$. Then, $a' \wedge b' \in x_1$ and $(a \otimes a') \vee (b \otimes b') \in z_2$. Now,

$$\begin{aligned}(a \vee b) \otimes (a' \wedge b') &= (a \otimes (a' \wedge b')) \vee (b \otimes (a' \wedge b')), && \text{by } LC_2,\ LC_4 \\ &\leq (a \otimes a') \vee (b \otimes b') \in z_2.\end{aligned}$$

Thus, $(a \vee b) \otimes (a' \wedge b') \in z_2$, since z_2 is an ideal, and therefore, $a \vee b \in J$. If $a \in J, a' \in x_1, a \otimes a' \in z_2$ and $b \leq a$, then, $b \otimes a' \leq a \otimes a'$ by Lemma 13.3(3), and $b \otimes a' \in z_2$ since z_2 is an ideal; hence, $b \in J$.

Assume that $J \cap y_1 \neq \emptyset$, say, $a \in J \cap y_1$. Let $a' \in x_1$ such that $a \otimes a' \in z_2$. Since $R(x, y, z)$, we have $x_1 \otimes_c y_1 \subseteq z_1$, and thus $a' \otimes a \in z_1$. LC_4 implies that $a \otimes a' \in z_1$, a contradiction. Thus, $\langle y_1, J \rangle$ is a filter–ideal pair which can be extended to a maximal pair u by Lemma 13.1. Let $a' \in x_1$ and $a \otimes a' \in z_2$; then, $a' \in J \subseteq u_2$, and thus, $Q(x, u, z)$. Consequently, FC_8 holds.

FC_{10}: Suppose that $Q(x, y, z)$. Observe that $R(x, y, u)$ for some $u \in L_X$ if and only if $x_1 \otimes_c y_1$ is contained in some proper filter of L, i.e. if it has the finite intersection property. Suppose that $a_1, \ldots a_n \in x_1, b_1, \ldots b_n \in y_1$. Then, $a := a_1 \wedge \ldots \wedge a_n \in x_1$ and $b := b_1 \wedge \ldots \wedge b_n \in y_1$, since x_1, y_1 are filters. If $a \otimes b = 0$, then $a \otimes b \in z_2$ and $Q(x, y, z)$ implies $b \in y_2$, contradicting $b \in y_1$. Thus,

$$0 \neq (a_1 \wedge \ldots \wedge a_n \in x_1) \otimes (b_1 \wedge \ldots \wedge b_n) \leq (a_1 \otimes b_1) \wedge \ldots \wedge (a_n \otimes b_n),$$

the latter by Lemma 13.3. Thus, $x_1 \otimes_c y_1$ can be extended to a proper filter F, and there is a maximal pair u such that $F \subseteq u_1$ by Lemma 13.1. It follows that $R(x, y, z)$. □

Theorem 13.4 *The complex algebra of a commutator frame is a commutator lattice.*

Proof Suppose that $W, Y, Z \in L_X$.

LC_1: Let $x \in W \otimes_X Y$ and assume $x \notin Y$. Then, $x \notin lr(Y)$, and there is some $x' \in X$ such that $x \leq_1 x'$ and $x' \in r(Y)$. Now, $x \in W \otimes_X Y$ and Lemma 13.5 imply that there are $u, v \in X$ such that $u \in W$, $Q(u, v, x')$ and $\uparrow_2 v \cap Y \neq \emptyset$. By FC_5, $x' \leq_2 v$, and the transitivity of $\leq_2$ and $\uparrow_2 v \cap Y \neq \emptyset$ imply that $\uparrow_2 x' \cap Y \neq \emptyset$. This contradicts $x' \in r(Y)$.

Next, assume that $x \notin W$. As above, there is some $x' \in X$ such that $x \leq_1 x'$ and $x' \in r(W)$; furthermore, there are $u, v \in X$ such that $u \in W$, $Q(u, v, x')$ and $\uparrow_2 v \cap Y \neq \emptyset$. By FC_{10}, there is some $x'' \in X$ such that $x' \leq_2 x''$ and $R(u, v, x'')$; by FC_6, we have $R(v, u, x'')$. From FC_8 we obtain some $u' \in X$ such that $u \leq_1 u'$ and $Q(v, u', x'')$, and $u \in W$ implies $u' \in W$. On the other hand, $x' \leq_2 x'' \leq_2 u'$ implies $x' \leq_2 u'$, which contradicts $x' \in r(W)$.

LC_2: It is straightforward, if somewhat tedious, to show that

$$W \circ_X (Y \vee_X Z) = (W \circ_X Y) \cap (W \circ_X Z), \tag{13.25}$$

see e.g. Lemma 22(i) of Düntsch et al. (2003). Now,

$$\begin{aligned}
W \otimes_X (Y \vee_X Z) &= l(W \circ_X (Y \vee_X Z)), \\
&= l((W \circ_X Y) \cap (W \circ_X Z)), && \text{by (13.25)}, \\
&= l(rl((W \circ_X Y)) \cap rl(W \circ_X Z)), && \text{by Lemma 13.4}, \\
&= lr(l(W \circ_X Y) \cup l(W \circ_X Z)), && \text{as } r \text{ is a sufficiency operator}, \\
&= lr((W \otimes_X Y) \cup (W \otimes_X Z)), && \text{by definition of } \otimes_X, \\
&= (W \otimes_X Y) \vee_X (W \otimes_X Z), && \text{by definition of } \vee_X .
\end{aligned}$$

LC_3: This is shown in Lemma 15.2.8 of Orłowska et al. (2015).

LC_4: We will show only $W \otimes_X Y \subseteq Y \otimes_X W$, since the other inclusion follows by symmetry. Let $x \in W \otimes_X Y$, and assume that $x \notin Y \otimes_X W$. Then,

$$(\forall x')[x \leq_1 x' \Rightarrow (\exists u, v)(Q(u, v, x') \text{ and } u \in W \text{ and } \uparrow_2 v \cap Y \neq \emptyset)], \tag{13.26}$$

$$(\exists x')[x \leq_1 x' \text{ and } (\forall s, t)(Q(s, t, x') \text{ and } s \in Y \Rightarrow t \in r(W))]. \tag{13.27}$$

Let x' be a witness for (13.27), and, for this x', let u, v be witnesses for (13.26), that is, the elements whose existence is asserted. Then, $Q(u, v, x')$, $u \in W$, and there is some w such that $v \leq_2 w$ and $w \in Y$. It follows from FC_3 that $Q(u, w, x')$. Now,

$$\begin{aligned}
Q(u, w, x') &\Rightarrow (\exists z)[x' \leq_2 z \text{ and } R(u, w, z), && \text{by } FC_{10},\\
&\Rightarrow (\exists z)[x' \leq_2 z \text{ and } R(w, u, z), && \text{by } FC_6,\\
&\Rightarrow (\exists u')[u \leq_1 u' \text{ and } Q(w, u', z), && \text{by } FC_8,\\
&\Rightarrow (\exists u')[u \leq_1 u' \text{ and } Q(w, u', x'), && \text{by } FC_3 \text{ and } x' \leq_2 z,\\
&\Rightarrow u' \in r(W), && \text{by (13.27) and } w \in Y,\\
&\Rightarrow u \notin lr(W) = W,
\end{aligned}$$

which contradicts $u \in W$. □

Theorem 13.5 *Let L be a commutator algebra and X be a commutator frame.*

1. *L can be embedded into the complex algebra of its canonical frame.*
2. *X can be weakly embedded into the canonical frame of its complex algebra.*

Proof 1. The mapping h of Theorem 13.1 is a lattice embedding. It furthermore preserves $\otimes$ and $\rightarrow$ by Orłowska et al. (2015, Lemma 15.2.12).

2. The mapping k of Theorem 13.2 embeds X as a lattice frame into X_{L_X}, and all that is left to show is that k preserves the ternary relations. Let $R(x, y, z)$; we will show that $R_{\otimes}(k(x), k(y), k(z))$.

Claim 13.1 *$R_{\otimes}(k(x), k(y), k(z))$ if and only if $z \in \uparrow_1 x \otimes \uparrow_1 y$.*

Proof First, observe that by definition of $R_{\otimes}$ in (13.19) and the definition of k,

$$\begin{aligned}
R_{\otimes}(k(x), k(y), k(z)) &\iff (\forall Y, Z \in L_X)[Y \in k_1(x), Z \in k_1(y) \Rightarrow Y \otimes Z \in k_1(z)],\\
&\iff (\forall Y, Z \in L_X)[x \in Y, y \in Z \Rightarrow z \in Y \otimes Z)].
\end{aligned}$$

By Lemma 13.2(1), $\uparrow_1 x$ and $\uparrow_1 y$ are stable, and thus, $R_{\otimes}(k(x), k(y), k(z))$ implies that $z \in \uparrow_1 x \otimes \uparrow_1 y$.

Conversely, suppose that $z \in \uparrow_1 x \otimes \uparrow_1 y$, and that $Y, Z \in L_X$ such that $x \in Y, y \in Z$. Our aim is to show that $z \in Y \otimes Z$. Since Y, Z are $\uparrow_1$–closed and $x \in Y, y \in Z$, we have $\uparrow_1 x \subseteq Y, \uparrow_1 y \subseteq Z$. By Lemma 13.3, $\otimes$ is isotone and therefore, we obtain $z \in \uparrow_1 x \otimes \uparrow_1 y \subseteq Y \otimes Z$. □

Claim 13.2 $z \in \uparrow_1 x \otimes \uparrow_1 y \iff (\forall t)[z \leq_1 t$ *implies* $(\exists v_t)(Q(x, v_t, t)$, *and* $y \leq_1 v_t)]$.

Proof This is an instance of Lemma 13.5. □

Let $R(x, y, z)$. In order to show that the right hand side of (13.2) holds, suppose that $z \leq_1 t$. Then, $R(x, y, t)$ by FC_1. By FC_8, there is some v_t such that $y \leq_1 v_t$ and $Q(x, v_t, t)$. From Claim 13.2 we obtain $z \in \uparrow_1 x \otimes \uparrow_2 y$, and thus, $R_{\otimes}(k(x), k(y), k(z))$ by Claim 13.1.

Next, we consider $Q(x, y, z)$. To show that $Q_{\otimes}(k(x), k(y), k(z))$ we will use (13.23) which becomes

$$Q_{\otimes}(F, G, H) \iff F_1 \otimes_c (-G_2) \subseteq -H_2.$$

Thus, we need to show that

$$k_1(x) \otimes_c -k_2(y) \subseteq -k_2(z),$$

i.e. for all $Y, Z \in L_X$,

$$Y \in k_1(x) \text{ and } Z \notin k_2(y) \Rightarrow Y \otimes Z \notin k_2(z).$$

Using Lemma 13.2, this becomes

$$x \in Y \text{ and } Z \not\subseteq [\leq_1](-\uparrow_2 y) \Rightarrow Y \otimes Z \not\subseteq [\leq_1](-\uparrow_2 z).$$

We can rewrite the right hand side by noting that

Claim 13.3 $Y \otimes Z \not\subseteq [\leq_1](-\uparrow_2 z) \iff \uparrow_2 z \cap Y \otimes Z \neq \emptyset$.

Proof "$\Rightarrow$": Let $u \in Y \otimes Z$ and $u \notin [\leq_1](-\uparrow_2 z)$. Then, there is some v such that $u \leq_1 v$ and $v \in \uparrow_2 z$, i.e. $z \leq_2 v$. Since $Y \otimes Z \in L_X$, it is $\leq_1$–closed, and thus, $v \in Y \otimes Z$.
"$\Leftarrow$": This follows from the reflexivity of $\leq_1$. □

Suppose that $x \in Y$ and $Z \not\subseteq [\leq_1](-\uparrow_2 y)$. Then, arguing as above, we obtain that $\uparrow_2 y \cap Z \neq \emptyset$; by FC_3, we may suppose w.l.o.g. that $y \in Z$.
Assume that $\uparrow_2 z \cap Y \otimes Z = \emptyset$. Then, for all $u \in \uparrow_2 z$,

$$(\exists t)[u \leq_1 t \text{ and } (\forall v, w)(v \in Y \text{ and } Q(v, w, t) \Rightarrow \uparrow_2 w \cap Z = \emptyset)]. \tag{13.28}$$

By FC_{10}, there is some u such that $z \leq_2 u$ and $R(x, y, u)$, and therefore, $u \in \uparrow_1 x \otimes \uparrow_1 y$ since k preserves R and the claim shown earlier. Now, by Lemma 13.5 and by FC_3,

$$u \in \uparrow_1 x \otimes \uparrow_1 y \iff (\forall t)[u \leq_1 t \Rightarrow (\exists w)(Q(x, w, t) \text{ and } \uparrow_2 w \cap \uparrow_1 y \neq \emptyset)], \tag{13.29}$$

Choose t as in (13.28), and for this t choose w as in (13.29). Since $x \in Y$ and $Q(x, w, t)$, we have $\uparrow_2 w \cap Z = \emptyset$ by (13.28). On the other hand, there is some $s \in \uparrow_2 w \cap \uparrow_1 y$. Since $y \in Z$ and Z is $\leq_1$–closed, we have $s \in Z$, a contradiction.
Similarly, one can show that $S(x, y, z)$ implies $S_\otimes(k(x), k(y), k(z))$. □

The mapping k is only a weak embedding, as the following example shows[2]:

Example 13.1 Let $X := \{w, x, y, z\}$, $\leq_1, \leq_2$ be given as in Fig. 13.1. The relations Q, R, S and $R_\otimes$ are shown in Table 13.1.
Then, $\langle X, \leq_1, \leq_2, R, S, Q\rangle$ is a commutator frame, where $S = Q$. We see from the table, that $R_\otimes(\uparrow_1 x, \uparrow_1 x, \uparrow_1 z)$, but not $R(x, x, z)$.

[2] The example was found by Mace4 (McCune 2010).

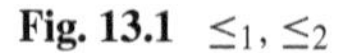

Fig. 13.1 $\leq_1, \leq_2$

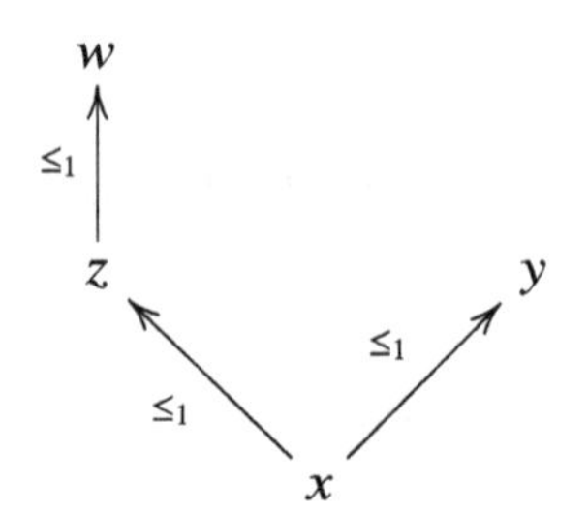

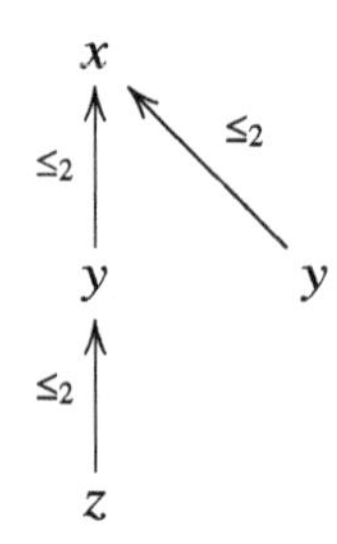

Table 13.1 The relations Q, R and $R_\otimes$

Q	R	$R_\otimes$
$Q(y, y, y)$	$R(y, y, y)$	$R_\otimes(\uparrow_1 y, \uparrow_1 y, \uparrow_1 y)$
$Q(y, y, z)$	$R(y, x, y)$	$R_\otimes(\uparrow_1 y, \uparrow_1 x, \uparrow_1 y)$
$Q(y, x, y)$	$R(x, y, y)$	$R_\otimes(\uparrow_1 x, \uparrow_1 y, \uparrow_1 y)$
$Q(y, x, z)$	$R(x, x, y)$	$R_\otimes(\uparrow_1 x, \uparrow_1 x, \uparrow_1 y)$
$Q(x, y, y)$	$R(x, x, w)$	$R_\otimes(\uparrow_1 x, \uparrow_1 x, \uparrow_1 w)$
$Q(x, y, z)$		$R_\otimes(\uparrow_1 x, \uparrow_1 x, \uparrow_1 z)$
$Q(x, x, y)$		
$Q(x, x, z)$		
$Q(x, x, w)$		

13.5 A Discrete Duality for DeMorgan Lattices

A *DeMorgan lattice* (DeM–lattice) is a structure $\langle L, \neg \rangle$, where L is a bounded lattice with natural order $\leq$, and $\neg$ is a unary operator on L where for all $a, b \in L$,

DeML$_1$. $a \leq \neg b \Rightarrow b \leq \neg a$.
DeML$_2$. $\neg\neg a = a$.

A representation theorem for DeMorgan lattices was presented by Dzik et al. (2006), see also Allwein and Dunn (1993) and Orłowska et al. (2015, Sect. 14.4). We briefly review the constructions.

A *DeM–frame* is a structure $\langle X, \leq_1, \leq_2, N \rangle$ where $\langle X, \leq_1, \leq_2 \rangle$ is a lattice frame, and for all $x, y \in X$,

M_1. N is an involution, i.e. $N(N(x)) = x$,
M_2. $x \leq_1 y$ implies $N(x) \leq_2 N(y)$,
M_3. $x \leq_2 y$ implies $N(x) \leq_1 N(y)$.

If L is a DeM–lattice, its *canonical frame* is the structure $\langle X_L, \leq_1, \leq_2, N_L \rangle$, where $\langle X_L, \leq_1, \leq_2 \rangle$ is the canonical frame of L as defined in Theorem 13.2, and $N_L(x) := \langle \neg x_2, \neg x_1 \rangle$, where—with some abuse of notation—for $W \subseteq L$, $\neg W := \{\neg a : a \in W\}$. Conversely, if $\langle X, \leq_1, \leq_2, N \rangle$ is a DeM–frame, its complex algebra is the structure $\langle L_X, \neg_X \rangle$, where L_X is the complex algebra of X, and $\neg_X Y := \{y \in X : N(y) \in r(Y)\}$.

Theorem 13.6 (Dzik et al. 2006)

1. *If L is a DeM–lattice, then X_L is a DeM–frame.*
2. *If X is a DeM–frame, then L_X is a DeM–lattice.*
3. *The lattice embedding $h : L \to L_{X_L}$ of Sect. 13.3 preserves $\neg$.*

To this we now add a frame representation.

Theorem 13.7 *The mapping $k : X \to X_{L_X}$ defined in Sect. 13.3 preserves N.*

Proof We need to show that

$$k(N(x)) = N_{L_X}(k(x)),$$

i.e. that $\langle k_1(N(x)), k_2(N(x))\rangle = \langle \neg_{L_X} k_2(N(x)), \neg_{L_X} k_1(x)\rangle$. Thus, we need to prove that

1. $k_1(N(x)) = \neg_{L_X} k_2(x)$,
2. $k_2(N(x)) = \neg_{L_X} k_1(x)$.

1. First, note that $Y \in \neg_{L_X} k_2(x)$ if and only if $\neg_{L_X} Y \in k_2(x)$, and that this holds if and only if $\neg_{L_X} Y \subseteq l(\uparrow_2 x)$ by Lemma 13.2(3). Since $\neg_{L_X} Y$ is $\leq_1$–increasing, by (13.7) this is equivalent to $\uparrow_2 x \subseteq r(\neg_{L_X} Y)$, i.e. $x \in r(\{y : N(y) \in r(Y)\})$, since $r(\{y : N(y) \in r(Y)\})$ is $\leq_2$–closed. Now,

$$\begin{aligned}
Y \in \neg_{L_X} k_2(x) &\iff x \in r(\{y : N(y) \in r(Y)\}), \\
&\iff \uparrow_2 x \cap \{y : N(y) \in r(Y)\} = \emptyset, \\
&\iff (\forall y)[x \leq_2 y \Rightarrow N(y) \notin r(Y)], \\
&\iff (\forall y)[N(x) \leq_1 N(y) \Rightarrow N(y) \notin r(Y)], \\
&\iff (\forall t)[N(x) \leq_1 t \Rightarrow t \notin r(Y)], && N \text{ is an involution,} \\
&\iff \uparrow_1 N(x) \cap r(Y) = \emptyset, \\
&\iff N(x) \in lr(Y), \\
&\iff N(x) \in Y, && \text{since } Y \text{ is stable,} \\
&\iff Y \in k_1(N(x)).
\end{aligned}$$

2. Similar to the previous step, we observe that

$$Y \in k_2(N(x)) \iff Y \subseteq l(\uparrow_2 N(x)) \iff N(x) \in r(Y).$$

Furthermore, a simple application of the definitions yields

$$Y \in \neg_{L_X} k_1(x) \iff \neg_{L_X} Y \in k_1(x) \iff x \in \neg_{L_X} Y \iff N(x) \in r(Y).$$

This finishes the proof. □

13.6 A Discrete Duality for Ortholattices

An *ortholattice* is a DeM–lattice $\langle L, \neg \rangle$ which satisfies

DeML3. $a \wedge \neg a = 0.$

The corresponding *orthoframe* is a DeM–frame $\langle X, \leq_1, \leq_2, N \rangle$ that additionally satisfies

M_4. $(\exists y)[x \leq_1 y$ and $N(x) \leq_2 y]$.

The motivating example of an ortholattice is the family of closed subspaces of a Hilbert space, see e.g. Kalmbach (1985). A discrete representation for ortholattices based on DeM–frames was presented by Dzik et al. (2006). They also show that the canonical frame of an ortholattice is an orthoframe and that the complex algebra of an orthoframe is an ortholattice. Since orthoframes are axiomatic extensions of DeM–frames, Theorem 13.7 also gives a discrete representation for orthoframes.

Using a different approach, Goldblatt (1974b) presented a relational semantics for an orthologic whose algebraic semantics is given in terms of ortholattices, see also Goldblatt (1974a, 1975). We shall now exhibit a discrete duality for ortholattices and their relational frames. Note that the duality does not use Urquhart's lattice representation.

An *O–frame* is a structure $\langle X, \perp \rangle$ where X is a nonempty set and $\perp$ is an irreflexive and symmetric relation on X, called an *orthogonality relation*. These are related to apartness frames which are O–frames with the additional condition of co–transitivity (Düntsch and Orłowska 2008).

Since $\perp$ is irreflexive we shall consider sufficiency operators. Irreflexivity and symmetry of a binary relation R can be expressed by its sufficiency operator $[[R]]$, see e.g. Düntsch and Orłowska (2001) or Goranko (1990):

Lemma 13.6 *Suppose that $\langle X, R \rangle$ is a frame.*

1. *R is irreflexive if and only if $[[R]](Y) \cap Y = \emptyset$ for all $Y \subseteq X$.*
2. *R is symmetric if and only if $Y \subseteq [[R]][[R]](Y)$ for all $Y \subseteq X$.*

Let $f_\perp : 2^X \to 2^X$ be defined by $f_\perp(Y) = [[\perp]][[\perp]](Y)$. We call $Y \subseteq X$ *closed* or *regular*, if $f_\perp(Y) \subseteq Y$. By Lemma 13.6(2), the closed sets are exactly those with $Y = f_\perp(Y)$. The complex algebra of an O–frame is the structure $\langle L_X, \cap, \vee_X, \emptyset, X, \neg_X \rangle$ where L_X is the set of closed subsets of X, and for $Y, Z \in L_X$,

$$Y \vee_X Z := f_\perp(Y \cup Z),$$
$$\neg_X Y := [[\perp]](Y).$$

It follows from the definition of $f_\perp$ and a sufficiency operator that

$$Y \vee_X Z = [[\perp]][[\perp]](Y \cup Z) = [[\perp]]([[\perp]](Y) \cap [[\perp]](Z)) = \neg_X(\neg_X Y \cap \neg_X Z).$$

Lemma 13.7 *Let $\langle X, \perp \rangle$ be an O–frame.*

1. *If Y, Z are closed, so are $Y \cap Z$, $Y \vee_X Z$, and $\neg_X Y$.*
2. *L_X is an ortholattice.*

Proof 1. follows immediately from the definition of $f_\perp$ and the operations. 2. follows from the fact that $\neg_X$ is a sufficiency operator and Lemma 13.6, see also Birkhoff (1967, Theorem 19, p.123) and Goldblatt (1975, Proposition 1). □

The canonical frame of an ortholattice is the structure $\langle X_L, \perp_L \rangle$ where X_L is the set of all proper filters of L and $\perp_L$ is the binary relation $\perp_X$ defined on X_L by $F \perp_L G$ if and only if there is some $a \in L$ such that $\neg a \in F$ and $a \in G$.

Theorem 13.8 (Goldblatt 1975, Proposition 2) *$\langle X_L, \perp_X \rangle$ is an O–frame, and the mapping $h : L \to L_{X_L}$ defined by $h(a) := \{F \in X_L : a \in F\}$ is an injective homomorphism of ortholattices.*

Goldblatt (1975) goes on to show that h maps L onto the collection of clopen regular subsets of a compact topology on X_L.

Theorem 13.9 *Each O–frame $\langle X, \perp \rangle$ can be embedded into the canonical frame of its complex algebra.*

Proof Define $k : X \to X_{L_X}$ by $k(x) := \{Y \in L_X : x \in Y\}$. We will show that k is a frame embedding. Clearly, $k(x)$ is a proper filter of L_X for every $x \in X$, and k is injective. Let $x \perp y$; we need to show that $x \perp y$ if and only if $k(x) \perp_{L_X} k(y)$, i.e. for all $x, y \in X$,

$$x \perp y \iff (\exists Y \in L_X)[x \in [[\perp]](Y) \text{ and } y \in Y] \iff (\exists Y \in L_X)[y \in Y \text{ and } (\forall z \in X)(z \in Y \Rightarrow x \perp z)]. \quad (13.30)$$

"$\Rightarrow$": Let $Y := \{t \in X : x \perp t\}$. Then, $Y \in L_X$ and $y \in Y$. Furthermore, $x \in [[\perp]](Y)$ by the definition of Y.
"$\Leftarrow$": Suppose that $k(x) \perp_{L_X} k(y)$ witnessed by Y; then, $x \in \neg_{L_X} Y$ and $y \in Y$, which implies $x \perp y$. □

An ortholattice L is called *orthomodular* if for all $a, b \in L$, $a \leq b$ implies $b \leq a \vee (b \wedge \neg a)$, equivalently, if $a \vee (\neg a \wedge (a \vee b)) = a \vee b$. Orthomodular lattices originate in the study of quantum mechanics. For their history and development we refer the reader to the monograph Kalmbach (1985), and for quantum logic to Dalla Chiara and Giuntini (2002) and Rédei (2009).

Goldblatt (1984) has exhibited two elementarily equivalent orthoframes $\langle X_1, \perp_1 \rangle$, $\langle X_2, \perp_2 \rangle$ such that the complex algebra of $\langle X_1, \perp_1 \rangle$ is orthomodular, while that of $\langle X_2, \perp_2 \rangle$ is not. This shows that there is no discrete duality for orthomodular lattices based solely on first order frame conditions.

13.7 A Discrete Duality for Relevant Structures

Urquhart (1996) presented a duality for algebras of relevant logics, based on the topological duality of Priestley (1970). In this section we present a discrete duality for these structures. Since algebras of relevant logic are distributive lattices, we use the representation going back to Stone (1937).

A *relevant algebra* (Urquhart 1996) is a structure

$$\langle L, \vee, \wedge, \circ, \rightarrow, \neg, 1', 0, 1\rangle$$

of type $\langle 2, 2, 2, 2, 1, 0, 0, 0\rangle$ such that

Ra$_1$. $\langle L, \vee, \wedge, 0, 1\rangle$ is a bounded distributive lattice,
Ra$_2$. $a \circ (b \vee c) = (a \circ b) \vee (a \circ c)$,
Ra$_3$. $(b \circ c) \vee a = (b \circ a) \vee (c \circ a)$,
Ra$_4$. $\neg(a \vee b) = \neg a \wedge \neg b$,
Ra$_5$. $\neg(a \wedge b) = \neg a \vee \neg b$,
Ra$_6$. $\neg 0 = 1, \neg 1 = 0$,
Ra$_7$. $a \circ 0 = 0 \circ a = 0$,
Ra$_8$. $1' \circ a = a$,
Ra$_9$. $a \circ b \leq c$ if and only if $a \leq b \rightarrow c$.

Let $X \neq \emptyset$, $R \subseteq X^3$, $f : X \rightarrow X$, $I \subseteq X$, and $\leq$ be a partial order on X. The tuple $\langle X, \leq, R, f, I\rangle$ is called a *relevant frame*, if for all $x, x', y, y', z, z' \in X$,

Rf$_1$. $R(x, y, z), x' \leq x, y' \leq y, z' \leq z$ imply $R(x', y', z')$,
Rf$_2$. f is antitone, i.e. $x \leq y$ implies $f(y) \leq f(x)$,
Rf$_3$. I is an order ideal, i.e. $x \in I$ and $x \geq y$ imply $y \in I$,
Rf$_4$. $y \leq z$ if and only if there is some $x \in I$ such that $R(x, y, z)$.

If L is a relevant algebra, its *canonical frame* is the structure $\langle X_L, \subseteq, R_L, f_L, I_L\rangle$ defined as follows:

1. X_L is the set of prime filters of L.
2. R_L is a ternary relation on X_L for which

$$R_L(F, G, H) \overset{\text{df}}{\Longleftrightarrow} F \circ_r G \subseteq H,$$

where

$$F \circ_r G := \{c \in L : (\exists a, b \in L)[a \in F, b \in G, a \circ b \leq c]\}.$$

3. $f_L : X \rightarrow X$ is a mapping defined by $f_L(F) := \{a \in F : \neg a \notin F\}$.
4. $I_L := \{F \in X_L : 1' \in F\}$.

The *complex algebra of the relevant frame* X is the structure

$$\langle L_X, \vee_X, \wedge_X, \circ_X, \rightarrow_X, \neg_X, 1'_X, 0_X, 1_X \rangle$$

of type $\langle 2, 2, 2, 2, 1, 0, 0, 0 \rangle$ such that $L_X := \{Y \subseteq X : Y = [\leq](Y)\}$, and for all $Y, Z \in L_X$,

$$Y \vee_X Z := [\leq](Y \cup Z), \tag{13.31}$$
$$Y \wedge_X Z := [\leq](Y \cap Z), \tag{13.32}$$
$$Y \circ_X Z := \{z \in X : (\exists x, y)[x \in Y \text{ and } y \in Z \text{ and } R(x, y, z)]\}, \tag{13.33}$$
$$Y \rightarrow_X Z := \{x \in X : (\forall y, z \in X)[R(x, y, z) \wedge y \in Y \Rightarrow z \in Z]\}, \tag{13.34}$$
$$\neg_X Y := \{x \in X : f(x) \notin Y\}, \tag{13.35}$$
$$1'_X := I, \ 0_X := \emptyset, \ 1_X := X. \tag{13.36}$$

Unlike Urquhart (1996) who uses clopen increasing sets of the Priestley topology, we use all increasing sets. A consequence of this is that L is not necessarily isomorphic to L_{X_L}, see the remarks in Orłowska et al. (2015, p. 34).

The following results establish the discrete duality:

Theorem 13.10 *The canonical frame of a relevant algebra L is a relevant frame.*

Proof Rf_1–Rf_3 are straightforward to prove. Suppose that G, H are prime filters of L and $G \subseteq H$. Let F be the principal filter of L generated by $1'$. Then, $F \circ G$ is a filter by Urquhart (1996, Lemma 2.1), and $G \subseteq H$ and Ra_8 imply that $F \circ G \subseteq H$. By Urquhart (1996, Lemma 2.2) there is some prime filter F' such that $F \subseteq F'$, and $F \circ G \subseteq$. This shows the $\Rightarrow$ direction of Rf_4. Conversely, suppose that $F, G, H \in X_L$ such that $1' \in F$ and $R_L(F, G, H)$. Then $F \circ G \subseteq H$, and $1' \in F$ implies $G = \{1'\} \circ G \subseteq F \circ G \subseteq H$. □

Theorem 13.11 *The complex algebra of a relevant frame X is a relevant algebra.*

Proof Our first task is to show that L_X is closed under the operations. Using the properties of the S4 modality $[\leq]$ this is straightforward for $\vee_X$ and $\wedge_X$. Let $Y, Z, W \in L_X$. Since $[\leq](Y) \subseteq Y$ by reflexivity of $\leq$, we shall only show the other inclusions.

Suppose that $z \in Y \circ_X Z$, and let $z \leq t$. Since $z \in Y \circ_X Z$, there are $x \in Y$, $y \in Z$ such that $R(x, y, z)$. By Rf_1 we have $R(x, y, t)$, and thus, $t \in Y \circ_X Z$. This shows that $Y \circ_X Z \subseteq [\leq](Y \circ_X Z)$.

Let $x \in Y \rightarrow_X Z$, $x \leq t$, and assume that $t \notin Y \rightarrow_X Z$. By definition of $\rightarrow_X$, there are y_t, z_t such that $y_t \in Y$, $z_t \notin Z$, and $R(t, y_t, z_t)$. Since $x \leq t$ and by Rf_1 this implies $R(x, y_t, z_t)$. However, $x \in Y \rightarrow_X Z$ and $y \in Y$ imply $z \in Z$, a contradiction. This shows that $Y \rightarrow_X Z \subseteq [\leq](Y \rightarrow_X Z)$.

Let $x \in \neg_X Y$; then, $f(x) \notin Y$. Let $x \leq t$; then, $f(t) \leq f(x)$ by Rf_2. Since $Y \in L_X$, this implies $f(t) \notin Y$, and therefore, $t \in \neg_X Y$. It follows that $x \in [\leq](\neg_X Y)$.

Let $x \in 1'_X$, i.e. $x \in I$, and suppose that $x \leq t$. Then, $t \in I$ by Rf_3, and it follows that $1'_X \subseteq [\leq](1'_X)$.

Thus, all operations are well defined. Next, we show that the axioms of a relevant algebra are fulfilled. Ra_1 is true because of the discrete duality for distributive lattices, see e.g. Orłowska et al. (2015, Sect. II). Ra_2–Ra_7 are straightforward to prove.

Ra_8: Let $z \in I \circ_X Y$; then, there are $x \in I, y \in Y$ such that $R(x, y, z)$. By the "$\Leftarrow$" part of Rf_4 we have $y \leq z$, and $y \in Y$ and $Y = [\leq](Y)$ imply $z \in Y$. Thus, $I \circ_X Y \subseteq Y$. Conversely, let $y \in Y$. Reflexivity of $\leq$ and Rf_4 imply that $R(x, y, y)$ for some $x \in X$. It follows that $y \in I \circ_X Y$.

Ra_9: Let $Y \circ_X Z \subseteq W$, and $x \in Y$. We need to show that $x \in Z \to_X W$, i.e. that $R(x, y, z)$ and $y \in Z$ imply $z \in W$. Thus, suppose that $R(x, y, z)$ and $y \in Z$. Then, $z \in Y \circ_X Z$, and the hypothesis implies that $z \in W$. The converse is shown similarly. □

Theorem 13.12 *1. L can be embedded into L_{X_L}.*
2. X can be embedded into X_{L_X}.

Proof 1. Let $h : L \to L_{X_L}$ be the Stone map, defined by $h(a) := \{F \in X_L : a \in F\}$. It is well known that h preserves the lattice operations and constants (Stone 1937), see also Orłowska et al. (2015, Sect. 2.5). To show that h preserves the additional operations and constants of a relevant algebra we can use Urquhart (1996, Theorem 3.3(2)).

2. For $x \in X$, let $k(x) := \{Y \in L_X : x \in Y\}$; since $\uparrow x$ is the smallest element of L_X containing x, $k(x)$ is the principal filter of L_X containing x. Suppose that $Y, Z \in L_X$ and $x \in Y \vee_X Z$; then, $\uparrow x \subseteq Y \cup Z$. If w.l.o.g. $x \in Y$, then $\uparrow x \subseteq Y$ since Y is $\leq$–increasing, and it follows that $Y \in k(x)$. Hence, $k(x)$ is prime, and k is well defined. Clearly, k is injective, and all that is left to show is that it is an embedding. Let $x, y, z \in X$ with $R(x, y, z)$. Below, we assume that $Y, Z, W \in L_X$. By the definitions,

$$\begin{aligned} R_{L_X}(k(x), k(y), k(z)) &\iff (\forall Y, Z, W)[x \in Y, y \in Z, Y \circ_X Z \subseteq W) \Rightarrow z \in W], \\ &\iff (\forall Y, Z)[x \in Y, y \in Z \Rightarrow z \in Y \circ_X Z], \\ &\iff z \in \uparrow x \ \circ_X \uparrow y, \\ &\iff (\exists u, v)[x \leq u, y \leq v, R(u, v, z)], \\ &\iff R(x, y, z), \quad \text{by } \mathrm{Rf}_1. \end{aligned}$$

Next we show that $k(f(x)) = f_{L_X}(k(x))$:

$$\begin{aligned} Y \in f_{L_X}(k(x)) \iff \neg_X Y \notin k(x) \iff x \notin \neg_X Y \\ \iff f(x) \in Y \iff Y \in k(f(x)). \end{aligned}$$

Finally,

$$I_{L_X} = \{F \in X_{L_X} : 1'_X \in F\} = \{F \in X_{L_X} : I \in F\} = \bigcap\{k(x) : x \in I\}.$$

This completes the proof. □

Acknowledgements We dedicate this article to Alasdair Urquhart, our friend and esteemed colleague, on the occasion of his 75th birthday. His work has been a valuable source of inspiration for us for many years. We also thank the second reader for her valuable comments. I. Düntsch gratefully acknowledges support by the National Natural Science Foundation of China, Grant No. 61976053.

References

Allwein, G., & Dunn, J. M. (1993). Kripke models for linear logic. *Journal of Symbolic Logic*, *58*, 514–545.

Birkhoff, G. (1967). *Lattice theory* (Vol. 25, 3rd Ed.). Providence: American Mathematical Society, Colloquium Publications.

Craig, A., & Haviar, M. (2014). Reconciliation of approaches to the construction of canonical extensions of bounded lattices. *Mathematica Slovaka*, *6*, 1335–1356.

Czelakowski, J. (2015). *The equationally-defined commutator, a study in equational logic and algebra*. Birkhäuser.

Dalla Chiara, M. L., & Giuntini, R. (2002). Quantum logics. In Gabbay, D., & Guenthner, F., (Eds.), *Handbook of philosophical logic* (Vol. 6, pp. 129–228). Kluwer.

Düntsch, I., & Gediga, G. (2018). Guttman algebras and a model checking procedure for Guttman scales. In Golińska-Pilarek, J., & Zawidzki, M., (Eds.), *Ewa Orłowska on relational methods in logic and computer science*. Outstanding contributions to logic (pp. 355–370). Berlin: Springer. MR3929609.

Düntsch, I., & Orłowska, E. (2001). Beyond modalities: Sufficiency and mixed algebras. In Orłowska, E., & Szałas, A., (Eds.), *Relational methods for computer science applications* (pp. 263–283). Heidelberg: Physica-Verlag. MR1858531.

Düntsch, I., & Orłowska, E. (2008). A discrete duality between apartness algebras and apartness frames. *Journal of Applied Non-Classical Logics*, *18*, 213–227. MR2462235.

Düntsch, I., & Orłowska, E. (2011). An algebraic approach to preference relations. In de Swart, H. C. M., (Ed.), *Proceedings of the 12th international conference on relational and algebraic methods in computer science (RAMiCS 12)*. Lecture notes in computer science (Vol. 6663, pp. 141–147). Berlin: Springer. MR2913845.

Düntsch, I., & Orłowska, E. (2019). A discrete representation of lattice frames. In Blackburn, P., Lorini, E., & Guo, M., (Eds.), *Logic, rationality, and interaction. LORI 2019*. Lecture notes in computer science (Vol. 11813). Berlin: Springer. MR4019594.

Düntsch, I., Orłowska, E., & Radzikowska, A. (2003). Lattice-based relation algebras and their representability. In H. de Swart, E. Orłowska, G. Schmidt, & M. Roubens (Eds.), *Theory and application of relational structures as knowledge instruments* (Vol. 2929, pp. 231–255)., Lecture notes in computer science Springer: Heidelberg.

Düntsch, I., Orłowska, E., & van Alten, C. (2016). Discrete dualities for n-potent MTL-algebras and 2-potent BL-algebras. *Fuzzy Sets and Systems*, *292*, 203–214. MR3471217.

Dzik, W., Orłowska, E., & van Alten, C. (2006). Relational representation theorems for general lattices with negations. In R. A. Schmidt (Ed.), *Relations and Kleene algebra in computer science. Lecture notes in computer science* (Vol. 4136, pp. 162–176). Berlin: Springer.

Freese, R., & McKenzie, R. (1987). *Commutator theory for congruence modular varieties*. Cambridge: Cambridge University Press.

Georgiev, D. (2006). An implementation of the algorithm SQEMA for computing first-order correspondences of modal formulas. Master's thesis, Sofia University, Faculty of Mathematics and Computer Science.

Goldblatt, R. (1974a). Metamathematics of modal logic. *Bulletin of the Australian Mathematical Society*, *10*, 479–480.

Goldblatt, R. (1974b). Semantic analysis of orthologic. *Journal of Philosophical Logic*, *3*, 19–35.
Goldblatt, R. (1975). The Stone space of an ortholattice. *Bulletin of the London Mathematical Society*, *7*, 45–48.
Goldblatt, R. (1984). Orthomodularity is not elementary. *The Journal of Symbolic Logic*, *49*(2), 401–404.
Goranko, V. (1990). Modal definability in enriched languages. *Notre Dame Journal of Formal Logic*, *31*(1), 81–105.
Hartonas, C. (2019). Discrete duality for lattices with modal operators. *Journal of Logic and Computation*, *29*(1), 71–89.
Hartonas, C., & Dunn, J. M. (1993). Duality theorems for partial orders, semilattices, Galois connections and lattices. Preprint IULG-93-26, Indiana University Logic Group.
Hartung, G. (1992). A topological representation of lattices. *Algebra Universalis*, *29*, 273–299.
Jónsson, B., & Tarski, A. (1951). Boolean algebras with operators I. *American Journal of Mathematics*, *73*, 891–939.
Kalmbach, G. (1985). *Orthomodular lattices*. London: Academic Press.
Kowalski, T., & Litak, T. (2008). Completions of GBL-algebras: Negative results. *Algebra Universalis*, *58*, 373–384.
McCune, W. (2005–2010). Prover9 and Mace4. http://www.cs.unm.edu/~mccune/prover9/.
Orłowska, E., & Rewitzky, I. (2007). Discrete duality and its applications to reasoning with incomplete information. *Lecture Notes in Artificial Intelligence*, *5785*, 51–56.
Orłowska, E., Rewitzky, I., & Radzikowska, A. (2015). *Dualities for structures of applied logics*. Studies in logic (Vol. 56). College Publications.
Priestley, H. A. (1970). Representation of distributive lattices by means of ordered Stone spaces. *Bulletin of the London Mathematical Society*, *2*, 186–190.
Rédei, M. (2009). The Birkhoff–von Neumann concept of quantum logic. In Engesser, K., Gabbay, D. M., & Lehmann, D., (Eds.), *Handbook of quantum logic and quantum structures: Quantum logic* (pp. 1–22). Elsevier.
Stone, M. (1937). Topological representations of distributive lattices and Brouwerian logics. *Časopis Pěst. Mat.*, *67*, 1–25.
Urquhart, A. (1978). A topological representation theorem for lattices. *Algebra Universalis*, *8*, 45–58.
Urquhart, A. (1996). Duality for algebras of relevant logics. *Studia Logica*, *56*, 263–276.
Urquhart, A. (2019). Relevant implication and ordered geometry. *The Australasian Journal of Logic*, *16*(8), 342–354.
van Benthem, J. (1984). Correspondence theory. In D. Gabbay & F. Guenthner (Eds.), *Handbook of philosophical logic* (Vol. III, pp. 325–408). Dordrecht: Reidel.

Chapter 14
Ockham Algebras—An Urquhart Legacy

T. S. Blyth and H. J. Silva

Second Reader
H.P. Sankappanavar
State University of New York at New Paltz

Abstract We highlight the fundamental influence that the work of Alasdair Urquhart has had in the area of distributive lattice-ordered algebras and in particular to the development of Ockham algebras, to which we attach some new results.

Keywords De Morgan algebra · Kleene algebra · Ockham algebra · Dual space · g-cycle · Subdirectly irreducible · Berman class · Urquhart class · Endomorphism

Since a Boolean algebra is a complemented distributive lattice, significant generalisations can be achieved by relaxing the distributivity and retaining the complementation (which is then no longer unique), or by retaining the distributivity and relaxing the complementation, or by relaxing the implication. Here, we shall be concerned uniquely with the second of these, for which a natural procedure is to consider a bounded distributive lattice with the complementation replaced by a dual endomorphism f, so that we have $f(0) = 1$ and $f(1) = 0$, together with the de Morgan type equalities $f(x \wedge y) = f(x) \vee f(y)$ and $f(x \vee y) = f(x) \wedge f(y)$. An early approach to such a consideration is to be found in the concept of a *de Morgan algebra* [to the above, add the property $f^2(x) = x$] introduced in 1935 by Moisil (1935) and known in the Polish school as 'quasi-Boolean algebras'. In 1958, they

T. S. Blyth (✉)
(cum saluere : Newburgh to Auchtermuchty!), Mathematical Institute,
University of St Andrews, St Andrews, Scotland
e-mail: tsblyth.prof@btinternet.com

H. J. Silva
Centro de Matemática e Aplicações and Departamento de Matemática,
Faculdade de Ciências e Tecnologia, Universidade Nova de Lisboa, Lisboa, Portugal
e-mail: hdjs@fct.unl.pt

I. Düntsch and E. Mares (eds.), *Alasdair Urquhart on Nonclassical and Algebraic Logic and Complexity of Proofs*, Outstanding Contributions to Logic 22,
https://doi.org/10.1007/978-3-030-71430-7_14

were investigated by Kalman (1958) who called them 'distributive i-lattices' and proved that the lattice of subvarieties of the variety $\mathbf{M}$ of de Morgan algebras is the 4-element chain $\boldsymbol{\omega} \subset \mathbf{B} \subset \mathbf{K} \subset \mathbf{M}$ where $\boldsymbol{\omega}$ is the trivial variety, $\mathbf{B}$ is the variety of Boolean algebras and $\mathbf{K}$ is the variety of Kleene algebras [defined by the identity $x \wedge f(x) \leqslant y \vee f(y)$]. In 1957, Bialynicki-Birula and Rasiowa (1957) proved a representation theorem for de Morgan algebras. For more on the early work on de Morgan algebras, we refer the reader to Balbes and Dwinger (1974) and to Rasiowa (1974).

It was in generalising de Morgan algebras by omitting the law of double negation that Berman began the study of Ockham algebras in a very deep paper in 1977 (Berman 1977). Two years later, Urquhart (1979) developed a topological duality theory for this type of algebra, gave a logical motivation for its study and introduced the name *Ockham lattice*. The stated justification for this etymology is 'the term Ockham lattice was chosen because the so-called de Morgan laws are due (at least in the case of propositional logic) to William of Ockham (1287–1347), an English Franciscan friar and scholastic philosopher'. The name *Ockham algebra* as applied to a bounded distributive lattice with a dual endomorphism was further promoted by Goldberg (1981, 1983) and has since become classical.

In order to appreciate the development of Ockham algebras, it is necessary to embrace the long history of duality theory in the context of distributive lattice-ordered algebras. It was in 1933 that Birkhoff (1933) established his famous representation theorem which provides a lattice isomorphism from a finite distributive lattice L to the lattice $\mathcal{O}\big(J(L)\big)$ of down-sets of the set $J(L)$ of join-irreducible elements of L. Three years later M. H. Stone (1936) developed a representation theory for arbitrary Boolean algebras using topological methods. Hilary Priestley (1972) provided an ingenious common generalisation of both these theories, thus allowing questions of a lattice-theoretic nature to be translated into the language of ordered topological spaces, and thereby be resolved more easily since the dual space is generally simpler and more tractable than the algebra itself.

Basically, an ordered topological space $(X; \tau, \leqslant)$ is said to be *totally order-disconnected* if, for $x, y \in X$ such that $x \nleqslant y$, there exists a clopen down-set U such that $y \in U$ and $x \notin U$. A *Priestley space* is then defined to be a compact totally order-disconnected space X. Of especial interest here is the family of clopen down-sets of X which is denoted by $\mathcal{O}(X)$.

If now L is a bounded distributive lattice and $I_p(L)$ is the set of prime ideals of L then the *dual space* of L is defined to be $\big(I_p(L); \tau, \subseteq\big)$ where a base for the topology τ consists of the sets $\{X \in I_p(L) \mid a \in X\}$ and $\{X \in I_p(L) \mid a \notin X\}$ for every $a \in L$. Then, a fundamental result due to Priestley (see Davey and Priestley 2002) is the following.

Theorem 14.1 *If L is a bounded distributive lattice then $X = \big(I_p(L); \tau, \subseteq\big)$ is a Priestley space and $L \simeq \mathcal{O}\big(I_p(L)\big)$ via $a \mapsto \{A \in I_p(L) \mid a \notin A\}$. Conversely, if P is a Priestley space then $\mathcal{O}(P)$ is a bounded distributive lattice and $P \simeq \big(I_p(\mathcal{O}(P)); \tau, \subseteq\big)$.*

In the language of category theory, if $\mathbf{D}_{01}$ denotes the category of bounded distributive lattices and 0, 1-preserving lattice morphisms, and if $\mathbf{P}$ is the category of

Priestley spaces and continuous isotone maps, then the above isomorphisms produce a dual equivalence between $\mathbf{D}_{01}$ and $\mathbf{P}$. Since Ockham algebras are bounded distributive lattices, they are dually equivalent to a suitable subcategory of $\mathbf{P}$. The identification of this is a fundamental result of Urquhart (1979). For this purpose, he defines an *Ockham space* to be a Priestley space X endowed with a continuous antitone mapping g. Then, as is established by the following theorem, the category $\mathbf{O}$ of Ockham algebras is dually equivalent to the category $\mathbf{Q}$ whose objects are Ockham spaces and whose morphisms are the continuous isotone mappings that commute with g.

Theorem 14.2 *If* $(X; g) \in \mathbf{Q}$ *then* $\big(\mathcal{O}(X); f\big) \in \mathbf{O}$ *where*

$$\big(\forall A \in \mathcal{O}(X)\big) \qquad f(A) = X \setminus g^{-1}(A).$$

Conversely, if $(L; f) \in \mathbf{O}$ *then* $\big(I_p(L); g\big) \in \mathbf{Q}$ *where*

$$\big(\forall A \in I_p(L)\big) \qquad g(A) = \{a \in L \mid f(a) \notin A\}.$$

Moreover, these constructions give a dual equivalence.

The power of duality theory is particularly evident in the study of Ockham congruences, a central component of which is the following notion. A subset Y of an Ockham space $(X; g)$ is a *g-subset* if $g(Y) \subseteq Y$. For every subset A of X there is a smallest g-subset that contains A, namely $g^{\omega}(A) = \{g^n(x) \mid n \geqslant 0,\ x \in A\}$. We denote the resulting lattice of closed g-subsets of X by $C_g(X)$. In particular, the g-subsets of the form $g^{\omega}\{x\}$ are called *monogenic*. For each $A \subseteq X$, we denote by $\overline{g^{\omega}(A)}$ the smallest closed g-subset of X that contains A. Then, the $\overline{g^{\omega}(x)}$ are the completely join-irreducible elements Grätzer (2011) of $C_g(X)$.

Fundamental results concerning congruences and subdirectly irreducible Ockham algebras (i.e. those which admit a smallest non-trivial congruence) are the following, also due to Urquhart (1979).

Theorem 14.3 *Let* $(X; g)$ *be the dual space of* $(L; f) \in \mathbf{O}$*. If* $Q \in C_g(X)$ *then the relation* ϑ_Q *defined on* $\mathcal{O}(X)$ *by*

$$(A, B) \in \vartheta_Q \iff A \cap Q = B \cap Q$$

is a congruence, and $Q \mapsto \vartheta_Q$ *is a dual lattice isomorphism from* $C_g(X)$ *to* Con L.

Theorem 14.4 *Let* $(X; g)$ *be an Ockham space. If* Y *is a* g*-subset of* X *then so is its closure* $\overline{Y}$*.*

Theorem 14.5 $(L; f) \in \mathbf{O}$ *is subdirectly irreducible if and only if, in the dual space* $(X; g)$,

$$\overline{\{x \in X \mid \overline{g^{\omega}\{x\}} \neq X\}} \neq X.$$

In particular, if $(L; f)$ *is finite then the topology 'evaporates' and*

(1) $(L; f)$ *is subdirectly irreducible* $\iff (\exists x \in X)\ g^{\omega}\{x\} = X$;

(2) $(L; f)$ *is simple* $\iff (\forall x \in X)\ g^{\omega}\{x\} = X$.

As to the subvarieties of $\mathbf{O}$, the following natural classification, also due to Urquhart, is closely related to the dual space. For $m > n \geqslant 0$ the *Urquhart class* $\mathbf{P}_{m,n}$ is defined as the subclass of $\mathbf{O}$ that is formed by those algebras whose dual space $(X; g)$ satisfies $g^m = g^n$. The dual space $(X; g)$ of a finite subdirectly irreducible Ockham algebra $(L; f)$ in $\mathbf{P}_{m,n}$ can be conveniently described as follows, in which the order on X is ignored and the action of g is indicated by the arrows:

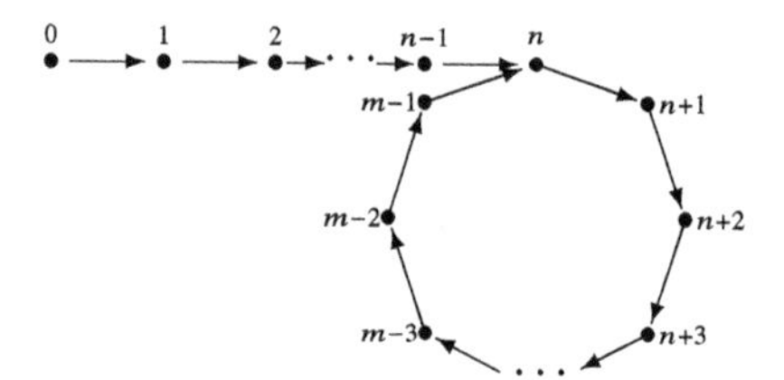

Here, we have $g^m(0) = g^n(0)$ so that $g^{\omega}\{0\} = X$ and Theorem 14.5(1) applies. The two subsets $\{0, 1, \ldots, n-1\}$ and $\{n, n+1, \ldots m-1\}$ are called the *tail* and *loop* of X, respectively. This Ockham space with the discrete order is denoted by m_n and its dual by $L_{m,n}$. Any order on m_n with respect to which g is antitone yields the dual space of a finite subdirectly irreducible Ockham algebra. Then, a finite Ockham algebra is subdirectly irreducible if and only if it is isomorphic to a subalgebra of $L_{m,n}$ for some m and n.

From the above representation, it is clear that the g-subsets of X are precisely the sets $\{k, k+1, \ldots, m-1\}$ where $0 \leqslant k \leqslant n$. Consequently, it follows from Theorem 14.3 that $\mathrm{Con}\, \mathrm{L}_{\mathrm{m,n}}$ is an $(n+2)$-element chain.

By way of illustration, the subdirectly irreducible algebras in $\mathbf{P}_{3,1}$ are the subalgebras of the algebra $L_{3,1}$ whose dual space $X = \{p, q, r\}$ has discrete order and on which g acts as follows:

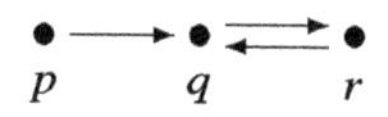

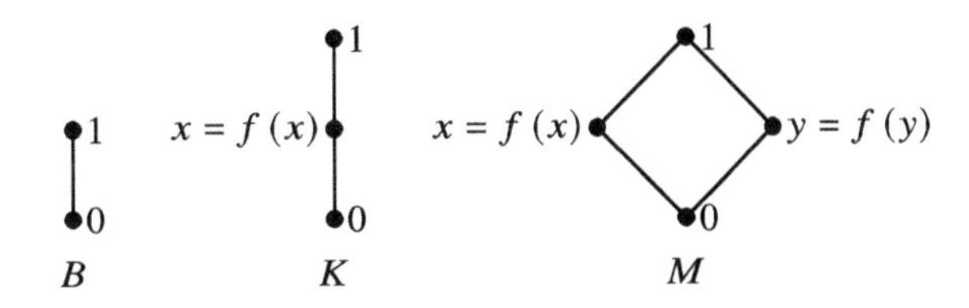

Fig. 14.1 The subdirectly irreducible algebras in **M**

Using Theorem 14.2, we see that the corresponding Ockham algebra $(L; f)$ is

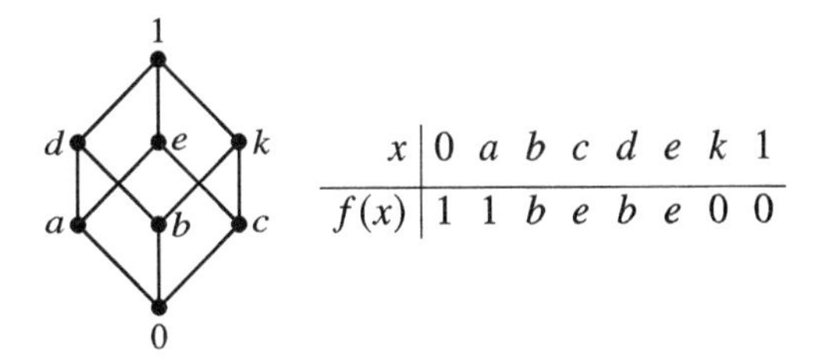

x	0	a	b	c	d	e	k	1
$f(x)$	1	1	b	e	b	e	0	0

By determining all orders on $(X; g)$ with respect to which g is antitone we can obtain the dual spaces of the subdirectly irreducible algebras in $\mathbf{P}_{3,1}$, whence all the Ockham subalgebras of $L_{3,1}$. There are 19 of these, and may be found tabulated together with their duals in Blyth and Varlet (1994).

In connection with this, we would add that the first description of the subdirectly irreducible algebras in the subvariety **M** of de Morgan algebras was obtained by Kalman (1958). There are precisely three, which are shown in Fig. 14.1.

Another subvariety of $\mathbf{P}_{3,1}$ of some interest is that of *MS-algebras* introduced in 1983 by Blyth and Varlet (1983a). These algebras essentially retain the properties that are common to de Morgan algebras and Stone algebras. To be more precise, in the variety **M** of de Morgan algebras $x \mapsto f(x)$ is a dual automorphism and $x \mapsto f^2(x)$ is the identity, whereas, in the variety **S** of Stone algebras, the pseudo-complementation $x \mapsto x^\star$ is a dual endomorphism and $x \mapsto x^{\star\star}$ is a closure. The class **MS** of MS-algebras is then defined as the subvariety of $\mathbf{P}_{3,1}$ obtained by adding the identity $x \wedge f^2(x) = x$. In **MS**, there are nine subdirectly irreducible algebras (Blyth and Varlet 1983a), and its lattice of subvarieties was determined in Blyth and Varlet (1983b).

More generally, a description of the subdirectly irreducible algebras in the variety $\mathbf{P}_{3,1}$ was obtained by Beazer (1984b) and, independently but not published earlier, by Sankappanavar (1985).

By comparing the tails and the loops in the representations of the dual spaces of subdirectly irreducible algebras, it is readily seen that the Urquhart classes may be ordered by

$$\mathbf{P}_{m,n} \subseteq \mathbf{P}_{m',n'} \iff m \leqslant m',\ n \leqslant n',\ m-n \mid m'-n'.$$

Whereas the Urquhart classes are defined in terms of the antitone mapping g on the dual space X, they may also be defined in terms of the dual endomorphism f on L. Indeed, more generally, the problem of translating from one language to the other was also solved by Urquhart (1981).

For this purpose, he defines a *term* to be a polynomial built from variables $a, b, c, \ldots$ and the constants 0, 1 by means of the operations $\wedge$, $\vee$ and f. A term is *atomic* if it is of the form $f^n(a)$ and is *even* or *odd* according to n. An *inequality* is then an expression of the form $A \leqslant B$ where A and B are terms, and is *basic* if A is a meet of atomic terms and B is a join of atomic terms. The occurrence of a variable a in an inequality $A \leqslant B$ is *positive* if a appears in an even term of A or in an odd term of B; and is *negative* if a appears in an odd term of A or in an even term of B. A basic inequality is said to be *simple* if each variable has precisely one positive and one negative appearance.

Two inequalities are *equivalent* if they determine the same equational class in $\mathbf{O}$. As shown by Urquhart, every basic inequality is equivalent to a simple inequality, with the latter being characterised as follows.

Theorem 14.6 *Let E be a finite subset of $\mathbb{N} \times \mathbb{N}$, say*

$$E = \{(p_1, q_1), (p_2, q_2), \ldots, (p_n, q_n)\}.$$

Let

$$A = \{f^{p_i}(x_i) \mid p_i \text{ even}\} \cup \{f^{q_i}(x_i) \mid q_i \text{ odd}\};$$
$$B = \{f^{p_i}(x_i) \mid p_i \text{ odd}\} \cup \{f^{q_i}(x_i) \mid q_i \text{ even}\},$$

where x_i and x_j are distinct variables for $(p_i, q_i) \neq (p_j, q_j)$ in E.

Then $(L; f)$ satisfies the simple inequality $\bigwedge A \leqslant \bigvee B$ if and only if its dual $(X; g)$ satisfies the disjunction

$$\bigvee_i \{g^{p_i} \geqslant g^{q_i} \mid (p_i, q_i) \in E\}.$$

An important consequence of this is that membership of the Urquhart classes may also be characterised in terms of f as follows.

Theorem 14.7 *Let $m > n \geqslant 0$ and $(L; f) \in \mathbf{O}$. Then*
(1) *when $m - n$ is even, $L \in \mathbf{P}_{m,n} \iff f^m = f^n$;*
(2) *when $m - n$ is odd,*

$$L \in \mathbf{P}_{m,n} \iff (\forall a \in L)\ f^m(a) \text{ and } f^n(a) \text{ are complementary.}$$

Proof (1) For each $x \in L$ the equality $f^m(x) = f^n(x)$ is the conjunction of $f^m(x) \leqslant f^n(x)$ and $f^n(x) \leqslant f^m(x)$. For m and n of the same parity, these are equivalent by Theorem 14.5 to $g^m \geqslant g^n$ and $g^n \geqslant g^m$, i.e. to $g^m = g^n$.

(2) Without loss of generality, we may assume that m is even and n is odd. Then, the equalities $f^m(x) \wedge f^n(x) = 0$ and $1 = f^m(x) \vee f^n(x)$ are equivalent, respectively, to $g^m \geqslant g^n$ and $g^n \geqslant g^m$, i.e. to $g^m = g^n$. □

The Urquhart classes of Ockham algebras extend the *Berman classes* $\mathbf{K}_{p,q}$ which were previously introduced in Berman (1977) and are given by

$$(L;\, f) \in \mathbf{K}_{p,q} \iff f^{2p+q} = f^{q}.$$

By Theorem 14.7(1), the connection is then that $\mathbf{K}_{p,q} = \mathbf{P}_{2p+q,q}$, and consequently

$$\mathbf{K}_{p,q} \subseteq \mathbf{K}_{p',q'} \iff q \leqslant q',\ p \mid p'.$$

It is noteworthy that every finite Ockham algebra belongs to a Berman class. Indeed, if $(L;\, f)$ is finite then the sets $\{f, f^3, f^5, \dots\}$ and $\{f^0, f^2, f^4, \dots\}$ of dual endomorphisms and endomorphisms are finite, so $f^q = f^{2p+q}$ for some p and q.

A subclass of $\mathbf{O}$ which contains all of the Berman classes is the class $\mathbf{K}_\omega$ where

$$(L;\, f) \in \mathbf{K}_\omega \iff (\forall x \in L)(\exists m, n \in \mathbb{N})(m \neq 0)\quad f^{m+n}(x) = f^{n}(x).$$

By its definition, the subclass $\mathbf{K}_\omega$ is closed under the formation of subalgebras, epimorphic images, and arbitrary direct powers. Notwithstanding the fact that it is not closed under arbitrary direct products, it was shown by Fang (1992) that $\mathbf{K}_\omega$ is closed under finite direct products and therefore forms a *generalised variety* in the sense of Ash (1985).

Congruences on Ockham algebras have been extensively investigated, the first fundamental result being the following, due to Berman, which describes for $a \leqslant b$ the smallest congruence $\vartheta(a, b)$ that identifies a and b.

Theorem 14.8 Berman (1977) *If* $(L;\, f) \in \mathbf{O}$ *and* $a \leqslant b$ *in* L*, then*

$$\vartheta(a, b) = \bigvee_{n \geqslant 0} \vartheta_{\mathrm{lat}}\big(f^n(a), f^n(b)\big).$$

For $(L;\, f) \in \mathbf{O}$, the relation Φ_n defined on L by

$$(x, y) \in \Phi_n \iff f^n(x) = f^n(y)$$

is a congruence on L, as is the relation $\Phi_\omega = \bigvee_{i \geqslant 0} \Phi_i$.

As a supplement to Theorem 14.3, the following result associates these basic congruences with corresponding closed g-sets in the dual space.

Theorem 14.9 (Blyth et al. (1991)) *Let* $(L;\, f) \in \mathbf{O}$ *and let* $(X;\, g)$ *be its dual space. Then for every* $n \geqslant 0$ *the congruence* Φ_n *is associated with the closed* g*-subset* $g^n(X)$*, and the congruence* Φ_ω *is associated with* $g^\omega(X) = \bigcap_{n \geqslant 0} g^n(X)$.

The subdirectly irreducible algebras in $\mathbf{K}_\omega$ have the following nice characterisation in which ω and ι denote, respectively, the trivial and the universal congruences.

Theorem 14.10 (Blyth et al. (1991)) *If* $(L; f)$ *belongs properly to* $\mathbf{K}_\omega$ *then* L *is subdirectly irreducible if and only if* Con L *reduces to the infinite chain*

$$\omega = \Phi_0 \prec \Phi_1 \prec \cdots < \Phi_\omega \prec \iota.$$

If $(L; f)$ *belongs to a Berman class, the smallest of which is* $\mathbf{K}_{p,q}$, *then* L *is subdirectly irreducible if and only if* Con L *reduces to the finite chain*

$$\omega = \Phi_0 \prec \Phi_1 \prec \cdots \prec \Phi_q = \Phi_\omega \prec \iota.$$

A celebrated theorem of Jónsson (1972) states that if $\mathbf{V}$ is a variety every algebra of which has a distributive congruence lattice then the lattice $\Lambda(\mathbf{V})$ of subvarieties of $\mathbf{V}$ is distributive. Applying this to the variety $\mathbf{O}$, Urquhart (1981) showed that $\Lambda(\mathbf{O})$ is uncountable and that a subvariety of $\mathbf{O}$ has finite height in $\Lambda(\mathbf{O})$ if and only if it is generated by a finite algebra. The most interesting part of $\Lambda(\mathbf{O})$ is the set $\Lambda_f(\mathbf{O})$ of subvarieties of $\mathbf{O}$ that are generated by finite algebras, this being mainly because it contains the well-known and extensively investigated subvarieties $\mathbf{B}$ of Boolean algebras, $\mathbf{K}$ of Kleene algebras, $\mathbf{M}$ of de Morgan algebras, and $\mathbf{S}$ of Stone algebras. These are all contained in the subvariety $\mathbf{P}_{3,1} = \mathbf{K}_{1,1}$ which may also be characterised by the fact that $(L; f) \in \mathbf{P}_{3,1}$ if and only if its *skeleton* $\Im f$ is a de Morgan algebra.

In Blyth and Varlet (1994, Chap. 5), there is developed a tabulation method for the purpose of reducing the number of atomic terms and variables in simple inequalities, thus easing the application of Urquhart's theorem. Using this method, it is established there that in $\mathbf{P}_{3,1}$ there can be defined 34 non-equivalent axioms, each involving at most 3 variables. Using a standard theorem of Davey (1979) from universal algebra, this leads to a successful description of the lattice $\Lambda(\mathbf{P}_{3,1})$.

In the years subsequent to the initial development of Ockham algebras, there have been many investigations into properties of specific subvarieties. Likewise, many generalisations have been inspired, promoting thereby a substantial literature. Whereas we make no attempt to list these, such a menu would certainly contain (Beazer 1984a; Fang 2006, 2011; Sankappanavar 1987, 2012, 2016) and, importantly, the many references therein.

Here our objective is to highlight some particular items that have been inspired by the Urquhart theorems. In this the main focus will be on summarising and extending some results on Ockham endomorphisms which have led to a description of the finite non-Boolean de Morgan algebras which have the endomorphism kernel property.

For this purpose, given an Ockham algebra $\mathcal{L} = (L; f)$, we denote by $\mathcal{S}(\mathcal{L}) = \big(I_p(L); g\big)$ the dual space of $\mathcal{L}$ where g is given by $g(P) = f^{-1}(L \setminus P)$. Correspondingly, given an Ockham space $\mathcal{X} = (X; g)$, we denote by $\widetilde{\mathcal{X}} = \big(O(X); f\big)$ the Ockham algebra where f is given by $f(A) = X \setminus g^{-1}(A)$.

An important observation is the following.

Theorem 14.11 *Let* $\mathcal{L} = (L; f)$ *be an Ockham algebra with dual space* $\mathcal{X} = (X; g)$. *The monoids* $\operatorname{End} \mathcal{X}$ *and* $\operatorname{End} \widetilde{\mathcal{X}} \simeq \operatorname{End} \mathcal{L}$ *are anti-isomorphic.*

Proof Consider the mapping $\Psi : \operatorname{End}\mathcal{X} \to \operatorname{End}\widetilde{\mathcal{X}}$ given by $\Psi(\alpha) = \widetilde{\alpha}$ where $\widetilde{\alpha} : \mathcal{O}(X) \to \mathcal{O}(X)$ is given by $\widetilde{\alpha}(A) = \alpha^{-1}(A)$.

Given $\alpha, \beta \in \operatorname{End}\mathcal{X}$, for every $A \in \mathcal{O}(X)$ we have

$$\Psi(\alpha \circ \beta)(A) = (\alpha \circ \beta)^{-1}(A) = \beta^{-1}\big(\alpha^{-1}(A)\big) = \big(\Psi(\beta) \circ \Psi(\alpha)\big)(A)$$

and consequently $\Psi(\alpha \circ \beta) = \Psi(\beta) \circ \Psi(\alpha)$. Thus, we have that Ψ is an anti-morphism of monoids.

As shown by Urquhart (1979), the Ockham spaces $\mathcal{X}$ and $\mathcal{S}(\widetilde{\mathcal{X}})$ are homeomorphic. Here, we make use of the isomorphism $\varepsilon : \mathcal{X} \to \mathcal{S}(\widetilde{\mathcal{X}})$ given by the assignment

$$\varepsilon(t) = \{A \in \mathcal{O}(X) \mid t \notin A\}.$$

For each $h \in \operatorname{End}\widetilde{\mathcal{X}}$ define $\overline{h} \in \operatorname{End}\mathcal{S}(\widetilde{\mathcal{X}})$ by $\overline{h}(P) = h^{-1}(P)$ and consider the mapping $\Phi : \operatorname{End}\widetilde{\mathcal{X}} \to \operatorname{End}\mathcal{X}$ given by $\Phi(h) = \varepsilon^{-1} \circ \overline{h} \circ \varepsilon$. For every $h \in \operatorname{End}\widetilde{\mathcal{X}}$ and every $A \in \mathcal{O}(X)$, we have

$$\begin{aligned}
[(\Psi \circ \Phi)(h)](A) &= (\varepsilon^{-1} \circ \overline{h} \circ \varepsilon)^{-1}(A) \\
&= \{x \in X \mid (\varepsilon^{-1} \circ \overline{h} \circ \varepsilon)(x) \in A\} \\
&= \{x \in X \mid A \notin \varepsilon\big((\varepsilon^{-1} \circ \overline{h} \circ \varepsilon)(x)\big)\} \\
&= \{x \in X \mid A \notin h^{-1}\big(\varepsilon(x)\big)\} \\
&= \{x \in X \mid h(A) \notin \varepsilon(x)\} \\
&= \{x \in X \mid x \in h(A)\} \\
&= h(A).
\end{aligned}$$

Thus, $(\Psi \circ \Phi)(h) = h$ and consequently $\Psi \circ \Phi = \mathrm{id}_{\operatorname{End}\widetilde{\mathcal{X}}}$.

Suppose now that $\alpha \in \operatorname{End}\mathcal{X}$ and $t \in X$. Given $A \in \mathcal{O}(X)$, we have

$$\begin{aligned}
A \in \overline{\widetilde{\alpha}}\big(\varepsilon(t)\big) \iff A \in \widetilde{\alpha}^{-1}\big(\varepsilon(t)\big) &\iff \widetilde{\alpha}(A) \in \varepsilon(t) \\
&\iff \alpha^{-1}(A) \in \varepsilon(t) \\
&\iff t \notin \alpha^{-1}(A) \\
&\iff \alpha(t) \notin A \\
&\iff A \in \varepsilon\big(\alpha(t)\big).
\end{aligned}$$

Thus, for every $t \in X$ we have that $\overline{\widetilde{\alpha}}\big(\varepsilon(t)\big) = \varepsilon\big(\alpha(t)\big)$ and consequently $\overline{\widetilde{\alpha}} \circ \varepsilon = \varepsilon \circ \alpha$. It follows that $\Phi\big(\Psi(\alpha)\big) = \Phi(\widetilde{\alpha}) = \varepsilon^{-1} \circ \overline{\widetilde{\alpha}} \circ \varepsilon = \varepsilon^{-1} \circ \varepsilon \circ \alpha = \alpha$. Hence $\Phi \circ \Psi = \mathrm{id}_{\operatorname{End}\mathcal{X}}$ and we conclude that Ψ is an anti-isomorphism. □

By Theorem 14.11, properties of $\operatorname{End}\mathcal{L}$ may be usefully investigated via $\operatorname{End}\mathcal{X}$.

Without loss of generality, if $\mathcal{L} = (L; f)$ is an Ockham algebra and $\mathcal{X} = (X; g)$ is its dual space we may identify L with the lattice $\mathcal{O}(X)$ of clopen down-sets of X. Then, by Theorem 14.2, for every $A \in \mathcal{O}(X)$ we have $f(A) = X \setminus g^{-1}(A)$, from which it follows that

(α) $(\forall i \in \mathbb{N})\ f^{2i}(A) = (g^{2i})^{-1}(A)$;

(β) $(\forall i \in \mathbb{N})\ f^{2i+1}(A) = X \setminus (g^{2i+1})^{-1}(A)$.

Since g is antitone, all even powers of g are isotone. However, it is possible for an odd power of g also to be isotone. This situation arises as follows.

Theorem 14.12 *If $(L; f)$ is an Ockham algebra with dual space $(X; g)$ then, for $n \in \mathbb{N}$, the following statements are equivalent*:

(1) g^{2n+1} *is isotone*;

(2) $f^{2n+1}(L)$ *is contained in the centre $Z(L)$ of L.*

Proof (1) $\Rightarrow$ (2): If $B \in f^{2n+1}(L)$ then, by (β), there exists $A \in \mathcal{O}(X)$ such that $B = f^{2n+1}(A) = X \setminus (g^{2n+1})^{-1}(A)$. But if g^{2n+1} is isotone then $(g^{2n+1})^{-1}(A) \in \mathcal{O}(X)$. Hence, B is a complemented element of $\mathcal{O}(X)$ and so belongs to $Z(L)$.

(2) $\Rightarrow$ (1): If (2) holds then for every $A \in \mathcal{O}(X) = L$ there exists $B \in \mathcal{O}(X)$ such that $f^{2n+1}(A) \cap B = \emptyset$ and $f^{2n+1}(A) \cup B = X$. Then $B = X \setminus f^{2n+1}(A)$ and therefore, by (β), $(g^{2n+1})^{-1}(A) = X \setminus f^{2n+1}(A) \in \mathcal{O}(X)$.

To see that g^{2n+1} is isotone, suppose by way of obtaining a contradiction that there exist $x, y \in X$ such that $x < y$ and $g^{2n+1}(x) \nleqslant g^{2n+1}(y)$. Since X is totally order-disconnected, there exists $U \in \mathcal{O}(X)$ with $g^{2n+1}(y) \in U$ and $g^{2n+1}(x) \notin U$; i.e. $y \in (g^{2n+1})^{-1}(U)$ and $x \notin (g^{2n+1})^{-1}(U)$. This contradicts the fact established above that $(g^{2n+1})^{-1}(U)$ is a down-set. □

In what follows we shall denote by σ the operation of complementation on the centre $Z(L)$ of L. Consider now the sets

$$\begin{aligned} \nabla(\mathcal{L}) &= \{f^k \mid k \text{ even}\} \cup \{\sigma f^k \mid k \text{ odd and } f^k(L) \subseteq Z(L)\}; \\ \Delta(\mathcal{X}) &= \{g^i \mid g^i \text{ isotone}\}. \end{aligned}$$

Clearly, $\nabla(\mathcal{L})$ is a submonoid of End $\mathcal{L}$, and $\Delta(\mathcal{X})$ is a submonoid of End $\mathcal{X}$.

Theorem 14.13 *The mapping $\varphi : \Delta(\mathcal{X}) \to \nabla(\mathcal{L})$ given by*

$$\varphi(g^k) = \begin{cases} f^k & \text{if } k \text{ is even;} \\ \sigma f^k & \text{if } k \text{ is odd,} \end{cases}$$

is a monoid isomorphism.

Proof Suppose now that $g^k \in \Delta(\mathcal{X})$ and let Ψ be the anti-isomorphism of Theorem 14.11. If k is even then it follows from (α) above that $\Psi(g^k) = f^k$. If k is odd then since g^k is isotone we have $(g^k)^{-1}(A) \in \mathcal{O}(X)$ and it follows from (β) above that $f^k(A) = X \setminus (g^k)^{-1}(A) \in Z\big(\mathcal{O}(X)\big)$. Consequently, $f^k\big(\mathcal{O}(X)\big) \subseteq Z\big(\mathcal{O}(X)\big)$ and $\sigma f^k = \Psi(g^k)$. In summary,

$$\Psi(g^k) = \begin{cases} f^k & \text{if } k \text{ is even;} \\ \sigma f^k & \text{if } k \text{ is odd,} \end{cases}$$

and $\Psi\big(\Delta(\mathcal{X})\big) \subseteq \nabla(\mathcal{L})$. The reverse inclusion is a consequence of Theorem 14.12. Finally, since $\Delta(\mathcal{X})$ is commutative, the isomorphism φ is now induced by Ψ. □

The significance of these submonoids is highlighted in the following description of the endomorphism semigroup of a finite subdirectly irreducible Ockham algebra.

Theorem 14.14 (Blyth et al. (2001b)) *If $\mathcal{L}$ is finite and subdirectly irreducible then* $\operatorname{End}\mathcal{L} = \nabla(\mathcal{L})$.

Proof In the particular case where $\mathcal{L}$ is finite and subdirectly irreducible, with dual space $\mathcal{X}$ as depicted previously by m_n, let $\vartheta \in \operatorname{End}\mathcal{X}$. Then, there exists i such that $\vartheta(0) = g^i(0)$. This gives $\vartheta(1) = \vartheta g(0) = g\vartheta(0) = g^{i+1}(0) = g^i g(0) = g^i(1)$. Suppose, by way of induction, that for $r > 1$ we have $\vartheta(r) = g^i(r)$. Then

$$\vartheta(r+1) = \vartheta g(r) = g\vartheta(r) = gg^i(r) = g^i g(r) = g^i(r+1).$$

Hence, $\vartheta(j) = g^i(j)$ for all j and so $\vartheta = g^i$. Since ϑ is isotone, we deduce that $\operatorname{End}\mathcal{X} \subseteq \Delta(\mathcal{X})$ and consequently $\operatorname{End}\mathcal{X} = \Delta(\mathcal{X})$.

Applying Theorem 14.11, we deduce that, correspondingly, $\operatorname{End}\mathcal{L} = \nabla(\mathcal{L})$. □

For a finite subdirectly irreducible Ockham algebra $\mathcal{L}$, the cardinality of $\operatorname{End}\mathcal{L}$ is known and may be summarised by the following result. For this we recall that for a finite Ockham algebra $\mathcal{L} = (L; f)$ the cardinality of its dual space is the number of $\vee$-irreducible elements of L, and equals the height $h(L)$ of L.

Theorem 14.15 (Blyth and Silva (1998/99)) *Let $\mathcal{L} = (L; f)$ be a finite subdirectly irreducible Ockham algebra with dual space $\mathcal{X} = (X; g)$. If the loop of X is not an antichain then*

$$|\operatorname{End}\mathcal{L}| = \left\lfloor \tfrac{1}{2}(\mathrm{h(L)} - 1) \right\rfloor + 1.$$

If the loop is an antichain and $i_f = \min\{k \in \mathbb{N} \mid k$ is odd and $f^k(L) \subseteq Z(L)\}$, then

$$|\operatorname{End}\mathcal{L}| = \begin{cases} \left\lfloor \tfrac{1}{2}(h(L) - 1) \right\rfloor + 1 & \text{if } i_f \geqslant h(L); \\ \left\lceil \tfrac{1}{2}(h(L) - 1) \right\rceil + \left\lfloor \tfrac{1}{2}(h(L) - 1 - i_f) \right\rfloor + 2 & \text{if } i_f < h(L). \end{cases}$$

It was first shown by Berman (1977) that every class $\mathbf{K}_{p,q}$ has only finitely many subdirectly irreducible algebras all of which are finite. For a general Urquhart class $\mathbf{P}_{m,n}$ with $m > n \geqslant 0$, we have that $\mathbf{P}_{m,n} \subseteq \mathbf{P}_{2m-n,n} = \mathbf{K}_{m-n,n}$ and a natural problem that arises is to determine those Urquhart classes that contain a *unique* subdirectly irreducible algebra. For this purpose, we recall that the dual space $(X; g)$ of a finite subdirectly irreducible algebra $(L; f)$ in $\mathbf{P}_{m,n}$ consists of a tail of order n and a loop of order $m - n$. Concerning this loop we require the following observations,

A *generalised crown* is an ordered set $C_{n.k}$ of cardinality $2n$ which is connected, is of height 1, and is such that all vertices of $C_{n.k}$ have the same degree k where $1 \leqslant k \leqslant n$. Their significance is exhibited in the following result.

Theorem 14.16 (Blyth and Varlet (1996)) *Every connected component of the dual space of a finite simple Ockham algebra is either a generalised crown or a singleton.*

Using this result, we may characterise the loop X° of $(X; g)$ as follows.

Theorem 14.17 *If the loop X° of $(X; g)$ is*
(1) *of odd order then it is an antichain;*
(2) *of even order then it is an antichain or a disjoint union of generalised crowns.*

Proof (1) The property is trivial for a loop of order 1. Suppose that X° is of order $2n + 1$ where $n \geqslant 1$, say $X^\circ = \{p_0, p_1, \ldots, p_{2n}\}$. If X° is not an antichain then there exist $p_i, p_j \in X^\circ$ such that $p_i \not\parallel p_j$. Since g acts cyclically on X°, there then exists $p_k \in X$ such that $p_0 \not\parallel p_k$. Suppose that $p_0 < p_k$. Then, since g^{2n+1} is antitone we have the contradiction $p_0 = g^{2n+1}(p_0) > g^{2n+1}(p_k) = p_k$. A similar contradiction obtains when $p_0 > p_k$. Consequently, X° is an antichain.

(2) Since g acts cyclically on X°, it follows by Theorem 14.5(2) that the corresponding dual algebra is simple. If X° is of even order then the result follows from Theorem 14.16. □

As a consequence of the above observations, we can determine the Urquhart classes $\mathbf{P}_{m,n}$ that contain a unique subdirectly irreducible algebra.

Theorem 14.18 $\mathbf{P}_{m,n}$ *properly contains a unique subdirectly irreducible algebra if and only if either $m = 1$ and $n = 0$, or $m - n$ is odd and greater than* 1.

Proof Let $\mathcal{X} = (X; g)$ be the dual space of a subdirectly irreducible Ockham algebra $\mathcal{L} = (L; f)$ that properly belongs to the Urquhart class $\mathbf{P}_{m,n}$.

$\Rightarrow$: Suppose that $\mathcal{L}$ is the only subdirectly irreducible algebra that properly belongs to $\mathbf{P}_{m,n}$. If $m - n$ is even then by Theorem 14.17(2) the loop X° of X is either an antichain, in which case so is X by the action of g, or is a disjoint union of generalised crowns, in which case the tail can have several different orders. More than one possibility therefore exists for the order on X and hence for $\mathcal{L}$. Thus, by the hypothesis, $m - n$ must be odd. Moreover, we must have $m - n \neq 1$ since otherwise the loop X° would consist of a g-fixed point and again the tail can have several different orders.

$\Leftarrow$: Conversely, suppose that the conditions hold. The case $m = 1, n = 0$ is clear. If now $m - n$ is odd and greater than 1 then by Theorem 14.17(1) the loop X° is an antichain. Clearly, by the action of g, this forces the tail also to be an antichain. Consequently, $\mathcal{L}$ is uniquely determined. □

Examples of the Urquhart classes in question are therefore $\mathbf{P}_{2n+1,0}$ $(n \geqslant 1)$ and $\mathbf{P}_{2n+k,1+k}$ $(n \geqslant 2,\ k \geqslant 0)$.

The Urquhart classes that are not Berman classes are those $\mathbf{P}_{m,n}$ for which $m - n$ is odd. To count the subdirectly irreducibles in these classes it suffices, in view of the above, to consider also the case where $m - n = 1$. Now the dual space of a subdirectly irreducible algebra in $\mathbf{P}_{k+1,k}$ with $k \geqslant 1$ is, ignoring the order, represented by

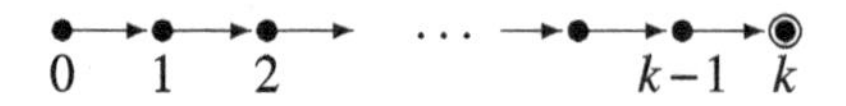

in which the arrows indicate the action of g and k is a g-fixed point.

A *valid order* on this is any order that produces a subdirectly irreducible algebra. Clearly, equality is a valid order, the resulting subdirectly irreducible algebra having a Boolean lattice reduct. In general, however, the problem of counting the subdirectly irreducible algebras that belong to the Urquhart class $\mathbf{P}_{k+1,k}$ remains open and appears to be a daunting task.

An Ockham algebra $\mathcal{L} = (L; f)$ is said to be *of Boolean shape* if its lattice reduct L is Boolean and f is not complementation. Such Ockham algebras, and in particular those that are subdirectly irreducible, are of particular interest in that they have a natural construction from any given monoid. As appropriate to the present discussion, the salient details of this are the following (see Blyth et al. 2001a).

Given a monoid M, let $\mathbf{2}^M$ denote its power set and, for each $c \in M$, consider the endomorphism $\varphi_c : \mathbf{2}^M \to \mathbf{2}^M$ given by the prescription

$$\varphi_c(A) = \{x \in M \mid xc \in A\}.$$

If σ denotes complementation in $\mathbf{2}^M$, define $f_c : \mathbf{2}^M \to \mathbf{2}^M$ by $f_c = \varphi_c\sigma = \sigma\varphi_c$. Then, we obtain an Ockham algebra $\mathcal{L}_c^M = (\mathbf{2}^M; f_c)$ whose lattice reduct is Boolean.

Immediate properties of the above mappings are

(α) $(\forall a, b \in M)\ \ a \neq b \implies \varphi_a \neq \varphi_b$;

(β) $(\forall a, b \in M)\ \ f_a\varphi_b = \varphi_a f_b = f_{ab}$ and $f_a f_b = \varphi_a\varphi_b = \varphi_{ab}$.

Theorem 14.19 *The Ockham algebra* $\mathcal{L}_c^M$

(1) *is of Boolean shape if and only if* $c \neq 1$;

(2) *belongs to the Urquhart class* $\mathbf{P}_{m,n}$ *if and only if* $c^m = c^n$.

Proof (1) Clearly, f_c coincides with the complementation σ if and only if φ_c coincides with the identity φ_1, and by (α) this holds if and only if $c = 1$. Thus $\mathcal{L}_c^M$ is of Boolean shape if and only if $c \neq 1$.

(2) There are two cases to consider:

$m - n$ even: By Theorem 14.7, $\mathcal{L}_c^M \in \mathbf{P}_{m,n}$ if and only if $f_c^m = f_c^n$. The result in this case then follows from the fact that

$$\begin{aligned} f_c^m = f_c^n &\iff \sigma^m f_c^m = \sigma^n f_c^n \\ &\iff (\sigma f_c)^m = (\sigma f_c)^n \\ &\iff \varphi_c^m = \varphi_c^n \\ &\iff \varphi_{c^m} = \varphi_{c^n} \quad \text{by } (\beta) \\ &\iff c^m = c^n \quad \text{by } (\alpha). \end{aligned}$$

$m - n$ odd: Here, we may assume that m is even and n is odd. In this case, again by Theorem 14.7, $\mathcal{L}_c^M \in \mathbf{P}_{m,n}$ if and only if, for every $a \in \mathcal{L}_c^M$, the elements $f_c^m(a)$ and $f_c^n(a)$ are complementary. But, under the assumption, it is clear that $f_c^m = \sigma f_c^n$ if and only if $\sigma^m f_c^m = \sigma^n f_c^n$, and precisely as above, this is equivalent to $c^m = c^n$. $\square$

The question of when, for a given monoid M and $c \in M$, the Ockham algebra $\mathcal{L}_c^M$ is subdirectly irreducible is settled as follows. In the case where c is the identity element of M it follows by Theorem 14.19(1) that f_c is the complementation, in which case subdirect irreducibility forces $\mathcal{L}_c^M$ to be the simple Boolean algebra **2** and then M is the trivial monoid. In contrast, the following obtains:

Theorem 14.20 (Blyth et al. (2001a)) *If M is a monoid and $c \in M$ with $c \neq 1_M$ then $\mathcal{L}_c^M$ is*

(1) *simple if and only if M is a finite cyclic group generated by c;*
(2) *subdirectly irreducible and not simple if and only if $M = \langle c \rangle \uplus \{1_M\}$.*

The above results served as a natural catalyst in the search for a first example of a (necessarily infinite) subdirectly irreducible Ockham algebra that does not belong to the generalised variety $\mathbf{K}_\omega$, the satisfactory outcome being the following.

Consider the monoid $(\mathbb{N}; +)$ of natural numbers and the corresponding Ockham algebra $\mathcal{L}_1^{\mathbb{N}}$ which, by Theorem 14.19(1), is of Boolean shape and, by Theorem 14.20(2), is subdirectly irreducible and not simple. We show as follows that $\mathcal{L}_1^{\mathbb{N}} \notin \mathbf{K}_\omega$.

Suppose, by way of obtaining a contradiction, that $\mathcal{L}_1^{\mathbb{N}} \in \mathbf{K}_\omega$. Then for the element $D = \{x^2 \mid x \in \mathbb{N}\}$ of $\mathcal{L}_1^{\mathbb{N}}$ there exist $m \geqslant 1$ and $n \geqslant 0$ such that $f_1^{m+n}(D) = f_1^n(D)$. Let $k, q \in \mathbb{N}$ be such that $k = qm \geqslant 1 + \sqrt{2n}$. Then

$$f_1^{2k+2n}(D) = f_1^{2(qm+n)}(D) = f_1^{2n}(D),$$

$$\text{i.e. } \{x \in \mathbb{N} \mid x + 2k + 2n \in D\} = \{x \in \mathbb{N} \mid x + 2n \in D\}.$$

Now $(k-1)^2 \geqslant 2n$ with $[(k-1)^2 - 2n] + 2n = (k-1)^2 \in D$ and so we have that

$$k^2 + 1 = [(k-1)^2 - 2n] + 2n + 2k \in D.$$

It follows that there exists $t \in \mathbb{N}$ such that $k^2 + 1 = t^2$ whence $t - k = 1 = t + k$. Since $k \geqslant 1$ we see that $1 = t + k \geqslant t + 1$ and so we must have $t = 0$ whence there follows the contradiction $k^2 + 1 = 0$. Consequently, $\mathcal{L}_1^{\mathbb{N}} \notin \mathbf{K}_\omega$.

More recent investigations into properties of endomorphisms of Ockham algebras have centred on the following general notion.

If $\mathcal{A}$ is an algebra then $\vartheta \in \operatorname{Con} \mathcal{A}$ is an *endomorphism kernel* if there exists $h \in \operatorname{End} \mathcal{A}$ such that $\vartheta = \operatorname{Ker} h$. We say that $\mathcal{A}$ has the *endomorphism kernel property* if every $\vartheta \in \operatorname{Con} \mathcal{A}$ different from the universal congruence $\iota_{\mathcal{A}}$ is an endomorphism kernel.

An alternative characterisation of the above property is given by

Theorem 14.21 (Blyth et al. (2004)) *An algebra $\mathcal{A}$ has the endomorphism kernel property if and only if every non-trivial epimorphic image of $\mathcal{A}$ is isomorphic to a subalgebra of $\mathcal{A}$.*

Theorem 14.22 *If $\mathcal{A}$ is an algebra with the endomorphism kernel property then any two finite simple epimorphic images of $\mathcal{A}$ are isomorphic.*

Proof Let $\mathcal{A}_1$ and $\mathcal{A}_2$ be finite simple epimorphic images of $\mathcal{A}$. Then, there exist $\vartheta, \varphi \in \mathrm{Con}\,\mathcal{A}$ such that $\vartheta, \varphi \neq \iota_{\mathcal{A}}$ with $\mathcal{A}_1 \simeq \mathcal{A}/\vartheta$ and $\mathcal{A}_2 \simeq \mathcal{A}/\varphi$. By Theorem 14.21, there is an injective morphism $\alpha : \mathcal{A}_1 \to \mathcal{A}$. Consider the natural epimorphism $\natural_\varphi : \mathcal{A} \to \mathcal{A}/\varphi$ and let

$$h = \natural_\varphi \circ \alpha : \mathcal{A}_1 \to \mathcal{A}/\varphi \simeq \mathcal{A}_2.$$

Since $\mathcal{A}_1$ is simple, h is injective and consequently there exists an injective morphism $\psi_1 : \mathcal{A}_1 \to \mathcal{A}_2$. Reversing the roles of $\mathcal{A}_1$ and $\mathcal{A}_2$, we obtain an injective morphism $\psi_2 : \mathcal{A}_2 \to \mathcal{A}_1$ and the conclusion follows from the assumption that both $\mathcal{A}_1$ and $\mathcal{A}_2$ are finite. □

Let $\mathcal{A}$ be an algebra and let $\vartheta \in \mathrm{Con}\,\mathcal{A}$. For $\gamma \in \mathrm{Con}\,\mathcal{A}$ such that $\vartheta \subseteq \gamma$ consider the congruence

$$\gamma/\vartheta = \big\{\big([a]_\vartheta, [b]_\vartheta\big) \mid (a, b) \in \gamma\big\}.$$

If $\vartheta \subseteq \gamma_1, \gamma_2$ and $\gamma_1 \circ \gamma_2 = \gamma_2 \circ \gamma_1$ then $(\gamma_1/\vartheta) \circ (\gamma_2/\vartheta) = (\gamma_2/\vartheta) \circ (\gamma_1/\vartheta)$. An immediate consequence of this, via the Correspondence Theorem, is that if $\mathcal{A}$ is an algebra that is congruence-permutable then, for every $\vartheta \in \mathrm{Con}\,\mathcal{A}$, so is the quotient algebra $\mathcal{A}/\vartheta$.

Theorem 14.23 *Let $\mathcal{A}$ be a congruence-permutable algebra such that the lattice* $\mathrm{Con}\,\mathcal{A}$ *is complemented. Then*

(1) *$\mathcal{A}$ is directly indecomposable if and only if it is simple;*

(2) *for every $\vartheta \in \mathrm{Con}\,\mathcal{A}$ the lattice $\mathrm{Con}\,(\mathcal{A}/\vartheta)$ is complemented.*

Proof (1) If $\mathcal{A}$ is directly indecomposable suppose, by way of obtaining a contradiction, that there exists $\vartheta \in \mathrm{Con}\,\mathcal{A}$ such that $\vartheta \notin \{\omega, \iota\}$. Then, $\mathrm{Con}\,\mathcal{A}$ being complemented, there exists $\vartheta^\star \in \mathrm{Con}\,\mathcal{A}$ such that $\vartheta \cap \vartheta^\star = \omega$ and $\vartheta \vee \vartheta^\star = \iota$. Since $\mathcal{A}$ is congruence-permutable, it follows that $\vartheta, \vartheta^\star$ is a non-trivial pair of factor congruences, contradicting the hypothesis that $\mathcal{A}$ is directly indecomposable. Hence $\mathrm{Con}\,\mathcal{A} = \{\omega, \iota\}$ and so $\mathcal{A}$ is simple. The converse is clear.

(2) The Correspondence Theorem gives $\mathrm{Con}\,(\mathcal{A}/\vartheta) \simeq [\vartheta, \iota]$ for every $\vartheta \in \mathrm{Con}\,\mathcal{A}$. We show as follows that the lattice $[\vartheta, \iota]$ is complemented. Let $\gamma \in \mathrm{Con}\,\mathcal{A}$ be such that $\vartheta \subseteq \gamma$. Since $\mathrm{Con}\,\mathcal{A}$ is complemented, there exists $\gamma^\star \in \mathrm{Con}\,\mathcal{A}$ such that $\gamma \cap \gamma^\star = \omega$ and $\gamma \vee \gamma^\star = \iota$. By a well-known result of Burris and Sankappanavar (1981, Theorem 5.10), the lattice $\mathrm{Con}\,\mathcal{A}$ is modular. Consequently,

$$\gamma \cap (\vartheta \vee \gamma^\star) = \vartheta \vee (\gamma \cap \gamma^\star) = \vartheta \vee \omega = \vartheta.$$

On the other hand, $\gamma \vee (\vartheta \vee \gamma^\star) = \vartheta \vee \iota = \iota$. Hence, $\vartheta \vee \gamma^\star$ is a complement of γ in $[\vartheta, \iota]$. □

When the algebra $\mathcal{A}$ is finite and has the endomorphism kernel property, the above observations lead to the following description of the structure of $\mathcal{A}$.

Theorem 14.24 *Let $\mathcal{A}$ be a finite congruence-permutable algebra such that the lattice* Con $\mathcal{A}$ *is complemented. If $\mathcal{A}$ has the endomorphism kernel property then $\mathcal{A}$ is isomorphic to a finite direct power of a finite simple algebra.*

Proof We use the well-known fact Burris and Sankappanavar (1981, Theorem 7.10) that we may write $\mathcal{A} = \times_{i=1}^{n} \mathcal{B}_i$ where each $\mathcal{B}_i$ is directly indecomposable.

Since $\mathcal{A}$ is finite by hypothesis, so also is every $\mathcal{B}_i$. Using Theorem 14.23(2) we see that each $\mathcal{B}_i$ is congruence-permutable with Con $\mathcal{B}_i$ a complemented lattice. It then follows by Theorem 14.23(1) that each $\mathcal{B}_i$ is a finite simple algebra. Invoking now Theorem 14.22, we see that $\mathcal{B}_i \simeq \mathcal{B}_1$ for each i, whence $\mathcal{A}$ is isomorphic to the direct power $\mathcal{B}_1^n$. □

We recall that an algebra $\mathcal{A}$ is said to have *factorisable congruences* if, whenever $\mathcal{A} = \times_{i=1}^{n} \mathcal{A}_i$, every $\vartheta \in \text{Con}\,\mathcal{A}$ is of the form $\times_{i=1}^{n} \vartheta_i$ with $\vartheta_i \in \text{Con}\,\mathcal{A}_i$ for each i. A variety $\mathcal{V}$ is said to have factorisable congruences if every member of $\mathcal{V}$ has factorisable congruences. In this situation, the following result is apposite.

Theorem 14.25 (Blyth et al. (2004)) *Let $\mathcal{V}$ be a variety that has factorisable congruences and let $\mathcal{A}_1, \ldots, \mathcal{A}_n$ be algebras in $\mathcal{V}$ such that* Mor $(\mathcal{A}_i, \mathcal{A}_j) \neq \emptyset$ *for $i \neq j$. Then if each $\mathcal{A}_i$ has the endomorphism kernel property so also does $\times_{i=1}^{n} \mathcal{A}_i$.*

The above general results may be applied, in the variety **O** of Ockham algebras, to algebras in $\mathbf{K}_{p,0} = \mathbf{P}_{2p,0}$ to obtain the following.

Theorem 14.26 *For a finite algebra $\mathcal{L} \in \mathbf{K}_{p,0}$ the following statements are equivalent:*

(1) *$\mathcal{L}$ is congruence-permutable and has the endomorphism kernel property;*

(2) *$\mathcal{L}$ is isomorphic to a finite direct power of a finite simple Ockham algebra.*

Proof (1) $\Rightarrow$ (2): Since by hypothesis $\mathcal{L} \in \mathbf{K}_{p,0}$ is finite, it follows by Blyth and Varlet (1994, Theorem 2.17) that Con $\mathcal{L}$ is a Boolean lattice. Consequently, if (1) holds then so does (2) by Theorem 14.24.

(2) $\Rightarrow$ (1): If (2) holds then since **O** has factorisable congruences it is readily seen that $\mathcal{L}$ is congruence-permutable. That $\mathcal{L}$ has the endomorphism kernel property is now a consequence of Theorem 14.25. □

Restricting our attention again to the class **O** of Ockham algebras, we now consider the problem of precisely when a finite subdirectly irreducible Ockham algebra has the endomorphism kernel property. This is settled as follows.

Theorem 14.27 *Let $\mathcal{L} = (L; f)$ be a finite subdirectly irreducible Ockham algebra and suppose that the smallest Berman class to which it belongs is $\mathbf{K}_{p,q}$.*

(1) *If $q = 0$ then $\mathcal{L}$ has the endomorphism kernel property;*

(2) *If $q \neq 0$ then $\mathcal{L}$ has the endomorphism kernel property if and only if either $q = 1$ or $f(L) \subseteq Z(L)$.*

Proof (1) If $q = 0$ then it follows from Theorem 14.10 that $\mathcal{L}$ is simple and therefore has the endomorphism kernel property.

(2) Suppose now that $q \neq 0$.

$\Rightarrow$: Suppose that $\mathcal{L}$ has the endomorphism kernel property. Then, since $\Phi_1 \neq \iota$, there exists $h \in \operatorname{End}\mathcal{L}$ such that $\Phi_1 = \operatorname{Ker h}$. By Theorem 14.14 and the definition of $\nabla(\mathcal{L})$, there are two cases to consider.

(*a*) *There exists an even* $r \in \mathbb{N}$ *such that* $h = f^r$.

In this case, we have $\Phi_1 = \operatorname{Ker h} = \Phi_{\mathrm{r}}$. Suppose, by way of obtaining a contradiction, that $q > 1$. Then, by Theorem 14.10, we have that $\Phi_0 \prec \Phi_1 \prec \Phi_2$. But since $r \geqslant 1$ and r is even, there arises the contradiction $\Phi_r = \Phi_1 \prec \Phi_2 \subseteq \Phi_r$.

(*b*) *There exists an odd* $k \in \mathbb{N}$ *such that* $f^k(L) \subseteq Z(L)$ and $h = \sigma f^k$.

In this case, we clearly have $\Phi_1 = \operatorname{Ker h} = \Phi_{\mathrm{k}}$. If now $q > 1$ then Theorem 14.10 gives $\Phi_0 \prec \Phi_1 \prec \Phi_2$. It follows that we must have $k = 1$ and therefore $f(L) \subseteq Z(L)$.

$\Leftarrow$: Suppose that $q = 1$. By Theorem 14.10, we have $\operatorname{Con}\mathcal{L} = \{\Phi_0, \Phi_1, \iota\}$. We deduce from this that $\Phi_1 = \Phi_2 = \operatorname{Ker f^2}$, whence $\mathcal{L}$ has the endomorphism kernel property.

Suppose that $f(L) \subseteq Z(L)$. For every $m \in \mathbb{N}$ we have $f^{2m+1}(L) \subseteq f(L) \subseteq Z(L)$. It follows from this that σf^{2m+1} is an endomorphism of $\mathcal{L}$ for every $m \in \mathbb{N}$. It now follows by Theorem 14.10 that $\mathcal{L}$ has the endomorphism kernel property. $\square$

Let $\mathcal{L} = (L; f)$ be an Ockham algebra with dual space $\mathcal{X} = (X; g)$. If $\alpha \in \operatorname{End}\mathcal{X}$ then by Theorem 14.3 the congruence $\vartheta_{\alpha(X)}$ is given by

$$\begin{aligned}(A, B) \in \vartheta_{\alpha(X)} &\iff A \cap \alpha(X) = B \cap \alpha(X)\\ &\iff \alpha^{-1}(A) = \alpha^{-1}(B),\end{aligned}$$

and consequently, by Theorem 14.11, $\vartheta_{\alpha(X)} = \operatorname{Ker}\widetilde{\alpha}$.

It follows that $\mathcal{L}$ has the endomorphism kernel property if and only if, for every $\emptyset \subset Q \in C_g(X)$, there exists $\alpha \in \operatorname{End}\mathcal{X}$ such that $\alpha(X) = Q$.

An important consequence of this is the following, in which a g-*cycle* is defined to be a g-subset of the form $Q = \{a, g(a), \ldots, g^{k-1}(a)\}$ with $g^k(a) = a$ and $|Q| = k$.

Theorem 14.28 (Blyth et al. (2004)) *Let* $\mathcal{L} = (L; f)$ *be an Ockham algebra with the endomorphism kernel property and let* P, Q *be* g*-cycles in the dual space* $\mathcal{X} = (X; g)$. *Then the Ockham spaces* $(P; g)$ *and* $(Q; g)$ *are isomorphic.*

With these observations to hand, we now concentrate on when the Ockham algebra $\mathcal{L} = (L; f)$ belongs to the variety $\mathbf{M} = \mathbf{K}_{1,0}$ of de Morgan algebras, and to its subclass $\mathbf{K}$ of Kleene algebras.

It follows from Theorem 14.28 that if $\mathcal{L}$ is a de Morgan algebra that enjoys the endomorphism kernel property then in the dual space $\mathcal{X}$ all g-cycles are equipotent. Consequently either every g-cycle consists of a g-fixed point (in which case $g = \mathrm{id}_X$ and $\mathcal{L}$ is Boolean), or every g-cycle is of the form $\{a, g(a)\}$. In the latter case we say that $\{a, g(a)\}$ is *connected* if $a \not\parallel g(a)$, and *disconnected* if $a \parallel g(a)$ and, again by Theorem 14.28, in the dual space $\mathcal{X}$ every g-cycle is connected or every g-cycle is disconnected.

Our main purpose now is to describe the structure of the finite non-Boolean de Morgan algebras (in particular, Kleene algebras) in which the endomorphism kernel property holds. Here we can only give a sketch whilst referring the reader to Blyth et al. (2004) for the complete details.

For this purpose, we recall that a *complete bipartite set* is an ordered set $K_{m,n}$ of height 1 having m maximal elements and n minimal elements, every minimal element being less than every maximal element. In the case where $m = n$ we say that such a set is *regular.*

In the regular complete bipartite set $K_{n,n}$ let

$$\max K_{n,n} = \{a_1, \dots, a_n\} \quad \text{and} \quad \min K_{n,n} = \{b_1, \dots, b_n\}.$$

By a *gate* we shall mean an ordered set that is obtained from $K_{n,n}$ by the adjunction of finite chains $I_1, \dots, I_n$ and $J_1, \dots, J_n$ such that $\min I_k = a_k$, $\max J_k = b_k$, $|I_k| = |J_k|$ for each k, with $I_r \parallel I_s$ and $J_r \parallel J_s$ for $r \neq s$.

By way of illustration, the following is a gate constructed from $K_{3,3}$:

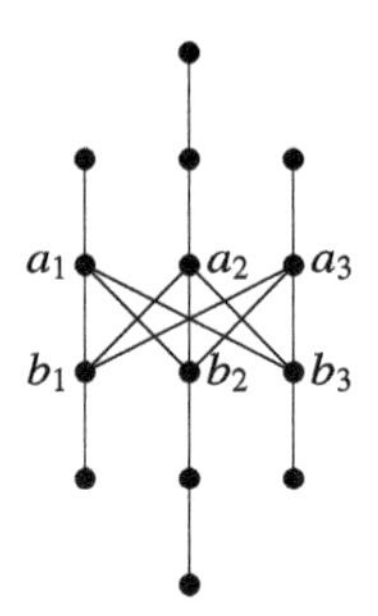

Taking g to be vertical reflection on this, it is readily seen that the corresponding Ockham algebra is the Kleene algebra

$$(L;\, f) \simeq \big((\mathbf{3}^2 \times \mathbf{4})\, \overline{\oplus}\, (\mathbf{3}^2 \times \mathbf{4});\, \Updownarrow\big)$$

where $\Updownarrow$ denotes the induced vertical reflection on the direct sum lattice.

That this algebra has the endomorphism kernel property is a particular case of the following result which highlights the notion of a gate.

Theorem 14.29 (Blyth et al. (2004)) *Let $\mathcal{L} = (L;\, f)$ be a finite de Morgan algebra that is not Boolean. Suppose that the dual space $\mathcal{X} = (X;\, g)$ of $\mathcal{L}$ is order connected. Then $\mathcal{L}$ has the endomorphism kernel property if and only if X is a gate.*

Let $\mathcal{L} = (L;\, f)$ be a finite de Morgan algebra with dual space $\mathcal{X} = (X;\, g)$ and let H be an order component of X. Then since $g^2 = \mathrm{id}_X$ it follows that $g(H)$ is also an order component of X, so that either $g(H) = H$ or $g(H) \parallel H$. In general, therefore, X can be expressed as the disjoint union of the g-sets $H_i \cup g(H_i)$. A deep analysis of these g-sets produces the following remarkable description of the structure of those finite de Morgan algebras that enjoy the endomorphism kernel property.

Theorem 14.30 (Blyth et al. (2004)) *Let $\mathcal{L} = (L; f)$ be a finite de Morgan algebra that is not Boolean and let $\mathcal{X} = (X; g)$ be its dual space.*

(1) *If $\mathcal{L}$ is not a Kleene algebra then the following are equivalent:*

(*a*) *$\mathcal{L}$ has the endomorphism kernel property;*

(*b*) *every order component of X is a chain that is not a g-subset;*

(*c*) $\mathcal{L} \simeq \bigtimes_{i=1}^{m} (C_i \times D_i\ ;\ \updownarrow)$ *in which C_i, D_i are chains with $|C_i| = |D_i|$ and $\updownarrow$ denotes vertical reflection on $C_i \times D_i$.*

(2) *If $\mathcal{L}$ is a Kleene algebra then the following are equivalent:*

(*d*) *$\mathcal{L}$ has the endomorphism kernel property;*

(*e*) *every order component of X is a gate that is a g-subset;*

(*f*) $\mathcal{L} \simeq \bigtimes_{i=1}^{m} \big((C_{i,1} \times \cdots \times C_{i,n_i})\overline{\oplus}(D_{i,1} \times \cdots \times D_{i,n_i})\ ;\ \Updownarrow\big)$ *in which $C_{i,j}$, $D_{i,j}$ are chains with $|C_{i,j}| = |D_{i,j}|$ and $\Updownarrow$ is the vertical reflection induced by the reflections on the chains $C_{i,j}\overline{\oplus}D_{i,j}$.*

* * * * *

This algebraic oak tree is one of many which, over the years, have risen from the acorns planted by Alasdair Urquhart.

Acknowledgements This work is partially supported by the Portuguese Foundation for Science and Technology, under the project UIDB/00297/2020 (Centro de Matemática e Aplicações). The authors are indebted to Professor Sankappanavar for his very helpful comments which have served to improve the original presentation.

References

Ash, C. J. (1985). Pseudovarieties, generalized varieties and similarly described classes. *Journal of Algebra*, *92*(1), 104–115.

Balbes, R., & Dwinger, P. (1974). *Distributive lattices*. Columbia, MO: University of Missouri Press.

Beazer, R. (1984a). Injectives in some small varieties of Ockham algebras. *Glasgow Mathematical Journal*, *25*(2), 183–191.

Beazer, R. (1984b). On some small varieties of distributive Ockham algebras. *Glasgow Mathematical Journal*, *25*(2), 175–181.

Berman, J. (1977). Distributive lattices with an additional unary operation. *Aequationes Mathematicae*, *16*(1–2), 165–171.

Bialynicki-Birula, A., Rasiowa, H. (1957). On the representation of quasi-Boolean algebras. *Bull. Acad. Polon. Sci. Cl. III*, *5*, 259–261, XXII.

Birkhoff, G. (1933). On the combination of subalgebras. *Proceedings of the Cambridge Philosophical Society*, *29*(4), 441–464.

Blyth, T. S., Fang, J., & Silva, H. J. (2004). The endomorphism kernel property in finite distributive lattices and De Morgan algebras. *Communications in Algebra, 32*(6), 2225–2242.
Blyth, T. S., Fang, J., & Varlet, J. C. (1991). Subdirectly irreducible Ockham algebras. *Contributions to general algebra, 7 Vienna (1990)* (pp. 37–48). Vienna: Hölder-Pichler-Tempsky.
Blyth, T. S., & Silva, H. J. (1998/1999). Singular antitone systems. *Order, 15*(3), 261–270.
Blyth, T. S., Silva, H. J., & Varlet, J. C. (2001a). Ockham algebras arising from monoids. *Algebra Colloquium, 8*(3), 315–326.
Blyth, T. S., Silva, H. J., & Varlet, J. C. (2001b). On the endomorphism monoid of a finite subdirectly irreducible Ockham algebra. *Unsolved problems on mathematics for the 21st century* (pp. 149–157). Amsterdam: IOS.
Blyth, T. S., & Varlet, J. C. (1983a). On a common abstraction of De Morgan algebras and Stone algebras. *Proceedings of the Royal Society of Edinburgh Section A: Mathematics, 94*(3–4), 301–308.
Blyth, T. S., & Varlet, J. C. (1983b). Subvarieties of the class of MS-algebras. *Proceedings of the Royal Society of Edinburgh Section A: Mathematics, 95*(1–2), 157–169.
Blyth, T. S., & Varlet, J. C. (1994). *Ockham algebras*. New York: The Clarendon Press, Oxford University Press, Oxford Science Publications.
Blyth, T. S., & Varlet, J. C. (1996). The dual space of a finite simple Ockham algebra. *Studia Logica, 56*(1–2), 3–21. Special issue on Priestley duality.
Burris, S., & Sankappanavar, H. P. (1981). *A course in universal algebra*, volume 78 of *Graduate Texts in Mathematics*. New York-Berlin: Springer.
Davey, B. A. (1979). On the lattice of subvarieties. *Houston Journal of Mathematics, 5*(2), 183–192.
Davey, B. A., & Priestley, H. A. (2002). *Introduction to lattices and order* (2nd ed.). New York: Cambridge University Press.
Fang, J. (1992). *Contributions to the theory of Ockham algebras*. Ph.D. Thesis, University of St. Andrews (United Kingdom), ProQuest LLC, Ann Arbor, MI.
Fang, J. (2006). Ockham algebras with double pseudocomplementation. *Algebra Universalis, 55*(2–3), 277–292. Special issue dedicated to Walter Taylor.
Fang, J. (2011). *Distributive lattices with unary operations*. Beijing: Science Press.
Goldberg, M. S. (1981). Distributive Ockham algebras: Free algebras and injectivity. *Bulletin of the Australian Mathematical Society, 24*(2), 161–203.
Goldberg, M. S. (1983). Topological duality for distributive Ockham algebras. *Studia Logica, 42*(1), 23–31.
Grätzer, G. (2011). *Lattice Theory: Foundation*. Basel: Birkhäuser/Springer Basel AG.
Jónsson, B. (1972). *Topics in universal algebra*. Lecture Notes in Mathematics (Vol. 250). Berlin-New York: Springer.
Kalman, J. A. (1958). Lattices with involution. *Transactions of the American Mathematical Society, 87*, 485–491.
Moisil, G. C. (1935). Recherches sur l'algèbre de la logique. *Annales scientifiques de l'Universitè de Jassy, 22*, 1–117.
Priestley, H. A. (1972). Ordered topological spaces and the representation of distributive lattices. *Proceedings of the London Mathematical Society, 3*(24), 507–530.
Rasiowa, H. (1974). *An algebraic approach to non-classical logics*. Studies in Logic and the Foundations of Mathematics (Vol. 78). Amsterdam-London: North-Holland Publishing Co.; New York: American Elsevier Publishing Co., Inc.
Sankappanavar, H. P. (1985). Distributive lattices with a dual endomorphism. *Z. Math. Logik Grundlag. Math., 31*(5), 385–392.
Sankappanavar, H. P. (1987). Semi-De Morgan algebras. *The Journal of Symbolic Logic, 52*(3), 712–724.
Sankappanavar, H. P. (2012). De Morgan algebras: New perspectives and applications. *Scientiae Mathematicae Japonicae, 75*(1), 21–50.
Sankappanavar, H. P. (2016). A note on regular De Morgan semi-Heyting algebras. *Demonstratio Mathematica, 49*(3), 252–265.

Stone, M. H. (1936). The theory of representations for Boolean algebras. *Transactions of the American Mathematical Society*, *40*(1), 37–111.
Urquhart, A. (1979). Distributive lattices with a dual homomorphic operation. *Studia Logica*, *38*(2), 201–209.
Urquhart, A. (1981). Distributive lattices with a dual homomorphic operation II. *Studia Logica*, *40*(4), 391–404.

Chapter 15
Temporal Logic of Minkowski Spacetime

Robin Hirsch and Brett McLean

Second Reader
J. Madarász
Alfréd Rényi Institute of Mathematics

Abstract We present the proof that the temporal logic of two-dimensional Minkowski spacetime is decidable, PSPACE-complete. The proof is based on a type of two-dimensional *mosaic*. Then, we present the modification of the proof so as to work for slower-than-light signals. Finally, a subframe of the slower-than-light Minkowski frame is used to prove the new result that the temporal logic of real intervals with *during* as the accessibility relation is also PSPACE-complete.

Keywords Temporal logic · Minkowski spacetime · Decidable · Filtration

15.1 Introduction

The theory of relativity raises several interesting issues for temporal reasoning. To mention a few: in special relativity, there can be no notion of absolute simultaneity and the separation of time from space adopted by conventional knowledge representation systems has to be rejected; in general, relativity fundamental theorems of computation become false, for example, the halting problem may be solvable when observers cross the event horizon of a black hole (Malament and Hogarth 1994).

R. Hirsch (✉)
Department of Computer Science, University College, London, UK
e-mail: r.hirsch@ucl.ac.uk

B. McLean
Laboratoire J. A. Dieudonné UMR CNRS 7351, Université Nice Sophia Antipolis, Nice, France
e-mail: brett.mclean@unice.fr

I. Düntsch and E. Mares (eds.), *Alasdair Urquhart on Nonclassical and Algebraic Logic and Complexity of Proofs*, Outstanding Contributions to Logic 22,
https://doi.org/10.1007/978-3-030-71430-7_15

Logical, axiomatic treatments of relativity have been developed, in the main part based on first-order, or in some cases second-order, logic; see Andréka et al. (2007) for references. However, a fundamental aspect of relativity theory is the abandonment of an absolute spacetime frame, and the requirement that all measurements and observations should be taken relative to a localised viewpoint or frame of reference. A modal treatment of relativity would therefore be more natural (in this respect) than a first-order one. A second connection between relativity theory and modal logic is to consider the points of spacetime as the worlds of a Kripke frame with an accessibility defined by the possibility of sending a signal from one spacetime point to another. This latter connection is the topic of this chapter.

According to relativity theory, it is impossible to send signals faster than the speed of light. However, the notion of a signal travelling at the speed of light is invariant under changes of reference frame. In contrast, travelling at 'half the speed of light', say, is not invariant—the most precise frame-invariant statement is that such a signal is travelling slower than lightspeed. So we may consider the points of spacetime as worlds in a Kripke frame where a spacetime point x is accessible from a spacetime point y if it is possible to send a signal from y to x, equivalently, if x lies in the future light cone of y.

The normal Minkowski spacetime considered in special relativity has three spatial dimensions and one time dimension, and if we use light-seconds for units of distance, a point (x, y, z, t) is accessible from (x', y', z', t') if $t \leq t'$ and $(x' - x)^2 + (y' - y)^2 + (z' - z)^2 \leq (t' - t)^2$. Goldblatt showed (Goldblatt 1980) that the *modal* validities over this frame are axiomatised by **S4.2**—reflexivity $\Box p \to p$, transitivity $\Box p \to \Box\Box p$, and confluence $\Diamond\Box p \to \Box\Diamond p$, along with instances of propositional tautologies, using modus ponens and generalisation as inference rules. The modal logic does not change if we vary the number of spatial dimensions, nor does it change if we require that signals are restricted to travel at strictly less than the speed of light, indeed the class of arbitrary reflexive, transitive, confluent frames has the same modal logic. The irreflexive version of the Minkowski spacetime frame has also been axiomatised; the modal logic is **OI.2** (Shapirovski and Shehtman 2002). The *exactly lightspeed* accessibility relation, however, yields an undecidable logic (Shapirovski 2010).

The topic of this chapter is the *temporal* logic of Minkowski spacetime: similar to the unimodal logic, but we use **F** instead of $\Diamond$ for 'sometime in the future' and we have a converse modality **P** to access spacetime points in the past. With the extra expressive power of the temporal language, the temporal validities of Minkowski spacetime now depend on the number of spatial dimensions (Hirsch and Reynolds 2018). We conjecture that the validities of Minkowski spacetime with at least three dimensions are not decidable, but this conjecture remains open. The main case we consider here is the temporal logic of two-dimensional Minkowski spacetime, with a single space dimension and one time dimension. A complete axiomatisation of the temporal validities of 2D Minkowski spacetime is not known. However, we will see that there is an algorithm to determine whether a given temporal formula is valid over 2D Minkowski spacetime or not.

In two dimensions it simplifies things if we change our coordinate system by rotating the axes anticlockwise by 45°, so the point (r, t) with spatial coordinate r and temporal coordinate t is represented as (x, y) where $x = \frac{1}{\sqrt{2}}(t + r)$, $y = \frac{1}{\sqrt{2}}(t - r)$. In this coordinate system, a point (x', y') is accessible from (x, y) if and only if $x \leq x'$ and $y \leq y'$. We write $(x, y) \leq (x', y')$ in this case. We can view this frame as a kind of product frame.

Let $\mathcal{K}_1=(W_1, R_1)$, $\mathcal{K}_2=(W_2, R_2)$ be two Kripke frames. Then the *product frame* $(W_1 \times W_2, R_1^*, R_2^*)$ has two accessibility relations, the first for accessing points horizontally: $(x, y)R_1^*(x', y') \iff (y = y' \wedge x R_1 y)$, and the second vertically: $(x, y)R_2^*(x', y') \iff (x = x' \wedge y R_2 y')$. These products have been studied intensely (Kurucz 2007; Gabbay et al. 2003). The logic of the product frame $(\mathbb{R}, \leq) \times (\mathbb{R}, \leq)$ is undecidable (Reynolds and Zakharyaschev 2001). But in 2D Minkowski spacetime, observers cannot tell whether a light signal arrives from the left or the right, since the map $(x, y) \mapsto (y, x)$ is an automorphism of $\mathbb{R}^2$ equipped with the up-to-speed-of-light accessibility relation. So here we consider a product frame $\mathcal{K}_1 \otimes \mathcal{K}_2$ with worlds $W_1 \times W_2$ and a single accessibility relation R given by $(x, y)R(x', y') \iff (x R_1 x' \wedge y R_2 y')$ (for reflexive relations, this is the composition of R_1^* and R_2^*). The 2D Minkowski frame we looked at above can be written as $(\mathbb{R}, \leq) \otimes (\mathbb{R}, \leq)$. The product of irreflexive frames $(\mathbb{R}, <) \otimes (\mathbb{R}, <)$ is the frame of two-dimensional spacetime points under the accessibility 'can send a signal to a distinct point at strictly less than the speed of light', which we will investigate later. The restriction of this latter frame to $\{(x, y) : x + y > 0\} \subseteq \mathbb{R}^2$ provides a frame for a temporal logic of the strict 'during' relation between intervals with real endpoints (Shapirovski and Shehtman 2002). This will be the third and final frame we will investigate.

Let $(\mathbb{R}, \leq) \otimes (\mathbb{R}, \leq)$ be the 2D frame of Minkowski spacetime we have described. In fact, we will work primarily with the frame M derived from $(\mathbb{R}, \leq) \otimes (\mathbb{R}, \leq)$ by making accessibility irreflexive (which is not the same as the frame $(\mathbb{R}, <) \otimes (\mathbb{R}, <)$). An M-*model* consists of the frame M and a map from propositional letters to subsets of $\mathbb{R}^2$. A *line* in M means a straight line. A line is *light-like* if it has the form $x = c$ or $y = c$ for some constant c. A line is *time-like* if is linearly ordered by accessibility. A line is *space-like* if it has strictly negative slope, equivalently no pair of distinct points from the line is ordered. For points $\mathsf{x} < \mathsf{y} \in \mathsf{M}$, we write $[\mathsf{x}, \mathsf{y}]$ for the closed interval in the poset M bounded by x and y. Geometrically, then, $[\mathsf{x}, \mathsf{y}]$ is a rectangle (modulo degenerate cases).

The problem we face, then, is to determine whether a given propositional temporal formula ϕ is valid over M. Rather than trying to find sound and complete axioms for this logic (which remains an unsolved problem) we construct a decision procedure based on an attempt to construct a certain type of finite description of a model, which we call a boundary map. The bulk of this work is based on Hirsch and Reynolds (2018), Hirsch and McLean (2018). These boundary maps can be thought of as a continuation of the research in *mosaics* pioneered by Németi and his group (Németi 1995, Sect. 4), and extended to the temporal logic of the reals with 'until' and 'since' in Reynolds (2011).

The idea is to start from basic building blocks and to synthesise our models by combining them together in various ways. The building blocks of Németi's mosaics are typically one-dimensional, whereas the ones we use here look like two-dimensional rectangles. We will use these rectangular blocks to produce an algorithm to determine validity of temporal formulas over M. From this, an algorithm for the reflexive frame $(\mathbb{R}, \leq) \otimes (\mathbb{R}, \leq)$ follows immediately, via the reduction given by applying $\mathbf{P}\psi \mapsto \psi \vee \mathbf{P}\psi$ and $\mathbf{F}\psi \mapsto \psi \vee \mathbf{F}\psi$ recursively to subformulas. Then, we will modify the algorithm to work for the frame $(\mathbb{R}, <) \otimes (\mathbb{R}, <)$. A new extension presented here is to modify the procedure used for slower-than-light accessibility to work for the temporal logic of real intervals, where the accessibility relation is 'during'. In each case, we can prove the decidability of the validity problem, and indeed these decision problems are all PSPACE-complete.

15.2 Temporal Filtration

A propositional temporal formula is either a propositional letter $p, q, r, \ldots$ or is built from smaller formulas using either propositional connectives $\neg, \vee$ or unary temporal operators $\mathbf{F}, \mathbf{P}$. We use standard propositional abbreviations $\wedge, \rightarrow$ and temporal abbreviations $\mathbf{G}\phi = \neg\mathbf{F}\neg\phi$, $\mathbf{H}\phi = \neg\mathbf{P}\neg\phi$ ('always in the future'/'always in the past').

Given a temporal formula ϕ, its closure set $\mathsf{Cl}(\phi)$ is the set of all subformulas and negated subformulas of ϕ. The cardinality of $\mathsf{Cl}(\phi)$ is linear in $|\phi|$.

Fixing a ϕ whose validity we want to decide, a *consistent set* is a subset of $\mathsf{Cl}(\phi)$ whose formulas can be (simultaneously) satisfied at some point in some (arbitrary) temporal frame. A *maximal consistent set* (abbreviated MCS) is a consistent set that is maximal with respect to set inclusion, i.e. for some point in a temporal frame it is *all* the formulas of $\mathsf{Cl}(\phi)$ holding there (So, in particular, an MCS includes one but not both of a subformula and its negation, for each subformula of ϕ.). We denote the set of all such MCSs by MCS.

It is well known that satisfiability in a temporal frame is PSPACE-complete (Spaan 1993). Hence, it can be checked in polynomial space whether any given subset of $\mathsf{Cl}(\phi)$ is an MCS. Such a subset can be stored in a number of bits linear in $|\phi|$, by recording its indicator function.

For $m, n \in \mathsf{MCS}$ we let

$$\begin{aligned} m \lesssim n \iff & \forall \mathbf{F}\psi \in \mathsf{Cl}(\phi)\ ((\psi \in n \rightarrow \mathbf{F}\psi \in m) \wedge (\mathbf{F}\psi \in n \rightarrow \mathbf{F}\psi \in m)) \\ & \wedge\ \forall \mathbf{P}\psi \in \mathsf{Cl}(\phi)\ ((\psi \in m \rightarrow \mathbf{P}\psi \in n) \wedge (\mathbf{P}\psi \in m \rightarrow \mathbf{P}\psi \in n)). \end{aligned}$$

The Kripke frame $(\mathsf{MCS}, \lesssim)$ will be finite and transitive. The set of MCSs on which $\lesssim$ is reflexive is partitioned into *clusters*—maximal sets of MCSs in which every pair of MCSs is in $\lesssim$. We extend the notation $\lesssim$ to clusters, thus we may write $m \lesssim c$ for an MCS m and a cluster c if $m \lesssim n$ for some (equivalently all) $n \in c$. For $\psi \in \mathsf{Cl}(\phi)$, we may say that ψ belongs to a cluster c if $\psi \in m$ for some $m \in c$. Observe that it is possible for both p and $\neg p$ to belong to a cluster, though a cluster cannot have both a temporal formula (e.g. $\mathbf{F}\psi$) and its negation.

A *defect* of an MCS m is simply a formula $\mathbf{F}\psi \in m$ (a future defect) or a formula $\mathbf{P}\psi \in m$ (a past defect). A formula $\mathbf{F}\psi$ is a defect of a cluster c if it belongs to c but ψ does not belong to c (with a similar definition for past defects). If $m \lesssim n$, the formula $\mathbf{F}\psi$ is a defect of m, and either $\psi \in n$ or $\mathbf{F}\psi \in n$ then we say that the defect $\mathbf{F}\psi$ of m is *passed up* to n (with a similar definition for clusters).

Any M-model defines a function $h : \mathbb{R}^2 \to \mathsf{MCS}$ recording the formulas in $\mathsf{Cl}(\phi)$ true at each point. Any two distinct vertical lines and two distinct horizontal lines define a closed rectangle in $\mathbb{R}^2$, and restricting h to this rectangle we obtain an instance of what we call a *closed rectangle model* (definition imminent). Similarly, we may define open (or semi-open) rectangle models by omitting all (or some) of the four boundary lines from the domain of h.

Definition 15.1 (*Rectangle model*) A *rectangle model* $h : R \to \mathsf{MCS}$ has a rectangle $R \subseteq \mathbb{R}^2$ (with edges parallel to the coordinate axes) as its domain and satisfies

coherence $\mathsf{x} \le \mathsf{y} \in R \to h(\mathsf{x}) \lesssim h(\mathsf{y})$;

no internal defects if $\mathbf{F}\psi \in h(\mathsf{x})$ then either there exists $\mathsf{y} \in R$ with $\mathsf{y} > \mathsf{x}$ and $\psi \in h(\mathsf{y})$ (no defect) or

- R includes the boundary point y due east of x and $\mathbf{F}\psi \in h(\mathsf{y})$ (defect passed east) or
- R includes the boundary point y due north of x and $\mathbf{F}\psi \in h(\mathsf{y})$ (defect passed north).

and (similarly) where all occurrences of $\mathbf{P}$ defects may be passed south or west.

It is straightforward to prove a 'truth lemma' stating that *open* rectangle models are equivalent to M-models on which a $\psi \in \mathsf{Cl}(\phi)$ holds at a point x precisely when $\psi \in h(\mathsf{x})$.

The plan is to provide a finite description of rectangle models by starting from simple rectangle models where a single cluster holds at all points in the interior of the rectangle. Then, we show how to combine these simple rectangle models in various ways to obtain more complex rectangle models, but still with finite descriptions. Finally, we will obtain a finite description of an open rectangle model. If we manage this, with ϕ mentioned during the construction, then we have a finite description of an M-model for ϕ.

Before continuing, it is worth noting that temporal formulas can describe quite complicated M-models. Consider the purely modal formula (not using $\mathbf{P}/\mathbf{H}$)

$$\phi = \mathbf{F}p_0 \wedge \mathbf{F}p_1 \wedge \mathbf{G} \bigwedge_{i \neq j \neq k < 3} (((\mathbf{F}p_i \wedge \mathbf{F}p_j) \to \mathbf{F}p_k) \wedge (p_i \to \mathbf{G}\neg p_j)).$$

This formula is satisfiable at $(0, 0)$ in an M-model with valuation v if $v(p_0)$, $v(p_1)$, $v(p_2)$ are disjoint subsets of the line $x + y = 1$ with a point in $v(p_i)$ between any pair of points in $v(p_j)$ and $v(p_k)$, where (i, j, k) is any permutation of $(0, 1, 2)$. And all M-models satisfying ϕ look roughly like that, i.e. there is a closed space-like line segment in the future where $v(p_0)$, $v(p_1)$, $v(p_2)$ cover disjoint non-empty subsets of

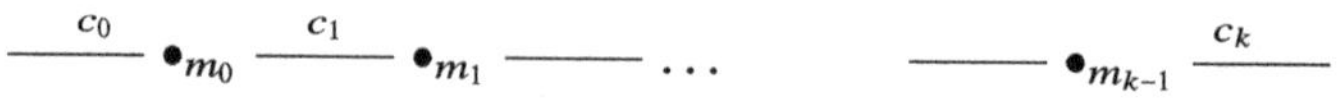

$$\overset{c_0}{\text{———}} \bullet_{m_0} \overset{c_1}{\text{———}} \bullet_{m_1} \text{———} \cdots \qquad \text{———} \bullet_{m_{k-1}} \overset{c_k}{\text{———}}$$

Fig. 15.1 A trace

the segment (they may also cover points off the spatial line), with the density property just described. We can even express that $v(p_0), v(p_1), v(p_2)$ covers the closed line segment (with no gaps) by the formula

$$\psi = \phi \wedge \mathbf{G}(\mathbf{F}p \vee \mathbf{P}p \vee p)$$

where p abbreviates $(p_0 \vee p_1 \vee p_2)$, though this property cannot be expressed by a purely modal formula. In an M-model of ψ, there is a segment of a space-like line l partitioned by $v(p_0), v(p_1), v(p_2)$ in which the convex subsets of $v(p_i)$ are closed subsegments of l for $i < 3$ and, by topological properties of $\mathbb{R}$, uncountably many of these closed segments are singleton sets (single points on the line)—a topological deduction that works in $(\mathbb{R}, \leq) \otimes (\mathbb{R}, \leq)$ but not in $(\mathbb{Q}, \leq) \otimes (\mathbb{Q}, \leq)$. Based on this observation, we can define temporal formulas satisfiable in the latter frame but not the former (Hirsch and Reynolds 2018, Fig. 14).

We have seen that the behaviour of the MCSs occurring along a space-like line can be quite unruly, making them hard to describe finitely. Time-like lines are much easier, since the function h is monotone with respect to $\leq$, so we may describe the MCSs holding at points along such a line by a *trace*: a $\lesssim$-ordered finite sequence $(c_0, m_0, c_1, m_1, \ldots, m_{k-1}, c_k)$ where the c_i are distinct clusters, and the m_i are MCSs (illustrated in Fig. 15.1). The clusters c_0 and c_k are called the *initial* and *final* cluster of the trace, respectively. If t, t' are traces, $m \in \mathsf{MCS}$, where $t \lesssim m \lesssim t'$ (i.e. each cluster/MCS of t is below m, which is below each cluster/MCS of t') we may join them together to form a single trace $t \oplus m \oplus t'$, by concatenation, with the proviso that if the final cluster of t equals the initial cluster of t' they are identified and m is omitted. A formula $\mathbf{F}\psi$ is a defect of a trace t if either it is a defect of the final cluster of t or it is a defect of some other MCS/cluster of t not passed up to the following cluster/MCS of t.

All pertinent information about a cluster c can be stored in a number of bits linear in $|\phi|$, for we only need record the indicator functions of

- a representative MCS $m \in c$ (for making $\lesssim$-comparisons),
- $\{\psi \in \mathsf{Cl}(\phi) \mid \exists m' \in c : \psi \in m'\}$ (to know the defects of c).

The maximal length of a chain of distinct clusters or irreflexive members of MCS is also linear in $|\phi|$ (because walking up such a chain, we permanently gain formulas of the form $\neg\mathbf{F}\psi$ or $\mathbf{P}\psi$ at a linear rate). Hence, any trace can be stored using a quadratic number of bits, and the number of traces is exponential in $|\phi|$.

As indicated, the rectangles we will use have edges parallel to the coordinate axes, so their edges are segments of light-like lines, hence time-like lines (at least while we consider the case where signals may be sent at the speed of light). Thus, we may describe the MCSs occurring along the perimeter of a rectangle R in M by

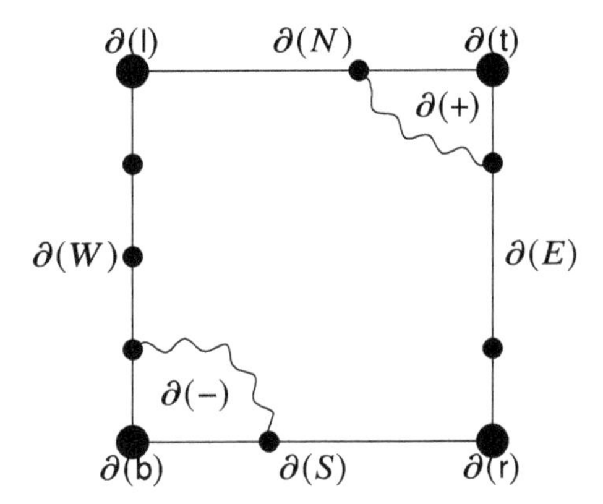

Fig. 15.2 A boundary map

four traces corresponding to the four open line segments bounding the rectangle and the four MCSs holding at the corners. The distribution of MCSs in the interior of the rectangle can be complex. However, the only information we need to record is the minimal cluster holding at interior points arbitrarily close to the bottom corner and the maximal cluster holding at points arbitrarily near the top, as illustrated in Fig. 15.2.

Definition 15.2 (*Closed boundary map*) A *closed boundary map* ∂ is a map from $\{N, S, E, W\} \cup \{\mathsf{b}, \mathsf{l}, \mathsf{r}, \mathsf{t}\} \cup \{-, +\}$ to traces (for the first four), MCSs (for the next four), and clusters (for the last two), satisfying the following conditions.

- Temporal ordering must be respected, i.e. $\partial(\mathsf{b}) \lesssim \partial(W) \lesssim \partial(\mathsf{l}) \lesssim \partial(N) \lesssim \partial(\mathsf{t})$. And (initial cluster of $\partial(W)$) $\lesssim \partial(-) \lesssim \partial(+) \lesssim$ (final cluster of $\partial(N)$).
- Future defects of $\partial(+)$ must be passed up to either the final cluster of $\partial(N)$ or the final cluster of $\partial(E)$. All future defects of $\partial(\mathsf{b})$ must be passed up to the initial cluster of $\partial(W)$ or the initial cluster of $\partial(S)$.
- Dual conditions obtained from those above by reflecting in either or both diagonals, i.e. by swapping future/past, up/down, t/b, $\lesssim/\gtrsim$, N/W, S/E, or by swapping N/E, S/W, l/r, or both, throughout.

Some data can be omitted if some of the bounding edges and/or corners are not themselves included in the rectangle.

Definition 15.3 (*Rectangular/rounded*) Let

$$\{-, +\} \subseteq D \subseteq \{-, +, N, S, E, W, \mathsf{b}, \mathsf{l}, \mathsf{r}, \mathsf{t}\}.$$

If a corner belongs to D if and only if the two adjacent edges are in D (e.g. $\mathsf{b} \in D \iff (S \in D \wedge W \in D)$) then we say that D is *rectangular*. If we have only a one-way implication (whenever a corner is in D then the two adjacent edges are also in D) we say that D is *rounded*.

The rectangular sets correspond to the different types of rectangles possible: open, partly open and closed rectangles, e.g. $\{-, +\}$ corresponds to an open rectangle, and $\{-, +, N, E, \mathsf{t}\}$ corresponds to a rectangle including its northern and eastern boundaries, but open to the south and west. We may define general boundary maps by restricting a closed boundary map to some rounded set D by modifying Definition 15.2. So, for example, a future defect of $\partial(+)$ must be passed up to either the final cluster of $\partial(N)$ or of $\partial(E)$ (provided N, E are in the set of ∂), but if neither N nor E is in the set then $\partial(+)$ can have no future defects. Mostly we consider only rectangular sets, but when we eventually discuss interval logic we will need rounded sets including S and W but *not* b.

As a notational convenience, we also define a *one-point boundary map* to be a constant map $\partial : \{\mathsf{b}, \mathsf{t}\} \to \mathsf{MCS}$, which may be identified with its image m. Boundary maps that are not one-point boundary maps are said to be *proper*.

In the simplest kind of proper boundary map we have $\partial(-) = \partial(+)$, corresponding to rectangle models where a single cluster holds over the entire interior of the rectangle. More complex boundary maps may be *fabricated* from simpler ones, by joins, limits and shuffles (see Fig. 15.3).

To first help with the intuition, given some propositional valuation, if R is a rectangle in the plane, and if R can be divided into two adjacent rectangles R_1, R_2 sharing a common edge, then the boundary map determined by R will be the join of the boundary map determined by R_1 and that determined by R_2 (in Fig. 15.3 we show a *northern* join).

If x, y_i are all incomparable ($i < \omega$), if the boundary map ∂_0 determined by the rectangle $[\mathsf{x} \wedge \mathsf{y}_i, \mathsf{x} \vee \mathsf{y}_i]$ is constant (as i varies), if the sequence y_i converges in a southeasterly direction to y, and if the maximal cluster of $[\mathsf{x} \wedge \mathsf{y}, \mathsf{x} \vee \mathsf{y}]$ extends all the way south to y and the minimal cluster all the way east to y, then the boundary map determined by $[\mathsf{x} \wedge \mathsf{y}, \mathsf{x} \vee \mathsf{y}]$ will be a southeastern limit of ∂_0. Intuitively, the region labelled with neither the maximal nor minimal cluster must be pinched at y, as sketched in the second section of Fig. 15.3.

To picture a boundary map ∂ that is a shuffle of closed boundary maps $\partial_0, \partial_1, \ldots$ (including at least one one-point boundary map m), imagine a rectangle in the plane with copies of rectangles for $\partial_0, \partial_1, \ldots$ disjointly but densely distributed in the manner of a Cantor set, along the diagonal from l to r, with gaps along this diagonal filled by the one-point boundary map m, with $\partial(-)$ covering the part of the rectangle below the diagonal and the copies of $\partial_0, \partial_1, \ldots$, and with $\partial(+)$ covering the area above—the paragraphs after Theorem 15.1 for a bit more detail of this construction.

Joins If ∂' fits to the north of ∂ (i.e. $\partial(N) = \partial'(S)$ is defined, all or none of $\partial(W)$, $\partial'(W)$, $\partial(\mathsf{l})$, $\partial'(\mathsf{b})$ are defined, and if defined $\partial(\mathsf{l}) = \partial'(\mathsf{b})$, and similarly for $\partial(E)$, $\partial'(E)$, $\partial(\mathsf{t})$, $\partial'(\mathsf{r})$), then we may form the join $\partial \oplus_N \partial'$ defined by

$$(\partial \oplus_N \partial')(-) = \partial(-) \qquad\qquad (\partial \oplus_N \partial')(+) = \partial'(+)$$

and when the right-hand sides are defined

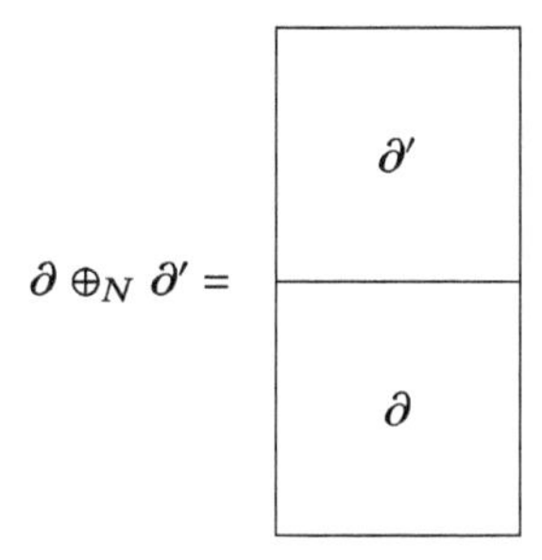

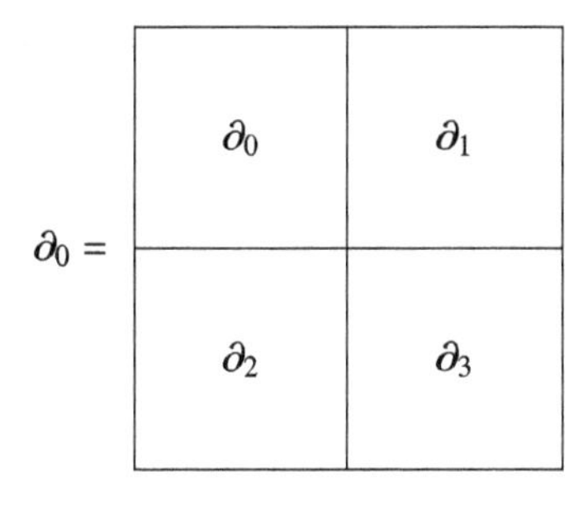

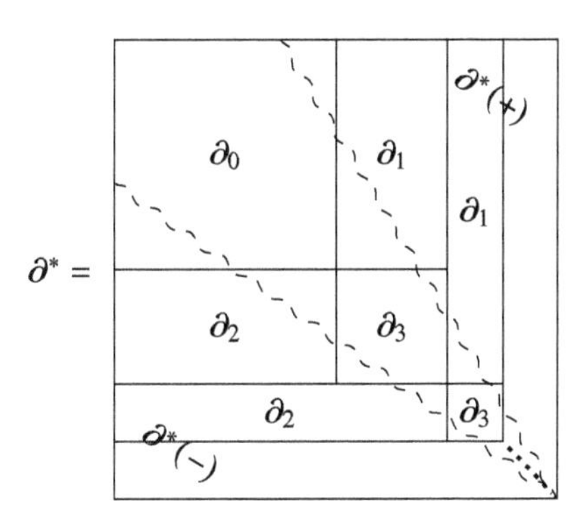

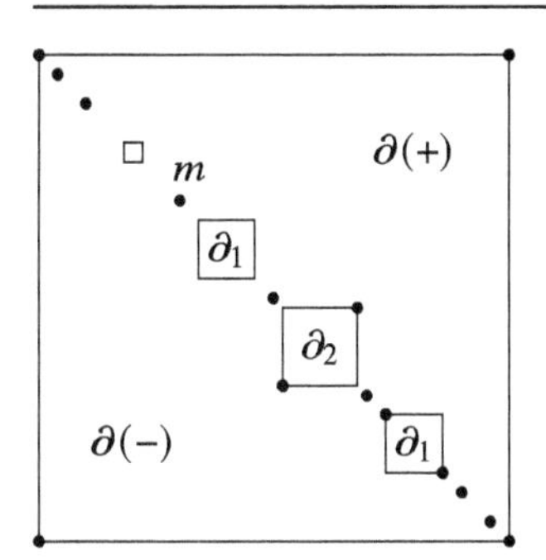

Fig. 15.3 A northern join, a southeastern limit and a shuffle of $\partial_1, \partial_2, m$

$$(\partial \oplus_N \partial')(N) = \partial'(N) \qquad (\partial \oplus_N \partial')(S) = \partial(S)$$
$$(\partial \oplus_N \partial')(W) = \partial(W) \oplus \partial'(W) \qquad (\partial \oplus_N \partial')(E) = \partial(E) \oplus \partial'(E)$$
$$(\partial \oplus_N \partial')(\mathsf{l}) = \partial'(\mathsf{l}) \qquad (\partial \oplus_N \partial')(\mathsf{t}) = \partial'(\mathsf{t})$$
$$(\partial \oplus_N \partial')(\mathsf{b}) = \partial(\mathsf{b}) \qquad (\partial \oplus_N \partial')(\mathsf{r}) = \partial(\mathsf{r})$$

(see the first part of Fig. 15.3). The join operator $\oplus_E$ is defined similarly.

Limits If

- $\partial_0 = (\partial_2 \oplus_E \partial_3) \oplus_N (\partial_0 \oplus_E \partial_1)$,
- ∂^* agrees with ∂_0 on $\{-, +, \mathsf{l}, W, N\}$ (if defined),
- $\partial_1(E)$ is the trace consisting of the single cluster $\partial^*(+)$, and $\partial_2(S)$ is the trace consisting of the single cluster $\partial^*(-)$,
- $\partial^*(-) = \partial_0(-)$ (necessarily equal to $\partial_2(-)$), and $\partial^*(+) = \partial_0(+)$ (necessarily equal to $\partial_2(+)$),
- if $\partial^*(S)$ is defined, every future defect is passed up to $\partial^*(-)$ or $\partial^*(\mathsf{r})$, and if $\partial^*(E)$ is defined, every past defect is passed down to $\partial^*(+)$ or $\partial^*(\mathsf{r})$,

then ∂^* is a *southeastern limit* of ∂_0 using $\partial_1, \partial_2, \partial_3$ (See the second part of Fig. 15.3, where the wavy lines indicate the implied boundaries of the regions labelled with $\partial^*(-)$ and $\partial^*(+)$.). Northeastern limits are defined similarly.

Shuffles If $\partial_1, \partial_2, \ldots, \partial_k$ are closed boundary maps including at least one one-point boundary map, every future defect of $\partial(-)$ is passed up to $\partial_i(\mathsf{b})$ for some $i \leq k$, every past defect of $\partial(+)$ is passed down to $\partial_i(\mathsf{t})$ (some $i \leq k$), every past defect of each $\partial_i(\mathsf{b})$ is passed down to $\partial(-)$, and every future defect of each $\partial_i(\mathsf{t})$ is passed up to $\partial(+)$ (all i), then ∂ is a shuffle of $\partial_1, \ldots, \partial_k$ (see the last part of Fig. 15.3).

Definition 15.4 (*Fabricated*) A boundary map ∂ is *fabricated* if it occurs in a sequence of distinct boundary maps each of which is either *simple* ($\partial(-) = \partial(+)$) or obtainable as the join, limit, or shuffle of earlier boundary maps in the sequence.

The length of such a sequence is bounded by the number of distinct boundary maps, an exponential function of $|\phi|$. And verifying whether the defining conditions of simple boundary maps, joins, limits and shuffles hold is straightforward. Hence, the set of fabricated boundary maps is decidable. We say that ϕ *occurs* in a fabricated boundary map ∂ if ϕ belongs to an MCS or cluster in one of the ∂-labels, or inductively ϕ occurs in one of the previous boundary maps in the sequence for ϕ used to fabricate ∂.

Proposition 15.1 *Let ϕ be a temporal formula. The following are equivalent.*

1. *ϕ occurs in an open, fabricated boundary map ∂.*
2. *ϕ is satisfiable in some M-model.*

An open boundary map is one summarising an open rectangle (of which $\mathbb{R}^2$ is a generic example), that is, a boundary map with domain $\{+, -\}$. Note that an open boundary map necessarily has no defects.

Once we establish the proposition, it follows that the validity of temporal formulas over M is decidable. In fact, (1) can be decided nondeterministically with polynomial space, by searching recursively for a decomposition of ∂ as a join, limit, or shuffle of fabricated boundary maps, using tail recursion where possible (Hirsch and Reynolds 2018, Lemma 5.1). The procedure to do this is shown in Algorithm 15.1.

Hence (1) is in NPSPACE, which by Savitch's theorem (Savitch 1970) is equal to PSPACE. Thus validity of temporal formulas over M is in PSPACE. As noted already,

Algorithm 15.1 Nondeterministic procedure to decide whether ∂ is fabricated

```
procedure FABRICATED(∂) choose either
  option 0
    check ∂ is simple
  option 1
    choose ∂1, ∂2; check they are boundaries
    check their join (choose some direction) is ∂; release ∂
    check FABRICATED(∂1); tail-call FABRICATED(∂2)
  option 2
    choose ∂1, ∂2, ∂3, ∂4; check they are boundaries
    check ∂ is the limit (choose direction) of ∂1 using ∂2, ∂3, ∂4; release ∂
    check FABRICATED(∂2), FABRICATED(∂3); release ∂2, ∂3
    check FABRICATED(∂4); tail-call FABRICATED(∂1)
  option 3
    choose k ∈ {1, ..., |φ|}, ∂1, ..., ∂k
    check they are boundaries including at least one one-point boundary
    check ∂ is the shuffle of ∂1, ..., ∂k; release ∂
    for i = 1, ..., k do
      check FABRICATED(∂i)
    end for
end procedure
```

the analogous result for $(\mathbb{R}, \leq) \otimes (\mathbb{R}, \leq)$ follows immediately, via the reduction given by applying $\mathbf{P}\psi \mapsto \psi \vee \mathbf{P}\psi$ and $\mathbf{F}\psi \mapsto \psi \vee \mathbf{F}\psi$ recursively to subformulas, and this validity problem is PSPACE-hard. Hence, we obtain the following theorem.

Theorem 15.1 (Hirsch and Reynolds 2018, Theorem 5.2) *The temporal logic of the frame* $(\mathbb{R}, \leq) \otimes (\mathbb{R}, \leq)$ *is PSPACE-complete.*

To prove Theorem 15.1, we prove the more general equivalence between (1′) ∂ being a fabricated boundary map and (2′) existence of a rectangle model h whose finite description is ∂.

The proof of (1′) $\Rightarrow$ (2′) is straightforward. We use dense valuations to construct rectangle models for simple boundary maps, and Fig. 15.3 shows how to synthesise rectangle models for joins, limits and shuffles of more primitive boundary maps. To be a bit more specific, a rectangle model for a simple boundary map may be constructed by using a dense valuation of clusters, that is, if a region (either the interior, or a segment of the boundary) is to be labelled by a cluster c and $m \in c$, then at any point in the region there must be points in the region above and below labelled by m. Joins and limits are straightforward (given that models may be variably dilated—in other words, reparameterised—horizontally/vertically, ensuring edges match up exactly). Given rectangle models $h_1, \ldots, h_{k-1}$ for closed boundary maps $\partial_1, \ldots, \partial_{k-1}$ and a one-point boundary map m, an open rectangle model for a shuffle ∂ of the ∂_i and m may be constructed over the base $R = (0, 1) \times (0, 1)$ in steps. Initially, there is a single open 'gap' $((0, 1), (1, 0))$ in the line $x + y = 1$, and no points are labelled. At each stage, using a fair schedule, a gap is chosen and an $i \leq k$. A copy of h_i is used to label the rectangle whose diagonal is the closed, central third of the gap. (If ∂_i is

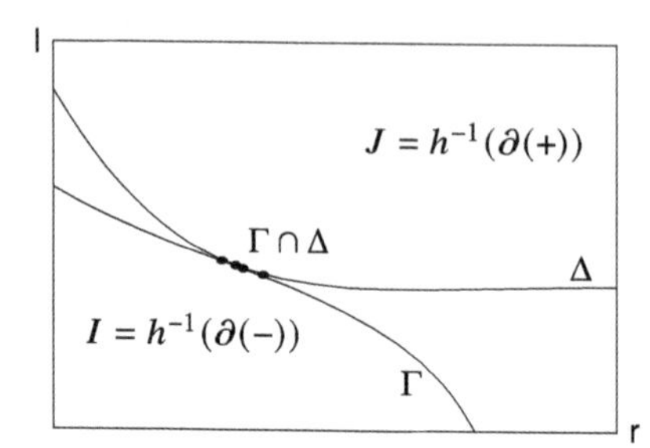

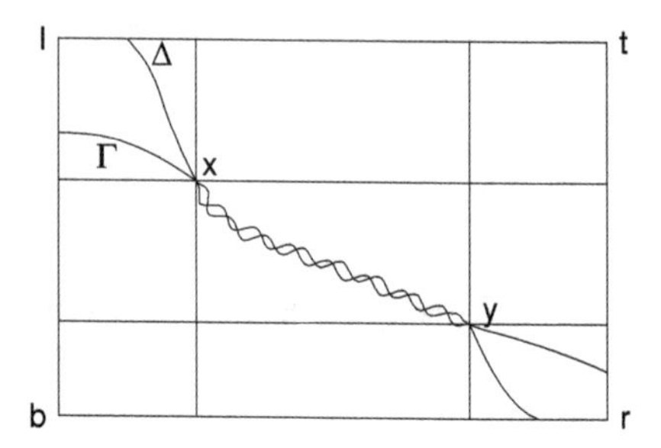

Fig. 15.4 Maximal and minimal clusters with boundaries Δ, Γ

a one-point boundary map, then only the single point on the diagonal in the exact centre of the gap is labelled.) The first and last third become new open gaps. This is repeated ω times. Finally, the Cantor set of unlabelled points on the line $x + y = 1$ are all labelled m, all remaining unlabelled points where $x + y < 1$ are given a label from $\partial(-)$ densely, and unlabelled points where $x + y > 1$ are given labels from $\partial(+)$ densely, completing the definition of the required rectangle model.

The converse implication is more intricate and only briefly outlined here (see Hirsch and Reynolds 2018, Lemma 4.1 for more details). Let h be a rectangle model. By reparameterising the coordinate axes, we may assume the rectangular domain of h is bounded. The proof that the rounded boundary map ∂_h defined by h is fabricated is by induction over the *height* of the rectangle—the maximum length of an ordered chain of distinct clusters from $\partial_h(-)$ to $\partial_h(+)$.[1] When the height is zero, $\partial_h(-) = \partial_h(+)$, so ∂_h is simple, hence fabricated. When the height is positive, we consider $I = h^{-1}(\partial_h(-))$, the set of points in the rectangle whose truth set belongs to $\partial_h(-)$—see the first diagram in Fig. 15.4.

Modulo points on the rectangle edges, I is a downward-closed set. The 'upper boundary', Γ, of I, is topologically equivalent (in the usual, open-ball topology on $\mathbb{R}^2$) to a closed line segment through the rectangle (Hirsch and Reynolds 2018, Lemma 2.11). Similarly, let Δ be the 'lower boundary' of $J = h^{-1}(\partial_h(+))$. If Γ and Δ are disjoint then, as they are closed, there is $\epsilon > 0$ such that each point in Γ is at least ϵ from each point of Δ. So ∂_h is a join of the boundary maps of finitely many rectangle models of side at most $\frac{\epsilon}{\sqrt{2}}$, of strictly smaller height, inductively fabricated; hence ∂_h is fabricated. Now suppose $\Gamma \cap \Delta$ is non-empty. The set $\Gamma \cap \Delta$ is closed, so consider the points x, y, where x is the nearest point in $\Gamma \cap \Delta$ to l, and y is the nearest point in $\Gamma \cap \Delta$ to r—see the second diagram in Fig. 15.4.

The rectangles $[\mathsf{b}, \mathsf{x}], [\mathsf{x}, \mathsf{t}], [\mathsf{b}, \mathsf{y}], [\mathsf{y}, \mathsf{t}]$ determine boundary maps of strictly smaller height, so they are fabricated. Consider the rectangle $[\mathsf{l} \wedge \mathsf{x}, \mathsf{l} \vee \mathsf{x}]$ with opposite corners l and x. There are only finitely many boundary maps, so let x_i be a sequence of points in this rectangle, strictly between Γ and Δ, converging to x such that the boundary map of the rectangle $[\mathsf{l} \wedge \mathsf{x}_i, \mathsf{l} \vee \mathsf{x}_i]$ is constant—call this boundary

[1] The proofs appearing in Hirsch and Reynolds (2018), Hirsch and McLean (2018) address only rectangular boundary maps, but the extension to rounded boundary maps is trivial.

map ∂_0. By the previous case (where Γ does not meet Δ) the map ∂_0 is fabricated. The boundary map of $[\mathsf{l} \wedge \mathsf{x}, \mathsf{l} \vee \mathsf{x}]$ is a southeastern limit of ∂_0, hence it is fabricated. Similarly, the boundary map of $[\mathsf{y} \wedge \mathsf{r}, \mathsf{y} \vee \mathsf{r}]$ is fabricated.

It remains to check that the boundary map of the rectangle $[\mathsf{x} \wedge \mathsf{y}, \mathsf{x} \vee \mathsf{y}]$ is fabricated. If $\mathsf{x} = \mathsf{y}$ this is trivial, so assume not. Let $\approx$ be the smallest equivalence relation over $\Gamma \cap \Delta$ including the successor relation (i.e. if $u, v \in \Gamma \cap \Delta$ and there is no $z \in \Gamma \cap \Delta$ strictly between u and v then $u \approx v$), including all pairs of points that differ in only one coordinate, and whose equivalence classes form closed sets (so if $\mathsf{x}_i \approx \mathsf{y}$ (all i) and x_i converges to x then $\mathsf{x} \approx \mathsf{y}$). The closed and bounded set $\Gamma \cap \Delta$ is partitioned by these equivalence classes, so by topological properties, either there is only one equivalence class, or uncountably many equivalence classes are singletons. Each $\approx$-equivalence class e also has a first and a last point $f(e)$ and $l(e)$, respectively. The rectangle $[f(e) \wedge l(e), f(e) \vee l(e)]$ defines a boundary map ∂_e (and in some cases this must be a one-point boundary map) that can be shown to be fabricated by considering joins and limits of boundary maps that, by previous cases, are fabricated. Hence each ∂_e is also fabricated (The argument here deviates from the proof of Hirsch and Reynolds (2018, Lemma 4.1), and is more similar to the proof of Hirsch and McLean (2018, Lemma 5.3.). If there is only one $\approx$-equivalence class, we are done. Otherwise, the proof of Theorem 15.1 is completed by showing that the boundary map of $[\mathsf{x} \wedge \mathsf{y}, \mathsf{x} \vee \mathsf{y}]$ is a shuffle of the boundary maps ∂_e.

15.3 Slower-Than-Light Signals

Now we consider the temporal logic of the frame $(\mathbb{R}, <) \otimes (\mathbb{R}, <)$, where the irreflexive accessibility relation is 'can send a signal at strictly less than the speed of light'. We can no longer give a direct description of the MCSs holding along the four light-lines bounding a rectangle, since distinct points on a light-line are now unordered. What we can do, nevertheless, is record, at each point on the light-line, the cluster of MCSs that hold arbitrarily soon in the future (the *upper cluster* of a point) and the cluster of MCSs that hold arbitrarily recently in the past (the *lower cluster* of a point). The function from points to upper cluster (or to lower cluster) is monotone with respect to the parametric ordering of the light-line.

Definition 15.5 (*Bi-trace*) A *bi-trace* consists of two sequences of clusters $c_0^+ \lesssim \dots \lesssim c_n^+$ and $c_0^- \lesssim \dots \lesssim c_n^-$, and one sequence of MCSs $b_1, \dots, b_n$ (for some n) such that for all i:

(i) $c_i^- \lesssim c_i^+$ and $(c_i^-, c_i^+) \neq (c_{i+1}^-, c_{i+1}^+)$,
(ii) $c_i^- \lesssim b_{i+1} \leq c_{i+1}^+$,
(iii) there exists $m_i \in \mathsf{MCS}$ such that: $c_i^- \lesssim m_i \lesssim c_i^+$, all future defects of m_i are passed up to c_i^+, and all past defects passed down to c_i^- (such an m_i is an *interpolant* of c_i^- and c_i^+),
(iv) all future defects of b_{i+1} are passed up to c_{i+1}^+ and all past defects of b_{i+1} are passed down to c_i^-.

Fig. 15.5 A bi-trace

See Fig. 15.5.

Intuitively, an interpolant of c_i^- and c_i^+ is an MCS that can appear *on* the light-line described by the bi-trace, on the segment between c_i^- and c_i^+.

A *bi-boundary* is like a boundary map, but uses bi-traces instead of traces.

Definition 15.6 (*Bi-boundary*) A *bi-boundary* ∂ is a map from a rounded subset of $\{-, +, N, S, E, W, \mathsf{b}, \mathsf{l}, \mathsf{r}, \mathsf{t}\}$ to clusters (for $-, +$), bi-traces (for N, S, E, W), and MCSs (for $\mathsf{b}, \mathsf{l}, \mathsf{r}, \mathsf{t}$) such that

- temporal ordering is respected, i.e. $\partial(+)$, the final lower cluster of $\partial(N)$, and the final lower cluster of $\partial(E)$ are all equal (when defined), and $\partial(+) \lesssim \partial(\mathsf{t})$ (when $\partial(\mathsf{t})$ is defined), plus a dual property for $\partial(-)$,
- future defects of $\partial(+)$ are passed up to either an interpolant of the final lower and upper clusters of $\partial(N)$ (provided $\partial(N)$ is defined), an interpolant of the final lower and upper clusters of $\partial(E)$ (if defined) or to t (if defined), plus a dual condition for past defects of $\partial(-)$.

The definitions of joins, limits, shuffles, fabricated, etc. above may be altered for bi-boundaries with no significant changes, hence the following theorem.

Theorem 15.2 (Hirsch and McLean 2018, Theorem 6.1) *The temporal logic of the frame* $(\mathbb{R}, <) \otimes (\mathbb{R}, <)$ *is PSPACE-complete.*

15.4 Real Intervals

There is a strong connection between these two-dimensional frames and interval logic, noted in Shapirovski and Shehtman (2002). Interval logics are based on frames consisting of strictly ordered pairs from a linear order. The full interval logic has modalities for all thirteen of Allen's interval relations (after, meets, overlaps, ends, during, starts, and their converses, and equals), the full logic is undecidable and remains undecidable if we consider the logic of intervals (x, y) with *real* endpoints $x < y$ (Halpern and Shoham 1986). A large body of research has investigated the decidability and complexity of the interval logic where the set of modalities is restricted (Bresolin et al. 2019), and over various linear orders. For example, the modal interval logic with the two modalities 'contained in' (i.e. during, starts, ends, or equals) and its converse, over real intervals, is known to be PSPACE-complete (Montanari et al. 2010). A problem that appears to remain open is the decidability and complexity of the logic of intervals with real endpoints with the two modalities, strict during and its converse (see the open problem 3 of Halpern and Shoham

1986). For comparison, the logic of intervals over *discrete* time with strict during and its converse is undecidable. With some modification, we will use the argument above to prove decidability of the interval logic with strict during and its converse as modalities, over the reals. With further analysis, the logic can be shown to be PSPACE-complete.

If we restrict the frame $(\mathbb{R}, <) \otimes (\mathbb{R}, <)$ to $\{(x, y) : x < y \in \mathbb{R}\}$ and treat its points as intervals, the accessibility relation is 'each endpoint is strictly earlier', equivalently, overlaps, meets, or before. Instead, we restrict $(\mathbb{R}, <) \otimes (\mathbb{R}, <)$ to $\{(x, y) : x + y > 0\}$ to obtain the open, upper triangular frame $\mathcal{T}$. A point (x, y) in this subframe represents the interval $(-x, y)$, more precisely, the map $(x, y) \mapsto (-x, y)$ is a frame isomorphism from $\mathcal{T}$ to the frame of intervals (x, y) with real endpoints $x < y$ under the during relation, i.e. $(x, y) \ \mathtt{dur} \ (x', y') \iff x' < x < y < y'$. The temporal logic of this frame differs from the temporal logic of $(\mathbb{R}, <) \otimes (\mathbb{R}, <)$. In particular, the past confluence axiom $\mathbf{PH}p \to \mathbf{HP}p$ is valid over $(\mathbb{R}, <) \otimes (\mathbb{R}, <)$ but not valid in $\mathcal{T}$.

Let $x + y > 0$. The *upper-closed triangle frame* $T[x, y]$ has worlds

$$\{(x', y') : x' + y' > 0, \ x' \leq x, \ y' \leq y\}$$

and accessibility inherited from $\mathcal{T}$; see the first diagram in Fig. 15.6. An *open* (or *semi-open*) triangle frame is obtained by requiring both (either) constraints $x' \leq x, \ y' \leq y$ to be strict. Note that points on the line $x' + y' = 0$ are excluded from the domain of triangle frames. If h is a propositional valuation over $T[x, y]$ and $x' < x, \ y' < y$ then h induces not only a valuation over $T[x', y']$ but also determines the upper cluster holding above but arbitrarily close to points along the northern and eastern perimeter of $T[x', y']$.

Definition 15.7 (*Closed triangle bi-boundary*) A *closed triangle bi-boundary* τ is a map from $\{+, N, E, \mathsf{t}\}$ to a cluster (for $+$), bi-traces (for N, E), and an MCS (for t) satisfying the following consistency constraints. (See the second diagram in Fig. 15.6.)

1. The cluster $\tau(+)$ equals the final lower cluster of $\tau(N)$ and of $\tau(E)$, which is below $\tau(\mathsf{t})$.
2. Future defects of $\tau(+)$ are passed up either to $\tau(\mathsf{t})$, or to an interpolant of $\tau(+)$ and the final upper cluster of $\tau(N)$, or to an interpolant of $\tau(+)$ and the final upper cluster of $\tau(E)$.
3. Past defects of $\tau(\mathsf{t})$ are passed down to $\tau(+)$.

Open and semi-open triangle bi-boundaries are similar. When τ is open, observe that none of the alternatives in (2) are possible, so $\tau(+)$ has no future defects. If h is a propositional valuation over a triangle frame and (x, y) is in the interior of the frame, we write $\tau_h[x, y]$ for the closed triangle bi-boundary induced by the triangle $T[x, y]$. The semi-open and open versions are denoted $\tau_h[x, y)$, $\tau_h(x, y]$, and $\tau_h(x, y)$.

A *simple* triangle bi-boundary is one where $\tau(+)$ has no past defects, and if the bi-traces $\tau(N)$ and/or $\tau(E)$ are defined then their lower clusters all equal $\tau(+)$.

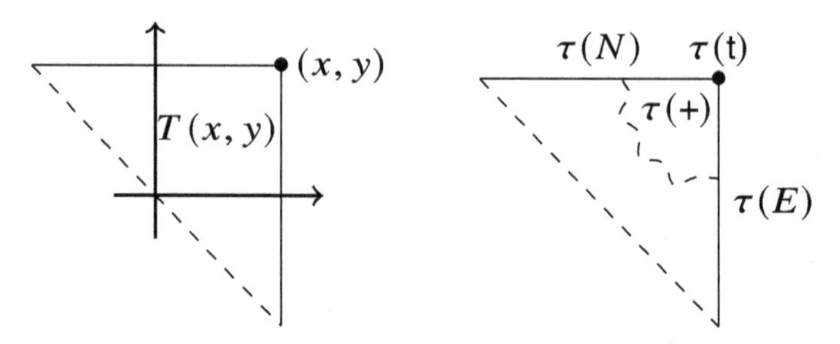

Fig. 15.6 The closed triangle frame $T[x, y]$ and a closed triangle bi-boundary

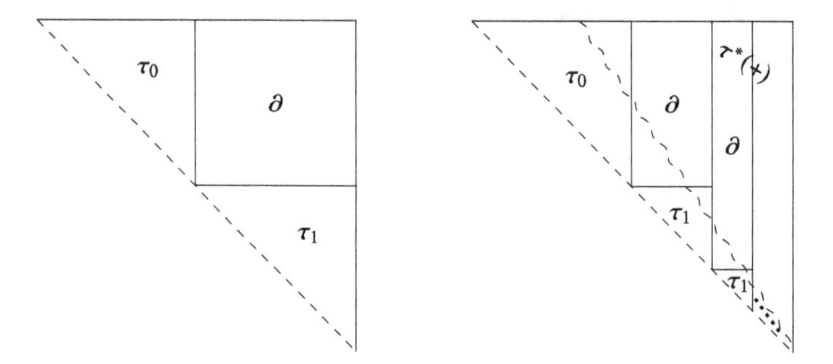

Fig. 15.7 The join $\tau_0 \oplus \partial \oplus \tau_1$ and a southeastern limit τ^* of τ_0

Given two closed triangle bi-boundaries τ_0, τ_1, and a bi-boundary ∂ with domain $\{-, +, N, S, E, W, \mathsf{l}, \mathsf{r}, \mathsf{t}\}$, if $\tau_0(E) = \partial(W)$, $\tau_0(\mathsf{t}) = \partial(\mathsf{l})$, $\partial(S) = \tau_1(N)$, $\partial(\mathsf{r}) = \tau_1(\mathsf{t})$, then we may form the closed triangle *join* $\tau_0 \oplus \partial \oplus \tau_1$ as illustrated in the first diagram of Fig. 15.7. Observe that the domain of ∂ omits b, so it is not rectangular, but it *is* rounded. Triangle joins that are open or semi-open (at the north and/or east) are defined similarly.

The triangle bi-boundary τ^* is a *southeastern limit* of τ_0 if

- $\tau_0 = \tau_0 \oplus \partial \oplus \tau_1$ for some ∂, τ_1,
- the lower cluster of the bi-trace $\partial(E)$ constantly $\tau^*(+)$,
- τ_0 agrees with τ^* over $\{+, N\}$ (allowing both to be undefined on N),
- $\tau^*(E)$, if defined, is a bi-trace where every lower cluster equals $\tau_0(+)$ (hence $\tau^*(+) = \tau_0(+) = \tau_1(+)$).

See the second diagram in Fig. 15.7, where the wavy line indicates the implied boundary of the region labelled with $\tau^*(+)$ *Northwestern* limits are similar.

If $\tau_1, \ldots, \tau_k$ are closed triangle bi-boundaries then τ is a *shuffle* of them if every past defect of $\tau(+)$ is passed down to $\tau_i(\mathsf{t})$ (some $i \leq k$) and every future defect of $\tau_i(\mathsf{t})$ is passed up to $\tau(+)$ (all $i \leq k$). Observe that a one-point triangle bi-boundary is not required for a shuffle since the diagonal boundary edge is excluded from the domain of the triangle, in contrast to the situation for shuffles of rounded bi-boundaries.

The set of *fabricated* triangle bi-boundaries is the closure of the set of simple triangle bi-boundaries under joins (using fabricated rounded bi-boundaries), limits and shuffles.

Proposition 15.2 *Let τ be a closed (respectively open, semi-open north, semi-open east) triangle bi-boundary map. The following are equivalent.*

1. *τ is fabricated.*
2. *There is a valuation h over a triangle frame T with $(1, 1)$ in the interior of T such that $\tau = \tau_h[1, 1]$ (respectively, $\tau = \tau_h(1, 1)$, $\tau_h(1, 1]$, $\tau_h[1, 1)$).*

As with rectangle models and boundary maps, it is simple to check that any fabricated triangle bi-boundary may be obtained as $\tau_h[x, y]$, $\tau_h[x, y)$, $\tau_h(x, y]$, or $\tau_h(x, y)$, for some valuation h (To be completely accurate, for any fabricated triangle bi-boundary τ there is an h such that the induced τ_h agrees with τ everywhere *except possibly* the upper clusters of $\tau(N)/\tau(E)$. Importantly, this technicality disappears in the case of open triangles—those corresponding to $\mathcal{T}$.)

Conversely, for any valuation h over some triangle frame T where $(1, 1)$ is in the interior, we must show that $\tau_h[1, 1]$, $\tau_h[1, 1)$, $\tau_h(1, 1]$, and $\tau_h(1, 1)$ are fabricated. We focus on $\tau_h[1, 1]$ below; the other cases are similar. The proof is by induction over the *size* of the triangle model—the number of distinct clusters witnessed in the open triangle $T(1, 1)$. Let λ be the line $\{(x, y) : x + y = 0\}$. For $(x, -x) \in \lambda$ and $x < y$, $-x < z$ we write $_*[(x, -x), (y, z)]$ for the 'rounded rectangle' $[(x, -x), (y, z)] \setminus \{(x, -x)\}$. By the bi-boundary equivalent of Theorem 15.1, the bi-boundary $\partial_h(_*[(x, -x), (y, z)])$ is fabricated.

For the base case, the single cluster $\tau_h(+)$ covers the entirety of $T(1, 1)$. The cluster $\tau_h(+)$ must have no past defects, and $\tau_h(N)$ and $\tau_h(E)$, if defined, must have all lower clusters equal to $\tau_h(+)$. Future defects of $\tau_h(+)$ must be passed up. Hence τ_h is simple, so fabricated.

Now suppose that more than one cluster is witnessed in the interior, so $J = h^{-1}(\tau_h(+)) \cap T(1, 1)$ is a proper, upward-closed subset of the interior of the triangle. Let Δ be the 'lower boundary' of J. Formally, Δ is given by first taking the boundary of J viewed as a subspace of $T(1, 1)$, then taking the closure of that within $\mathbb{R}^2$. For topological reasons, Δ is topologically equivalent to a line segment, closed by definition.

If Δ does not meet λ then since the sets are closed, and Δ is bounded, there is $\epsilon > 0$ such that all points in Δ are at least ϵ from any point in λ. The triangle $T[1, 1]$ (whose orthogonal sides are of length 2) may be divided into at most $t = \lceil \frac{2\sqrt{2}}{\epsilon} \rceil$ triangles $T(x, y)$, with (x, y) below Δ, of side at most $\frac{\epsilon}{\sqrt{2}}$ and at most $t - 1$ rectangles. Then since each triangle has strictly smaller size, by induction and by the result for bi-boundaries of rectangles, τ_h is fabricated.

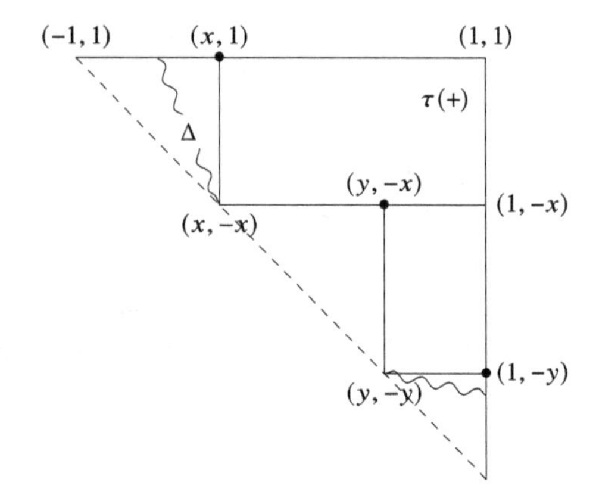

Fig. 15.8 A decomposition of $\tau = \tau[1, 1]$ as $\tau_h[x, 1] \otimes \partial_h({}_*[(x, -x), (1, 1)]) \otimes (\tau_h[y, -x] \otimes \partial_h({}_*[(y, -y), (1, -x)]) \otimes \tau_h[1, -y])$. The wavy line indicates part of the lower boundary Δ of $\tau(+)$

So assume $\Delta \cap \lambda$ is non-empty. Let $x \leq y$ be the infimum and supremum respectively of $\{z : (z, -z) \in \Delta\}$—see Fig. 15.8. Since $\Delta \cap \lambda$ is closed, we know $(x, -x)$, $(y, -y) \in \Delta \cap \lambda$. Suppose $x \neq -1$. Then first we wish to fabricate $\tau_h[x, 1]$. We may assume there are no points of Δ directly north of $(x, -x)$ (otherwise we could fabricate $\tau_h[x, 1]$ using a triangle join and the $\Delta \cap \lambda = \emptyset$ cases, for triangles and rectangles). By this assumption, and as there are only finitely many triangle bi-boundaries, there is an increasing sequence x_i converging to x such that: the sequences $\tau_h[x_{i+1}, -x_i]$, $\tau_h[x_i, 1]$, and $\partial_*[(x_i, -x_i), (x_{i+1}, 1)]$ are all constant, and the bi-boundary $\partial_*[(x_i, -x_i), (x_{i+1}, 1)](E)$ is the single cluster $\tau_h(+)$. Since Δ is bounded away from λ in $\tau_h[x_i, 1]$, by the $\Delta \cap \lambda = \emptyset$ cases all three of $\tau_h[x_{i+1}, -x_i]$, $\tau_h[x_i, 1]$, and $\partial_*[(x_i, -x_i), (x_{i+1}, 1)]$ are fabricated. Then $\tau_h[x, 1]$ is a southeastern limit of the constant triangle bi-boundary $\tau_h[x_i, 1]$. Hence $\tau_h[x, 1]$ is fabricated.

If $y = x$ then τ is the fabricated join $\tau_h[x, 1] \otimes \partial_h({}_*[(x, -x), (1, 1)]) \otimes \tau_h[1, -x]$. So assume $x < y$. Then τ is the join $\tau_h[x, 1] \otimes \partial_h({}_*[(x, -x), (1, 1)]) \otimes (\tau_h[y, -x] \otimes \partial_h({}_*[(y, -y), (1, -x)]) \otimes \tau_h[1, -y])$, as shown in Fig. 15.8. So it remains to show that $\tau_h[y, -x]$ is fabricated.

Let $\lhd$ be the northwest-to-southeast ordering on $\Delta \cap \lambda$. Let $\approx$ be the smallest equivalence relation over $\Delta \cap \lambda$ such that

(a) $\approx$ includes the successor relation over $(\Delta \cap \lambda, \lhd)$ (i.e. if $(x', -x'), (y', -y') \in \Delta \cap \lambda$ and there is no z strictly between x' and y' such that $(z, -z) \in \Delta \cap \lambda$ then $(x', -x') \approx (y', -y')$),
(b) each $\approx$-equivalence class is closed in $\mathbb{R}^2$ (so if y_i is a sequence converging to y' and $(x, -x) \approx (y_i, -y_i)$ (all i) then $(x', -x') \approx (y', -y')$).

By minimality of $\approx$, the equivalence classes are convex with respect to $\lhd$. It follows that, as with rounded boundary maps, there is either a single $\approx$-equivalence class or an uncountable number of them.

Each (non-singleton) $\approx$-equivalence class e determines an upper-closed triangle $T[\bigvee e, -\bigwedge e]$ where $\bigvee e$ and $\bigwedge e$ are the supremum and infimum, respectively, of the x-coordinates of e. As with rounded boundary maps, by considering joins and limits of fabricated triangle bi-boundaries, the bi-boundary of each $T[\bigvee e, -\bigwedge e]$ can be shown to be fabricated. If there is a single $\approx$-equivalence class, we are done. Otherwise, the proof of Theorem 15.2 is completed by showing that $\tau_h[y, -x]$ is a shuffle of triangle bi-boundaries $\tau_h[\bigvee e, -\bigwedge e]$ where e ranges over all non-singleton $\approx$-equivalence classes.

Theorem 15.3 *The temporal logic of the frame of intervals* $\{(x, y) \in \mathbb{R}^2 \mid x < y\}$ *under the* during *relation,* $(x, y)\ \mathtt{dur}\ (x', y') \iff x' < x < y < y'$, *is PSPACE-complete.*

PSPACE-hardness of the temporal logic follows from the PSPACE-hardness of the modal logic of $(\mathbb{R}, <) \otimes (\mathbb{R}, <)$ (Shapirovski 2004). By Theorem 15.2, we now know that the temporal formulas satisfiable in the *during* frame of intervals is decidable—it suffices to check if a given formula ϕ occurs in an open, fabricated triangle bi-boundary. To obtain a polynomial space upper bound on the complexity, note first that just like for boundary maps and bi-boundary maps, a triangle bi-boundary can be stored in polynomial space, in terms of $|\phi|$. Then just as for Theorems 15.1 and 15.2, whether a triangle bi-boundary map is fabricated can be decided nondeterministically with polynomial space by searching recursively for a decomposition as a join, limit, or shuffle, using tail recursion where possible. The procedure to do this is shown in Algorithm 15.2. In Algorithm 15.2, a call of the form FABRICATED(∂) indicates a call to the bi-boundary analogue of Algorithm 15.1 needed to prove Theorem 15.2, which is Hirsch and McLean (2018, Algorithm 1).

15.5 Open Problems

1. Find an axiomatisation of the valid temporal formulas over $(\mathbb{R}, \leq) \otimes (\mathbb{R}, \leq)$.
2. Generalise the material above to find more general conditions where the temporal logic of $(W, R) \otimes (W', R')$ is decidable.
3. Determine the decidability of the temporal logic of Minkowski spacetime of higher dimensions. We conjecture this logic is undecidable for three or more spacetime dimensions.

Algorithm 15.2 Nondeterministic procedure to decide whether τ is fabricated

procedure FABRICATED(τ) **choose** either
  option 0
    check τ is simple
  option 1
    choose τ_0, ∂, τ_1
    check τ_0, τ_1 are triangle bi-boundaries; **check** ∂ is a bi-boundary
    check $\tau = \tau_0 \oplus \partial \oplus \tau_1$; **release** τ
    check FABRICATED(τ_0), FABRICATED(τ_1); **tail-call** FABRICATED(∂)
  option 2
    choose τ_0, ∂, τ_1
    check τ_0, τ_1 are triangle bi-boundaries; **check** ∂ is a bi-boundary
    check τ is the limit (**choose** direction) of τ_0; **release** τ
    check FABRICATED(τ_1); **release** τ_1;
    check FABRICATED(τ_0); **tail-call** FABRICATED(∂)
  option 3
    choose $k \in \{1, \ldots, |\phi|\}$, $\tau_1, \ldots, \tau_k$; **check** they are bi-boundaries
    check τ is the shuffle of $\tau_1, \ldots, \tau_k$; **release** τ
    for $i = 1, \ldots, k$ **do**
      check FABRICATED(τ_i)
    end for
end procedure

Acknowledgements The authors would like to thank our reviewer, Judit Madarász, for constructive suggestions—we believe the chapter is considerably improved!

References

Andréka, H., Madarász, J. X., & Németi, I. (2007). Logic of space-time and relativity theory. In M. Aiello, I. Pratt-Hartmann, & J. V. Benthem (Eds.), *Handbook of spatial logics* (pp. 607–711). Berlin: Springer.

Bresolin, D., Monica, D. D., Montanari, A., Sala, P., & Sciavicco, G. (2019). Decidability and complexity of the fragments of the modal logic of Allen's relations over the rationals. *Information and Computation, 266,* 97–125.

Gabbay, D. M., Kurucz, A., Wolter, F., & Zakharyaschev, M. (2003). *Many-dimensional modal logics: Theory and applications*. Studies in logic and the foundations of mathematics. Amsterdam: Elsevier Science.

Goldblatt, R. (1980). Diodorean modality in Minkowski space-time. *Studia Logica, 39,* 219–236.

Halpern, J., & Shoham, Y. (1986). *A propositional modal logic of time intervals*. In *1st International Symposium on Logic in Computer Science*. Boston: IEEE.

Hirsch, R., & McLean, B. (2018). The temporal logic of two dimensional Minkowski spacetime with slower-than-light accessibility is decidable. *Advances in Modal Logic, 12,* 347–366.

Hirsch, R., & Reynolds, M. (2018). The temporal logic of two-dimensional Minkowski spacetime is decidable. *The Journal of Symbolic Logic, 83*(3), 829–867.

Kurucz, A. (2007). Combining modal logics. In P. Blackburn, J. V. Benthem, F. Wolter (Eds.), *Handbook of modal logic* (Vol. 3, pp. 869–924). Studies in logic and practical reasoning. Amsterdam: Elsevier.

Malament, D., & Hogarth, M. (1994). Non-Turing computers and non-Turing computability. *Journal of the Philosophy of Science Association, 1,* 126–138.
Montanari, A., Pratt-Hartmann, I., & Sala, P. (2010). Decidability of the logics of the reflexive sub-interval and super-interval relations over finite linear orders. In *17th International Symposium on Temporal Representation and Reasoning* (pp. 27–34).
Németi, I. (1995). Decidable versions of first-order logic and cylindric-relativized set algebras. In L. Csirmaz, D. Gabbay, & M. D. Rijke (Eds.), *Logic colloquium'92, Studies in logic, language and computation* (pp. 177–241). Stanford: CSLI Publications & FoLLI.
Reynolds, M. (2011). A tableau for until and since over linear time. *18th International Symposium on Temporal Representation and Reasoning* (pp. 41–48).
Reynolds, M., & Zakharyaschev, M. (2001). On the products of linear modal logics. *Journal of Logic and Computation, 11*(6), 909–931.
Savitch, W. J. (1970). Relationships between non-deterministic and deterministic tape complexities. *Journal of Computer and System Science, 4,* 177–192.
Shapirovski, I. (2004). On PSPACE decidability in transitive modal logics. *Advances in Modal Logic, 5,* 269–287.
Shapirovski, I. (2010). Simulations of two dimensions in unimodal logics. *Advances in Modal Logic, 8,* 373–391.
Shapirovski, I., & Shehtman, V. (2002). Chronological future modality in Minkowski spacetime. *Advances in Modal Logic, 4,* 437–459.
Spaan, E. (1993). The complexity of propositional tense logics. In M. de Rijke (Ed.), *Diamonds and defaults, studies in logic, language and information* (pp. 287–307). Amsterdam: Kluwer.

Chapter 16
Dynamic Contact Algebras with a Predicate of Actual Existence: Snapshot Representation and Topological Duality

Dimiter Vakarelov

Second Reader
I. Pratt-Hartmann
Manchester University

Abstract The paper is in the field of region-based theory of space and time (RBTST). This is an extension of the region-based theory of space (RBTS) with time. Its origin goes back to some ideas of Whitehead, De Laguna, and Tarski and is related to the problem of how to build the theory of space without the use of the notion of point. The notion of *contact algebra* (CA) presents an algebraic formulation of RBTS. CA is an extension of Boolean algebra, considered as an algebra of spatial regions with an additional relation of *contact*. *Dynamic contact algebra* (DCA) considered as an algebraic formulation of RBTST is an extension of CA aiming to study regions changing in time. In this paper, we study a version of DCA incorporating an explicit predicate AE of *actual existence*. We first develop the representation theory of such DCAs by means of the so-called *snapshot models*. Second, we introduce topological models of DCA and develop the corresponding topological representation and duality theory.

Keywords Mereotopology · (Stone-type)-duality · Regular closed set · Spacetime · Temporal relation · Ultrafilter

2020 Mathematics Subject Classification: 03B44 · 03G05 · 08A02 · 18A23 · 54D10 · 54D30 · 54H10

D. Vakarelov (✉)
Faculty of Mathematics and Informatics, Department of Mathematical Logic and Applications, Sofia University "St. Kliment Ohridski", Sofia, Bulgaria
e-mail: dvak@fmi.uni-sofia.bg

I. Düntsch and E. Mares (eds.), *Alasdair Urquhart on Nonclassical and Algebraic Logic and Complexity of Proofs*, Outstanding Contributions to Logic 22,
https://doi.org/10.1007/978-3-030-71430-7_16

16.1 Introduction

In his pioneering work (Urquhart 1978), Alasdair Urquhart shows how to extend the famous Stone duality for Boolean algebras and Priestley duality for distributive lattices to the case of general lattices. Since then duality theory become an established area of mathematics studying relations between different classes of algebras and classes of topological spaces. The presented paper also contributes to this area. More precisely the paper is in the field of region-based theory of space and time (RBTST). This is an extension of the region-based theory of space (RBTS) with time.

Let us discuss in some detail RBTS and RBTST. RBTS is a kind of point-free theory of space based on the notion of *region*. Another name of RBTS is *mereotopology*, because it combines notions and methods of mereology and topology (Simons 1987). The origin of this theory goes back to some ideas of Whitehead (1929), de Laguna (1922) and Tarski (1956) related to the problem of how to build the theory of space without the use of the notion of point. The reason is that the notion of point does not have a separate existence in reality and hence should not be put as a primitive notion in geometry. This does not mean that this notion should be disregarded at all, it is an useful notion and it is doubtful if geometry can be developed without this concept. What is required is that in an axiomatic development of the theory the notion of point should be defined by means of the other primitive notions. More information on RBTS, mereotopology and their applications can be found in Vakarelov (2007), Bennett and Düntsch (2007), Hahmann and Grüninger (2012), Pratt-Hartmann (2007).

One algebraic version of RBTS is the notion of contact algebra (CA) Dimov and Vakarelov (2006) which is an extension of Boolean algebra with an additional relation C called contact. The elements of the Boolean algebra are called *regions* considered as abstractions of material or geometrical bodies, and Boolean operations are considered as operations of constructing new regions by means of given ones. Boolean algebra in this interpretation is considered as a formal explication of *mereology*. This is a suggestion by Tarski who considered mereology as a (complete) Boolean algebra with zero deleted (see, for instance, Simons 1987 page 25 for this view). The reason for deleting the zero is because mereology, as a part of ontology is a science of the existent and the zero element does not designate any existing object. We, however, assume the zero element just as a designation of "non-existing object" and use it to define a certain predicate of ontological existence of an object: "a ontologically exists", in symbols $E(a)$ iff $a \neq 0$. The predicate of ontological existence is very important for RBTST and for studying objects changing in time. Just in presence of time we may say for a given object a when it *begins to exist*, *ceases to exist*, *always exists* (like God), *exists simultaneously* with some other object b, *exists before* (or after) b. Thus, there are various spatio-temporal relations between changing objects which formalization needs the existence predicate. We used the predicate $E(a)$ and some of the mentioned relations in the representation theory of *dynamic contact algebra* (DCA) studied in our papers (Vakarelov 2010, 2012, 2014, 2016a, b, 2020) and considered as a formal algebraic explication of RBTST. We obtained DCA as an abstraction from a special *dynamic model of space and time* (DMST), called also *snapshot* or *cinematographic* model. DMST is constructed from a series of contact algebras considered as snapshots of changing regions. We found, however, that the predicate $E(a) \Leftrightarrow_{\text{def}} a \neq 0$, although expressing existence, is quite weak and possesses some disadvantages: there are too many existing objects and only one non-existing—the zero 0. We discussed this and other questions related to the existence predicate in our papers (Vakarelov 2017a, b). As a result, we obtain a new axiomatic definition of existence predicate formalizing the intuitive situation for an object a for which we can say that it *actually exists* (in symbols $AE(a)$) as a special mode of existence, existence at the actual (current) moment of time. We also proposed a corresponding contact relation between objects called *actual contact*. The obtained generalization of CA is called shortly *AE-algebra with actual contact*. These algebras are more natural than CAs because they can distinguish from actual and non-actual ontological existence.

One of the main purposes of the present paper is to introduce a new version of DCA based on the existence predicate AE and to study this more realistic version with corresponding models and representation theory. We first develop the representation theory of such DCAs by means of snapshot models which motivates the axioms of the abstract definition. The second aim is to introduce topological models of DCAs which present a new view on the nature of space and time and show what happens if we are abstracting from their metric properties. We develop for these models the expected topological representation and duality theory, generalizing in a certain sense the well-known Stone duality for Boolean algebras (Stone 1936). Due to these models, DCA can also be called a *dynamic mereotopology* considering CA as a *static mereotopology*.

The paper is organized as follows: Sect. 16.2 contains all needed technical information on contact and precontact algebras, their models and representation theory. Section 16.2 has also a preliminary character. It discusses in more detail results in Vakarelov (2017a, b) on predicates of actual existence and actual contact: the formal definitions and their motivations, relational and topological models and representation theory. Section 16.4 starts the study of the new version of DCA by introducing first a concrete model called the *dynamic model of space and time based on snapshot construction* and shows how to construct a special DCA in this model called *standard DCA*. This is a concrete point-based model of space and time which can be considered in some sense as a formal analog of a description of an area of changing regions by means of making a video film. This film is a series of snapshots of the area describing in this way the picture of change. The model is used as a source of true facts which then are taken as axioms of the formal definition of DCA. Similar models are considered in philosophy of spacetime as the so-called *Block Universe*. Section 16.5 presents the abstract definition of DCA and studies some technical constructions and various kinds of abstract points. Section 16.6 is devoted to the representation theorem of DCA by means of the snapshot model. It is shown in this section how to define in DCA in a canonical way a structure of snapshot model and to prove that DCA is isomorphically embedded in the standard DCA associated with this model. In Sect. 16.7, we introduce special topological model of DCA called dynamic mereotopological space (DMS). DMS is a special kind of a topological mathematical structure which is a generalization of mereotopological spaces introduced in Goldblatt and Grice (2016). The main result here is the topological representation theorem for DCA. Topological models of DCA present another view on the nature of space and time and show what they look like if we are abstracting from their metric properties. In Sect. 16.8, we consider the classes of all DCAs and all DMSs as categories with a special morphisms and prove that they are dually isomorphic.

16.2 Preliminaries

We assume basic knowledge of Boolean algebras, filters, ideals, ultrafilters, and Stone representation theory. We will use the following signature for Boolean algebras: $(B, \leq, 0, 1, ., +, *)$ where $0 \neq 1$ and a^* is the complement operation in B. If $A \subseteq B$ then $Ult(A)$ is the set of all ultrafilters contained in A.

We consider Sikorski (1964), Engelking (1977), and Mac Lane (1998) as standard reference books correspondingly for Boolean algebras, topology, and category theory.

16.2.1 Contact and Precontact Algebras

Definition 16.1 (*Contact and precontact relations*) Let $(B, 0, 1, \leq, +, ., *)$ be a Boolean algebra and C be a binary relation in B. C is called a **contact relation** in B if the following axioms are satisfied:

(C1) If aCb, then $a \neq 0$ and $b \neq 0$, **existentiality axiom**
(C2) If aCb and $a \leq a'$ and $b \leq b'$ then $a'Cb'$, **monotonicity axiom**,
(C3') If $aC(b + c)$ then aCb or aCc, (C3") If $(a + b)Cc$ then aCc or bCc, **distributivity axioms**
(C4) If aCb then bCa, **symmetry axiom**,
(C5) If $a.b \neq 0$, then aCb, (or, equivalently (C5') if $a \neq 0$, then aCa) **reflexivity axiom**.

We write $\overline{C}$ for the complement of C. If the axioms (C4) and (C5) are omitted then C is called a **precontact relation**. A precontact relation C satisfying (C4) (respectively, (C5)) is called a **symmetric precontact** (respectively, **reflexive precontact**). Note that if C is a symmetric precontact, then only one of the axioms (C3') and (C3") is needed.

If C is a contact (respectively, precontact) relation in B, then the pair (B, C) is called a **contact algebra** (respectively, **precontact algebra**).

Remark 16.1 Let us note that later on in this paper we will be interested in Boolean algebras containing several precontact relations sometimes satisfying some interacting axioms.

We will be interested also in contact and precontact algebras satisfying the following additional axiom:

(CEf) If $a\overline{C}b$, then $(\exists c)(a\overline{C}c$ and $c^*\overline{C}b)$ and

The axiom (CEf) is called sometimes **Efremovich axiom**, because it is used in the definition of Efremovich proximity spaces (Naimpally and Warrack 1970). In

Vakarelov (1997), this axiom is called transitivity axiom (see Lemma 16.2 for a motivation of this name).

Contact algebras are introduced in Dimov and Vakarelov (2006) and precontact algebras (under the name of Boolean proximity algebras) are introduced in Düntsch and Vakarelov (2007).

In the present context, we treat the Boolean part of the contact algebra as its *mereological component* and the contact relation—as its *mereotopological component.* In our treating of mereology, we consider the zero element 0 as a *non-existing region* and this can be used to define the ontological predicate of existence $E(a)$—*a "ontologically exists"* in the following way: $E(a)$ iff $a \neq 0$ (instead of $E(a)$ we will write sometimes $a \in E$). We will discuss and generalize this predicate in Sect. 16.3. Note that on the base of the predicate E the axioms (C1) and C5) can be presented as follows:

(C1) If aCb, then $a \in E$ and $b \in E$,

(C5) If $a.b \in E$, then aCb.

The above presentation of (C1) motivates its name as *existence axiom.*

The definitions of mereological relations *"part-of"* and *"overlap"* are the following:

• a is part of b iff $a \leq b$, i.e., part of is just the Boolean ordering,

• a overlaps b (in symbols aOb) iff there exists a region $c \neq 0$ such that $c \leq a$ and $c \leq b$ iff $a.b \neq 0$ iff $a.b \in E$.

Note that by the definition of overlap the axiom (C5) can be presented thus

(C5) aOb implies aCb.

Remark 16.2 It is easy to see that the relation O of overlap satisfies all axioms of contact relation and by axiom (C5) it can be considered as the smallest contact in B. It will be denoted also by C^{min}_{con}. Non-degenerate Boolean algebras have also another contact C^{max}_{con} definable by "$a \neq 0$ and $b \neq 0$". It follows by axiom (C1) that this is the largest contact in B.

Lemma 16.1 *The contact relation C^{max}_{con} satisfies the axiom of Efremovich ($C^{max}_{con} Ef$).*

Proof Suppose $a\overline{C^{max}_{con}}b$ and define c as follows:

$$c = \begin{cases} 1, & \text{if } b \neq 0 \\ 0, & \text{if } b = 0 \end{cases}$$

It is easy to see that thus defined c satisfies $a\overline{C^{max}_{con}}c$ and $c^*\overline{C^{max}_{con}}b$.

16.2.2 Examples of Contact and Precontact Algebras

Topological example of contact algebra. The intended example of contact algebra is a topological one and can be defined in the following way. Let X be a topological space and Cl and Int be the operations of closure and interior of a subset of X.

A set $a \subseteq X$ is called *regular closed* if $a = Cl(Int(a))$. The set $RC(X)$ of regular closed subsets of X is a Boolean algebra with respect to the following operations: $0 = \emptyset$, $1 = X$, $a + b = a \cup b$, $a.b = Cl(Int(a \cap b))$, $a^* = Cl(X \setminus a) = Cl(-a)$. The algebra $RC(X)$ becomes a contact algebra with respect to the following contact relation $C_X : aC_Xb$ iff $a \cap b \neq \emptyset$, i.e., if a and b have a common point, which is the intended meaning of contact relation. Obviously any Boolean subalgebra of $RC(X)$ under the same definition of C_X is also a contact algebra. Topological contact algebra will be considered as the standard example of contact algebra meaning that every contact algebra can be represented as a contact algebra of such a kind.

Relational examples of precontact and contact algebras. Let X be a non-empty set and R be a binary relation in X, called adjacency relation in X. The pair (X, R) is called *adjacency space* (see Düntsch and Vakarelov 2007 for the origin of this name and the intuitive meaning of adjacency spaces).

One can construct a precontact algebra from an adjacency space (X, R) as follows: let $B(X)$ be the Boolean algebra of all subsets under the set theoretical operations $a + b = a \cup b$, $a.b = a \cap b$ and complement $a^* = X \setminus a = -a$ and define C_R between two members of B as follows: aC_Rb iff there exist $x \in a$ and $y \in b$ such that xRy. It can easily be verified that all axioms of precontact are satisfied, so $(B(X), C_X)$ is a precontact algebra. Obviously, every Boolean subalgebra of $B(X)$ with the same definition of C_R is also a precontact algebra, called the discrete (or relational) precontact algebra. We will consider this kind of precontact algebra as a typical one, because every precontact algebra can be represented as such an algebra.

Lemma 16.2 (Düntsch and Vakarelov (2007)) *Let (X, R) be an adjacency space and $(B(X), C_R)$ be the precontact algebra over all subsets of X. Then, the following conditions hold:*

(i) R is a symmetric relation in X iff $(B(X), C_R)$ satisfies the symmetry axiom (C4),
(ii) R is reflexive relation in X iff $(B(X), C_R)$ satisfies the reflexivity axiom (C5),
(iii) R is a transitive relation in X iff $(B(X), C_R)$ satisfies the Efremovic axiom (CEf).

As a consequence of the above lemma, we see that the precontact algebra over a reflexive and symmetric adjacency space is a contact algebra. Such contact algebras are called discrete or relational. They also can be considered as typical examples of contact algebras, because every contact algebra can be represented as an algebra of such a kind.

16.2.3 Discrete (Relational) Representation of Contact and Precontact Algebras

Representation theory of precontact and contact algebras over adjacency spaces is considered in Düntsch and Vakarelov (2007). We will remind some definitions and constructions from Düntsch and Vakarelov (2007) which will be used later on in this paper.

Definition 16.2 (*Canonical adjacency space of a precontact algebra*) Let (B, C) be a precontact algebra. The adjacency space (X, R) where $X = Ult(B)$ and $R = R_C$ is called the canonical adjacency space for (B, C) if the relation R_C has the following definition—for arbitrary $U, V \in Ult(B)$:

UR_CV iff $(\forall a, b \in B)(a \in U$ and $b \in V \to aCb)$

We associate to this space the Stone embedding $s(a) = \{U \in Ult(B) : a \in U\}$ and the precontact algebra $(B(X), C_R)$ over (X, R), called the canonical relational precontact algebra of (B, C).

Note that the definition of R_C is meaningful not only between ultrafilters, but also between arbitrary subsets of B and especially between filters. Then, the following lemma is true.

Lemma 16.3 (Düntsch and Vakarelov (2007) R_C-extension Lemma) *Let C be a precontact relation in a Boolean algebra B. Then,*

(i) For $a \in B$, the set $[a) = \{b \in B : a \leq b\}$ is the smallest filter in B containing a.
(ii) aCb iff $[a)R_C[b)$,
(iii) Let U_0 and V_0 be filters in a precontact algebra (B, C) and let $U_0R_CV_0$. Then, there exist ultrafilters U and V such that $U_0 \subseteq U$, $V_0 \subseteq V$ and UR_CV.

Lemma 16.4 *(i) aCb iff there exist ultrafilters U, V such that UR_CV, $a \in U$ and $b \in V$.*

(ii) aCb iff $s(a)C_Rs(b)$.

Proof Apply Lemma 16.3.

Lemma 16.5 (Düntsch and Vakarelov (2007)) *Let $Ult(B)$ be the set of ultrafilters of (B, C). Then,*

(i) R_C is a symmetric relation in $Ult(B)$ iff (B, C) satisfies the axiom (C4).
(ii) R_C is a reflexive relation in $Ult(B)$ iff (B, C) satisfies the axiom (C5).
(iii) R_C is a transitive relation in $Ult(B)$ iff (B, C) satisfies the axiom (CEf).

The above lemmas imply the following representation theorem:

Theorem 16.1 (Relational representation theorem for precontact and contact algebras Düntsch and Vakarelov (2007)) *Let (B, C) be a precontact algebra, $(Ult(B), R_C)$ be the canonical adjacency space over (B, C), and s be the Stone embedding. Then,*

(i) s is an embedding of (B, C) into the precontact algebra over the canonical adjacency space $(Ult(B), R_C)$.
(ii) If (B, C) is a contact algebra, then the precontact algebra over the canonical adjacency space over (B, C) is a contact algebra.

Proof The idea of the proof is very simple. By the well-known properties of Stone embedding s embeds isomorphically the Boolean algebra B into the Boolean algebra of all subsets of $Ult(B)$. Then, the proof follows by applying Lemmas 16.4 (ii) and 16.5 (i) and (ii).

The above representation theorem of contact algebras is not the intended one because the contact is not of Whiteheadean type, namely to have a common point. In the next section, we will describe another representation of contact algebras using topology, which presents an Whiteheadean type contact between regions.

16.2.4 Topological Representation of Contact Algebras. Grills and Clans

All results of this section are taken from Dimov and Vakarelov (2006). We start by introducing another kind of abstract points in contact algebras called clans.

Definition 16.3 (*Clans and grills*) Let (B, C) be a contact algebra. A non-empty subset $\Gamma \subseteq B$ is called a **clan** in (B, C) if it satisfies the following conditions:

(i) $0 \notin \Gamma$, **existentiality condition,**
(ii) if $a \in \Gamma$ and $a \leq b$, then $b \in \Gamma$, **monotonicity condition,**
(iii) if $a + b \in \Gamma$, then $a \in \Gamma$ or $b \in \Gamma$, **distributivity condition,**
(iv) If $a, b \in \Gamma$ then aCb, **contact-closed condition.**
If we omit the condition (iv) Γ is called a **grill**. Γ is a *maximal clan* (maximal grill) if it a maximal set under the set inclusion.

If Γ is a clan (grill) then we denote by $Ult(\Gamma)$ the set of all ultrafilters contained in Γ and by $Clans(B)$ $(grills(B)$)—the set of all clans (grills) of (B, C).

If B contains several contact relations then the clans corresponding to each relation C are called C-clans and their set is denoted by C-clans(B).

The above definitions of clan and grill are algebraic abstractions from analogous notions in the proximity theory (for the origin of these notions see Thron (1973), from where we adopt these names).

Lemma 16.6 (Grill Lemma Thron (1973)) *If $F subseteq\, G$, is a grill then there exists an ultrafilter U such that $F \subseteq U \subseteq G$.*

Lemma 16.7 (Properties of grills)

(i) Every ultrafilter is a grill.
(ii) A non-empty union of ultrafilters is a grill.
(iii) If G is a grill then $G = \bigsqcup_{U \in Ult(G)} U$.
(iv) The complement of a grill is a proper ideal in B.
(v) If Γ is a grill, $a^*.b \notin \Gamma$ and $b \in \Gamma$, then $a \in \Gamma$.

Proof (i) and (2) are easy consequences from the definition of grill. For (iii) suppose $a \in G$. Then, the filter $[a)$ generated by a is contained in G and by the Grill Lemma there exists an ultrafilter U such that $[a) \subseteq U \subseteq G$. Consequently $a \in \bigsqcup_{U \in Ult(G)}$ and hence $G \subseteq \bigsqcup_{U \in Ult(G)} U$. The converse inclusion is obvious. (iv) follows by a direct verification.

(v) Suppose that Γ is a grill, $a^*.b \notin \Gamma$ and $b \in \Gamma$. We have $b = 1.b = (a + a^*).b = a.b + a^*.b$. Then $a^*.b \notin \Gamma$ and $b \in \Gamma$ imply $a.b \in \Gamma$ and since $a.b \leq a$, then $a \in \Gamma$.

Let us note that ultrafilters are clans, but there are other clans and they can be obtained by the following construction.

Let Σ be a non-empty set of ultrafilters of (B, C) such that if $U, V \in \sum$ then $U R_C V$, where R_C is the canonical adjacency relation in the set of ultrafilters of (B, C). Such sets of ultrafilters are called R_C-*cliques*. An R_C-clique is maximal, if it is a maximal set under the set inclusion. By the axiom of choice every R_C-clique is contained in a maximal R_C-clique. Let $\Gamma(\Sigma)$ be the union of all ultrafilters from Σ. Then, it can be verified that $\Gamma(\Sigma)$ is a clan. Moreover, every clan can be obtained by this construction from an R_C-clique and there is a correspondence between maximal cliques and maximal clans. All these facts about clans are contained in the following technical lemma:

Lemma 16.8 *(i) Every clan is contained in a maximal clan.*
(ii) Let Σ be an R-clique and $\Gamma(\Sigma) = \bigcup_{\Gamma \in \Sigma} \Gamma$. Then, $\Gamma(\Sigma)$ is a clan.
(iii) If $U, V \in Ult(\Gamma)$ then $U R_C V$, so $Ult(\Gamma)$ is an R_C-clique.
(iv) If Γ is a clan and $a \in \Gamma$ then there is an ultrafilter $U \in Ult(\Gamma)$ such that $a \in U$.
(v) Let Γ be a clan and $\sum$ be the R_C-clique $Ult(\Gamma)$. Then $\Gamma = \Gamma(\Sigma)$, so every clan can be defined by an R_C-clique as in (ii).
(vi) If Σ is a maximal R_C-clique, then $\Gamma(\Sigma)$ is a maximal clan.
(vii) If Γ is a maximal clan then $Ult(\Gamma)$ is a maximal R_C-clique.
(viii) For all ultrafilters U, V: $U R_C V$ iff there exists a (maximal) clan Γ such that $U, V \in Ult(\Gamma)$.
(ix) For all $a, b \in B$: aCb iff there exists a (maximal) clan Γ such that $a, b \in \Gamma$.
(x) For all $a, b \in B$: $a \not\leq b$ iff there exists an ultrafilter (clan) Γ such that $a \in \Gamma$ and $b \notin \Gamma$.
(xi) $a \neq 0$ iff there exists a clan Γ containing a.

The topological representation theory of contact algebras is based on the following construction. Let (B, C) be a contact algebra and let $X = Clans(B)$ and for $a \in B$, define $g(a) =_{\text{def}} \{\Gamma \in Clans(B) : a \in \Gamma\}$. We introduce a topology in X by taking the set $\mathbf{B} = \{g(a) : a \in B\}$ as the base of closed sets in X. The obtained topological space X is called the **canonical topological** space of (B, C). Then, we have the following lemma

Lemma 16.9 *(i)* $g(0) = \varnothing$, $g(1) = X$.
(ii) $g(a + b) = g(a) \cup g(b)$.

(iii) $g(a^*) = Cl_X(X \setminus a)$.
(iv) $g(a)$ *is a regular closed set of* X.
(v) aCb *iff* $g(a) \cap g(b) \neq \varnothing$.
(vi) $a \leq b$ *iff* $g(a) \subseteq g(b)$.

The above lemma implies the following representation theorem:

Theorem 16.2 (Topological representation theorem of contact algebras) *Let* (B, C) *be a contact algebra and* X *be its canonical topological space. Then, the mapping* g *from Lemma 16.9 is an embedding from* (B, C) *into the contact algebra* $RC(X)$.

In fact in Dimov and Vakarelov (2006) the corresponding representation theorem is stronger, stating additional properties of the canonical topological space $X = Clans(B)$: compactness, T_0 separation property and semiregularity (X is semiregular if it has a base for closed sets a family of regular closed sets), $g(B)$ is a dense subspace of$X$, but for the purposes of the present section these properties are not important.

Remark 16.3 Let us note that the clans corresponding to the least contact C^{max}_{con} (overlap) in B are just the proper filters of B and the maximal C^{max}_{con}-clans are just the ultrafilters. Analogously C^{max}_{con}-clans are just the grills in B. There exists only one maximal C^{max}_{con}- clan in B (the maximal grill) which is just the union of all ultrafilters in B.

16.2.5 Contact Relations Satisfying Efremovich Axiom (CEf). Clusters.

We show in this section that if the contact relation C of a contact algebra (B, C) satisfies the Efremovich axiom (CEf) we can introduce a new kind of abstract points in the algebra, called clusters. Our definition is an algebraic abstraction of the analogous notion used in the compactification theory of proximity spaces (see for instance Naimpally and Warrack 1970).

Definition 16.4 (*Clusters.* Dimov and Vakarelov (2006)) Let (B, C) be a contact algebra. A subset $\Gamma \subseteq B$ is called a **cluster** in (B, C) if it is a C-clan satisfying the following condition:

(Cluster) If $a \notin \Gamma$ then there exists $b \in \Gamma$ such that $a\overline{C}b$.

The set of clusters of (B, C) is denoted by Clusters(B).

If B contains several contact relations satisfying the Efremowich axiom then the clusters for each relation C are called C-clusters.

Lemma 16.10 *Let* (B, C) *be a contact algebra satisfying the Efremovich axiom (CEf). Then,* Γ *is a cluster in* (B, C) *iff* Γ *is a maximal clan in* (B, C).

Proof ($\Rightarrow$) Let Γ be a cluster in (B, C) and suppose that Γ is not a maximal clan. Then, there exists a clan Δ such that $\Gamma \subseteq \Delta$, $a \in \Delta$ and $a \notin \Gamma$. These facts imply that $a \in \Delta$ and that there exists $b \in \Gamma$ such that $a\overline{C}b$. But $a \in \Delta$ and $b \in \Delta$ imply aCb which is a contradiction.

($\Leftarrow$) Let Γ be a maximal clan and let $\Sigma = Ult(\Gamma)$. Then, by Lemma 16.8, the set Σ is a maximal R_C-clique and $\Gamma = \bigcup\{U : U \in \Sigma\}$. Let us note that in contact algebras satisfying the Efremovich axiom (CE) the canonical relation R_C is an equivalence relation (by Lemma 16.5). Note that maximal R_C-cliques for equivalence relations are just the equivalence classes generated by R_C. So the R_C-clique Σ is an R_C-equivalence class and can be represented as $\Sigma = |U_0|_R = \{U : UR_CU_0\}$ by some ultrafilter $U_0 \in \Sigma$. Since Γ is a clan, to show that Γ is a cluster we have to verify the condition (Cluster). Suppose for the sake of contradiction that this condition is not true, i.e., that there exists some $a \notin \Gamma$ such that for all $b \in \Gamma$ we have aCb. Since $U_0 \subseteq \Gamma$, then for all $b \in U_0$ we have aCb. This implies that $[a)R_CU_0$ where $[a) = \{c : a \leq c\}$ is the filter generated by a. Then, by Lemma 16.3, there exists an ultrafilter U such that $[a) \subseteq U$ and URU_0. Hence $a \in U$, $U \in \Sigma$, $U \subseteq \Gamma$ and consequently $a \in \Gamma$—a contradiction.

Let us note that the above lemma is a lattice-theoretic version of a result of Leader about clusters in proximity spaces mentioned in Thron (1973).

Lemma 16.11 *Let (B, C) be a contact algebra satisfying the Efremovich axiom (CEf). Then for any $a, b \in B$: aCb iff there is a cluster Γ containing a and b.*

Proof aCb iff (by Lemma 16.8) there exists a maximal clan Γ containing a and b. By Lemma 16.10 Γ is a cluster.

Note that we cannot prove a representation theorem for contact algebras satisfying the Efremovich axiom as subalgebras of regular closed sets using only clusters as abstract points, because in general we can not distinguish different regions by means of clusters. Ultrafilters can distinguish different regions, but in general they are not clusters.

The following lemma states how we can distinguish clusters.

Lemma 16.12 *Let (B, C) be a contact algebra satisfying the Efremovich axiom (CEf) and let Γ, Δ be clusters. Then, the following conditions are equivalent:*

(i) $\Gamma \neq \Delta$,
(ii) *there exist $a \in \Gamma$ and $b \in \Delta$ such that $a\overline{C}b$,*
(iii) *there exists $c \in B$ such that $c \notin \Gamma$ and $c^* \notin \Delta$.*

Proof $(i) \Rightarrow (ii)$ Suppose $\Gamma \neq \Delta$, then, since they are maximal clans, there exists $a \in \Delta$ and $a \notin \Gamma$. Consequently, there exists $b \in \Gamma$ such that $a\overline{C}b$, so (ii) is fulfilled.

$(ii) \Rightarrow (iii)$ Suppose that there exist $a \in \Gamma$ and $b \in \Delta$ such that $a\overline{C}b$. From $a\overline{C}b$ we obtain by the Efremovich axiom (CEf) that there exists c such that $a\overline{C}c$ and $c^*\overline{C}b$. Conditions $a \in \Gamma$ and $a\overline{C}c$ imply $c \notin \Gamma$. Similarly $b \in \Delta$ and $c^*\overline{C}b$ imply $c^* \notin \Delta$.

$(iii) \Rightarrow (i)$ Suppose that there exists $c \in B$ such that $c \notin \Gamma$ and $c^* \notin \Delta$ and for the sake of contradiction that $\Gamma = \Delta$. Since $c + c^* = 1$ then either $c \in \Gamma$ or $c^* \in \Delta$—a contradiction.

Remark 16.4 As we have mentioned the largest contact C^{max}_{con} of (B, C) satisfies the Efremowich axiom $(C^{max}_{con} Ef)$ (see Lemma 16.1) and C^{max}_{con}-clans are just the grills. Because there is only one maximal grill G^{max} (the union of all ultrafilters) it is the only C^{max}_{con}-cluster in B.

16.2.6 Boolean Algebras with Several Precontact Relations

We study in this section Boolean algebras with several precontact relations which may satisfy some additional axioms and some interacting axioms between them. The proofs which follow from a routine verification of some conditions are omitted and left to the reader.

Lemma 16.13 *(i) Let X be a non-empty set an R be a binary relation in X and let $(B(X), C_R)$ be the precontact algebra over all subsets of X determined by R. Then, C_R is a non-empty precontact relation iff R is a non-empty relation.*
(ii) Let B be a Boolean algebra and C be a precontact relation in B. Then, the following conditions are equivalent:
(iia) C is a non-empty relation,
(iib) $1C1$,
(iic) the canonical relation R_C is non-empty.

Lemma 16.14 *(i) Let $\underline{X} = (X, R_1, R_2)$ be a set with two binary relations and let $(B(X), C_{R_1}, C_{R_2})$ be the corresponding Boolean algebra with two precontacts over the set of all subsets of X. Then, the following condition holds:*
$C_1 \subseteq C_2$ iff $(C_{R_1} \subseteq C_{R_2})$.
(ii) Let $\underline{B} = (B, C_1, C_2)$ be a Boolean algebra with two precontact relations and let R_{C_1} and R_{C_2} be the corresponding canonical relations of C_1 and C_2. Then,
$C_1 \subseteq C_2$ iff $R_{C_1} \subseteq R_{C_2}$.

Definition 16.5 (i) Let R and R' be a binary relations in X. R is left (respectively, right) quasi-reflexive with respect to R' if it satisfies the following condition:
If xRy, then $xR'x$ (respectively, $yR'y$).
R is quasi-reflexive with respect to R' if it is left and right quasi-reflexive with respect to R'. R is quasi-reflexive if it is quasi-reflexive to itself.
(ii) Let C and C' be precontact relations in a Boolean algebra B. We say that C is left (respectively, right) quasi-reflexive with respect to C' if it satisfies the condition
If aCb, then $aC'a$ (respectively, $bC'b$).
C is quasi-reflexive with respect to C' if it is left and right quasi-reflexive with respect to R'. R is quasi-reflexive if it is quasi-reflexive to itself.

Lemma 16.15 *(i) Let R and R' be two binary relations in a non-empty set X and let $(B(X), C_R, C_{R'})$ be the Boolean algebra of all subsets of X with the corresponding precontact relations C_R and $C_{R'}$. Then,*

(ia) *C_R is left (right) quasi-reflexive with respect to $C_{R'}$ iff R is left (right) quasi-reflexive with respect to R'.*
(ib) *C_R is quasi-reflexive iff R is quasi-reflexive.*
(ii) *Let (B, C, C') be Boolean algebra with precontact relations C and C' and let R_C and $R_{C'}$ be their canonical relations. Then,*
(iia) *C is left (right) quasi-left reflexive with respect to C' iff R_C is left (right) quasi-reflexive with respect to $R_{C'}$.*
(iib) *C is quasi-reflexive iff R_C is quasi-reflexive.*

Proof (i) is easy.
(iia) $\Rightarrow$ Suppose that C is left quasi-reflexive with respect to C' and suppose also that for the ultrafilters U, V we have $U R_C V$. We have to show $U R_{C'} U$. To do this suppose first that $a, b \in U$. Then, $a.b \in U$ and $1 \in V$ and by $U R_C V$ we get $a.bC1$. By left quasi-reflexivity of C with respect to C' $a.bC1$ implies $a.bC'a.b$. Since $a.b \leq a$ and $a.b \leq b$, then by monotonicity we get $aC'b$ which by the definition of canonical relation (see 16.2) shows that $U R_{C'} U$.
$\Leftarrow$ Let the relation R_C be left quasi-reflexive with respect to $R_{C'}$ and let aCb. We have to show $aC'a$. By Lemma 16.4 there are Ultrafilters U, V such that $a \in U, b \in V$ and $U R_C V$. By the assumption we get $U R_{C'} U$. Since $a \in U$ and $b \in V$, then, by the definition of canonical relation we get $aC'a$. In a similar way, we consider the case for right quasi-reflexivity
The proof of (iib) is similar.

Lemma 16.16 *Let C be a non-empty left (or right) quasi-reflexive precontact relation in a Boolean algebra B and let $G = \{a \in B : aCa\}$. Then,*

(i) *G is a grill.*
(ii) *$U \in Ult(G)$ iff $U R_C U$, so ultrafilters contained in G are just the reflexive elements of the canonical relation R_C.*

Proof We will consider only the case of left quasi-reflexivity.

(i) The non-trivial part of (i) is the verification of the distributivity condition for grills. Let $a + b \in G$. Then, we have the following implications: $(a + b)C(a + b)$ $\Rightarrow$ (by the distributivity axiom for C) aCa or aCb or bCa or bCb $\Rightarrow$ (by left quasi-reflexivity of C) aCa or aCa or bCb or bCb $\Rightarrow$ aCa or bCb $\Rightarrow$ $a \in G$ or $b \in G$.
(ii) $\Rightarrow$ Let $U \in Ult(G)$ and suppose that $U\overline{R_C}U$. Then, there are exist $a \in U$ and $b \in U$ (and hence $a.b \in U$)) such that $a\overline{C}b$. But $U \subseteq G$, so $a.b \in G$, $(a.b)C(a.b)$ which implies aCb—a contradiction.
$\Leftarrow$ Suppose that U is an ultrafilter such that $U R_C U$ and let a be an arbitrary element in B such that $a \in U$. This implies aCa, so $a \in G$ and hence $U \subseteq G$ and $U \in Ult(G)$.

Definition 16.6 Let X be a non-empty set and R, R', R'' be binary relations in X. The following condition connecting R, R', R'' and denoted by $(Gtr\,RR'R'')$ is called general transitivity:

If xRy and $yR'z$, then $xR''z$.

In the case $R = R' = R''$ this is the standard transitivity of R.

Lemma 16.17 *Let R, R', R'' be binary relations in a non-empty set X and let $(B(X), C_R, C_{R'}, C_{R''})$ be the Boolean algebra of all subsets of X with corresponding precontact relations C_R, $C_{R'}$ and $C_{R''}$. Then, the following conditions are equivalent:*

(i) $(GtrRR'R'')$ For all $x, y, z \in X$ If xRy and $yR'z$, then $xR''z$.

(ii) $(C_RC_{R'}C_{R''}Ef)$ For all $a, b \in B(X)$: If $a\overline{C_{R''}}b$, then $(\exists c \in B(X)(a\overline{C_R}c$ and $c^*\overline{C_{R'}}b)$.

Proof $(i) \Rightarrow (ii)$ Suppose (i) and let $a\overline{C_{R''}}b$. Put $c = \{y \in X : (\exists z \in X)(yR'z$ and $z \in b\})$. It can easily be shown that $a\overline{C_R}c$ and $c^*\overline{C_{R'}}b$.

$(ii) \Rightarrow (i)$. Suppose (ii) and for the sake of contradiction suppose that (i) is not true, i.e., for some $x, y, z \in X$ we have xRy, $yR'z$ but $x\overline{R''}z$. From here we get $\{x\}\overline{C_{R''}}\{z\}$ and applying $(C_RC_{R'}C_{R''}Ef)$ we obtain that for some $c \in B(X)$ we have $\{x\}\overline{C_R}c$ and $c^*\overline{C_{R'}}\{z\}$.

Case 1: $y \in c$. Since xRy we get $\{x\}C_Rc$—a contradiction.

Case 2: $y \in c^*$. Since $yR'z$ we get $c^*C_{R'}\{z\}$—again a contradiction.

Remark 16.5 The condition $(C_RC_{R'}C_{R''}Ef)$ is called the *generalized Efremowich axiom*. For $R = R' = R''$ this is the Efremowich axiom. Later on in this paper we will use some special cases of this axiom and some special cases of Lemma 16.17.

The next technical lemma from Vakarelov (2010) will be used several times in this paper.

Lemma 16.18 *Let U, V be filters in a Boolean algebra and C be a precontact relation in B. Define the following sets:*

$F_I^C(U) =_{\text{def}} \{b \in B : (\exists a \in U)(a\overline{C}b^*)\}$,

$F_{II}^C(V) =_{\text{def}} \{a \in B : (\exists b \in V)(a^*\overline{C}b)\}$.

Then, (i) $F_I^C(U)$ and $F_{II}^C(V)$ are filters.

(ii) If V is an ultrafilter then UR_CV iff $F_I^C(U) \subseteq V$.

(iii) If U is an ultrafilter, then UR_CV iff $F_{II}^C(V) \subseteq U$.

Lemma 16.19 *Let B be a Boolean algebra and C, C', C'' be precontact relations in B. Then, the following conditions are equivalent:*

(i) $(CC'C''Ef)$ if $a\overline{C''}b$, then $(\exists c)(a\overline{C}c$ and $c^*\overline{C'}b)$,

(ii) $(GtrC_RC_{R'}C_{R''})$ For all $U, V, W \in Ult(B)$ If UR_CV and $VR_{C'}W$, then $UR_{C''}W$.

Proof $(i) \Rightarrow (ii)$ Suppose that (i) holds and for the sake of contradiction that (ii) does not hold, i.e., that for some ultrafilters U, V, W we have UR_CV, $VR_{C'}W$ but $U\overline{R_{C''}}W$. Then, there exist $a \in U$ and $b \in W$ such that $a\overline{C''}b$. This implies by $(CC'C''Ef)$ that there exists some c such that $a\overline{C}c$ and $c^*\overline{C'}b$. There are two cases for c:

Case 1: $c \in V$. Since $a \in U$ and UR_CV, this implies aCb—a contradiction.

Case 2: $c \notin V$. Then $c^* \in V$ and since $b \in W$ and $V R_{C'} W$, this implies $c^* C' b$—again a contradiction. Consequently (ii) holds.

$(ii) \Rightarrow (i)$. Suppose that (ii) holds and that (i) does not hold. Then for some $a, b \in B$ we have $a\overline{C''}b$ but there is no c such that $a\overline{C}c$ and $c^*\overline{C'}b$. We search for a contradiction. The strategy is to find ultrafilters U, V, W such that $a \in U$, $b \in W$ and $U R_C V$ and $V R_{C'} W$. This by generalized transitivity implies $U R_{C''} W$ and $a \in U$ and $b \in W$ implies $aC''b$ which is the desired contradiction.

To realize this strategy start with $[a)$ and $[b)$—the filters generated by a and b. Let $F_I^C([a))$ and $F_{II}^{C'}([b))$ be the filters defined in Lemma 16.18 (U is replaced by $[a)$ and V is replaced by $[b)$) and denote by $F_I^C([a)) \oplus F_{II}^{C'}([b))$ the smallest filter containing both of them. We will show that it is a proper filter. Suppose not: then there exists $c \in B$ such that

(1) $c^* \in F_I^C([a))$ and (2) $c \in F_{II}^{C'}([b))$.

Condition (1) implies that $\exists a' \in [a)$ (so $a \leq a'$) and $a'\overline{C}c^{**}$. This by monotonicity implies $a\overline{C}c$. By a similar reasoning condition (2) implies $c^*\overline{C'}b$. Thus, $(\exists c \in B)(a\overline{C}c$ and $c^*\overline{C'}b)$—a contradiction. Hence $F_I^C([a)) \oplus F_{II}^{C'}([b))$ is a proper filter. Now extend it to an ultrafilter V. Then, we get

(3) $F_I^C([a)) \subseteq V$ and

(4) $F_{II}^{C'}([b)) \subseteq V$.

By Lemma 16.18 (ii) (3) implies $[a) R_C V$. Applying Lemma 16.3 extend $[a)$ to an ultrafilter U such that $U R_C V$. Analogously by Lemma 16.18 (iii), (4) implies $V R_{C'} [b)$. Again by Lemma 16.3 extend $[b)$ to an ultrafilter W such that $V R_{C'} W$. Thus, we have obtained

(5) $U R_C V$, $V R_{C'} W$, $a \in U$ and $b \in W$—the strategy is realized.

As a corollary of Lemma 16.17, we obtain a new proof of Lemma 16.2 (iii).

16.3 Boolean Algebras with Predicates of Actual Existence and Actual Contact

One view of mereology adopted from Tarski is to identify it with complete Boolean algebra without zero 0 (see Simons 1987 page 25). The reason is that the zero elements do not correspond to existing objects. In our treatment we identify mereology with Boolean algebra including zero and use it to define a predicate of ontological existence $E(a) \leftrightarrow_{\text{def}} a \neq 0$ (see Sect. 16.2.1). In some sense, this can be considered as an approximate definition of the existence predicate and we used this in our papers (Vakarelov 2012, 2014, 2016a, b). This definition, however, possesses the following unpleasant property: there are too many existing regions and only one non-existing region—0. In fact, there are things (regions) different from 0 which are not actually (factually) existing. For instance, if we see something we intuitively conclude that it actually exists (exists at the moment of observation). But in general this is not true for things far from the observer, due to the finite velocity of light. For instance, if we see in the sky a star it is quite possible that this star is not existing at the moment of observation because it ceased to exist billions of years ago and now we see only the last light emitted from it. So, the observed star had been existing in the past but now it actually does not exist. In fact there is something related to the star which still exists—this is its last emitting of light. Of course the observed star may be actually existing at the moment of observation, if it is, for instance, the Sun. We conclude this fact, because the light travels from the Sun to the Earth several minutes and it is not possible for the Sun to cease to exist in such a short time. Another example: suppose we see a pregnant woman. We can not say for the child which she is expecting that it is actually existing now—it will exist only after its birth. But nevertheless the expected child exists now in some incomplete way, which may be phrased as possible or incomplete existing. There are things for which there are alternate periods of time in their life history with states of existence and states of non-existence. For example, a country that had been conquered alternately for some periods of time and after of any such period it had been free.

The above examples say that we have different modes of existence: actual existence (strong or factual existence) which stated for a region a will be denoted by $AE(a)$ or $a \in AE$, and the predicate $E(a) \Leftrightarrow a \neq 0$ which will be treated either as an existence for which we do not have a full information, or as a kind of possible, weak, or incomplete existence.

The predicates of existence E and actual existence AE have been discussed in our papers (Vakarelov 2017a, b) where we propose some axiomatic definition of AE and corresponding modification of the predicate of contact connected naturally to AE and called *actual contact*. In the next sections, we will remind some definitions and facts from these papers which will be used later on. Proofs will be given only for new statements not included in Vakarelov (2017a, b).

16.3.1 Actual Existence

First, we will characterize the predicate AE of actual existence in the context of Boolean algebras by a set of reasonable axioms. The problem is how to find these "reasonable axioms". In the next lemma we give abstract characterization of the predicate $E(a) \Leftrightarrow_{\text{def}} a \neq 0$ and from this characterization we extract the axioms for AE.

Lemma 16.20 *Let $\underline{B}$ be a Boolean algebra and let $E(a) \leftrightarrow_{\text{def}} a \neq 0$. Then,*
(i) $E(a)$ satisfies the following first-order conditions:

(AE1) $E(1)$ and $\overline{E}(0)$,

(AE2) If $E(a)$ and $a \leq b$, then $E(b)$,

(AE3) If $E(a + b)$, then $E(a)$ or $E(b)$.

(ii) E is the largest (under inclusion) predicate satisfying the axioms (AE1), (AE2), and (AE3).

Lemma 16.20 suggests the following definition.

Definition 16.7 Let B be a Boolean algebra and AE be a subset of B considered as a one-place predicate. We call AE a predicate of **actual existence** if it satisfies the axioms (AE1), (AE2), and (AE3) from Lemma 16.20. If an element $a \in B$ satisfies AE this will be denoted by $a \in AE$ or by $AE(a)$. The complement of AE will be denoted by $\overline{AE}$. The pair (B, AE) is called Boolean algebra with predicate of actual existence or shortly Boolean AE-algebra. An ultrafilter U included in AE is called **actual ultrafilter** (A-ultrafilter for short). We denote by $Ult(AE)$ the set of ultrafilters of B contained in AE.

Remark 16.6 (i) The formal definition of AE shows that $AE(a)$ implies $E(a)$, so EA is stronger than E and by Lemma 16.20 that E is the weakest predicate of existence and in the presence of EA can be treated as a predicate of non-actual existence.

(ii) Comparing with the definition of a grill (see Definition 16.3) we see that AE is a fixed grill in the Boolean algebra B and by Lemma 16.7 AE is union of a non-empty set of ultrafilters. Because ultrafilters are grills, then every ultrafilter is also a predicate of existence and these are the minimal predicates of actual existence. The predicate $E(a) \Leftrightarrow_{\text{def}} a \neq 0$ is the largest predicate of existence and it coincides with the union of all ultrafilters of B.

(iii) By means of the predicate of actual existence we may define an "actual" version of the mereological predicate overlap called **actual overlap** AO as follows: $a(AO)b \Leftrightarrow_{\text{def}} a.b \in AE$.

16.3.2 Actual Contact and Actual Precontact

Let us remind the reader that in the definition of contact relation two of its axioms use the definable existence predicate $E(a) \Leftrightarrow_{\text{def}} a \neq 0$: in (C1) aCb implies $E(a)$ and $E(b)$, and in (C5) $E(a.b)$ implies aCb. An "actualized" version of contact, called *actual contact* can be obtained by replacing E in these axioms by AE. In a similar way we obtain the definition of actual precontact. So we introduce the following formal definition.

Definition 16.8 Let (B, AE) be Boolean AE-algebra. A precontact relation C in B is called **actual contact**, shortly **A-contact**, if it satisfies the following additional axioms:

(AC1) If aCb, then $a, b \in AE$.
(AC2) If $a.b \in AE$, then aCb (or, equivalently (AC2′) If $a \in AE$, then aCa).
(AC3) If aCb, then bCa.

If we omit the axioms (AC2) and (AC3), C is called **actual precontact** (A-precontact). If C is an A-contact in (B, AE) then (B, AE, C) is called Boolean AE-algebra with A-contact C.

Although A-contact relations in general are not contact relations we put in their name the part "contact" because, as we will see later on, they behave almost as contact relations.

Lemma 16.21 *Let (B, AE, C) be Boolean AE-algebra and C be a binary relation in B. Then, the following conditions are equivalent:*

(i) C is an actual contact in (B, AE).
(ii) C is a non-empty quasi-reflexive and symmetric precontact relation in B such that $AE = \{a \in B : aCa\}$ (see Definition 16.5 for quasi-reflexive and symmetric precontact).

Proof (i)⇒(ii). By $(AC1)$, aCb implies $a, b \in AE$ which by (AC2′) implies aCa and bCb. This shows quasi-reflexivity of C. Since $1 \in AE$ we get that $1C1$, so C is a non-empty relation. By (AC1) and (AC2′) we obtain that $AE = \{a \in B : aCa\}$.

(ii)⇒(i) Quasi-reflexivity of C and the condition $AE = \{a \in B : aCa\}$ imply axioms (AC1) and (AC2′).

Lemma 16.22 *Let (B, AE, C) be Boolean AE-algebra with actual contact C. Then,*

(i) AE is definable by C as follows $AE = \{a \in B : aCa\}$.
(ii) Let $aC^{min}_{act}b \Leftrightarrow_{\text{def}} a.b \in AE$, i.e., C^{min}_{act} is the actual overlap in B. Then, C^{min}_{act} is the least actual contact in B.
(iii) Let $C^{max}_{act} \Leftrightarrow_{\text{def}} a \in AE$ and $b \in AE$. Then,
(1) C^{max}_{act} is the largest actual contact in (A, AE, C).
(2) C^{max}_{act} together with and $C^{max}con$ (see Remark 16.2) satisfy the following extended Efremovich axiom (which implies trivially the Efremovich axiom $(C^{max}_{act}Ef)$).
$(C^{max}_{act}C^{max}_{con}Ef+)$: $a\overline{C^{max}_{act}}b \Rightarrow (\exists c)(a\overline{C^{max}_{act}}c$ and $c^*\overline{C^{max}_{con}}c$ and $c^*\overline{C^{max}_{act}}b)$.

Proof The proofs of (i), (ii) and (iii) (1) can be obtained by a direct verification. We give the proof of (iii) (2). Suppose $a\overline{C^{max}_{act}}b$ and define c as follows:

$$c = \begin{cases} 1, & \text{if } b \in AE \\ 0, & \text{if } b \notin AE \end{cases}$$

It can be verified that thus defined c satisfies the remaining part of the extended Efremowich axiom.

Remark 16.7 Let (B, C) be a Boolean algebra and C be a non-empty quasi-reflexive and symmetric precontact relation in B. By Lemma 16.16 the set $G = \{a \in B : aCa\}$ is a grill. If we define $AE = G$ then by Lemma 16.21 C is an actual contact in the Boolean AE-algebra (B, AE, C). This means that the notion of Boolean algebra with actual existence predicate AE and actual contact C can be identified with Boolean algebra (B, C) with quasi-reflexive and symmetric precontact relation C and definable actual existence relation AE by C. We, however, will keep AE into the signature (B, AE, C) because of the separate importance of the predicate AE.

Lemma 16.23 *Let (B, AE, C) be Boolean AE-algebra with A-contact C. Let R_C be the canonical relation of C in the set $Ult(B)$ (see Definition 16.2). Then $U \in Ult(AE)$ iff $U R_C U$ (in other words: actual ultrafilters of B are just the reflexive ultrafilters with respect to R_C).*

Proof The proof follows by Lemma 16.16 from the following facts: C is a non-empty quasi-reflexive and symmetric precontact and that $AC = \{a \in B : aCa\}$.

Topological models of Boolean AE-algebras with A-contact. Since by Remark 16.7 a Boolean AE-algebra (B, AE, C) with A-contact C can be identified with the precontact algebra (B, C) with non-empty quasi-symmetric and reflexive precontact C, Lemmas 16.15 and 16.13 give relational examples of such precontact algebras.

Definition 16.9 (*Topological construction of actual contact and actual existence predicate*) A topological model of a Boolean AE-algebra (B, AE, C) with A-contact C can be defined as follows. By a topological space with actual points we mean any pair $\underline{X} = (X, X^a)$ such that X is a topological space and X^a is a non-empty subset of X whose elements are called **actual points**. Let $RC(X)$ be the Boolean algebra of regular closed subsets of X. For $a, b \in RC(X)$ define actual contact, denoted $C_{(X,X^a)}$, by $aC_{(X,X^a)}b \leftrightarrow_{\text{def}} a \cap b \cap X^a \neq \varnothing$, and the predicate of actual existence, denoted $AE_{(X,X^a)}$, by $AE_{(X,X^a)} =_{\text{def}} \{a \in RC(X) : a \cap X^a \neq \varnothing\}$.

It is easy to see that the system $(RC(X), AC_{(X,X^a)}, C_{(X,X^a)})$ is a Boolean AE-algebra and $C_{(X,X^a)}$ is an actual contact in it. This algebra is called the canonical topological AE-algebra with actual contact over (X, X^a).

16.3.3 Actual Clans

Definition 16.10 Let (B, AE, C) be Boolean AE-algebra with an A-contact C. The notion of a clan with respect to C, called now **actual clan** (A-clan for short) is the same as the definition of clan with respect to contact relations (see Definition 16.3), namely, a subset Γ of B is called **actual clan** of B if it is a grill such that aCb holds for all $a, b \in \Gamma$. Γ is a maximal A-clan if it is a maximal element in the set A-clans(B) of all A-clans of B. If Γ is an A-clan we denote by $Ult(\Gamma)$ the set of all ultrafilters included in Γ.

If in B there are several A-contact or contact relations, then for each C the corresponding clan will be called C-clan and from the type of C one can conclude if it is an A-clan or simply a clan.

Note that although the definition of A-clan is the same as the definition of clan for contact relations, A-clans are slightly different from clans because of the difference between contact and A-contact. In order to simplify the terminology, we will use the term "clan" for both type of clans.

Lemma 16.24 *Let (B, AE) be Boolean AE-algebra and C be an A-contact in B and let R_C be the canonical relation of C. Then,* *(i)* *Every A-clan can be extended into a maximal A-clan.*

(ii) *Every A-ultrafilter is an A-clan.*

(iii) *If* Γ *is an A-clan and* $U \in Ult(\Gamma)$, *then* U *is an A-ultrafilter. If* $U, V \in Ult(\Gamma)$, *then* $U R_C V$.

(iv) *If* $\sum$ *is a set of A-ultrafilters such that for every* $U, V \in \sum$ *we have* $U R_C V$, *then* $\Gamma = \bigsqcup_{U \in \sum} U$ *is an A-clan.*

(v) *If* Γ *is an A-clan, then* $\Gamma = \bigsqcup_{U \in Ult(\Gamma)} U$.

(vi) Γ *is an A-clan iff* Γ *is a grill and for all* $U, V \in Ult(\Gamma)$ *we have* $U R_C V$. *An ultrafilter* U *is an A-clan iff* U *is an A-ultrafilter.*

(vii) *For all* $a, b \in B$: aCb *iff there exists a (maximal) A-clan* Γ *such that* $a, b \in \Gamma$.

Proof The proofs of most of the statements of the lemma can be found in Vakarelov (2017a). Let us prove (vi). First we consider the first part of (vi).

$\Rightarrow$ Let Γ be an A-clan. Then by the definition of A-clan Γ is a grill. Suppose $U, V \in Ult(\Gamma)$. Then by (iii) we get $U R_C V$.

$\Leftarrow$. Suppose Γ is a grill and for all $U, V \in \Gamma : U R_C V$. We have to show that for all $a, b \in \Gamma$ we have aCb. Suppose $a, b \in \Gamma$. By the properties of grills there are $U, V \in Ult(\Gamma)$ such that $a \in U$ and $b \in V$. But we have $U R_C V$. Then, by the definition of the canonical relation (Definition 16.2)), we obtain aCb.

Let us prove the second part of (vi).

$\Rightarrow$ let U be an ultrafilter and let U be an A-clan. Then $Ult(U) = \{U\}$. By the first part of (vi) we have $U R_C U$. Then, by Lemma 16.23 we have $U \in AE$, so U is an A-ultrafilter.

$\Leftarrow$. This part follows from (ii).

Remark 16.8 It is natural to ask what are the clans corresponding to least and largest A-contacts in B. We mention them below without proofs.

(i) By Lemma 16.22 the least A-contact denoted by C^{min}_{act} is the actual overlap in B, $aO^a b \Leftrightarrow_{\text{def}} a.b \in AE$. Its actual clans are just A-ultrafilters. They also are the maximal actual C^{min}_{act}-clans.

(ii) Again by Lemma 16.22 the largest A-contact in B denoted by C^{max}_{act} is just $AE \times AE$. Its actual clans are just all grills included in AE. The only maximal actual C^{max}_{act}-clan coincides with AE.

16.3.4 Topological Representation of Boolean AE-Algebras with A-Contact

Remark 16.9 For contact algebras clans considered as abstract points were enough to prove a topological representation theorem. The crucial fact for this is that every ultrafilter is a clan, which makes possible to characterize the Boolean ordering by clans as follows: $a \nleq b$ iff there is a clan Γ such that $a \in \Gamma$ and $b \notin \Gamma$—here Γ is an ultrafilter and this characterization guarantees the isomorphic embedding. However, A-clans are not enough to characterize Boolean ordering in this way. So we need to use an additional definable contact relation in B. By means of A-contact C we may define a contact relation in B denoted here by $\widetilde{C}$ as follows: $a\widetilde{C}b \Leftrightarrow_{\text{def}} aCb$ or $a.b \neq 0$. The set of clans with respect to $\widetilde{C}$ include all A-clans (with respect to the A-contact C) which makes it possible to prove in Vakarelov (2017a) a topological representation theorem for Boolean AE-algebras with A-contact C. We outline here another, more natural topological representation theorem for such algebras using the definable contact relation $C^{max}_{con} \Leftrightarrow a \neq 0$ and $b \neq 0$. As we have mentioned in Remark 16.3 C^{max}_{con}-clans are just grills of B.

Let (B, AE, C) be Boolean AE-algebra with A-contact C. We say that the topological space with actual points (X, X^a) is the canonical topological space of B with actual points if $X = Grills(B)$ and $X^a = A\text{-}clans(B)$. Since A-clans are grills, obviously $X^a \subseteq X$. We define a topology in X by the function $g(b) = \{\Gamma \in Grills(B) : b \in \Gamma\}$ considering the set $\{g(a) : a \in B\}$ as the base of closed subsets of X. Let $(RC(X), AE_{(X,X^a)}, C_{(X,X^a)})$ be the Boolean AE-algebra with A-contact over the canonical topological space (X, X^a) of (B, AE, C). It can be proved that the mapping g is an isomorphic embedding of (B, AE, C) into the canonical Boolean AE-algebra over the canonical space of (B, AE, C). The proof can be done in a similar way as the topological representation theorem for contact algebras using grills and A-clans in the place of clans.

16.3.5 Actual Contacts Satisfying the Axiom of Efremovich. Actual Clusters

Let (B, AE, C) be Boolean AE-algebra with A-contact C satisfying the axiom of Efremovich (CEf) (see Remark 16.1).

Definition 16.11 (*Definition of actual clusters*) Let (B, AE, C) be a Boolean AE-algebra with A-contact C. The notion of a cluster with respect to C called now **actual cluster** (A-cluster for short) has the same definition as for cluster with respect to a contact relation. Namely, an A-clan Γ is an A-cluster if for every $a \notin \Gamma$ there exist $b \in \Gamma$ such that $a\overline{C}b$. The set of A-clusters of (B, AE, C) is denoted by A-clusters(B). If B contains several contact or A-contact relations satisfying the Efremowich axiom then the clusters for each relation C are called C-clusters.

The properties of A-clusters are almost the same as the properties of clusters with respect to contact relations in the presence of the Efremovich axiom (CEf) for C. Namely, we have the following lemma.

Lemma 16.25 *Let (B, AE, C) be a Boolean AE-algebra a with an A-contact C satisfying the Efremovich axiom (CEf). Then, the following equivalence is true:*
Γ is an A-cluster in (B, AE, C) iff Γ is a maximal A-clan in (B, AE, C).

Proof The proof is the same as the proof of Lemma 16.10 using the fact that axiom (CEf) induces transitivity of the canonical relation R_C (see Lemma 16.5 (iii)) which makes it an equivalence relation on the set $Ult(AE)$.

Lemma 16.26 *Let (B, AE, C) be a Boolean AE-algebra with A-contact C satisfying the Efremovich axiom (CEf). Then for any $a, b \in B$: aCb iff there is an A-cluster Γ containing a and b.*

Lemma 16.27 *Let (B, AE, C) be a Boolean AE-algebra with A-contact satisfying the Efremovich axiom (CEf) and let Γ, Δ be A-clusters. Then, the following conditions are equivalent:*

(i) $\Gamma \neq \Delta$,
(ii) *there exist $a \in \Gamma$ and $b \in \Delta$ such that $a\overline{C}b$,*
(iii) *there exists $c \in B$ such that $c \notin \Gamma$ and $c^* \notin \Delta$.*

Remark 16.10 By Lemma 16.22 (iii) the largest A-contact C_{act}^{max} in B satisfies the Efremowich axiom and for this A-contact there exist only one maximal C_{act}^{max}-clan, namely AE, so AE is the only C_{act}^{max}-cluster in B.

16.3.6 Semi-actual Contact Relations

Late on we will need the following modification of actual precontact.

Definition 16.12 Let (B, AE) be a Boolean AE-algebra and let C be a precontact relation in (B, AE). C is called a **semi-actual contact** if it satisfies the following axioms:

(SemiA 1) If aCb, then $a \in AE$ and $b \neq 0$
(SemiA 2) If $a \neq 0$, then $1Ca$,
(SemiA 3) If $a \in AE$, then aCa.

If C is a semi-actual contact in (B, AE), then the triple (B, AE, C) is called Boolean AE-algebra with semi-actual contact.

The following lemma is obvious.

Lemma 16.28 *The following conditions are equivalent:*

(i) *(B, AE, C) is Boolean AE-algebra with semi-actual contact C.*
(ii) *C is a left quasi-reflexive relation in (B, AE) satisfying (SemiA 2) and $a \in AE$ iff aCa.*

The following lemma gives an example of a semi-actual precontact in B.

Lemma 16.29 *Let $aC_{semi}^{max}b \Leftrightarrow_{\text{def}} a \in AE$ and $b \neq 0$. Then,*

(i) *C_{semi}^{max} is the largest semi-actual precontact in B.*

(ii) Together with the relation C^{max}_{con} *(see Lemma 16.22)* C^{max}_{semi} *satisfy the following version of generalized Efremovich axiom:*
$(C^{max}_{semi}C^{max}_{con}C^{max}_{semi}Ef)\ a\overline{C^{max}_{semi}}b \Rightarrow (\exists c)(a\overline{C^{max}_{semi}}c$ *and* $c^*C^{max}_{con}b)$.

Proof We will verify only (ii). Suppose $a\overline{C^{max}_{semi}}b$ and define c as follows:

$$c = \begin{cases} 1, & \text{if } a \notin AE \\ 0, & \text{if } a \in AE \end{cases}$$

It is easy that thus defined c satisfies the remaining conditions for c: $a\overline{C^{max}_{semi}}c$ and $c^*C^{max}_{con}b$.

16.4 A Dynamic Model of Space and Time Based on Snapshot Construction

In this section, we give a specific point-based spacetime structure called dynamic model of space and time (DMST) built by a special construction called **snapshot construction**. The development follows the general line of Vakarelov (2014, 2016a, b). The difference with Vakarelov (2014, 2016a, b) is that now we base this construction on the notion of Boolean AE-algebras with A-contact C which makes the model much more realistic, because there is a way to differ actual existence from non-actual existence. Since the notion of **time structure** is one of the base ingredients of the construction we start with this notion.

16.4.1 Time Structures

Classical physics describes changing objects by presenting their main features as functions of time. So it presupposes that the time is given by its sets of time points identifying them with real or rational numbers with their specific arithmetic structure. This structure of the set of time points is not obligatory for all situations where we have to describe change. Very often time structures have the form of an abstract relational system of the form $(T, \prec)$, where T is a non-empty set of time points (instants of time, moments) and $\prec$ is a binary relation on T such that $m \prec n$ means that *m is before n*. We also suppose that T is supplied with the standard notion of equality denoted as usual by $=$. Such systems are used in temporal logic (see for instance van Benthem 1983) to characterize temporal modalities. We do not presuppose in advance any fixed set of conditions for the relation $\prec$. One possible list of first-order conditions for $\prec$ which are called **time conditions** are listed in the following table. These conditions are typical for some systems of temporal logic. We describe them with their specific names and notations which will be used in this paper.

The list of time conditions

- **(RS)** *Right seriality* $(\forall m)(\exists n)(m \prec n)$,
- **(LS)** *Left seriality* $(\forall m)(\exists n)(n \prec m)$,
- **(Up Dir)** *Updirectedness* $(\forall i, j)(\exists k)(i \prec k \text{ and } j \prec k)$,
- **(Down Dir)** *Downdirectedness* $(\forall i, j)(\exists k)(k \prec i \text{ and } k \prec j)$,
- **(Circ)** *Circularity* $(\forall i, j)(i \prec j \rightarrow (\exists k)(j \prec k \text{ and } k \prec i))$
- **(Dens)** *Density* $i \prec j \rightarrow (\exists k)(i \prec k \text{ and } k \prec j)$,
- **(Ref)** *Reflexivity* $(\forall m)(m \prec m)$,
- **(Irr)** *Irreflexivity* $(\forall m)(m \not\prec m)$,
- **(Lin)** *Linearity* $(\forall m, n)(m \prec n \text{ or } n \prec m)$,
- **(Tri)** *Trichotomy* $(\forall m, n)(m = n \text{ or } m \prec n \text{ or } n \prec m)$,
- **(Tr)** *Transitivity* $(\forall ijk)(i \prec j \text{ and } j \prec k \rightarrow i \prec k)$.

We call the set of formulas (RS), (LS), (Up Dir), (Down Dir), (Circ), (Dens), (Ref), (Irr), (Lin), (Tri), (Tr) *time conditions*. If the relation $\prec$ satisfies the condition (Irr) it will be called "strict". If $\prec$ satisfies (Ref) the reading of $i \prec j$ should be, more precisely: "i is equal or before j".

Note that the above-listed conditions for time ordering are not independent. Taking some meaningful subsets of them we obtain various notions of time order. Of course, this list is not absolute and is open for extensions but in this paper we will consider only these 11 conditions.

16.4.2 Dynamic Model of Space and Time

Now we want to present a specific dynamic model of space and time based on a given time structure $(T, \prec)$ by means of the so called *snapshot construction*. The intuition based on this construction is the following. Suppose that we want to describe a dynamic environment consisting of a regions changing in time. First we suppose that we are given a time structure $(T, \prec)$ and want to know what the spatial configuration of regions at each moment of time $m \in T$ is. We assume that for each $m \in T$ the spatial configuration of the regions forms an AE-contact algebra (B_m, AE_m, C_m) with an A-contact C_m, or in more details—$(B_m, 0_m, 1_m, \leq_m, +_m, \cdot_m, *_m, AE_m, C_m)$. In other words (B_m, AE_m, C_m) is a "snapshot" of this configuration. We identify a given changing region a with the series $< a_m >_{m \in T}$ of snapshots and call such a series a dynamic region. In a sense this series can be considered also as a "trajectory" or "time history" of a. We denote by $\mathbb{B}$ the set of all dynamic regions. If $a = \langle a_m \rangle_{m \in T}$ is a given dynamic region, then a_m can be considered as the m-th coordinate of a, or *a at the time point m*. For instance the expression $a_m \in AE_m$ means that *a actually exists at the moment m*, $a_m \neq 0_m$ that a_m satisfies the weakest predicate of existence E_m with the meaning of weak or non-actual existence, and $a_m C_m b_m$ means that a and b are in an *actual contact* at the moment m. Thus, (B_m, AE_m, C_m) contains all m-th coordinates of the changing regions. We assume that the set $\mathbb{B}$ is a Boolean algebra with Boolean constants and operations defined as follows: $1 =< 1_m >_{m\in T}$, $0 =< 0_m >_{m\in T}$, Boolean ordering $a \leq b$ iff $(\forall m \in T)(a_m \leq_m b_m)$ and Boolean operations are defined "coordinatewise": $a + b =_{\text{def}}< a_m +_m b_m >_{m\in T}$, $a.b =_{\text{def}}< a_m \cdot_m b_m >_{m\in T}$, $a^* =_{\text{def}}< a_m^{*_m} >_{m\in T}$.

Let us define the Cartesian product (direct product) $\mathbf{B}$ of the coordinate Boolean algebras B_i, $i \in T$, namely $\mathbf{B} = \prod_{i\in T} B_i$. Obviously, $\mathbb{B}$ is a subalgebra of $\mathbf{B}$. Now we introduce the following important definition.

Definition 16.13 By a **dynamic model of space and time** (DMST) we understand the system $\mathcal{M} = \langle (T, \prec), \{(B_i, AE_i, C_i) : i \in T\}, \mathbb{B}, \mathbf{B}\rangle$. We say that $\mathcal{M}$ is a **full model** if $\mathbb{B} = \mathbf{B}$, and that $\mathcal{M}$ is a **rich model** if $\mathbb{B}$ contains all regions $a = \langle a_i \rangle_{i\in T}$ such that for all $i \in T$ either $a_i = 0_i$, or $a_i = 1_i$. (obviously every full model is a rich model).

Dynamic model of space and time we sometimes call "snapshot model" or "cinematographic model".

Let us note that DMST is a very expressive model of space and time with the main component the Boolean algebra $\mathbb{B}$ of dynamic regions which can be supplied with additional structure by various ways using the other components of the model. In this model time is presented by the point-based notion of time structure and space is presented by the coordinate algebras which are point-free. But having in mind their intended topological representation they can be substituted in an obvious way by the corresponding topological spaces with actual points (X_i, X_i^a) (called coordinate spaces) and instead of abstract coordinate algebras to consider the corresponding Boolean AE-algebras with A-contact over them. So in this model we do not have a single space but a series of spaces corresponding to each moment of time. This model can be considered also as a version of the so-called **block universe of reality** discussed in the contemporary philosophy of physics.

16.4.3 Standard Dynamic Contact Algebras

In this section, we supply the Boolean algebra $\mathbb{B}$ of dynamic regions with several spatio-temporal relations.

- **Actual existence**: $a \in AE$ iff $(\exists i \in T)(a_i \in AE_i)$
- **Actual space contact** (space A-contact): $aC^s b$ iff $(\exists i \in T)(a_i C_i b_i)$.
- **Actual time contact** (time A-contact): $aC^t b$ iff $(\exists i \in T)(a_i \in AE_i$ and $b_i \in AE_i)$.
- **Semi-actual time contact** (time SA-contact): $a \overrightarrow{C}^t b$ iff $(\exists i \in T)(a_i \in AE_i$ and $b_i \in E_i$. We define $a \overleftarrow{C}^t b \Leftrightarrow_{\text{def}} b \overrightarrow{C}^t a$.
- **Possible time contact**: $a\widehat{C}^t b$ iff $(\exists i \in T)(a_i \in E_i$ and $b_i \in E_i)$.
- **Actual precedence** (simply **Precedence**): $aC^{\prec} b$ iff $(\exists i, j \in T)(i \prec j$ and $a_i \in AE_i$ and $b_j \in AE_j)$.

Remark 16.11 The introduced above spatio-temporal relations are not the only possible ones, for instance in Vakarelov (2014, 2016a, b) we considered special classes of regions called "time representatives". But in this paper we are concentrating only on the above list of relations, which we call "the basic spatio-temporal relations".

The meaning of existence predicates AE and E was discussed in Sect. 16.3. The relations C^t, $\overrightarrow{C}^t$, $a\overleftarrow{C}^t b$ and $\widehat{C}^t$ express all possible kinds of simultaneous existence between dynamic regions at a given moment of time definable by the actual existence predicate AE and the definable predicate E. In a sense, they realize different kinds of contact in time which motivates their names.

Note that if we consider the special case when AE=E, then the relations C^t and $\overrightarrow{C}^t$ will collapse to $\widehat{C}^t$. This is just the case which we studied in Vakarelov (2014, 2016a, b, 2020).

Intuitively a is in a local precedence relation with b (in words *a precedes b*) means that there is a time point in which a actually exists which is before a time point in which b actually exists, which motivates the name of $C^{\prec}$ as a (local) precedence relation. Note the following similarity between the relations C^t and $C^{\prec}$: if in the definition of $C^{\prec}$ we replace the relation $\prec$ with $=$, then we obtain just the definition of C^t.

Definition 16.14 Let $\mathcal{M} = \langle (T, \prec), \{(B_i, AE_i, C_i) : i \in T\}, \mathbf{B}, \mathbb{B}\rangle$ be a DMST. The Boolean algebra $\mathbb{B}$ of dynamic regions supplied with the above defined spatio-temporal relations is called **standard dynamic contact algebra** (standard DCA) over $\mathcal{M}$. If $\mathcal{M}$ is a full (rich) DMST, then $\mathbb{B}$ is also called full (rich) standard DCA.

The notion of standard DCA is used in this paper as a source of true sentences to be used as axioms for the abstract notion of DCA. In the next lemma, we collect a set of such sentences.

Lemma 16.30 *Let* $(\mathbb{B}, AE, C^s, C^t, \overrightarrow{C}^t, \widehat{C}^t)$ *be a standard DCA over a rich DMST* $\mathcal{M} =< (T, \prec), \{(B_i, C_i) : i \in T\}, \mathbb{B}, \mathbf{B}\rangle$. *Then, the set* AE *and the relations* C^s, C^t, $\widehat{C}^t$ *and* $C^{\prec}$ *satisfy the following abstract conditions which are called DCA-axioms:*

(DCA AE) AE *is a predicate of actual existence in* $\mathbb{B}$.

(DCA C^s*)* C^s *is an actual contact relation in* $\mathbb{B}$,

(DCA C^t*)* C^t *is an actual contact relation in* $\mathbb{B}$ *such that* $C^s \subseteq C^t$.

(DCA $\overrightarrow{C}^t$*)* $\overrightarrow{C}^t$ *is a semi-actual precontact relation such that* $C^t \subseteq \overrightarrow{C}^t$.

(DCA $\widehat{C}^t$*)* $\widehat{C}^t$ *is a contact relation in* $\mathbb{B}$. *It satisfies the following additional axioms: the inclusion axiom:*

$\overrightarrow{C}^t \subseteq \widehat{C}^t$, *the Efremovich axiom:*

$(\widehat{C}^t Ef)$ $a\overline{\widehat{C}^t}b \Rightarrow (\exists c)(a\overline{\widehat{C}^t}c$ *and* $c^*\overline{\widehat{C}^t}b)$, *the extended Efremovich axiom (mentioned in Lemma 16.22):*

$(C^t\widehat{C}^t Ef+)$ $a\overline{C^t}b \Rightarrow (\exists c)(c^*\overline{C^t}c$ *and* $c^*\overline{\widehat{C}^t}c$ *and* $c^*\overline{C^t}b)$, *and a version of the generalized Efremowich axiom:*

$(\overrightarrow{C}^t\widehat{C}^t\overrightarrow{C}^t Ef)$ $a\overline{\overrightarrow{C}^t}b \Rightarrow (\exists c)(a\overline{\overrightarrow{C}^t}c$ *and* $c^*\overline{\widehat{C}^t}b)$

(DCA $C^{\prec}$*)* $C^{\prec}$ *is an actual precontact relation in* $\mathbb{B}$ *satisfying the following two versions of the generalized Efremovich axioms (see Remark 16.5)*

$(C^{\prec}C^tC^{\prec}Ef)$ $a\overline{C^{\prec}}b \Rightarrow (\exists c)(a\overline{C^t}c$ *and* $c^*\overline{C^{\prec}}b)$, *and*

$(C^tC^{\prec}C^{\prec}Ef)$ $a\overline{C^{\prec}}b \Rightarrow (\exists c)(a\overline{C^{\prec}}c$ *and* $c^*\overline{C^t}b)$.

Proof We will give proofs only for the corresponding Efremovich-like axioms. Since they are existential first-order conditions only for their proofs the assumption that the algebra is rich will be used. We omit the proofs for the other statements, because they are based on simple verifications of the corresponding definitions.

Proof of ($\widehat{C}^t Ef$). Suppose $a\overline{\widehat{C}^t}b$ and define c coordinatewise as follows:

$$c_i = \begin{cases} 1_i, & \text{if } a_i = 0_i \\ 0_i, & \text{if } a_i \neq 0_i \end{cases}$$

Note that $c \in \mathbb{B}$ because the algebra is rich. We will show that c satisfies the remaining part of $\widehat{C}^t Ef$: $a\overline{\widehat{C}^t}c$ and $c^*\overline{\widehat{C}^t}b$. Suppose for the sake of contradiction that $a\widehat{C}^t c$. Then, there exists $i \in T$ such that $a_i \neq 0_i$ and $c_i \neq 0_i$. Then, by the definition of c, we get $c_i = 0_i$—a contradiction. Suppose now that $c^*\widehat{C}^t b$. Then there exists $i \in T$ such that $c_i^* \neq 0_i$. Then, by the definition of c we get $c_i^* \neq 0_i$ and hence $c_i \neq 1_i$, which implies $c_i = 0_i$ and consequently $a_i \neq 0_i$. But $a_i \neq 0_i$ and $b_i \neq 0_i$ implies $a\widehat{C}^t b$—in a contradiction with the assumption $a\overline{\widehat{C}^t}b$.

Proof of ($C^t\widehat{C}^t$**Ef+**). Suppose $a\overline{C}^t b$ and define c coordinatewise as follows:

$$c_i = \begin{cases} 1_i, & \text{if } a_i \notin AE_i \\ 0_i, & \text{if } a_i \in AE_i \end{cases}$$

$c \in \mathbb{B}$ because the algebra is rich. We will show that c satisfies the remaining part of ($\widehat{C}^t$Ef+): ($a\overline{C}^t c$ and $c^*\overline{\widehat{C}^t}c$ and $c^*\overline{C}^t b$). Suppose for the sake of contradiction that $aC^t c$. Then, there exists $i \in T$ such that $a_i \in AE_i$ and $c_i \in AE_i$. From here we get $c_i = 0_i$ and consequently $0_i \in AE_i$—a contradiction. Suppose now $c^*\widehat{C}c$. Then, there exists $i \in T$ such that $c_i^* \neq 0_i$ and $c_i \neq 0_i$. The last condition implies that $c_i = 1_i$ and consequently $c_i^* = 0_i$—a contradiction with $c^* \neq 0_i$. To verify $c^*\overline{C}^t b$ suppose that it is not true. Then, there exists $i \in T$ such that $c_i^* \in AE_i$ and $b_i \in AE_i$. Then $c_i^* \neq 0_i$ and by the definition of c we conclude that $c_i = 0_i$ which implies that $a_i \in AE_i$. But $a_i \in AE_i$ and $b_i \in AE_i$ imply $aC^t b$ which contradicts the initial assumption $a\overline{C}^t b$.

Proof of ($\overrightarrow{C}^t\widehat{C}^t\overrightarrow{C}^t$**Ef**)) Suppose $a\overline{\overrightarrow{C}^t}b$ and define c as follows;

$$c_i = \begin{cases} 1_i, & \text{if } a_i \notin AE_i \\ 0_i, & \text{if } a_i \in AE_i \end{cases}$$

The remaining part of the proof is as in the previous cases.

Proof of ($C^{\prec}C^t C^{\prec}$**Ef**)). Suppose $a\overline{C^{\prec}}b$ and define c as follows:

$$c_i = \begin{cases} 1_i, & \text{if } a_i \notin AE_i \\ 0_i, & \text{if } a_i \in AE_i \end{cases}$$

As in the above proofs, one can show that c satisfies the remaining part of $C^{\prec}C^{t}C^{\prec}$Ef).

Proof of ($C^{t}C^{\prec}C^{\prec}$**Ef**). The proof is similar to the previous one using the following definition of c:

$$c_i = \begin{cases} 1_i, & \text{if } b_i \in AE_i \\ 0_i, & \text{if } b_i \notin AE_i \end{cases}$$

16.4.4 Time Conditions and Time Axioms

We do not presuppose in the formal definition of dynamic model of space that the time structure $(T, \prec)$ satisfies some of the abstract properties of the precedence relation introduced in Sect. 16.4.1 and named *time conditions*. In this section, we shall see that these time conditions are in an exact correlation with some special conditions on dynamic regions called *time axioms*. The correlation can be seen in the table below and the meaning of the correlation is stated in the Correlation Lemma after the table.

Correlation table

(RS) *Right seriality* $(\forall i)(\exists j)(i \prec j) \Longleftrightarrow$
 (rs) $a \in AE \rightarrow aC^{\prec}1$,
(LS) *Left seriality* $(\forall i)(\exists j)(j \prec i) \Longleftrightarrow$
 (ls) $a \in AE \rightarrow 1C^{\prec}a$,
(Up Dir) *Updirectedness* $(\forall i, j)(\exists k)(i \prec k$ and $j \prec k) \Longleftrightarrow$
 (up dir) $a \in AE$ and $b \in AE \rightarrow aC^{\prec}p$ or $bC^{\prec}p^*$,
(Down Dir) *Downdirectedness* $(\forall i, j)(\exists k)(k \prec i$ and $k \prec j) \Longleftrightarrow$
 (down dir) $a \in AE$ and $b \in AE \rightarrow pC^{\prec}a$ or $p^*C^{\prec}b$,
(Circ) *circularity* $i \prec j \rightarrow (\exists k)(k \prec i$ and $j \prec k) \Longleftrightarrow$
 (cirk) $aC^{\prec}b \rightarrow bC^{\prec}p$ or $p^*C^{\prec}a$
(Dens) *Density* $i \prec j \rightarrow (\exists k)(i \prec k \wedge k \prec j) \Longleftrightarrow$
 (dens) $aC^{\prec}b \rightarrow aC^{\prec}p$ or $p^*C^{\prec}b$,
(Ref) *Reflexivity* $(\forall i)(i \prec i) \Longleftrightarrow$
 (ref) $aC^{t}b \rightarrow aC^{\prec}b$,
(Irr) *Irreflexivity* $(\forall i)(i \not\prec i) \Longleftrightarrow$
 (irr) $aC^{\prec}a \rightarrow (\exists c, d)(aC^{t}c$ and $aC^{t}d$ and $c\overline{C}^{t}d)$,
(Lin) *Linearity* $(\forall i, j)(i \prec j$ or $j \prec i) \Longleftrightarrow$
 (lin) $a \in AE$ and $b \in AE \rightarrow aC^{\prec}b$ or $bC^{\prec}a$,
(Tri) *Trichotomy* $(\forall i, j)(i = j$ or $i \prec j$ or $j \prec i) \Longleftrightarrow$
 (tri) $a \in AE$ and $b \in AE \rightarrow aC^{t}b$ or $aC^{\prec}b$ or $bC^{\prec}a$,
(Tr) *Transitivity* $i \prec j$ and $j \prec k \rightarrow i \prec k \Longleftrightarrow$
 (tr) $a\overline{C^{\prec}}b \rightarrow (\exists c)(a\overline{C^{\prec}}c$ and $c^*\overline{C^{\prec}}b)$.

Definition 16.15 The formulas in the list (rs), (ls), (up dir), (down dir), (circ), (dens), (ref), (irr), (lin), (tri), (tr) which are taken from the Correspondence table, are called **time axioms**.

Lemma 16.31 (Correspondence Lemma) *Let* $(\mathbb{B}, AE, C^s, C^t, \widehat{C}^t$ *be a rich standard DCA over DMST* $\mathcal{M} =< (T, \prec), \{(B_i, C_i) : i \in T\}, \mathbb{B}, \mathbf{B}(T)\rangle$. *Then, all the correspondences in the above table are true in the following sense: the left site of a given equivalence is true in* $(T, \prec)$ *iff the right site is true in* $\mathbb{B}$.

Proof We will demonstrate the proofs only for three examples. The other proofs can be obtained in a similar way.

Proof of (RS) $\Longleftrightarrow$ **(rs).** $\Rightarrow$ Suppose that (RS) holds in $(T, \prec)$ and $a \in AE$. Then, there exists $i \in T$ such that $a_i \in AE_i$. By (RS) there exist $j \in T$ such that $i \prec j$. We have that $1_j \in AE_j$. All this implies $aC^{\prec}1$ which sows that (rs) holds in the algebra $\mathbb{B}$.

$\Leftarrow$ Suppose now that (rs) holds in the algebra $\mathbb{B}$ and suppose that (RS) does not hold in $(T, \prec)$, i.e., there exists $i_0 \in T$ such that $(\forall j \in T)(i_0 \not\prec j)$. Define a coordinatewise as follows:

$$a_i = \begin{cases} 1_i, & \text{if } i = i_0 \\ 0_i, & \text{if } i \neq i_0 \end{cases}$$

Since the algebra is rich, $a \in \mathbb{B}$. This implies that $a_{i_0} = 1_0 \in AE_{i_0}$ and hence $a \in AE$. This by (rs) implies $aC^{\prec}1$. This means that there exist i, j such that $i \prec j$, $a_i \in AE_i$ and $1_j \in AEj$. Condition $a_i \in AE_i$ implies that $a_i \neq 0_i$, which by the definition of a implies that $a_i = 1_i$ and consequently $i = i_0$. Substituting this in $i \prec j$ we get $i_0 \prec j$—a contradiction.

Proof of (Dense) $\Longleftrightarrow$ **(dense).** $\Rightarrow$ Suppose that (Dense) holds and suppose $aC^{\prec}b$ and that $p \in \mathbb{B}$. Then, there exist $i, j \in T$ such that $i \prec j$, $a_i \in AE_i$ and $b_j \in AE_j$. By (Dense) there exist $k \in T$ such that $i \prec k \prec j$. There are two cases for p_k:

Case 1: $p_k \in AE_k$. This case together with $a_i \in AE_i$ and $i \prec k$, implies $aC^{\prec}p$.

Case 2: $p_k \notin AE_k$. Then $p_k^* \in AE_k$. This, together with $b_j \in AE_j$ and $k \prec j$ implies $p^*C^{\prec}b$, which shows that (dense) holds in $\mathbb{B}$.

$\Leftarrow$ Suppose that (dense) holds in $\mathbb{B}$ but that (Dense) does not hold in $(T, \prec)$, i.e., there exist $i_0, j_0 \in T$ such that $i_0 \prec j_0$ and that it is not true that $(\exists k \in T)(i_0 \prec k \prec j_0)$. Define a, b, p as follows:

$$a_i = \begin{cases} 1_i, & \text{if } i = i_0 \\ 0_i, & \text{if } i \neq i_0 \end{cases}, b_j = \begin{cases} 1_j, & \text{if } j = j_0 \\ 0_i, & \text{if } j \neq j_0 \end{cases}, p_k = \begin{cases} 1_k, & \text{if } k \prec j_0 \\ 0_i, & \text{if } k \not\prec j_0 \end{cases}$$

From here we obtain $a_{i_0} = 1_{i_0} \in AE_{i_0}$ and $b_{j_0} = 1_{j_0} \in AE_{j_0}$ which by $i_0 \prec j_0$ implies $aC^{\prec}b$. Then, by (dense), we get $aC^{\prec}p$ or $p^*C^{\prec}b$. We will show that both cases imply contradiction.

Case 1: $aC^{\prec}p$. Then, there exist i, k such that $i \prec k$, $a_i \in AE_i$ and $p_k \in AE_k$. The condition $a_i \in AE_i$ implies $a_i \neq 0_i$, so $a_i = 1_i$, so $i = i_0$ and hence $i_0 \prec k$.

With a similar reasoning the condition $p_k \in AE_k$ implies $p_k = 1_k$ and consequently $k \prec j_0$, which together with $i_0 \prec k$ imply a contradiction with the assumption.

Case 2: $p^*C^{\prec}b$. Then, there exist $k, j \in T$ such that $k \prec j$, $p_k^* \in AE_k$ and $b_j \in AE_j$. The condition $p_k^* \in AE_k$ implies that $p_k^* \neq 0_k$, hence $p_k \neq 1_k$, and finally—$p_k = 0_k$. This implies $k \not\prec j_0$. With a similar reasoning condition $b_j \in AE_j$ implies that $b_j = 1_j$ and consequently $j = j_0$. Substituting this in $k \prec j$ we get $k \prec j_0$—which contradicts $k \not\prec j_0$.

Proof of (Irr) $\Longleftrightarrow$ (irr). $\Rightarrow$ Suppose (Irr) and let $aC^{\prec}a$. Then, for some $i, j \in T$, we have $i \prec j$, $a_i \in AE_i$ and $a_j \in AE_j$. Define c, d as follows:

$$c_m = \begin{cases} 1_m, & \text{if } m = i \\ 0_m, & \text{if } m \neq i \end{cases}, \quad d_n = \begin{cases} 1_n, & \text{if } n = j \\ 0_n, & \text{if } n \neq j \end{cases}$$

It is easy to see that thus defined c, d satisfy the conditions $aC^t c$, $aC^t d$ and $c\overline{C}^t d$ so (irr) is fulfilled.

$\Leftarrow$ For the converse implication suppose (irr) and for the sake of contradiction that there exists $i \in T$ such that $i \prec i$. Define a as follows:

$$a_k = \begin{cases} 1_k, & \text{if } k = i \\ 0_k, & \text{if } k \neq i \end{cases}$$

Obviously $a_i = 1_i \in AE_i$ and since $i \prec i$, then $aC^{\prec}a$. Then, by (irr), there exist c, d such that $aC^t c$, $aC^t d$ and $c\overline{C}^t d$. Condition $aC^t c$ implies that there exists $k \in T$ such that $a_k \in AE_k$ and $c_k \in AE_k$. Also $a_k \in AE_k$ implies that $a_k \neq 0_k$ and by the definition of a this yields $k = i$, so $c_i \in AE_i$. In a similar way the condition $aC^t d$ implies that $d_i \in AE_i$ which together with $c_i \in AE_i$ both imply $cC^t d$, which contradicts $c\overline{C}^t d$.

Proof of (Tr) $\Longleftrightarrow$ (tr). $\Rightarrow$ Suppose that (Tr) holds in $(T, \prec)$ and let $a\overline{C^{\prec}}b$. Define c as follows:

$$c_j = \begin{cases} 1_j, & \text{if } (\exists k)(j \prec k \text{ and } b_k \in AE_k) \\ 0_j, & \text{otherwise} \end{cases}$$

We will show that $a\overline{C^{\prec}}c$ and $c^*\overline{C^{\prec}}b$.

The proof of $a\overline{C^{\prec}}c$. Suppose for the contrary that $aC^{\prec}c$. Then, there exist i, j such that $i \prec j$, $a_i \in AE_i$ and $c_j \in AE_j$. The condition $c_j \in AE_j$ implies that $c_j \neq 0_j$, so $c_j = 1_j$, which implies $(\exists k)(j \prec k$ and $b_k \in AE_k)$. The conditions $i \prec j$ and $j \prec k$ imply by (Tr) $i \prec k$, which together with $a_i \in AE_i$ and $b_k \in AE_k$ imply $aC^{\prec}b$—contradiction with the assumptions.

For the proof of $c^*\overline{C^{\prec}}b$ suppose again the contrary: $c^*C^{\prec}b$. Then, there exist $j, k \in T$ such that $j \prec k$, $c^* \in AE_j$ and $b_k \in AE_k$. The conditions $j \prec k$ and $b_k \in AE_k$ yield $c_j = 1_j$ and hence $c_j^* = 0_j$. This together with $c^* \in AE_j$ implies $0_j \in AE_i$—a contradiction.

$\Leftarrow$ Suppose that (tr) holds in $\mathbb{B}$ but (Tr) does not hold in $(T, \prec)$, Then, there exist $i_0, j_0 \in T$ such that $i_0 \prec j_0$, $j_0 \prec k_0$, but $i_0 \not\prec k_0$. Define a, b as follows:

$$a_i = \begin{cases} 1_i, & \text{if } i = i_0 \\ 0_i, & \text{if } i \neq i_0 \end{cases}, b_j = \begin{cases} 1_j, & \text{if } j = k_0 \\ 0_i, & \text{if } j \neq k_0 \end{cases}$$

We get from here that $a_{i_0} = 1_{i_0} \in AE_{i_0}$ (and hence $a_{i_0} \in AE_{i_0}$) and $b_{k_0} = 1_{k_0} \in AE_{k_0}$ (and hence $b_{k_0} \in AE_{k_0}$) and $a\overline{C^{\prec}}b$. It follows by (tr) that $(\exists c)(a\overline{C^{\prec}}c$ and $c^*\overline{C^{\prec}}b)$. There two cases for c_{j_0}.

Case 1: $c_{j_0} \in AE_{j_0}$. This together with $i_0 \prec j_0$ and $a_{i_0} \in AE_{i_0}$ imply $aC^{\prec}c$—in a contradiction with $a\overline{C^{\prec}}c$.

Case 2: $c_{j_0} \notin AE_{j_0}$. This implies $c^* \in AE_{i_0}$, which together with $j_0 \prec k_0$ and $b_{k_0} \in AE_{k_0}$ yield $c^*C^{\prec}b$—in a contradiction with $c^*\overline{C^{\prec}}b$.

Remark 16.12 The Correlation Lemma 1 is very important because it states that the *time conditions* , which are reasonable abstract properties of the time structures are determined correspondingly by the *time axioms* which are *point-free* in the sense that they do not contain time points and are formulated using only variables for for dynamic regions by means of some of the spatio-temporal relations AE, C^t and $C^{\prec}$. This correlation suggests considering (abstract) DCA-s which will be introduced in the next section satisfying some reasonable collections of the time axioms.

16.5 Dynamic Contact Algebra (DCA)

The aims of this section are three:

(1) To introduce an abstract notion of *dynamic contact algebra*, DCA for short. This can be done by taking the signature of standard DCA introduced in Sect. 16.4.3 and taking as axioms the conditions satisfying by standard DCA formulated in Lemma 16.30.

(2) To study some formal properties of DCA-s needed in (3).

(3) To proof a representation theorem for DCA-s representing them in standard DCA-s.

The above aims will be realized in several subsections.

16.5.1 Abstract Definition of DCA

Definition 16.16 The system $A = (B, AE, C^s, C^t, \overrightarrow{C^t}\widehat{C^t}, C^{\prec})$ is called **dynamic contact algebra** (DCA), if B is a Boolean algebra and the following conditions, taken from Lemma 16.30 are satisfied:

(DCA AE) AE is the predicate of actual existence in A.

(DCA C^s) C^s is an actual contact relation in A.

(DCA C^t) C^t is an actual contact relation in A such that $C^s \subseteq C^t$.

(DCA $\overrightarrow{C^t}$) $\overrightarrow{C^t}$ is a semi-actual precontact relation such that $C^t \subseteq \overrightarrow{C^t}$.

(DCA $\widehat{C^t}$) $\widehat{C^t}$ is a contact relation in B. It satisfies the following additional axioms: the inclusion axiom:

$\overrightarrow{C^t} \subseteq \widehat{C^t}$, the Efremovich axiom:

$(\widehat{C^t}Ef)$ $a\overline{\widehat{C^t}}b \Rightarrow (\exists c)(a\overline{\widehat{C^t}}c$ and $c^*\overline{\widehat{C^t}}b)$, the extended Efremovich axiom (mentioned in Lemma 16.22):

$(C^t\widehat{C^t}$Ef+$)$ $a\overline{C^t}b \Rightarrow (\exists c)(c^*\overline{C^t}c$ and $c^*\overline{\widehat{C^t}}c$ and $c^*\overline{C^t}b)$, and a version of the generalized Efremowich axiom:

$(\overrightarrow{C^t}\widehat{C^t}\overrightarrow{C^t}Ef)$ $a\overline{\overrightarrow{C^t}}b \Rightarrow (\exists c)(\overline{\overrightarrow{C^t}}c$ and $c^*\overline{\widehat{C^t}}b)$.

(DCA $C^{\prec}$) $C^{\prec}$ is an actual precontact relation in A satisfying the following two versions of the generalized Efremovich axioms (see Remark 16.5):

$(C^{\prec}C^tC^{\prec}$Ef$)$ $a\overline{C^{\prec}}b \Rightarrow (\exists c)(a\overline{C^t}c$ and $c^*\overline{C^{\prec}}b)$, and

$(C^tC^{\prec}C^{\prec}$Ef$)$ $a\overline{C^{\prec}}b \Rightarrow (\exists c)(a\overline{C^{\prec}}c$ and $c^*\overline{C^t}b)$.

It is assumed also that A may satisfy also some of the **time axioms** (see Definition 16.15) except the axiom (irr) (the reasons why we will not consider (irr) will be given later on).

Remark 16.13 (i) The name "dynamic contact algebra" (DCA) is used in the papers (Vakarelov 2010, 2012, 2014, 2016a, b, 2020) as an integral name for point-free theories of space and time with different definitions in different papers. This is just for the economy of names. In this paper DCA is just the DCA defined in the above definition.

(ii) Examples of DCA-s are the standard DCA-s from Lemma 16.30, so DCA-s exist. More simple example is stated in the Lemma 16.33 below.

Lemma 16.32 *The following conditions are consequences from the axioms of DCA:*

(i) the Efremovich axiom (C^tEf) and the generalized Efremovich axioms $(C^t\overrightarrow{C^t}\overrightarrow{C^t}Ef)$ and $(C^t\widehat{C^t}\overrightarrow{C^t}Ef)$.

(ii) $a \in AE$ iff aC^sa iff aC^ta iff $a\overrightarrow{C}^ta$.

Proof (i) The condition (C^tEf) is a direct consequence from the extended Efremovich axiom $(C^t\widehat{C^t}$Ef+$)$. The conditions $(C^t\overrightarrow{C^t}\overrightarrow{C^t}Ef)$ and

$(C^t\widehat{C^t}\overrightarrow{C^t}Ef)$ follow from the axiom $(\overrightarrow{C^t}\widehat{C^t}\overrightarrow{C^t}Ef)$ and inclusion axioms $C^t \subseteq \overrightarrow{C^t}$ and $\overrightarrow{C^t} \subseteq \widehat{C^t}$.

(ii) is true because C^s and C^t are actual contact relations and that $\overrightarrow{C^t}$ is a semi-actual relation.

The following lemma gives a simple example of DCA and shows that DCA is a generalization of Boolean AE-algebra with A-contact.

Lemma 16.33 *Let $A = (B, AE, C)$ be a Boolean AE-algebra with an A-contact C. Let $C^t = AE \times AE = C^{max}_{act}$ (see Remark 16.2), $\widehat{C^t} = C^{max}_{con}$ (see Remark 16.2) and $C^{\prec} = C^{max}_{act}$. Then A is a DCA.*

Proof The non-trivial part of the proof follows from Lemmas 16.32, 16.22 and 16.29.

16.5.2 Ultrafilters, Clans Clusters and Extended Canonical Relations in DCA

We suppose in this section that A is a fixed DCA and let B_A be the underlying Boolean algebra of A.

Notation 1 *We adopt the following notations and terminology in DCA.*

Clans with respect to C^s, C^t and $\widehat{C^t}$ are called, respectively, C^s-clans, C^t-clans, $\widehat{C^t}$-clans and similarly for clusters for C^t and $\widehat{C^t}$.

Note that C^s-clans and C^t-clans (respectively, clusters) are actual clans (respectively, actual clusters) so they are unions of actual ultrafilters, while $\widehat{C^t}$-clans are clans with respect to a contact relation and they are unions of ultrafilters.

We denote the set of all C^s-clans of A by C^s-clans(A) and similarly for other clans (clusters).

Lemma 16.34 *(i) C^s-clans$(A) \subseteq C^t$-clans$(A) \subseteq \widehat{C^t}$-clans$(A)$. Every actual ultrafilter is a C^s-clan and a C^t-clan.*
(ii) C^t-clusters$(A) \subseteq \widehat{C^t}$-clusters$(A)$.
(iii) Every C^t-clan Γ can be extended into a unique C^t-cluster Δ.
(iv) Every $\widehat{C^t}$-clan Γ can be extended into a unique $\widehat{C^t}$- cluster Δ.

Proof (i) The proof of the first part follows from the inclusion axioms of DCA. The second part follows from the fact that C^s-clans are actual clans and that all actual clans are unions of R_{C^t}-related actual filters. The proof of (ii) is similar.
(iii) Applying Zorn's Lemma every C^t-clan Γ can be extended into a maximal C^t-clan Δ which is an actual C^t-clan. Since the A-contact C^t satisfies the Efremowich axiom (C^tEf), then by Lemma 16.25 every maximal actual C^t-clan is a C^t-cluster (see Lemma 16.25). To show that Δ is a unique C^t-cluster in which Γ is included suppose that $\Delta_1 \neq \Delta_2$ be C^t-clusters and that $\Gamma \subseteq \Delta_1$ and $\Gamma \subseteq \Delta_2$. Then, by Lemma 16.27, there exists $c \in B_A$ such that $c \notin \Delta_1$ and $c^* \notin \Delta_2$. Then $c \notin \Gamma$ and $c^* \notin \Gamma$, hence $1 = c + c^* \notin \Gamma$—a contradiction.
(iv) The proof is similar to the proof of (iii).

Lemma 16.35 *The canonical relations between ultrafilters for the basic relations in A satisfy the following conditions:*

(i) R_{C^s} is a non-empty quasi-reflexive and symmetric relation which is included in R_{C^t}.
(ii) R_{C^t} is a non-empty quasi-reflexive, symmetric and transitive relation included in $R_{\overrightarrow{C^t}}$.
(iii) $R_{\overrightarrow{C^t}}$ is a non-empty left quasi-reflexive relation included in $R_{\widehat{C^t}}$ and satisfying the following generalized transitivity condition: $(Gtr R_{C^t} R_{\overrightarrow{C^t}} R_{\overrightarrow{C^t}})$.
(iv) $R_{\widehat{C^t}}$ is an equivalence relation satisfying the following generalized transitivity condition: $(Gtr R_{\overrightarrow{C^t}} R_{\widehat{C^t}} R_{\overrightarrow{C^t}})$.
(v) $R_{C^{\prec}}$ is quasi-reflexive relation with respect to R_{C^t} satisfying the following generalized transitivity conditions: $(Gtr R_{C^{\prec}} R_{C^t} R_{C^{\prec}})$, $(Gtr R_{C^t} R_{C^{\prec}} R_{C^{\prec}})$.

Proof Each condition is guaranteed by a corresponding axiom of DCA and Lemma 16.32. For the proofs see Sect. 16.2.6.

Definition 16.17 Let $\Gamma, \Delta \subseteq B_A$ and let $C \in \{C^s, C^t, \widehat{C^t}, C^{\prec}\}$. Define $\Gamma R_C \Delta \Leftrightarrow (\forall a \in \Gamma$ and $\forall b \in \Delta)(aCb)$. When Γ, Δ are clans or clusters of a given type the relation is called the canonical relation between corresponding clans and clusters.

Lemma 16.36 *This lemma lists some facts for C^t-clans and C^t-clusters and separately for $\widehat{C^t}$-clans and $\widehat{C^t}$-clusters concerning the canonical relations R_{C^t} and $R_{\widehat{C^t}}$ introduced by Definition 16.17.*

(A) The case for C^t-clans and C^t-clusters.

(I) Let Γ, Δ be C^t-clans. Then, the following conditions are equivalent:

(i) $\Gamma R_{C^t} \Delta$,

(ii) There exists an ultrafilter $U_0 \subseteq \Gamma$ and an ultrafilter $V_0 \subseteq \Delta$ such that $U_0 R^t_C V_0$,

(iii) For all $U \in Ult(\Gamma)$ and for all $V \in Ult(\Delta)$, $U R^t_C V$.

Let $\Gamma, \Gamma', \Delta, \Delta'$ be C^t-clans. Then,

(II) (i) The relation R_{C^t} considered in the set of all C^t-clans is an equivalence relation.

(ii) $\Gamma R_{C^t} \Delta$ and $\Gamma R_{C^t} \Delta'$, then $\Delta \cup \Delta'$ is a C^t-clan and $\Gamma R_{C^t} (\Delta \cup \Delta')$.

(iii) If $\Gamma \subseteq \Delta$ then $\Gamma R_{C^t} \Delta$.

(iv) Let Γ, Δ be C^t-clans and let $\Gamma R_{C^t} \Delta$. Then, Γ, Δ can be extended correspondingly to C^t-clusters Γ', Δ' such that $\Gamma' R_{C^t} \Delta'$.

(v) $aC^t b$ iff there exists a C^t-clan (cluster) Γ such that $a, b \in \Gamma$.

(III) (i) If $\Gamma \subseteq \Gamma'$, $\Gamma \subseteq \Delta$ and Δ is a C^t-cluster, then $\Gamma' \subseteq \Delta$.

(ii) If Δ is a C^t-cluster and $\Gamma R_{C^t} \Delta$, then $\Gamma \subseteq \Delta$. If in addition Γ is a C^t-cluster, then $\Gamma = \Delta$.

(B) The case of $\widehat{C^t}$-clans and $\widehat{C^t}$-clusters. All statements of this part can be formulated by a simple replacement of C^t with $\widehat{C^t}$.

(C) In this case we list some combined properties.

(i) If Γ, Δ are C^t-clans and $\Gamma C^t \Delta$, then $\Gamma \widehat{C^t} \Delta$.

(ii) If Γ is a C^t-cluster, Δ is a $\widehat{C^t}$-clan, Θ is a C^t-clan, $\Theta \subseteq \Delta$, and $\Gamma \subseteq \Delta$, then $\Theta \subseteq \Gamma$.

Proof *Proof of the case (A)* (I) The implications (i)⇒(ii) and (iii)⇒(i) are obvious.

(ii)⇒(iii) Let (ii) holds, i.e., $U_0 R_{C^t} V_0$ and suppose $U \in Ult(\Gamma)$ $V \in Ult(\Delta)$. Since every two ultrafilters in an A-clan are R_{C^t}-related we get $U R_{C^t} U_0$ and $V_0 R_{C^t} V$. But R_{C^t} is a transitive relation (Lemma 16.35 (ii)), so we get $U R_{C^t} V$.

(II) can be proved by a direct verification of the corresponding definitions using (I).

(III)(i) Let the assumptions of (III)(i) are fulfilled. From the assumptions $\Gamma \subseteq \Gamma'$ and $\Gamma \subseteq \Delta$ we obtain by (II)(iii) that $\Gamma R_{C^t} \Delta$ and $\Gamma R_{C^t} \Delta'$ and by (II)(ii) we get $\Gamma R_{C^t} (\Delta \cup \Delta')$. Again by (II)(iii) we get that $\Delta \cup \Delta'$ is a C^t-clan and that $\Delta \subseteq \Delta \cup \Delta'$. But Δ is a C^t-cluster and consequently it is a maximal C^t-clan which implies that $\Delta = \Delta \cup \Delta'$ and consequently $\Delta' \subseteq \Delta$.

The following is another proof of III(i). Suppose Γ, Γ' are C^t-clans, $\Gamma \subseteq \Gamma'$, $\Gamma \subseteq \Delta$ and Δ is a C^t-cluster and for the sake of contradiction that $\Gamma' \nsubseteq \Delta$. Then, there exists $a \in \Gamma'$ and $a \notin \Delta$. By the Efremovich Axiom $(C^t Ef)$ there exists c such that $a\overline{C^t}c$ and $c^*\overline{C^t}b$. Condition $c^*\overline{C^t}b$ implies that $c^* \notin \Delta$ and consequently $c^* \notin \Gamma$. Condition $a\overline{C^t}c$ implies that $c \notin \Gamma'$ and consequently $c \notin \Gamma$, so $c^* \in \Gamma$—a contradiction with $c^* \notin \Gamma$.

III(ii) Suppose Γ is a C^t clan, Δ is a C^t-cluster, $\Gamma R_{C^t} \Delta$ and for the sake of contradiction suppose that $\Gamma \nsubseteq \Delta$. Then, there exists $a \in \Gamma$ and $a \notin \Delta$. Since Δ is a C^t cluster, then there exists $b \in \Delta$ such that $a\overline{C}^t b$. But $a \in \Gamma$, $b \in \Delta$ and $\Gamma R_{C^t} \Delta$ imply by (I) that $aC^t b$—a contradiction. The second part of (ii) follows from the first part.

Proof of the case (B). The proofs of the cases (I)–(III) are similar to that of the case (A) using the corresponding facts and axioms for the relation $\widehat{C}^t$.

Proof of the case (C) (i) follows from the inclusion axiom $(C^t \subseteq \widehat{C}^t)$.

(C)(ii) Suppose Γ is a C^t-cluster, Δ is a $\widehat{C}^t$-clan, Θ is a C^t-clan, $\Theta \subseteq \Delta$, and $\Gamma \subseteq \Delta$ and for the sake of contradiction that $\Theta \nsubseteq \Gamma$. Then, there exists $a \in \Theta$ and $a \notin \Gamma$. Because Γ is a C^t-cluster, there is $b \in \Gamma$ such that $a\overline{C^t}b$. Now by the axiom $(C^t\widehat{C}^t\text{Ef+})$ there exists c such that $a\overline{C^t}c$, $c^*\overline{\widehat{C}^t}c$ and $c^*\overline{C^t}b$. Conditions $a \in \Theta$ and $a\overline{C^t}c$ imply that $c \notin \Theta$, so $c^* \in \Theta$ and by $\Theta \subseteq \Delta$ we get $c^* \in \Delta$. Analogously $b \in \Gamma$ and $c^*\overline{C^t}b$ imply $c^* \notin \Gamma$, so $c \in \Gamma$ which together with $\Gamma \subseteq \Delta$ imply $c \in \Delta$. Because Δ is a $\widehat{Ct}$-clan, $c^* \in \Delta$ and $c \in \Delta$ imply $c^*\widehat{C}^t c$ which contradicts $c^*\overline{\widehat{C}^t}c$.

Lemma 16.37 *(I) Let Γ be a C^t-clan, Δ be $\widehat{C}^t$-clan and $R_{\overrightarrow{C^t}}$ be the canonical relation corresponding to $\overrightarrow{C^t}$ between C^t-clans and $\widehat{C}^t$-clans. Then, the following conditions are equivalent:*

(i) $\Gamma R_{\overrightarrow{C^t}} \Delta$,

(ii) There exists an ultrafilter $U_0 \subseteq \Gamma$ and an ultrafilter $V_0 \subseteq \Delta$ such that $U_0 R_{\overrightarrow{C^t}} V_0$,

(iii) For all $U \in Ult(\Gamma)$ and for all $V \in Ult(\Delta)$, $U R_{\overrightarrow{C^t}} V$.

Let Γ, Γ' be C^t-clans and Δ, Δ' be $\widehat{C}^t$-clans.

(II)(i) If $\Gamma R_{C^t} \Gamma'$ and $\Gamma' R_{\overrightarrow{C^t}} \Delta'$ and $\Delta' R_{\widehat{C^t}} \Delta$, then $\Gamma R_{\overrightarrow{C^t}} \Delta$.

(ii) $\Gamma \overrightarrow{C^t} \Gamma$.

(iii) If $\Gamma R_{\overrightarrow{C^t}} \Delta$, then $\Gamma R_{\widehat{C^t}} \Delta$.

(iv) If $\Gamma \subseteq \Delta$, then $\Gamma R_{\overrightarrow{C^t}} \Delta$.

(v) If $\Gamma R_{\overrightarrow{C^t}} \Gamma'$ and $\Gamma R_{\overrightarrow{C^t}} \Delta$, then $\Gamma' \cup \Delta$ is a $\widehat{C}^t$-clan and $\Gamma R_{\overrightarrow{C^t}} \Gamma' \cup \Delta$.

(vi) $\Gamma R_{\overrightarrow{C^t}} \Delta$ and Δ is a $\widehat{C}^t$-cluster, then $\Gamma \subseteq \Delta$.

(vii) If $\Gamma R_{\overrightarrow{C^t}} \Delta$, then Γ can be extended into a C^t-cluster Γ' and Δ can be extended into a $\widehat{C}^t$-cluster Δ' such that $\Gamma' R_{\overrightarrow{C^t}} \Delta'$.

(viii) $a\overrightarrow{C^t}b$ iff there exists a C^t-clan (cluster) Γ and a $\widehat{C}^t$-clan (cluster) Δ such that $\Gamma R_{\overrightarrow{C^t}} \Delta$ and $a \in \Gamma$ and $b \in \Delta$.

Proof (I) The proof is similar to the proof of Lemma 16.36 combining cases (A)(I) and (B)(I) with the use of generalized transitivity conditions

(Gtr$R_{\overrightarrow{C^t}} R_{\widehat{C^t}} R_{\overrightarrow{C^t}}$) and (Gtr$R_{\overrightarrow{C^t}} R_{\widehat{C^t}} R_{\overrightarrow{C^t}}$) from Lemma 16.35 (iii) and (iv) which are guaranteed by the corresponding axioms in DCA for the relation $\overrightarrow{C}^t$.

(II) The proof of (II) follows from (I) and Lemma 16.36 (A) and (B) with applications of inclusion axioms ($C^t \subseteq \overrightarrow{C^t}$) and ($\overrightarrow{C^t} \subseteq \widehat{C^t}$). As an example we will demonstrate the cases (iv) and (vii).

(II)(iv) Suppose $\Gamma \subseteq \Delta$. Then, by Lemma 16.36 (A)(II)(iii), we get $\Gamma R_{C^t} \Delta$ which by monotonicity axiom ($C^t \subseteq \overrightarrow{C^t}$) implies b $\Gamma R_{\overrightarrow{C^t}} \Delta$.

(II)(viii) Let Δ be a $\widehat{C^t}$-cluster and $\Gamma R_{\overrightarrow{C^t}} \Delta$. Then, by (II)(iv), we get $\Gamma R_{\widehat{C^t}} \Delta$. Then, applying Lemma 16.36 (B)(III)(ii), we obtain $\Gamma \subseteq \Delta$.

Lemma 16.38 *Let Γ, Δ be C^t-clans and $R_{C^\prec}$ be the canonical relation between C^t-clans corresponding to $C^\prec$. Then,*

(I) The following conditions are equivalent:
(i) $\Gamma R_{C^\prec} \Delta$.
(ii) $U_0 R_{C^\prec} V_0$ for some ultrafilters $U_0 \subseteq \Gamma$ and $V_0 \subseteq \Delta$.
(iii) $U R_{C^\prec} V$ for every ultrafilter $U \subseteq \Gamma$ and $V \subseteq \Delta$.
(II) If Γ', Δ' are C^t-clans and $\Gamma' R_{C^\prec} \Delta'$, then there exist C^t-clusters Γ, Δ such that $\Gamma' \subseteq \Gamma$, $\Delta' \subseteq \Delta$ and $\Gamma R_{C^\prec} \Delta$.
(III) $aC^\prec b$ iff there exist C^t-clusters Γ, Δ such that $a \in \Gamma$, $b \in \Delta$ and $\Gamma R_{C^\prec} \Delta$.

Proof The proof is similar to the proof of Lemma 16.36. The proof of (I) is based of the following generalized transitivity conditions for the canonical relations R_{C^t} and $R_{C^\prec}$ in the set of ultrafilters: (Gtr$R_{C^\prec} R_{C^t} R_{C^\prec}$) and (Gtr$R_{C^t} R_{C^\prec} R_{C^\prec}$), and the fact that $R^\prec$ is quasi-reflexive relation with respect to R^t (which means that if $U R_{C^\prec} V$, then the ultrafilters U, V are actual C^t-clans). This is used in the proofs of (II) and (III).

16.5.3 Clan Pairs

There is no notion of a clan corresponding to the relation $\overrightarrow{C^t}$ because it is asymmetric one. However, using ordered pairs of sets we can find an analog of the notion of clan, maximal clan, and cluster for the relation $\overrightarrow{C^t}$.

Definition 16.18 (*Clan pairs*) The pair $\langle\Gamma, \Delta\rangle$ is called a clan pair if Γ is a C^t-clan, Δ is a $\widehat{C^t}$-clan and $\Gamma \subseteq \Delta$. We say that $\langle\Gamma, \Delta\rangle$ is included in the clan pair $\langle\Gamma', \Delta'\rangle$ if $\Gamma \subseteq \Gamma'$ and $\Delta \subseteq \Delta'$. $\langle\Gamma, \Delta\rangle$ is maximal if it is a maximal element in the set of clan pairs under inclusion. $\langle\Gamma, \Delta\rangle$ is called a cluster pair if Γ is a C^t-cluster and Δ is a $\widehat{C^t}$-cluster.

Clan pairs and cluster pairs will be used to give a special characterization of the basic relations in DCA.

Lemma 16.39 *(i) Every clan pair can be extended to a maximal clan pair.*

(ii) *$\langle \Gamma, \Delta \rangle$ is a maximal clan pair iff $\langle \Gamma, \Delta \rangle$ is a cluster pair.*

Proof (i) The proof can be obtained by the Zorn's Lemma.

(ii) The implication right to left is obvious. For the implication from left to right suppose that $\langle \Gamma, \Delta \rangle$ is a maximal clan pair. First we will show that Δ is a maximal $\widehat{C^t}$-clan. Suppose the contrary, then there exists a $\widehat{C^t}$-clan $\Delta' \neq \Delta$ which extends Δ. Then, obviously $\langle \Gamma, \Delta' \rangle$ is a clan pair which extends $\langle \Gamma, \Delta \rangle$- a contradiction with the maximality of $\langle \Gamma, \Delta \rangle$. So Δ is a maximal $\widehat{C^t}$-clan which implies that Δ is a $\widehat{C^t}$-cluster. Now we will show that Γ is a maximal C^t-clan. Again suppose the contrary, so there exists a C^t-clan $\Gamma' \neq \Gamma$ which extends Γ. So we have $\Gamma \subseteq \Gamma'$ and $\Gamma \subseteq \Delta$. Then by Lemma 16.36 (B)(ii) $\Gamma' \subseteq \Delta$ which sows that $\langle \Gamma', \Delta \rangle$ is a clan pair extending $\langle \Gamma, \Delta \rangle$—again a contradiction with the maximality of $\langle \Gamma, \Delta \rangle$. This proves that Γ is a maximal C^t-cluster, which shows that $\langle \Gamma, \Delta \rangle$ is a cluster pair.

Lemma 16.40 *(i) Let Γ be a C^t-clan, Δ be a $\widehat{C^t}$-clan and $\Gamma R_{\overrightarrow{C^t}} \Delta$. Then $\langle \Gamma, \Gamma \rangle$ and $\langle \Gamma, \Gamma \cup \Delta \rangle$ are clan pairs.*

(ii) Let Γ be C^t-cluster and Δ b e a $\widehat{C^t}$-cluster. Then, $\langle \Gamma, \Delta \rangle$ is a cluster pair iff $\Gamma R_{\overrightarrow{C^t}} \Delta$.

(iii) If Γ is a C^t-cluster, then there exists a unique $\widehat{C^t}$-cluster Δ such that $\langle \Gamma, \Delta \rangle$ is a cluster pair.

(iv) If Δ is a $\widehat{C^t}$-cluster then there exists a unique C^t-cluster Γ such that $\langle \Gamma, \Delta \rangle$ is a cluster pair.

(v) Let $\langle \Gamma, \Delta \rangle$ a cluster pair. Then, for every C^t-clan Θ: $\Theta \subseteq \Gamma$ iff $\Theta \subseteq \Delta$.

Proof (i) The first part follows by Lemma 16.37(II)(ii) and the second part from (II)(v) and the first part.

(ii) follows from Lemma 16.37 (II)(iv) and (vi).

(iii) Let Γ be a C^t-cluster. Then, by Lemma 16.34 (i) Γ is also a $\widehat{C^t}$-clan. Then, by Lemma 16.34 (iii) Γ can be extended into a unique cluster Δ. So, $\langle \Gamma, \Delta \rangle$ is a cluster pair.

(iv) Let Δ be a $\widehat{C^t}$-cluster. Take any ultrafilter $V \in Ult(\Delta)$. Since $\overrightarrow{C^t}$ is a semi-actual precontact it satisfies the axiom (SemiA 2) If $a \neq 0$, then $1\overrightarrow{C^t}a$ (see Sect. 16.3.6). It follows by this axiom that $\{1\}R_{\overrightarrow{C^t}} V$. Indeed, $\{1\}$ is the unit filter, and for all $b \in V$ we have $b \neq 0$. Then, by axiom (SemiA 2), we get $1\overrightarrow{C^t}b$ which shows $\{1\}R_{\overrightarrow{C^t}} V$. Now extend $\{1\}$ into an ultrafilter U (see Lemma 16.3) such that $U\overrightarrow{C^t}V$. We get from this that $a\overrightarrow{C^t}1$ for any element $a \in U$. From axiom (SemiA 1) we obtain that $a \in AE$ which shows that U is an A-ultrafilter and hence a C^t-clan. Now extend U into a C^t-cluster Γ. Since $U \in \Gamma$, $V \in \Delta$ and $U\overrightarrow{C^t}V$ we get (by Lema 16.37 (I) that $\Gamma R_{\overrightarrow{C^t}} \Delta$. Then, by (ii), we obtain that $\langle \Gamma, \Delta \rangle$ is a cluster pair.

To show the uniqueness of Γ, suppose that $\langle \Gamma_1, \Delta \rangle$ and $\langle \Gamma_2, \Delta \rangle$ are cluster pairs. Then, $\Gamma_1 \subseteq \Delta$ and $\Gamma_2 \subseteq \Delta$, Γ_1, Γ_2, are C^t-clusters and Δ is a $\widehat{C^t}$-clan. By Lemma 16.36 (B)(III)(iii), we obtain $\Gamma_1 \subseteq \Gamma_2$ and $\Gamma_2 \subseteq \Gamma_1$ which shows that $\Gamma_1 = \Gamma_2$.

(v) Assume that $\langle \Gamma, \Delta \rangle$ is a cluster pair and Θ is a C^t-clan. The implication from left to the right is obvious. For the converse implication suppose that $\Theta \subseteq \Delta$. By Lemma 16.14 Θ is a C^t-clan. So we have $\Gamma, \Theta \subseteq \Delta$, Γ is a C^t-cluster, Θ is a C^t-clan and Δ is a $\widehat{C^t}$-clan. Then, by lemma 16.36 (B)(III)(iii), we obtain $\Theta \subseteq \Gamma$.

Notation 2 *Let $\langle \Gamma, \Delta \rangle$ be a cluster pair. We see by Lemma 16.40 that any of the components of the pair uniquely determines the other one. So we may introduce the following one-one-functions: if Γ is a C^t-cluster we denote by Γ^+ the unique $\widehat{C^t}$-cluster such that $\langle \Gamma, \Gamma^+ \rangle$ is a cluster pair. Similarly, if Δ is a $\widehat{C^t}$-cluster we denote by Δ^- the unique C^t-cluster such that $\langle \Delta^-, \Delta \rangle$ is a cluster pair. So $()^+$ is a one-one mapping from the set of C^t-clusters into the set of $\widehat{C^t}$ clusters of a given DCA B with the property $\Gamma \subseteq \Gamma^+$ and $()^-$ is its converse mapping from the set of $\widehat{C^t}$-clusters into the set of C^t-clusters. Consequently, the following conditions are satisfying:*

(i) For Γ a C^t-cluster: $(\Gamma^+)^- = \Gamma$ and $\Gamma \subseteq \Gamma^+$.
(ii) For Δ a $\widehat{C^t}^t$-cluster: $(\Delta^-)^+ = \Delta$ and $\Delta^- \subseteq \Delta$.

Lemma 16.41 (Characterization of the basic relations in DCA by cluster pairs) *Let B be a DCA. Then, the following conditions hold:*

(i) $a \neq 0$ iff there exists a cluster pair $\langle \Gamma, \Delta \rangle$ such that $a \in \Delta$.
(ii) $a \in AE$ iff there exists a cluster pair $\langle \Gamma, \Delta \rangle$ such that $a \in \Gamma$.
(iii) $aC^s b$ iff there exists a cluster pair $\langle \Gamma, \Delta \rangle$ and an C^s-clan $\Theta \subseteq \Gamma$ such that $a, b \in \Theta$.
(iv) $aC^t b$ iff there exists a cluster pair $\langle \Gamma, \Delta \rangle$ such that $a, b \in \Gamma$.
(v) $a\widehat{C^t}b$ iff there exists a cluster pair $\langle \Gamma, \Delta \rangle$ such that $a, b \in \Delta$.
(vi) $a\overrightarrow{C}^t b$ iff there exists a cluster pair $\langle \Gamma, \Delta \rangle$ such that $a \in \Gamma$ and $b \in \Delta$.

Proof (i) $a \neq 0$ iff there exists an ultrafilter Θ (which is also a $\widehat{C^t}$-clan) such that $a \in \Theta$. In addition to the last condition there exists an $\widehat{C^t}$-cluster Δ such that $\Theta \subseteq \Delta$, and consequently $a \in \Delta$, So, the cluster pair is $\langle \Delta^-, \Delta \rangle$ and $a \in \Delta$. The converse is obvious.

(ii) $a \in AE$ iff there exists an actual ultrafilter Θ (which is also a C^t-clan) such that $a \in \Theta$. In addition to this there exists a C^t-cluster Γ, such that $\Theta \subseteq \Gamma$. The required cluster pair is $\langle \Gamma, \Gamma^+ \rangle$. For the converse let $\langle \Gamma, \Delta \rangle$ be a cluster pair and let $a \in \Gamma$. Because Γ is actual clan then $a \in AE$.

(iii) $aC^s b$ iff there exists a C^s-clan Θ (which is an actual clan) such that $a, b \in \Theta$ (Lemma 16.24 (vii). Because Θ is also a C^t-clan (by the inclusion axiom $(C^s \subseteq C^t)$, then there exists a C^t-cluster Γ such that $\Theta \subseteq \Gamma$. The required cluster pair is $\langle \Gamma, \Gamma^+ \rangle$. The converse is obvious because $a, b \in \Theta$ and Θ is an actual C^s-clan.

(iv) $aC^t b$ iff there exists a C^t-cluster Γ such that $a, b \in \Gamma$ (Lemma 16.36 (A)(II)(v). The required cluster pair is $\langle \Gamma, \Gamma^+ \rangle$.

(v) $a\widehat{C^t}b$ iff there exists a $\widehat{C^t}$-cluster Δ such that $a, b \in \Delta$ (Lemma 16.36 (B)(II)(v). The required cluster pair is $\langle \Delta^-, \Delta \rangle$.

(vi) $a\overrightarrow{C^t}b$ iff there exist a C^t-cluster Γ and a $\widehat{C^t}$-cluster Δ such that $a \in \Gamma$, $b \in \Delta$ and $\Gamma R_{\overrightarrow{C^t}} \Delta$ (Lemma 16.37 (II)(viii)). Then, by Lemma 16.37 (III)(i), we obtain that $\Gamma \subseteq \Delta$ and hence the required cluster pair is $\langle \Gamma, \Delta \rangle$.

The lemma shows the importance of cluster pairs. Since every cluster pair is determined by any of it components the lemma can be reformulated by means of $\widehat{C^t}^t$-clusters, or by means of C^t-clusters and using the operations + and –.

16.6 Representation Theorem of DCA by Means of Snapshot Models

In this section we will show that every DCA A can be isomorphically embedded into a standard DCA over a dynamic model of space and time. The strategy is the following. First we will associate to each DCA A a canonical time structure $\mathbf{T}_A = (T_A, \prec_A)$. Second, to each element $i \in T_A$ we associate in a canonical way an AE-Boolean algebra (B_i, AE_i, C_i) with actual contact C_i as a coordinate algebra corresponding to i. And, third, we will define by means of $\mathbf{T}_A$ and the set of coordinate algebras $B_i, i \in T_A$ the corresponding full standard DCA $\mathbf{B}$ and an embedding f from A into $\mathbf{B}$ which will imply the representation theorem for DCA as a subalgebra of a full standard DCA.

16.6.1 *Canonical Time Structure*

Definition 16.19 Let A be a DCA. The **canonical time structure** of A, $\mathbf{T}_A = (T_A, \prec_A)$ is defined as follows: $T_A = C^t$-clusters(A) and $\prec_A = R_{C^\prec}$, where $R_{C^\prec}$ acts on C^t-clusters (equality is standard). We will consider also a **preliminary canonical time structure** $\mathbf{T}'_A = (T'_A, \prec'_A, =')$, where T'_A is the set of actual ultrafilters, and $\prec'_A = R_{C^\prec}$ considered on actual ultrafilters. We have here a non-standard equality $='$ which is the relation R_{C^t}. The interpretation of time conditions from Sect. 16.4.1 in these two time structures is obvious. Running variables for C^t-clusters will be $\Gamma, \Delta, \Theta, \ldots$ and for actual ultrafilters—$U, V, W, \ldots$. We will use the same short names for time conditions but the names for their translations into the language of preliminary time structure will be surrounded by $\langle , \rangle$, for instance $\langle$Up Dir$\rangle$.

Lemma 16.42 (Axiomatic correspondence) *Let (Ψ) be any formula from the list of time conditions (RS), (LS), (Up Dir), (Down Dir),(Circ), (Dens), (Ref), (Irr), (Lin), (Tri), (Tr) translated in the language of time structure $\mathbf{T}_B$, $\langle A \rangle$ be its translation in the preliminary time structure $\mathbf{T}'_B$ and (ψ) be the corresponding formula from the list of time axioms (rs), (ls), (up dir), (down dir), (circ), (dens), (ref), (lin), (tri), (tr) (see Definition 16.15). Then, the following conditions are equivalent:*

(i) (ψ) is true in the algebra A,

(ii) $\langle\Psi\rangle$ *is true in the structure* $\mathbf{T}'_A$,
(iii) (Ψ) *is true in the time structure* $\mathbf{T}_A$.

Proof We will demonstrate the proof by two typical cases.

• **Case:** (i) (rs) $a \in AE \Rightarrow aC^{\prec}1$, (ii) $\langle RS\rangle$ $(\forall U)(\exists V)(U R_{C^{\prec}} V)$,
(iii) (RS) $(\forall\Gamma)(\exists\Delta)(\Gamma R_{C^{\prec}}\Delta)$.

(i)⇒(ii) Suppose (i) and let U be an A-ultrafilter. We show that $U R_{C^{\prec}}[1)$ Suppose $a \in U$. Because U is an A-ultrafilter, then $U \subseteq AE$ and consequently $a \in AE$. By (i) we get $aC^{\prec}1$ and hence $U R_{C^{\prec}}[1)$. Then, extend $[1)$ to an ultrafilter V such that $U R_{C^{\prec}} V$ (Lemma 16.3). Because $R_{C^{\prec}}$ is a quasi-reflexive relation with respect to R_{C^t} we have that $V R_{C^t} V$ and consequently V is an A-ultrafilter (Lemma 16.23).

(ii)⇒(iii) Suppose (ii) and let Γ be a C^t-cluster and let U be any ultrafilter included in Γ. Because Γ is a C^t-cluster, then U is an A-ultrafilter. By (ii) there exists an A-ultrafilter V (which is also a C^t-clan) such that $U R_{C^{\prec}} V$. Now extend V into a C^t-cluster Δ such that $\Gamma R_{C^{\prec}}\Delta$ (Lemma 16.37).

(iii)⇒(i) Suppose (iii) and let $a \in AE$. Then, there exists a C^t-cluster Γ such that $a \in \Gamma$ (Lemma 16.41). By (iii) there exists a C^t-cluster Δ such that $\Gamma R_{C^{\prec}}\Delta$. But $1 \in \Delta, a \in \Gamma$, so $aC^{\prec}1$.

In a similar way, one can treat the case of (RS). The proofs of (Ref), (Lin), and (Tri) are much more direct.

• **Case:** (i)(up dir) $a \in AE$ and $b \in AE$, then $aC^{\prec}c$ or $bC^{\prec}c^*$, (ii) $\langle$Up Dir$\rangle$ $(\forall U, V)(\exists W)(U R_{C^{\prec}} W$ and $V R_{C^{\prec}} W)$,
(iii) (Up Dir) $(\forall\Gamma, \Delta)(\exists\Theta)(\Gamma R_{C^{\prec}}\Theta$ and $\Delta R_{C^{\prec}}\Theta)$.

(i)⇒(ii) Suppose (i) and let U, V be A-ultrafilters. To show that there exists an A-ultrafilter W such that $U R_{C^{\prec}} W$ and $V R_{C^{\prec}} W$ we will need a construction based on the Lemma 16.18. Consider the filters $F_I^{C^{\prec}}(U) = \{b : (\exists a \in U)(a\overline{C}^{\prec} b^*)\}$ and $F_{II}^{C^{\prec}}(V) = \{a : (\exists b \in V)(a^*\overline{C^{\prec}}b)\}$ and let $F_I^{C^{\prec}}(U) \oplus F_{II}^{C^{\prec}}(V)$ be the smallest filter containing them. We will show that it is a proper filter. Assume otherwise, then there exists c such that $c^* \in F_I^{C^{\prec}}(U)$ and $c \in F_{II}^{C^{\prec}}(V)$ which implies

(♯) $a\overline{C}^{\prec}c$ and $b\overline{C}^{\prec}c^*$.

But $a \in U$ and $b \in V$ imply that $a \in AE$ and $b \in AE$ and by (i) that for every c we have $aC^{\prec}c$ or $bC^{\prec}c^*$, which contradicts (♯). So $F_I^{C^{\prec}}(U) \oplus F_{II}^{C^{\prec}}(V)$ is a proper filter and it can be extended into an ultrafilter W. By Lemma 16.18 we obtain $U R_{C^{\prec}} W$ and $V R_{C^{\prec}} W$. Since $R_{C^{\prec}}$ is a quasi-reflexive relation with respect to R_{C^t} we get $W R_{C^t} W$ which implies that W is an A-ultrafilter (Lemma 16.23).

(ii)⇒(iii) Suppose (ii) and let Γ, Δ be C^t-clusters. Let U, V be ultrafilters such that $U \subseteq \Gamma$, $V \subseteq \Delta$. Then, U, V are A-ultrafilters and by (ii) there exists an A-ultrafilter W such that $U R_{C^{\prec}} W$ and $V R_{C^{\prec}} W$. Then, W can be extended into a unique C^t-cluster Θ and by Lemma 16.37 we obtain $\Gamma R_{C^{\prec}}\Theta$ and $\Delta R_{C^{\prec}}\Theta$.

(iii)⇒(i) Suppose (iii) and let $a, b \in AE$. There exist C^t clusters Γ, Δ such that $a \in \Gamma$ and $b \in \Delta$. Let c be an arbitrary element of B. There are two cases for c.

Case 1: $c \in \Theta$. This implies $aC^{\prec}c$.

Case 2: $c \notin \Theta$. Then $c^* \in \Theta$ and this implies $bC^{\prec}c^*$.

In a similar way one can treat the cases (Down Dir), (Circ), (Dens). For the proof of case of (Tr) use Lemma 16.17.

Remark 16.14 Note that the above lemma does not treat the case of (Irr). It is easy to see that (Irr) implies (irr) but the problem is the converse implication. In standard DCA the two conditions are equivalent but the proof of the implication (irr)⇒(Irr) uses the fact that the algebra is rich. However, the abstract DCA-s do no have enough syntax to imitate richness. In the version of DCA developed in Vakarelov (2014, 2016a, b) richness is imitated by a class of special regions called "time representatives" which make possible to prove the equivalence (irr)⇔(Irr). It is possible to introduce analogs of "time representatives" in the present version and to solve the above difficulty but this will make the system complicated. Another solution is to assume a second-order axiom by means of clusters, but this is not an aesthetic solution.

16.6.2 Canonical Standard DCA and the Main Representation Theorem

First we will introduce a canonical construction for coordinate algebras. Let A be a DCA and Γ be a fixed C^t-cluster (which now we consider as a time point in the canonical time structure) and let Γ^+ be the $\widehat{C^t}$-cluster (see Notation 2). Because Γ^+ is also a grill, the complement $\overline{\Gamma^+}$ is a proper ideal in the underline Boolean algebra $B = B_A$ (see Lemma 16.7). It is a well-known fact in the theory of Boolean algebras that every ideal determines the following congruence relation in B (which we denote by $\equiv_\Gamma$):

$a \equiv_\Gamma \Leftrightarrow_{\text{def}} a.b^* + a^*.b \in \overline{\Gamma^+}$.

The factor Boolean algebra $B \backslash \equiv_\Gamma$ with respect to $\equiv_\Gamma$ will be denoted by B_Γ. The elements of B_Γ are the equivalence classes with respect to $\equiv_\Gamma$ and denoted by $|a|_\Gamma =_{\text{def}} \{b \in B : a \equiv_\Gamma b\}$. The operations and constants in B_Γ are defined and denoted as usual: $|1|_\Gamma$, $|0|_\Gamma$, $|a|_\Gamma + |b|_\Gamma = |a+b|_\Gamma$, $|a|_\Gamma.|b|_\Gamma = |a.b|_\Gamma$ and $|a|_\Gamma^* = |a^*|_\Gamma$.

Note that $|a|_\Gamma \leq_\Gamma |b|_\Gamma$ iff $a.b^* \notin \Gamma^+$. We introduce in B_Γ the following set and a relation:

$AE_\Gamma = \{|a|_\Gamma : a \in \Gamma\}$,

$|a|_\Gamma C_\Gamma |b|_\Gamma$ iff there exists a C^s-clan $\Theta \subseteq \Gamma$ such that $a, b \in \Theta$.

Lemma 16.43 *The system* $(B_\Gamma, AE_\Gamma, C_\Gamma)$ *is a Boolean AE-algebra with A-contact* C_Γ.

Proof First we show that the definition of AE_Γ is correct namely, if $a \equiv_\Gamma b$ then $|a|_\Gamma \in AE_\Gamma$ iff $|b|_\Gamma \in AE_\Gamma$. And indeed suppose first $|a|_\Gamma \in AE_\Gamma$ (i.e., $a \in \Gamma$) and $a \equiv_\Gamma b$ (i.e., $a.b^* + a^*.b \in \overline{\Gamma^+}$). The last condition together with the fact that $\Gamma \subseteq \Gamma^+$ (see Notation 2), imply $a.b^* \notin \Gamma$ and $a^*.b \notin \Gamma$. But conditions $a \in \Gamma$ and $a.b^* \notin \Gamma$ imply by Lemma 16.7 (v) that $b \in \Gamma$ and consequently $|b|_\Gamma \in AE_\Gamma$. In the same way we verify the converse implication.

Now we show that AE_Γ is an actual existence predicate, i.e., that AE_Γ is a grill. The non-trivial part is the monotonicity axiom. Suppose $|a|_\Gamma \in AE_\Gamma$ and $|a|_\Gamma \leq_\Gamma |b|_\Gamma$.

Then, we have $a \in \Gamma$ and $a.b^* \notin \Gamma^+$ and consequently $a.b^* \notin \Gamma$ (because $\Gamma \subseteq \Gamma^+$). Then, again by Lemma 16.7 (v), we get $b \in \Gamma$ and equivalently $|b|_\Gamma \in AE_\Gamma$. Thus, (B_Γ, AE_Γ) is a Boolean AE-algebra.

The proof of the correctness of the definition of C_Γ is similar to the proof of the correctness of AE_Γ with several applications of Lemma 16.7 (v). The non-trivial part of the proof that C_Γ is an A-contact is the verification of the axiom:

(AC2′) If $|a|_\Gamma \in AE_\Gamma$, then $|a|_\Gamma C|a|_\Gamma$.

Suppose $|a|_\Gamma \in AE_\Gamma$, i.e., $a \in \Gamma$. Then, there exists an actual ultrafilter $U \subseteq \Gamma$ such that $a \in U$. But actual ultrafilters are C^s-clans (Lemma 16.34), and this implies $|a|_\Gamma C|a|_\Gamma$. Thus, $(B_\Gamma, AE_\Gamma, C_\Gamma)$ is an AE-algebra with A-contact C_Γ.

Definition 16.20 Let A be a DCA and B_A (denoted simply by B) be the underlying Boolean algebra of A. We associate in a canonical way a **standard DCA over** A as follows. By means of the canonical time structure $\mathbf{T}_B = (T_A, \prec_A)$ of A and the set of canonical coordinate algebras B_Γ, $\Gamma \in T_A$ one can construct as in Definition 16.14 the full standard DCA $(\mathbf{B}, AE_\mathbf{B}, C^s_\mathbf{B}, C^t_\mathbf{B}, \overrightarrow{C}^t_\mathbf{B}, \widehat{C}^t_\mathbf{B})$ where $\mathbf{B}$ is the Cartesian product $\prod_{\Gamma \in T_B} B_\Gamma$ of the coordinate Boolean algebras. We define a mapping $f : B \mapsto \mathbf{B}$, called a **canonical mapping** from B into $\mathbf{B}$ as follows: $f(a) = \langle |a|_\Gamma \rangle_{\Gamma \in T_A}$, or coordinatewise by $f(a)_\Gamma = |a|_\Gamma$.

Lemma 16.44 (Embedding Lemma) *The mapping f satisfies the following properties:*

(i) *f preserves the Boolean constants and operations.*
(ii) *$a \neq 0$ iff $\exists \Gamma \in T_A$: $a \in \Gamma^+$ iff $\exists \Gamma \in T_A$: $|a|_\Gamma \neq |0|_\Gamma$ iff $f(a) \neq 0$.*
(iii) *$a \in AE$ iff $\exists \Gamma \in T_A$: $|a|_\Gamma \in AE_\Gamma$ iff $f(a) \in AE_\mathbf{B}$.*
(iv) *$aC^s b$ iff $\exists \Gamma \in T_A$: $|a|_\Gamma C_\Gamma |b|_\Gamma$ iff $f(a) C^s_\mathbf{B} f(b)$.*
(v) *$aC^t b$ iff $\exists \Gamma \in T_A$: $|a|_\Gamma \in AE_\Gamma$ and $|b|_\Gamma \in AE_\Gamma$ iff $f(a) C^t_\mathbf{B} f(b)$.*
(vi) *$a\overrightarrow{C}^t b$ iff $\exists \Gamma \in T_A$: $|a|_\Gamma \in AE_\Gamma$ and $|b|_\Gamma \neq |0|_\Gamma$ iff $f(a)\overrightarrow{C}^t_\mathbf{B} f(b)$.*
(vii) *$a\widehat{C}^t b$ iff $\exists \Gamma \in T_A$: $|a|_\Gamma \neq |0|_\Gamma$ and $|b|_\Gamma \neq |0|_\Gamma$ iff $f(a)\widehat{C}^t_\mathbf{B} f(b)$.*

Proof (i) is obvious.

(ii) Suppose $a \neq 0$. By Lemma 16.41 this is equivalent to the following: there exists a cluster pair $\langle \Gamma, \Delta \rangle$ such that $a \in \Delta$ and because $\Delta = \Gamma^+$, i.e., $a \in \Gamma^+$ which is equivalent to $a \notin \overline{\Gamma^+}$. By the definition of $\equiv_\Gamma$ this is just $a \not\equiv_\Gamma 0$, i.e., $|a|_\Gamma \neq |0|_\Gamma$, which is equivalent to $f(a) \neq 0$.

(iii) $a \in AE$ iff (by Lemma 16.41) there exist a C^t-cluster Γ such that $a \in \Gamma$ iff (by the proof of Lemma 16.43) there exists a C^t-cluster Γ such that $|a|_\Gamma \in AE_\Gamma$ iff $f(a) \in AE_\mathbf{A}$.

(iv) $aC^s b$ iff (by Lemma 16.41) there exist a C^t-cluster Γ and a C^s-clan $\Theta \subseteq \Gamma$ such that $a, b \in \Theta$ iff (by Lemma 16.43) there exists a C^t-cluster Γ such that $|a|_\Gamma C_\Gamma |b|_\Gamma$ iff $f(a) C^s f(b)$.

(v) $aC^t b$ iff (by Lemma 16.41) there exist a C^t-cluster Γ such that $a, b \in \Gamma$ iff (by the proof of Lemma 16.43) there exists a C^t-cluster Γ such that $|a|_\Gamma \in AE_\Gamma$ and $|b|_\Gamma \in AE_\Gamma$ iff $f(a) C^t_\mathbf{B} f(b)$.

(vi) $a\overrightarrow{C}^t b$ iff (by Lemma 16.41) there exist a cluster pair $\langle\Gamma, \Gamma^+\rangle$ such that $a \in \Gamma$ and $b \in \Gamma^+$ iff $\exists\Gamma \in T_A$: $|a|_\Gamma \in AE_\Gamma$ and $|b|_\Gamma \neq |0|_\Gamma$ iff $f(a)\overrightarrow{C}^t f(b)$.

(vii) $a\widehat{C}^t b$ iff (by Lemma 16.41) there exist a cluster pair $\langle\Gamma, \Gamma^+\rangle$ such that $a, b \in \Gamma^+$ iff $\exists\Gamma \in T_A$: $|a|_\Gamma \neq |0|_\Gamma$ and $|b|_\Gamma \neq |0|_\Gamma$ iff $f(a)\widehat{C}^t_{\mathbf{B}} f(b)$.

The next statement is one of the main theorems of this paper.

Theorem 16.3 (Representation theorem for DCA by means of snapshot models) *Let A be a DCA,* **B** *be the standard DCA associated to A and $f : B_A \mapsto \mathbf{B}$ be the canonical mapping introduced in Definition 16.20. Then, f is an embedding from A into* **B**.

More over, let (ψ) be any of the time axioms from the list (rs), (ls), (up dir), (down dir), (circ), (dens), (ref), (lin), (tri), (tr) . Then, (ψ) is true in B iff (ψ) is true in **B**.

Proof (i) follows from Lemma 16.44.

(ii) Let (ψ) be any of the time axioms from the list (rs), (ls), (up dir), (down dir), (circ), (dens), (ref), (lin), (tri), (tr) and denote by (Ψ) the corresponding time condition from the list (RS), (LS), (Up Dir), (Down Dir), (Circ), (Dens), (Ref), (Lin), (Tri), (Tr). Then, by Lemma 16.42, (ψ) is true in A iff (Ψ) is true in the canonical time structure $(T_A, \prec_A)$ of A. Because the standard DCA **B** is full then it is rich. Then, by Lemma 16.31, (Ψ) is true in the time structure $(T_A, \prec_A)$ iff (ψ) is true in the standard DCA **B**, which finishes the proof.

16.7 Topological Models for DCA

16.7.1 What Kind of Topological Models for DCA We Need?

In this section, we will present a topological model and topological representation theory for DCA. This model should combine the topological models studied before for contact relations, actual contact relations, and precontact relations. So the Boolean algebra of the intended topological space X will be the algebra $RC(X)$ of regular closed sets of X and $RC(X)$ should have analogs of the relations C^s, C^t, $\widehat{C}^t$, $\overrightarrow{C}^t$, $C^\prec$. The first three C^s, C^t, $\widehat{C}^t$ are contact like, so X must have several kinds of points characterizing the corresponding contact between regions as sharing a common point of the corresponding kind. The last two relations $\overrightarrow{C}^t$ and $C^\prec$ are precontact relations, so there must be two binary relations in X to characterize the corresponding precontact between regions as in the representation theory in precontact algebras. Analogs of the points in X will be the clans and clusters of DCA.

When proving embedding theorems it is important to know some conditions for the pair (A_1, A_2) where A_2 is a Boolean algebra with a number of precontact relations and A_1 is a Boolean subalgebra of A_2 ensuring the following property: A_1 satisfies a number of additional axioms for the precontact relations if and only if A_2 satisfies the same axioms. Such conditions will be called lifting conditions and this is the subject of the next section.

16.7.2 Lifting Conditions

Let $A_i = (B_i, AE_i, C_i^s, C_i^t, \overrightarrow{C^t}_i, \widehat{C^t}_i, C_i^{\prec}), i = 1, 2$ be two Boolean algebras with the signature of DCA such that AE_i is a predicate of actual existence and for the other relations it is assumed that they are precontact relations and that $\widehat{C^t}_i$ is assumed to be a contact relation. We assume also that A_1 is a subalgebra of A_2. This means that B_1 is a Boolean subalgebra of B_2 and the precontact relations of B_1 are restrictions of the corresponding relations of B_2 to B_1.

We need some abstract "lifting" conditions guarantying that A_1 satisfies the remaining axioms of DCA and possibly some time axioms from the list *time axioms* **(rs)**, **(ls)**, **(up dir)**, **(down dir)**, **(circ)**, **(dens)**, **(ref)**, **(lin)**, **(tri)**, **(tr)** iff A_2 satisfies the same axioms. The conditions are given in the next definition and generalize similar conditions considered in Vakarelov (2007) (pages 283-4) only for contact algebras. For convenience the elements from the set B_{A_i} are denoted correspondingly by a_i, b_i, c_i, ... etc.

Definition 16.21 (*Lifting conditions*) Having in mind the above notations the Boolean subalgebra A_1 is said to be a Boolean **dense subalgebra** of A_2 if the following condition is satisfied:

(Dense) $(\forall a_2)(a_2 \neq 0 \Rightarrow (\exists a_1)(a_1 \neq 0 \text{ and } a_1 \leq a_2)$,

and to be a **co-dense subalgebra** of A_2 if

(Co-dense) $(\forall a_2)(a_2 \neq 1 \Rightarrow (\exists a_1)(a_1 \neq 1 \text{ and } a_2 \leq a_1)$.

(Lift AE) $(\forall a_2)(a_2 \notin AE) \Rightarrow (\exists a_1)(a_1 \notin AE) \text{ and } a_2 \leq a_1)$.

Let C be any of the relations $C^s_{A_2}$, $C^t_{A_2}$, $\mathcal{B}_{A_2}$ and its restriction to B_{A_1} to be denoted also by C. We say that A_1 is a C-separable subalgebra of A_2 if the following condition is satisfied:

(C-separation) $(\forall a_2, b_2))(a_2\overline{C}b_2 \Rightarrow (\exists a_1, b_1)(a_1\overline{C}b_1 \text{ and } a_2 \leq a_1 \text{ and } b_2 \leq b_1)$.

If all lifting conditions are satisfied, then A_1 is said to be a **stable subalgebra** of A_2.

If g is an isomorphic embedding of A_1 into A_2, then g is said to be a **dense** (**co-dense**) embedding provided that $g(A_1)$ is a dense (co-dense) subalgebra of A_2. We say that g is a C-separable embedding if $g(A_1)$ is a C-separable subalgebra of A_2. If all lifting conditions are satisfied, then g is called a **stable embedding** of A_1 into A_2.

It is easy to see that (Dense) is equivalent to (Co-dense).

Under the assumptions of the above definition, the following lemma is true.

Lemma 16.45 *Let A_1 be a stable subalgebra of A_2. Then, the following conditions are satisfied:*

(i) *A_1 is DCA iff A_2 is DCA.*
(ii) *Let Ax be a time axiom (see Definition 16.15). Then, Ax is true in A_1 iff Ax is true in A_2.*

Proof Because we assumed that the relations in A_i, $i = 1, 2$ are precontact, then we have to verify the statement for all other axioms for DCA (for instance that $\widehat{C^t}$ satisfies the contact axioms). Let us note that the proofs for all axioms go uniformly that is why we will consider only two informative examples from (i).

• The axiom $(C^s \subseteq C^t)$ $aC^sb \Rightarrow aC^tb$.

(From A_1 to A_2) Suppose that $(C^s \subseteq C^t)$ is true in A_1 and for the sake of contradiction that it is not true in A_2. Then, for some a_2, b_2, we have $a_2C^sb_2$ and $a_2\overline{C}^t b_2$. Then, by the condition (C^t-separation), we obtain there exist a_1, b_1, such that $a_2 \leq a_1$, $b_2 \leq b_1$ and $a_1\overline{C}^t b_1$. From here and $a_2C^sb_2$ we get $a_1C^sb_1$ which by $a_1\overline{C}^t b_1$ shows that the axiom $(C^s \subseteq C^t)$ is not true in A_1—a contradiction.

(From A_2 to A_1). Suppose now that the axiom is true in A_2. Since it is an universal formula, then it is trivially true in A_1.

• The axiom (AC^s1) If aC^sb, then $a, b \in AE$.

(From A_1 to A_2). Suppose that the axiom is true in A_1 but not in A_2. So for some a_2, b_2 we have $a_2C^sb_2$ but $a_2 \notin AE$. Then, by (**Lift AE**), there exists a_1 such that $a_2 \leq a_1$ and $a_1 \notin AE$. From here and the obvious condition $b_2 \leq 1$ we obtain a_1C^s1, which implies $a_1 \in AE$—a contradiction.

(From A_2 to A_1) This case is obvious because the axiom is a first-order condition.

• The axiom $(C^t\widehat{C^t}$Ef+) $a\overline{C^t}b \Rightarrow (\exists c)(a\overline{C^t}c$ and $c^*\overline{\widehat{C^t}}c$ and $c^*\overline{C^t}b)$,

(From A_1 to A_2). Suppose that the axiom is true in A_1 and let $a_2\overline{C^t}b_2$. Then, there exist a_1, b_1 such that $a_2 \leq a_1$, $b_2 \leq b_1$ and $a_1\overline{C^t}b_1$. Then, there exists c_1 such that $(a_1\overline{C^t}c_1$ and $c_1^*\overline{\widehat{C^t}}c_1$ and $c_1^*\overline{C^t}b_1)$. Using monotonicity and $a_2 \leq a_1$ and $b_2 \leq b_1$ we get $(a_2\overline{C^t}c_1$ and $c_1^*\overline{\widehat{C^t}}c_1$ and $c_1^*\overline{C^t}b_2)$. Since c_1 is also in B_2 this shows that the axiom is true in B_2.

(From A_2 to A_1). Suppose that the axiom is true in A_2 and let $a_1\overline{C^t}b_1$. Since a_1, b_1 are also in B_2, then there exists c_2 such that

(1) $(a_1\overline{C^t}c_2$ and
(2) $c_2^*\overline{\widehat{C^t}}c_2$ and
(3) $c_2^*\overline{C^t}b_1)$. Condition (1) implies that there exists c_1 such that
(4) $c_2 \leq c_1$ and
(5) $a_1\overline{C^t}c_1$. Condition (2) implies that there exist e_1, f_1 such that
(6) $c_2^* \leq e1$ and
(7) $c_2 \leq f_1$ and
(8) $e_1\overline{\widehat{C^t}}f_1$. From (4) and (6) we obtain $1 = c_2 + c_2^* = c_1 + f_1$ which implies
(9) $c_1 + f_1 = 1$. Analogously (6) and (7) imply
(10) $e_1 + f_1 = 1$. Since $\widehat{C^t}$ is a contact relation, condition (8) implies
(11) $e_1.f_1 = 0$. Conditions (10) and (11) imply $e_1 = f_1^*$ which substituted in (9) and (8) gives
(12) $f_1 \leq c_1$ and

(13) $f_1^* \overline{\widehat{C^t}} f_1$. Now from (12) and (8) we get by monotonicity
(14) $a_1 \overline{C^t} f_1$. Condition (3) implies by (C-separability) that there exists d_1 such that
(15) $c_2^* \leq d_1$ and
(16) $d_1 \overrightarrow{C^t} b_1$. Now (7) and (15) imply $d_1 + f_1 = 1$ which gives
(17) $f_1^* \leq d_1$. From (16) and (17) we get
(18) $f_1^* \overrightarrow{C^t} b_1$.
Now conditions (14), (13), and (18) show that the axiom ($C^t \widehat{C}^t$Ef+) is true in the algebra A_1).

16.7.3 Dynamic Mereotopological Space

In this section, we will introduce a topological model of DCA called *dynamic mereotopological space* (DMS, DM-space for short), which can be considered as a generalization of the notion *mereotopological space* from Goldblatt and Grice (2016) adapted for the case of DCA. DMS is a multi-sorted topological structure with the signature

$$S = (X_S^s, X_S^t, \widehat{X^t}_S, T_S, \widehat{T}_S, \overrightarrow{R}_S, \prec_S, M_S),$$

where

• $\widehat{X^t}_S$ is a topological space and M_S is a subalgebra of the algebra $RC(\widehat{X^t}_S)$ of regular closed subsets of $\widehat{X^t}_S$. $\widehat{X^t}_S$ has several sorts of points and relations between them described as follows:

• $X_S^s, X_S^t, \widehat{X^t}_S, T_S, \widehat{T}_S$ are non-empty sets satisfying the following inclusions: $X_S^s \subseteq X_S^t \subseteq \widehat{X^t}_S$, $T_S \subseteq X_S^t$, $\widehat{T}_S \subseteq \widehat{X^t}_S$,

• $\overrightarrow{R}_S \subseteq X_S^t \times \widehat{X^t}_S$ and $\prec_S \subseteq X_S^t \times X_S^t$. The system S is called a **dynamic mereotopological space** if the following axioms are satisfied:

The axioms of DMS:

(S1) M_S is considered as a closed base of the topology of $\widehat{X^t}_S$.

Before the formulation of the other axioms, we introduce some terminology and

Some definitions:

• The elements of X_S^s are called **space points**.

• The elements of X_S^t and T_S are called **actual time points**. The elements of X_S^t are considered as **approximate or partial** actual time points while the elements of the subset T_S are considered as **exact** actual time points.

• The relation $\prec_S$ between the elements of X_S^t is called "before-after" relation and the substructure $(T_S, \prec_S)$ is called the **time structure of** S.

• The elements of the sets $\widehat{X^t}_S$ and $\widehat{T}_S$ are called time points of S. The elements of $\widehat{X^t}_S$ are considered as **approximate or partial** time points while the elements of the subset $\widehat{T}_S$ are considered as **exact** time points. Compared to the points of the set X_S^t the elements of $\widehat{X^t}_S$ are considered as **possibly actual** or **non-actual** points.

• The relation $\overrightarrow{R}_S$ with the source X^t_S and co-source $\widehat{X^t}_S$ is called **semi-actual simultaneity**. If $x \in X^t_S$ and $y \in \widehat{X^t}_S$ and $x\overrightarrow{R}_S y$, then we say that the actual time point x is **simultaneous** with the time point y.

We define some relations and a subset in the Boolean algebra $RC(\widehat{X^t}_S)$: $a, b \in RC(\widehat{X^t}_S$

• **Actual existence** $a \in AE_S$ iff $a \cap X^t_S \neq \varnothing$—the region a actually exists if it has at least one actual time point.

• **Space contact**: $aC^s_S b$ iff $a \cap b \cap X^s_S \neq \varnothing$, a and b have a common space point.

• **Actual time contact**: $aC^t_S b$ iff $a \cap b \cap X^t_S \neq \varnothing$—$a$ and b have a common actual time point.

• **Possibly actual or non-actual time contact**: $a\widehat{C^t}_S b$ iff $a \cap b \neq \varnothing$, a and b are in the standard topological contact in $RC(\widehat{X^t}_S)$.

• **Semi-actual time contact**: $a\overrightarrow{C^t}_S b$ iff $\exists x \in X^t_S$, $\exists y \in \widehat{X^t}_S$ such that $x \in a$, $y \in b$ and $x\overrightarrow{R}_S y$.

• **Precedence**: $aC^{\prec}_S b$ iff $\exists x, y \in X^t_S$ such that $x \in a$, $y \in b$ and $x \prec_S y$.

• **The regular-set algebra**: $RC(S) =_{\text{def}} (RC(\widehat{X^t}_S), AE_S, C^s_S, C^t_S, \overrightarrow{C^t}_S, \widehat{C^t}_S, C^{\prec}_S)$. It can easily be shown the following fact for $RC(S)$:

• **Fact 1:** AE_S is indeed a predicate of actual existence in $RC(\widehat{X^t}_S)$, C^s_S and C^t_S are actual contacts, $C^{\prec}_S$ is an actual precontact, that $\overrightarrow{C^t}_S$ is a semi-actual contact and that $\widehat{C^t}_S$ is the standard topological contact relation in $RC(\widehat{X^t}_S)$.

• **The dual of** S, $S^+ =_{\text{def}} (M_S, AE_S, C^s_S, C^t_S, \overrightarrow{C^t}_S, \widehat{C^t}_S, C^{\prec}_S)$ where the set AE_S and the relations $C^s_S, C^t_S, \overrightarrow{C^t}_S, \widehat{C^t}_S, C^{\prec}_S$ are restricted to the set M_S.

• We define for $x \in \widehat{X^t}_S$ the following function to subsets of M_S
$\rho_S(x) =_{\text{def}} \{a \in M_S : x \in a\}$.

• **Approximation ordering** for $x, y \in \widehat{X^t}_S$: $x \sqsubseteq_S y \Leftrightarrow_{\text{def}} \rho_S(x) \subseteq \rho_S(y)$.

The next axioms for DMS are the following.

(S2) S^+ is a DCA. It is called also the canonical DCA of S.

It can easily be proved the following fact for $\rho_S(x)$:

• **Fact 2:** Let $x \in X^t_S$ and $y \in \widehat{X^t}_S$. Then $\rho_S(x)$ is a C^t-clan, called a point C^t-clan, and $\rho_S(y)$ is a $\widehat{C^t}$-clan, called a point $\widehat{C^t}$-clan in S^+.

(S3) Let $x \in T_S$, and $y \in \widehat{T_S}$. Then $\rho_S(x)$ is a C^t-cluster in S^+, and $\rho_S(y)$ is a $\widehat{C^t}$-cluster in S^+. $\rho_S(x)$ is called a point $\widehat{C^t}$-cluster and $\rho_S(y)$ is called a point $\widehat{C^t}$-cluster in S^+.

(S4) For $x \in X^t_S$ and $y \in \widehat{X^t}_S$: $x\overrightarrow{R}_S y$ iff $\rho_S(x) R_{\overrightarrow{C^t}_S} \rho_S(y)$ where $R_{\overrightarrow{C^t}_S}$ is the canonical relation between C^t-clans and $\widehat{C^t}$-clans in S^+.

For $x, y \in X^t_S$: $x \prec_S y$ iff $\rho_S(x) R_{C^{\prec}_S} \rho_S(y)$ where $R_{C^{\prec}_S}$ is the canonical relation between C^t-clans corresponding to the relation $C^{\prec}_S$ in S^+.

Let us note that Fact 2 and axioms (S3) and (S4) show that ρ_S maps the points of S into the corresponding point-clans and point-clusters in S^+ and transfers the relations $\overrightarrow{R}_S$ and $\prec_S$ into the corresponding clan relations in S^+.

We say that S is a $T0$ space if $\widehat{X^t}_S$ is a $T0$ space.

Let $\widehat{Ax}$ be a subset of the time conditions from the list (RS), (LS), (Up Dir), (Down Dir), (Circ), (Dens), (Ref), (Lin), (Tri), (Tr). We say that S satisfies the conditions from the list $\widehat{Ax}$ if the time structure $(T_S, \prec_S)$ satisfies these conditions.

16.7.4 *T*0 and DMS-Compactness Property of DMS

In this section, we will study first some consequences of $T0$ property of DMS. The next lemma states an equivalent definition in terms of approximate ordering $\sqsubseteq$.

Lemma 16.46 *(i) S is a $T0$ space iff the relation $\sqsubseteq_S$ is a partial ordering in $\widehat{X^t_S}$.*
(ii) ρ_S is an injective mapping from $\widehat{X^t}_S$ into the set of all $\widehat{C^t}_S$-clans of S^+.

Proof (i) ($\Rightarrow$) Suppose that S is a $T0$ space. It is obvious that the relation $\sqsubseteq_S$ is a reflexive and transitive relation. Suppose that $x \sqsubseteq_S y$ and $y \sqsubseteq_S x$. Then $\rho_S(x) = \rho_S(y)$. We shall show that $x = y$. Suppose for the contrary that $x \neq y$. Then, by the $T0$ property, there is a closed set $A \subseteq \widehat{X}_S$ containing one of the points and not the other. Consider the case $x \in A$ and $y \notin A$. Because M_S is a closed base of the space the condition $y \notin A$ implies that there exists $a \in M_S$ such that $A \subseteq a$ and $y \notin a$, so $a \notin \rho_S(y)$. Because $x \in A$ and $A \subseteq a$ we get $x \in a$ and hence $a \in \rho_S(x)$. This implies $\rho_S(x) \neq \rho_S(y)$—a contradiction.

($\Rightarrow$) Suppose that the relation $\sqsubseteq_S$ is a partial ordering. We will show the $T0$ property of S. Suppose that the $T0$ property is not true, then there exist $x \neq y$ such that for all closed subsets $A \subseteq \widehat{X^t}_S$ we have $x \in A$ iff $y \in A$. The elements of M_S are regular closed subsets of $\widehat{X^t}_S$ and consequently: for all $a \in M_S$, $x \in a$ iff $y \in a$. This implies $\rho_S(x) = \rho_S(y)$ which gives $x \sqsubseteq_S y$ and $y \sqsubseteq_S x$. Then, because $\sqsubseteq_S$ is a partial ordering we get $x = y$ which is a contradiction.

(ii) Suppose $\rho_S(x) = \rho_S(y)$. Then by (i) $x = y$.

Lemma 16.47 *Let S possesses the $T0$ property. Then, the following properties of $\sqsubseteq_S$ are true:*

(i) If $x \in X^t_S$, then there exists at most one $y \in T_S$ such that $x \sqsubseteq_S y$.
(ii) If $x \in \widehat{X^t}_S$, then there exists at most one $y \in \widehat{T}_S$ such that $x \sqsubseteq_S y$.
(iii) For $x \in T_S$ and $y \in \widehat{T}_S$: $x \sqsubseteq_S y$ iff $x \overrightarrow{R}_S y$.

Proof (i) Suppose $x \in X^t_S$ and that there are $y, z \in \widehat{T}_S$ such that $x \sqsubseteq y$, $x \sqsubseteq z$ and that $x \neq z$. This implies either $y \not\sqsubseteq z$ or $z \not\sqsubseteq y$. In both cases we get $\rho_S(y) \neq \rho_S(z)$. By axiom (S3) $\rho_S(y)$ and $\rho_S(z)$ are C^t-clusters in S^+, so by Lemma 16.27 there exists $c \in M_S$ such that $c \notin \rho_S(y)$ and $c^* \notin \rho_S(z)$. Because $x \sqsubseteq y$ and $x \sqsubseteq z$ this implies $c \notin \rho_S(x)$ and $c^* \notin \rho_S(x)$, which is impossible because by Fact 1 from Sect. 16.7.3 $\rho_S(x)$ is a C^t-clan.

The proof of (ii) is analogous to that of (i) and uses Lemma 16.12.

The proof of (iii) is similar to the above and uses Lemma 16.40 (ii).

Let us note that in Lemma 16.47 condition (i)(and similarly (ii)) is a statement for the uniqueness of y and in general we can not prove that such an y exists. Now we will introduce a special compactness property for S which will imply the existence of such an y.

Definition 16.22 We say that a DMS S possesses a **DMS-compactness property** (S is DMS-compact or S is a DM-compact space for short) if the following conditions are satisfied:

(i) If Γ is a C^s-clan in M_S^+ then there exists $x \in X_S^s$ such that $\Gamma = \rho_S(x)$.
(ii) If Γ is a C^t-clan in M_S^+ then there exists $x \in X_S^t$ such that $\Gamma = \rho_S(x)$.
(iii) If Γ is a C^t-cluster in M_S^+ then there exists $x \in T_S$ such that $\Gamma = \rho_S(x)$.
(iv) If Γ is a $\widehat{C^t}$-clan in M_S^+ then there exists $x \in \widehat{X^s}_S$ such that $\Gamma = \rho_S(x)$.
(v) If Γ is a $\widehat{C^t}$-cluster in M_S^+ then there exists $x \in \widehat{X^t}_S$ such that $\Gamma = \rho_S(x)$.
Shortly: all types of clans and clusters in M_S^+ are correspondingly point-clans and clusters taking the points from the corresponding set of points of S. Or, equivalently: the function ρ_S maps surjectively the sets X_S^s, X_S^t, $\widehat{X^t}_S$, T_S, $\widehat{T}_S$ onto the sets of the corresponding type of clans and clusters in M_S^+.

Note that DMS-compactness is a modification of the notion of *mereocompactness* from Goldblatt and Grice (2016) adapted for the case of DMS.

Lemma 16.48 *Let S possesses the $T0$ and DMS-compactness properties. Then, the following properties of $\sqsubseteq_S$ are true:*

(i) If $x \in X_S^t$, then there exists exactly one $y \in T_S$ such that $x \sqsubseteq_S y$.
(ii) If $x \in \widehat{X^t}_S$, then there exists exactly one $y \in \widehat{T}_S$ such that $x \sqsubseteq_S y$.
(iii) If $x \in T_S$ then there exists exactly one $y \in \widehat{T}_S$ such that $x \sqsubseteq_S y$.
(iv) If $y \in \widehat{T}_S$ then there exists exactly one $x \in T_S$ such that $x \sqsubseteq_S y$.

Proof The proofs follow from $T0$ and DMS-compactness properties by using Lemmas 16.47 and 16.40.

Corollary 16.1 *Let S be a DMS. Then, the following conditions are equivalent:*

(i) S is $T0$ and DM-compact:
(ii) ρ_S is a one-one mapping from $\widehat{X^t}_S$ to the set of all $\widehat{C^t}_S$-clans of S^+ such that the following additional properties are true:
(1) ρ_S is a surjection from X_S^s onto all C_S^s-clans of S^+.
(2) ρ_S is a surjection from X_S^t onto all C_S^t-clans of S^+.
(3) ρ_S is a surjection from T_S onto all C_S^t-clusters of S^+.
(4) ρ_S is a surjection from $\widehat{T}_S$ onto all $\widehat{C^t}_S$-clusters of S^+.

Proof The proof follows from Lemma 16.46 and DM-compactness.

Lemma 16.49 *If S is a DM-compact space then $\widehat{X^t}_S$ is a compact space.*

Proof The proof is similar to the proof of Theorem 4.2. (2) of Goldblatt and Grice (2016).

In the next section, we will derive other important consequences from $T0$ and DMS-compactness properties of S.

16.7.5 Canonical Filters

We assume in this section that S is a DM-compact space. Our aim is to introduce a technical notion—*canonical filter*, generalizing a similar notion from Vakarelov (2007). By means of canonical filters and the assumption of DM-compactness of a given S, we will establish that the algebra S^+ is a stable subalgebra of $RC(S)$ in the sense of Definition 16.21 which fact implies several important consequences.

Definition 16.23 Let $\alpha \in RC(X^t_S)$. Then, the set $F_\alpha =_{\text{def}} \{a \in M_S : \alpha \subseteq a\}$ is called a canonical filter of S^+.

Lemma 16.50 *Let* $\alpha, \beta \in RC(\widehat{X^t}_S)$. *Then,*

(i) F_α *is a filter in* S^+.
(ii) $\forall x \in \widehat{X^t}_S$: $x \in \alpha$ *iff* $F_\alpha \subseteq \rho_S(x)$.
(iii) $\alpha \neq \widehat{X^t}_S$ *iff there exists* $a \in M_S$ *such that* $\alpha \subseteq a$ *and* $a \neq \widehat{X^t}_S$.
Let C *denote any of the basic relations* C^s_S, C^t_S, $\overrightarrow{C^t}_S$, $\widehat{C^t}_S$, $C^{\prec}_S$ *of the algebra* $RC(S)$ *and let* R_C *be the canonical relation between ultrafilters in* S^+ *corresponding to* C *from* S^+.
(iv) *The following conditions are equivalent for each* C *of* $RC(S)$:
(1.1) $\alpha C^s_S \beta$ *(1.2)* $F_\alpha R_{C^s_S} F_\beta$ *(1.3)* $\alpha \cap \beta \cap T_S \neq \varnothing$.
(2.1) $\alpha C^t_S \beta$ *(2.2)* $F_\alpha R_{C^t} F_\beta$ *(2.3)* $\alpha \cap \beta \cap X^t_S \neq \varnothing$.
(3.1) $\alpha \overrightarrow{C^t}_S \beta$, *(3.2)* $F_\alpha R_{\overrightarrow{C^t}} F_\beta$, *(3.3)* $(\exists x \in \alpha \cap X^t_S)(\exists y \in \beta \cap \widehat{X^t}_S)(x \overrightarrow{R}_S y)$.
(4.1) $\alpha \widehat{C^t}_S \beta$, *(4.2)* $F_\alpha R_{\widehat{C^t}} F_\beta$, *(4.3)* $\alpha \cap \beta \cap \widehat{T}_S \neq \varnothing$.
(5.1) $\alpha C^{\prec}_S \beta$, *(5.2)* $F_\alpha R_{C^{\prec}} F_\beta$, *(5.3)* $(\exists x \in \alpha \cap T_S)(\exists y \in \beta \cap T_S)(x \prec_S y)$.
(6.1) $\alpha \in AE_S$, *(6.2)* $F_\alpha \subseteq AE_{S^+}$, *(6.3)* $\alpha \cap T_S \neq \varnothing$.

Proof (i) The proof is by a direct checking the corresponding definitions.

(ii) The implication from left to right is by straightforward checking. For the converse direction, we will reason by contraposition. Suppose $x \notin \alpha$. Now we will apply the fact that M_S is a closed base of the topology of X. Because α is a regular closed set then α is a closed set and then there exists $a \in M_S$ such that $\alpha \subseteq a$ and $x \notin a$. Then, $a \in F_\alpha$ and $a \notin \rho_S(x)$, so $F_\alpha \nsubseteq \rho_S(x)$.

(iii) can be derived by direct application of (ii).

(iv) We will consider with more details two typical examples: one for contact like relation and one for precontact relation. First consider the case $C = C^t_S$.

(1.1)⇒(1.2). Suppose $\alpha C^t_S \beta$. Then, there is a point $x \in X^t_S$ such that $x \in \alpha$ and $x \in \beta$. By (ii) this implies

(1) $F_\alpha \subseteq \rho_S(x)$ and

(2) $F_\beta \subseteq \rho_S(x)$.

In order to show $F_\alpha R_{C^t_S} F_\beta$ suppose $a \in F_\alpha$ and $b \in F_\beta$ and proceed to show $a C^t_S b$ in S^+. Then, by (1) and (2) we get $a \in \rho_S(x)$ and hence $x \in a$, and $b \in \rho_S(x)$ and hence $x \in b$, which shows $a \cap b \cap X^t_S \neq \varnothing$. So, $a C^t_S b$ in S^+. Thus, by the definition of the canonical relation $R_{C^t_S}$, we get $F_\alpha R_{C^t_S} F_\beta$.

(1.2)⇒(1.3). Suppose $F_\alpha R^t F_\beta$. By Lemma 16.3 there exist ultrafilters U, V such that $F_\alpha \subseteq U$, $F_\beta \subseteq V$ and $U R_{C^t_S} V$. Let $\Gamma = U \cup V$. Obviously $F_\alpha \subseteq \Gamma$ and $F_\beta \subseteq \Gamma$. By Lemma 16.8 Γ as a union of $R_{C^t_S}$-related ultrafilters is a C^t-clan in S^+. Now extend Γ into a C^t-cluster Γ'. Then, by DM-compactness, there is $x \in T_S$ such that $\Gamma = \rho_s(x)$. Hence $F_\alpha \subseteq \rho_s(x)$ and $F_\beta \subseteq \rho_s(x)$. By (ii) $x \in \alpha$ and $x \in \beta$ hence $\alpha \cap \beta \cap T_S \neq \varnothing$.

(1.3)⇒(1.1). Because $T_S \subseteq X^t_S$ we get from (1.3) $\alpha \cap \beta \cap X^t_S \neq \varnothing$ which by the definition of C^t_S in $RC(S)$ we obtain $\alpha C^t_S \beta$.

Let us consider the case $C = C^{\prec}_S$.

(5.1)⇒(5.2). Suppose $\alpha C^{\prec}_S \beta$. Then, there exists a points $x, y \in X^t_S$ such that $x \in \alpha$ and $y \in \beta$ and $x \prec_S y$. By (ii) this implies

(3) $F_\alpha \subseteq \rho_S(x)$ and

(4) $F_\beta \subseteq \rho_S(y)$ and by axiom (S4) of DMS

(5) $\rho_S(x) R_{C^{\prec}_S} \rho_S(y)$.

Then by (3), (4), and (5) we get $F_\alpha R_{C^{\prec}_S} F_\beta$.

(5.2)⇒(5.3). Suppose (5.2): $F_\alpha R_{C^{\prec}_S} F_\beta$. Then, by Lemma 16.3, there are ultrafilters U, V in S^+ such that $F_\alpha \subseteq U$, $F_\beta \subseteq V$ and $U R_{C^{\prec}_S} V$. Extend them to C^t_S-clusters Γ, Δ in S^+ such that $\Gamma R_{C^t_S} \Delta$. By DM-compactness, there exist $x', y' \in T_S$ such that $\Gamma = \rho_S(x')$ and $\Delta = \rho_S(y')$. From here we get

(6) $F_\alpha \subseteq \Gamma = \rho_S(x')$,

(7) $F_\beta \subseteq \Delta = \rho_S(y')$,

(8) $\rho_S(x') R_{C^{\prec}_S} \rho_S(y')$.

From (6) we obtain $x' \in \alpha \cap T_S$ (by (ii)), analogously from (7)—$y' \in \beta \cap T_S$ and from (8) we obtain by DMS axiom (S4) that $x' \prec_S y'$. This is just (5.3).

(5.3)⇒(5.1). Suppose (5.3). Because $T_S \subseteq X^t_S$ we get easily (5.1).

Lemma 16.51 *The following conditions are true for S:*

(i) The algebra S^+ is a stable Boolean subalgebra of $RC(S)$.
(ii) $RC(S)$ is a DCA.

Proof (i) We first show that S^+ satisfies the lifting conditions (see Definition 16.21) and then (i) is a corollary of Lemma 16.45. First, we verify the lifting condition (co-dense). Suppose $A \in RC(X^t_S)$ and $\alpha \neq \widehat{X^t}_S$. Then, by Lemma 16.23(iii), there exists $a \neq M_S$ such that $a \neq X^t_S$ and $\alpha \subseteq a$ which is just the dense condition. We do not treat (dense) because it is equivalent to (co-dense).

To verify the condition (C-separation) for C from the list C^s_S, C^t_S, $\overrightarrow{C^t}_S$, $\widehat{C^t}_S$, $C^{\prec}_S$ we proceed as follows. Looking at the conditions (iv) of Lemma 16.23, we see that they have the following common form. Let R_C be the canonical relation between filters corresponding to the relation C. Then for any $\alpha, \beta \in RC(\widehat{X^t}_S)$: $\alpha C \beta$ iff $F_A R_C F_B$. Taking the negation in both sides we obtain: $\alpha \overline{C} \beta$ iff $F_A \overline{R}_{C_C} F_B$ iff there exists $a, b \in M_S$ such that $a \in F_A$, $b \in F_B$ and $a \overline{C} b$ iff there exists $a, b \in M_S$ such that $A \subseteq a$, $B \subseteq b$ and $a \overline{C} b$. Thus, $F_A \overline{R} F_B$ implies that for some $a, b \in M_S$, $A \subseteq a$, $B \subseteq b$ and $a \overline{C} b$ which is just the (C-separation) condition. Note that just this implication needed DM-compactness in Lemma 16.23.

(ii) is a corollary of (i) and the fact that S^+ is a DCA, so by Lemma 16.45 the axioms of DCA are lifted from S^+ to $RC(S)$.

Lemma 16.52 *Let (φ) be any of the time axioms:* ***(rs)****,* ***(ls)****,* ***(up dir)****,* ***(down dir)****,* ***(circ)****,* ***(dens)****,* ***(ref)****,* ***(lin)****,* ***(tri)****,* ***(tr)*** *Then, the following conditions are equivalent:*

(i) (φ) is true in the algebra S^+.
(ii) (φ) is true in the algebra $RC(S)$.

Proof The proof follows from Lemmas 16.51 (i) and 16.45.

Lemma 16.53 *Let S be DM-compact DMS, $RC(S)$ be its regular-sets algebra, $(T_S, \prec_S)$ be its time structure and let $(T_{S^+}, \prec_{S^+})$ be the canonical time structure of S^+ (see Definition 16.19). Let (Φ) be the time condition from the list (RS), (LS), (Up Dir), (Down Dir), (Circ), (Dens), (Ref), (Irr), (Lin), (Tr) (condition (Tri) is excluded). Then, the following conditions are true:*

(i) (Φ) is true in $(T_S, \prec_S)$ iff (Φ) is true in $(T_{S^+}, \prec_{S^+})$.
(ii) If S is $T0$ DMS, then (Tri) is true in $(T_S, \prec_S)$ iff (Tri) is true in $(T_{S^+}, \prec_{S^+})$.

Proof (i) Let us remind that the members of T_{S^+} are C^t-clusters of S^+ , which we will denote by $\Gamma, \Delta, \Theta, \ldots$. We will demonstrate the proof considering the case (Dense) the proofs for the other cases can be done in the same manner.

(Dense) $(\forall i, j)(i \prec j \Rightarrow (\exists k)(i \prec k$ and $k \prec j)$.

($\Rightarrow$) Suppose (Dense) is true in $(T_S, \prec_S)$ and let $\Gamma, \Delta \in T_{S^+}$ and $\Gamma \prec_{S^+} \Delta$. Then, by DM-compactness there exist $x, y \in T_S$ such that $\Gamma = \rho_S(x)$, and $\Delta = \rho_S(y)$, so $\rho_S(x) \prec_{S^+} \rho_S(y)$ which by axiom (S4) implies $x \prec_S y$. Then, by (Dens) there exists $z \in T_S$ such that $x \prec_S z \prec_S y$. Again by axiom (S4) we obtain $\rho_S(x) \prec_{S^+} \rho_S(z) \prec_{S^+} \rho_S(y)$. Because $\rho_S(z)$ is a cluster in S_+ we put $\Theta = \rho_S(z)$ and obtain $\Gamma \prec_{S^+} \Theta \prec_{S^+} \Delta$ which shows that (Dense) is true in $(T_{S^+}, \prec_{S^+})$.

($\Leftarrow$) Suppose (Dense) is true in $(T_{S^+}, \prec_{S^+}), x, y \in T_S$ and $x \prec_S y$. Then $\rho_S(x) \prec_{S^+} \rho_{S^+}(y)$. By (Dense) there exists a cluster Θ (hence there exists $z \in T_S$ with $\rho_S(z) = \Theta$) such that $\rho_S(x) \prec_{S^+} \rho_S(z) \prec_{S^+} \rho_{S^+}(y)$. This implies $x \prec_S z \prec_S y$ which shows that (Dens) is true in $(T_S, \prec_S)$.

(ii) The case of (Tri) $(\forall i, j)(i = j$ or $i \prec j$ or $j \prec i$.

($\Rightarrow$) The proof of this implication is straightforward and requires neither DM-compactness nor $T0$ property.

($\Leftarrow$) Suppose (Tri) is true in $(T_{S^+}, \prec_{S^+})$ and let $x, y \in T_S$. Then, $\rho_S(x), \rho_S(y)$ are clusters in S^+. Then, by (Tri), we have $\rho_S(x) = \rho_S(y)$ or $\rho_S(x) \prec_{S^+} \rho_S(y)$ or $\rho_S(y) \prec_{S^+} \rho_S(x)$. **Case 1**: $\rho_S(x) = \rho_S(y)$. Since $\rho_S(x)$ and $\rho_S(y)$ are also t-clans then by the assumption that S is a $T0$ space case 1 implies $x = y$ (by Lemma 16.46).

Case 2: $\rho_S(x) \prec_{S^+} \rho_S(y)$. By DMS axiom (S4) this implies $x \prec_S y$.

Case 3: $\rho_S(y) \prec_{S^+} \rho_S(x)$. Again by (S4) we get $y \prec_S x$. Thus, (Tri) is fulfilled in the time structure $(T_S, \prec_S)$.

Lemma 16.54 (Topological definability) *Let $(T_S, \prec_S)$ be the time structure of S, (Φ) be the time condition from the list (RS), (LS), (Up Dir), (Down Dir), (Circ),*

(Dens), (Ref), (Lin), (Tri) (Tr), and (φ) be the corresponding time axiom from the list **(rs), (ls), (up dir), (down dir), (circ), (dens), (ref), (lin), (tri), (tr)**. *Then, the following conditions are equivalent (for the case of (Tri) we assume also that S is T0):*

(i) (Φ) *is true in* $(T_S, \prec_S)$.
(ii) (φ) *is true in* $(RC)(S)$.

Proof (Φ) is true in $(T_S, \prec_S)$ iff (by Lemma 16.53) (Φ) is true in the canonical time structure of S^+, $(T_{S^+}, \prec_{S^+})$ iff (by Lemma 16.42 (φ) is true in S^+ iff (by Lemma 16.52) (φ) is true in the algebra $RC(S)$.

16.7.6 Canonical DMS for DCA and Topological Representation Theorem for DCA

Let $A = (B_A, AE_A, C^s_A, C^t_A, \overrightarrow{C^t}_A \widehat{C^t}_A, C^{\prec}_A)$ be a DCA. We associate to A in a canonical way a DM-space denoted by A_+ and called the **canonical DMS of A** or the **dual DMS of** A as follows:

• $A_+ =_{\text{def}} (X^s_A, X^t_A, \widehat{X^t}_A, T_A, \widehat{T_A}, \overrightarrow{R}_A, \prec_A, M_A)$, where

• X^s_A are the C^s-clans of A, X^t_A are the C^t-clans of A, $\widehat{X^t}_A$ are the $\widehat{C^t}$-clans of A, T_A are the C^t-clusters of A, $\widehat{T_A}$ are the $\widehat{C^t}$-clusters of A.

• $\overrightarrow{R}_A \subseteq X^t_A \times \widehat{X^t}_A$ defined by $\overrightarrow{R}_A = R_{\overrightarrow{C}_A}$ where $R_{\overrightarrow{C}_A}$ is the canonical relation between clans corresponding to the relation $\overrightarrow{C^t}_A$ (see Definition 16.17 and Lemma 16.37).

• $\prec_A \subseteq X^t_A \times X^t_A$ is the before-after relation in the set X^t_A defined by $\prec_A = R_{C^{\prec}_A}$, where $R_{C^{\prec}_A}$ is the canonical relation between clans corresponding to $C^{\prec}_A$ (see Definition 16.17 and Lemma 16.38).

• The structure $(T_A, \prec_A)$ - the time structure of A is now the time structure of A_+.

M_A is defined as follows and is used to introduce a topology in the set X^t_A considering it as a basis of the closed sets in the topology:

• For $a \in B_A$ let $g_A(a) = \{\Gamma \in \text{t-}Clans(A) : a \in \Gamma\}$ and put

• $M_A = \{g_A(a) : a \in B_A\}$.

By the topological representation theory of contact algebras (see Sect. 16.2.4) the set $\{g_A(a) : a \in B_A\}$ defines a topology in the set $\widehat{X^t}_A$ and g_A is an isomorphic embedding of B_A into the algebra $RC(\widehat{X^t}_A)$ and M_A is a Boolean subalgebra of $RC(\widehat{X^t}_A)$ isomorphic to B_A.

We define the algebra $(A_+)^+$—the dual of A_+ as follows.

• $(A_+)^+ =_{\text{def}} (M_A, AE_{A_+}, C^s_{A_+}, C^t_{A_+}, \overrightarrow{C^t}_{A_+}, \widehat{C^t}_{A_+}, C^{\prec}_{A_+})$, where the set

AE_{A_+} and the relations $C^s_{A_+}, C^t_{A_+}, \overrightarrow{C^t}_{A_+}, \widehat{C^t}_{A_+}, C^{\prec}_{A_+}$ have their standard definitions by means of the elements of M_A which are sets of the form $g_A(a)$ as in the corresponding representation theory by clans. These definitions are in accordance with the corresponding topological definition of the algebra $(A_+)^+$ from the space A_+.

Having in mind the topological representation theory of contact algebras (see Sect. 16.2.4) and AE-algebras with actual contact (see Sect. 16.3.4) it can be seen that g_A is also an isomorphism from A onto $(A_+)^+$, so $(A_+)^+ = g_A(A)$ which proves the following lemma.

Lemma 16.55 *A is isomorphic to $(A_+)^+$ and hence $(A_+)^+$ is a DCA.*

By definition, we have $\rho_{A_+} =_{\rm def} \{g_A(a) \in M_A : \Gamma \in g_A(a)\} = \{g_A(a) \in M_A : a \in \Gamma\}$.

Lemma 16.56 *(i) For any $\Gamma \in X^s_A$ $\rho_{A_+}(\Gamma)$ is a $C^s_{(A_+)^+}$-clan in $(A_+)^+$.*
(ii) For any $\Gamma \in X^t_A$ $\rho_{A_+}(\Gamma)$ is a $C^t_{(A_+)^+}$-clan in $(A_+)^+$.
(iii) For any $\Gamma \in \widehat{X^t}_A$ $\rho_{A_+}(\Gamma)$ is a $\widehat{C^t}_{(A_+)^+}$-clan in $(A_+)^+$.
(iv) For any $\Gamma \in T_A$ $\rho_{A_+}(\Gamma)$ is a $C^t_{(A_+)^+}$-cluster in $(A_+)^+$.
(v) For any $\Gamma \in \widehat{T}_A$ $\rho_{A_+}(\Gamma)$ is a $\widehat{C^t}_{(A_+)^+}$-cluster in $(A_+)^+$.

Proof The proof is by a routine verification of the corresponding definitions. As an example, we will demonstrate the proof of (iv)

Let $\Gamma \in T_A$. Then Γ is a C^t_A-cluster in A, so Γ is a C^t_A-clan in A. By (ii) $\rho_{A_+}(\Gamma)$ is a $C^t_{(A_+)^+}$-clan in $(A_+)^+$. We will show that $\rho_{A_+}(\Gamma)$ is a $C^t_{(A_+)^+}$-cluster in $(A_+)^+$. Suppose that for some $a \in B_A$, $g_A(a) \notin \rho_{A_+}(\Gamma)$. Then $\Gamma \notin g_A(a)$, so $a \notin \Gamma$. Then, there exists $b \in B_A$ such that $b \in \Gamma$ and $a\overline{C^t}_A b$. Then, $g_A(b) \in \rho_{A_+}(\Gamma)$ and $g_A(a) \cap g_A(b) \cap X^t_A = \varnothing$, so $g_A(a)\overline{C^t}_{A_+} g_A(b)$. Note that (iv) and (v) verify the DMS axiom (S4) for A_+.

Lemma 16.57 *(i) Let Γ be C^t_A-clan in A and Δ be a $\widehat{C^t}_A$-clan in A. Then, $\Gamma R_{\overrightarrow{C^t}_A} \Delta$ iff $\rho_{A_+}(\Gamma) R_{\overrightarrow{C^t}_{A_+}} \rho_{A_+}(\Delta)$.*
(ii) Let Γ, Δ be C^t_A-clans. Then, $\Gamma R_{C^{\prec}_A} \Delta$ iff $\rho_{A_+}(\Gamma) R_{C^{\prec}_{A_+}} \rho_{A_+}(\Delta)$.

Proof (i) Let Γ be C^t_A-clan in A and Δ be a $\widehat{C^t}_A$-clan in A. Then by Definition 16.17 for the left side of (i) we have $\Gamma R_{\overrightarrow{C^t}_A} \Delta$ iff $(\forall a \in \Gamma)(\forall b \in \Delta)(aC^{\prec}b)$.

Analogously for the right side of (i) we have $\rho_{A_+}(\Gamma) R_{\overrightarrow{C^t}_{A_+}} \rho_{A_+}(\Delta)$ iff $(\forall g_A(a) \in \rho_{A_+}(\Gamma))(\forall g_A(b) \in \rho_{A_+}(\Delta))(g_A(a)C^{\prec}_{A_+} g_A(b))$ iff $(\forall a \in \Gamma)(\forall b \in \Delta)$ $(aC^{\prec}_A b)$—we have obtained the same result as for the left side of (i) which proves (i). The proof of (ii) is similar.

Lemma 16.58 *A_+ is a DMS.*

Proof The proof follows from Lemmas 16.56, 16.57, and 16.55 which establish the DMS axioms (S3) and (S4). The other axioms are obviously true.

Lemma 16.59 *Let A be a DCA and $\Gamma \subseteq M_A$. Define $\widehat{\Gamma} =_{\rm def} \{a \in B_A : g_A(a) \in \Gamma\}$. Then, the following conditions are true:*

(i) If Γ is a $C^s_{(A_+)^+}$-clan in $(A_+)^+$, then $\widehat{\Gamma}$ is a C^s_A-clan in A and $\rho_{A_+}(\widehat{\Gamma} = \Gamma$.
(ii) If Γ is a $C^t_{(A_+)^+}$-clan in $(A_+)^+$, then $\widehat{\Gamma}$ is a C^t_A- clan in A and $\rho_{A_+}(\widehat{\Gamma}) = \Gamma$.

(iii) *If* Γ *is a* $\widehat{C^t}_{(A_+)^+}$*-clan in* $(A_+)^+$, *then* $\widehat{\Gamma}$ *is a* $\widehat{C^t}$*-clan in* A *and* $\rho_{A_+}(\widehat{\Gamma}) = \Gamma$.
(iv) *If* Γ *is a* $C^t_{(A_+)^+}$*-cluster in* $(A_+)^+$, *then* $\widehat{\Gamma}$ *is a* C^t_A*-cluster in* A *and* $\rho_{A_+}(\widehat{\Gamma}) = \Gamma$.
(v) *If* Γ *is a* $\widehat{C^t}_{(A_+)^+}$*-cluster in* $(A_+)^+$, *then* $\widehat{\Gamma}$ *is a* $\widehat{C^t}_A$*-cluster in* A *and* $\rho_{A_+}(\widehat{\Gamma}) = \Gamma$.

Proof The proof is almost the same for all cases. We will consider two examples.

(ii) Let Γ be a $C^t_{(A_+)^+}$-clan in $(A_+)^+$. The grill properties of $\widehat{\Gamma}$ follow from the grill properties of Γ. For the C^t_A-clan property suppose $a, b \in \widehat{\Gamma}$. Then, $g_A(a), g_A(b) \in \Gamma$. Because Γ is a $C^t_{(A_+)^+}$-clan, then $g_A(a) C^t_{(A_+)^+} g_A(b)$. By the definition of the relation $C^t_{(A_+)^+}$ we get $g(a) \cap g(b) \cap X^t_A \neq \varnothing$. This sows that there exists $\Delta \in X^t_A$ (and hence Δ is a C^t_A-clan), $\Delta \in g(a)$ and $\Delta \in g(b)$. This implies $a, b \in \Delta$ and consequently $a C^t_A b$ which shows that $\widehat{\Gamma}$ is a C^t_A-clan.

Let us show that $\rho_{A_+}(\widehat{\Gamma}) = \Gamma$. The following sequence of equivalencies proves this:

$g_A(a) \in \rho_{A_+}(\widehat{\Gamma})$ iff $\widehat{\Gamma} \in g_A(a)$ iff $a \in \{b \in B_A : g_A(b) \in \Gamma\}$ iff $g_A(a) \in \Gamma$.

We use here the that $\widehat{\Gamma}$ is a C^t_A-clan, so $\widehat{\Gamma} \in X^t_A \subseteq \widehat{X^t}_A$ and hence is the domain of ρ_{A_+}.

(iv) Let Γ be a $C^t_{(A_+)^+}$-cluster in $(A_+)^+$. We will show that $\widehat{\Gamma}$ is a C^t_A-cluster in A. The verification that $\widehat{\Gamma}$ is a C^t_A-clan in A is as in the above proof. For the cluster property of $\widehat{\Gamma}$ suppose that $a \notin \widehat{\Gamma}$. Then $g_A(a) \notin \Gamma$. Since Γ is a $C^t_{(A_+)^+}$-cluster in $(A_+)^+$, there exists $g_A(b) \in \Gamma$ (and hence $b \in \widehat{\Gamma}$) such that $g_A(a) \overline{C^t}_{(A_+)^+} g_A(b)$. This is equivalent to $g_A(a) \cap g_A(b) \cap X^t_A \neq \varnothing$. As in the above proof this implies $a \overline{C^t}_A b$, which shows that $\widehat{\Gamma}$ is a cluster in A. The proof of the equality $\rho_{A_+}(\widehat{\Gamma}) = \Gamma$ is the same as above.

The following theorem is important.

Theorem 16.4 *A_+ is T0 and DM-compact DMS.*

Proof The $T0$ property of A_+ follows from Lemma 16.46 and DM-compactness follows from Lemma 16.59.

Theorem 16.5 (Topological representation theorem for DCA) *Let A be a DCA. Then, the following conditions for A are true:*

(i) $(A_+)^+$ *is a stable subalgebra of the algebra* $RC(A_+)$.
(ii) *The algebra* $RC(A_+)$ *is a DCA.*
(iii) *The function* g_A *is a stable isomorphic embedding of* A *into* $RC(A_+)$.
(iv) *Let* (θ) *be a time axiom from the list (rs), (ls), (up dir), (down dir), (circ), (dens), (ref), (lin), (tri), (tr). Then,* (θ) *is true in* A *iff* (θ) *is true in* $RC(A_+)$.

Proof (i) By Theorem 16.4 A_+ is a DM-compact DMS and hence by Lemma 16.51 (i) $(A_+)^+$ is a stable Boolean subalgebra of $RC(A_+)$.

(ii) follows from (i) and Lemma 16.51 (ii).

(iii) By Lemma 16.55 g_A is an isomorphism from A onto $(A_+)^+$ and hence by (i) g_A is a stable isomorphic embedding of A into $RC(A_+)$.

(iv) follows from Theorem 16.4 and Lemma 16.52.

16.8 Topological Duality Theory for DCA

In this section, we extend the topological representation of DCAs to a topological duality theory of DCAs in terms of DMSes. We assume basic knowledge of category theory: categories, morphisms, functors, and natural isomorphisms (see, for instance, Chap. I from Mac Lane 1998). Since DCA is a generalization of contact algebra, and DMS is a generalization of mereotopological space, the developed duality theory in this section will generalize the duality theory for contact algebras and mereotopological spaces presented by Goldblatt and Grice (2016). This makes that some proofs below will be the same (or almost the same) as in Goldblatt and Grice (2016). Other topological duality theorems for contact and precontact algebras are presented in Dimov et al. (2017) and it is possible to generalize them for DCAs, but in this paper we follow the scheme of Goldblatt and Grice (2016) for two purposes: first, because the corresponding notion of DMS fits quite well to the topological representation theory for DCS-s, and second, because the proofs in this case are more short.

16.8.1 The Categories DCA and DMS

Definition 16.24 The category **DCA** consists of the class of all DCAs supplied with the following morphisms, called DCA-morphisms.

Let A and A' Then $f : A \longrightarrow A'$ is a DCA-morphism if it is a mapping $f : B_A \longrightarrow B_{A'}$ which satisfies the following conditions:

(f 1) f is a Boolean homomorphism from B_A into $B_{A'}$.

For all $a, b \in B_A$:

(f2) If $f(a) \in AE_{A'}$, then $a \in AE_A$.

Let C_A be any of the relations C^s_A, C^t_A, $\overrightarrow{C^t_A}$, $\widehat{C^t_A}$, $C^{\prec}_A$ and similarly for $C_{A'}$. Then

(f 3) if $f(a)C_{A'}f(b)$, then aC_Ab,

A is the domain of f and A' the codomain of f.

We define $f_+ =_{\text{def}} f^{-1}$ acting on t-clans of A_2 as follows: for $\Gamma \in t\text{-}Clans(A')$, $f^{-1}(\Gamma) =_{\text{def}} \{a \in B_A : f(a) \in \Gamma\}$.

A DCA-morphism $f : A \longrightarrow A'$ is a DCA-isomorphism (in the sense of category theory) if there is a DCA-morphism $g : A' \longrightarrow A$ such that the compositions $f \circ g$ and $g \circ f$ are the identity morphism of their domains. It is a well-known fact that this definition is equivalent to the standard algebraic definition of isomorphism in universal algebra.

Definition 16.25 The category **DMS** consists of the class of all DMSes equipped with suitable morphisms called DMS-morphism. The definition is as follows. θ is a DMS-morphism from S into S' if it is a mapping $\theta : \widehat{X^t}_S \rightarrow \widehat{X^t}_{S'}$ which satisfies the following conditions:

(θ 1) If $x \in \widehat{X^s}_S$, then $\theta(x) \in X^s_{S'}$.

(θ 2) If $x \in \widehat{X^t}_S$, then $\theta(x) \in X^t_{S'}$.

(θ 3) If $x \overrightarrow{R}_{S} y$ then $\theta(x) \overrightarrow{R}_{S'} \theta(y)$.

(θ 4) If $x \prec_S y$, then $\theta(x) \prec_{S'} \theta(y)$.

Let $a \subseteq \widehat{X^t}_{S'}$ and $\theta^{-1}(a) =_{\text{def}} \{x \in \widehat{X^t}_S : \theta(x) \in a\}$. We define $\theta^+ =_{\text{def}} \theta^{-1}$. The next two requirements for θ are the following:

(θ 5) If $a \in M_{S'}$ then $\theta^{-1}(a) \in M_S$ and

(θ 6) the map $\theta^{-1} : M_{S'} \longrightarrow M_S$ is a Boolean algebra homomorphism from (M') into (M).

Note that in M_S the join operation is a set theoretical union of regular closed sets. Since the meet in a Boolean algebra is definable by the join and the complement $*$, for the condition (θ 6) it is sufficient to assume that θ^{-1} preserves complement (note that a^* is not a set-complement of a).

A DMS-morphism $\theta : S \longrightarrow S'$ is a DMS-isomorphism if there exists a converse DMS-morphism $\eta : S' \longrightarrow S$ such that the compositions $\theta \circ \eta$ and $\eta \circ \theta$ are identity morphisms in the corresponding domains.

The following lemma states an equivalent definition of DMS-isomorphism. Similar statement for mereotopological isomorphism is Theorem 2.2 from Goldblatt and Grice (2016).

Lemma 16.60 *Let S, S' be DM-spaces and $\theta : S \to S'$ be DMS-morphism from S into S'. Let $a \subseteq X^t_S$ and define $\theta[a] = \{\theta(x) : x \in a\}$. Then the following two conditions are equivalent:*

(i) θ is a DMS-isomorphism from S onto S'.

(ii) θ is a bijection from $\widehat{X^t}_S$ onto $\widehat{X^t}_{S'}$ satisfying the following conditions:

(1) If $\theta(x) \in X^s_{S'}$, then $x \in X^s_S$.

(2) If $\theta(x) \in X^t_{S'}$, then $x \in X^t_S$.

(3) If $\theta(x) \overrightarrow{R}_{S'} \theta(y)$, then $x \overrightarrow{R}_S y$.

(4) If $\theta(x) \prec_{S'} \theta(y)$, then $x \prec_S y$.

(5) If $a \in M_S$, then $\theta[a] \in M_{S'}$.

Proof (i)$\Rightarrow$(ii) Suppose that θ is a DMS isomorphism from S onto S'. Then obviously θ is a bijection with converse η such that θ is a DMS-morphisms from S onto S' and η is a DMS-morphism from S' onto S such that the composition $\theta \circ \eta$ is the identity in S' and $\eta \circ \theta$ is the identity in S. To show (1) let $\theta(x) \in X^s_{S'}$. Then $x = \eta(\theta(x)) \in X^s_S$, because η is a DMS-morphism from S' onto S. In a similar way we show (2), (3) and (4).

To show (5) let $a \in M_S$. Then $\eta^{-1}(a) \in M_{S'}$, because η is a DMS-morphism from S' onto S. This means that for any $x \in \widehat{X^t}_{S'}$ and $a \in M_S$ the following holds: $x \in \eta^{-1}(a)$ iff $\eta(x) \in a$ iff (by the definition of $\theta[a]$) $\theta(\eta(x)) \in \theta[a]$ iff (because $\theta(\eta(x)) = x$) $x \in \theta[a]$. This shows that $\theta[a] = \eta^{-1}(a)$, which proves that $\theta[a] \in M_{S'}$.

(i)$\Leftarrow$(ii) Suppose that θ is a DMS-morphism from S into S' and that (ii) is true. Conditions (1), (2), (3), and (4) imply that η satisfy conditions (θ1), (θ2) and (θ3) for DMS-morphism. Since θ is a DMS-morphism, it follows that the map $a \mapsto \theta^{-1}(a)$

is a Boolean homomorphism from $M_{S'}$ to M_S. Because θ is a bijection, it follows that for its converse η, the map $a \mapsto \eta^{-1}(a)$ is a Boolean homomorphism from M_S to $M_{S'}$, which shows that the condition (θ4) is also fulfilled. So η is a DMS-morphism from S' to S. Because θ and η are converses to each other, their compositions are the identity mappings in the corresponding domains. So, θ is a DMS-isomorphism from S onto S'.

Let $f : A_1 \longrightarrow A_2$ and $g : A_2 \longrightarrow A_3$ be two DCA-morphisms. The composition $h = f \circ g$ is a mapping $h : B_{A_1} \longrightarrow B_{A_3}$ acting as follows; for $a \in B_{A_1}$: $h(a) = g(f(a))$. In a similar way, we define composition for DMS-morphisms.

The following lemma has an easy proof.

Lemma 16.61 *(i) The composition of two DCA-morphisms is a DCA-morphism. The identity mapping 1_A on each DCA A is a DCA-morphism. Hence **DCA** is indeed a category.*
*(ii) The composition of two DMS-morphisms is a DMS-morphism. The identity mapping 1_S on each DMS S is a DMS-morphism. Hence **DMS** is indeed a category.*

It follows from Lemma 16.61 that **DCA** and **DMS** are indeed categories.

We denote by **DMS*** the full subcategory of **DMS** of all T0 and DM-compact DMSes.

We introduce two contravariant functors

Φ: **DCA**$\mapsto$**DMS**, and Ψ: **DMS**$\mapsto$**DCA** as follows:

(I) For a given DCA A we put $\Phi(A) = A_+$ and for a DCA-morphism $f : A \longrightarrow A'$ we put $\Phi(f) = f_+$ and prove that f_+ is a DMS-morphism from $(A')_+$ into A.

(II) For a given DMS S we put $\Psi(S) = S^+$ and for a DMS-morphism $\theta : S \longrightarrow S'$ we put $\Psi(\theta) = \theta^+$ and prove that θ^+ : is a DMS morphism from $(S')^+$ into S.

(III) We show that for each DCA A the mapping $g_A(a) = \{\Gamma \in t\text{-}Clans(A) : a \in \Gamma\}$, $a \in B_A$ is a natural isomorphism from A to $\Psi(\Phi(A)) = (A_+)^+$. (For natural isomorphisms see Mac Lane 1998 Chap. I, 4.)

(IV) We show that for each $T0$ and DM-compact DMS S the mapping $\rho_S(x) = \{a \in M_S : x \in a\}$, $x \in X^t_S$ is a natural isomorphism from S to $\Phi(\Psi(S) = (S^+)_+$.

All this shows that the category **DCA** is dually equivalent to the category **DMS*** of T0 an DM-compact DMS. The realization of (I)–(IV) is given in the next subsection.

16.8.2 Facts for DCA-Morphisms and DMS-Morphisms

Lemma 16.62 *Every DMS-morphism is a continuous mapping.*

Proof Let $\theta : S \longrightarrow S'$ be a DMS-morphism. Since θ^{-1} maps $M_{S'}$ (which is the closed basis of the topology of S') into M_S, then θ is continuous.

Lemma 16.63 *Let $f : A \longrightarrow A'$ be a DCA-morphism, let C_A be any of the relations $C^s_{A'}$, $C^t_{A'}$, and $\widehat{C^t}_A$ and similarly for A'. If Γ is a $C_{A'}$-clan in A' then $f^{-1}(\Gamma) =_{\text{def}} \{a \in B_A : f(a) \in \Gamma\}$ is a C_A-clan in A.*

Proof The proof consists of a routine check of the corresponding definitions of clan.

Lemma 16.64 *(i) Let A, A' be two DCAs and $f : A \longrightarrow A'$ be a DCA-morphism. Then, f_+ is a DMS-morphism from $(A')_+$ to A_+.*

(ii) The mapping $g_A(a) = \{\Gamma \in t\text{-}Clans(A) : a \in \Gamma\}$, $a \in B_A$ is a natural DCA-isomorphism of A onto $\Psi(\Phi(A)) = (A_+)^+$.

Proof (i) Remind that $(A')_+ = (C^s - Clans(A'), C^t - Clans(A'), \widehat{C^t} - clans(A'), C^t - Clusters(A'), \widehat{C^t} - Clusters(A'), \overrightarrow{R}_{A'}, \prec_{A'}, M_{A'})$ where $\overrightarrow{R}_{A'} = R_{\overrightarrow{C^t_A}}$ and $\prec_{A'} = R_{C^{\prec}_A}$ and similarly for A_+. If $\Gamma \in C^s - Clans(A')$, then by Lemma 16.63 $f^{-1}(\Gamma)$ is a C^s-clan of A and similarly for the other types of clans. This shows that the conditions $(\theta 1)$–$(\theta 3)$ for DMS-morphisms are fulfilled.

For the condition $(\theta 4)$ let $\Gamma R_{\overrightarrow{C^t}_{A'}} \Delta$, Γ is a $C^t_{A'}$-clan and Δ is a $\widehat{C^t}_{A'}$-Clan in A'. We have to show that

$f^{-1}(\Gamma) R_{\overrightarrow{C^t}_A} f^{-1}(\Delta)$. By the definition of $R_{\overrightarrow{C^t}_A}$ for clans (see Lemma 16.37) this means the following. Let $a \in f^{-1}(\Gamma)$, $b \in f^{-1}(\Delta)$. Then $f(a) \in \Gamma$ and $f(b) \in \Delta$. But $\Gamma R_{\overrightarrow{C^t}_{A'}} \Delta$, so $f(a) \overrightarrow{C^t}_{A'} f(b)$, which by (f3) implies $a \overrightarrow{C^t}_A b$. This shows that $\Gamma R_{\overrightarrow{C^t}_A} \Delta$.

The next step is to verify the condition $(\theta 5)$ of DMS-morphisms, namely that $(f_+)^+$ maps the members of $M_{A'}$ into the members of M_A. Note that the members of M_A are of the form $g_A(a)$ for $a \in B_A$ and that $g_A(a) = \{\Gamma \in \widehat{C^t} - Clans(A) : a \in \Gamma\}$ and similarly for the members of $M_{A'}$. In order to verify $(\theta 5)$, we will show that for any $a \in B_A$ the following equality holds which indeed shows that $(f_+)^+$ maps M_A into $M_{A'}$:

$$(f_+)^+(g_A(a)) = g_{A'}(f(a)) \tag{16.1}$$

To show (16.1) note that $(f_+)^+(g_A(a))$ is a subset of $\widehat{C^t}$-Clans(A'). So let Γ be $\widehat{C^t}$-clan in A'. Then, the following sequence of equivalences proves (16.1):

$\Gamma \in (f_+)^+(g_A(a))$ iff $\Gamma \in (f^{-1})^{-1}(g_A(a))$ iff $f^{-1}(\Gamma) \in g_A(a)$ iff $a \in f^{-1}(\Gamma)$ iff $f(a) \in \Gamma$ iff $\Gamma \in g_{A'}(f(a))$.

Now we verify the condition $(\theta 6)$ of DMS-morphisms: $(f_+)^+$ preserves the Boolean complement. We show this by applying (16.1) and the facts that f and $g_{A'}$ acts as Boolean homomorphisms:

$(f^+)_+((g_A(a))^*)$=$(f^+)_+(g_A(a^*))$=$g_{A'} f(a^*)$=$(f^+)_+(g_A(a^*))$= $((f^+)_+(g_A(a)))^*$.

(ii) The statement that g_A is a natural isomorphism in the sense of category theory means the following: first, that g_A is indeed an isomorphism from A onto A_+ (this is the Theorem 16.55) and second, that for any DCA-morphism $f : A \longrightarrow A'$, the following equality should be true: $g_{A'} \circ f = (f_+)^+ \circ g_A$. By the definition of the composition $\circ$ for DCA-morphisms, this equality is equivalent to the following: for any $a \in B_A$ the following holds:

$g_{A'}(f(a)) = (f_+)^+(g_A(a))$, which is just (16.1).

Lemma 16.65 *Let S, S' be two DMS-s and $\theta : S \longrightarrow S'$ be a DMS-morphism from S to S'. Then, θ^+ is a DCA-morphism from $(S')^+$ to S^+.*

Proof We have to verify that $\theta^+ = \theta^{-1}$ satisfies the conditions (f1)–(f3) for DCA-morphism. Condition (f1) is fulfilled by the condition $(\theta 6)$ for DMS-morphisms. For the condition (f2) suppose that $f(a) \in AE_{A'}$. Then, because $C^s_{A'}$ is actual contact, then $f(a)C^s_{A'}f(a)$ which by $f(3)$ implies that $aC^s_A a$ and hence $a \in AE_A$.

For condition (f3) suppose that for some $a, b \in M_{S'}, \theta^{-1}(a)C^s\theta^{-1}(b)$ and proceed to show $aC^a_{S'}b$. Then, there exists $x \in X^s_S$ such that $x \in \theta^{-1}(a)$ and $x \in \theta^{-1}(b)$. From here we obtain $\theta(x) \in a$, $\theta(x) \in b$ and $\theta(x) \in X^s_{S'}$ (by condition $(\theta 1)$ for DMS-morphism) which yields $aC^t_{S'}b$. In a similar way, one can verify condition (f3) for the other relations.

Before the formulation of the next statement let us remind what is $(S^+)_+$ for a DMS S. S^+ is the dual of S which is a DCA algebra and $(S^+)_+$ is just the dual space of the algebra S^+.

Lemma 16.66 *(i) Let S be a DMS. Then ρ_S is a DMS-morphism from S to $(S^+)_+$.*

(ii) Let S be a DM-compact DMS and let for $a \subseteq \widehat{X^t}_S$, $\rho_S[a] =_{\mathrm{def}} \{\rho_S(x) : x \in a\}$. Then for $a \in M_S$: $\rho_S[a] = g_{S^+}(a)$ (for the function g_A for a DCA A see Sect. 16.7.6).

(iii) If S is $T0$ and DM-compact, then ρ_S is a DMS-isomorphism from S onto $(S^+)_+$.

(iv) If S is a $T0$ and DM-compact DMS, then ρ_S is a natural isomorphism from S to $\Phi(\Psi(S)) = (S^+)_+$.

Proof (i) We have to verify whether ρ_S satisfies the conditions $(\theta 1)$–$(\theta 6)$ for DMS-morphisms. Note that $\rho_S(x)$ is a C^s-clan in S^+ for $x \in X^s_S$, such that for $x \in X^s_S$ it is a C^t_S-clan in S^+ and for $x \in X^t_S$ it is a C^t_S-clan in S^+. This verifies the conditions $(\theta 1)$ and $(\theta 2)$ for DMS-morphisms. Conditions $(\theta 3)$ and $(\theta 4)$ follow from axiom (S4) for DMS. For the condition $(\theta 5)$, we have to show that $(\rho_S)^{-1}$ transforms the members from M_{S^+} into the members from M_S (recall that the members of M_{S^+} are of the form $g_{S^+}(a)$, $a \in M_S$). This can be seen from the following equality

$$(\rho_S)^{-1}(g_{S^+}(a)) = a \tag{16.2}$$

Indeed, for $x \in \widehat{X^t}_S$ we have

$x \in (\rho_S)^{-1}(g_{S^+}(a))$ iff $\rho_S(x) \in g_{S^+}(a)$ iff $a \in \rho_S(x)$ iff $x \in a$.

For the condition $(\theta 6)$, we have to show that $(\rho_S)^{-1}$ preserves Boolean complement. The following sequence of equalities proves this

$(\rho_S)^{-1}(g_{S^+}(a^*)) = a^* = ((\rho_S)^{-1}(g_{S^+}(a)))^*$, which is true on the base of (16.2).

(ii) Suppose $a \in M_S$ and let us show first $\rho_S[a] \subseteq g_{S^+}(a)$:

$\rho_S(x) \in \rho_S[a] \Rightarrow x \in a \Rightarrow a \in \rho_S(x) \Rightarrow \rho_S(x) \in g_{S^+}(a)$ (because $\rho_S(x)$ is a $\widehat{C}^t$-clan in the DCA algebra S^+).

For the converse inclusion, let Γ be a $\widehat{C}^t{}_{S^+}$-clan in S^+. Then, by DM-compactness, there exists $x \in \widehat{X^t}_S$ such that $\Gamma = \rho_S(x)$. Then for $a \in M_S$:

$\Gamma \in g_{S^+}(a) \Rightarrow a \in \Gamma \Rightarrow a \in \rho_S(x)$ and $x \in a \Rightarrow \rho_S(x) \in \rho_S[a] \Rightarrow \Gamma \in \rho_S[a]$, which completes the proof.

(iii) Let S be $T0$ and DM-compact. Then, by Lemma 16.1, ρ_S is a one-one mapping from $\widehat{X^t}_S$ onto the set of all $\widehat{C^t}$-clans of S^+, which are the points of $(S^+)_+$. By (i) ρ_S is a DMS-morphism from S to $(S^+)_+$. So in order to show that ρ_S is a DMS-isomorphism from S onto $(S^+)_+$ we have to see if ρ_S satisfies the conditions (1)–(5) of Lemma 16.60 (ii).

For condition (1) suppose $\rho_S(x) \in X^s_{S^+}$. Then $\rho_S(x)$ is a t-clan in M_S. By DM-compactness there exists $y \in X^s_S$ such that $\rho_S(x) = \rho_S(y)$. By $T0$ condition this implies $x = y$, so $x \in X^s_S$. In a similar way, we verify condition (2) of the lemma.

For condition (3) suppose $x \in X^t_S$, $y \in \widehat{X^t}_S$ and $\rho_S(x) R_{\overrightarrow{C^t}_{S^+}} \rho_S(y)$. Then, by axiom (S4) of DMS we get for DMS we obtain $x \overrightarrow{R}_S y$. In a similar way, we verify condition (4)

For condition (5) suppose $a \in M_S$ and proceed to show that $\theta[a] \in M_{(S^+)_+}$. By (ii) $\theta[a] = g_{S^+}(a)$ and since $g_{S^+}(a) \in M_{(S^+)_+}$ we get $\theta[a] \in M_{(S^+)_+}$.

Thus, the conditions of (1)–(5) are fulfilled which proves that ρ_S is a DMS-isomorphism from S onto $(S^+)_+$.

(iv) Let S be a T0 and DM-compact DMS. In order ρ_S to be a natural isomorphism from S to $(S^+)_+$ it has to satisfy the following two conditions: first, ρ_S have to be a DMS-isomorphism—this is guaranteed by (iii), and second, for every DMS-morphism $\theta : S \Rightarrow S'$: the following equality should be true: $\theta \circ \rho_{S'} = \rho_S \circ (\theta^+)_+$. This equality is equivalent to the following condition: for $x \in X^t_S$

$$(\theta^+)_+(\rho_S(x) = \rho_{S'}(\theta(x)) \tag{16.3}$$

The following sequence of equivalencies proves (16.3). For $a \in M_{S'}$:

$a \in (\theta^+)_+(\rho_S(x))$ iff $a \in (\theta^+)^{-1}(\rho_S(x))$ iff $\Theta^+(a) \in \rho_S(x)$ iff $x \in \theta^+(a)$ iff $x \in \theta^{-1}(a)$ $\theta(x) \in a$ iff $a \in \rho_{S'}(\theta(x))$.

As applications of the developed theory, we establish some isomorphism correspondences between the objects of the two categories. The isomorphism between two objects will be denoted by the symbol $\cong$.

Lemma 16.67 *Let A, A' be two DCAs. Then, the following conditions are equivalent:*

(i) $A \cong A'$,
(ii) $A_+ \cong (A')_+$,
(iii) $(A_+)^+ \cong ((A')_+)^+$.

Proof **(i)⇔(iii)**. By Lemma 16.55, we have $A \cong (A_+)^+$ and $A' \cong (A'_+)^+$. This makes obvious the equivalence (i)⇔(iii).

(i)⇒(ii) Suppose $A \cong A'$, then there exists a on-one mapping f from A onto A' with a converse mapping h such that $f : A \mapsto A'$ is a DCA-morphism from A onto A' and $h : A' \mapsto A$ is a DCA- morphism from A' onto A such that the composition $f \circ h$ is the identity mapping in A' and the composition $h \circ f$ is the identity mapping in A. Then, by Lemma 16.64, f_+ is a DMS-morphism from A'_+ onto A_+ and h_+ is a DMS-morphism from A_+ onto A'_+.

We shall show the following:

(I) The composition $f_+ \circ h_+$ is the identity in A'_+, and

(II) The composition $h_+ \circ f_+$ is the identity in A_+.

Then, by the definition of DMS-isomorphism, this will imply that both f_+ and h_+ are DMS-isomorphisms in the corresponding directions.

Note that the members of A_+ are the t-clans of A and similarly for A_+.

To show (I) let Γ be a point of the space A'_+, i.e., Γ is a $\widehat{C^t}$-clan in A'. We shall show that $(f_+ \circ h_+)(\Gamma) = \Gamma$ which will prove (I). This is seen from the following sequence of equivalencies where a is an arbitrary element of $B_{A'}$:

$a \in (f_+ \circ h_+)(\Gamma\ )$ iff $a \in (f_+(h_+(\Gamma))$ iff $a \in f^{-1}(h_+(\Gamma))$ iff $f(a) \in h_+(\Gamma)$ iff $f(a) \in h^{-1}(\Gamma)$ iff $h(f(a)) \in \Gamma$ iff $a \in \Gamma$.

Here, we use that $h(f(a)) = a$ for $a \in B_{A'}$ because h is the converse of the one-one mapping f from B_A onto $B_{A'}$.

In a similar way we show (II).

(ii)⇒(iii) The proof is similar to the above one. Suppose $A_+ \cong (A')_+$, then there exists a one-one mapping θ and its converse η such that θ is a DMS-morphism from A_+ onto $(A')_+$ and η is a DMS-morphism from $(A')_+$ onto A_+. Then, by Lemma 16.65, θ^+ is a DCA-morphism from $(A'_+)^+$ into $(A_+)^+$ and $(\eta^+$ is a DCA-morphism from $(A'_+)^+$ into $(A_+)^+$. We shall show that both θ^+ and η^+ are DCA-isomorphisms in the corresponding directions by showing that their compositions are identities in the corresponding domains. Let us note that the domain of θ^+ are the members of the algebra $(A'_+)^+$ which are of the form $g_{A'}(a)$, $a \in_{B_{A'}}$, and similarly for the members of $(A_+)^+$. Namely, we will show the following two things:

(III) $(\theta^+ \circ \eta^+)(g_{A'}(a)) = g_{A'}(a)$ for any $a \in B_{A'}$,

(IV) $(\eta^+ \circ \theta^+)(g_{A'}(a)) = g_{A'}(a)$ for any $a \in B_A$.

To show (III) note that $g_{A'}(a) = \{\Gamma \in \widehat{C^t}clans(A') : a \in \Gamma$. So let Γ be a $\widehat{C^t}$-clan A'. Then, the following sequence of equivalencies proves (III):

$\Gamma \in (\theta^+ \circ \eta^+)(g_{A'}(a))$ iff $\Gamma \in (\theta^+(eta^+(g_{A'}(a))))$ iff $\Gamma \in (\theta^{-1}(eta^+(g_{A'}(a))))$ iff $\theta(\Gamma) \in (eta^+(g_{A'}(a)))$ iff $\theta(\Gamma) \in (eta^{-1}(g_{A'}(a)))$ iff $\eta(\theta(\Gamma)) \in g_{A'}(a)$ iff $\Gamma \in g_{A'}(a)$.

We have just used that $\eta(\theta(\Gamma)) = \Gamma$, because η is the converse of the one-one mapping θ. The proof of (IV) is similar.

Lemma 16.68 *Let S, S' be two DMSes. Then, the following conditions are equivalent:*

(i) $S \cong S'$,
(ii) $S^+ \cong (S')^+$,
(iii) $(S^+)_+ \cong ((S')^+)_+$.

Proof The proof is analogous to the proof of Lemma 16.67

As a corollary from Lemmas 16.67 and 16.68, we obtain the following addition to the topological representation theorem for DCAs.

Corollary 16.2 *There exists a bijective correspondence between the class of all, up to DCA-isomorphism DCAs, and the class of all, up to DMS-isomorphism DMSes; namely, for every DCA algebra A the corresponding DMS of A is A_+—the canonical algebra of A; and for every DMS S the corresponding DCA of S is S^+—the canonical algebra of S.*

16.8.3 Topological Duality Theorem for DCAs

In this section, we prove the third important theorem of this paper.

Theorem 16.6 (Topological duality theorem for DCAs) *The category* ***DCA*** *of all dynamic contact algebras is dually equivalent to the category* $\mathbf{DMS}^*$ *of all $T0$ and DM-compact DMSes.*

Proof The proof follows from Lemmas 16.64, 16.65, and 16.66.

The above theorem has several consequences to some important subcategories of **DCA** and **DMS**. The first example is the following. Let Ax be a subset of the set of temporal axioms (rs), (ls), (up dir), (down dir), (circ), (dens), (ref), (lin), (tri), (tr). Consider the class of all DCAs satisfying the axioms from Ax. It is easy to see that this class forms a full subcategory of the category of all DCAs under the DCA-morphism. Denote this subcategory by **DCA(Ax)**. Let $\widehat{Ax}$ be the subset of the corresponding to Ax time condition from the list (RS), (LS), (Up Dir), (Down Dir), (Circ), (Dens), (Ref), (Lin), (Tri), (Tr). Consider the class of all $T0$ and DM-compact DMSes which satisfy the axioms $\widehat{Ax}$. It is easy to see that this class is a full subcategory of the category $\mathbf{DMS}^*$ of all T0 and DM-compact dynamic mereotopological spaces. Denote this subcategory by $\mathbf{DMS}(\widehat{Ax})^*$

Theorem 16.7 *The category* ***DCA(Ax)*** *of all dynamic contact algebras satisfying Ax is dually equivalent to the category* $\mathbf{DMS}(\widehat{Ax})^*$ *of all $T0$ and DM-compact DMSes satisfying $\widehat{Ax}$.*

Proof Let S be a $T0$ and DM-compact DMS. It follows by Lemma 16.54 that S satisfies $\widehat{Ax}$ iff S^+ satisfies Ax. Now the theorem is a corollary of Theorem 16.6.

Acknowledgements The author is sponsored by Contract DN02/15/19.12.2016 with the Bulgarian NSF, project title *Space, Time and Modality: Relational, Algebraic and Topological Models*. Thanks are due to my colleagues Georgi Dimov, Tinko Tinchev, Philippe Balbiani, and Ivo Düntsch for the collaboration. I am very much indebted to Ian Pratt-Hartmann for carefully reading the manuscript and for helpful suggestions to improve the quality of the text.

References

Aiello, M., Pratt-Hartmann, I., & van Benthem, J. (Eds.). (2007). *Handbook of Spatial Logics*. Dordrecht: Springer.

Bennett, B., & Düntsch, I. (2007). Axioms, algebras and topology. In Aiello, M., (Ed) *Pratt-Hartmann and van Benthem* (pp. 99–159).
de Laguna, T. (1922). Point, line and surface as sets of solids. The Journal of Philosophy, 19:449–461.
Dimov, G., Ivanova-Dimova, E., & Vakarelov, D. (2017). A generalization of the Stone duality theorem. *Topology Appl.*, *221*, 237–261.
Dimov, G. and Vakarelov, D. (2006). Contact algebras and region-based theory of space: A proximity approach - I, II. Fundamenta Informaticae, 74:209–282.
Düntsch, I. and Vakarelov, D. (2007). Region-based theory of discrete spaces: A proximity approach. Ann. Math. Artif. Intell., 49(1–4):5–14.
Engelking, R. (1977). *General topology*. PWN—Polish Scientific Publishers, Warsaw. Translated from the Polish by the author, Monografie Matematyczne, Tom 60. [Mathematical Monographs, Vol. 60].
Goldblatt, R., & Grice, M. (2016). Mereocompactness and duality for mereotopological spaces. In *Journal of Michael Dunn on information based logics*. Outstanding contributions to logic (Vol. 8, pp. 313–330). Cham: Springer.
Hahmann, T., & Grüninger, M. (2012). Region-based theories of space: Mereotopology and beyond. In Hazarika, S., (Ed.), *Qualitative spatio-temporal representation and reasoning: Trends and future directions* (pp. 1–62). IGI Global.
Mac Lane, S. (1998). *Categories for the working mathematician*. Graduate texts in mathematics (Vol. 5, 2nd Ed.). New York: Springer.
Naimpally, S. A., & Warrack, B. D. (1970). *Proximity spaces*. Cambridge tracts in mathematics and mathematical physics (Vol. 59). London: Cambridge University Press.
Pratt-Hartmann, I. (2007). First-order mereotopology. In In Aiello, M., (Ed) *Pratt-Hartmann and van Benthem* (pp. 13–97).
Sikorski, R. (1964). *Boolean algebras* (2nd Ed.). Ergebnisse der Mathematik und ihrer Grenzgebiete, NeueFolge, Band 25. New York: Academic Inc.; Berlin: Springer.
Simons, P. (1987). *Parts. A study in ontology*. (Oxford: Clarendon Press).
Stone, M. (1936). The theory of representations for Boolean algebras. *Trans. Amer. Math. Soc.*, *40*, 37–111.
Tarski, A. (1956). Foundation of the geometry of solids. In J. H. Woodger (Ed.), *Logic, semantics, metamathematics* (pp. 24–29). Oxford: Clarendon Press. Translation of the summary of an address given by A. Tarski to the First Polish Mathematical Congress, Lwów, 1927.
Thron, W. (1973). Proximity structures and grills. Math. Ann., 206:35–62.
Urquhart, A. (1978). A topological representation theorem for lattices. Algebra Universalis, 8:45–58.
Vakarelov, D. (1997). Proximity modal logic. In *Proceedings of the 11th Amsterdam colloquium* (pp. 301–306).
Vakarelov, D. (2007). Region-based theory of space: Algebras of regions, representation theory, and logics. In *Mathematical problems from applied logic. II*. International mathematical series (New York) (Vol. 5, pp. 267–348). Springer, New York.
Vakarelov, D. (2010). Dynamic mereotopology: A point-free theory of changing regions. I. Stable and unstable mereotopological relations. *Fund Inform*, *100*(1–4), 159–180.
Vakarelov, D. (2012). Dynamic mereotopology II: Axiomatizing some Whiteheadean type space-time logics. *Advances in modal logic* (Vol. 9, pp. 538–558). London: Coll. Publ.
Vakarelov, D. (2014). Dynamic mereotopology. III. Whiteheadean type of integrated point-free theories of space and time. I. *Algebra and Logic*, *53*(3), 191–205.
Vakarelov, D. (2016a). Dynamic mereotopology. III. Whiteheadian type of integrated point-free theories of space and time. II. *Algebra and Logic*, *55*(1), 9–23.
Vakarelov, D. (2016b). Dynamic mereotopology. III. Whiteheadian type of integrated point-free theories of space and time. III. *Algebra and Logic*, *55*(3), 181–197.
Vakarelov, D. (2017a). Actual existence predicate in mereology and mereotopology (extended abstract). In L. Polkowski, Y. Yao, P. Artiemjew, D. Ciucci, D. Liu, D. Ślęzak, & B. Zielosko,

(Eds.), *Rough sets. International joint conference, IJCRS 2017, Proceedings Part II*. LNAI (Vol. 10314, pp. 138–157).
Vakarelov, D. (2017b). Mereotopologies with predicates of actual existence and actual contact. Fund. Inform., 156(3–4):413–432.
Vakarelov, D. (2020). Point-free theories of space and time. *Journal of Applied Logics - IFCoLog Journal of Logics and their Aapplications, 7*(6), 1243–1321 (2020).
van Benthem, J. (1983). *The logic of time*. Reidel.
Whitehead, A. N. (1929). *Process and reality: An essay in cosmology*. New York: MacMillan. Revised Edition, edited by D.R. Griffin and D.W. Sherburne (1978).

Chapter 17
Substitution and Propositional Proof Complexity

Sam Buss

Second Reader
W. Dzik
University of Silesia

Abstract We discuss substitution rules that allow the substitution of formulas for formula variables. A substitution rule was first introduced by Frege. More recently, substitution is studied in the setting of propositional logic. We state theorems of Urquhart's giving lower bounds on the number of steps in the substitution Frege system for propositional logic. We give the first superlinear lower bounds on the number of symbols in substitution Frege and multi-substitution Frege proofs.

Keywords Substitution · Proof complexity · Lower bounds · Multisubstitution · Frege proof · Frege · Begriffsschrift · Grundgesetze

> "The length of a proof ought not to be measured by the yard. It is easy to make a proof look short on paper by skipping over many links in the chain of inference and merely indicating large parts of it. Generally people are satisfied if every step in the proof is evidently correct, and this is permissable if one merely wishes to be persuaded that the proposition to be proved is true. But if it is a matter of gaining an insight into the nature of this 'being evident', this procedure does not suffice; we must put down all the intermediate steps, that the full light of consciousness may fall upon them."
>
> [G. Frege, *Grundgesetze*, 1893; translation by M. Furth]

[1]We do not concern ourselves with the much more common use of substitution allowing replacing (first-order) variables with terms.

S. Buss (✉)
Department of Mathematics, University of California, San Diego,
La Jolla CA 92093-0112, USA
e-mail: sbuss@ucsd.edu

I. Düntsch and E. Mares (eds.), *Alasdair Urquhart on Nonclassical and Algebraic Logic and Complexity of Proofs*, Outstanding Contributions to Logic 22,
https://doi.org/10.1007/978-3-030-71430-7_17

17.1 Introduction

The present article concentrates on the substitution rule allowing the substitution of *formulas* for formula variables.[1] The substitution rule has long pedigree as it was a rule of inference in Frege's *Begriffsschrift* (Frege 1879) and *Grundgesetze* (Frege 1903), which contained the first in-depth formalization of logical foundations for mathematics. Since then, substitution has become much less important for the foundations of mathematics, being subsumed by comprehension axioms. Indeed, the substitution of formulas for formula variables, when it is even permitted, is generally viewed as being merely a derived rule of inference.

Interest in substitution rules was revived in the 1970s, however, by Cook (1975), Cook and Reckhow (1979), Reckhow (1976) working with propositional proof systems. Motivated by logical questions arising out of the P versus NP question, they were interested in the logical and computational strength of propositional proof systems (called "Frege systems") including propositional proof systems in which both modus ponens and substitution rule are allowed as inferences. In the propositional setting, if $\varphi(p)$ is a derived formula with p a propositional variable, then the substitution rule allows inferring $\varphi(\psi/p)$, namely by replacing every use of the variable p with the formula ψ.

The present article was instigated by the opportunity to write a contribution for a volume honoring Alasdair Urquhart. A large part of Urquhart's work concerns the proof complexity of propositional proof systems, especially relatively weak proof systems. Two of his papers give lower bounds for substitution Frege proof length. In addition, Urquhart's work also addresses Russell's use of substitution for the foundations of mathematics based on Russell's efforts to fix the paradoxes present in Frege's work. Quite apart from the inherent interest of the substitution rule, this makes it an appropriate topic for the present collection of papers.

Section 17.2 will briefly discuss the substitution rule in second-order logic. The substitution was used already in Frege's work introducing formal methods for mathematical reasons, encompassing both first- and second-order arithmetic. Section 17.2 discusses the well-known equivalence of substitution and comprehension in second-order logic, and touches very lightly on later attempts of Russell to use substitution as the foundation for logicism.

Our principal interest in the substitution rule is in the setting of propositional logic, using the so-called Frege proof systems augmented with the substitution rule; this topic is discussed in Sects. 17.3 and 17.4, which form the main parts of this paper. We are particularly interested in general bounds on the size of substitution Frege proofs. Section 17.3 gives the main definitions and discusses connections with substitution Frege systems and quantified propositional logic. It also discusses different forms of substitution including renaming substitution and $\top/\bot$-substitution. Section 17.4 states lower bounds on the number of inferences in substitution Frege proofs due to Urquhart; we extend these to obtain new lower bounds on symbol length as well.

17.2 Substitution in Metamathematics

In preparing this article, I (the present author) took the opportunity to read the core work of Frege (in English translation) presented in his *Begriffsschrift* (Frege 1879) and volume 1 of his *Grundgesetze* (Frege 1903; Furth 1964). This was an eye-opening experience. Although it is at times hard to read Frege as his less formal, philosophical discussions do not always correspond exactly to his formal system, in the end, Frege gives more or less complete formal definitions, and overall, Frege's formulations of quantification theory are remarkably well-developed and advanced, and very clearly elucidated.[2] Frege's formal system encompasses both first-order and second-order logic. Frege's second-order objects are functions, defined in terms of their "course-of-values" or "value ranges" ("Werthverlauf"); this together with his Basic Law V provides a general comprehension axiom. Frege's pictograph representation of assertions may look awkward by modern standards when first encountered, but his proof system includes propositional logic and first- and second-order universal and existential quantifiers in a full and modern form. It even contains the sequent calculus as a special case.[3] Frege uses sophisticated methods for reasoning about inductive properties in a general way. He defines the non-negative integers in terms of equinumerous classes ("Gleichzahligkeit") starting in Section 38 of the *Grundgesetze* (Frege 1903; Furth 1964). The *Grundgesetze* also contains some rudimentary type theory in the sense of first- and second-level functions ("erster und zweiter Stufe"); however, for Frege, everything collapsed to the first-level functions. The notation, $\grave{\epsilon}$, used for defining a function in terms of its course-of-values is a precursor to lambda notation. Frege also used a definite article ("bestimmter Artikel") similar to Russell's upside-down iota symbol ℩ and a precursor to the Hilbert epsilon symbol. Furthermore, Frege used a formal method of introducing definitions, allowing the introduction of new pictographs to represent more complex formulas.

The substitution rule is given in Section 48 of the *Grundgesetze*, where it is presented as inference rule 9, allowing "Replacement of Roman letters" ("Ersatz der lateinischen Buchstaben"). In short, it allowed any first-order function ("Funktion erster Stufe") to be substituted for a free variable as long as the number of argument places matched up. In the earlier *Begriffsschrift*, substitution is introduced without much fanfare or explanation starting in the derivation of (2) in Section 13. Later parts of the *Begriffsschrift* use increasingly strong substitution principles, culminating in

[2] Indeed, Urquhart (1999) mentions "Frege's limpid clarity," comparing it favorably with the work of Russell.

[3] The sequent calculus is included in the sense that a figure of the form (for instance)

corresponds to the sequent $\Pi, \Lambda, \Delta \to \Gamma$. The *Begriffsschrift* and *Grundgesetze* spend an unexpectedly long time discussing things that correspond to the structural rules, the cut rule, and the negation rules of the sequent calculus.

derivations of (97), (98) and (110). (See Boolos (1985), for more on the use of substitution in the *Begriffsschrift*.)

We can state a version of the substitution rule in modern terms as follows. We work in second-order logic. Unlike Frege, we use sets as second-order objects instead of functions. Let $F(x)$ be a second-order object; that is, $F(a)$ takes on Boolean values for arbitrary first-order objects a. Let $\Phi(x)$ and Ψ be an arbitrary second-order formulas. We write $\Psi(\Phi/F)$ for the formula that results from replacing every instance $F(s)$ in Ψ with $\Phi(s/x)$, where $\Phi(s/x)$ means the formula obtained from Φ by replacing free occurrences of x with the term s.[4] The substitution rule allows us to infer

$$\frac{\Psi}{\Psi(\Phi/F)}$$

This form of the substitution rule is equivalent to comprehension; a fact first noticed by von Neumann (1927) and again by Henkin (1953). The fact that substitution implies comprehension can be proved as follows (see Boolos 1985). Let $\Phi(x)$ be an arbitrary formula, and let A and X be unary predicate symbols. By substituting Φ for A in the valid sequent $\exists X\, \forall x(X(x) \leftrightarrow A(x))$, we obtain $\exists X\, \forall x(X(x) \leftrightarrow \Phi(x))$. This is just the comprehension principle for Φ. The converse, that substitution follows from comprehension, is similarly easy to prove.

We hasten to add however that the substitution rule in the *Grundgesetze* was not as powerful as the substitution in second-order logic; instead, it served more as a convenience method to shorten proofs by being able to replace variables with arbitrary formulas. In particular, the substitution rule was not the mechanism used by Frege to establish comprehension. That was done instead of using course-of-values syntax and the Basic Law V of the *Grundgesetze*.

As is well known, the *Grundgesetze* proof system is inconsistent due to Russell's paradox. As Frege himself maintains in the appendix to volume 2 of Frege (1903), the Basic Law V is the problematic axiom. In fact, the problematic part is often referred to as "Basic Law Vb",[5] which we can restate in the language of second-order logic as

$$X = Y \;\rightarrow\; \forall x(X(x) \leftrightarrow Y(x)).$$

Stated in this form, Basic Law Vb presents as an instance of an equality axiom, and thus as completely unproblematic. However, it was a crucial axiom for the *Grundgesetze*. For Russell's paradox, one posits the existence of a set of all sets which do not contain themselves as a member. In the *Grundgesetze*, and taking some liberties in notation, the property of a set a not being a member of itself is expressed as

$$(\exists b)(b = a \wedge a \notin b).$$

It requires the use of an equality axiom to conclude from this that $a \notin a$. My own take on this is that the real problem does not lie in Basic Law Vb. That law, stated as an

[4] The usual conditions on 'substitutability' must hold of course. These can always be enforced by renaming bound variables as necessary.

[5] See Frege (1903, Sect. 52).

equality axiom, seems completely true. Instead, the root cause of the inconsistency is the fact that the *Grundgesetze* system allows unrestricted use of the course-of-values notation $\grave{\epsilon}$ to introduce functions.

Frege of course was devastated by Russell's paradox and understood very clearly the problems it raised. He wrote in part,[6] "And even now I do not see how arithmetic can be scientifically founded, how numbers can be conceived as logical objects and brought under study, unless we are allowed—at least conditionally—the transition from a concept to its extension. Is it always permissable to speak of the extension of a concept, of a class? And if not, how do we recognize the exceptional cases?" Furth (1964, p. 127).

After discovering the paradoxes, Russell took up the effort of recasting Frege's theories into a consistent theory for the foundations of mathematics, including the foundations of arithmetic. He made a strong effort, in both published and unpublished works, to develop a "substitutional theory" in which substitution played a leading role. In those theories, the notation $p\frac{a}{x}!q$ indicated that the result of substituting x for every appearance of a in p yields q. This substitution notation was not just a syntactic construction; instead, "$p\frac{a}{x}!q$" served as a formula and indeed p, x, a, q could be quantified over as variables. Unfortunately, the substitutional theory also suffered from paradoxes, and Russell abandoned it favor of the type system of Whitehead and Russell's *Principia Mathematica*.

Russell's substitutional theory is not particularly relevant to the main topics of the present paper, nor is it not really about the syntactic operation of substitution. Furthermore, the present author is not particularly knowledgeable about it. We therefore do not consider it further. The interested reader is referred instead to Landini (1998) and to the articles (Grattan-Guiness 1974; Hylton 1980; Pelham and Urquhart 1995; Stevens 2003; Urquhart 1999).

There is a great deal of work on Frege's formal systems for the foundations of mathematics. As a start, some that I have consulted include (Bentzen 2020; Boolos 1985; Heck 2011, 2012; Zalta 2018).

17.3 Substitution and Propositional Proofs

We now turn to the substitution rule in the setting of propositional proof systems. Propositional proof systems are much weaker than the second-order systems discussed above but are still of more-than-considerable interest. The first reason for our interest is the connection to fundamental open questions in computational complexity such as the P versus NP question, or especially the NP versus coNP question, a connection first discovered by Cook (1975). A second reason is that propositional proof systems form the basis for many computerized verification and theorem-proving systems. Of course, a third reason is the intrinsic interest of the proof systems. The present section will discuss Frege proof systems, extended Frege proof systems, and

[6] Translation by Furth of the appendix to volume 2 of the *Grundgesetze*.

several forms of substitution Frege proof systems. It will also explain the connection between the substitution rule and quantified propositional logic. The next section will discuss upper and lower bounds on the lengths of substitution Frege proofs.

A *propositional language* is a finite set L of propositional connectives, e.g., $L = \{\neg, \wedge, \vee, \rightarrow, \leftrightarrow\}$. We usually require that L is a complete set of Boolean connectives in that any Boolean function can be represented by an L-formula. A *propositional proof system* for L is a (total) polynomial time computable function f mapping $\{0, 1\}^*$ onto the set of L-tautologies (Cook and Reckhow 1979). A traditional proof system P can be viewed as a propositional proof system by defining the function f_P so that when w encodes a valid F-proof, $f_P(w)$ is equal to the formula proved by w, and for other w, $f_P(w)$ is equal to some arbitrary L-tautology.

We write $|w|$ and $|\varphi|$ to denote the length of a string w and a propositional formula φ. By "length," we mean the number of symbols in the string or formula. When P is a traditional proof in one of the inference systems defined below, and π is a P-proof, we write $|\pi|$ for the number of symbols in π, namely the sum of the lengths of the distinct formulas appearing in P. We call $|\pi|$ the *length* of π. The number of distinct formulas in π is called the *step length* of π. Since we only count distinct formulas, proofs are implicitly dag-like, not tree-like.

Suppose f and g are proof systems for L. We say that f *polynomially simulates* g provided there is a polynomial $p(n)$ so that for any g-proof w of a formula φ, there is an f-proof v of φ with $|v| \leq p(|w|)$. If $p(n)$ is linear, then we say f *linearly simulates* g. We call f and g *polynomially equivalent* if they polynomially simulate each other.

In propositional logic, the substitution rule is generally used as an augmentation of a Frege proof system. A Frege proof system is a propositional proof system with a finite number of axiom schemes and inference schemes which is implicationally sound and implicationally complete.[7] Without loss of generality, propositional formulas are formed using the connectives $\neg$, $\wedge$, $\vee$, $\rightarrow$ and $\leftrightarrow$ and the only inference rule is *modus ponens*. The finitely many axiom schemes typically include schemes such as $A \rightarrow (B \rightarrow A)$ where any formulas may be substituted for A and B. It is known that all Frege systems polynomially simulate each other (Cook and Reckhow 1979), so the exact choice of connectives, axioms, and inference rules is not particularly important. In addition, if two Frege systems use the same language, then they linearly simulate each other.

An *extended Frege proof* is allowed to use the *extension rule* (Tseitin 1968) which permits inferring a formula

$$x \leftrightarrow C$$

where x is a new variable that does not appear in the formula C, or in the proof so far, or in the final line of the proof. The idea is that the new variable x serves as an abbreviation for the formula C. In principle, this may allow extended Frege proofs to be shorter than Frege proofs. However, it is open how much speedup of

[7] We only briefly describe Frege proof systems here. For more background see Buss (1999) or Krajicek (2019).

proof length extended Frege proofs provide over Frege proofs. This seems to be an extremely hard question, as it is related to the question of whether Boolean circuits can be represented by polynomial-size formulas.

Definition 17.1 The *substitution rule* for Frege systems allows inferences of the form

$$\frac{A}{A(B/p)}$$

where the notation "$A(B/p)$" means the result of replacing every occurrence of the variable p in A with the formula B.

The substitution rule is not implicationally sound since the hypothesis A may not logically imply the conclusion $A(B/p)$. However, it is sound, since the conclusion is valid whenever the hypothesis is valid.

The extension rule and substitution rule provide two ways to (apparently) add substantial strength to a Frege proof system. An *extended Frege* proof system is defined to be a Frege system augmented with the extension rule. Likewise, a *substitution Frege* proof system is a Frege proof system augmented with the substitution rule. It is common to use $\mathcal{F}$ to denote a particular Frege proof system. Then $e\mathcal{F}$ and $s\mathcal{F}$ denote the associated extended Frege and substitution Frege proof systems obtained by added the extension rule and the substitution rule, respectively, to $\mathcal{F}$.

It is also possible, although not nearly as common, to define a substitution rule that allows multiple substitutions in parallel.

Definition 17.2 The *multi-substitution rule* for Frege systems allows inferences of the form

$$\frac{A}{A(B_1/p_1, \ldots, B_k/p_k)} \tag{17.1}$$

where $p_1, \ldots, p_k$ are distinct variables, and the notation "$A(B_1/p_1, \ldots, B_k/p_k)$" means the result of replacing every occurrence of each variable p_i in A with the formula B_i. The *multi-substitution Frege* proof system, $ms\mathcal{F}$, is obtained by adding the multi-substitution rule to a Frege system $\mathcal{F}$.

It is clear that a $s\mathcal{F}$-proof is also an $ms\mathcal{F}$-proof, so $ms\mathcal{F}$ trivially polynomially simulates $s\mathcal{F}$. Conversely, the action of a multi-substitution inference as shown above can be simulated by (at most) $2k - 1$ substitution inferences. Namely, $k - 1$ substitution inferences are used to replace $p_2, \ldots, p_k$ with new variables $p'_2, \ldots, p'_k$ and then k substitution inferences are used to replace $p_1, p'_2, \ldots, p'_k$ with $B_1, \ldots, B_k$. (The first $k - 1$ inferences are used in case any p_j occurs in any B_i.) This shows that the (single) substitution Frege system $s\mathcal{F}$ polynomially simulates the multi-substitution system $ms\mathcal{F}$.

When working with a $s\mathcal{F}$ or $ms\mathcal{F}$ proof π, we can view the formulas in π as being implicitly universally quantified. That is, if a formula A has been proved in π, it means the same as $(\forall p)A$. Indeed, the substitution rule can be simulated in quantified propositional logic with the inferences

$$\dfrac{\forall\text{-intro}\ \dfrac{A}{(\forall p)A} \qquad (\forall p)A \to A(B/p)}{A(B/p)}\ \text{modus ponens}$$

The next theorem gives a central result about Frege systems.

Theorem 17.1 (Cook and Reckhow 1979; Dowd 1985; Krajíček and Pudlák 1989) *The extended Frege proof systems and substitution Frege proof systems are polynomially equivalent.*

It is open whether Frege systems can polynomially simulate the extended Frege and substitution Frege systems.

The fact that $s\mathcal{F}$ polynomially simulates $e\mathcal{F}$ systems as was proved by Cook and Reckhow (1979), and we sketch the proof to give an example of the power of substitution. Suppose π is an extended Frege proof of a formula A. Enumerate the uses of the extension rule in π as $x_i \leftrightarrow C_i$ for $i = 1, 2, \ldots, \ell$. We assume these extension axioms are given in the order in which they appear in π, and thus, the condition that each x_i is a new variable implies that x_i does not appear in C_j for any $j < i$. Applying the deduction theorem ℓ times, there is a Frege proof of the formula

$$(x_\ell \leftrightarrow C_\ell) \to ((x_{\ell-1} \leftrightarrow C_{\ell-1}) \to (\cdots((x_2 \leftrightarrow C_2) \to ((x_1 \leftrightarrow C_1) \to A))\cdots)). \tag{17.2}$$

We use the substitution rule to replace x_ℓ with C_ℓ; note that x_ℓ does not appear anywhere in the formula (17.2) other than where it is indicated. We then prove the tautology $C_\ell \leftrightarrow C_\ell$ (with a proof of length polynomial in $|C_\ell|$), and use modus ponens to infer

$$(x_{\ell-1} \leftrightarrow C_{\ell-1}) \to (\cdots((x_2 \leftrightarrow C_2) \to ((x_1 \leftrightarrow C_1) \to A))\cdots).$$

This process is repeated $\ell - 1$ many more times until a derivation of A is obtained. This gives an $s\mathcal{F}$ proof of A with length polynomially bounded by the length of $|\pi|$.

The fact that $e\mathcal{F}$ polynomially simulates $s\mathcal{F}$ was proved by Dowd (1985) in unpublished work, and then by Krajíček and Pudlák (1989). Dowd gave a proof based on proving the soundness of $s\mathcal{F}$ in the bounded arithmetic theory PV; Krajíček and Pudlák describe that proof in the setting of S^1_2, and also give an explicit simulation of $s\mathcal{F}$ by $e\mathcal{F}$. The reader should refer to Krajíček and Pudlák (1989) for details.

We conclude this section with two restrictions on the (multi-)substitution rule from Buss (1995) which have turned out to be as strong as unrestricted substitution. We henceforth assume that the language L contains the two constant symbols $\top$ and $\bot$ denoting the constants *True* and *False*, respectively.

Definition 17.3 A $\top/\bot$ *substitution inference* is a multi-substitution inference of the form (17.1) in which each formula B_i is either $\top$ or $\bot$.

Definition 17.4 A *variable renaming inference* is a multi-substitution inference of the form (17.1) in which each formula B_i is a variable.

Definition 17.5 A *permutation substitution inference* is a multi-substitution inference of the form (17.1) in which each B_i is a variable, and the mapping $p_i \mapsto B_i$ is a permutation of $\{p_1, \ldots, p_k\}$. It is permitted that some p_i's do not appear in A.

The difference between a variable renaming inference and a permutation substitution is that the former can replace two variables with the same variable. Namely, a variable renaming inference may have some B_i and B_j equal to each other, and thus, the renaming inference causes the two variables p_i and p_j to be mapped to the same variable. This is not permitted in a permutation substitution inference. Since it may be that not all the p_i's actually appear in A, a permutation substitution inference should be viewed as a permutation acting on all variables, not just on the variables appearing in A.

A $\top/\bot$- *substitution Frege* proof system is a Frege system augmented with the $\top/\bot$ substitution inference rule. A *renaming Frege* proof system is a Frege system augmented with the variable renaming rule. These are known to be equivalent to (multi-)substitution Frege:

Theorem 17.2 (Buss 1995) *$\top/\bot$-substitution Frege and renaming Frege are both polynomially equivalent to $s\mathcal{F}$ (and hence to $ms\mathcal{F}$ and $e\mathcal{F}$).*

However, it is open whether permutation Frege proof systems are polynomially equivalent to extended Frege systems. It is also open whether a Frege system can polynomially simulate a permutation Frege proof system.

17.4 Bounds on Substitution Frege Proof Length

This section discusses the best-known lower bounds on the lengths and step lengths of substitution Frege proofs. This work was initiated by Urquhart (1997, 2005)) who proved linear lower bounds on the step length of substitution Frege proofs. Urquhart was in turn motivated by the results of Buss (1995) giving a quadratic lower bound on the numbers of symbols in Frege proofs and extended Frege proofs for certain tautologies. Theorem 17.5 below gives new lower bounds on the lengths of (multi-)substitution Frege proofs. These kinds of lower bounds are of interest because of the connections between propositional proof length and the question of whether NP $=$ coNP. In particular, if there exists a proof system P (in the sense of Cook and Reckhow) such that all tautologies have polynomial length P-proofs, then NP $=$ coNP (Cook and Reckhow 1979). Thus, it is interesting to prove non-trivial lower bounds on proof length even for specific systems such as Frege, extended Frege, substitution Frege, etc.

For Frege and extended Frege systems, we have (a weaker bound for Frege was proved earlier by Tseitin and Choubarian 1975):

Theorem 17.3 (Buss 1995) *There is an infinite family of tautologies φ_n for which the shortest extended Frege proofs have length $\Omega(|\varphi_n|^2)$. In addition, the shortest extended Frege proofs have step length $\Omega(|\varphi_n|)$.*

Of course, this implies the same lower bounds for Frege proofs.

We henceforth make the (inessential) assumption that the propositional language contains the symbols $\top$, $\bot$, $\neg$, $\wedge$, $\vee$ and $\rightarrow$. The proof of Theorem 17.3 in Buss (1995) used formulas φ_n of the form

$$\bot \vee (\bot \vee (\bot \vee (\cdots (\bot \vee \top) \cdots))), \qquad (17.3)$$

where there are n many $\bot$'s. Suppose π is an (extended) Frege proof of φ_n. A formula B appearing in π is defined to be *active* in π provided that there is an axiom or an inference in π which involves an occurrence of B, and the validity of the inference depends on the presence of the principal connective of B. For example, in an axiom $D \rightarrow (C \rightarrow D)$, the two formulas $D \rightarrow (C \rightarrow D)$ and $C \rightarrow D$ are active; however, C and D and their subformulas are not. Similarly, in a modus ponens inference inferring D from C and $C \rightarrow D$, only the formula $C \rightarrow D$ is active.

It is a simple observation, that if a formula B is not active in an (extended) Frege proof π, then the result of replacing every appearance of B in π uniformly with another formula B' results in a valid (extended) Frege proof π'. It follows that every subformula of φ_n must be active in π. This is because otherwise, we could replace that subformula by the constant $\bot$, thereby obtaining a valid proof of a false formula.

The proof of Theorem 17.3 is now almost immediate. An axiom or inference in π has only $O(1)$ many active formulas. Since every subformula of φ_n must be active in π, there must be $\Omega(n)$ many lines in π, so the step length of π is $\Omega(|\varphi_n|)$. Any active occurrence of a formula in π can be a subformula of only $O(1)$ many other occurrences of active formulas. Therefore, the number of symbols in π is bounded by $\Omega(s)$ where s is the sum of the sizes of the subformulas in φ_n. Clearly, $s \geq n^2$, so $|\pi| = \Omega(n^2) = \Omega(|\varphi_n|^2)$. This completes the proof sketch for Theorem 17.3.

Theorem 17.3 does not say anything about the lengths of substitution Frege proofs. In fact, the formulas φ_n have $s\mathcal{F}$-proofs of length $O(n)$ and step length. Urquhart (1997, 2005) addressed this by proving the following:

Theorem 17.4 *There are tautologies ψ_n such that any $ms\mathcal{F}$-proof of ψ_n requires step length $\Omega(n/\log n)$. There are tautologies χ_n such that any $s\mathcal{F}$-proof of χ_n requires step length $\Omega(n)$.*

The second part of Theorem 17.4 is proved in Urquhart (2005). The formulas χ_n are formed by letting $n = 2^N$, and letting χ_n be a balanced conjunction of the formulas $p_i \rightarrow p_i$ for $i = 1, \ldots, n$. The $\Omega(n)$ lower bound is proved by extending the notion of "active" formulas to include also any formula B_i used in a (multi-)substitution inference (17.1), and then arguing that for every i, either p_i or $p_i \rightarrow p_i$ is active. As there can be only $O(1)$ many active formulas per inference in an $s\mathcal{F}$-proof, this gives the $\Omega(n)$ step length lower bound for $s\mathcal{F}$. This argument fails for $ms\mathcal{F}$-proofs however. Indeed, as Urquhart shows, there are $ms\mathcal{F}$-proofs of χ_n of step length $O(\log n)$.

The proof of the first part of Theorem 17.4, giving lower bounds on the step length of $ms\mathcal{F}$-proofs, uses formulas similar to, but more complicated than the φ_n's used for Theorem 17.3. The idea is to encode a binary string w into a propositional formula Ψ_w.

Inductively define Ψ_w for $w \in \{0, 1\}^*$ as follows. For w the empty string, let Ψ_w equal just $\top$. Further let Ψ_{0w} be the formula $(\bot \vee \Psi_w)$, and let Ψ_{1w} be the formula $(\top \rightarrow \Psi_w)$. For instance, Ψ_{0101} is the formula

$$(\bot \vee (\top \rightarrow (\bot \vee (\top \rightarrow \top)))).$$

Urquhart (1997) gives an information-theoretic/counting proof that there is a w of length n such that any $ms\mathcal{F}$-proof of Ψ_w requires step length $\Omega(n/\log n)$. The basic idea is to encode $ms\mathcal{F}$-proofs of step length m by a binary string of length $O(m \log m)$ using a "condensed detachment" inference—in essence this construction shows that an $ms\mathcal{F}$-proof can be specified by stating, for each line B in proof, what axiom or inference was used to derive B and which earlier lines (if any) were used as hypotheses for the inference deriving B. This description sets up a unification problem that can be solved to find the "most general" desired proof. The end result gives a non-constructive proof that there are strings w such that Ψ_w requires $ms\mathcal{F}$-proofs of step length $\Omega(n/\log n)$. For each n, ψ_n is set (non-constructively) be one of the Ψ_w's with $|w| = n$ that require $ms\mathcal{F}$-Frege proofs of step length $\Omega(n/\log n)$.

We now extend Theorem 17.4 to give a lower bound on the (symbol) length of $ms\mathcal{F}$-proofs.

Theorem 17.5 *There are tautologies ψ_n of length $|\psi_n| = \theta(n)$ such that any $ms\mathcal{F}$-proof of ψ_n has length $\Omega(n \log n)$.*

Theorem 17.5 will be proved using a version of Urquhart's formulas Ψ_w, together with an extension of the concept of "active" formula. First, we extend the notation Ψ_w somewhat, and let $\Psi_w \circ B$ denote the result of replacing the final $\top$ symbol in Ψ_w with the formula B. For w the empty string, $\Psi_w \circ B$ is just the formula B. Then, for any w, $\Psi_{0w} \circ B$ is $(\bot \vee \Psi_w \circ B)$ and $\Psi_{1w} \circ B$ is $(\top \rightarrow \Psi_w \circ B)$. Note that $\Psi_w \circ \top$ is the same as Ψ_w.

We also use the symbol "$\circ$" to denote string concatenation: for $v, w \in \{0, 1\}^*$, $v \circ w$ denotes the concatenation of v and w. Clearly, $\Psi_v \circ (\Psi_w \circ B)$ is equal to $\Psi_{v \circ w} \circ B$. We write $v \sqsubseteq w$ to indicate that v is a substring of w. We write $w[i, j]$ for the substring of w starting with the $(i + 1)$st symbol of w and ending with the jth symbol of w. Thus, $w[0, i]$ denotes the prefix of w containing the first i symbols of w; and $w[i, |w|]$ denotes the suffix of length $|w| - i$.

Before proving Theorem 17.5, we show the results are optimal for our Ψ_w formulas, by explicitly constructing $s\mathcal{F}$-proofs that have length $O(n \log n)$ and step length $O(n)$ for $n = |w|$. The $s\mathcal{F}$-proof proceeds by proving the tautologies

$$p \rightarrow (\Psi_v \circ p) \tag{17.4}$$

for longer and longer $v \sqsubseteq w$. First consider strings v of length one. Here $\Psi_0 \circ p$ is $(\bot \vee p)$ and $(\Psi_1 \circ p)$ is $(\top \rightarrow p)$, and the formulas

$$p \to (\bot \vee p) \qquad \text{and} \qquad p \to (\top \to p)$$

have constant size Frege proofs. For $|v| > 1$, express v as $v = u_1 \circ u_2$, where $|u_1| = \lceil \frac{1}{2}|v| \rceil$ and $|u_2| = \lfloor \frac{1}{2}|v| \rfloor$. Suppose that $p \to (\Psi_{u_1} \circ p)$ and $p \to (\Psi_{u_2} \circ p)$ have already been derived. From these, we derive

$$\dfrac{p \to (\Psi_{u_2} \circ p) \qquad \dfrac{p \to (\Psi_{u_1} \circ p)}{(\Psi_{u_2} \circ p) \to (\Psi_{u_1} \circ (\Psi_{u_2} \circ p))}\text{ substitution}}{p \to (\Psi_{u_1} \circ (\Psi_{u_2} \circ p))}$$

The upper inference is a substitution replacing p with $(\Psi_{u_2} \circ p)$. The double line above the last step indicates that some steps (may) have been omitted. The last line follows tautologically as an instance of the rule "from $A \to B$ and $B \to C$ deduce $A \to C$". Since the Frege system is implicationally complete, this has a schematic derivation with $O(1)$ steps, and with $O(|v|)$ many symbols. Since $(\Psi_{u_1} \circ (\Psi_{u_2} \circ p))$ is the same as $(\Psi_v \circ p)$, this completes the desired derivation for $p \to (\Psi_v \circ p)$.

The final step of the $s\mathcal{F}$-proof applies substitution to $p \to (\Psi_w \circ p)$ to obtain $\top \to (\Psi_w \circ \top)$. From this, $\Psi_w \circ \top$ is derived with $O(1)$ more steps without further use of substitution. The $s\mathcal{F}$-proof does not need to prove the tautologies (17.4) for all v, only the ones that are needed to prove $p \to (\Psi_w \circ p)$. This gives a divide-and-conquer recursion. By inspection, the resulting $s\mathcal{F}$-proof has $O(n)$ many steps, and $O(n \log n)$ many symbols.

Theorem 17.5 will be proved using formulas Ψ_w. We just sketched how to form an $s\mathcal{F}$-proof of Ψ_w with length $O(|w| \log |w|)$ and step length $O(n)$. Thus, for these formulas at least, the length lower bound in Theorem 17.5 cannot be improved. However, the proof of Theorem 17.5 needs an additional assumption about w, since Ψ_w does have much shorter proofs for some w's. In particular, the formulas φ_n used for Theorem 17.3 have the form Ψ_{0^n}. For these formulas, the $s\mathcal{F}$-proofs just constructed have length $O(n)$ and step length $O(\log n)$. The reason for this shorter length and step length is that all the substrings v of w have the form 0^i, so there are only $O(\log n)$ many distinct tautologies $p \to (\Psi_v \circ p)$ needed for the $s\mathcal{F}$-proof of Ψ_0^n.

The needed additional assumption is that all the substrings v of w of length $K = \lceil 2 \log n \rceil$ are distinct. In other words, for $i \leq |w| - K$, the substrings $w[i, i + K]$ are distinct. It is easy to give a non-constructive proof of the existence of such a w; namely, a randomly chosen binary string w of length n has all its length K substrings distinct with probability approximately $\frac{1}{2}$. For $i < j$, the probability that two substrings of w, $w[i, i + K]$ and $w[j, j + K]$ are identical is equal to 2^{-K}. There are $\binom{n-K+1}{2} < n^2/2$ many ways to choose $i < j < n$. Thus, a union bound probability argument implies that most w's have all of their length K substrings distinct.[8]

[8] A constructive way to find w with all length K substrings disjoint is as follows. Let $J = \lfloor \frac{1}{2} K \rfloor$. For $0 \leq i < n$, let $b_i \in \{0, 1\}^J$ be the binary representation of the integer i padded with leading zeros as needed to make it have length J. Form the concatenation $b_0 \circ b_1 \circ b_2 \circ \cdots \circ b_{n-1}$. It can be shown that all length $2L$ substrings of w' are distinct. Let w be w' truncated to length n.

Proof (*Of Theorem* 17.5) Let $n > 0$ and $K = \lceil 2 \log n \rceil$. Let $w \in \{0, 1\}^n$ such that all of w's length K substrings are distinct. Suppose π is a $ms\mathcal{F}$-proof of Ψ_w. The goal is to give a lower bound on the length of π. Recall the definition from Buss (1995) of "active occurrence" of a formula that was given above in the proof of Theorem 17.3. We shall modify that definition to define what it means for a Ψ_v to be "s-active" in π. The main point of "s-active" is to take into account substitution inferences in deciding what parts of what formulas are essential for the correctness of π as an $s\mathcal{F}$-proof. (This is different from Urquhart's notion of active formulas in $s\mathcal{F}$ proofs in Urquhart (2005).)

Let v be a substring of w. If one of the following situations hold, then we say Ψ_v is *s-active* in π. In each situation, we express v as a (non-trivial) concatenation $v = v_1 \circ v_2$.

(a) Suppose a substitution inference in π derives $A(B_1/p_1, \ldots, B_\ell/p_\ell)$ from A. Also suppose that for some $j \leq \ell$, A contains $\Psi_{v_1} \circ p_j$ as a subformula and that B_j has the form $\Psi_{v_2} \circ C$ for some v_2 and C. Further suppose that $v = v_1 \circ v_2$ and that neither v_1 nor v_2 is empty. Then Ψ_v is s-active in that inference and thus in π.
(b) Suppose that an inference I in π has an active occurrence of $\Psi_v \circ B$ for some B. Then Ψ_v is s-active in this inference and thus in π. Let v_2 be the maximal length suffix of v such that $\Phi_{v_2} \circ B$ is not active in I (if such a v_2 exists). Then v_1 is the prefix of v such that $v = v_1 \circ v_2$.

In situation (b), we wish to have v_2 exist and be non-empty. This can be arranged by noting that for any particular Frege system $\mathcal{F}$, there is an upper bound K_0 on the length of v_1. This is because the Frege system is schematic, and the finitely many axiom schemes and inference schemes only nest connectives to a fixed depth. (In fact, we can take K_0 equal to 2 in the most common axiomatization for Frege systems.) We shall only consider whether Ψ_v is s-active when $|v| > K_0$. Then, when Ψ_v is s-active due to condition (b) holding for an active occurrence of $\Psi_b \circ B$, it must be that $|v_1| \leq K_0$ and hence that v_2 is non-empty.

The condition $|v| > K_0$ will automatically be satisfied in our construction below if $K + K_0 \leq K^2$; this holds for n sufficiently large. In fact, $n \geq 2$ will suffice if $K_0 = 2$.

Lemma 17.1 *Let π be an $ms\mathcal{F}$-proof of Ψ_w and suppose $v \sqsubseteq w$ is non-empty. Then Ψ_v must be s-active in π.*

Proof Suppose for sake of contradiction, that Ψ_v is not s-active in π. We shall modify π so that it remains a syntactically correct $ms\mathcal{F}$ proof, but ends with a formula which is not a tautology. This will be a contradiction.

Modify π as follows. Let v' be the substring of v containing all but the first symbol of v. Find every occurrence in π of a subformula of the form $\Psi_v \circ B$. Such a subformula has one of the forms $(\bot \vee (\Psi_{v'} \circ B))$ or $(\top \to (\Psi_{v'} \circ B))$ depending on whether v's first symbol is a 0 or a 1. In either event, replace this formula with

$$(\bot \wedge (\Psi_{v'} \circ B)).$$

By inspection, the transformed π remains a correct $ms\mathcal{F}$ proof after this transformation, since otherwise Ψ_v would have been s-active in π. And, the final line, $\Psi_w \circ \top$, has been transformed into a false formula because of the presence of "$(\bot \wedge \cdots)$" in the transformed $\Psi_w \circ \top$. This gives the desired contradiction.

The proof of Theorem 17.5 will be based on a dynamic process searching for a v with $|v| \geq K^2$ such that Ψ_v is not s-active in the given proof π. Of course, by the lemma, there is no such v. However, the process of searching for a non-s-active v will identify appearances of formulas $\Psi_v \circ B$ in π which jointly contain $\Omega(n \log n)$ symbols. The search process will maintain two sets, Q and P, of strings $v \sqsubseteq w$: strings in Q are called "queued" and strings in P are called "processed''. Initially, Q contains only w, and P is empty. Strings in Q will be processed one at a time, and then moved to P, possibly adding additional strings to Q. The following two invariants i. and ii. will be maintained throughout the process.

i. The strings in Q all have length at least K^2 and are substrings of w. Since $K^2 \geq K$, this means each $v \in Q$ corresponds to a unique substring location in w, namely v can be uniquely expressed as $v = w[i, j]$ (with $j = i + |v|$). We call this the "w-location" of v. The w-locations of the queued v's are disjoint (non-overlapping) substrings of w.
ii. Each substring v in P will have earlier been in Q; hence $|v| \geq K^2$ and $v \sqsubseteq w$. For any $v_1 \neq v_2 \in P$, one the following holds: (a) $v_1 \sqsubseteq v_2$, (b) $v_2 \sqsubseteq v_1$, or (c) the w-locations of v_1 and v_2 are disjoint substrings of w. Furthermore, each $v \in P$ will be associated with an occurrence of a subformula $\sigma(v)$ somewhere in π. The subformula $\sigma(v)$ will have the form $\Psi_v \circ B$.

The process runs as follows. Pick an arbitrary $v \in Q$. By Lemma 17.1, v must be s-active in π. This means that at least one of the following situations hold. It may be that there are multiple ways that (a) and (b) hold; but in this case, the process just picks arbitrarily one way they hold, so that v gets processed and put into P only once.

(a) Case (a) of the definition of s-active holds for Ψ_v. There are v_1 and v_2 such that $v = v_1 \circ v_2$ and a substitution inference deriving the formula $A(B_1/p_1, \ldots, B_k/p_k)$ from A. The formula A contains a subformula $\Psi_{v_1} \circ p_j$ and B_j has the form $\Psi_{v_2} \circ C$. This introduces a new subformula of the form $\Psi_v \circ C$. We move v from Q to P, and let $\sigma(v)$ equal one of the subformulas $\Psi_v \circ C$ introduced in D by the substitution of B_j for p_j. We add v_1 and v_2 to Q unless they have length $< K^2$.
(b) Case (b) of the definition of s-active holds for Ψ_v. There is an inference I in π in which $\Psi_v \circ B$ is active. For such an I and active occurrence of $\Psi_v \circ B$, we can express v as $v = v_1 \circ v_2$ where $|v_1| \leq K_0$ and v_2 is the maximal suffix of v such that $\Psi_{v_2} \circ B$ is not active in the inference I. If there is more than one way to choose I and an active occurrence of a formula $\Psi_v \circ B$, then choose them so as to maximize the length of v_1. We move v from Q into P, and let the associated formula $\sigma(v)$ be the chosen active occurrence of $\Psi_v \circ B$. We add v_2 to Q if it has

length $\geq K^2$. Since $|v_1| \leq K_0 < K^2$ (for n sufficiently large), v_1 is not added to Q.

The process stops when Q becomes empty.

It is not hard to see that the invariants i. and ii. hold throughout the process. Every v in P or Q has length $K^2 > K$ and thus has a unique w-location as $v = w[i, j]$. Since the process acts by splitting strings v into two substrings as $v = v_1 \circ v_2$, moving v to P and possibly adding v_1 and v_2 to Q, it is clear that invariants i. and ii. hold.

To finish the proof of Theorem 17.5, we shall show that the subformulas $\sigma(v)$ for $v \in P$ contribute $\Omega(n \log n)$ symbols to the length of π.

It is helpful to view strings in P as being vertices of a tree in which each node has degree at most 2. The node w is the root, since w is the first string put in Q, and thus the first string put in P. And, if $v \in P$, the children (if any) of v are the $\sqsubseteq$-maximal $v' \in P$ such that $v' \sqsubset v$. Namely, v' is a child of v iff $v' \sqsubset v$ and there is no $v'' \in P$ such that $v' \sqsubset v'' \sqsubset v$. The uniqueness of the w-locations, the invariant ii., and the fact that the process always splits strings v into at most two substrings means this gives a tree in which each node has at most two children. Any v in P which is a leaf vertex has length $|v| < 2K^2 - 1$; otherwise, either case (a) or (b) would act to give at least one child of v.

A $v \in P$ will called "type (a)" or "type (b)" depending whether the process used case (a) or case (b) to add v to P.

It would be nice if we could argue that the subformulas $\sigma(v)$ for $v \in P$ were all disjoint and non-overlapping; however, we have not been able to do this. Instead, we will show three things: First, Lemma 17.2 will limit how much the w-locations of any two v's in P of type (a) can overlap, and thereby identify a way to avoid double-counting symbols in the formulas $\sigma(v)$ for v's of type (a). Second, Lemma 17.3 will show that the subformulas $\sigma(v)$ for $v \in P$ of type (b) are disjoint and do not overlap. Third, Lemma 17.4 shows that for $v \in P$ of type (a), $\sigma(v)$ overlaps with $\sigma(v')$ for at most one $v' \in P$ of type (b).

For $v \in P$, the formula $\sigma(v)$ has the form $\Psi_v \circ B$ for some B. The first $|v|$ binary connectives are $\vee$'s and $\rightarrow$'s according to the symbols 0 and 1 in v. We call these the *top binary connectives* of $\sigma(v)$. The terminology "top" is since we think of the formula $\sigma(v)$ as a tree with root at the top: the "top" part is the part above B. There are $|v|$ many top binary connectives in $\sigma(v)$.

Lemma 17.2 *Suppose $v \neq v' \in P$ and both v and v' are type (a). Then the subformulas $\sigma(v)$ and $\sigma(v')$ have less than K top binary connectives $\vee$ and $\rightarrow$ in common.*

Because of the linear (non-branching) structure of the formula Ψ_v and $\Psi_{v'}$, Lemma 17.2 means that the overlapping top binary connectives of $\sigma(v)$ and $\sigma(v')$ are connectives corresponding to the right end of v and the left end of v', or vice-versa. Our lower bound on the length of the $ms\mathcal{F}$-proof π will be obtained by arguing that each $v \in P$ contributes $\geq |v| - K$ many symbols to π (subject to the multiplicative reduction required by the next two lemmas).

Proof (*Of Lemma* 17.2) Suppose $\sigma(v)$ and $\sigma(v')$ contain K or more top binary connectives ($\vee$'s and $\rightarrow$'s) in common. These are formulas $\Psi_v \circ B$ and $\Psi_{v'} \circ B'$.

Without loss of generality, $\sigma(v')$ is a subformula of $\sigma(v)$. Therefore, one of the following situations hold: (1) $v' \sqsubset v$ and $v = u_1 \circ v' \circ u_3$ for some u_1, u_3, or (2) $v = u_1 \circ u_2$ and $v' = u_2 \circ u_3$ for some u_1, u_2, u_3 with u_2 non-empty.

We claim that case (1), $v' \sqsubset v$, is impossible. Since v is type (a), the process found a multi-substitution inference J, and v_1 and v_2, such that $v = v_1 \circ v_2$ and the multi-substitution inference created $\Psi_{v_1 \circ v_2} \circ B = \Psi_v \circ B$ by substituting $\Psi_{v_2} \circ B$ for a variable p_j. Exactly the same holds for v' for some substrings v'_1 and v'_2 of v', with the *same* substitution inference J. (This is because $\sigma(v)$ and $\sigma(v')$ overlap and are both type (a).) Since $v' \sqsubset v$, the tree properties for the strings in P means that either $v' \sqsubseteq v_1$ or $v' \sqsubseteq v_2$. As $\sigma(v')$ is a subformula of $\sigma(v)$, this means that it is impossible for the same multi-substitution inference to have been used to process both v and v', as the decomposition $v = v_1 \circ v_2$ would have to split v somewhere outside of v', and the decomposition $v' = v'_1 \circ v'_2$ has to do the split inside v'.

Now consider case (2), $v = u_1 \circ u_2$ and $v' = u_2 \circ u_3$. The invariant ii. implies that the w-locations for v and v' are disjoint. The fact that there are no repeated substrings of length K in w thus means that $|u_2| < K$, so v and v' overlap in $< K$ symbols. Thus, $\sigma(v)$ and $\sigma(v')$ have $< K$ top binary connectives in common.

Lemma 17.3 *Let v and v' be of type (b) in P. Then $\sigma(v)$ and $\sigma(v')$ are disjoint subformulas in π.*

lema

Proof Suppose $\sigma(v)$ and $\sigma(v')$ not disjoint, Because of the "linear" structure of the formulas Ψ_v and Ψ'_v, one of $\sigma(v)$ and $\sigma(v')$ is a subformula of the other. The strings v and v' were processed using case (b), using inferences I and I', respectively, and expressing $v = v_1 \circ v_2$ and $v' = v'_1 \circ v'_2$ and finding active formulas $\Psi_{v_1} \circ \Psi_{v_2} \circ B$ and $\Psi_{v'_1} \circ \Psi_{v'_2} \circ B'$. By the fact that we choose I and I' and the active formulas $\Psi_{v_1} \circ \Psi_{v_2} \circ B$ and $\Psi_{v'_1} \circ \Psi_{v'_2} \circ B'$ so as to maximize $|v_1|$ and $|v'_1|$, it must be that $\Psi_{v_2} \circ B$ and $\Psi_{v'_2} \circ B'$ are maximal non-active subformulas and thus are are exactly the same subformula. (In fact, we may assume that I and I' are the same inference.) Thus, either v_2 is a prefix of v'_2 or vice-versa.

From $|v_1|, |v'_1| \le K_0$ and $|v|, |v'| \ge K^2$, we have that both $|v_2|$ and $|v'_2|$ are $\ge K^2 - K_0 \ge K$. So v and v' share a common substring of length $\ge K$. Hence, their w-locations overlap, and by the tree properties for members of P, either $v \sqsubset v'$ or $v' \sqsubset v$. W.l.o.g., $v' \sqsubset v$. Since v is type (b), v_1 did not get added to Q, so $v' \sqsubseteq v_2$.

We have $|v'| \le |v_2|$ and $|v'_1| \ge 1$; thus $|v'_2| < |v_2|$ and v'_2 is a proper prefix of v_2. Recall however that $|v'_2| \ge K$. This means that v'_2 appears at two places in v_2 as a substring: once since v'_2 is a prefix of v_2, and once since $v' = v'_1 \circ v'_2 \sqsubseteq v_2$. (Possibly the places overlap.) This violates the uniqueness property for w-locations for strings of length K.

Lemma 17.4 *Suppose $v \in P$ is type (a). Then $\sigma(v)$ overlaps with $\sigma(v')$ for at most one v' of type (b).*

Proof Suppose $v, v', v'\prime \in P$ have types (a), (b) and (b), respectively; also suppose $\sigma(v)$ overlaps with both $\sigma(v')$ and $\sigma(v'\prime)$. Now $\sigma(v)$ has the form $\Psi_v \circ C$, and $|v| \ge$

$K^2 > K_0$. Each of $\sigma(v')$ and $\sigma(v'\prime)$, must either contain $\sigma(v)$ or be nested no more than K_0 levels deep inside $\sigma(v)$. Because of the "linear" structure of Ψ_v, this implies that $\sigma(v')$ must overlap with $\sigma(v'\prime)$, contradicting Lemma 17.3.

Based on three lemmas, we can lower bound the number of connectives $\vee$ and $\rightarrow$ appearing in π by

$$\frac{1}{2}\sum_{v\in P}(|v| - K).$$

This counts all but the K topmost (leftmost) top binary connectives in $\sigma(v)$ for $v \in P$. By Lemma 17.2, this avoids double counting symbols in $\sigma(v)$'s with v of type (a). The multiplicative factor of $\frac{1}{2}$ takes into account Lemma 17.4 so that we do not double count connectives that appear both in a $\sigma(v)$ and a $\sigma(v')$ with v of type (a) and v' of type (b).

To finish the proof of Theorem 17.5 it suffices to prove that $\sum_{v\in P}(|v| - K)$ is $\Omega(n \log n)$. This is a straightforward, albeit a bit detailed, computation, which we now carry out. Let

$$f(m) \;=\; m\log(m/K^3) + K,$$

where the logarithm is base 2. Recall that $K = \lceil 2\log n \rceil$.

Lemma 17.5 *For $v \in P$, let P_v be the set $\{v' \in P : v' \sqsubseteq v\}$. Then*

$$f(|v|) \;\le\; \sum_{v'\in P_v}(|v'| - K). \tag{17.5}$$

Note that P_v is the set of strings v' in the subtree rooted at v.

Proof We may assume n is sufficiently large. In particular, it is convenient to require at least $n \ge 4$ and $K \ge 4$, and $K^2 \ge K + K_0$. The proof is by induction on $m = |v|$. First, suppose v is a leaf in the tree of members of P. Since v is a leaf member of P, we have $K^2 \le m < 2K^2$. In fact, we need only that $K^2 \le m \le K^3$. Since $P_v = \{v\}$, the inequality (17.5) becomes

$$m\log(m/K^3) + K \;\le\; m - K. \tag{17.6}$$

From $m \le K^3$, we have $\log(m/K^3) \le 0$, so it will suffice to show $K \le m - K$. This holds as $m \ge K^2 \ge 2K$.

There are two induction cases to consider. The first is when v has two children v_1 and v_2 in P (with $|v_1|, |v_2| \ge K^2$). Let $m_1 = |v_1|$ and $m_2 = |v_2|$, so $m = m_1 + m_2$. Since P_v is the union of $\{v\}$, P_{v_1} and P_{v_2}, and by the induction hypothesis applied to v_1 and v_2, it suffices to show that

$$f(m) \;\le\; (m - K) + f(m_1) + f(m_2). \tag{17.7}$$

We claim that $f(m)$ is concave up. This is easy to check by noting that the first derivative

$$f'(m) \;=\; \log(m/K^3) + 1, \tag{17.8}$$

is an increasing function. Therefore, by convexity and since $m_1 + m_2 = m$, it suffices to prove that $f(m) \le (m - K) + 2f(m/2)$. In other words,

$$m\log(m/K^3) + K \;\le\; (m - K) + 2\Big(\frac{m}{2}\log(m/2K^3) + K\Big).$$

In fact, the two sides are equal.

The other induction step is when v has a single child. This arises in case (b), and also in case (a) when v_1 or v_2 has length $< K^2$. Let's assume $|v_1| < K^2$; the other case is dual. We have $m_1 = |v_1| < K^2$ and $m_2 = m - m_1 > m - K^2$, and P_v is $\{v\} \cup P_{v_2}$. If $m \le K^3$, the desired inequality (17.5) follows from (17.6); recall that (17.6) was proved using only the hypothesis $K^2 \le m \le K^3$. So we may assume $m > K^3$. By the induction hypothesis for v_2 it suffices to show

$$f(m) \;\le\; (m - K) + f(m_2). \tag{17.9}$$

We claim that the derivative (17.8) is positive for $m \ge K^3 - K^2$. To see this, note that $(K^3 - K^2)/K^3 \ge 3/4$ since $K \ge 4$, so for $m \ge K^3 - K^2$, we have $\log m/K^3 > -1$. Therefore, to establish (17.9), it suffices to show

$$f(m) \;\le\; (m - K) + f(m - K^2).$$

since $m_2 \ge m - K^2$. In other words,

$$m(\log m - \log K^3) + K \;\le\; m - K + (m - K^2)(\log(m - K^2) - \log K^3) + K.$$

We have $\log m - \log(m - K^2) \le (K^2/(m - K^2))/\ln 2 \le \frac{3}{2}(K^2/(m - K^2))$ from a first-order approximation to $\log x$ at $x = m - K^2$. So, regrouping and canceling terms, it will suffice to show

$$K^2\log(m - K^2) + K \;\le\; m\Big(1 - \frac{3K^2}{2(m - K^2)}\Big) + K^2\log K^3,$$

With $m \ge K^3$ and $K \ge 4$, we have $\frac{3}{2}(K^2/(m - K^2)) \le 1/2$, so it will further suffice to show

$$K^2\log m + K \;\le\; \frac{m}{2} + K^2\log K^3, \tag{17.10}$$

When we set $m = K^3$ in (17.10), it becomes, after cancellation, $K \le K^3/2$. So (17.10) holds for $m = K^3$. To handle $m > K^3$, let LHS and RHS be the left- and righthand sides of (17.10). Their derivatives are

$$\frac{\partial}{\partial m}(\text{LHS}) = \frac{K^2}{m} \quad \text{and} \quad \frac{\partial}{\partial m}(\text{RHS}) = \frac{1}{2}.$$

Thus, $\frac{\partial}{\partial m}(\text{LHS}) \leq \frac{\partial}{\partial m}(\text{RHS})$ for $m \geq K^3$, since $K \geq 2$. It follows that (17.10) holds for all $m \geq K^3$. That proves Lemma 17.5.

We can now finish the proof of Theorem 17.5. We have Ψ_w is a tautology of length $n = |w|$. Any $ms\mathcal{F}$-proof of Ψ_w must have at least $\frac{1}{2}f(n)$ many occurrences of binary connectives. Furthermore, since $K = \lceil 2\log n \rceil$,

$$f(n) = n\log n - n\log K^3 + K$$

has growth rate $\Omega(n\log n)$.

Theorem 17.5 bounds the length of $ms\mathcal{F}$ proofs with length measured as the number of symbols in the proof. Urquhart's theorem gives a $\Omega(n/\log n)$ lower bound on the number of steps in an sF-proof of (a randomly chosen version of) the same formulas. We conjecture that for suitable w of length n, the correct lower bound for the step length of an $s\mathcal{F}$-proof of Ψ_w is $\Omega(n)$. In fact, we even conjecture an $\Omega(n)$ lower bound for the step length of $ms\mathcal{F}$-proofs of these formulas.

Of course, if the commonly accepted conjectures about NP are true, then there are formulas that require exponential size $ms\mathcal{F}$-proofs. But obtaining such lower bounds is at present well out of reach.

Acknowledgements We thank Jeremy Avigad, Bruno Bentzen, Wojciech Dzik, and Alasdair Urquhart for very useful comments and feedback on an earlier draft of this paper.

References

Bentzen, B. (2020). Frege's theory of types. https://arxiv.org/abs/2006.16453.

Boolos, G. (1985). Reading the Begriffsschrift. *Mind*, (New series) *94*(375), 331–344.

Buss, S. R. (1995). Some remarks on lengths of propositional proofs. *Archive for Mathematical Logic, 34*, 377–394.

Buss, S. R. (1999). Propositional proof complexity: An introduction. In U. Berger & H. Schwichtenberg (Eds.), *Computational Logic* (pp. 127–178). Berlin: Springer.

Cook, S. A. (1975). Feasibly constructive proofs and the propositional calculus. In *Proceedings of the seventh annual ACM symposium on theory of computing* (pp. 83–97). Association for Computing Machinery.

Cook, S. A., & Reckhow, R. A. (1979). The relative efficiency of propositional proof systems. *Journal of Symbolic Logic, 44*, 36–50.

Dowd, M. (1985). Model-theoretic aspects of $P \neq NP$. Typewritten manuscript.

Ebert, P. A., & Rossberg, M. (2013). *Basic laws of arithmetic*. Oxford: Oxford University Press.

Frege, G. (1879). *Begriffsschrift, eine der arithmetischen nachgebildete Formelsprache des reinen Denkens*. Halle. English translation by Stefan Bauer-MengelBerg, with an introduction by van Heijenoort, in [29] (pp. 1–82).

Frege, G. (1893/1903). *Grundgesetze der Arithmetik*. Verlag Hermann Pohle. Two volumes. English translation in [8]; partial translation of volume 1 in [11].

Furth, M. (1964). *The basic laws of arithmetic*. California: University of California Press.
Grattan-Guiness, I. (1974). The Russell archives: Some new light on Russell's logicism. *Annals of Science, 31*(5), 387–406.
Heck, R. G. (2011). *Frege's theorem*. Oxford: Oxford University Press.
Heck, R. G. (2012). *Reading Frege's Grundgesetze*. Oxford: Oxford University Press.
Henkin, L. (1953). Banishing the rule of substitution for functional variables. *Journal of Symbolic Logic, 18*(3), 201–208.
Hylton, P. (1980). Russell's substitutional theory. *Synthese, 45,* 1–31.
Krajíček, J. (2019). *Proof complexity*. Cambridge: Cambridge University Press.
Krajíček, J., & Pudlák, P. (1989). Propositional proof systems, the consistency of first-order theories and the complexity of computations. *Journal of Symbolic Logic, 54,* 1063–1079.
Landini, G. (1998). *Russell's hidden substitutional theory*. Oxford: Oxford University Press.
Pelham, J., & Urquhart, A. (1995). Russellian propositions. In *Proceedings of the logic, methodology and philosophy of science IX*. Studies in logic and foundations of mathematics (Vol. 134, pp. 307–326). Elsevier.
Reckhow, R. A. (1976). *On the lengths of proofs in the propositional calculus*. Ph.D. thesis, Department of Computer Science, University of Toronto. Technical Report #87.
Siekmann, J., & Wrightson, G. (1983). *Automation of reasoning* (Vols. 1 & 2). Berlin: Springer.
Stevens, G. (2003). Substitution and the theory of types: Review of Landini, "Russell's Hidden Substitutional Theory". *Russell, The Journal of Bertrand Russell Studies, 23*(2), 161–176.
Tseitin, G. S., & Choubarian, A. (1975). On some bounds to the lengths of logical proofs in classical propositional calculus (Russian). Trudy Vyčisl. *Centra AN ArmSSR i Erevanskovo Univ., 8,* 57–64.
Tseitin, G. S. (1968). On the complexity of derivation in propositional logic. *Studies in Constructive Mathematics and Mathematical Logic*, 2, 115–125. Reprinted in: [22, vol 2], pp. 466–483.
Urquhart, A. (1997). The number of lines in Frege proofs with substitution. *Archive for Mathematical Logic, 37,* 15–19.
Urquhart, A. (1999). Review of G. Landini, "Russell's Hidden Substitutional Theory". *Journal of Symbolic Logic, 64*(3), 1370–1371.
Urquhart, A. (2005). The complexity of propositional proofs with the substitution rule. *Logic Journal of the IGPL, 13*(3), 287–291.
van Heijenoort, J., (Ed.), (1967). *From Frege to Gödel: A source book in mathematical logic (1879–1931)*. Harvard University Press.
von Neumann, J. (1927). *Zur Hibertschen Beweistheorie. Mathematische Zeitschrift, 26,* 1–46.
Zalta, E. N. (1998, revised 2018). Frege's theorem and foundations for arithmetic. Stanford Encyclopedia of Philosophy. https://plato.stanford.edu/entries/frege-theorem. Retrieved July 26(2020.

Chapter 18
Reflections on Proof Complexity and Counting Principles

Noah Fleming and Toniann Pitassi

Second Reader
P. Beame
University of Washington

Abstract This paper surveys the development of propositional proof complexity and the seminal contributions of Alasdair Urquhart. We focus on the central role of counting principles, and in particular Tseitin's graph tautologies, to most of the key advances in lower bounds in proof complexity. We reflect on a couple of key ideas that Urquhart pioneered: (i) graph expansion as a tool for distinguishing between easy and hard principles and (ii) "reductive" lower bound arguments, proving via a simulation theorem that an optimal proof cannot bypass the obvious (inefficient) one.

Keywords Theory of computation · Complexity theory · Propositional proof complexity · Counting principles · Tseitin tautologies

18.1 Introduction

One of the most basic questions of logic is the following: Given a universally true statement (tautology) what is the length of the shortest proof of the statement in some standard axiomatic proof system? The propositional logic version of this question is particularly important in computer science for both theorem proving and complexity theory. An important related algorithmic question is whether there is an efficient algorithm that will produce a proof of any tautology? Such questions of theorem

---

N. Fleming (✉) · T. Pitassi
University of Toronto, Toronto, Canada
e-mail: noahfleming@cs.toronto.edu

T. Pitassi
e-mail: toni@cs.toronto.edu

I. Düntsch and E. Mares (eds.), *Alasdair Urquhart on Nonclassical and Algebraic Logic and Complexity of Proofs*, Outstanding Contributions to Logic 22,
https://doi.org/10.1007/978-3-030-71430-7_18

proving and complexity inspired Cook's seminal paper on NP-completeness notably entitled "The complexity of theorem-proving procedures" (Cook and Reckhow 1979) and were contemplated even earlier by Gödel in his now well-known letter to von Neumann (see Sipser 1992)[1].

These questions have fundamental implications for complexity theory. As formalized by Cook and Reckhow (1979), there exists a propositional proof system giving rise to short (polynomial size) proofs of all tautologies if and only if NP equals co-NP. Cook and Reckhow were the first to propose a program of research aimed at attacking the NP versus co-NP problem by systematically studying and proving strong lower bounds for standard proof systems of increasing complexity.

The second motivation concerns automated theorem proving. The main goal is to investigate the efficiency of heuristics for testing satisfiability and to give some theoretical justification for them. The third and perhaps the most compelling reason to study the complexity of propositional proof systems is as a principled way to understand the limitations of current algorithmic approaches for solving NP-hard problems. Almost all algorithms that implement a deterministic or randomized procedure for solving an NP-hard optimization problem are based on a standard propositional proof system, and thus upper and lower bounds on these systems shed light on the inherent complexity of any theorem-proving system based upon it. The most striking example is Resolution on which almost all propositional theorem provers (and even first-order theorem provers) are based.

Cook and Reckhow's program has led to many beautiful results in the last twenty years, including strong connections to circuit lower bounds and celebrated exponential lower bounds on proof size for a variety of important and well-studied proof systems (see the following surveys Beame and Pitassi 1998; Urquhart 1995; Razborov 2016; Segerlind 2007).

Most of these breakthroughs were established for counting principles such as the propositional pigeonhole principle and a special family of mod p counting principles, introduced by Tseitin and called *Tseitin's graph tautologies*. Indeed, the pigeonhole principle and the Tseitin tautologies are the most well-studied structured hard instances in propositional proof complexity. A Tseitin instance $\mathrm{TS}(G, l)$ is defined relative to an undirected graph $G = (V, E)$, and a labelling $l : V \to \{0, 1\}$ of the vertices of G. The variables correspond to the edges of G, and for each $v \in V$, we have a constraint $\bigoplus_{e:v\in e} x_e = l(v)$ asserting that the parity of the edge variables incident with vertex v must agree with the label $l(v)$. By the handshake principle, for any connected graph G, $\mathrm{TS}(G, l)$ is unsatisfiable if and only if the sum of all labels is odd (see Fig. 18.1 for an example).

Tseitin was the first to study the optimal size of propositional proofs and in particular the optimal size of Resolution proofs. In his landmark 1968 paper (Tseitin 1968), Tseitin introduced his now famous Tseitin formulas and proved lower bounds on the length of *regular* Resolution refutations of the Tseitin formulas on the grid graph.

[1] However, this letter was not discovered until after Cook's paper.

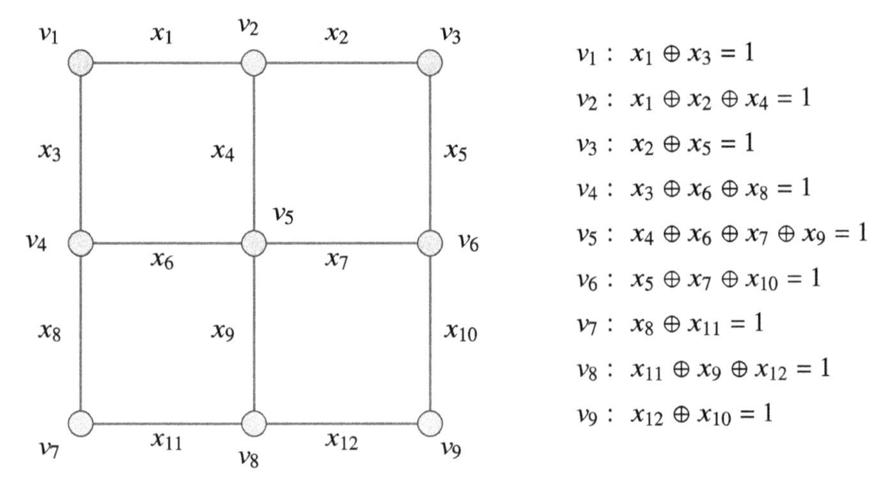

Fig. 18.1 An unsatisfiable instance of the Tseitin formulas on a 3×3 grid graph with $l(v) = 1$ for all $v \in V$

Subsequently, these formulas have been central to nearly every result in propositional proof complexity. Super-polynomial lower bounds for the Tseitin formulas have been established for many well-studied proof systems, beginning with the seminal paper by Urquhart (1987), who proved that Tseitin formulas require exponential-size Resolution refutations building on Haken's (1985) sub-exponential lower bound on the propositional pigeonhole principle. In the last thirty years, exponential lower bounds for Tseitin formulas were established for stronger proof systems, including including Nullstellensatz (1998), the Polynomial Calculus (Buss et al. 2001), Sum-of-Squares (Grigoriev 2001; Schoenebeck 2008), and bounded-depth Frege (Ben-Sasson 2002; Pitassi et al. 2016; Håstad 2017).

In this paper, we survey the landscape of results on the proof complexity of Tseitin formulas and the important role they have played in our understanding of the proof complexity of stronger systems, as well as the complexity of large families of algorithms based on these proof systems. In Section 3, we present Urquhart's seminal result proving truly exponential lower bound on the length of Resolution refutations for Tseitin formulas on any constant-degree expander graph. We highlight two central concepts that have remained quite important in nearly all subsequent lower bounds. The first is the role of graph expansion as the key combinatorial property underlying the lower bound. Over a highly expanding graph (which behaves like a random graph with respect to expansion), when viewing the graph *locally*, by looking at a small subset of the graph, there is a partial assignment to the edges of this subset which satisfies the Tseitin constraints on the vertices, whereas *globally* there is no satisfying assignment. Thus, graph expansion is a crucial *pseudorandom* property used to show that weak proof systems that reason locally (such as Resolution) cannot reason about properties where there is a big distinction between the local versus global behaviour of the property. Secondly, we highlight the *reductive* nature of the lower bound:

the proof not only rules out Resolution refutations of sub-exponential length, but it actually shows that *any* Resolution refutation must essentially mimic the obvious upper bound strategy that corresponds to Gaussian elimination.

In Section 4, we survey some of the subsequent important lower bounds in proof complexity. In these breakthrough results, we will see that the proofs, following Urquhart and Haken, show reductive lower bounds using graph expansion as the underlying pseudorandom property. Finally, in Section 5, we present a new and very surprising result due to Dadush and Tiwari (2020), who showed that Tseitin formulas are easy for Cutting Planes proofs, thereby refuting a widely believed conjecture. We conclude in Section 6 with some open problems and potential barriers to future progress.

18.2 Preliminaries

18.2.1 Resolution

Using the standard reduction from SAT to 3-SAT, one can take an arbitrary propositional formula F and convert it to a CNF or 3-CNF formula in such a way that it has only polynomially larger size and is unsatisfiable if the original formula was a tautology. To do this, one adds new variables x_A to stand for each of its subformulas A and clauses to specify that the value at each connective is computed correctly, as well as one clause of the form $\neg x_F$. In this way, one can consider any sound and complete system that produces refutations for CNF formulas as a general propositional proof system.

A *literal* is a propositional variable x or its negation $\neg x$. A *clause* is a disjunction of literals. The Resolution refutation system has a single inference rule:

$$\frac{A \vee x,\ \ B \vee \neg x}{A \vee B}.$$

The Resolution rule says that if A and B are clauses and x is a variable, then any assignment that satisfies both of the clauses $A \vee x$ and $B \vee \neg x$ also satisfies $A \vee B$. The clause $A \vee B$ is said to be a *resolvent* of the clauses $A \vee x$ and $B \vee \neg x$ derived by *resolving on* the variable x. A *Resolution derivation* of a clause C from a CNF formula F consists of a sequence of clauses in which each clause is either a clause of F, or a resolvent of two previous clauses, and C is the last clause in the sequence; it is a *refutation* of F if C is the empty clause Λ. The *size* of a refutation is the number of resolvents in it. We can represent it as a directed acyclic graph (dag) where the nodes are the clauses in the refutation, each clause of F has out-degree 0, and any other clause has two incoming arcs from the two clauses that produced it. The arcs pointing from $C \vee x$ and $D \vee \neg x$ to $C \vee D$ are labelled with the literals x and $\neg x$, respectively. It is well known that Resolution is a *sound* and *complete* propositional

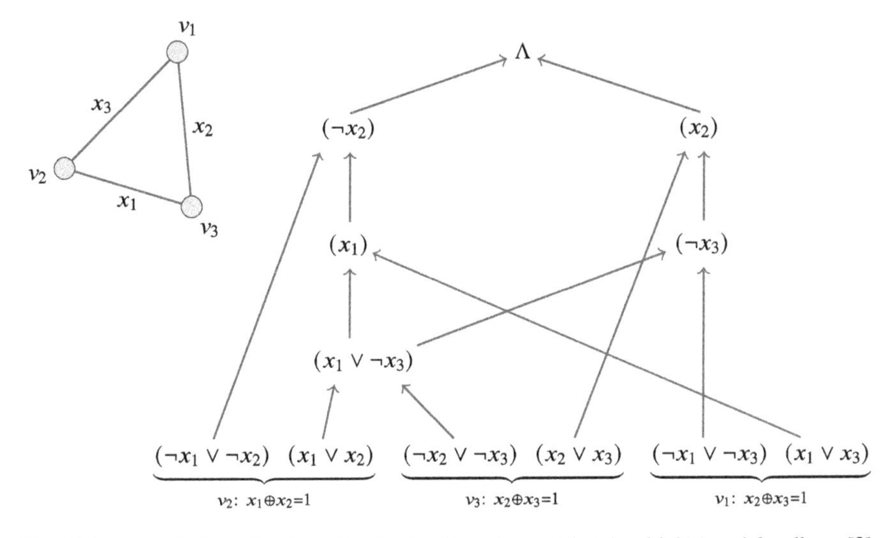

Fig. 18.2 A Resolution refutation of the Tseitin formula on a triangle with $l(v) = 1$ for all $v \in [3]$.

proof system, i.e. a formula F is unsatisfiable if and only if there is a Resolution refutation for F.

History of the Complexity of Resolution Refutations
Resolution was pre-dated by two systems known as Davis–Putnam procedures which are still the most widely used in propositional theorem proving. The general idea of these procedures is to convert a problem on n variables to problems on $n - 1$ variables by eliminating all references to some variable. The former Davis and Putnam (1960) which we call DP does this by applying all possible uses of the Resolution rule on a given variable to eliminate it. The latter Davis et al. (1962), which we call DLL and is the form used today, branches based on the possible truth assignments to a given variable; although at first, this does not look like Resolution, it is an easy argument to show that this second form is equivalent to the special class of tree-like Resolution proofs. As a proof system, Resolution is strictly stronger than DP (Goerdt 1992) and DLL (Urquhart 1995). The reasons for DLL's popularity are related to its proof search properties which we discuss below.

A more general but still restricted form of Resolution is called *regular* Resolution, and was first introduced by Tseitin in a ground-breaking article (Tseitin 1968), the published version of a talk given in 1966 at a Leningrad seminar. A regular Resolution refutation is a Resolution refutation whose underlying directed acyclic graph has the property that along each path from the root (empty clause) to a leaf (initial clause), each variable is resolved upon at most once. Observe that DP refutations are automatically regular. If refutations are represented as trees, rather than directed acyclic graphs, then minimal-size refutations are regular, as can be proved by a simple pruning argument (Urquhart 1995, p. 436).

Tseitin (1968) established the first super-polynomial lower bounds on the size of regular Resolution refutations of the Tseitin formulas. Interestingly, obtaining an improvement of this bound to an exponential one by Galil (1977) was a driving force behind some of the early work in the development of the theory of expander graphs (Gabber and Galil 1981; Hoory et al. 2006).

There was a 15+ year gap before the first super-polynomial lower bound for proofs in general Resolution was obtained by Haken (1985) who showed exponential lower bounds for the pigeonhole principle. Subsequently, Urquhart proved the first truly exponential bounds for Resolution refutations of the Tseitin formulas (Urquhart 1987).

Urquhart's original proof used the technique known as *bottleneck counting* due to Haken. In this method, one views the proof as a directed acyclic graph of clauses and views the truth assignments as flowing from the root of the directed acyclic graph to a leaf, where an assignment flows through a clause C if and only if: (i) it flows through the parent clause of C and (ii) the assignment falsifies C. Each assignment can be seen to flow through a unique path in any Resolution refutation. The idea is to show that for the formula in question, there must exist a large set of truth assignments with the property that each must pass through a wide clause (a clause containing many literals). Since a wide clause cannot be falsified by too many assignments, this implies that there must exist many wide clauses and hence the proof must be large.

An essential lemma in any bottleneck counting argument is to show that any Resolution refutation of F must involve a wide clause (a bottleneck). An important paper by Ben-Sasson and Wigderson (2001), using ideas from Clegg et al. (1996), shows that this lemma is *sufficient*; namely, they prove that any Resolution refutation of small size can be converted into a refutation with no wide clauses. This result is important since it reduces the more difficult problem of proving Resolution *size* lower bounds to the easier problem of proving Resolution *width* lower bounds.

Lower bounds for the Tseitin formulas have paved the way to proving lower bounds for *random* unsatisfiable instances. Indeed, there is a strong link between Tseitin formulas and random CNF formulas. By varying the underlying odd labelling, and d-regular graph, we get precisely a uniform distribution on d-XOR instances, where each variable occurs in exactly two equations. Because of this connection, lower bounds for Tseitin formulas has been a precursor to understanding lower bounds for random instances, such as random k-XOR and random k-SAT. Urquhart's lower bound for Tseitin was the precursor to Chvátal and Szemerédi's exponential lower bounds for Resolution refutations of random kCNF formulas (Chvátal and Szemerédi 1988), and similarly for other proof systems including Nullstellensatz, Polynomial Calculus, and SOS (Grigoriev 1998, 2001; Pitassi et al. 2016; Håstad 2017).

18.3 Urquhart's Resolution Lower Bound

In this section, we present the main ideas behind Urquhart's exponential lower bound for Resolution refutations of the Tseitin formulas

Theorem 18.1 *Let G be a d-regular odd-charged graph on n vertices with expansion* $e(G) = \Omega(n)$. *Then any resolution refutation of TS*(G) *has size* $2^{\Omega(n)}$.

As previously mentioned, Urquhart's original proof used the bottleneck counting method due to Haken. Here we give a simpler presentation of his argument, using Ben-Sasson and Wigderson's size-width theorem for Resolution (Ben-Sasson and Wigderson 2001).

We start with some intuition behind the proof. Without loss of generality, we will consider a d-regular graph $G = (V, E)$ on n vertices, where n is odd and such that all charges are odd. The variables of TS(G) are $x_{i,j}$ where $(i, j) \in E$. Each vertex $i \in V$ corresponds to a constraint which says that the mod-2 sum of the edges incident to i is odd:

$$\sum_{(i,j)\in E} x_{i,j} = 1 \ (mod\ 2).$$

The central intuition behind the proof is to relate the combinatorial notion of *expansion* of the underlying graph G to the complexity of refuting TS(G)

Definition 18.1 (*Graph Expansion*) Let G be an undirected graph with n vertices. The expansion of G, $e(G)$, is $\min\{|E(V', V - V')| : V' \subseteq V, n/3 \leq |V'| \leq 2n/3\}$.

For any odd-charged graph G, the equations in TS(G) form a set of mod-2 constraints with the property that every variable occurs in exactly two constraints. Thus, if G is a random d-regular graph with a random odd-charged labelling, then TS(G) can be viewed as a random XOR formula, where each mod-2 constraint contains $d - 1$ variables, such that each variable occurs in exactly 2 equations. The most obvious way to obtain a contradiction is iteratively deriving new mod-2 constraints. For instance, if $x_{1,2} + x_{1,3} + x_{1,4} = 1$ and $x_{1,3} + x_{2,5} = 1$ have been derived, then we can derive their sum mod-2: $x_{1,2} + x_{1,4} + x_{2,5} = 0$. This simple addition rule is sometimes called the Gaussian rule. A Gaussian refutation consists of a sequence of mod-2 equations where each equation is either an initial one or obtained by two previous equations by the Gaussian rule, and such that the final equation is $0 = 1$. For a Gaussian refutation Π of TS(G), let $width(\Pi)$ be the maximum number of variables that occur in any equation in Π. Now it follows fairly easily from the definition of expansion that if G is a connected expanding graph (i.e. $e(G) = \Omega(n)$), then any Gaussian refutation of G must have width $\Omega(n)$.

Let's see what happens when we try to mimic a Gaussian using Resolution. Since Resolution can only express *disjunctions* of literals, a mod-2 constraint involving k variables translates into an equivalent conjunction of 2^{k-1} clauses. Now if G is expanding, any Gaussian refutation has linear width, and therefore, translating this refutation to a Resolution refutation will lead to a huge blowup in size—the proof will have size exponential in n.

The difficult step in proving Resolution lower bounds for TS(G) is, therefore, to prove that Resolution can do no better—that is, that the optimal size Resolution refutation for TS(G) is that obtained by mimicking a Gaussian refutation.

We now proceed to the proof. An *assignment* for a formula F (sometimes we call it also a *restriction*) is a Boolean assignment to some of the variables in the formula; the assignment is *total* if all the variables in the formula are assigned values. If C is a clause, and σ an assignment, then we write $C\lceil\sigma$ for the result of applying the assignment to C, that is, $C\lceil\sigma = 1$ if $\sigma(l) = 1$ for some literal l in C, otherwise, $C\lceil\sigma$ is the result of removing all literals set to 0 by σ from C (with the convention that the empty clause is identified with the Boolean value 0). If F is a CNF formula, then $F\lceil\sigma$ is the conjunction of all the clauses $C\lceil\sigma$, C a clause in F. If $R = C_1, \ldots, C_k$ is a Resolution derivation from a formula F, and σ an assignment to the variables in F, then we write $R\lceil\sigma$ for the sequence $C_1\lceil\sigma, \ldots, C_k\lceil\sigma$.

Ben-Sasson and Wigderson (2001) proved the following relationship between Resolution width and Resolution proof size.

Lemma 18.1 (Size-Width Lemma) *Let F be an unsatisfiable k-CNF formula over n variables, with a Resolution refutation of size s. Then F also has a refutation of width $O(\sqrt{n \log s}) + k$.*

Thus, sufficiently strong lower bounds on *width* imply super-polynomial or even exponential lower bounds on proof *size*. For tree-like Resolution proofs, they obtained a similar result, proving that tree-like refutations of size S imply refutations of width $O(\log S)$.

We first reproduce Ben-Sasson and Widerson's proof of the Size-Width Lemma, which uses the following Lemma.

Lemma 18.2 (Lemma 3.2 in Ben-Sasson and Wigderson (2001)) *Let F be an unsatisfiable k-CNF formula and let ℓ be a literal appearing in F. If $F\lceil(\ell = 1)$ has a refutation of width $w - 1$ and $F\lceil(\ell = 0)$ have a refutation of width at most w, then F has a width-w refutation.*

Proof (*Proof sketch*) Let $\pi\lceil(\ell = 1)$ be a width $w - 1$ refutation of $F\lceil(\ell = 1)$. By adding ℓ and $\neg\ell$ back to all initial clauses of F, $\pi\lceil(\ell = 1)$ becomes a derivation of the clause $\neg\ell$ of width w. We can then resolve the derived clause $\neg\ell$ with all clauses in F to derive $F\lceil(\ell = 0)$ in width w. Then by assumption that $F\lceil(\ell = 0)$ has a width w refutation, we can refute F in width w.

Proof (*Proof of Size-Width Lemma*) Let F be an unsatisfiable k-CNF over n variables, and let π be a size s resolution refutation of F. Let π^* denote the set of wide clauses in π, where a wide clause is one that contains at least $w := \sqrt{2n \log s}$ literals. We prove by induction on b and n that if $|\pi^*| < a^b$ then F has a refutation of width at most $w + k + b$, where $a = (1 - w/(2n))^{-1}$. The base case ($b = 0$) is trivially true. For the induction step, by an averaging argument, there must exist a literal, say

ℓ, appearing in at least the average number of wide clauses, which is $|\pi^*| \cdot w/(2n)$. Restricting the entire proof π by setting $\ell = 1$ gives a refutation of $F\lceil(\ell = 1)$ with at most $(1 - w/2n)|\pi^*| < a^{b-1}$ wide clauses, which by induction on b has a refutation of width at most $w + k + b - 1$. On the other hand, setting $\ell = 0$ gives a refutation of $F\lceil(\ell = 0)$ of width at most $w + k + b$, by induction on n. Applying Lemma 18.2 completes the proof.

The next lemma shows that as long as G is a connected graph with good expansion, then any resolution refutation of TS(G) must have linear width. This combined with the Size-Width Lemma completes the proof of Theorem 18.1.

Lemma 18.3 (Tseitin Width Lower Bound) *Let G be a connected d-regular odd-charged graph with n vertices and linear expansion, i.e. $e(G) = \Omega(n)$. Then any Resolution refutation of TS(G) requires width $\Omega(n)$.*

Proof Let π be a Resolution refutation of TS(G). Let A be the set of clauses of TS(G), let $A(v)$ denote the clauses associated with vertex v, and for $V^* \subseteq V$ let $A(V^*) := \bigcup_{v \in V^*} A(v)$. We define the following complexity measure $\mu(C)$ on clauses C over the variables of TS(G).

$$\mu(C) = min\{|V'| \mid V' \subseteq V,\ A(V') \implies C^*\}.$$

That is, $\mu(C)$ is the size of the minimal set of vertices $V' \subseteq V$ such that the clauses associated with V' imply C. Since Resolution is a sound procedure, μ is subadditive. That is, if C is derived from the Resolution rule applied to clauses C_1 and C_2, then $\mu(C) \leq \mu(C_1) + \mu(C_2)$. Note that for an initial clause, $C \in A$, $\mu(C) = 1$, and for the final empty clause of Π, $\mu(\Lambda) = |A|$, because if one of the clauses of A is left out then A becomes satisfiable. Therefore, by subadditivity, there exists a clause C^* in π such that $|A|/3 \leq \mu(C^*) \leq 2|A|/3$.

Let V' denote a minimal set of vertices such that $A(V') \implies C$. By expansion of G, the size of the boundary $|E(V', V \setminus V')| = \Omega(n)$. We will argue that $width(C) = \Omega(n)$ since all literals associated with the edges in $E(V', V \setminus V')$ must occur in C. Let $x_i \in E(V', V \setminus V')$ and suppose that x_i does not occur in C. We will construct an assignment that falsifies C but satisfies $A(V')$, contradicting that $A(V') \implies C$. Because $x_i \in E(V', V \setminus V')$, there is exactly one vertex $v \in V'$ incident to x_i. Pick an assignment α that falsifies $A(v)$, falsifies C, and satisfies $A(v')$ for all $v' \in V' \setminus \{v\}$; the existence of such an assignment follows because $A(V') \implies C$ and V' is minimal. Let $\alpha^{\oplus i}$ be the assignment obtained from α by flipping the ith bit, i.e. $\alpha_j^{\oplus i} = \alpha_j$ for $j \neq i$ and $\alpha_i^{\oplus i} = 1 - \alpha_i$. Then $\alpha^{\oplus i}$ falsifies $A(v')$, and therefore, $A(V')$, however, it satisfies C because C does not depend on x_i; this contradicts that $A(V') \implies C$. Therefore, C must depend on all of the variables in $|E(V', V \setminus V')|$. Altogether, we have shown that any Resolution refutation of TS(G) has width $\Omega(n)$.

18.4 Subsequent Lower Bounds for Tseitin Formulas

18.4.1 Bounded-Depth Frege

A Frege system is a propositional proof system where the underlying lines in a proof are Boolean formulas over the basis $\wedge$, $\vee$ and $\neg$. There are a large number of axiomatizations of Frege systems, and by the foundational results of Cook and Reckhow (1979), they are all known to be polynomially equivalent, meaning that the minimum proof length of any tautology remains the same to within a polynomial factor. Obtaining even super-linear lower bounds for Frege proofs for any family of tautologies remains one of the most important open problems in proof complexity.

A well-known restricted proof system is *bounded-depth* Frege, where the rules remain the same, but we impose the restriction that every formula in the proof has depth at most d. (In order for this to remain a complete system, we measure depth of a formula as the number of alternations of $\vee$ and $\wedge$ connectives in the formula, or equivalently, we can generalize the connectives $\wedge$ and $\vee$ and associated rules to have unbounded-fan-in, and then the depth is simply the number of alternations of unbounded-fan-in $\wedge$ and $\vee$ gates.

In a breakthrough result, Ajtai (1988) established super-polynomial lower bounds on the length of bounded-depth Frege proofs of the propositional pigeon hole principle. In this tour-de-force paper, he actually proved the existence of a nonstandard model for an axiomatic system of arithmetic ($I\Delta_0$) that corresponds to bounded-depth Frege, where the pigeonhole principle is false. His ingenious construction of a nonstandard model is obtained by showing, through a combinatorial switching lemma, that any small bounded-depth Frege proof of the pigeonhole principle actually implies the existence of a small, simpler Resolution proof. Thus, the proof again establishes a reductive lower bound by reducing a more complex bounded-depth Frege proof to a simpler Resolution proof. Subsequently, Bellantoni et al. (1991) obtained somewhat stronger bounds, via a purely combinatorial analysis of Ajtai's proof. Exponential lower bounds for bounded-depth Frege proofs of the pigeonhole principle were established in Pitassi et al. (1993), Krajícek et al. (1994).

Urquhart and Fu (1996) gave a more streamlined presentation of the lower bounds mentioned above and extended the argument to prove super-polynomial lower bounds for Tseitin's formulas on the complete graph. Shortly thereafter and independently, Ben-Sasson (2002) managed to prove lower bounds for the Tseitin formulas over any expander graph by a clever reduction to the pigeonhole lower bound.

The above lower bounds left open the question of proving *optimal* lower bounds. Due to the difficulty of proving a switching lemma in this context, the state-of-the-art lower bounds for depth-d Frege proofs of the pigeonhole principle (as well as for Tseitin) are exponential in $n^{1/2^d}$, whereas the best possible upper bounds are exponential in $n^{1/d}$, leaving a large gap. Improving the lower bound to matching the upper bound appeared to be much more difficult, and closing this gap remained open for over twenty years. The first progress was made by Pitassi et al. (2016), who were able to close this gap, but only for a certain range of parameters, by establishing a switching lemma following some of the high-level ideas in a related breakthrough result (Håstad et al. 2017). In a remarkable recent paper, Håstad (2017) finally obtained near-optimal lower bounds for Tseitin formulas and grid graph (the original graph used by Tseitin to prove super-polynomial lower bounds on regular Resolution proofs). Recently, Galesi et al. (2019) extended Håstad's result to every graph, obtaining a lower bound of $2^{tw(G)^{\Omega(1/d)}}$, where $tw(G)$ is the *tree-width* of G. Furthermore, they show that this is tight up to a multiplicative constant in the top exponent.

18.4.2 Algebraic Proof Systems

Algebraic proof systems, first defined in Beame et al. (1994), are aimed at proving the unsolvability of a family of polynomial equalities or inequalities over an underlying field, and as a special case, they are refutation systems for unsatisfiable CNF formulas. Given an unsatisfiable k-CNF formula over n variables, by a standard translation we can convert the formula into a family of degree-k polynomial equations $\mathcal{P} = \{p_1 = 0, \ldots, p_m = 0\}$ over variables $x_1, \ldots, x_n$ such that the polynomial equations are satisfiable over $\{0, 1\}$ if and only if the formula is satisfiable. A Nullstellensatz refutation is a set of polynomials $q_1, \ldots, q_m$ such that $\sum_i p_i q_i = 1$—that is, the polynomials q_i witness the fact that 1 is in the ideal generated by the p_i's, and therefore, they are not simultaneously satisfiable. The *degree* of a refutation is the maximal degree of the q_i's. The Polynomial Calculus (PC) is a dynamic version of the Nullstellensatz refutation system allowing proofs of potentially lower degree. The *semialgebraic* systems, Sherali-Adams (SA) and Sum-of-Squares (SoS), are further extensions which refute families of polynomial inequalities.

There is a long history of degree lower bounds for all of these systems, again with the Tseitin formulas being the prototypical hard instance. The sequence of papers (Grigoriev 1998; Buss et al. 2001; Grigoriev 2001; Schoenebeck 2008) culminated in a linear degree lower bound for Tseitin formulas over expander graphs. In Beame et al. (2007), and later Göös and Pitassi (2014), lower bounds for more general *lifted* Tseitin formulas were proven by a reduction to communication complexity lower bounds, which, in turn, implies lower bounds for more general dynamic SoS systems as well as certain extensions of them. (See, e.g. Fleming et al. 2019 (Chap. 5) for a simplified presentation of the SoS lower bound, as well as a survey of related results.)

By fairly standard low-degree reductions, lower bounds, as well as integrality gaps, have been proven for a variety of other problems. An integrality gap is aimed at proving lower bounds for approximation algorithms for NP-hard optimization algorithms, which is more general than proving lower bounds for solving the problem exactly. At a high level, these SoS integrality gap lower bounds show that no efficient SoS-based algorithm can approximate Max-3SAT any better than the trivial algorithm that achieves a 7/8-approximation factor. SoS-based algorithms capture a natural family of linear and semidefinite programming algorithms, and thus SoS lower bounds rule out a large and natural family of algorithms for approximating NP-hard optimization problems.

Extended Formulations of Linear Programs
In a beautiful line of work, the above-mentioned SoS lower bounds have been central to proving strong lower bounds for *extended formulations* of linear programs. More specifically, Sherali-Adams degree bounds for Tseitin formulas were used to prove lower bounds for LP (linear programming) extended formulations (Chan et al. 2016; Kothari et al. 2017), and similarly, SoS bounds were shown to imply lower bounds for SDP (semidefinite programming) extended formulations. These lower bounds on extension complexity are again reductive—they show that, given an extended formulation for the optimization problem, the algorithm implies a much simpler SA-based algorithm for the same problem. That is, they prove in a very constructive sense that an algorithm coming from the larger family of polynomial size extended formulations, actually implies the much simpler SA-based algorithm.

We cannot begin to do justice to this fascinating topic and developments, but refer the reader to Fleming et al. (2019) for a comprehensive treatment.

18.5 Cutting Planes and Tseitin Formulas

The method of using cutting planes for inference in the study of polytopes in integer programming was first described by Gomory (2010), modified and shown to be complete by Chvátal (1973), and first analyzed for its efficiency as a proof system in Cook et al. (1987).

Cutting Planes proofs manipulate integer linear inequalities. A CNF formula $F = (C_1, \ldots, C_m)$ is translated into a system of linear inequalities as follows: for each variable x_i add the inequalities $x_i \geq 0$ and $x_i \leq 1$. For each clause $C_i = \bigvee_{i \in P} x_i \vee \bigvee_{i \in N} \neg x_i$, add

$$\sum_{i \in P} x_i + \sum_{i \in N} (1 - x_i) \geq 1.$$

It can be checked that F is satisfiable if and only if there exists an assignment in $\{0,1\}^n$ satisfying this system of inequalities.

Given a system of linear inequalities $Ax \geq b$, a Cutting Planes (CP) derivation of $Ax \geq b$ is a sequence of inequalities $\{c_i x \geq d_i\}_{i \in [t]}$, where $c_i \in \mathbb{Z}^n$ is a vector and $d_i \in \mathbb{Z}$, such that every $c_i x \geq d_i$ either belongs to $Ax \geq b$, or is obtained from earlier inequalities by a *Chvátal-Gomory cut* (CG cut). A CG cut consists of two steps:

- *Linear Combination*: From previously derived inequalities

$$(c_{i_1} x \geq d_{i_1}), \ldots, (c_{i_k} x \geq d_{i_k}) \text{ and } \lambda_{i_1}, \ldots, \lambda_{i_k} \geq 0$$

  let $ax \geq b$ be such that $a = \sum_{j=1}^{k} \lambda_{i_j} c_{i_j}$ and $b = \sum_{j=1}^{k} \lambda_{i_j} d_{i_j}$.
- *division*: Derive $ax \geq \lceil b \rceil$.

A CP derivation is a *refutation* if the final inequality is $0 \geq 1$.

The *size* of the refutation is the number of lines (inequalities) in the refutation (which is polynomially related to the bit-size complexity Cook et al. 1987). Associated with any CP refutation is a directed acyclic graph labeled with the inequalities in the refutation, such that (i) each leaf is an inequality from $Ax \geq b$; (ii) intermediate nodes follow from their children by a CG cut; (iii) the root is $0 \geq 1$. A CP refutation is *tree-like* if the graph is a tree.

History of the Complexity of Cutting Planes Refutations

There is a long history of lower bounds for CP, beginning with a paper by Impagliazzo et al. (1994), who proved lower bounds on tree-like CP proofs by a reduction to the communication complexity of an associated search problem. The first lower bounds for general CP were established by Pudlák (1997) and independently by Bonet (1997) (for the case of bounded coefficients) using the method of feasible interpolation (Krajícek 1997). However, the formulas for which these lower bounds were obtained had to be specially tailored to the method of feasible interpolation, and it remained a longstanding open problem to resolve the complexity of the Tseitin formulas, as well as random instances, for CP.

Recently, Fleming et al. (2017) and Hrubes and Pudlák (2017) proved super-polynomial lower bounds on the size of Cutting Planes refutations for random k-CNF formulas, for $k = O(\log n)$. This is the first example of a proof system for which lower bounds on random formulas did not follow from lower bounds for the Tseitin formulas. Following this, Garg et al. (2018) showed that Urquhart's lower bound for Resolution implied lower bounds on the size of CP refutation of *lifted* Tseitin formulas by establishing a general lifting theorem from Resolution lower bounds to CP lower bounds. Recently, Dadush and Tiwari (2020) showed that CP has quasi-polynomial size proofs of the Tseitin formulas.

Lower Bounds via Communication Complexity

Here we sketch the main idea behind all of the aforementioned lower bounds. At their core, all of these lower bounds are reductions to the communication complexity[2] of an associated search problem.

Definition 18.2 (*Canonical Search Problem*) Let $F = (C_1, \ldots, C_m)$ be a CNF formula and (X, Y) be a partition of its variables. The associated search problem $Search_{X,Y}(F) \subseteq \{0, 1\}^X \times \{0, 1\}^Y \times [m]$ asks, given $(x, y) \in \{0, 1\}^X \times \{0, 1\}^Y$ to find the index of a clause $i \in [m]$ that is violated by (x, y), i.e. $C_i(x, y) = 0$.

The use of communication complexity to obtain proof complexity lower bounds was pioneered in the work of Impagliazzo et al. (1994). We illustrate their main technique in Lemma 18.4 for the case of *low-weight* CP refutations in which the sum of the magnitude of the coefficients of each line require at most $t = O(\log n)$ bits to express.

[2] We refer the reader to the excellent book (Rao and Yehudayoff 2020) for definitions and a rigorous treatment of communication complexity.

Lemma 18.4 (Impagliazzo et al. (1994)) *Let F be an unsatisfiable formula and (X, Y) be any partition of the variables. If there is a tree-like CP refutation of F of size s in which every line can be expressed in t bits, then the communication complexity of solving $Search_{X,Y}(F)$ is $O(t \log s)$.*

Proof First, note that with at most a polynomial increase in the size, we can assume that the graph of the refutation has fan-in at most 2; let s' be the size of the fan-in 2 proof. Let the input to Alice and Bob be $x \in \{0, 1\}^X$ and $y \in \{0, 1\}^Y$ respectively. The proof is by induction on s'. Viewing the refutation π as a tree, there exists an intermediate node l such that the number of nodes above l is between $s'/3$ and $2s'/3$. Denote by π_1 the subtree rooted at l, and let the remainder $\pi \setminus \pi_1$ be denoted by π_0.

Alice and Bob will evaluate $l := ax + by \geq d$. To do so, Alice evaluates $ax - d$ and sends the result to Bob in $O(\log n)$ bits, who can then evaluate whether $ax + by \geq d$. If l is falsified under (x, y) then Alice and Bob proceed on the subtree π_1. Otherwise, they recurse on π_0. Both π_0 and π_1 have size at most $2s'/3$ and thus by induction they have communication protocols of size $t \log(2s'/3)$. Therefore, in total the number of bits communicated by the protocol solving the search problem on π is at most $t + t \log(2s'/3) \leq t \log s' = O(t \log s)$.

The correctness of the protocol follows by the soundness of the refutation: if l is falsified by (x, y), then at least one child of l must also be falsified by (x, y). Therefore, this procedure is guaranteed to arrive at a clause of F which is falsified by (x, y).

Using this reduction together with the monotone formula lower bounds of Raz and Wigderson (1990), Impagliazzo et al. (1994) obtained lower bounds for tree-like CP. The issue of low coefficients is avoided by using randomized or real models of communication, both of which have short protocols for computing integer linear inequalities. Lower bounds on general CP size can be obtained by switching to an appropriate *dag-like* model of communication (Razborov 1995; Sokolov 2017).

18.5.1 A Warmup to an Upper Bounds on the Tseitin Formulas

Reductions to communication complexity form the backbone of all of the known lower bounds for CP. This presented a significant barrier against obtaining lower bounds on the Tseitin formulas: the search problem associated with the Tseitin formulas has a short communication protocol. We will first give a general strategy for finding a violated constraint in $\mathrm{TS}(G, l)$ which will be useful in the following section. In Lemma 18.5, we will show that this strategy can be implemented with a short communication protocol, and in Sect. 18.5.2, we will describe how this upper bound strategy can be implemented in CP.

Fix an assignment $(x, y) \in X \times Y$. For $U \subset V$ let $E[U] := E[U, V \setminus U]$ and let $l(U) := \bigoplus_{u \in U} l(u)$ be the parity of the total labelling on U. At each recursive round, we will maintain a subset of $U \subseteq V$ and a value $\delta \in \{0, 1\}$ such that $\delta \neq l(U)$ (initially $U = V, \delta = 0$) such that we have determined that $\sum_{e \in E[U]} x_e = \delta \pmod 2$. This will ensure that U contains a vertex whose constraint is falsified by (x, y). Indeed, suppose the constraints of U are satisfiable, then

$$l(U) = \sum_{u \in U} \sum_{e: u \in e} x_e \pmod 2 = \sum_{e=(u,v): u,v \in U} 2x_e + \sum_{e \in E[U]} x_e \pmod 2 = \delta,$$

which contradicts $\delta \neq l(U)$.

At each round, perform the following:

1. Partition U into two halves, U_1 and U_2.
2. Determine $\delta_1 = \sum_{e \in E[U_1]} x_e \pmod 2$ and $\delta_2 = \sum_{e \in [U_2]} x_e \pmod 2$.
3. Recurse on U_1 if $\delta_1 \neq l(U_1)$ and otherwise on U_2 when $\delta_2 \neq l(U_2)$.

The recursion halts when $|U| = 1$, at which point we have found a violated constraint.

Lemma 18.5 *Fix a graph G, an odd labelling l, and let (X, Y) be any partition of the variables of $TS(G, l)$. There is a $O(\log n)$ communication protocol solving $Search_{X,Y}(TS(G, l))$.*

Proof Given $x \in X$ and $y \in Y$ respectively, Alice and Bob will implement the aforementioned upper bound strategy. To do so, we must show that they are able to perform step (2) while communicating $O(1)$ bits. Under the partition (X, Y) of the edges E, the first sum in (2) can be written as $\sum_{e \in E[U_1], e \in X} x_e + \sum_{e \in E[U_1], e \in Y} y_e$. Alice evaluates $\sum_{e \in E[U_1], e \in Y} y_e$ and sends the answer (a single bit) to Bob, who is then able to determine δ_1 and send the answer to Alice. Similarly, they are able to compute δ_2 in 2 bits of communication.

Each recursive round halves the size of the set U. Thus, there are $\log n$ rounds, each costing 4 bits of communication.

18.5.2 Tseitin Formulas are Easy For Cutting Planes

In a surprising breakthrough result, Dadush and Tiwari (2020) refuted the conjecture that the Tseitin formulas are hard for CP.

Theorem 18.2 (Dadush-Tiwari (2020)) *For any graph G and odd labelling l there is a quasi-polynomial size CP refutation of $TS(G, l)$.*

Their proof shows that a known refutation of the Tseitin formulas in the stronger Stabbing Planes proof system (introduced by Beame et al. 2018) could be simulated in CP. In the remainder, we will describe the proof of this upper bound.

Definition 18.3 (*Stabbing Planes*) Let F be an unsatisfiable system of linear inequalities. A Stabbing Planes (SP) refutation of F is a directed binary tree in which each edge is labelled with a linear integral inequality satisfying the following conditions:

- *Internal Nodes*: For any internal node u, if the right outgoing edge is labelled with $ax \geq b$, then the left outgoing edge is labelled with $ax \leq b-1$.
- *Leaves*: Each leaf node u is labelled with a non-negative linear combination of inequalities in F with inequalities along the path leading to u that yield $0 \geq 1$.

Associated with every node u in the SP proof is a polytope P_u formed by the intersection of the inequalities in F together with the inequalities labelling the root-to-u path. The polytopes labelling the leaves are empty. The pair of inequalities $(ax \leq b-1,\ ax \geq b)$ is called the *query* corresponding to the node. The *slab* corresponding to the query is $\{x \in \mathbb{R}^n \mid b-1 < ax < b\}$. The *size* of a refutation is the number of queries in the tree, which is polynomially equivalent to the bit-length needed to encode a description of the entire proof tree (Dadush and Tiwari 2020).

Lemma 18.6 (Beame et al. (2018)) *For any G and odd labelling l there is a quasi-polynomial size SP refutation of $TS(G, l)$.*

Proof (*Proof Sketch*) We show that the upper bound strategy from the previous section can be implemented in SP. However, this implementation will be lossy because SP cannot reason about mod 2 equations directly. Instead of maintaining that we have some $\delta \in \{0, 1\}$ such that $\sum_{e\in E[U]} x_e = \delta$ and $\delta \neq l(U)$, we will strengthen our invariant to require that we have determined $\sum_{e\in E[U]} x_e$ exactly. That is, we have determined that $\sum_{e\in E[U]} x_e = \delta$ for $\delta \in \{0, \dots, |E[U]|\}$.

At each round we will perform the following:

1. Partition U into two halves, U_1 and U_2.
2. Determine $\delta_1 = \sum_{e\in E[U_1]} x_e$ and $\delta_2 = \sum_{e\in E[U_2]} x_e$.
3. Recurse on U_1 if $l(U_1) \neq \delta_1$ (mod2) and otherwise on U_2 when $l(U_2) \neq \delta_2$ (mod 2). If neither holds then we have a contradiction to our inductive assumption and we can derive $0 \geq 1$.

The recursion stops when $|U| = 1$, at which point, we obtain a contradiction between δ and the constraint of the single vertex in U.

It remains to show how to perform step (2) in SP: First, query

$$\Big(\sum_{e\in E[U_1]} x_e \leq \delta_1 - 1,\ \sum_{e\in E[U_1]} x_e \geq \delta_1\Big) \qquad \text{for } \delta_1 = 1, \dots, |E[U_1]|,$$

where the ith query is attached to the right child of the $(i-1)$st query (see Fig. 18.3). Each leaf of this tree has determined that $\sum_{e\in E[U_1]} x_e = \delta_1$ (for the edge cases $\delta_1 = 0, \delta_1 = |E[U_1]|$ we use the axioms $x_i \geq 0$ and $x_i \leq 1$). At each leaf of this tree, we query

$$\Big(\sum_{e\in E[U_2]} x_e \leq \delta_2 - 1,\ \sum_{e\in E[U_2]} x_e \geq \delta_2\Big) \qquad \text{for } \delta_2 = 1, \dots, |E[U_2]|,$$

where again the ith query is attached to the right child of the $(i-1)$st. This completes the simulation of (2).

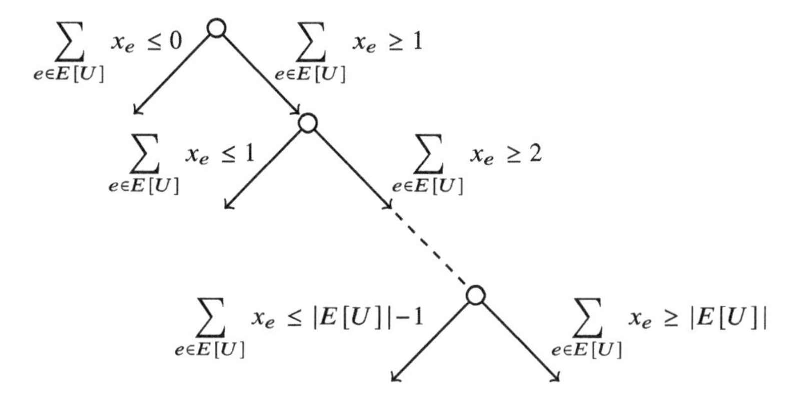

Fig. 18.3 The first tree of SP queries in a recursive step

Each recursive step halves the size of the set U under consideration and converges in $O(\log n)$ recursive steps. Each recursive step can be implemented in $O(|E|^2)$ queries. Thus, the total proof size is quasi-polynomial.

Simulating the Tseitin Refutation in CP

Dadush and Tiwari showed that Lemma 18.6 could be converted into a CP refutation. Let us first recall some basic facts about polytopes and the geometry of CP proofs. A hyperplane $ax \geq b$ is *valid* for a polytope P if $ax \geq b$ for every $x \in P$. $F := P \cap \{x : ax = b\}$ is *face* of P if at least one of $ax \geq b$, $ax \leq b$ is valid for P. We can view a CP refutation of a set of linear inequalities F as a sequence of polytopes $F = P_0, P_1, \ldots, P_t = \emptyset$ where P_i is derived from P_{i-1} by a Chvátal-Gomory (CG) cut, which corresponds to taking a hyperplane $ax \geq b$ that is valid for P_{i-1} and shifting it to the nearest integral point (see Fig. 18.5). The principal difference between CP and SP is that SP can cut *within* the current polytope by removing a slab from it, while CP can only cut on the boundary of the current polytope.

The key observation is that the SP refutation from Lemma 18.6 works from the boundary of the polytope $P = \mathrm{TS}(G, l)$ inwards. Indeed, if the vertices U are partitioned into U_1 and U_2, then the first branch (see Fig. 18.3) corresponds to refuting the face $\sum_{e \in E[U_1]} x_e = 0$, the second to refuting $\sum_{e \in E[U_1]} x_e = 1$, and so on (see Fig. 18.4). More generally, for every query $(ax \leq b, ax \geq b+1)$ corresponding to some node u in the SP refutation of $\mathrm{TS}(G, l)$, $ax \geq b$ will be valid for the current polytope P_u

- If $b = 0$ this follows from the axioms $x_i \geq 0$.
- If $b > 0$ then this follows because the SP proof queries $(ax \leq b, ax \geq b+1)$ sequentially from $b = 1, 2, \ldots$ and so $ax \geq b$ is one of the defining inequalities for P.

Therefore, the SP proof refutes $TS(G, l)$ by recursively cutting away at the sides of the polytope.

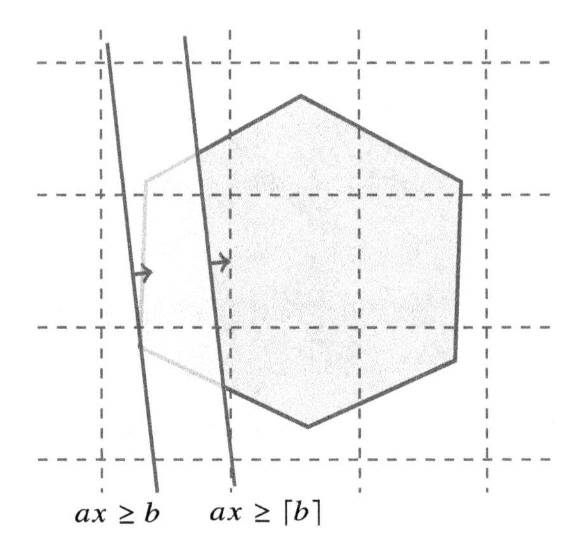

Fig. 18.4 A CG cut $ax \geq \lceil b \rceil$, the shift of a valid halfspace $ax \geq b$ to the nearest integer point. The grid intersection points represent integer points

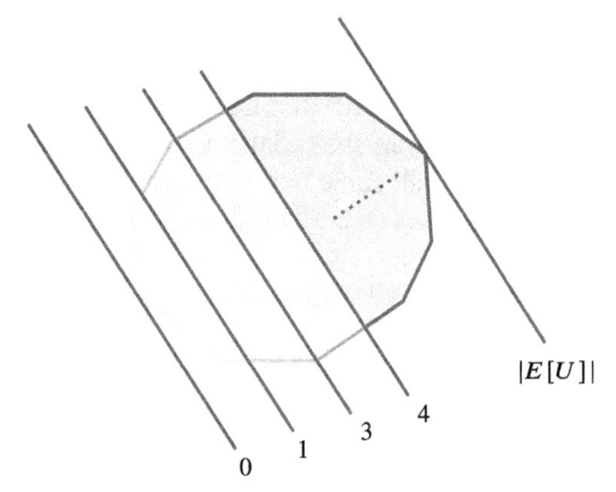

Fig. 18.5 The faces $\sum_{e \in E[U]} x_e = \delta$ for $\delta \in [E[U]|]$ of the $\mathrm{TS}(G, l)$ polytope refuted sequentially in the SP refutation

A second important observation is that if we have a CP derivation of a polytope P' from P such that $P' \cap \{x : ax = b\} = \emptyset$ (i.e. we have refuted the face $\{x : ax = b\}$) then CP can drive $ax \geq b + 1$ from P'. Indeed, there must be some $\varepsilon \in (0, 1)$ such that $ax \geq b + \varepsilon$ is valid for P' and so $ax \geq \lceil b + \varepsilon \rceil$ is a GC cut from P'.

Together, these observations show that each recursive round of the SP proof looks locally like a CP proof. The challenge is to show that we can unroll the recursion. We would like to show that if we have CP refutations of $P \cap \{x : \sum_{e \in E[U]} x_e = \delta\}$ for all $\delta = 0, \ldots, |E[U]|$ then we can glue these refutations together to form a refutation P. The following Lemma due to Schrijver (1980) will allow us to do this.

Lemma 18.7 (Schrijver (1980)) *Let P be a polytope and F be a face of P. If F' is obtained from F by a CG cut then there is a polytope P' obtained by a CG cut from P such that $P' \cap F \subseteq F'$.*

The high-level idea of the proof is as follows: Let $P = \{Ax \leq b\}$, $F = \{A_0x = b_0, A_1x \leq b_1\}$ and $ax \geq b$ be the CG cut which obtains F' from F. Observe that shifting $ax \geq b$ by factors of $A_0x \leq b_0$ does not change its effect on F. Since we care only about the resulting polytope when restricted to the face F, we can shift $ax \geq b$ by $A_0x \leq b_0$ so that it no longer depends on $A_0x \geq b_0$ and is therefore a valid cut from P.

Repeated application of this lemma allows us to simulate a refutation of a face F on P itself.

Corollary 18.1 *Suppose that $ax \geq b$ is valid for P, $a \in \mathbb{Z}^n, b \in \mathbb{Z}$, and let $F := P \cap \{x : ax = b\}$. Let $F = F_0, \ldots, F_k = \emptyset$ be a CP refutation of F, then there is a CP derivation $P = P_0, \ldots, P_k, P_{k+1}$ such that $P_{k+1} \subseteq P \cap \{x : ax \geq b + 1\}$.*

Proof For $i = 1, \ldots, k$, apply Lemma 18.7 to obtain P_i from F_i and P_{i-1} such that $P_i \cap F \subset F_i$ and therefore $P_k \cap F = \emptyset$. Because $ax \geq b$ is valid for P and $P_k \cap \{x : ax = b\} = \emptyset$ it follow that there exists $0 < \varepsilon \leq 1$ such that $ax \geq b + \varepsilon$ is valid for P. Therefore, $P_{k+1} := P_k \cap \{x : ax \geq b + 1\}$ is a CG cut from P_k.

We now sketch the proof of the CP refutation of the Tseitin formulas by Dadush and Tiwari.

Proof (*Proof Sketch of Theorem* 18.2) The proof is a post-order traversal of the SP refutation of $\mathrm{TS}(G, l)$. Consider some node in the SP refutation corresponding to a query $(ax \leq b,\ ax \geq b + 1)$, and let P be the polytope associated with this node. Suppose that we have CP refutations of the left and right children $P \cap \{x : ax = b\}$ and $P \cap \{x : ax = b + 1\}$. We will construct a CP refutation of P as follows:

1. Apply Corollary 18.1 to the refutation of the left child in order to obtain a CP derivation $P = P_0, \ldots, P_{t+1}$ such that $P_{t+1} \subseteq P \cap \{x : ax \geq b + 1\}$.
2. Append the CP refutation of the right child in order to refute P_{t+1}.

Since the root of the SP refutation corresponds to the polytope $TS(G, l)$, this procedure will produce a CP refutation of $TS(G, l)$. Observe that this simulation preserves the size of the SP refutation.

In a follow-up work, Fleming et al. (2021) gave an alternative proof of Theorem 18.2 by showing that CP can quasi-polynomially simulate any SP proof provided that the coefficients of the proof are quasi-polynomially bounded.

18.6 Concluding Remarks

We end with several related open problems.

Are Optimal Resolution Refutations of Tseitin Formulas Regular?

In his paper (Tseitin 1968), Tseitin makes the following remarks about the heuristic interpretation of the regularity restriction:

> The regularity condition can be interpreted as a requirement for not proving intermediate results in a form stronger than that in which they are later used (if A and B are disjunctions such that $A \subseteq B$, then A may be considered to be the stronger assertion of the two); if the derivation of a disjunction containing a variable ξ involves the annihilation of the latter, then we can avoid this annihilation, some of the disjunctions in the derivation being replaced by "weaker" disjunctions containing ξ.

These heuristic remarks of Tseitin suggest that there is always a regular Resolution refutation of minimal size, as in the case of tree Resolution. Consequently, some authors tried to extend Tseitin's results to general Resolution by showing that regular Resolution can simulate general Resolution efficiently. The results in Huang and Yu (1987), Goerdt (1993), Alekhnovich et al. (2007), Urquhart (2011) show that these attempts were doomed to failure. Despite this negative result showing that regular Resolution can be substantially weaker than general Resolution in the worst case, it remains open for natural examples such as counting principles. In particular, Urquhart (1987) conjectured that the minimal-size Resolution refutations for Tseitin are always regular.

There has been significant progress towards resolving this conjecture for constant-degree graphs. For constant-degree graphs, In this case, the results in Galesi et al. (2018) and Alekhnovich and Razborov (2011) imply a regular Resolution proof of size $2^{O(tw(G))} \log |V|$, where $tw(G)$ is the tree-width of G. As observed by Itsykson et al. (2019), the technique of Galesi et al. (2019) implies a $2^{\Omega(tw(G)^{1/10})}$ lower bound for Resolution. Recently Itsykson et al. (2019) proved near-optimal $2^{\tilde{\Omega}(tw(G))}$ lower bounds for regular Resolution.

Further Lower Bounds for Bounded-depth Frege

It is still an open problem to prove optimal lower bounds for the propositional pigeonhole principle. As mentioned above, the best known lower bound is exponential in $n^{1/2^{O(d)}}$ for depth-d Frege proofs, whereas the best known upper bound is exponential in $n^{1/d}$. Another longstanding open problem is to prove lower bounds for bounded-depth Frege proofs for k-CNF random formulas; in this case, there are no nontrivial lower bounds known for $d > 2$. Lower bounds for the Tseitin formulas have typically paved the way for obtaining lower bounds on random formulas. Indeed, k-XOR instances can be viewed as k-CNF formulas with additional clauses, and thus random k-XOR lower bounds imply lower bounds on random k-CNFs. While near-optimal lower bounds on the Tseitin formulas are known, these lower bounds rely on a certain structure of the underlying graph that is not available for random k-XOR instances.

Finally, the most longstanding open problem concerning bounded-depth Frege systems is to prove lower bounds for bounded-depth Frege systems over the basis which includes mod-p gates, for any prime $p \geq 2$. It is conjectured that the Tseitin formulas (mod 2) should be hard for bounded-depth Frege systems over the basis which includes mod-p gates, for any $p \neq 2$.

Unprovability of $\mathsf{P} \neq \mathsf{NP}$

In Razborov (1995), Razborov showed that if one assumes the existence of strong pseudorandom generators, then certain systems of Bounded Arithmetic cannot prove circuit lower bounds, thus ruling out any approach for proving $\mathsf{NP} \not\subseteq \mathsf{P/poly}$ that could be implemented in these systems. As well, this implies the same result for any propositional proof system which admits feasible interpolation by monotone circuits (equivalently, any proof system that can be simulated by the RCC_1 proof system Fleming et al. 2017). In several follow-up works, Razborov established unconditionally that the Polynomial Calculus (Razborov 1998), Resolution (Razborov 2004), and Resolution over $o(\log n)$-DNFs (Razborov 2015) do not possess short proofs of $\mathsf{NP} \not\subseteq \mathsf{P/poly}$. It remains an open problem to extend these lower bounds to stronger systems such as bounded-depth Frege.

Acknowledgements Toniann Pitassi would like to express her gratitude to Alasdair Urquhart and acknowledge his enormous influence on her academic and intellectual development. He is a true scholar, and has been a source of great ideas and inspiration. The authors thank Paul Beame (the second reader) for many excellent comments and suggestions that greatly improved this paper.

References

Ajtai, M. (1988). The complexity of the pigeonhole principle. In *29th annual symposium on foundations of computer science, White Plains, New York, USA, 24–26 October 1988* (pp. 346–355). IEEE Computer Society.

Alekhnovich, M., Johannsen, J., Pitassi, T., & Urquhart, A. (2007). An exponential separation between regular and general resolution. *Theory of Computing, 3*(1), 81–102.

Alekhnovich, M., & Razborov, A. A. (2011). Satisfiability, branch-width and tseitin tautologies. *Computer Complex, 20*(4), 649–678.

Beame, P., Fleming, N., Impagliazzo, R., Kolokolova, A., Pankratov, D., Pitassi, T., & Robere, R. (2018). Stabbing planes. In *9th innovations in theoretical computer science conference, ITCS 2018, January 11–14 (2018, Cambridge, MA, USA* (pp. 10:1–10:20).

Beame, P., Impagliazzo, R., Krajícek, J., Pitassi, T., & Pudlák, P. (1994). Lower bound on hilbert's nullstellensatz and propositional proofs. In *35th annual symposium on foundations of computer science, Santa Fe, New Mexico, USA (20–22 November 1994* (pp. 794–806). IEEE Computer Society.

Beame, P., & Pitassi, T. (1998). Propositional proof complexity: Past, present and future. *Electronic Colloquium on Computational Complexity (ECCC), 5*(67).

Beame, P., Pitassi, T., & Segerlind, N. (2007). Lower bounds for lov[a-acute]sz-schrijver systems and beyond follow from multiparty communication complexity. *SIAM Journal of Computational, 37*(3), 845–869.

Bellantoni, S., Pitassi, T., & Urquhart, A. (1991). Approximation and small depth frege proofs. In *Proceedings of the sixth annual structure in complexity theory conference, Chicago, Illinois, USA, June 30 - July 3 (1991* (pp. 367–390). IEEE Computer Society.

Ben-Sasson, E. (2002). Hard examples for the bounded depth frege proof system. *Computer Complex, 11*(3–4), 109–136.

Ben-Sasson, E., & Wigderson, A. (2001). Short proofs are narrow - resolution made simple. *Journal of the ACM, 48*(2), 149–169.

Bonet, M. L., Pitassi, T., & Raz, R. (1997). Lower bounds for cutting planes proofs with small coefficients. *Journal of Symbolic Logic, 62*(3), 708–728.

Buss, S. R., Grigoriev, D., Impagliazzo, R., & Pitassi, T. (2001). Linear gaps between degrees for the polynomial calculus modulo distinct primes. *Journal of Computer and System Sciences, 62*(2), 267–289.

Chan, S. O., Lee, J. R., Raghavendra, P., & Steurer, D. (2016). Approximate constraint satisfaction requires large LP relaxations. *Journal of the ACM, 63*(4), 34:1–34:22.

Chvátal, V. (1973). Edmonds polytopes and a hierarchy of combinatorial problems. *Discrete Mathematics, 4*(4), 305–337.

Chvátal, V., & Szemerédi, E. (1988). Many hard examples for resolution. *Journal of the ACM, 35*(4), 759–768.

Clegg, M., Edmonds, J., & Impagliazzo, R. (1996). Using the groebner basis algorithm to find proofs of unsatisfiability. In Miller, G. L., (Ed.), *Proceedings of the twenty-eighth annual ACM symposium on the theory of computing, Philadelphia, Pennsylvania, USA, May 22–24 (1996* (pp. 174–183). ACM.

Cook, S. A., & Reckhow, R. A. (1979). The relative efficiency of propositional proof systems. *Journal of Symbolic Logic, 6,* 169–184.

Cook, W. J., Coullard, C. R., & Turán, G. (1987). On the complexity of cutting-plane proofs. *Discrete Applied Mathematics, 18*(1), 25–38.

Dadush, D. & Tiwari, S. (2020). On the complexity of branching proofs. *CoRR,* arXiv:abs/2006.04124.

Davis, M., Logemann, G., & Loveland, D. (1962). A machine program for theorem proving. *Communications of the Association for Computing Machinery, 5,* 394–397.

Davis, M., & Putnam, H. (1960). A computing procedure for quantification theory. *Journal of the ACM, 7*(3), 201–215.

Fleming, N., Göös, M., Impagliazzo, R., Pitassi, T., Robere, R., Tan, L.-Y., & Wigderson, A. (2021). On the Power and Limitations of Branch and Cut. *Electronic Colloquium Computational Complexity, 28,* 12. https://eccc.weizmann.ac.il/report/2021/012.

Fleming, N., Kothari, P., & Pitassi, T. (2019). Semialgebraic proofs and efficient algorithm design. *Foundations and Trends in Theoretical Computer Science, 14*(1–2), 1–221.

Fleming, N., Pankratov, D., Pitassi, T., & Robere, R. (2017). Random (log n)-cnfs are hard for cutting planes. In *58th IEEE annual symposium on foundations of computer science, FOCS 2017, Berkeley, CA, USA, October 15–17 (2017* (pp. 109–120).

Gabber, O., & Galil, Z. (1981). Explicit constructions of linear-sized superconcentrators. *Journal of Computer and System Sciences, 22*(3), 407–420.

Galesi, N., Itsykson, D., Riazanov, A., & Sofronova, A. (2019). Bounded-depth frege complexity of tseitin formulas for all graphs. In P. Rossmanith, P. Heggernes, & J. Katoen (Eds.), *44th international symposium on mathematical foundations of computer science, MFCS 2019, August 26–30 (2019, Aachen, Germany, volume 138 of LIPIcs* (pp. 49:1–49:15). Schloss Dagstuhl - Leibniz-Zentrum für Informatik.

Galesi, N., Talebanfard, N., & Torán, J. (2018). Cops-robber games and the resolution of tseitin formulas. In Beyersdorff, O., & Wintersteiger, C. M., (Eds.), *Theory and applications of satisfiability testing - SAT 2018 - 21st international conference, SAT 2018, held as part of the federated logic conference, FloC 2018, Oxford, UK, July 9–12 (2018, Proceedings.* Lecture notes in computer science (vol. 10929, pp. 311–326). Springer.

Galil, Z. (1977). On the complexity of regular resolution and the davis-putnam procedure. *Theoretical Computer Science, 4*(1), 23–46.

Garg, A., Göös, M., Kamath, P., & Sokolov, D. (2018). Monotone circuit lower bounds from resolution. In Diakonikolas, I., Kempe, D., & Henzinger, M., (Eds.), *Proceedings of the 50th annual ACM SIGACT symposium on theory of computing, STOC 2018, Los Angeles, CA, USA, June 25–29 (2018* (pp. 902–911). ACM.

Goerdt, A. (1992). Davis-Putnam resolution versus unrestricted resolution. *Annals of Mathematics and Artificial Intelligence, 6,* 1–3.

Goerdt, A. (1993). Regular resolution versus unrestricted resolution. *SIAM Journal of the Computer, 22*(4), 661–683.

Gomory, R. E. (2010). Outline of an algorithm for integer solutions to linear programs and an algorithm for the mixed integer problem. In Jünger, M., Liebling, T. M., Naddef, D., Nemhauser,

G. L., Pulleyblank, W. R., Reinelt, G., Rinaldi, G., and Wolsey, L. A., (Eds.), *50 years of integer programming 1958–2008 - From the early years to the state-of-the-art* (pp. 77–103). Springer.
Göös, M., Jain, R., & Watson, T. (2016). Extension complexity of independent set polytopes. *Electronic Colloquium on Computational Complexity (ECCC), 23,* 70.
Göös, M., & Pitassi, T. (2014). Communication lower bounds via critical block sensitivity. In Shmoys, D. B., (Ed.), *Symposium on theory of computing, STOC 2014, New York, NY, USA, May 31 - June 03(2014* (pp. 847–856). ACM.
Grigoriev, D. (1998). Tseitin's tautologies and lower bounds for nullstellensatz proofs. In *39th annual symposium on foundations of computer science, FOCS '98, November 8–11 (1998, Palo Alto, California, USA* (pp. 648–652). IEEE Computer Society.
Grigoriev, D. (2001). Linear lower bound on degrees of positivstellensatz calculus proofs for the parity. *Theoretical Computer Science, 259*(1–2), 613–622.
Haken, A. (1985). The intractability of resolution. *Theoretical Computer Science, 39,* 297–308.
Håstad, J. (2017). On small-depth frege proofs for tseitin for grids. In Umans, C., (Ed.), *58th IEEE annual symposium on foundations of computer science, FOCS 2017, Berkeley, CA, USA, October 15–17 (2017* (pp. 97–108). IEEE Computer Society.
Håstad, J., Rossman, B., Servedio, R. A., & Tan, L. (2017). An average-case depth hierarchy theorem for boolean circuits. *Journal of the ACM, 64*(5), 35:1–35:27.
Hoory, S., Linial, N., & Wigderson, A. (2006). Expander graphs and their applications. *Bulletin of the American Mathematical Society, 43*(4), 439–561.
Hrubes, P., & Pudlák, P. (2017). Random formulas, monotone circuits, and interpolation. In *58th IEEE annual symposium on foundations of computer science, FOCS 2017, Berkeley, CA, USA, October 15–17 (2017* (pp. 121–131).
Huang, W., & Yu, X. (1987). A DNF without regular shortest consensus path. *SIAM Journal of the Computer, 16*(5), 836–840.
Impagliazzo, R., Pitassi, T., & Urquhart, A. (1994). Upper and lower bounds for tree-like cutting planes proofs. In *Proceedings of the ninth annual symposium on logic in computer science (LICS '94), Paris, France, July 4–7 (1994* (pp. 220–228). IEEE Computer Society.
Itsykson, D., Riazanov, A., Sagunov, D., & Smirnov, P. (2019). Almost tight lower bounds on regular resolution refutations of tseitin formulas for all constant-degree graphs. *Electronic Colloquium on Computational Complexity (ECCC), 26,* 178.
Kothari, P. K., Meka, R., & Raghavendra, P. (2017). Approximating rectangles by juntas and weakly-exponential lower bounds for LP relaxations of csps. In Hatami, H., McKenzie, P., & King, V., (Eds.), *Proceedings of the 49th annual ACM SIGACT symposium on theory of computing, STOC 2017, Montreal, QC, Canada, June 19–23 (2017* (pp. 590–603). ACM.
Krajícek, J. (1997). Interpolation theorems, lower bounds for proof systems, and independence results for bounded arithmetic. *Journal of Symbolic Logic, 62*(2), 457–486.
Krajícek, J., Pudlák, P., & Woods, A. R. (1994). An exponential lower bound to the size of bounded depth frege proofs of the pigeonhole principle. *Electronic Colloquium on Computational Complexity (ECCC), 1*(18),
Pitassi, T., Beame, P., & Impagliazzo, R. (1993). Exponential lower bounds for the pigeonhole principle. *Computer Complex, 3,* 97–140.
Pitassi, T., Rossman, B., Servedio, R. A., & Tan, L. (2016). Poly-logarithmic frege depth lower bounds via an expander switching lemma. In *Proceedings of the 48th annual ACM SIGACT symposium on theory of computing, STOC 2016, Cambridge, MA, USA, June 18–21 (2016* (pp. 644–657).
Pudlák, P. (1997). Lower bounds for resolution and cutting plane proofs and monotone computations. *Journal of Symbolic Logic, 62*(3), 981–998.
Rao, A., & Yehudayoff, A. (2020). *Communication complexity and applications*. Cambridge: Cambridge University Press.
Raz, R., & Wigderson, A. (1990). Monotone circuits for matching require linear depth. In Ortiz, H., (Ed.), *Proceedings of the 22nd annual ACM symposium on theory of computing, May 13–17 (1990, Baltimore, Maryland, USA* (pp. 287–292). ACM.

Razborov, A. (1995). Unprovability of lower bounds on circuit size in certain fragments of bounded arithmetic. *Izvestiya Mathematics, 59*(1), 205–227.

Razborov, A. A. (1998). Lower bounds for the polynomial calculus. *Computer Complex, 7*(4), 291–324.

Razborov, A. A. (2004). Resolution lower bounds for perfect matching principles. *Journal of Computer and System Sciences, 69*(1), 3–27.

Razborov, A. A. (2015). Pseudorandom generators hard for k-dnf resolution and polynomial calculus resolution. *Annals of Mathematics*, 415–472.

Razborov, A. A. (2016). Guest column: Proof complexity and beyond. *SIGACT News, 47*(2), 66–86.

Schoenebeck, G. (2008). Linear level lasserre lower bounds for certain k-csps. In *49th annual IEEE symposium on foundations of computer science, FOCS 2008, October 25–28 (2008, Philadelphia, PA, USA* (pp. 593–602). IEEE Computer Society.

Schrijver, A. (1980). On cutting planes. In Hammer, P. L., (Ed.), *Combinatorics 79.* Annals of discrete mathematics (Vol. 9, pp. 291 – 296). Elsevier.

Segerlind, N. (2007). The complexity of propositional proofs. *Bulletin of Symbolic Logic, 13*(4), 417–481.

Sipser, M. (1992). The history and status of the P versus NP question. In Kosaraju, S. R., Fellows, M., Wigderson, A., and Ellis, J. A., (Eds.), *Proceedings of the 24th annual ACM symposium on theory of computing, May 4–6 (1992, Victoria, British Columbia, Canada* (pp. 603–618). ACM.

Sokolov, D. (2017). Dag-like communication and its applications. *ECCC TR16-202.*

Tseitin, G. (1968). On the complexity of derivation in propositional calculus. In *Studies in constructive mathematics and mathematical logic, part 2, Consultants Bureau, New York-London* (pp. 115–125).

Urquhart, A. (1987). Hard examples for resolution. *Journal of the ACM, 34*(1), 209–219.

Urquhart, A. (1995). The complexity of propositional proofs. *Bulletin of Symbolic Logic, 1*(4), 425–467.

Urquhart, A. (2011). A near-optimal separation of regular and general resolution. *SIAM Journal of the Computer, 40*(1), 107–121.

Urquhart, A., & Fu, X. (1996). Simplified lower bounds for propositional proofs. *Notre Dame Journal of Formal Logic, 37*(4), 523–544.

Chapter 19
Satisfiability, Lattices, Temporal Logic and Constraint Logic Programming on Intervals

André Vellino

Second Reader
I. Düntsch
Fujian Normal University & Brock University

Abstract This essay narrates some of the influences that Alasdair Urquhart has had on computer science at the intersection of automated theorem proving, temporal logic and lattice theory—topics that have no obvious relationship to one another. I illustrate this by showing how Allen's temporal relations are represented in a system for constraint logic programming over intervals and how the combination of a linear-resolution theorem prover and an interval constraint satisfaction system are connected via a lattice-theoretical semantic model. I also speculate on whether some of Thich Nhat Hanh's meditation teachings might have been one of the sources of Alasdair's creative inspirations and preternatural capacity for lateral thinking.

Keywords Constraint logic programming · Interval arithmetic · Temporal logic · Lattice-theoretic semantics · Proof complexity · Thich Nhat Hanh

19.1 Introduction

My aim, in this essay, is to narrate some of the many and long-lasting influences that Alasdair Urquhart has had on me over the last forty years of my career in computer science and information studies. I hope it succeeds in illustrating the unique talents and eclectic brilliance with which Alasdair approaches problems and to illustrate some of the ways in which he achieved deep insights into unsolved problems. I hope also that this narrative serves to acknowledge the many debts of gratitude that I owe Alasdair, both intellectual and personal.

A. Vellino (✉)
School of Information Studies, University of Ottawa, Ottawa, Canada
e-mail: avellino@uottawa.ca

I. Düntsch and E. Mares (eds.), *Alasdair Urquhart on Nonclassical and Algebraic Logic and Complexity of Proofs*, Outstanding Contributions to Logic 22,
https://doi.org/10.1007/978-3-030-71430-7_19

The opportunity to contribute to this volume affords a perspective of hindsight that I did not anticipate at the outset: many of Alasdair's preoccupations—starting with his work on temporal logic, but also lattice theory and the complexity of propositional theorem proving—have found their way into a variety of endeavours in my own work—particularly as it concerns constraint logic programming over intervals, but also in automated theorem proving and logic programming more generally. This is what I focus on primarily.

But there is also another touchpoint in our personal friendship that may, in a subtle way, be related to our intellectual one: a common and long-standing relationship to the Zen master Thich Nhat Hanh. This Vietnamese Buddhist monk is one of the preeminent teachers of Zen in the twentieth and twenty-first century and second in renown only to His Holiness the Dalai Lama. He seems to attract the attention of intellectuals who dominant activity is the process of thought. I believe Thich Nhat Hanh's meditation teaching is one of the sources of Alasdair's creative inspirations and I endeavour to explain why.

19.2 Non-Classical Logics

When I first met Alasdair in 1979—by a photocopy machine, as I recall—I had only just joined the University of Toronto philosophy department with a view to continuing the kind of philosophy of science into which I had been indoctrinated at the London School of Economics (LSE). My time at the LSE had given me a very narrow and simplistic view of logic: my experience with the Popperians with which I was acquainted at the time is that they had little patience for anything other than Boolean semantics for classical logic—propositional and first-order. Logics for modelling human reasoning such as non-monotonic logics or relevance logics, while known and understood by the LSE school, were not taught to students, perhaps because they believed it served no function in supporting the (hypothetico-deductive) scientific method. Hence, in my early philosophical education, I had never been exposed to the broad range of alternative logical models to which students like Alasdair had been exposed at Pittsburgh: modal logics, many valued logics, relevance logics—all these were either foreign or dangerous or both.

At the time when I was debating with myself about whether or not to cross the Atlantic to move to Toronto, the then chair of the LSE Philosophy department John Watkins took me out for a drink at the LSE pub and advised me to "go west, young man". If Watkins had known that my going west would result my having a supervisor who would soon write the chapter on Many-Valued logics for the Handbook of Philosophical Logic, he might not have been so encouraging.

Yet west I went, and this is how I had the great fortune of being associated for decades with an impish Scottish logician with a penchant for tackling and a genial talent for solving impossible problems. Most notable among these problems is the undecidability of propositional relevance logic and ticket entailment Urquhart (1984), a proof I feel sure I will never understand. I will, therefore, say little more about it except to note that it illustrates this remarkable capacity that Alasdair has for lateral thinking and making connections between multiple and varied fields.

In this case, the key inspiration for this proof was an analogy he saw between the axioms of projective geometry and the axioms of relevance logic and ticket entailment. Alasdair himself offers an account of his discovery of the association between models for KR and the axioms of projective geometry in a lesser-known but more intuitive paper Urquhart (1983) written after but published before the final proof came out in the JSL. It is in that paper that he refers to those logicians who "fulminate" against relevance logics—a good characterization of the attitude taken by my friends at the LSE!

In this paper Alasdair writes "...it is extremely easy to 'get a picture' of R model structures. In a literal sense, these models have been staring us in the face for a long time." Truth, once discovered is obvious, perhaps. And yet only a polymath with an extraordinary ability to see deeply into the inter-relatedness of abstract ideas and with great intuitive imagination could have drawn these connections.

19.3 Theorem Proving

After I had fulfilled my graduate course requirements and finished a difficult comprehensive exam (one question on this exam was "Is Logic Empirical?—Answer with more than one word") I went to see Alasdair looking for a supervisor and a thesis topic. I inquired about what he was working on at the time and as I recall, he briefly explained the $\mathcal{P} = \mathcal{NP}$ problem and how the complexity of proofs in the propositional calculus related to it. I didn't fully understand the notion of a non-deterministic polynomial time ($\mathcal{NP}$) Turing machine nor why the $\mathcal{P} = \mathcal{NP}$ question was of interest in computer science, but I was able to accept on faith that the propositional satisfiability problem was $\mathcal{NP}$-complete, i.e. characteristic of the entire class of $\mathcal{NP}$ Turing machines.

The satisfiability/tautology problem for the propositional calculus is important because if one can show that there is a decision procedure that can determine whether any formula is a tautology in a number of steps that is a polynomial function of the length of that formula, i.e. that $\mathcal{P} = \mathcal{NP}$, it follows that $\mathcal{NP}$ = co-$\mathcal{NP}$. Consequently, if there is no polynomially bounded proof procedure for the tautologies, then $\mathcal{P} \neq \mathcal{NP}$. It was this strategy of exploring propositional proof-complexity for solving this major and still open question in computer science that Cook and his students took as the starting point for tackling this problem (Cook 1971).

As Alasdair points out in his later survey paper on the complexity of proofs "the classical propositional calculus has an undeserved reputation among logicians as being essentially trivial" (Urquhart 1995). At the time that I approached him for a thesis topic, I too thought, naively, that a basic understanding of Normal Forms and Boolean semantics for the propositional calculus had equipped me to explore its proof-complexity. I therefore embarked enthusiastically on writing propositional theorem provers of various sorts—in Lisp, of course, since theorem proving was

"AI", and on a time-sharing PDP-10 (using 18-bit addresses and 256 kilobytes of RAM!)—with the aim of showing that the minimal proof-length complexity for these theorem provers was, to within a polynomial factor, equivalent—a slightly less ambitious project than tackling $\mathcal{P} = \mathcal{NP}$.

Alasdair urged me to write these computer programs to develop an experimental feel for how theorem provers—mostly variants on Resolution such as the Davis-Putnam Procedure—behave with different kinds of contradictions as inputs. Even though proofs about the minimal proof length of theorem provers—or any other topic for that matter—ought to be analytical, the inspiration and understanding that the (mathematical) world is as it is, comes from experience. In this case, the key to understanding minimal proof lengths was in the structure of the propositional formulae that were used as inputs to the theorem provers and why that structure made them "hard" for the proof method. Alasdair thought it was important to have an empirical, experimental "feel" for how things worked in practice before reasoning analytically about them.

The analytical results that researchers in the efficiency of proof systems were seeking and writing about were dominated by the notion of polynomial simulation. Informally speaking, if any proof of a propositional tautology T in proof system A can be transformed into a proof of T in system B such that the length of this proof is a polynomial function of the length of T then system B *p-simulates* system A. Conversely, if one can show that proof system A cannot *p-simulate* system B by showing that there are tautologies that are hard-to-prove for A are but not hard-to-prove for B it follows that B is strictly more powerful than A.

In his Ph.D. thesis, Reckhow (1975, Chap. 5) draws a diagram of what was known at the time about the relative complexity of theorem proving methods, from Frege systems to Gentzen systems and variants of Resolution, including the Davis-Putnam Procedure, Tree Resolution and Tableau methods. A version of this diagram, updated by Alasdair in his 1995 survey is reproduced in Fig. 19.1 (the relation *p-simulates* is shown as an arrow and *not p-simulates* is shown as a broken arrow).

Some of the results in my own thesis, such as the *p-simulation* results for the clash restricted improved analytic tableau method and SL-resolution (Vellino 1993), filled out some of the details of this map. While that filling out does not directly contribute to a solution to the $\mathcal{P} = \mathcal{NP}$ problem, Alasdair's general proof that there are hard examples for resolution (Urquhart 1987) did serve to firm up the boundary (dotted line in Fig. 19.1) up the hierarchy of increasingly powerful proof methods. In that paper, he shows that it is possible to construct sets of propositional formulae of length $O(n)$ that require their resolution proofs to be of length c^n for some constant $c > 1$. A similar result had been proven earlier by Haken (1985). He had shown that so-called "pigeon hole clauses" that encode the proposition that $n + 1$ objects can fit into n holes require superpolynomial-size resolution refutations.

In comparison to Haken's, Alasdair's proof is simpler and more elegant. It relies on insights about the tautologies encoded by Tseitin graphs (Tseitin 1970 reprinted in Siekmann and Wrightson 1983) and a method for measuring minimal proof-lengths by counting edge-deletions. The general idea is that propositional formulae can be constructed from finite, connected undirected graphs, whose edges are labelled

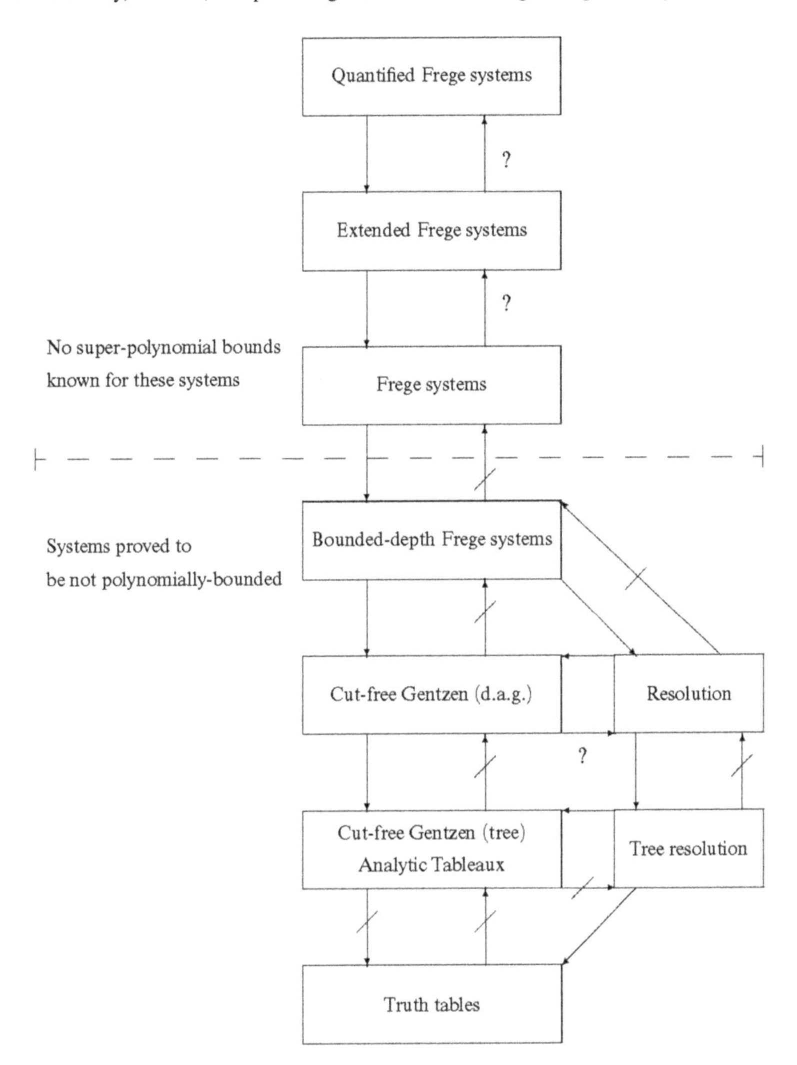

Fig. 19.1 Urquhart's map of not-polynomially bounded propositional proof systems

by propositional variables and whose nodes have a *charge* (0 or 1). The formulae constructed from these graphs are essentially assertions of propositional equivalence relations between the edge-labels.

For example, consider a graph with two nodes (X and Y) and three edges $\{a, b, c\}$ and a charge of 1 on node X and a charge of 0 on node Y, shown in Fig. 19.2 below. The propositional formula corresponding to vertex X is the formula in Disjunctive Normal Form that corresponds to an even (odd) number of negations for a vertex with charge 1 (0). In the graph, the formula for vertex X are:

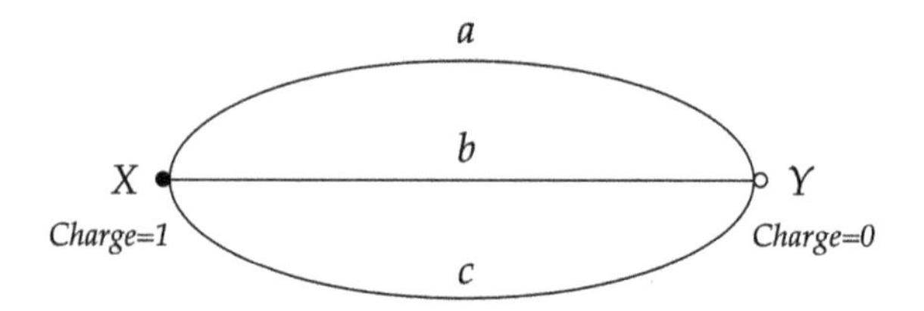

Fig. 19.2 Simple Tseitin Graph

$$(a \vee b \vee c) \wedge (a \vee \neg b \vee \neg c) \wedge (\neg a \vee b \vee \neg c) \wedge (\neg a \vee \neg b \vee c) \tag{19.1}$$

and the propositional formulae corresponding to vertex Y are:

$$(a \vee b \vee \neg c) \wedge (a \vee \neg b \vee c) \wedge (\neg a \vee b \vee c) \wedge (\neg a \vee \neg b \vee \neg c), \tag{19.2}$$

which in turn are equivalent to

$$(a \equiv (b \equiv c)) \tag{19.3}$$

$$\neg(a \equiv (b \equiv c)) \tag{19.4}$$

which, when taken together, is obviously a contradiction.

Alasdair's hard examples for resolution are generated from Tseitin graphs, using the idea of a critical truth-value assignment (cta) for a clause (disjunction) C in a set S of inconsistent clauses. A cta is essentially a set of truth-value assignments that make all the formulae in S true except for C.

Alasdair saw that one could easily count the ctas for such formulae constructed from Tseitin graphs and that the number of ctas are exponentially large in formulae generated from Tseitin graphs. In Urquhart (1987), Alasdair used this counting method to show that clauses generated from bipartite expander graphs have exponentially long resolution proofs. I recently stumbled upon a short letter (reproduced in the appendix) in which he communicated this insight to me while on summer holidays in 1984. Even though he now thinks this insight was obvious—hindsight being 20/20—it clearly wasn't at the time and I do remember his (and my) excitement about it.

I think that if we had more such correspondence that kept track of Alasdair's other insights—be they on the axioms of projective geometry or others that led him to his discoveries, we would find both the same kind of uncanny flavour of his ability to bring together multiple fields—in this case, graph theory and combinatorics—as well as his humility and excitement.

19.4 Lattice Theory and Constraint Logic Programming

While the strongest and most obvious influence Alasdair had on me was in the area of proof-complexity, another equally strong one was in equipping me for my career in Constraint Logic Programming. As a graduate student who had accepted his fate in the tutelage of this radical logician, I decided on a deep-immersion approach and took his course on Lattice Theory. The pre-requisite was a familiarity with set theory and I had devoured Halmos' Naive Set Theory at LSE. How hard could George Grätzer's book on Lattice Theory (Grätzer 1978) be? As gruelling as it was—I am not sure that even the professors who were attending this seminar got all the proofs—I am grateful for this hardship. It came in handy some years later when I landed my first real job in the private sector.

After a period of being steeped in Prolog (Programming in Logic) at the University of Georgia, as a post-doc, I was hired by Bell-Northern Research in 1987 to work on a Constraint Programming language known as BNR Prolog and sometimes referred to as CLP(BNR). The unique feature of this language was the way it processes constraints on intervals on the real line as an integral part of a Prolog engine. It is the combination of these two elements—a linear-resolution theorem prover and an interval constraint satisfaction system—that are connected via a lattice-theoretical semantic model that accounts for their common properties.

One of the key features (indeed also a significant flaw) of Prolog as a programming language is its built-in depth-first search of the solution tree (Covington et al. 1996). All the building blocks of a practical programming language must be linear-time in their execution and the built-in resolution theorem proving algorithm in Prolog achieves that by being a resolution theorem prover for Horn clauses, i.e. a disjunction of literals with at most one positive literal (Dowling and Gallier 1984). This also forces the sequential program execution to have a specific order—the order in which the clauses are written.

The major downside of using such an algorithm is that it breaks the semantic purity of meta-computations performed on logic programs (van Emden and Kowalski 1976). In particular, Prolog programs may not terminate as a result of the order in which the horn clauses are written. The non-termination of recursive Prolog programs has long been a pedagogical stumbling block for students as well. Hence, Prolog is not a "logical" language—it has lost the declarative nature of logic: the order in which clauses are written matters for a program's execution—the literals $[p \wedge q]$ and $[q \wedge p]$ do not necessarily produce the same result. This flaw was noticed and addressed in the declarative logic programming language Gödel (Hill et al. 1994).

The non-logical nature of Prolog execution is especially remarkable when it comes to computing arithmetic, which, in Prolog, is a functional rather than a relational operation: arithmetic operations in Prolog typically assign to a free variable the value that results from computing a function on arithmetically known quantities. It is impossible to express the assertion that arithmetic relations between variables must be true.

As we noted in Older and Vellino (1993), the variables in a purely declarative logic program ought to obey the following three properties:

- narrowing (the answer is a substitution of the question)
- idempotence (executing the answer adds nothing) and
- monotonicity (the more specific the question is, the more specific the answer will be).

These properties should also apply to a system of declarative relational arithmetic. To integrate a system of interval arithmetic that had a logical semantics into a declarative language was a suggestion first made by Cleary (1987). Its first implementation was developed at Bell-Northern Research (BNR) in 1987 and has since been imitated several times by various constraint satisfaction systems, notably IBM's Ilog Solver and Prolog III among others. Recently BNR Prolog has been re-implemented as open-source.[1]

The key feature of interval computations in this constraint programming system is that answers to questions can only narrow the interval variables. Intervals themselves are represented as pairs on the real line and the process of narrowing a variable X whose initial values are $[Xl_0, XI_0]$ means that subsequent values 1..n for $[Xl_{1..n}, XI_{1..n}]$ must be such that, for every j < k

$$Xl_j \leq XI_k \tag{19.5}$$

$$XU_j \geq XU_k \tag{19.6}$$

An interval constraint network can be thought of as a collection of nodes (interval variables) related to one another by edges that represent primitive constraint operations. As the process of computation proceeds forward, the only thing that happens to the interval variables is a narrowing of the bounds. To satisfy all the constraints on intervals in the network of interval relations, the constraint satisfaction algorithm performs successive narrowing operations on all the intervals in the network until it reaches a fixed-point.

The relational nature of interval arithmetic in a logic programming system (van Emden 2010) does not have obvious semantics, at least not as an extension of the standard semantics for logic programming languages (van Emden and Kowalski 1976). Lattice theory enters the picture here because it turns out to provide an effective framework for characterizing the properties of "logically" well-behaved constraint interval relations and provides a much simpler way of representing the semantics of logic programs generally. Just as intervals in a constraint network are partially ordered by set inclusion, so the constraint network's set of computational states can also be treated as a partially ordered set. The meet semi-lattice ℓ defined by the partial order $\preceq$ and a meet (computational state intersection) denoted by $\wedge$. Its bottom element $\bot$ corresponds to Prolog failure and the top element $\top$ is the largest possible state of the interval network in which all intervals range from $+\infty$ to $-\infty$. If we define the

[1] https://github.com/ridgeworks/clpBNR_pl.

join operation $\vee$ on a collection of states as the meet of all the elements larger than each item in the collection, ℓ becomes a complete lattice.

Analyzing the primitive operations in the constraint network in lattice-theoretical terms proved to be key—both practically and theoretically. In particular, the properties of narrowing, idempotence and monotonicity are central to providing a logical interpretation of interval arithmetic in a logic programming environment in a lattice-theoretical semantic model. The following examples illustrate this.

19.4.1 Narrowing

The narrowing property that applies to logic programs—wherein the answer is a substitution of the question—applies also to interval constraints. Consider, for example, the simple equality constraint on two intervals: if two intervals have an intersection then the equality constraint narrows both of them to the intersection. If the intervals have no intersection at the outset then imposing the constraint fails (the bottom element of the lattice) and the interval variable cannot have a value.

Constraints on intervals and the execution of logic programs, in general, share a narrowing semantics. As computations proceed forward, the degree of "groundedness" of the variables in the execution of the program becomes smaller. If P is the operator that maps one state of a computation to the next, and $\subseteq$ defines the partial ordering and Z is the state of the set of constrained intervals, the narrowing constraint is expressed as

$$P(Z) \subseteq Z \tag{19.7}$$

19.4.2 Idempotence

If you perform the computation P on the state of a computation X and doing so produces the result $P(X)$, idempotence means performing the computation P on this state again (i.e. $P(P(X))$) introduces no further change, i.e.

$$P(P(X)) = P(X) \tag{19.8}$$

Here it applies also to interval operations and is equivalent to the meet operation in the meet semi-lattice ℓ. In so-called "pure" logic programs—those that have no side-effects like printing a result to a screen or writing to a file—executing the program and then executing it again also changes nothing. Applied to interval computations, idempotence means that the intervals in a constraint network do not narrow further if the same constraint is re-applied.

19.4.3 *Monotonicity*

Applied to Prolog variables, monotonicity means that the more instantiated the Prolog variables are before the computation begins, the more instantiated the Prolog variables are after the computation is performed. For a lattice of interval states Y and Z, the monotonicity requirement can be expressed as

$$Y \subseteq Z \implies P(Y) \subseteq P(Z) \tag{19.9}$$

Applied to a network of interval constraints, this means that smaller initial intervals yield smaller final intervals once the constraints are propagated in the network.

In Older and Vellino (1993), we show how these and other properties (commutativity, associativity, composition of interval primitives, etc.) make interval constraint networks isomorphic to a meet semi-lattice and form a coherent model for BNR Prolog as a whole.

19.5 Temporal Logic

The success of BNR Prolog's system of constraint satisfaction over intervals encouraged me and my colleague, William Older, to consider what applications it may have besides the purely arithmetic constraint solving for which it was designed. One of the questions we were curious about was whether this system could express Allen's interval-based representation of temporal constraints (Allen 1981). I was not even aware at the time that Alasdair had co-authored a book on temporal logic (Rescher and Urquhart 1971).

The context in which Allen was seeking a representation of temporal events was to reason about events that have duration in natural language discourse. Two representations of time can be used: the instant-based model and the interval-based model. The instant-based model of time E@T—event E occurred at instant T—only allows for the expression of precedence relations. The interval representation allows for the idea that events with durations may overlap or that one event may occur inside another.

The representation of interval temporal relations proposed by Allen was designed to be independent of an absolute scale of time and to permit imprecision and uncertainty. It may be that "the exact relationship between two time intervals is not known, but some constraints are known about how they could be related" (Allen 1981). Like System R described in Rescher and Urquhart (1971), Allen wanted to be able to express the notion of "now" and permit the expression of persistence ("true henceforth").

The six primitive interval relations that Allen proposed are: the ability to assert whether one event interval occurs strictly earlier (later) than another event (before/after), whether one event immediately precedes (succeeds) another

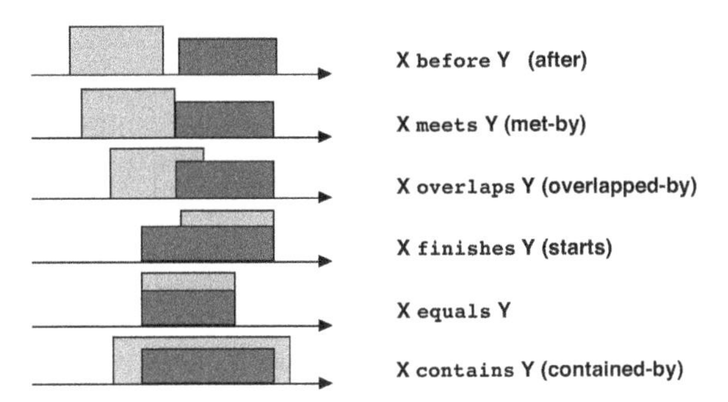

Fig. 19.3 Allen's temporal relations

(meets/met-by), whether two event interval have a partial intersection in time (overlaps/overlapped-by), whether two event interval have the same end (start) points (finishes/starts), whether two event intervals occur in the same time period (equal), or have a complete intersection (contains/contained-by), as shown in Fig. 19.3. Note that the relational algebra generated by these relations has an intimate relationship with the corresponding relations for qualitative spatial reasoning as well as the region connection calculus (RCC) (Düntsch 2005).

The question of whether or not the Allen model of temporal logic could be expressed in BNR Prolog to reason usefully about time has a mixed answer. There are two ways to interpret what it is that an interval variable refers to in BNR Prolog: the existential and the universal interpretation. In the existential interpretation, an interval represents an (unknown) point on the real number line that lies somewhere between the upper and lower bounds. Under the universal interpretation, an interval is a set of *all* the points between its upper and lower bounds. When dealing with temporal events that have duration, the universal interpretation is the most natural one. Furthermore, the narrowing semantics of constraint-bound arithmetic intervals has no obvious philosophical temporal interpretation, so it would seem that BNR Prolog would not be of much use to reason about time intervals.

However, this computational approach with arithmetic constraints does provide the ability to quantitatively preserve some of Allen's qualitative relations on interval bounds. For example, one can assert that two event intervals occur at the same time (equal) or occur before one another; that one event immediately precedes or succeeds another (meets) and that one event overlaps (contains) another. The existential interpretation of BNR Prolog intervals and their narrowing semantics is what makes the behaviour of such constraint relations somewhat unusual.

The simplest relation—equals—is interpretable either existentially or universally. It is expressed as (X == Y) and constrains the intervals X and Y to have the same start and finish points. Its negation (Prolog's "negation by failure") not(X == Y) succeeds only if X and Y have no overlap.

A trickier relation is the relation "before". The relation "before" is expressed as X =< Y and constrains the interval X to be before interval Y in the sense that X *starts* before or at the same time as Y starts and X *finishes* before Y finishes, but allows the two intervals to overlap. The strong sense of before ("serial before"), in which X finishes before Y starts, cannot be imposed directly as a constraint although it can be tested for by using not(X >= Y). The inability to constrain a serial relation is due to the inability to constrain non-overlap, which is due to the need to enforce an arbitrary point of separation.

Similarly, the temporal overlap relation between two intervals X and Y cannot be expressed directly in this language. However, introducing a new interval T (to represent the overlapping time period) and adding the condition that Y =< T and T =< X has the effect that if X and Y ever narrow such that they no longer overlap, then T becomes empty and forces a Prolog "failure".

In summary, constraint logic programming with interval arithmetic provides a logical, declarative and executable language in which to express temporal constraint relations embedded in a language that operates with a resolution theorem prover and has a lattice-theoretic semantic model. This knits together at least three of Alasdair's life-long intellectual passions—temporal logic, theorem proving and lattice theory.

19.6 Zen Master

Without a doubt Alasdair's preternatural abilities to see the hidden connections between logical truths, projective geometry, lattice theory and complexity theory stand out as evidence of his intellectual accomplishments. Yet I think that his personal interest in Zen master Thich Nhat Hanh is germane to understanding them as well. One of Thich Nhat Hanh's important practical teachings is the practice of arresting thought: to let go of the thinking mind and cultivate simply "being". In the Buddhist worldview, not all knowledge is mediated by thought and, indeed, thinking often obscures insight into truth.

One of the key teachings of Thich Nhat Hanh is the realization of Interbeing (Hanh 2010). One simple way of formulating the teaching is that there a causal connection between all things (interdependent co-arising) and I find it particularly fitting that this doctrine contains exactly the flavour of causality espoused by another great Scottish philosopher David Hume. But Interbeing also has another interpretation—that every individuated object does not have an identity that is separate from the elements that object is composed of. Thich Nhat Hanh has often used the phrase—"This flower is made of non-flower elements". The core idea being that the "identity"—including personal identity—is a conventional illusion—also a Humean idea. In the context of Alasdair's research in logic, one could also say—"This logical puzzle is made of non-logical elements".

The practice of Zen begins with "stopping" with busyness and doing and offering the mind an opportunity to rest. Over the years I have bumped into Alasdair and his wife Patricia many times at Thich Nhat Hanh's Zen centers, where I have never seen them much more at ease in a state of non-busyness. The practice of mindfulness is well-known as an antidote to the stresses and strains of cerebral work and paradoxically also a wellspring for creative inspiration. I think it may well be that at least some of Alasdair's remarkable leaps of creativity have something to do with his learned ability to not-do and not-think. In any case, there is definitely a Yoda-like quality to Alasdair—both in his gentle personal demeanour and in his natural wisdom.

Acknowledgements I would like to acknowledge the long-time influence of my former colleague William Older at Bell-Northern Research, who, next to Alasdair, was a prominent mentor and showed me by example that one can attend to the abstract mathematics of a problem and without losing track of the practical details of the assembler code that implements its solution. I am very grateful to Ivo Düntsch and Alasdair Urquhart for their detailed comments, suggestions and corrections to drafts of this paper as well as to the editors for their Herculean efforts in putting this whole volume together—in the midst of a global pandemic, no less.

Appendix

Alasdair wrote the letter shown in Figs. 19.4, 19.5, 19.6, 19.7 in the summer of 1984 as I was nearing a first draft of my thesis. He had been studying contradictions generated by Tseitin graphs (discussed in Sect. 19.3) and was looking for a simple way of counting the resolution steps needed to eliminate a variable (edge in a Tseitin graph) to show that resolution proofs for them must grow exponentially. He eventually proved his conjecture (stated on p. 3 of the letter) using his insight that the critical truth- value assignments are an exponential function of the cyclomatic number of Tseitin graphs.

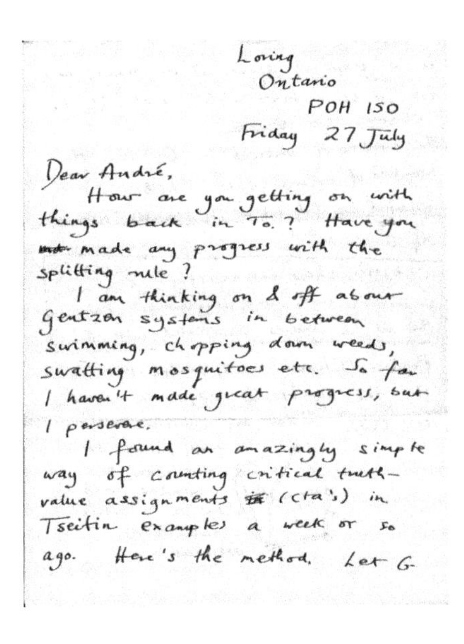

Loring
Ontario
POH 150
Friday 27 July

Dear André,
How are you getting on with things back in To.? Have you made any progress with the splitting rule?
I am thinking on & off about Gentzen systems in between swimming, chopping down weeds, swatting mosquitoes etc. So far I haven't made great progress, but I persevere.
I found an amazingly simple way of counting critical truth-value assignments (cta's) in Tseitin examples a week or so ago. Here's the method. Let G

Fig. 19.4 Page 1

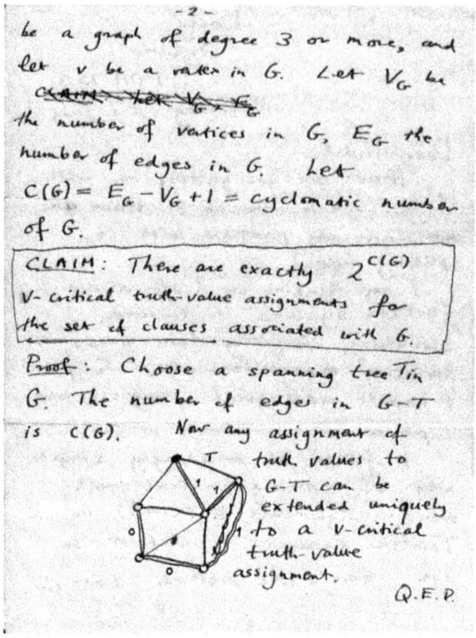

-2-

be a graph of degree 3 or more, and let v be a vertex in G. Let V_G be the number of vertices in G, E_G the number of edges in G. Let $C(G) = E_G - V_G + 1$ = cyclomatic number of G.

CLAIM: There are exactly $2^{C(G)}$ V-critical truth-value assignments for the set of clauses associated with G.

Proof: Choose a spanning tree T in G. The number of edges in G−T is C(G). Now any assignment of truth values to G−T can be extended uniquely to a V-critical truth-value assignment.

Q.E.D.

Fig. 19.5 Page 2

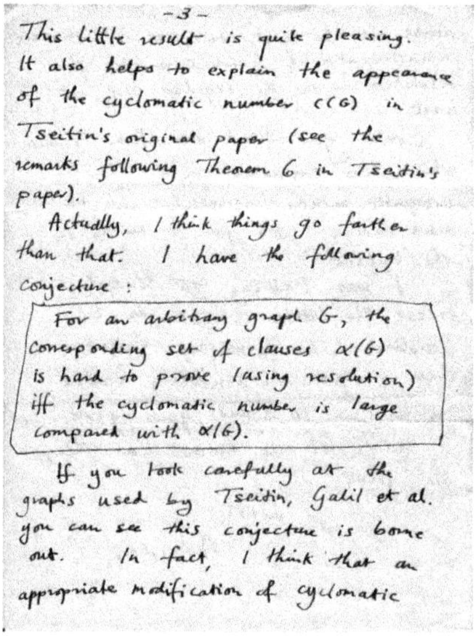

-3-

This little result is quite pleasing. It also helps to explain the appearance of the cyclomatic number C(G) in Tseitin's original paper (see the remarks following Theorem 6 in Tseitin's paper).

Actually, I think things go farther than that. I have the following conjecture:

For an arbitrary graph G, the corresponding set of clauses $\alpha(G)$ is hard to prove (using resolution) iff the cyclomatic number is large compared with $\alpha(G)$.

If you look carefully at the graphs used by Tseitin, Galil et al. you can see this conjecture is borne out. In fact, I think that an appropriate modification of cyclomatic

Fig. 19.6 Page 3

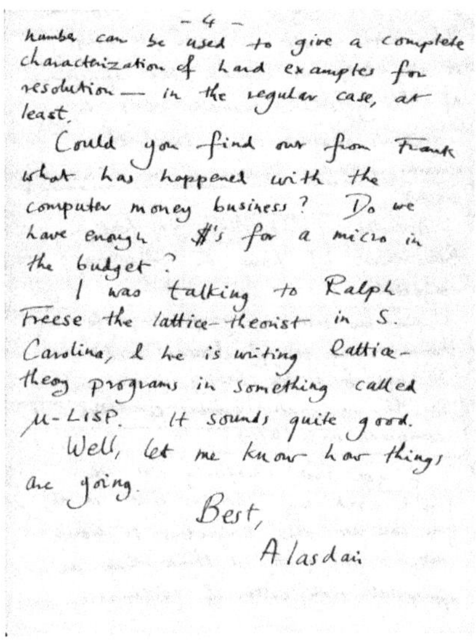

– 4 –

number can be used to give a complete characterization of hard examples for resolution — in the regular case, at least.

Could you find out from Frank what has happened with the computer money business? Do we have enough $'s for a micro in the budget?

I was talking to Ralph Freese the lattice-theorist in S. Carolina, & he is writing lattice-theory programs in something called μ-LISP. It sounds quite good.

Well, let me know how things are going.

Best,

Alasdair

Fig. 19.7 Page 4

References

Allen, J. F. (1981). An interval-based representation of temporal knowledge. *IJCAI*, *81*, 221–226.

Cleary, J. G. (1987). Logical arithmetic. *Future Computing Systems*, *2*(2), 125–149.

Cook, S. A. (1971). The complexity of theorem-proving procedures. *Proceedings of the third annual ACM symposium on Theory of computing* (pp. 151–158).

Covington, M. A., Nute, D., & Vellino, A. (1996). *Prolog Programming in Depth*. Hoboken: Prentice-Hall, Inc.

Dowling, W. F., & Gallier, J. H. (1984). Linear-time algorithms for testing the satisfiability of propositional horn formulae. *The Journal of Logic Programming*, *1*(3), 267–284.

Düntsch, I. (2005). Relation algebras and their application in temporal and spatial reasoning. *Artificial Intelligence Review*, *23*(4), 315–357.

Grätzer, G. (1978). *General Lattice Theory*. Basel: Birkhäuser.

Haken, A. (1985). The intractability of resolution. *Theoretical Computer Science*, *39*, 297–308.

Hanh, T. N. (2010). *The Sun My Heart: Reflections on Mindfulness, Concentration, and Insight* (2nd ed.). Berkeley: Parallax Press.

Hill, P., Lloyd, J., & Lloyd, J. W. (1994). *The Gödel Programming Language*. Cambridge: MIT press.

Older, W. J., & Vellino, A. (1993). Constraint arithmetic on real intervals. In Benhamou, F. & Colmerauer, A. (Eds.), *Constraint Logic Programming: Selected Research* (pp. 175–195). Cambridge: MIT Press.

Reckhow, R. A. (1975). *On the lengths of proofs in the propositional calculus*. Ph.D. thesis, University of Toronto.

Rescher, N., & Urquhart, A. (1971). *Temporal Logic* (Vol. 3). Berlin: Springer Science & Business Media.

Siekmann, J., & Wrightson, G. (1983). *Automation of Reasoning Vol. 2: Classical Papers on Computational Logic 1967–1970*. Berlin, Heidelberg: Springer.

Tseitin, G. S. (1970). On the complexity of derivation in propositional calculus. In Siekmann, J., Wrightson, G. (Eds.), *Automation of Reasoning Vol. 2: Classical Papers on Computational Logic 1967–1970* (pp. 466–483). Berlin: Springer.

Urquhart, A. (1983). Relevant implication and projective geometry. *Logique et Analyse*, *26*(103/104), 345–357.

Urquhart, A. (1984). The undecidability of entailment and relevant implication. *Journal of Symbolic Logic*, *49*(4), 1059–1073.

Urquhart, A. (1987). Hard examples for resolution. *Journal of the ACM (JACM)*, *34*(1), 209–219.

Urquhart, A. (1995). The complexity of propositional proofs. *Bulletin of Symbolic Logic*, *1*(4), 425–467.

van Emden, M. H. (2010). Integrating interval constraints into logic programming. Preprint, https://arxiv.org/abs/1002.1422.

van Emden, M. H., & Kowalski, R. A. (1976). The semantics of predicate logic as a programming language. *Journal of the ACM (JACM)*, *23*(4), 733–742.

Vellino, A. (1993). The relative complexity of analytic tableaux and sl-resolution. *Studia Logica*, *52*(2), 323–337.

Chapter 20
Russellian Propositions in *Principia Mathematica*

Bernard Linsky

Second Reader
E. Mares
Victoria University

Abstract As Alasdair Urquhart has noted, Bertrand Russell asserted that developing the theory of definite descriptions from 1905 was the first step towards solving the paradoxes that were finally resolved after 1908 in *Principia Mathematica* with the theory of types. I extend Urquhart's suggestion that Russell was referring to the use of the notion of *incomplete symbol* in his solution to the paradoxes in his doomed theory "substitutional theory" of "Russellian propositions" in 1906. The Introduction to PM states that expressions for propositions are incomplete symbols. This paper assesses the status of propositions in PM and connects the theory of types with the theory of descriptions.

Keywords Russell · Definite descriptions

20.1 The First Step

As has been noted by Alasdair Urquhart, Bertrand Russell described his theory of definite descriptions in 1905 as

> ...the first step towards overcoming the difficulties which baffled me for so long.[1]

[1]Russell (1967, p. 152). Urquhart notes this in Urquhart and Lewis (1994, p. xxxv). As well, Urquhart finds Russell describing the theory of denoting in this way as early as 1906 in a letter to Philip Jourdain, before Russell had adopted the theory of types. Urquhart and Lewis (1994, p. xxxiii)).

B. Linsky (✉)
Department of Philosophy, University of Alberta, Edmonton, Alberta T6G 2E7, Canada
e-mail: blinsky@ualberta.ca

I. Düntsch and E. Mares (eds.), *Alasdair Urquhart on Nonclassical and Algebraic Logic and Complexity of Proofs*, Outstanding Contributions to Logic 22,
https://doi.org/10.1007/978-3-030-71430-7_20

These difficulties were the paradoxes Russell had begun to collect in 1901, following the discovery of the "Russell Paradox"—the contradiction of the class of all classes that are not members of themselves. This paradox is announced in print in Chapter X of *The Principles of Mathematics* (Russell 1903), and about which he wrote to Frege in his famous letter of June, 1902.[2]

It might be thought that this is only a first step in the temporal sense. The theory of descriptions in "On Denoting" from 1905 (Russell 1905), it would seem, only introduced the notion of "incomplete symbols" as a first conceptual step in a series of different developments that resulted in a solution to the paradoxes, only to be completed in 1908 (Russell 1908) with the introduction of the theory of logical types. According to the theory of descriptions, an occurrence of a definite description in the position of a singular term, as in "The ϕ is ψ", can be defined by presenting a sentence which provides a *contextual definition* , namely, "There is one and only one ϕ and it is ψ'

$$*14 \cdot 01 \ \ [\imath x\phi x]\ \psi\ \imath x\phi x =_{df} (\exists x)(y) \colon \phi y\ .\equiv .\ y = x\ \wedge\ \psi x^{3}$$

Similarly, the apparent use of the class expression "the class of ψs" in the sentence "f is true of the class of ϕs", can be replaced by the higher-order sentence "There is a (predicative) function ϕ true of exactly the same individuals as ψ, and f is true of ϕ"

$$*20 \cdot 01 \ \ f\{x|\psi x\} =_{df} \exists\phi \colon (x)\ .\ \phi!x \equiv \psi x\ .\wedge .\ f\phi^{4}$$

This much of the theory in PM is clearly inspired by the theory of definite descriptions. However, the essential restriction that allows one to bar the "class of classes not members of themselves" comes from the restriction on types of the functions that are introduced by the contextual definition of classes. In the definition $*20 \cdot 01$, the notion of membership in a class ($x \in \{x|\psi x\}$)is reduced to that of membership in a function ($x \in \psi$), and that in turn is defined in terms of the application of that function

$$*20 \cdot 02 \ \ x \in \psi =_{df} \psi x$$

As no function can apply to another of its own type, an expression of the form $\{x|\psi x\} \in \{x|\psi x\}$ as well as $\sim\{x|\psi x\} \in \{x|\psi x\}$ are prohibited by the theory of types. Thus, it would seem, while Russell's analysis of class terms as incomplete symbols

[2] See van Heijenoort (1967, pp. 124–125)

[3] I follow Russell's notation for quantifiers and some other notation, but not all, as in the use of '$\wedge$' for conjunction, and a modified version of Peano's use of dots ('.' and ':') for punctuation in which dots sometimes appear in pairs like opening and closing parentheses: '(' and ')' and also as flanking principal connectives, as in '$. \equiv .$'. The expression $[\imath x\phi x]$ indicates the *scope* of the description. The difference between the two readings of 'The present King of France is not bald' in Russell (1905) depends on the relative scope of the negation and description operator, as in 'The present King of France is not–bald' as distinguished from 'It is not the case that the present King of France is bald'.

[4] The exclamation mark in $\phi!$ indicates that ϕ is a *predicative* function, a complication required by the theory of types. At Whitehead and Russell (1910, p. 78), it is indicated that a convention about the relative scope of the class operator obviates the need for explicit scope indicators as with definite descriptions.

may have indeed been a step "towards the overcoming" of the paradoxes, it was only the first step.

Urquhart's suggestion is that the route to the crucial second step towards the solution to the paradoxes, the theory of types, began in the use of incomplete symbols in the so-called "substitutional theory of 1906.[5] This system, based on the notion of one constituent being replaced by another in a "Russellian proposition", made Russell think that he could solve the paradoxes simply by eliminating the class expressions with contextual definitions involving the result of the substitution of objects for each other in propositions. It was only after the collapse of this system, Urquhart argues, that Russell saw that he must add a theory of types of propositions, and consequently of propositional functions according to the types of their values, in order to resolve the paradoxes in PM.

20.2 Propositions in PM

It is, however, not obvious that one can see PM as a system which distinguishes types of Russellian propositions. The problem of determining the status of propositions in PM is presented clearly in the "Introduction"

> ...what we call a 'proposition' (in the sense in which it is distinguished from the phrase expressing it) is not a single entity at all. That is to say, the phrase expressing a proposition is what we call an 'incomplete' symbol ..."[6]

The logical system of *Principia Mathematica* is a higher-order predicate logic that allows quantification over intensional *propositional functions*, where the types are *ramified*. This means that co-extensive predicates are not identified, and predicates true of the members of a given type of argument are further divided into types distinguished by the *order* of the propositions that are their values. The range of arguments for which a function is true or false form a *type*, in the strict sense. The propositions that result from a given argument will differ by the totalities over which their quantifiers range, making the functions differ by *order*. There are, however, (almost) no cases in which propositional variables are bound by quantifiers.[7]

The aim of this paper is to explain how Russell could say that the symbol for a proposition is in some way an "incomplete symbol" despite having based the system

---

[5] Urquhart and Lewis (1994, pp. xxxv–xxxvi).

[6] Whitehead and Russell (1910, p. 44). Hereafter PM. In his review of PM in the *Times Literary Supplement* of September 7, 1911, G. H. Hardy remarks on this passage that while propositions "...must be regarded, for the main purposes of the book, as ultimate and unanalysable." After citing this passage, he complains that "It would have been better, we think, if the authors had made up their minds to 'face the music,' and had begun with a definitely philosophical excursus—a mere ha'porth of tar in the outfit of such a leviathan." Hardy (1911, p. 321).

[7] See Urquhart (2003, pp. 293–297) for a presentation of the "Ramified Theory of Types".

of PM on a theory of types of propositions.[8] There are no contextual definitions in PM that make it possible to eliminate terms expressing propositions. Indeed, there are several occasions on which bound variables ranging over propositions do occur, and elsewhere all that is said is that such use is to be "avoided".

Are there propositions in PM or not? This seemingly simple question is not easy to answer. In the commentary on ∗14·3, which is a theorem that asserts that the scope of *proper* definite descriptions is irrelevant in extensional contexts, there is an explicit use of just such bound variables ranging over propositions:

> In this proposition, however, the use of propositions as apparent variables involves an apparatus not required elsewhere, and we have therefore not used this proposition in subsequent proofs.[9]

$$*14\cdot 3\ \ (p)(q).\ fp \equiv fq\ .\wedge .\ E!\,\imath x\phi x\ :\supset :\ f.\,[\imath\, x\phi x]\ \chi\ \imath\, x\phi x\ . = [\imath\, x\phi x]\ .f\{\chi\ \imath\, x\phi x\}.$$[10]

Even though there are no appearances of bound variables p, q, etc., after ∗14, there are several references to variables ranging over propositions in the Introduction.[11]

The "Prefatory Statement of Symbolic Conventions" at the beginning of Volume II of PM discusses the notions that result from considering expressions of different types, some occurring even in the same proposition. Thus a *homogenous* cardinal is one that contains a number of pairs, each pair of objects of the same type, rather, for example, between pairs consisting of an individual and a class. This prefatory statement begins with a discussion of three hierarchies:

(1) the functional hierarchy
(2) the propositional hierarchy.
(3) the extensional hierarchy.[12]

Yet symbols for classes and relations in extension are called "incomplete symbols" and only the last are given contextual definitions. The philosophical motivation for the theory of types includes the "epistemological" paradoxes such as the Epimenides paradox of the assertion that all propositions asserted by Epimenides are false.[13] The solution to this paradox involves the theory of types. The propositions that are asserted by Epimenides that are asserted to be false are all of a lower type than the proposition that they are false.

---

[8] I take the Introduction to be Russell's work and describe the material in the body as the work of "Whitehead and Russell". See below for an account of the Introduction.

[9] Whitehead and Russell (1910, p. 185).

[10] The notation $E!\,\imath x\phi x$ indicates that the description is *proper*, that is, there is one and only one ϕ. The difference between the two readings of "The present King of France is not bald" does lead to a difference of truth value in this case, as there is no present King of France and so the description "The present King of France" is not proper.

[11] See Whitehead and Russell (1910), pages 5, 4, 42, 62, 129, and 185.

[12] Whitehead and Russell (1912, p. ix).

[13] The Epimenides paradox and its resolution in the theory of types is sketched in the Introduction to Whitehead and Russell (1910, pp. 60–62).

Only variables over propositional functions are required for the results for the foundations of logic and mathematics that are in the body of *Principia Mathematica*. The three volumes on mathematics do not need any account of intensional notions such as belief or judgment (that are propositional attitudes directed towards propositions which are their objects).

Thus, quantification over propositions is not needed for the technical work of reducing mathematics which occupies the bulk of the work of PM after the first half of Volume I. Even the "functional hierarchy" of possibly intensional propositional functions is supplanted after ∗30 by the fully extensional "descriptive functions" and "relations" that are used to define further mathematical notions.

In "Comparison of Russell's Resolutions of the Semantical Antinomies with that of Tarski", Church (1976) Alonzo Church says that his own account takes propositions as values of propositional variables, despite the text of the Introduction. In the original article, Church (1976), the footnote containing this, remarking on the quote about propositions as incomplete symbols he continues as follows:

> They seem to be aware that this fragmenting of propositions requires a similar fragmenting of propositional functions. But the contextual definitions that are implicitly promised by the "incomplete symbol" characterization are never fully supplied, and it is in particular not clear how they would explain away the use of bound propositional and functional variables.[14]

In addition to pointing out that there are no contextual definitions for proposition or propositional function expressions, Church points out that such definitions would not be able to explain bound variables where the instances would be mere symbols. Church may have been pessimistic about a purely "substitutional" interpretation of type theory, by which the higher-order variables range over linguistic substitution instances, while the bound individual variables remain interpreted with standard objectual quantification. Such a substitutional interpretation of the quantifiers would allow one to interpret quantification over propositions and propositional functions as making assertions about all substitution instances of those quantifiers.[15]

This is one way of interpreting the talk of symbols for propositions (and propositional functions) as "incomplete symbols". Indeed, Church modifies this footnote in the later version of his paper, eliminating the implied criticism of this "substitutional" interpretation of the logic.[16]

---

[14] Church (1976, p. 748, n.4).

[15] Such an interpretation of the first edition of *Principia Mathematica* is presented in Hazen and Davoren (2000). Landini (1996) has a very different interpretation of this approach. He terms his substitutional semantics for bound higher- order variables a "nominalist" interpretation. On this account, the syntax of the language for PM, however, has no higher-order free variables. Such letters are treated schematically. Among other consequences of his view, expressions of the form "$\phi\hat{x}$" for propositional functions are not to be treated as functional abstracts. See Landini (1998) for his interpretation of the relation between the 1906 substitutional theory and PM.

[16] Urquhart cites Parsons (1971) for this notion of substitutional interpretation of quantification.

> This seems to mark the beginning of Russell's long search for a substitute for propositions, or other way to be rid of them. It is probably a late addition to the *Introduction* of PM, as no trace of it appears in (Russell 1908) or in the main text of PM.[17]

What's more, if expressions for propositional functions are also incomplete symbols, they surely could not be eliminated, as the notion of propositional function is basic and certainly could not be defined in terms of more basic notions. There could not be a contextual definition of a primitive notion of the logic that is modeled on the contextual definitions of definite descriptions in $*14$, or classes in $*20$. Quantification over classes is defined in $*20$ with the use of special Greek variables, with contextual definitions as in $(\exists\alpha)$ and (α) at $*20 \cdot 07$ and $*20 \cdot 01$. This consequence still does not forbid the use of bound variables for propositions, if they can be interpreted as predicate variables, and so bound by higher type quantifiers. The use of variables for propositions is adopted by Church (1976) as variables over propositions as treated simply as variables over 0-place propositional functions.

There are yet more logical consequences of adopting the view that symbols for propositions are "incomplete symbols", and perhaps that it is not, as Church suggests, simply a promissory note that cannot be redeemed. Contrary to what he says above, we will see in what follows that there are in fact "traces" of the notion of incomplete symbols in what follows in $*9$. In fact, the notion of propositions, and functions, as incomplete symbols, is relevant to the application of the theory of types as the solution to the paradoxes.

Church suggests that the sections of the Introduction that conclude with the assertion that symbols for propositions are incomplete symbols are "late addition" to *Principia Mathematica*. In fact, as has been observed recently by James Levine, those pages in the Introduction are literally a late addition to a paper called "The Theory of Logical Types" (Russell 1910), which was then inserted into the Introduction. In that manuscript, there is an assertion between pages 14 and 15, labeled "14a" through "14g".[18] The passage inserted into the manuscript begins with Russell's familiar ontology of *facts* that was to be the center of his logical atomist writing and extends to the end of Section III "Definition and Systematic Ambiguity of Truth and Falsehood" of Chapter II of the Introduction on "The Theory of Logical Types":

> The universe consists of objects having various qualities and standing in various relations. Some of the objects which occur in the world are complex. When an object is complex, it consists of interrelated parts. Let us consider a complex object composed of two parts a and b standing in the relation R. The complex object "a-in-the-relation-R" may be capable of being *perceived*; when perceived, it is perceived as one object. Attention may show that it is complex; we then *judge* that a and b stand in the relation R. Such a judgment, being derived from perception by mere attention, may be called a "judgment of perception". This judgment of perception, considered as an actual occurrence, is a relation of four terms, namely a and

[17] Church (1984b, p. 291, n.4). Church repeats this with slightly different wording in Church (1984a) as marking: "...the beginning of Russell's long attempt to do away with or modify the notion of proposition, but it has fortunately not affected the main text of 'Principia'."

[18] These facts were to be found all along in a condensed textual note buried at Slater and Frohmann (1992, 503) and only recently appreciated by Levine. This led him to the Bertrand Russell Archives, where the insertion of pages into the manuscript of Russell (1910) is readily apparent.

> *b* and *R* and the percipient". ...in fact, we may define *truth*, where such judgments are concerned, as consisting in the fact that there is a complex *corresponding* to the discursive thought which is the judgment.[19]

Russell's new talk of *facts* which are objects of perception, and the intended targets of belief, enters Russell's writing in 1910 and is most explicitly developed only in 1918 in the *Philosophy of Logical Atomism* lectures.[20] While perception is veridical, that is, one can only perceive a fact that obtains, we can believe falsehoods. In this new ontology, Russell abandoned the representation of belief as a relation to a proposition, which can be either true or false. When someone judges falsely, even though there is no fact that is believed, there is still the true occurrence of the judgment, which is itself the fact that someone believes such and such. One might think that this late addition to the Introduction marks the adoption of this replacement of an ontology of true and false propositions with the new ontology of facts, which can only *obtain*, and so simply do not exist and cannot be the objects of attitudes in the case of false beliefs. Church seems to hold that there is no way that this new ontology can be reflected in the previously worked out technical sections in the body of *Principia Mathematica*, and that the talk of symbols for propositions as "incomplete symbols" cannot be taken at face value.[21] How are they to be interpreted?

20.3 Symbols for Propositions Are Incomplete

An explanation of these puzzling remarks about propositions and incomplete symbols, and their relation to the theory of types will come from an examination of some work of Urquhart on Russell's collected logical papers, and of Pelham and Urquhart 1994 on Russellian propositions. My interpretation of these remarks involves an examination of another late addition to PM, chapter "$*9$ Extension of the Theory of Deduction from Lower to Higher Types of Propositions". The demonstrations in $*9$ are intended to show that the theory of quantified logic can be based on the propositional logic of $*1$ to $*5$, in which the propositions are to be interpreted as *elementary*, that is, quantifier free atomic sentences in various propositional combinations of $\sim$ and $\vee$, preceded by a string of existential and universal quantifiers. This is the "prenex normal form" of contemporary logic. (As there are no predicate or individual constants in the system, the atomic sentences will consist only of predicate and individual variables.) Russell's new ideas in $*9$ are a technical expression of why he wants to avoid the use of quantifiers over propositions where possible.[22]

---

19 Whitehead and Russell (1910, p. 43).

20 See Slater (1986, pp. 171–173) . Russell also stresses the special status of entities for which there is a "simple symbol".

21 For a substitutional semantics of the sort in Hazen and Davoren (2000) that explicitly uses the notion of fact see Mares (2007).

22 See Linsky (2013) for examples showing that Russell did not hold that all "incomplete symbols" are eliminable with contextual definitions.

Russell's association of "incomplete symbols" with expressions that should not be represented by a single variable but instead by a complex quantificational expression originated with his theory of descriptions. In his introduction to Urquhart and Lewis (1994), Urquhart quotes from a letter to Couturat of 23 October, 1905, almost immediately after Russell had written "On Denoting":

> As for denoting functions, here are the main ideas. I find that to avoid contradictions, and make the elements of mathematics rigorous, it is absolutely necessary not to employ a single letter, such as ϕ or f, for a variable which cannot become an arbitrary entity, but which is really a *dependent* variable. Suppose we wish to say e.g.
>
> $(\phi, f) : \phi ! f`x$ (A)
>
> The values of ϕ and f in question are not the same as the values of x in $(x) . \phi ! x$. Now we can always reduce propositions such as (A) to another form which does not contain this other type of variability. All that the theory of denoting functions does is to replace the variability such as is possessed by f by the variability such as is possessed by ϕ; that is the first stage. ...Instead of $\phi ! x$ we can put $p \frac{x}{a}$, which is to mean "the result of substituting x for a in p"; if a does not occur in p, $p \frac{x}{a} = p$ ...Thus we shall have only one type of independent variable ...I think once more that the solution of the contradictions is to be found in maintaining that there are no classes or relations.[23]

Here an "incomplete symbol" for something cannot be straightforwardly expressed by a single variable and is instead expressed using quantification that is legitimate. This is the sense in which the analysis of definite descriptions, such as "the present King of France" can be said to "maintain" that there is no present King of France. The expression "the King of France" does not occur as a singular term would, subject to quantification. From $(x)(\exists y) y = x$ one cannot infer "$(\exists y)$ the present King of France $= y$". In general, however, if a definite description is *proper*, then the description can be treated as though it were a singular term

$*14{\cdot}204\ E! \,\imath x \phi x \equiv (\exists y)\ y = \imath x \phi x$

The solution to the Russell paradox is not simply to say that there is no "class of all classes that are not members of themselves" simply because there are *no* classes whatever. Instead, it is to show how expressions for classes are expressed using quantifiers over propositional functions and then how the paradox is avoided by some feature of those expressions.

The theory that treats class expressions as incomplete symbols which Russell describes in his letter to Couturat is the "substitutional theory" that he explored tentatively in the years 1906 and 1907. The theory was based on the notion of substituting individuals in propositions. These "Russellian Propositions" have individuals as constituents that can be replaced, or "substituted" for each other yielding new propositions.[24] The theory does not distinguish propositions from other objects by

---

[23] Urquhart and Lewis (1994, p. xxxvii).

[24] Pelham and Urquhart (1994, p. 307), Kaplan (1986) for the use of the term "Russellian propositions" in the theory of direct reference, although they note that the expression circulated in manuscripts before that. See Church (1984b) for an account of Russell's theory of propositions in Russell (1903).

type, but instead defines individuals and propositional functions in terms of the result of various substitutions in propositions. The distinctive logical notion of this theory of substituting one individual for another, "the result of replacing a for b in p is q", symbolized as "$p\frac{a}{b}!q$".

The notion of a propositional function is then defined as an "incomplete symbol". It is not a constituent of a proposition that can be substituted, but rather is given a contextual definition in terms of the result of substitutions of objects for each other in propositions. In one of the unpublished papers from this period, in a discussion of the expression "$(s)s$" used in the account of negation in *The Principles of Mathematics*, Russell discusses a variation on this view which appears to allow substitution of both functions and individuals[25]:

> We can't get any form of $(s)s$ in $*7$ because we don't yet know any single particular function. The first is $p\frac{x}{a}!q$ and this will have to be
>
> $$(\phi y)\frac{x}{a}!(\psi z).$$
>
> 26

and further:

> Thus propositions, like classes, are to be replaced by *two* variables. But now we take entities and *functions*, not entities and *propositions*, as our fundamentals. It is to be understood that a proposition is not an entity, and is not to be expressed by a single letter, but always by two, ϕ and x.[27]

However, later in the paper, we have this qualification:

> N.B. We have to be careful not to let functions creep back into being. We must now have functions, but they must be quite distinct from entities.
>
> If ϕa is significant, $\phi(\psi a)$ is to be nonsense. Similarly if ϕa is significant, $(\phi a)\frac{\psi x}{b}$ is to be nonsense.[28]

When later Russell holds that variables for propositions and perhaps also propositional functions are *both* incomplete symbols, it is something like this that is behind his view. Neither propositions nor propositional functions can be substituted for genuine individuals, only now because the theory of types forbids it. These restrictions from the theory of types formerly emerged from the substitutions in propositions.

We can see the sense in which a propositional function is an "incomplete symbol" in the details of the substitutional theory. However, as Pelham and Urquhart demonstrate the substitutional theory is itself subject to a paradox similar to that in Russell (1903, Appendix B), which this time led Russell to adopt a theory of types

---

[25] The assertion of $(s)s$, amounting to "all propositions are true", is used as to define negation, as $p \supset (s)s$, as later could be expressed using the *falsum* ($\perp$) as: $p \supset \perp$.

[26] Russell (1906, p. 264).

[27] Russell (1906, p. 265).

[28] Russell (1906, p. 265).

for propositions.[29] As they say traces of the substitutional theory still appear in the first presentation of the theory of types in Russell (1908), but also, I will suggest, in *Principia Mathematica* as well.

Even though Russell's foundational account of judgment was indeed changing even while he was composing the Introduction, this still does not mean that there is no place for propositions in the logical theory of PM. Propositions figure prominently in the Introduction:

> The paradoxes of symbolic logic concern various sorts of objects: propositions, classes, cardinal and ordinal numbers ...[30]

The paradoxes that are used to motivate the theory of types are listed in Section VII of the Introduction, "The Contradictions", which follows the inserted material.[31]

This would seem to support the view that what appears to be a "recursive definition" of truth is really about the truth of propositions, not an anticipation of Tarski's hierarchy of truth-predicates of sentences in different meta-languages:[32]

> ...that the words "true" and "false" have different meanings, according to the kind of propositions to which they are applied ...[33]

and, on the next page:

> ..."or" and "not" adapt themselves to propositions of any *order*.[34]

Then appears the denial that propositions are entities:

> ...the phrase which expresses a proposition is what we call an "incomplete symbol"[35]

After the insertion of the material explaining the "multiple relation theory of judgment", we have:

> Let us call the sort of truth which is applicable to ϕa "*first truth*. ...Consider now the proposition $(x).\ \phi x$. If this has the sort of truth appropriate to it, this will mean that every value ϕx has *first truth*. Thus if we call the sort of truth that is appropriate to $(x)\phi x$ "second truth" we may define $\{(x).\ \phi x\}$ has second truth as meaning "every value for $\phi\hat{x}$ has first truth, i.e. "$(x).\ \phi x$ has first truth".[36]

A quantified proposition is not a member of the totality of its instances, it is hence of a higher type by the vicious circle principle. But what does it mean to say that "the phrase expressing a proposition is what we call an 'incomplete' symbol..."?

---

[29] See Church (1984b) for a formalization of the argument.

[30] Whitehead and Russell (1910, p. 38).

[31] Whitehead and Russell (1910, pp. 60–65).

[32] See Landini (1996, p. 316). Landini reads this as just such a definition of truth. The notion of the hierarchy of meta-languages is presented in Tarski (1956, Sect. 2).

[33] Whitehead and Russell (1910, p. 42).

[34] Whitehead and Russell (1910, p. 43).

[35] Whitehead and Russell (1910, p. 44).

[36] Whitehead and Russell (1910, p. 42).

20.4 Propositions in $*9$

Progress towards making sense of these puzzling remarks about propositions in PM involves looking at $*9$, which is alluded to in the "late addition" to the Introduction. $*9$ is itself also a late addition to the technical body of *Principia Mathematica*, that was originally to begin quantificational logic with $*10$.

Russell states the project of the chapter:

> In what follows the single letters p and q will represent *elementary* propositions, and so will "ϕx", "ψy", etc. We shall show how, assuming the primitive ideas and propositions of $*1$ as applied to elementary propositions, we can define and prove analogous propositions as applied to propositions of the forms $(x).\phi x$ and $(\exists x).\phi x$. By mere repetition of the analogous process, it will then follow that analogous ideas and propositions can be defined and proved for propositions of any order; whence further, it follows that, in all that concerns disjunction and negation, so long as propositions do not appear as apparent variables, we may wholly ignore the distinction between different types of propositions and between different meanings of negation and disjunction. Since we never have occasion, in practice, to consider propositions as apparent variables, it follows that the hierarchy of propositions (as opposed to the hierarchy of propositional functions) will never be relevant in practice after the present number.[37]

The argument of $*9$ is intended to prove that all the axioms of propositional logic in the first chapters will still hold for substitution instances of the variables p, q, etc. for quantified expressions "$(x)\phi x$" and "$(\exists y)\psi y$". The method involves showing that those quantified expressions can be defined in terms of expressions in prenex normal form, in which the matrices can be derived from the early chapters. This is because the "Russellian propositions" in the matrices will be formulas of the form of ϕx or $\phi x \vee \psi y$ where those are variables for propositional functions and individuals. In the logic, those elementary propositions will occur only when bound by variables. This is the technical way in which propositions appear as "incomplete symbols". They occur only in the context of quantified expressions and are not themselves separate entities that can occur in the subject position of an atomic Russellian proposition, albeit one of a higher order. This is because a quantified proposition is not a member of the totality of its instances. It is hence of a higher type by the vicious circle principle.[38]

It is clear from the series of previous theorems used in later chapters that $*9$ of *Principia Mathematica* is another "late addition".[39] The numbering system begins with quantificational logic at $*10$, and in fact, all of the later theorems cite $*10$. Indeed the technical failings of $*9$ support the suggestion that it was a late addition,

---

[37] Whitehead and Russell (1910, p. 129) In $*9$ Russell uses the expression "first order" to refer to a formula with only one quantifier with an elementary matrix, and "higher order" to refer to those with a second, or further, number of quantifiers. In this place there seems to be a more fine-grained notion of types than elsewhere.

[38] See the discussion in Chapter II of the Introduction, Whitehead and Russell (1910, pp. 37–41).

[39] There is a draft of the material in $*9$ in Moore (2014, pp. 632–634), titled "Deduction of Theory of Propositions of Higher Type from That of Those of Lower Type" and dated by the editor with some other fragments as written in 1908. This is still likely later than $*10$ and the following numbers of Volume I.

and so it is most likely to be solely the work of Russell, with Whitehead having been content with the original formulation based on $*10$.[40] The only technical role of $*9$ is to prove what had been the primitive propositions of $*10$ as theorems based on an alternative foundation. The theorems of $*10$ allow for quantified expressions to occur as subformulas of sentences. The use of any inferences of the "propositional logic" of $*1$ to $*5$ would thus have to be typically ambiguous with regard to the types of the propositional variables, and also the number of quantifiers they include. With the addition of $*9$ it becomes possible to interpret the propositional variables p and q, etc., as standing now for "elementary propositions". An elementary proposition is one that contains no variables. It will therefore be of the form ϕx where x is an individual, and ϕ a "matrix", or predicate which does not contain a quantifier. The logic of elementary propositions is then extended in $*9$ to apply to propositions with quantifiers by first quantifying the variables in the elementary propositions with the quantifiers in prenex position, and then moving them into the formula by the definitions in $*9$. These definitions are stated with the variable p to stand for elementary propositions.

The result is that the propositional inferences of $*1$ to $*5$ can be used with arbitrary formulas, however, many quantifiers may be involved. Furthermore, these are defined propositions and the connectives $\sim$, $\vee$, $\supset$, are in fact "typically ambiguous" and are of higher orders (in the sense of "order" special to $*9$). Because of the definitions of $*9$, this extended use of the theorems of $*1 - *5$ is legitimated, so that for practical purposes, the typical ambiguity and the use of multiple quantifiers of a given type can be ignored. As will be repeated in $*14$, we see a remark that variables for propositions will not be used later, as a matter of practice only:

> Since we never have occasion, in practice, to consider propositions as apparent variables, it follows that the hierarchy of propositions (as opposed to the hierarchy of propositional functions) will never be relevant in practice after the present number.[41]

Landini (2000) points out that the version of $*9$ that was published is incomplete, in the sense that not all valid first order formulas are provable. Indeed, the system is not complete for its purported purpose of showing that the axioms of the system are all provable when presented in prenex normal form. The problem comes from having an insufficient list of primitive propositions. An examination of the details of $*9$ will help to understand its role in the account of propositions in PM.

The definitions of negation are:

$$*9 \cdot 01 \sim\{(x)\phi x\} =_{df} (\exists x) \sim\phi x$$
$$*9 \cdot 02 \sim\{(\exists x)\phi x\} =_{df} (x) \sim\phi x$$

Then:

> *Definition of Disjunction.* To define disjunction when one or both of the propositions concerned is of the first order, we have to distinguish six cases:

[40] In a letter to the editor Whitehead (1926, p. 120) of *Mind*, Whitehead says that of the material in the first edition relevant to the second, $*10$ was *not* primarily Russell's work, and by implication that $*9$ *was* Russell's work alone.

[41] Whitehead and Russell (1910, p. 129).

$*9 \cdot 03\ (x)\phi x \vee p =_{df} (x).\ \phi x \vee p.$

$*9 \cdot 04\ p \vee (x)\phi x =_{df} (x).\ p \vee \phi x.$

$*9 \cdot 05\ (\exists x)\phi x \vee p =_{df} (\exists x).\ \phi x \vee p.$

$*9 \cdot 06\ p \vee (\exists x)\phi x =_{df} (\exists x).\ p \vee \phi x.$

$*9 \cdot 07\ (x)\phi x \vee (\exists y)\psi y =_{df} (x)(\exists y).\ \phi x \vee \psi y.$

$*9 \cdot 08\ (\exists y)\psi y \vee (x)\phi x =_{df} (x)(\exists y).\ \psi y \vee \phi x.$

(The definitions $*9 \cdot 07 \cdot 08$ are to apply also when ϕ and ψ are not both elementary functions).

The connective $\supset$ is defined, even though it, notoriously, occurs in the primitive propositions of the system of *Principia Mathematica*:

$*1 \cdot 01\ p \supset q =_{df} \sim p \vee q$

The rule of *Modus Ponens* ($*1 \cdot 11$) is used as well as a "Primitive proposition" (Pp) that amounts to a rule universal generalization (where "real variables" are free and "apparent" variables are bound by a quantifier):

$*9 \cdot 13$. In any assertion containing a real variable, this real variable may be turned into an apparent variable of which all possible values are asserted to satisfy the question in question.

In addition to this series of definitions only two axioms, or "primitive propositions" are needed:

$*9 \cdot 1\ \phi x \supset (\exists x)\phi x$ Pp

and

$*9 \cdot 11\ \phi x \vee \psi y \supset (\exists z)\phi z$ Pp

Landini (2000) points out that there is an error in the proof of the case for existential quantified elementary propositions of the very first propositional axiom:

$*1 \cdot 2\ \ p \vee p\ .\supset .\ p$ Pp

The error occurs in the second proof of two cases for quantified instances of p:

$*9 \cdot 3\ \ (x)\phi x \vee (x)\phi x\ .\supset .\ (x)\phi x$

and

$*9 \cdot 31\ (\exists x)\phi x \vee (\exists x)\phi x\ .\supset .\ (\exists x)\phi x$

It might seem that this Existential Generalization axiom is all that is needed for a system of quantificational logic, given the interdefinability of the existential and universal quantifiers, so that axioms such as:

$*10 \cdot 1\ (x)\phi x\ .\supset .\ \phi y$

can be proved as theorems from $*9$. However, of axiom $*9 \cdot 11$, Russell says:

> The second of the above primitive propositions is only used once, in proving, $(\exists x)\phi x \vee (\exists x)\phi x \supset (\exists x)\phi x$ …The effect of this primitive proposition is to emphasize the ambiguity of the z required in order to secure $\exists z \phi z$.

This description of the "effect" of the axiom is not very helpful in understanding this non-trivial theorem. The complications involved are not obvious.[42]

The faulty proof of $*9 \cdot 31$ has five lines (those numbered are indicated in parentheses on the left of the lines), and as is almost universal in *Principia Mathematica*, compresses a number of steps into each line. The error occurs in the last step of this expanded version:

1.	$\phi x \vee \phi y \supset (\exists z)\phi z$	$*9{\cdot}11$ (Pp)
2. (1)	$(y) . \phi x \vee \phi y \supset (\exists z)\phi z .$	$*9{\cdot}13$ (U.G.)
3.	$(y) . \sim(\phi x \vee \phi y) \vee (\exists z)\phi z .$	$*1{\cdot}01$ (Def. $\supset$)
4.	$(y) \sim(\phi x \vee \phi y) . \vee . (\exists z)\phi z$	$*9 \cdot 03$ (Def.)
5.	$\sim(\exists y)(\phi x \vee \phi y) . \vee . (\exists z)\phi z$	$*9{\cdot}02$ (Def.)
6. (2)	$(\exists y)(\phi x \vee \phi y) . \supset . (\exists z)\phi z$	$*1{\cdot}01$ (Def. $\supset$)
7. (3)	$(x) . (\exists y)(\phi x \vee \phi y) \supset (\exists z)\phi z .$	$*9{\cdot}13$ (U.G.)
8.	$(x) . \sim (\exists y)(\phi x \vee \phi y) \vee (\exists z)\phi z .$	$*1{\cdot}01$ (Def. $\supset$)
9.	$(x) \sim(\exists y)(\phi x \vee \phi y) . \vee . (\exists z)\phi z$	$*9{\cdot}03$ (Def.)
10.	$\sim(\exists x)(\exists y)(\phi x \vee \phi y) . \vee . (\exists z)\phi z$	$*9{\cdot}02$ (Def.)
11. (4)	$(\exists x)(\exists y)(\phi x \vee \phi y) . \supset . (\exists z)\phi z$	$*1{\cdot}01$ (Def. $\supset$)
12.	$(\exists x)(\phi x \vee (\exists y)\phi y) . \supset . (\exists z)\phi z$	$*9 \cdot 06$ (Def.)
13.	$(\exists x)\phi x \vee (\exists y)\phi y . \supset . (\exists z)\phi z$	$*9 \cdot 05$ (Def.)

The error is in the very last step, from 12 to 13, an illegitimate use of $*9 \cdot 05$ because it takes $(\exists y)\phi y$ as an instance of p. As is said on the preceding page the variables p, q, etc. are restricted to elementary propositions. In the proof above, p must be elementary, and $(\exists y)\phi y$ is not elementary, containing as it does a single quantifier and an elementary matrix, ϕy.

---

[42] As Urquhart points out this is "…perhaps the first appearance of the *contraction rule* in the foundations of logic …". Urquhart (2013, p. 17). Urquhart says, in a discussion of the influence of $*9$ on Kurt Gödel's "*Dialectica* interpretation", that this case is the most difficult axiom to verify. Urquhart (1976, p. 508).

The error can be repaired by adding one more axiom that allows for defining expressions with more than one sentential component that is not elementary. The system with this one repair is not complete for first-order semantics.[43]

Indeed, earlier in the Introduction, on the last page of the "late addition", there is a discussion of precisely this principle.[44] After a discussion of the definition of negation for quantified expressions, as given by $*9 \cdot 01$ and $*9 \cdot 02$, we find:

> An analogous explanation will apply to disjunction. Consider the statement "either p, or ϕx always." We will denote the disjunction of two propositions, p, q by "$p \vee q$." Then our statement is "$p \,.\vee .\, (x)\phi x$" we will suppose that p is an elementary proposition, and that ϕx is always an elementary proposition. We take the disjunction of two elementary propositions as a primitive idea, and we wish to *define* the disjunction
>
> "$p \,.\vee .\, (x)\phi x$"
>
> This may be defined as "$(x).\, p \,.\vee .\, \phi x$", *i.e.* "either p is true or ϕx is always true" is to mean " 'p or ϕx' is always true." Similarly we will define
>
> "$p \,.\vee .\, (\exists x)\phi x$"
>
> as meaning "$(\exists x).\, p \vee \phi x$ ", *i.e.* "either p is true or there is an x for which ϕx is true" as meaning "there is an x for which either p or ϕx is true."[45]

This much is straightforwardly carried out in $*9{\cdot}04$ and $*9{\cdot}06$. But, then Russell goes on to say that there are definitions that are needed, these are the very definitions with two quantified expressions that are missing from the list in $*9$:

> Similarly we can define the disjunction of two universal propositions: "$(x).\, \phi x \vee (y).\, \psi y$" will be defined as meaning $(x, y).\, \phi x \vee \psi y$", *i.e.* "either ϕx is always true or ψy is always true" is to mean " 'ϕx or ψy' is always true." By this method we obtain definitions of disjunctions of elementary propositions of the form $(x)\phi x$ or $(\exists x)\phi x$ in terms of disjunctions of elementary propositions; but the meaning of "disjunction" is not the same for propositions of the forms $(x)\phi x$, $(\exists x)\phi x$, as it was for elementary propositions.[46]

There are, however, no such definitions to be found in $*9$, either for the existential or universally quantified sentences "$(x).\, \phi x \vee (y).\, \psi y$ " or "$(\exists x).\, \phi x \vee (\exists y).\, \psi y$". The latter is what is needed at step 12 of the faulty demonstration of $*9 \cdot 31$.[47]

---

[43] See Landini (2000) for details of one selection of additional axioms borrowed from $*8$ in the second edition, and a proof that the resulting system provides the axioms of a complete system of quantificational logic. The attention given above to the details of the error that Landini has spotted is intended to change his focus on $*9$ as an unsuccessful attempt to produce a complete system of quantificational logic there as well. The discussion here is limited to proving that theorems of the system of $*10$ can all be presented in prenex normal form and the consequent connection between expressions for propositions and the theory of types.

[44] It is on 14*h* of the manuscript. See Urquhart and Lewis (1994, p. 679).

[45] Whitehead and Russell (1910, p. 47).

[46] Whitehead and Russell (1910, p. 47).

[47] Landini (2000) cites this passage at page 47 of the Introduction to PM, but indicates it as compounding the failings of $*9$, and not as indicating that omitting these axioms is simply an error.

In the "late addition" to PM Russell claims both that symbols for propositions are "incomplete symbols" and includes this description of what is needed for $*9$. It seems, then, that the purpose of presenting propositions in prenex normal form in $*9$ is to show how the symbols for propositions are to be understood as incomplete symbols. The variables for propositions p, q, etc., are to be replaced by expressions such as "$(x).\ \phi x \vee (y).\ \psi y$", which is in turn defined by "$(x, y).\ \phi x \vee \psi y$" in which "$\phi x$" and "$\psi y$" are "elementary" propositions, but rather propositions involving variables that are quantified, and so are not Russellian propositions with particular individuals as constituents at these quantified places. These expressions, I claim, can be termed "incomplete symbols", as they only have meaning in the context of a quantifier binding the variable, or an assertion in which the variable is to be left free and interpreted as ranging over "anything", as a symbol that is only meaningful in a particular syntactic context, and not on its own as a genuine term.

In this respect the symbols for propositions are worse off than expressions replacing variables over classes. In chapter $*20$ we are shown how we *can* quantify over classes, despite the symbols for classes being "incomplete symbols":

$*20{\cdot}07\ (\alpha)\ f\alpha\ .=_{df}\ .\ (\phi)\ f\{\hat{z}(\phi!z)\}$

$*20{\cdot}071\ (\exists\alpha)\ f\alpha\ .=_{df}\ .\ (\exists\phi) f\{\hat{z}(\phi!z)\}$

This certainly *appears* to be a case where *Principia Mathematica* does allow quantification over things that are not "terms", namely, the referents of "incomplete symbols". Indeed, in the Introduction, Russell says we are to:

> ...think of classes as "quasi-things", capable of immediate representation by a single name.[48]

There is no "single name", constant or variable x, y ..., that is to represent classes. It is still possible to quantify over classes with variables α, β, ..., provided they are understood as subject to the definitions $*20{\cdot}07$ and $*20{\cdot}071$.

20.5 Conclusion

To what extent is it possible to give an account of the status of Russellian propositions in *Principia Mathematica* given the variety of seemingly inconsistent things said about them? A summary of the results of this paper can be organized around a list of some obscure and apparently incorrect assertions Russell makes:

1 The theory of definite descriptions was the first step towards overcoming the difficulties ...

The theory of descriptions led Russell to the notion of an "incomplete symbol", one to be defined in terms of a complex quantificational analysis. Russell's first solution to the paradoxes was the substitutional theory explored in Pelham and Urquhart

[48] Whitehead and Russell (1910, p. 81).

(1994), but it was abandoned and ultimately replaced by a theory of types of propositions and, consequently, of propositional functions. The project of $*9$ reveals a connection between the theory of definite descriptions and other incomplete symbols, and the theory of types. The expressions for propositions and even propositional functions will, in *Principia Mathematica*, always be an inseparable constituent of a quantified expression. And as Russell explains the effects of the vicious circle principle formulated as "whatever involves a totality cannot be a member of that totality", an expression quantifying over some domain must define a proposition or function of a higher type. The propositions in PM are all quantified Russellian propositions. That is, they are expressible with a string of quantifiers applied to a matrix which is an elementary proposition. Expressions for propositions are, therefore, "incomplete symbols" without being eliminable by contextual definitions. The role of quantifiers is key to the understanding of the theory of types.

2 ...what we call a "proposition" ...is not a single entity at all ....

Russell's metaphysics for logic replaces the analysis of propositions of belief and other attitudes as entities with the "multiple relation theory of judgment".

3 ...the phrase expressing a proposition is what we call an "incomplete" symbol ....

The notion of "incomplete symbol" applies to a range of different cases, from the theory of definite descriptions and the "no-classes" theory of classes. The case of propositions is different again. There can be no definitions which allow the elimination of symbols for propositions from contexts in which they occur, and so no definitions of the occurrences of bound propositional variables. Yet, these expressions count as "incomplete symbols" in an extended sense of the term, because of their similarity with the paradigm case of the theory of descriptions.

4 The use of bound variables for propositions "involves an apparatus not required elsewhere" and is therefore "avoided".

The upshot of $*9$ is to show propositions to be even less "things" than the "quasi-things" that classes are. The technical project of $*9$ is to show that all propositions in *Principia Mathematica* might be represented by an expression in the prenex form:

> A string of $n \geq 0$ quantifiers Q_i that are either existential or universal, (x_i) or $(\exists x_i)$ for each $i \leq n$, preceding an *elementary* matrix Φ: $Q_1 \ldots Q_n . \Phi(x_1, \ldots x_n)$.

Without additional metalinguistic resources, this result is unstable in the language of PM, even with the very different resources of the contextual definitions used for the cases of definite descriptions or classes.

5 ...a proposition can not be "expressed by a single letter".

Church's practice of treating propositions as 0-place predicates would treat symbols for propositions as "single letters". However, it involves an approach to assigning types to variables that Russell did not accept. Church (1976) only distinguishes "*r-types*" for variables and not connectives. The restrictions of the vicious circle principle, by which a function defined in terms of quantification over a given r-type must be of a higher r-type are enforced by determining the r-types of the variables in the comprehension schema. So, to conclude that there is a function ϕ true of just those things expressed by a formula Φ, we must observe the restriction that the bound variables of Φ are all of "order" less than the order of ϕ.[49] As a result, it makes no sense to speak of the type of a connective such as $\sim$ or $\vee$, and so to worry, as Russell did, about the meaning of those connectives when flanked by propositions of various types. Russell could not accept this strategy, I suggest, because he thought it would obscure the quantificational essence of propositions that determines their type. He is ambivalent about propositional variables, as we have seen, in his both making use of them in some places, and in saying that his avoidance of them, in general, is simply a matter of their not being "required" for his purposes in PM. The only reason, then, that he did not consider symbols for propositions to be incomplete, in the way that the Greek letters α, etc. for classes were incomplete, was the impossibility of providing a contextual definition. Perhaps he should not have suggested that there could not be variables ranging over propositions in PM, but just that they would not be used in the body of the work.

Urquhart concludes his discussion of the influence that $*9$ had on later logic with this:

> The interesting but slightly odd developments in $*9$ are often overlooked in discussions of *Principia Mathematica*. Yet ...they suggested new ideas to both Gödel and Herbrand. In the work of these later logicians, however, the developments of $*9$, inspired by philosophical scruples, are transformed into technical tools.[50]

I think that more can be said for $*9$. Perhaps that these "philosophical scruples" about the proper expressions for propositions in the theory of types are in fact a link between Russell's discovery of "incomplete symbols" with the theory of definite descriptions in 1905 and his final resolution of the paradoxes in *Principia Mathematica* in 1910.

Acknowledgements The author wishes to acknowledge the crucial assistance of the second reader who made me appreciate more fully what Alasdair Urquhart has written about the semantics of the theory of types in *Principia Mathematica*.

[49] See Church (1976, p. 750).

[50] See Urquhart (1976, p. 509).

References

Church, A. (1976). Comparison of Russell's resolution of the semantical antinomies with that of Tarski. The Journal of Symbolic Logic, 41:747–760.

Church, A. (1984a). Comparison of Russell's resolution of the semantical antinomies with that of Tarski. In R. L. Martin (Ed.), *Recent Essays on Truth and the Liar Paradox* (pp. 289–306). Oxford: Oxford University Press.

Church, A. (1984b). Russell's theory of identity of propositions. Philosophia Naturalis, 21:513–522.

Hardy, G. H. (1911). The new symbolic logic. Times Literary Supplement, 504:321–322.

Hazen, A., & Davoren, J. (2000). Russell's 1925 logic. *Australas. J. Philos.*, *78*(4), 534–556.

Kaplan, D. (1986). Opacity. In Hahn, L. E. and Schilpp, P. A., editors, The Philosophy of W. V. Quine, pages 229–289. Open Court, La Salle.

Landini, G. (1996). Will the real Principia Mathematica please stand up? Reflections on the formal logic of the Principia. In R. Monk & A. Palmer (Eds.), *Bertrand Russell and the Origins of Analytical Philosophy* (pp. 287–330). Bristol: Thoemmes.

Landini, G. (1998). *Russell's Hidden Substitutional Theory*. Oxford: Oxford University Press.

Landini, G. (2000). Quantification theory in *9 of Principia Mathematica. *Hist. Philos. Logic*, *21*, 57–78.

Linsky, B. (2013). Russell's theory of descriptions and the idea of logical construction. In M. Beaney (Ed.), *The Oxford Handbook of the History of Analytic Philosophy* (pp. 407–429). Oxford: Oxford University Press.

Mares, E. D. (2007). The fact semantics for ramified type theory and the axiom of reducibility. Notre Dame Journal of Formal Logic, 48(2), 237–251.

Moore, G. H., (Ed.) (2014). *Toward "Principia Mathematica" 1905–08*. The collected papers of Bertrand Russell (Vol. 5). London: Routledge.

Parsons, C. (1971). A plea for substitutional quantification. Journal of Philosophy, 68:231–237.

Pelham, J., & Urquhart, A. (1994). Russellian propositions. In D. Prawitz, B. Skyrms, D. Westerståhl (Eds.), *Logic, methodology and philosophy of science IX* (pp. 307–326). Dordrecht: Elsevier Science B.V.

Russell, B. (1903). *The Principles of Mathematics*. Cambridge: Cambridge University Press.

Russell, B. (1905). On denoting. *Mind*, *14*, 479–493. Reprinted in [22], 415–427.

Russell, B. (1906). Logic in which propositions are not entities. In *[12]* (pp. 262–267).

Russell, B. (1908). Mathematical logic as based on the theory of types. *American Journal of Mathematics*, *30*, 222–262. Reprinted in Russell 2014b, 585–625.

Russell, B. (1910). The theory of logical types. In *[22]* (pp. 3–31).

Russell, B. (1967). *The autobiography of Bertrand Russell Vol.I: 1872–1914*. London: George Allen and Unwin.

Slater, J. G., (Ed.) (1986). *The philosophy of logical atomism and other essays 1914–19*. The collected papers of Bertrand Rusell (Vol. 8). London: Routledge.

Slater, J. G., & Frohmann, B., (Eds.) (1992). *Logical and philosophical papers 1909–13*. The collected papers of Bertrand Russell (Vol. 6). London: Routledge.

Tarski, A. (1956). The concept of truth in formalized languages. *Logic, Semantics, Meta-Mathematics* (pp. 152–278). Oxford: Clarendon Press.

Urquhart, A. (1976). Russell and Gödel. *Bull. Symb. Log.*, *22*, 504–520.

Urquhart, A. (2003). The theory of types. In N. Griffin (Ed.), *The Cambridge companion to Bertrand Russell* (pp. 286–309). Cambridge: Cambridge University Press.

Urquhart, A. (2013). Principia Mathematica: The first 100 years. In N. Griffin & B. Linsky (Eds.), *The Palgrave Centenary Companion to Principia Mathematica* (pp. 3–20). New York: Palgrave.

Urquhart, A., & Lewis, A. C., (Eds.) (1994). *Foundations of logic 1903–05*. The collected papers of Bertrand Russell (Vol. 4). London: Routledge.

van Heijenoort, J. (1967). *From Frege to Gödel: A Source Book in Mathematical Logic, 1879–1931*. Cambridge: Harvard University Press.

Whitehead, A., & Russell, B. (1910). *Principia mathematica* (Vol. I, p. 1925). Cambridge: Cambridge University Press. (Page references to the second edition).

Whitehead, A., & Russell, B. (1912). *Principia mathematica* (Vol. II, p. 1927). Cambridge: Cambridge University Press. (Page references to the second edition).

Whitehead, A. N. (1926). Principia Mathematica. Mind, 35:120.

Chapter 21
Some Lessons Learned About Adding Conditionals to Certain Many-Valued Logics

Allen P. Hazen and Francis Jeffry Pelletier

Second Reader
L. Humberstone
Monash University

Abstract There are good reasons to want logics, including many-valued logics, to have *usable conditionals*, and we have explored this in certain logics. However, it turns out that we "accidentally" chose some favourable logics. In this paper, we look at some of the unfavourable logics and describe where usable conditionals can be added and where it is not possible.

Keywords First-degree entailment · Many-valued logics · Conditionals · Intuitionistic logic · Disjunctions · Substitutivity in many-valued logics

---

Dedication Alasdair Urquhart has dealt with many-valued logics at a number of places in his illustrious career, starting at least as early as his 1971 "Interpretation of many-valued logic" (Urquhart 1971). Possibly the best known of these places is his 2001 "Basic Many-Valued Logic" entry in the 2nd Edition of the *Handbook of Philosophical Logic* (Urquhart 2001). Alasdair's 1973 dissertation was *The Semantics of Entailment* and his supervisors were Nuel D. Belnap and Alan Ross Anderson. We hope that our discussion of the Anderson-Belnap logic FDE brings back some happy thoughts from those early, heady days.

A. P. Hazen · F. J. Pelletier (✉)
Department of Philosophy, University of Alberta, Edmonton, Alberta T6G 2E1, Canada
e-mail: francisp@ualberta.ca

A. P. Hazen
e-mail: aphazen@ualberta.ca

I. Düntsch and E. Mares (eds.), *Alasdair Urquhart on Nonclassical and Algebraic Logic and Complexity of Proofs*, Outstanding Contributions to Logic 22,
https://doi.org/10.1007/978-3-030-71430-7_21

21.1 Introduction

In a series of papers, we have considered adding a "useful conditional" connective to a number of many-valued logics (Hazen and Pelletier 2018a, b, 2019; Sutcliffe et al. 2018). We called this conditional a "classical material implication", or "a cmi conditional" for short (and for even shorter we used $\rightarrow_{\text{cmi}}$ or simply $\rightarrow$). Since we were investigating different many-valued logics, the details of the conditional would differ in the different logics, but the underlying sameness was this: if the antecedent of the conditional took a designated value, then the conditional would take the value of the consequent; and if the antecedent took an undesignated value, then the conditional would take the "highest" or "most true" truth value in the semantics. The 3- and 4-valued logics we investigated were all members of the so-called FDE family (see Belnap 1992, and, for a modern survey, Omori and Wansing 2017). These logics are were all introduced with no conditional operator, other than one defined by employing their $\vee$, $\wedge$, $\neg$ connectives. But in some of the logics, these definitions do not validate modus ponens, while in others conditional introduction can't be carried out. And, therefore, the unaugmented logics do not allow chaining of arguments, conditional proof, and various other logical properties that make conditionals useful—perhaps even indispensable—for any logical applications. Our articles investigated the logical properties, such as their relative strength, completeness, etc., of the logics when augmented with the conditional we advocated. But as we show in the present paper, our choice of logics in which to add "a real conditional" was in a sense quite fortunate, for there are very many other many-valued logics in connection with which a conditional operator like ours seems likely to be less useful. We here canvas the landscape of logical options available for a useful conditional.

21.2 Our Previous Work

In his master's thesis (Tedder 2014), Andrew Tedder considered the problem of axiomatizing the true sentences of some of the inconsistent ("paraconsistent") models of arithmetic described semantically by Priest (1997, 2000, 2006). A major difficulty was that, although typical axioms of mathematical systems contain occurrences of a conditional connective, the appropriate logic for Priest's semantically described arithmetics is his three-valued logic LP ("Logic of Paradox": like Kleene's "strong" three-valued logic (Kleene 1952), but with the middle value counted as designated), and no useful implicational connective is definable in this logic. (One can, of course, introduce a new connective by one of the classical definitions of material implication, $\neg(\phi \wedge \neg\psi)$ or $(\neg\phi \vee \psi)$, but *modus ponens* is not valid for the connectives so defined, and no other defined connective works either.) Tedder's solution to this problem was to show that a new implicational connective, not definable in terms of LP's "native" connectives, having most of the familiar properties of the material implication connective of Classical Logic, can be added conservatively to LP, and then (after giving pleasantly familiar-looking and usable axiomatic and sequent calculus systems for the augmented logic) formulating his axioms in the enriched formal language.[1]

Inspired by his example, we tried to see whether a "material" conditional could be added—conservatively!—to several related logics: Kleene's (strong) three-valued logic, K3 (like LP but with the middle value not counted as designated), the Belnap-Dunn four-valued logic FDE (two "intermediate" values, **B** and **N**, with **B** designated and **N** not, conjunction and disjunction treated lattice-wise, so that $(\mathbf{B} \wedge \mathbf{N})$ is **F** and $(\mathbf{B} \vee \mathbf{N})$ is **T**, and truth tables otherwise mimicking K3 over the values **F**, **N**, **T** and LP over the values **F**, **B**, **T**, and universal and existential quantifiers treated as infinitary conjunctions and disjunctions), and the logic obtained by adding the "mingle" inference, $(\phi \wedge \neg\phi) \vdash (\psi \vee \neg\psi)$, to the Belnap-Dunn logic. (This last might be called a "three and a half valued" logic: it has the four values of FDE, but in any model at most one of the intermediate values gets used.) For each of these logics, we defined a material conditional that in a way generalizes Tedder's definition for LP

- if ϕ has one of the non-designated values, $(\phi \to \psi)$ takes the top value, **T**, and
- if ϕ has a designated value, $(\phi \to \psi)$ has the same value as ψ.

The logics we are now describing are located as in this "kite" diagram, where the strongest logic, CL ("classical logic") is at the top and the lines descending from CL indicate weaker logics, until we get to FDE. The similar diagram holds when we add our $\to_{\mathrm{cmi}}$. (Of course, CL and $\mathrm{CL}^{\to}$ are the same logics.)

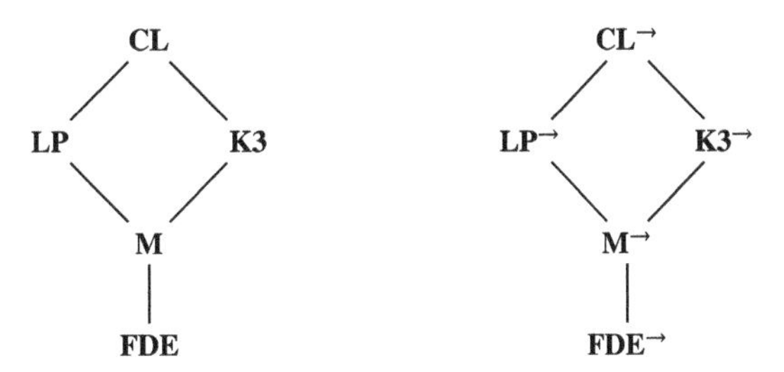

Fig. 21.1 Left: The FDE family of logics. Right: The FDE logics each augmented with $\to_{\mathrm{cmi}}$

Somewhat to our surprise, we obtained quite pleasing results. The conservativity of the extensions is immediate, since the semantic interpretation of formulas not containing the new $\to$ connective were left unchanged. (Thus, the versions of these logics with the added conditional can be used as "auxiliary" logics for, e.g., the problem of axiomatization of theories in one of the conditional-deprived logics.) On the other hand, the sets of valid formulas (and valid inferences) of the pure $\to$ fragments, and indeed of the positive ($\neg$ free) fragments, are exactly those of the corresponding fragments of classical logic. (Thus, the intuitions of the classically

[1] Unfortunately, the version published (Tedder 2015) did not contain the sequent calculus formulation of his Tedder (2014).

trained axiomatizer won't lead to *too* many mistakes.) The enriched logics all turned out to have reassuringly familiar-looking natural deduction systems, easily shown sound and complete: for each logic without a $\rightarrow$ we have the classical rules for $\wedge, \vee$ and the quantifiers (though the "restriction" principle, $\forall x(\phi \vee \psi(x)) \vdash (\phi \vee \forall x\psi(x))$, is no longer derivable and has to be added as a primitive rule), together with double-negation and De Morgan rules (and quantificational analogues of De Morgan rules) for their interactions with negation, and (for the logics stronger than FDE) some special rule (*ex falso quodlibet* for K3, excluded middle for LP, the mingle inference for the logic we call M) for negation. To these we add rules giving $\rightarrow_{\text{cmi}}$ the properties of classical material implication: conditionalization, *modus ponens*, De Morgan-ish rules making $\neg(\phi \rightarrow_{\text{cmi}} \psi)$ equivalent to $(\phi \wedge \neg\psi)$, and something to yield the (classically but not intuitionistically valid) principle $\vdash (\phi \vee (\phi \rightarrow_{\text{cmi}} \psi))$. Crucially (the system wouldn't be nearly as usable otherwise!), the "native" rules and the new implicational rules can each be used inside each others' hypothetical deductions: rules for $\rightarrow_{\text{cmi}}$ can be used in, e.g., the hypothetical subproofs of $\vee$ Elimination, and all the rules for $\forall, \exists, \wedge, \vee$ and $\neg$ can be used in the subproofs for conditionalization.

And, if one prefers, (multiple succedent) sequent calculi in the spirit of Tedder's can also easily be shown sound and complete (and to have *Hauptsätze*).

The systems we defined are analogous, to a greater or lesser degree, to many other logics known in the literature. Perhaps the most interesting comparison is with the logics defined by Nelson (1949, 1959). These papers were concerned with axiomatic theories of arithmetic, but abstracting from the application to arithmetic we can see him as presenting two First Order logics which have since been dubbed N3 (in Nelson 1949) and N4 (in Nelson 1959). N3 combines the positive fragment of Heyting's Intuitionistic logic with a new ("constructible": subsequent writers have often called it "strong negation") negation operator. The new negation's interactions with the Boolean operators are governed by Kleene's (strong) 3-valued logic, K3, and the "Kleene negation" of a conditional asserts the truth of the antecedent and falsity of the consequent: $\neg(\phi \rightarrow \psi)$ is equivalent to $(\phi \wedge \neg\psi)$. The combination is quite a natural one: we all know that Intuitionistic Logic is not a 3-valued logic, but it resembles 3-valued logics in allowing for truth value "gaps." It is thus natural to ask whether intuitionistic ideas can be expressed in a language dealing with truth value gaps in the manner of K3, and Nelson's work showed that the combination works very smoothly.[2] Nelson (1949), gives a variant form of realizability for a system of Intuitionistic arithmetic formulated with the "Kleene" negation instead of the usual, Heyting, negation. With a slight change of perspective, we can think of N3 as the logic produced when we supplement K3 with an Intuitionistic implication connective: it is, so to speak, the "K3 plus Intuitionistic implication" corresponding to our "K3 plus Classical material implication". The analogy to our system extends to details: Nelson (1959) observes phenomena that also appear in our cmi systems, notably the failure of contraposition and the failure of contraction for a defined contrapos-

[2] Nelson was a Ph.D. student of Kleene, and his dissertation was devoted to working out details in Kleene's realizability interpretation of Intuitionistic arithmetic: cf. his Nelson (1947).

able conditional ($\phi \Rightarrow \psi$) defined as $((\phi \rightarrow \psi) \wedge (\neg\psi \rightarrow \neg\phi))$. In Nelson (1959), he similarly grafted an Intuitionistic conditional to the gap-and-glut tolerating logic FDE. (The numerals in the designations N3 and N4 would seem to refer to the numbers of truth values in the related, conditional free, logics: hence the lower number designates the stronger logic, contrary to what a reader familiar with the Lewis systems of modal logic might expect.) Nelson's systems have been extensively treated by later authors: to cite particularly rich recent work, see Spinks and Veroff (2018) and their extensive references. They have studied the close relationships between N4 and a variety of other logics, in particular relevance logics. There is, however, a fundamental difference: the conditionals of Nelson's logics are Intuitionistic, and the systems compared to them by Spinks & Veroff all have non-classical conditionals. (Few, if any, of the logics considered by Spinks & Veroff are many-valued in the sense of being determined by finite matrices.) Our aim, in contrast, was to see how a *classical* implication operator could be added (conservatively!) to logics in which such connectives were not definable. Since, however, some of the systems we considered are proper extensions of systems they consider, some of our results, such as the synonymy of $LP^{\rightarrow}$ and RM3, can be seen as simple corollaries of results of Spinks & Veroff, though our proofs, because of the simplicity of the logics treated, are much simpler than their more general arguments.

21.3 About Generalizing Our Previous Work

We were quite pleased with our results, and—since the logics suggested semantically by various kinds of consideration don't always have convenient and easily used conditional operators—thought that the technique of adding (conservatively) a new (and basically classical) implication operator to other logics might be a useful one. So, we wondered, how well can our results be extended to other logics? Here, alas, we have to report mixed (though not exclusively negative) results: to some degree, we seem to have been lucky in choosing the FDE family of logics to look at first.

Let us begin the report with a description of the general problem and a few positive results. Starting with an arbitrary many-valued logic, what properties are assumed of it in our definition of an auxiliary material implication operator?

- There is a set of truth values, of which a non-empty proper subset is designated.
- We need to use one of the designated values—in the most general case, we might choose one at random—to be our "top" value: the value taken by conditionals with non-designated antecedents.
- Conditionals with designated antecedents (either the "top" value or another if there are multiple designated values) are then defined as taking the value of their consequents.

Given these features, our auxiliary material conditional can be given a semantic definition in any such logic; it remains to be seen whether it will be useful. The restriction to many-valued logics still includes a broad range of interesting logics, much broader, since we allow infinite "matrices", than what are usually thought of as "many-valued" logics. We furthermore allow *sets* of matrices, as we illustrate shortly. The notion of *consequence* we assume—on which a conclusion is implied by a set (and not a multi-set) of premisses, and is implied by a set if it is implied by one of its subsets—might seem to rule out many substructural logics. Even there, however, there can be a role for something like our classical kind of implication. Thus, for example, the primary notion of implication in Relevant logics is substructural (in that {A,B} does not have to imply C if {A} does), but there is also the notion of a conclusion being implied by a conjunction of premisses, which might be captured by a supplementary material conditional added in our fashion to the logic (cf. Anderson and Belnap 1975, pp. 261–262).

The recipe we described for adding our cmi connective to a many-valued logic is semantic: adding a "truth table" for the cmi connective to the matrix characterizing the logic. It applies, however, a bit more widely: we can add cmi to a logic characterized, not by a single matrix, but by a set of matrices, where validity is defined as what holds in all the matrices of the set. (Such logics, then, can be seen as the intersections of a set of many-valued logics.) We treated an example of such a logic (without describing it in those terms) in our earlier papers: the logic M (= FDE $+((\varphi \wedge \neg\varphi) \vdash \neg(\psi \vee \neg\psi)))$ cannot be characterized by a single matrix (as pointed out in Humberstone 2011, p. 214), but is the intersection of two 3-valued logics, K3 and LP. The recipe for dealing with such logics is simply to add an interpretation of $\rightarrow_{\text{cmi}}$ to each of the matrices in the set by the above recipe and then take the set of supplemented matrices as the set of matrices for the supplemented logic.

Unless further properties are assumed of the logic, the obvious notion of a valid rule of inference is that of a rule which never takes us from premisses, all of which have designated values, to a conclusion having a non-designated value. (There is also the notion of a validity preserving rule: one which never leads from premises, each of which is a valid formula, to a conclusion which is not. Any rule which is valid in the obvious sense is also validity preserving, but not conversely: think about the rule of substitution for propositional variables, or the rule of Necessitation in modal logics.) It is immediate that *modus ponens* is valid in this sense: if ϕ has a designated value, $(\phi \rightarrow \psi)$ will have the same value as ψ, so if ϕ and $(\phi \rightarrow \psi)$ both have designated values, ψ must too. (At least as far as axiomatizing the valid formulas of a logic are concerned, then, once we have the auxiliary implication operator, *modus ponens* can be taken as the sole rule of inference: any other rule of inference can be replaced by an implicational axiom.)

We can also, extending this notion of validity to *suppositional* rules in the obvious way, show that the rule of conditionalization ($\rightarrow$ Introduction) is sound. Suppose that ψ is derived from the hypothesis ϕ, together, perhaps, with additional, undischarged, premisses $\chi, \theta, \ldots$, and that in the hypothetical deduction only valid rules of inference are used. Then the inference of $(\phi \rightarrow \psi)$ from the undischarged premisses is valid. If one or more of the undischarged premisses has an undesignated value, there is nothing to prove. Otherwise, suppose ϕ has a designated value: then, from the validity of the rules used in the hypothetical deduction, ψ must also have a designated value, and this is the value of $(\phi \rightarrow \psi)$. If, on the other hand, ϕ has an undesignated value, $(\phi \rightarrow \psi)$ automatically takes the designated "top" value.

Modus ponens and conditionalization by themselves, of course, yield only the intuitionistic logic of $\rightarrow$. Our classical material implication logics were therefore formulated with an additional, classicalizing, rule, which can be formulated without appeal to disjunction: a formula χ may be asserted if it can be derived (validly) both from the hypothesis ϕ and from the hypothesis $(\phi \rightarrow \psi)$. It is easy to see that this rule will be valid: if ϕ has a designated value, the derivability of χ from ϕ guarantees that χ is designated, and if ϕ is not designated, $(\phi \rightarrow \psi)$ will have the top, designated, value, so again the derivability of χ guarantees that its value will be designated. (Alternatively, we could adopt the "rule" that instances of Peirce's Law can be taken as axioms, the validity of Peirce's Law being easily shown. But we think our rule involving hypothetical derivations is more in the spirit of a natural deduction system, and likely to be easier to use in practice.)

These are reassuring results, but notice that nothing has as yet been said about the interaction of the rules for $\rightarrow$ with suppositional rules of the underlying logic. ...

Note also that the underlying logic may have an implicational structure that is not captured by our rather minimal sense of valid rule. For example, relevance logics do not require that all valid formulas (relevantly) imply each other, even though valid formulas take exclusively designated values in their (infinite) matrices. In this case, however, we don't necessarily see a problem: $(\phi \rightarrow \psi)$, where ϕ and ψ are both valid formulas and $\rightarrow$ is our material implication, will always take a designated value (the value of ψ, in fact) and so count as itself valid, but so what? Material implication and relevant implication are different things, as has been emphasized in the literature of relevance logic from the outset. But mismatches between the implicational structure of the underlying logic and that of our auxiliary material implication may in some cases make it less helpful to use the latter.

21.4 Substitution

One nice feature of $\rightarrow$ in classical logic is that

$$(\star) \qquad ((\phi \rightarrow \psi) \wedge (\psi \rightarrow \phi))$$

licenses the intersubstitution of ϕ and ψ in other formulas. Let us call two formulas *substitution equivalent* if each can be substituted for the other in arbitrary contexts: *eadem sunt quae sibi mutuo substitui possunt*.[3] Then, we can say that $(\star)$, in classical logic, expresses the substitution equivalence of ϕ and ψ. This fails in the logics obtained by adding our auxiliary implication to FDE, etc. (If $(\star)$ takes a designated value, ϕ and ψ will either both have designated values or both undesignated values,

[3] Being substitution equivalent is not (necessarily) a logical truth about the formulas: two formulas can be substitution equivalent on one assignment of values to the variables but not on others, so there is an interesting sense in which the truth of one formula can imply the substitution equivalence of two others.

but they can have different designated values in logics with more than one designated value or different undesignated values in logics with more than one undesignated.) This seemed unfortunate: one of the main advantages to having a conditional operator in a logic is that of making it possible to represent arguments otherwise formulable only in a metalanguage by object language deductions. Given the centrality, in the metatheory of various logics, of concepts related to substitution equivalence, failure to express this notion is surely a severe limitation on the utility of the material implication connective.

For the particular logics we considered, however, there was a work-around: substitution equivalence was expressed by a slightly more complex formula,

$$(\star\star) \qquad (((\phi \rightarrow \psi) \wedge (\psi \rightarrow \phi)) \wedge ((\neg\psi \rightarrow \neg\phi) \wedge (\neg\phi \rightarrow \neg\psi)))$$

In each of the logics we considered ($\star\star$) takes a designated value only when ϕ and ψ have the same value. (The key to seeing this is that these logics have at most two designated and at most two undesignated values, so if ϕ and ψ have different ones, one must have a "classical" value (Top or Bottom), and at least one of the four conjuncts of ($\star\star$) will take an undesignated value.) This is obviously not a general solution to the problem—for one thing, it depends on the presence of the $\neg$ operator—but one might hope it applied a bit more widely. Alas, it fails in very simple cases. Consider the logic with a chain of four values—1, $\frac{2}{3}$, $\frac{1}{3}$, 0—with only 1 designated, conjunctions (disjunctions) taking the minimum (maximum) value of their conjuncts (disjuncts), and the value of a negation being found by subtracting the value of the negated formula from 1. (This is, of course, the matrix for the $\wedge, \vee, \neg$ fragment of the four-valued Łukasiewicz logic, $Ł_4$.) If ϕ and ψ have the two intermediate values, $\frac{2}{3}$ and $\frac{1}{3}$, ($\star\star$) will have the value 1 (because the antecedents of all four conjuncts have undesignated values), so ($\star\star$) fails to distinguish the two undesignated values. Worse, no other formula will do: we can show, by induction on the complexity of formulas, that changing the value of one variable from $\frac{1}{3}$ to $\frac{2}{3}$, or vice versa, will never change the value of a formula containing it from a designated one to an undesignated one or vice versa. Thus, no formula in the $\wedge, \vee, \neg, \rightarrow$ language can have the desired property of taking a designated value when and only when a specified pair of its subformulas have the same value. This unfortunate result extends easily to all "larger" (more truth values) Łukasiewicz logics.

There is a further disappointing corollary to this. When our classical material implication is added to the $\wedge, \vee, \neg$ fragment of Łukasiewicz's three-valued logic $Ł_3$, Łukasiewicz's implication operator is definable (by $(\phi \rightarrow \psi) \wedge (\neg\psi \rightarrow \neg\phi)$). However, substitution equivalence is expressible in all Łukasiewicz logics (by ($\star$), now interpreting the $\rightarrow$ as Łukasiewicz implication), so it follows from the above result that Łukasiewicz implication is not definable from our classical material implication in any Łukasiewcz logic with four or more values. Classical material implication is definable in terms of Łukasiewicz implication in all finitely valued Łukasiewicz logics [by the nested implication

$$(\phi \rightarrow (\phi \rightarrow (\phi \rightarrow \ldots (\phi \rightarrow \psi) \ldots))$$

with $(n - 1)$ ϕ antecedents in n-valued Łukasiewicz logic]; so it follows that, for $n > 3$, the n-valued Łukasiewicz logic is properly more expressive than the logic obtained by adding our classical material implication, $\rightarrow_{\text{cmi}}$, to its $\wedge$, $\vee$, $\neg$ fragment.

21.5 Defined Versus Native Disjunctions

In Classical Logic, it is possible to define disjunction in terms of the material implication connective, as $((\phi \rightarrow \psi) \rightarrow \psi)$. (The same definiens, with the $\rightarrow$ interpreted as Łukasiewicz implication, works in all Łukasiewicz logics, but there are other logics—intuitionistic logic, for example—with nice implication connectives in which it gives something that does not have the properties of a disjunction operator: see Humberstone 2011, pp. 1320 and 1068). As a consequence of the classical behaviour of our cmi connective, the connective defined in terms of it has disjunction-like properties: $((\phi \rightarrow \psi) \rightarrow \psi)$ follows from each of ϕ and ψ, and anything derivable both from ϕ and from ψ is derivable from it. Let us call it *cmi disjunction*, and symbolize it as $\ddot{\vee}$ (as in Humberstone 2011, p. 555[4]). On the other hand, the systems of truth values of many interesting logics have a lattice structure allowing the (semantic) definition of (conjunction and) disjunction connectives: the value of a disjunction is the lattice join of the values of the disjuncts. Call these connectives the *native disjunctions* of the logics. So the question arises: how do the native and cmi disjunctions relate to each other in various logics? In the logics we initially studied—those obtained by adding a cmi connective to LP, K3, M, or FDE—$(\phi \ddot{\vee} \psi)$ is equivalent to $(\phi \vee \psi)$ (with the native disjunction)...but only in the sense of $(\star)$: one will have a designated value if and only if the other does. They are *not* equivalent in the stronger sense of $(\star\star)$. Thus, in K3, let ϕ have the middle value and ψ the bottom value. Since the middle value is not designated, $(\phi \rightarrow_{\text{cmi}} \psi)$ has the top value, and so the cmi disjunction defined as $((\phi \rightarrow_{\text{cmi}} \psi) \rightarrow_{\text{cmi}} \psi)$ has the bottom value; but the native disjunction takes the middle value: thus the two "disjunctions" can take different undesignated values. Similarly, in LP have ϕ take the top value and ψ the middle value. Since ϕ is designated, $(\phi \rightarrow_{\text{cmi}} \psi)$ has the same, middle, value as ψ, and since the middle value is designated in LP, $((\phi \rightarrow_{\text{cmi}} \psi) \rightarrow_{\text{cmi}} \psi)$ will also have the middle value, but the native disjunction takes the top value: the two "disjunctions" can take different designated values. And, of course, in FDE both things can happen.

This turns out to be a widespread phenomenon: native and cmi disjunctions fail to be equivalent in a wide variety of interesting and well-motivated logics. For example, *Heyting's intuitionistic logic*. This is easily seen by considering a five-valued matrix which satisfies this logic (and so characterizes an extension of it): an inverted kite-shaped matrix, with the designated value at the top and a diamond of undesignated values below it.

[4] Abbreviating it as $\vee_{cmi}$ would be in keeping with our $\rightarrow_{cmi}$, but the $\ddot{\vee}$ is established in the literature.

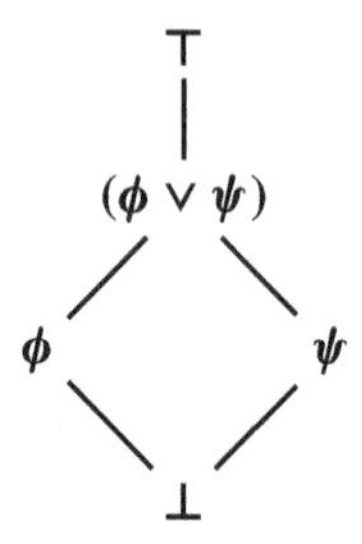

This is, of course, the lattice of "propositions" in a three-world Kripke model in which neither of the possible "futures" comes after the other.

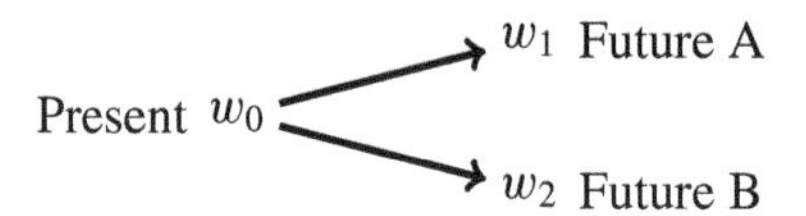

Let ϕ and ψ have the two side-by-side values at the next-to-bottom level. Their native disjunction will have the lattice join of these values, here, the highest of the undesignated values, but since ϕ has an undesignated value, $\phi \to \psi$ will take the designated, top, value, and the cmi disjunction, therefore, the same value as ψ: the native and cmi disjunctions take different (though in both cases undesignated) values.

Things are even worse with some other logics. For example, *orthomodular quantum logic*. This is easily seen by considering a six-valued matrix which satisfies this logic (and so characterizes an extension of it): a top value (the only designated one), a bottom value, and a four-element antichain between them, with conjunction and (native) disjunction interpreted as lattice meet and join (and—not that we need this to make the current point—orthonegation pairing the top and bottom values and also pairing two disjoint pairs of the intermediate values).

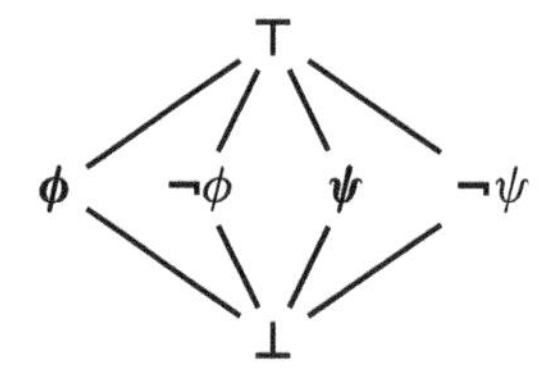

Let ϕ and ψ have two of the intermediate values. Their native disjunction will have the top value, but since ϕ has an undesignated value, $(\phi \to \psi)$ has the top value, and $(\phi \to \psi) \to \psi$ therefore has the same value as ψ. Since in this case the native disjunction has the designated value and the cmi disjunction one of the undesignated ones, the two disjunctions in this logic aren't even ($\star$) equivalent! The inference from the native disjunction to the cmi disjunction wouldn't, in the sense of validity we have been looking at, be valid.

One might hope that this, more serious, kind of inequivalence between the two disjunctions would only arise in seriously weird (i.e., non-distributive) logics, but not so: it arises even in an almost-classical context. Consider, as a system of values, a Boolean Algebra with more than two elements, with only the top one counted as designated. (This kind of structure was suggested by Carnap 1943, as giving a non-standard interpretation of classical logic, one validating all the classical logical truths and rules of inference, but in which, a true disjunction may not have a true disjunct. Similar structures arise in connection with the notion of *supervaluations* (van Fraassen 1966, 1969). Let ϕ have some undesignated value other than the bottom, and let the value of ψ be its Boolean complement. Their native disjunction will have the top, designated, value, but their cmi disjunction will have the same undesignated value as ψ.

21.6 More Disheartening Difficulties

This is all very disheartening, all the more so since the underlying logics (intuitionistic, quantum, and "Carnap-van Fraassen classical") all have very nice natural deduction systems. (Natural deduction for intuitionistic and classical logics go back to Gentzen and Jaśkowski. The possibility of a natural deduction formulation of orthomodular logic was noted by Dummett (1978, p. xiv), by one of the present authors in unpublished work from the late 1970s, and has been presented in a textbook (Gibbins 1987). It involves restricting the use of auxiliary premises in the hypothetical deductions of rules like Disjunction Elimination.) Naïvely combining the rules for the native disjunction with the rules for cmi, however, allows the derivation of clearly invalid results. In either the orthomodular example or the Boolean algebraic example, the native disjunction $(\phi \vee \psi)$ and the cmi conditional $(\phi \rightarrow \psi)$ will both have the top value. (Since ϕ is not designated, $(\phi \rightarrow_{\text{cmi}} \psi)$ will be; and the value of $(\phi \vee \psi)$ will be the join of ϕ and ψ, namely the designated $\top$.) But in the combined system we would have the derivation

1	$\phi \vee \psi$	Premise
2	$\phi \rightarrow \psi$	Premise
3	$\quad \phi$	Hypothesis
4	$\quad \phi \rightarrow \psi$	2, Reiterate
5	$\quad \psi$	3,4 modus ponens
6	$\quad \psi$	Hypothesis
7	$\quad \psi$	6, Repeat
8	ψ	1, 3–5, 6–7, $\vee$-elimination

We have derived ψ, whose value is undesignated, from premises which both have the top value!

We get, similarly, disastrous results when we merge our cmi rules with intuitionistic logic. The conjunction $(\phi \wedge \psi)$ takes the bottom value $\bot$ in our inverted kite model (equivalently, it isn't true at any of the worlds in the corresponding Kripke model), so $\neg(\phi \wedge \psi)$ takes the top value $\top$ (is true at the "present" world): like $(\phi \rightarrow \psi)$, it has a designated value. But then we have

1	$\neg(\phi \wedge \psi)$	Premise
2	$\phi \rightarrow \psi$	Premise
3	$\quad \phi$	Hypothesis
4	$\quad \phi \rightarrow \psi$	2, Reiterate
5	$\quad \psi$	3, 4 modus ponens
6	$\quad (\phi \wedge \psi)$	3, 5 $\wedge$-introduction
7	$\quad \neg(\phi \wedge \psi)$	1, Reiterate
8	$\quad (\phi \wedge \psi) \wedge \neg(\phi \wedge \psi)$	6, 7 $\wedge$-introduction
9	$\neg\phi$	3–8 $\neg$-introduction

Again, both premisses have the top value, but the conclusion has an undesignated value (it has the same value as ψ).

What has gone wrong? The relationship between our Boolean-valued logic and supervaluations provides a clue: it is a familiar part of the lore of supervaluations that, though they satisfy all the classically valid formulas and even the classical relation of logical consequence, they don't allow unrestricted use of the classical natural deduction rules. (Thus, for example, in a supervaluational treatment of the Liar, $(\mathcal{L} \vee \neg\mathcal{L})$ (where $\mathcal{L}$ is the Liar sentence) is true, as an instance of the Law of Excluded Middle; but we can't derive a contradiction from it because we can't use the rule of $\vee$ Elimination.) Our problematic derivations both involve hypothetical derivations (for $\vee$ Elimination in the first, for $\neg$ Introduction in the second) with *modus ponens* for the cmi $\rightarrow$ employed in them.

In the logics of the examples considered in this and the previous section, rules for the "native" connectives (native disjunction, intuitionistic negation) are not only valid in our sense of never taking designated premisses to an undesignated conclusion, but they also have the property—let's call it L-validity—of never yielding a conclusion with a value properly lower than that of the (lattice meet of the) premiss(es). (On the natural assumption that the designated values form a filter on the lattice of values, every L-valid rule is also valid in our sense, but the converse doesn't always hold.) We can only expect good results from the hypothesis-discharging rules of the underlying logic if all the rules used in the hypothetical derivations are L-valid. Modus ponens for cmi, alas, is valid but not L-valid.

Summing up: we are led to the disappointing conclusion that adding a cmi connective to some logic is likely to be more misleading than helpful if we are interested in L-validity.

21.7 Fallback: Strict Implication

If we don't insist on having exactly a classical material conditional, but are willing to make use of something more like *strict* implication, the technique of adding a new conditional to a logic may still occasionally be useful even when we are interested in L-validity. For example, full orthomodular quantum logic has a "native" conditional, the so-called *Sasaki Hook*, but an interesting subsystem, Weak Orthologic, does not. It is, therefore, impossible to give a Hilbert-style axiomatic formulation of it: hence the use, in Goldblatt (1974), of something more in the spirit of a sequent calculus. We can, however, enrich the logic with a strict conditional, $\Rightarrow$, not definable in terms of the other operators of Weak Orthologic, interpreting it in any ortholattice by the condition that $(\phi \Rightarrow \psi)$ takes the top value of the lattice if and only if ϕ takes a value equal to or below the value of ψ, and takes the bottom value of the lattice otherwise. It is then simple to turn Goldblatt's system into something looking more Hilbert-stylish.

Acknowledgements We gratefully acknowledge the very helpful (and learned!) comments of the second reader.

References

Anderson, A., & Belnap, N. (1975). *Entailment: The Logic of Relevance and Necessity* (Vol. I). Princeton, NJ: Princeton UP.

Belnap, N. (1992). A useful four-valued logic: How a computer should think. In A. Anderson, N. Belnap, & J. Dunn, (Eds.), *Entailment: The logic of relevance and necessity, Volume II* (pp. 506–541). Princeton UP, Princeton. First appeared as "A Useful Four-valued Logic" *Modern uses of multiple-valued logic* J.M. Dunn & G. Epstein (Eds.) (pp. 3–37); Dordrecht: D. Reidel, 1977; and "How a Computer Should Think" *Contemporary aspects of philosophy* G. Ryle (Ed.) (pp. 30–56); Oriel Press, 1977.

Carnap, R. (1943). *Formalization of Logic*. Cambridge, MA: Harvard University Press.

Dummett, M. (1978). *Truth and Other Enigmas*. Cambridge, MA: Harvard University Press.

Gibbins, P. (1987). *Particles and Paradoxes: The Limits of Quantum Logic*. Cambridge, UK: Cambridge UP.

Goldblatt, R. (1974). Semantic analysis of orthologic. *Journal of Philosophical Logic*, *3*, 19–35.

Hazen, A. P. and Pelletier, F. J. (2018a). Pecularities of some three- and four-valued second order logics. Logica Universalis, 12:493–509.

Hazen, A. P. and Pelletier, F. J. (2018b). Second-order logic of paradox. Notre Dame Journal of Formal Logic, 59:547–558.

Hazen, A. P. and Pelletier, F. J. (2019). K3, Ł3, RM3, A3, FDE, M: How to make many-valued logics work for you. In Omori, H. and Wansing, H., editors, New Essays on Belnap-Dunn Logic, pages 201–235. Springer, Berlin.
Humberstone, L. (2011). *The Connectives*. Cambridge, MA: MIT Press.
Kleene, S. (1952). *Introduction to Metamathematics*. Amsterdam: North-Holland.
Nelson, D. (1947). Recursive functions and intuitionistic number theory. *Transactions of the American Mathematical Society, 61*, 307–368. See Errata, ibid., p. 556.
Nelson, D. (1949). Constructible falsity. Journal of Symbolic Logic, 14:16–26.
Nelson, D. (1959). Negation and separation of concepts in constructive systems. In A. Heyting (Ed.), *Constructivity in mathematics: Proceedings of the colloquium held at Amsterdam, 1957* (pp. 208–225). Amsterdam: North Holland.
Omori, H. and Wansing, H. (2017). 40 years of FDE: An introductory overview. Studia Logica, 105:1021–1049.
Priest, G. (1997). Inconsistent models of arithmetic, I: Finite models. *Journal of Philosophical Logic*, 223–235.
Priest, G. (2000). Inconsistent models of arithmetic, II: The general case. Journal of Symbolic Logic, 65:1519–1529.
Priest, G. (2006). *In Contradiction: A Study of the Transconsistent* (2nd ed.). Oxford: Oxford University Press.
Spinks, M., & Veroff, R. (2018). Paraconsistent constructive logic with strong negation as a contraction-free relevant logic. In J. Czelakowski (Ed.), *Don Pigozzi on Abstract Algebraic Logic, Universal Algebra, and Computer Science* (pp. 323–379). Berlin: Springer.
Sutcliffe, G., Pelletier, F. J., & Hazen, A. P. (2018). Making Belnap's 'useful 4-valued logic' useful. In *Proceedings of the thirty-first international Florida artificial intelligence research society conference (FLAIRS-31)*. Association for the advancement of artificial intelligence.
Tedder, A. (2014). Paraconsistent logic for dialethic arithmetics. Master's thesis, University of Alberta, Philosophy Department, Edmonton, Alberta, Canada. Available at https://www.library.ualberta.ca/catalog/6796277.
Tedder, A. (2015). Axioms for finite collapse models of arithmetic. Review of Symbolic Logic, 8:529–539.
Urquhart, A. (1971). An interpretation of many-valued logic. Zeitschrift für mathematische Logik und Grundlagen der Mathematik, 19:111–114.
Urquhart, A. (2001). Basic many-valued logic. In F. Guenthner & D. Gabbay (Eds.), *Handbook of Philosophical Logic* (2nd ed., Vol. 2, pp. 249–294). Dordrecht: Kluwer.
van Fraassen, B. (1966). Singular terms, truth-value gaps, and free logic. Journal of Philosophy, 63:481–495.
van Fraassen, B. (1969). Presuppositions, supervaluations and free logic. In K. Lambert (Ed.), *The Logical Way of Doing Things* (pp. 67–92). New Haven: Yale UP.

Chapter 22
Comments on the Contributions

Alasdair Urquhart

Acknowledgment and thanks. I am deeply grateful to everybody who has contributed to the volume, and wish to express my heartfelt thanks to my old friends and colleagues Ivo Düntsch and Ed Mares who have worked so hard to produce a volume in my honour.

I've done research in a lot of areas in logic, and the selection of authors provides a good cross-section of my preoccupations. I am particularly pleased that some of the authors have taken the opportunity to publish new results, often extending my own ideas, or solving open problems that I have posed. Thank you to all my friends and colleagues in logic and computer science who have contributed!

David Makinson: Relevance-sensitive truth trees. David Makinson gives us an ingenious suggestion for modifying the classical notion of truth trees (also known as "semantic tableaux") to produce a formulation of logic that respects relevance. The basic idea is as follows. Given an application of the negated implication rule

$$\neg(\alpha \to \beta) \vdash \alpha, \neg\beta$$

on a branch of the tree, the nodes labelled with α and $\neg\beta$ are dubbed a *critical pair*. Makinson's restriction on the trees is this: if a branch of the tree terminates with a pair $\zeta, \neg\zeta$ (dubbed a "crash-pair"), then provided one of a critical pair is in the trace of ζ (equivalently, the trace of $\neg\zeta$), the other is as well. (Here the "trace" of a node is the cumulative record of the nodes from which it was derived.) The formulas α that can be proved by truth trees with $\neg\alpha$ at the root satisfying this restriction, Makinson calls "directly acceptable."

A. Urquhart (✉)
Department of Philosophy, University of Toronto, Toronto, Canada
e-mail: urquhart@cs.toronto.edu

I. Düntsch and E. Mares (eds.), *Alasdair Urquhart on Nonclassical and Algebraic Logic and Complexity of Proofs*, Outstanding Contributions to Logic 22,
https://doi.org/10.1007/978-3-030-71430-7_22

This restriction rules out things like *ex falso quodlibet*, $(\alpha \wedge \neg\alpha) \rightarrow \beta$, also known as *explosion*, as well as other "baddies" such as disjunctive syllogism. What is more, all of the axioms of **R** are directly acceptable, so the suggestion seems to be along the right lines. On the other hand, it is rather difficult to form a clear picture of the set of directly acceptable formulas. As Makinson explains in his Sect. 1.7.1, the notion of direct acceptability has several drawbacks. It is not closed under *modus ponens*, so there are theorems of **R** that are not directly acceptable.

Makinson corrects these drawbacks, expanding the notion of acceptability by closing the directly acceptable formulas iteratively under that rule and adjunction. Unfortunately, he is unable to show that there are classical tautologies that are not acceptable, leaving this question as an open problem. Makinson aims "to articulate a clear rationale for relevance-sensitive propositional logic." However, given the open problem above, it seems that more work is needed before we can conclude that acceptability is a fully satisfactory replacement for the Anderson-Belnap approach.

It is curious what strong feelings are aroused by relevance logic and entailment. In spite of Quine's persistent hostility, modal logic now seems to be an accepted part of the logical landscape. However, relevance logic still bears the brunt of fairly frequent attacks. Both modal logic and relevance logic, though, both owe their origins to dissatisfaction with material implication. Lewis (1913) thought it absurd that a false proposition implied an arbitrary proposition, and that a true proposition was implied by any proposition, and this led him to his formulations of modal logic.

Ackermann (1956) formulated a narrower concept of strict implication, and formulated a calculus in which, for example, *ex falso quodlibet* is no longer a theorem. Anderson and Belnap's favourite system **E** is equivalent to Ackermann's. They explain the motivation for the system as follows. They wish to avoid the fallacies of modality and relevance, but on the other hand, they aim to adhere to Quine's maxim (Quine 1986, p. 7) of "minimum mutilation" (Anderson et al. 1992, p. 507). Thus, in avoiding the fallacies, they want to preserve as much of traditional classical logic as possible.

The two desiderata listed above certainly don't lead to an unambiguous solution, but they seem quite clear as a guide to constructing a system free of the fallacies of modality and relevance. Jean-Yves Girard subjects their outlook to blistering attacks in his monograph (Girard 2011, p. 203). Quite a few of the features of Girard's linear logic, however, are surprisingly prefigured in the relevance logic tradition, in spite of the considerable differences in motivation. Girard grudgingly admits these facts, though in the context of further insults (Girard 2011, pp. 184–185). Perhaps this is "the rage of Caliban at seeing his own face in the glass" (Wilde 1891, Preface).

I have always felt that the system defined by the semilattice semantics is the most natural extension of Church's theory of weak implication (Church 1951) to include conjunction and disjunction. In the systems **E** and **R**, the distribution law (required by the minimum mutilation maxim) does not follow from the introduction and elimination rules and must be put in "by hand"—this seems a serious flaw. Girard is not worried by this—he puts cut elimination in a central place in his philosophy of logic and so he simply drops the distribution rule. However, as Makinson mentions in his Appendix A.2, Anderson and Belnap (1975, p. 348) briefly consider a modification to the disjunction rules, that justify distribution directly. The rule is: if we have proved the formula $(\alpha \vee \beta)_a$, then we can split the derivation into two branches, one beginning with α_a, the other with β_a.

This last rule corresponds exactly with the semilattice semantics, as I observed in (Urquhart 1989). Dag Prawitz (1965, Chap. VII) postulated an equivalent form of the rule. The equivalence of the two versions was proved by Charlwood in his doctoral thesis (Charlwood 1978). This rule seems very natural, though it does not mesh well with the quasi-classical treatment of negation from the Anderson-Belnap tradition. However, if you interpret the elements of a semilattice model as pieces of information, the constructive reading of relevant implication fits well.

Roger D. Maddux: Tarskian classical relevant logic. As I mentioned in my autobiographical notes, while an undergraduate I was very fond of reading old bound volumes of the *Journal of Symbolic Logic*. It was in one of these that I discovered the enchanting article (Tarski 1941) by Alfred Tarski on the calculus of relations. I was familiar with the calculus of relations as it appears in *Principia Mathematica*. However, Whitehead and Russell adopt a rather cool attitude towards the Schröder calculus, even though there are substantial parts of their *magnum opus* (for example $*23$ and $*25$) that are almost unadulterated relation algebra.

Tarski's finely written and seductive article aroused my interest in the calculus of relations. Tarski concludes his exposition with the remark: "The calculus of relations has an intrinsic charm and beauty which makes it a source of intellectual delight to all who become acquainted with it" (Tarski 1941, p. 89). Tarski lists some open problems at the end of his article that intrigued me. One of them was what may be called the "condensation problem" for the first-order calculus of relations.

The condensation problem can be explained as follows. Let $L(R)$ be the first-order language with identity containing one relation symbol Rxy. Some formulas of $L(R)$ are equivalent to equations in the Schröder calculus; for example, transitivity of R can be expressed as $R; R \subseteq R$ (a formulation frequently employed by Whitehead and Russell). Similarly, density of R is expressible as $R \subseteq R; R$, a formulation employed in Jacob Garber's contribution. However, Alwin Korselt proved that there are formulas of $L(R)$ that are not expressible in this way. Tarski asks whether there is a decision method to decide whether this holds for a given formula. This question fascinated me as an undergraduate, though I never worked seriously on it. On one of the first occasions when I met Roger Maddux at a conference, I asked him about this problem; he told me that, in fact, it had been solved by a student of Tarski.

It was only after I discovered the construction of models for the logic **KR** and the resulting undecidability results for a family of relevant logics that I realized the closeness of the connection between the mathematics of these logics and that of relation algebras. The construction of (Urquhart 1984b) that produces models for **KR** from projective spaces is essentially identical with an old method of Lyndon (1961) that constructs non-representable relation algebras from geometries.

Maddux's contribution to this volume provides a deep investigation of the relation between the two fields. Among the highlights are: a penetrating discussion of the relation between the semantics of relevant logics and logics with only four variables,

and a sequent calculus for these logics. Maddux focuses on the Tarskian classical relevant logic **TR** by interpreting Meyer and Routley's classical relevant logic **CR***
in the language of relation algebras. The result is a logic stronger than **CR*** that Maddux shows not to be finitely axiomatizable.

Maddux's chapter contains a rich collection of results, including theorems characterizing **TR** and **KR** in terms of logic in four variables. He concludes with a remarkable section giving both exact numbers and asymptotic formulas for the number of **TR**-frames and and **KR**-frames on a finite universe. As Roger mentions in his Sect. 2.21, I was impressed by the number and variety of finite models that can be constructed from finite projective spaces—these are the models that I used in (Urquhart 1984b) to show that a large variety of logics determined by finite frames are not axiomatizable.

However, the models derived from projective spaces are only a *very small* selection from the set of finite frames. The amazing fact, proved here, is that we can construct such frames by a random process; if we choose triples from a finite universe $\{1, \ldots, n\}$ at random, there is a very high probability (approaching 1 asymptotically) that the result is a frame of the required kind! This leads to a lower bound of the form $2^{\Omega(n^3)}$ for the number of isomorphism types of such frames, a truly astounding result!

Willem Conradie and Valentin Goranko: Algorithmic Correspondence for Relevance Logics I. The Algorithm PEARL. In a paper)Urquhart 1996) published in 1996, I remarked that

> Correspondence theory in the case of modal and intuitionistic logic has been extensively studied, but the analogous theory for the case of relevant logics is surprisingly neglected.

I gave a fairly general result in that paper from which many of the better known correspondence results for relevant logics follow, but observed that a good deal remained to be done.

Conradie and Goranko have gone a considerable distance in fulfilling the challenge in my article. They provide a calculus *PEARL* that computes first-order equivalents for many formulas of relevant logic. They define the class of inductive formulas, extending the family of Sahlqvist formulas defined by earlier researchers, and show that *PEARL* works correctly for all of this class. This fine result goes a considerable distance in fulfilling the problem that I sketched in 1996.

Jacob Garber: Beth Definability in the logic KR. In my paper, written in memory of Helena Rasiowa (Urquhart 1999), I expounded a very simple proof that Beth's theorem fails in the logics between $\mathbf{B} + A22$ and **R**, where $A22$ is the transitivity axiom

$$[(A \to B) \wedge (B \to C)] \to (A \to C).$$

Blok and Hoogland later improved this result, replacing $\mathbf{B} + A22$ with the basic logic **B**. My little piece was inspired by Ralph Freese's brilliant paper (Freese 1979), whose main result is the theorem that the variety of modular lattices is not generated by its finite members, a precursor to his great article (Freese 1980) showing that the word problem for the free modular lattices on five generators is unsolvable. In addition to his main theorem, Freese proves a supplementary result Freese (Freese 1979, Theorem 3.3) to the effect that epimorphisms are not necessarily surjective in the variety of modular lattices. The proof of this theorem relies on the fact that relative complements in distributive lattices are unique.

Since relevant logics eschew classical negation, I realized that this last observation could be used directly to refute the Beth definability property for many such systems. All that was needed was to find a De Morgan monoid containing an element that has a relative complement that is not explicitly definable. A quick search in the monograph (Thistlewaite et al. 1988) revealed that the well known CRYSTAL lattice has this property, and my paper was basically complete.

This easy proof does not work in logics such as **KR** that contain classical negation. However, I conjectured in (Urquhart 1999) that it might be possible to adapt Freese's result on epimorphisms to extend the failure of the Beth definability property to such logics, and repeated the conjecture in my article on the geometry of relevant implication (Urquhart 2017). I worked on this idea for a little while and corresponded with Roger Maddux in 2018 about it, as I hoped it might serve for a talk in Toruń. [1] However, I didn't make much headway, so I talked about something else in Toruń.

I am delighted that Jacob Garber has managed to carry out my plan of 1999, based in part on results of Steve Givant in the area of relation algebras. I congratulate him on his excellent work!

Greg Restall: Geometric Models for Relevant Logics. Greg Restall gives an illuminating survey of the semilattice semantics, and its generalization to the ternary Routley–Meyer model theory. He then introduces the elegant theory of collection frames that provide a common generalization of both approaches. Restall shows that not only does this theory generalize the Routley–Meyer approach, but in appropriate spaces, the compositional relation takes on natural geometric meanings, providing insight into the models based on affine and projective spaces.

Shawn Standefer: Revisiting Semilattice Semantics. I am grateful to Shawn Standefer for providing an excellent and very scholarly history and survey of the semilattice semantics and its later evolution. I was always attracted by the simplicity and elegance of the semantics, and found the ternary relation models of Routley and Meyer much less intuitive—I felt I couldn't form a picture of them. Of course, I changed my mind completely after discovering the models based on projective spaces where you can quite literally form pictures of them.

It's good to hear that people are still working on the semilattice system—I learned some new results from Standefer's contribution. The theory of set frames provides a common framework to discuss both ternary relational models and the semilattice models, which turn out to be equivalent to functional set models.

Shay Allen Logan: The Universal Theory Tool Building Toolkit is Substructural. Shay Logan's contribution arouses feelings of nostalgia in me, as it reminds me of my 25-year-old self as I worked out the semantical analysis of relevant logics. As

[1] I was saddened to hear from him about Steve Givant's passing, a splendid mathematician and a fine person as I remarked to Roger.

a graduate student, I learned the basics of the natural deduction systems in courses by my teachers Alan Anderson and Nuel Belnap. From these, it is only a short jump to the semilattice semantics—all you have to do is interpret a subscripted formula A_x as a statement that the formula A is true at the semilattice element x, and the truth condition for implications emerges immediately. I interpreted the elements of the semilattice as "pieces of information."

Of course, Anderson and Belnap wanted to include modality as well, as the subtitle of their famous monograph (Anderson and Belnap 1975) indicates. I was acquainted with semantic tableaux, and had studied Kripke's completeness proofs for modal logics. So, I interpreted the subproofs in the Fitch-style natural deduction system for **E** (Anderson and Belnap's favourite system) as possible worlds in the style of Kripke. This gives my semantical analysis of the implicational connective in **E**, where the possible worlds play the role of "background information." I felt that this produced an intuitively satisfying picture of these logics—or at least the implication-conjunction fragments of them.

In thinking of the semilattice elements as pieces of information, I was inspired by Kripke's semantics for intuitionistic logic. In this picture, pieces of information can be both inconsistent and incomplete. Kit Fine, on the other hand, considered the points of his models to be theories. His closure operation tu on theories t and u is given by the same definition as Logan employs in Sect. 8.6 of his contribution. As Logan points out, this operation does not satisfy the usual properties of a closure operation. In fact, its properties depend on what implications we assume for the theory t, so there is a kind of circularity involved here. The interaction between the closure operation and theorems of a given logic is explained both by Fine, and by Routley and Meyer (1973) in their calculus of intensional theories.

Anderson and Belnap promoted **E** as their favourite logic. Bob Meyer's affections, on the other hand, centered on the logic **R**. Richard Routley, however, was fond of very weak systems, and even accused Bob and his students of driving around in a "dilapidated vehicle." If pressed, I would have to admit that I am at heart a classical logician, but I enjoy investigating alternative logics, and to some extent take a pragmatic view of the matter. The later sections of Logan's dialogue seem to take a similarly pragmatic view of logic, so as a "traditional logician," I am satisfied.

Marcus Kracht: More on the power of a constant. Marcus Kracht provides an elegant and concise proof of the surprising fact that the addition of a single constant to a monomodal logic can increase the number of Post-complete extensions from two to continuum many.

While I am on the subject of propositional constants, I might mention that I am not in agreement with David Makinson in Appendix A.1 of his contribution, where he says that the constant t is "not in the spirit of relevance logic." If we interpret t as the conjunction of all logical truths (Anderson and Belnap 1975, p. 342), then this seems to me perfectly natural, and not at all in conflict with the motivation of Anderson and Belnap.

Philip Kremer: Strong completeness of S4 for the real line. As Philip Kremer mentions at the beginning of his contribution, my course on relevance logic was the first logic course in his undergraduate career, so that he learned entailment and relevance logic before he encountered the classical systems! This would surely have pleased Bob Meyer, for whom **R** was always the One True Logic. Phil and I are in fact doctoral brothers, since we share the same *Doktorvater*, Nuel Belnap.

Some of Kremer's early work was in the area of relevance logic, particularly versions of the systems with propositional quantifiers. He proved various results showing that when you add propositional quantifiers, then you produce systems of monstrous complexity.

Lately, he has become a major figure in the area of topological models for modal logics. Early work in the mathematics of modal logic, such as that of McKinsey and Tarski described in Kremer's essay, used topological spaces as the foundation. However, the great success of Kripke's model theory in the 1950s and 1960s tended to put this earlier research in the shade. Lately, there has been a revival of this stream of investigation, in which logicians from Georgia have played a starring part. Kremer's contribution provides a very elegant strong completeness proof for **S4** for the real line.

Robert Goldblatt: Modal Logics of Some Hereditarily Irresolvable Spaces. I am grateful to Rob Goldblatt for his generous assessment of my contributions to various areas in non-classical logics. There is one item here where I have to disclaim much in the way of originality. My paper (Urquhart 2015) solves a problem that I first encountered in a talk by Max Cresswell at the meeting of the *Society for Exact Philosophy* at McMaster University in May 2015. It was only after it was published that I discovered that its main result had already appeared as part of Theorem 9 in Kit Fine's brilliant paper (Fine 1978) of 1978.

As a graduate student in Pittsburgh, I was captivated by the way in which properties of the accessibility relation in Kripke models were reflected (and sometimes *not* reflected) in formulas of modal logic. In the same way, we can find properties of topological spaces reflected in modal axioms in the older approach of McKinsey and Tarski. Goldblatt shows with his usual elegance the topological properties that characterize the logics $K4\mathbb{C}_n$, where the possibility modality $\Diamond$ is interpreted as the derived set (of limit points) operation, as well as extensions of these logics and the extension of K4 by the McKinsey axiom.

Katalin Bimbó and J. Michael Dunn: St. Alasdair on lattices everywhere. I am grateful to Katalin Bimbó and J. Michael Dunn for providing an excellent overview of my work as it relates to lattice theory. My early efforts in the area were all variations on the theme of distributive lattices and their representation theory, in particular, the work of Hilary Priestley. This was, in turn, an outgrowth of my fascination with the model theory of non-classical logic, particularly the ideas of Saul Kripke.

Whenever I encountered a problem in lattice theory, I would immediately try to translate it into a dual form, since I was impressed by how apparently complicated questions could be made transparent by this process. My representation theory for general lattices fits in to this framework, since I was trying to generalize Priestley's work. I was disappointed with the results, for several reasons. First, I failed to solve the two famous problems that inspired me, as I explain in my autobiographical essay. Second, the representation theory does not include a satisfactory duality for homomorphisms.

Nevertheless, I am very happy that later authors have been inspired by my efforts, even though I gave up on my own work. Bimbó and Dunn provide a splendid summary of this later research.

Readers who are not familiar with the Logicians Liberation League may be puzzled by the title of this contribution. The League was founded by Bob Meyer (aka the Maximum Leader) in the fall of 1969, who read out a manifesto at the close of a philosophy talk given by Paul Eisenberg at Indiana University. The Manifesto itself is archived on the web site of the Australasian Association for Logic, together with a list of current members of the LLL, such as the Boss of Bloomington (J. Michael Dunn) and the Queen of Combinators (Katalin Bimbó). I myself was reduced to canonical form in a Bull issued by the Maximum Leader in 1984, appended here.

LOGICIANS LIBERATION LEAGUE
BULL#1 1 September 1984

We, the members of
THE LOGICIANS LIBERATION LEAGUE
Having allocated all posts of temporal honour and glory,
SUCH AS IT IS,
Do now announce the reduction of our esteemed Leader and Teacher,
ALASDAIR URQUHART,
To canonical form. The case having long been studied in the Holy Office,
(Coombs Building 2203, Tel #49-2156, We deliver),
And the said Alasdair Urquhart having performed 2 authentic MIRACLES, to wit,
THE REVELATION OF A RELEVANT SEMANTICS,
And
THE REFUTATION OF RECURSIVENESS IN RELEVANT LOGICS,
The first by taking 2 subscripts and 5 relevance numerals and transforming them into
PIECES OF INFORMATION,
And the second by transubstantiation of the system KR into
MODULAR LATTICE THEORY,
Which for reasons that tend to escape us is also
PROJECTIVE GEOMETRY,
And the same Alasdair Urquhart having lived moreover
A SPOTLESS AND BLAMELESS LIFE,
By the lights of LLL criteria for judging such things, and moreover having
SEEN VISIONS,
HEARD VOICES,
FOUND ENLIGHTENMENT,
With the assistance of such devoted comrades as
S. GIAMBRONE AND C.E. MORTENSEN,
Masters of Inductive

TRANQUILITY,
We accordingly declare, by the powers invested in us, and with the full authority of the Holy Office,
SUCH AS IT IS,
That the aforementioned is to be known henceforth as
ST. ALASDAIR THE BLESSED,
And that all members of the League are henceforth encouraged to
DRAW IN THE SWEET INCENSE OF HIS SACRED CANDLES
MAKE PILGRIMAGES TO TORONTO TO WORSHIP AT HIS SHRINE,
PRAY FOR HIS AID TO SECURE RELEASE FROM
THE PURGATORY OF UNSOLVED PROBLEMS.
St. Alasdair is moreover declared
THE PATRON SAINT OF TYPEWRITERS.
(All Temporal Power remains in the Hands of the Maximum Leader)
All my love, MAX

Ivo Düntsch and Ewa Orłowska: Application of Urquhart's Representation of Lattices to Some Non-classical Logics. After I failed to solve the two famous problems of lattice theory with my representation theory for general lattices, I abandoned it. Perhaps I should have persevered longer, but I felt frustrated and wanted to move on to areas where I might have more success. However that may be, I am delighted that other researchers have been inspired by my efforts, and have used the ideas in their own work, extending my own earlier analyses considerably. The contribution by Ivo Düntsch and Ewa Orłowska is an excellent example of this.

The results in their contribution are all examples of discrete dualities, roughly corresponding to completeness theorems in non-classical logics. They prove such results for a variety of non-classical systems, such as commutator algebras; this latter case is a very pleasing application of my representation theorem. The theory of commutator algebras bears a resemblance to the $\{\vee, \wedge, \otimes, \rightarrow, 0, 1\}$ fragment of affine linear logic (linear logic with weakening). I was happy that Gerard Allwein and J. Michael Dunn used my representation theorem for general lattices in their semantical analysis of linear logic (Allwein and Dunn 1993).

T.S. Blyth and H.J. Silva: OCKHAM ALGEBRAS—an Urquhart legacy. I am delighted and honoured that Blyth and Silva have contributed such a fine essay on Ockham algebras. They provide an excellent survey of my own work on these structures, and then go on to inform the reader about further work that has been done since I left the field in the early 1980s.

As a student of Anderson and Belnap, I was of course familiar with De Morgan algebras and their basic theory. So, when my old friend Joel Berman sent me a preprint of his paper (Berman 1977) [2] generalizing these structures by omitting the law of double negation, I was immediately taken by this very natural idea. I was pleased to find that I could extend Berman's results in a number of different ways.

[2] This was of course before the days of the internet!

As usual in my approach to algebraic logic, I started by dualizing everything in sight, inspired by the representation theory for quasi-Boolean algebras.

Berman applied the rather colourless designation "$\mathcal{K}$" to this class of algebras. Since I was aware that the so-called "De Morgan laws" were originally stated by the great mediaeval logician William of Ockham, I decided to honour him by naming them "Ockham algebras." I am very happy that my terminology has stood the test of time! My first paper gives the basic duality theory; the second gives what I think is a much deeper analysis of these structures, including what I think are my best results in this area, the fact that all of the varieties of Ockham algebras are determined by their finite members, while the lattice of varieties is uncountable.

I continued working in algebraic logic for a few years longer, and published a couple of papers in the area of distributive pseudo-complemented lattices, but after my visit to Canberra, I moved on to the area of complexity theory.

I had the pleasure of meeting Tom Blyth for lunch when I attended a logic meeting in 2000 in St. Andrews on analytic tableaux and related methods organized by the late Roy Dyckhoff. I had not realized that our birthplaces are less than five miles apart—Ockham algebras seem to be a specialty of Fife!

Robin Hirsch and Brett McLean: Temporal Logic of Minkowski Spacetime. Shortly before the monograph *Temporal Logic* that I co-authored with Nicholas Rescher appeared, Gerald Massey published an article with the provocative title "Tense logic! Why Bother?". Massey argued that the tense logic program of Arthur Prior and others was unviable, partly because it was grounded in bad physics (Massey 1969, p. 31), largely ignoring the special theory of relativity.

Since the main applications of temporal logic have been in computer science, where a Newtonian view of time is adequate, this criticism may not matter too much in practice. However, since the early 1970s, authors such as Goldblatt, Shapirovski and Shehtmann have made deep investigations of the logic of Minkowski space-time. Hirsch and McLean provide an elegant treatment of the two-dimensional version of this logic, showing that it is decidable and PSPACE-complete.

Dimiter Vakarelov: Dynamic Contact Algebras With a Predicate of Actual Existence: Snapshot Representation and Topological Duality. As an undergraduate, and later as a graduate student, I was fascinated by the ideas that I found in Whitehead and Russell about logical constructions of abstract objects out of more primitive entities such as sense-data or extended physical objects. Russell wrote quite a lot in this vein, for example, in Lecture IV of his Lowell Lectures at Harvard (Russell 1914). In that lecture, he explains:

> The following illustrative method, simplified so as to be easily manipulated, has been invented by Dr. Whitehead for the purpose of showing how points might be manufactured from sense-data (Russell 1914, p. 114).

Although Russell is at pains to attribute the method to Whitehead, the latter was annoyed by what he considered the premature publication of his ideas and complained to Russell in a letter of January 8 1917 (Russell 1968, pp. 100–101). Whitehead had planned to publish his method of Extensive Abstraction in Volume IV of *Principia*

Mathematica—one of the most renowned in the catalogue of non-existent books. Russell himself continued to expound the method in his philosophical writings, such as *The Analysis of Matter* (Russell 1927), and in fact his last paper on mathematical logic is an analysis (Russell 1936) of the assumptions that we need to make if we are to construct moments in time from events.

Following in the tradition of Whitehead, De Laguna and Tarski, Dimiter Vakarelov's essay gives an excellent exposition of the Region Based Theory of Space (RBTS), extending it to include the idea of regions changing in time, formalized as the theory of Dynamic Contact Algebras (DCA). In addition, he adds a predicate of *actual existence*. The zero, or empty region, is a problem for mereological theories, since if we count it as a region, it is both everywhere and nowhere. The actual existence predicate is intended to deal with this problem.

Sam Buss: Substitution and Propositional Proof Complexity. My first job as a teaching assistant at the University of Pittsburgh was as a tutorial leader and grader for a logic course taught by Storrs McCall. Storrs was a devotee of Polish (prefix) notation in the propositional calculus, so of course I became quite familiar with this old notation. (One of the questions on the final examination was: "Explain three advantages of Polish notation.") I knew a little about it already, having studied Tarski's classic collection (Tarski 1956) of logic papers. While at Pittsburgh, I studied Arthur Prior's idiosyncratic logic textbook (Prior 1967), where Polish notation is used throughout. This text has an unusual appendix where Prior explains C.A. Meredith's rule of condensed detachment, that permits succinct representations of propositional derivations – also represented in Polish notation!

In 1996, it occurred to me that Meredith's device of condensed detachment could be used to prove lower bounds on the number of lines in Frege proofs with substitution. This resulted in a short paper (Urquhart 1997) published in 1997. In it, I also posed the problem of improving this result by proving stronger lower bounds on the number of symbols in $ms\mathcal{F}$ proofs. An $ms\mathcal{F}$ proof is one in a Frege system with the multiple substitution rule. I published a lower bound for systems with the singular substitution rule in 2005 (Urquhart 2005), showing that we obtain a speedup with the multiple substitution rule. However, this is a fairly easy result, not much more difficult than the corresponding result for Frege systems. I worked for a little while on the problem of proving a length lower bound for $ms\mathcal{F}$ proofs, but found it discouragingly difficult, so I abandoned the attempt.

I am delighted that Sam Buss has risen to the challenge and proved an $\Omega(n \log n)$ lower bound on the number of symbols in $ms\mathcal{F}$ proofs of tautologies with length $\Theta(n)$. Although his proof makes use of some of the ideas I used in my two papers, it is much more intricate and difficult, and conveys an idea of just how challenging these problems are. I congratulate him on his fine result.

Noah Fleming and Toniann Pitassi: Reflections on proof complexity and counting principles. I am grateful to Noah Fleming and Toni Pitassi for providing a splendid survey of the Tseitin graph clauses and their uses in proof complexity. When I started working in the field of proof complexity in 1983, Tseitin's lower bounds for regular resolution were among the very few substantial results in the area. One of the things that attracted me to his approach was the fact that I could visualize a regular resolution refutation as a graph decomposition process. I tried to visualize irregular refutations in the same fashion, but ultimately my efforts failed completely. However, when Armin Haken produced his lower bounds for general resolution refutations of the pigeonhole clauses using the probabilistic method, my experience with graph-based examples allowed me to adapt his bottle-neck counting techniques to produce a truly exponential lower bound. I am very happy that my efforts have inspired other researchers to many new results.

I was always fond of the graph-based clauses, not only because of their visual aspects, but also because they are remarkably flexible. Fleming and Pitassi have provided a splendid account of their importance in later work. I also learned quite a number of new results from their excellent contribution.

My collaboration with Toni Pitassi has been among the most fruitful of my career. I am delighted that she has become a star of the Toronto computer science department, and a major figure in theoretical computer science.

André Vellino: Satisfiability, Lattices, Temporal Logic and Constraint Logic Programming on Intervals. André Vellino's contribution brings back to me the early days of my work on propositional proof complexity and my collaboration with André. It is surprising to recall just how little was known in the area in the early 1980s. When I started work on the subject in Toronto in 1983, there were no good lower bounds known even for such a simple system as resolution (though Tseitin had already proved a strong lower bound for regular resolution in 1966). This explains why I concentrated initially on this problem, though, as I explained in my autobiographical account, I made no progress until the breakthrough by Armin Haken in early 1984. Even then, there were still plenty of open problems remaining concerning weaker proof systems, so it was a good topic for a doctoral thesis.

When I started programming in 1983, I used a time-sharing system on a PDP-10 (the workhorse of the Canberra group led by Bob Meyer). Later, when personal computers began to appear on people's desktops in the 1980s, André and I wrote little programs in Turbo Pascal. We had a lot of fun together playing with computers and programs—I recall writing a little routine with André to generate configurations in Conway's Game of Life. It was a revelation to see gliders, blinkers and other Life objects emerging from a random soup of pixels.

André went on to a distinguished career in industry and academia, making his mark initially in the field of logic programming. In the latter part of his essay, André gives a lucid introduction to constraint logic programming and interval arithmetic. As he points out, these areas provide a neat combination of my interests in temporal logic, theorem proving and lattice theory!

Bernard Linsky: Russellian Propositions in *Principia Mathematica*. Whitehead and Russell's great treatise is often hard to interpret. It is not a formal system of logic in the modern sense, though parts of it (for example, $*$**1** to $*$**5** on propositional logic) are close to a modern rigorous axiomatic system. However, a great deal of the foundational material in the first volume leaves a lot to be desired from the point of view of rigour. In particular, the presentation of the basic material on the ramified theory of types is often extremely obscure.

A notorious crux for interpreters is the nature of propositions. From his earliest writings on logic, such as the 1903 *Principles of Mathematics* (Russell 1937), Russell had used quantification over propositions extensively. This held true up until the demise of Russell's substitutional theory on which he worked extensively from 1905 to 1907. In that theory, classes are defined contextually in terms of primitive notions of propositions and substitution. The 1908 theory of types (Russell 1908) emerged from the substitutional theory as a result of the introduction of type distinctions when it became clear that the paradoxes could not be avoided in an untyped theory of propositions.

Although in his 1908 paper, Russell begins from a hierarchy of propositions, he then proceeds to what he claims is the more convenient hierarchy of functions introduced initially through the procedure of substitution. *Principia Mathematica* avoids quantification over propositions almost completely, though as Linsky notes, there is one proposition in it, $*14.3$, where it appears. In addition, quantification over propositions is needed for the analysis of some of the antinomies, for example, the Epimenides paradox (Whitehead and Russell 1927, p. 62). In fact, it is inevitable that *Principia Mathematica* admits categories of propositions of various types. If there is at least one individual x (as is required for the validity of $*10.25$), and ϕ is a first-order function, then ϕx is a first-order proposition.

The major interpretive difficulty, explained clearly by Linsky, lies in making sense of the claim (Whitehead and Russell 1927, pp. 44) that propositions are "incomplete symbols." Russell derives this claim from a metaphysical theory of judgement (Whitehead and Russell 1927, pp. 43–44) that propositions express judgments:

> Thus, the proposition "Socrates is human" uses "Socrates is human"in a way which requires a supplement of some kind before it acquires a complete meaning; but when I judge "Socrates is human" the meaning is completed by the act of judging, and we no longer have an incomplete symbol.

Clearly, we are here in a realm quite far removed from the formal developments of the main text. In my survey article on the theory of types (Urquhart 2003), I followed Church (Church 1976) in ignoring this as a philosophical excrescence that does not affect the main symbolic development. As Church surmised, and James Levine confirmed by archival research, these passages on propositions as incomplete symbols are a late addition to the text of *Principia Mathematica*.

Linsky, however, does not adopt a completely dismissive attitude to the passage from the introduction, but makes an interesting suggestion that we can make some sense of the idea of propositions as incomplete symbols by examining the developments in $*\mathbf{9}$. He suggests that Russell's method of extending the axioms of propositional logic to higher types in $*\mathbf{9}$ provides a technical way to interpret propositions as "incomplete symbols."

Allen P. Hazen and Francis Jeffrey Pelletier: Some Lessons Learned About Adding Conditionals to Certain Many-Valued Logics. My involvement with many-valued logics has been somewhat peripheral, even though I have written two versions of a survey of the field. My first paper on the subject (Urquhart 1973) was prompted by the discovery that the conditional in the many-valued logics of Łukasiewicz could be given a semantical reading akin to that of the conditional in the semilattice semantics for relevant implications. Scott (1974) had the same idea independently, and worked it out in greater detail than I did.

My little paper, of 1971, led to the invitation from Gabbay and Guenthner to contribute the chapter (Urquhart 1984a) on many-valued logic to the first edition of the *Handbook of Philosophical Logic*. I wrote a good deal of it when in Australia, in 1982, but I was rather bored with the project and distracted by the numerous pleasures of Canberra and my interactions with the fine group of logicians there. Consequently, the chapter contains several serious errors. Fortunately, I was able to correct these in the second version (Urquhart 2001).

Hazen and Pelletier address the important question of adding a conditional to certain logics. Their earlier work achieved success in several cases. Here, they exhibit some problematic examples.

References

Ackermann, W. (1956). Begründung einer strengen Implikation. *Journal of Symbolic Logic*, *21*, 113–128.

Allwein, G., & Dunn, J. M. (1993). Kripke models for linear logic. *Journal of Symbolic Logic*, *58*, 514–545.

Anderson, A. R., & Belnap, N. D. (1975). *Entailment* (Vol. 1). Princeton, NJ: Princeton University Press.

Anderson, A. R, Jr., & N. D. B., and Dunn, J. M., (1992). *Entailment* (Vol. 2). Princeton, NJ: Princeton University Press.

Berman, J. (1977). Distributive lattices with an additional unary operation. *Aequationes Mathematicae*, *16*, 165–171.

Charlwood, G. (1978). *Representations of semilattice relevance logic*. Ph.D. thesis, University of Toronto.

Church, A. (1951). The weak theory of implication. In A. Menne, A. Wilhelmy, (Eds.), *Kontrolliertes Denken, Untersuchungen zum Logikkalkül und der Logik der Einzelwissenschaften* (pp. 22–37). Munich: Kommissions-Verlag Karl Alber.

Church, A. (1976). Comparison of Russell's resolution of the semantical antinomies with that of Tarski. *Journal of Symbolic Logic*, *41*, 747–760.

Fine, K. (1978). Model theory for modal logic part I: The "de re/de dicto" distinction. *Journal of Philosophical Logic*, *7*, 125–156.

Freese, R. (1979). The variety of modular lattices is not generated by its finite members. *Transactions of the American Mathematical Society*, *255*, 277–300.

Freese, R. (1980). Free modular lattices. *Transactions of the American Mathematical Society*, *261*, 81–91.

Girard, J. -Y. (2011). *The Blind Spot: Lectures on Logic*. Zürich: European Mathematical Society.

Lewis, C. I. (1913). A new algebra of implication and some consequences. *The Journal of Philosophy, Psychology and Scientific Methods*, *10*, 428–438.

Lyndon, R. C. (1961). Relation algebras and projective geometries. *Michigan Mathematical Journal, 8*, 21–28.
Massey, G. J. (1969). Tense logic! why bother? *Noûs, 3*, 17–32.
Prawitz, D. (1965). *Natural Deduction. A Proof-theoretical study*. Reprinted by Dover books 2006. Stockholm: Almqvist and Wiksell.
Prior, A. (1967). *Past Present and Future*. Oxford: Oxford Univerity Press.
Quine, W. (1986). Autobiography of W.V. Quine. In L. E. Hahn, & P. A. Schilpp (Eds.), *The Philosophy of W. V. Quine (The Library of Living Philosophers Volume XVIII)* (pp. 1–48). Chicago: Open Court Publishing Company.
Routley, R., & Meyer, R. K. (1973). Semantics of entailment. In H. Leblanc, (Ed.), *Proceedings of the Temple University Conference on Alternative Semantics Truth Syntax and Modality* (pp. 199–243). Amsterdam: North-Holland Publishing Company.
Russell, B. (1908). Mathematical logic as based on the theory of types. *American Journal of Mathematics, 30*, 222–262. Reprinted in [42].
Russell, B. (1914). *Our Knowledge of the External World as a field for scientific method in philosophy*. Chicago: The Open Court Publishing Company.
Russell, B. (1927). *The Analysis of Matter*. Brace: Harcourt.
Russell, B. (1936). On order in time. *Proceedings of the Cambridge Philosophical Society, 32*, 216–228.
Russell, B. (1937). The Principles of Mathematics. George Allen and Unwin. Second impression with a new introduction: first edition 1903.
Russell, B. (1968). *The Autobiography of Bertrand Russell 1914–1944* (Vol. II). Boston: An Atlantic Monthly Press Book/Little, Brown and Company.
Scott, D. (1974). Completeness and axiomatizability in many-valued logic. In *Proceedings of the Tarski Symposium: Symposia in Pure Mathematics* (Vol. 25). Providence: American Mathematical Society.
Tarski, A. (1941). On the calculus of relations. *Journal of Symbolic Logic, 6*, 73–89.
Tarski, A. (1956). *Logic, Semantics, Metamathematics: Papers from 1923 to 1938* (J. H. Woodger Trans.). Oxford: Oxford University Press.
Thistlewaite, P., McRobbie, M., & Meyer, R. (1988). *Automated theorem-proving in non-classical logics*. London: Pitman.
Urquhart, A. (1973). An interpretation of many-valued logic. *Zeitschrift für Mathematische Logik und Grundlagen der Mathematik, 19*, 212–219.
Urquhart, A. (1984a). Many-valued Logic. In D. Gabbay & F. Guenthner (Eds.), *Handbook of Philosophical Logic* (Vol. III). Dordrecht: D. Reidel Publishing Company.
Urquhart, A. (1984b). The undecidability of entailment and relevant implication. *Journal of Symbolic Logic, 49*, 1059–1073.
Urquhart, A. (1989). What is relevant implication? In J. Norman, & R. Sylvan (Eds.), *Directions in Relevant Logic* (pp. 167–174). Alphen aan den Rijn: Kluwer.
Urquhart, A. (1996). Duality for algebras of relevant logics. *Studia Logica, 56*, 263–276.
Urquhart, A. (1997). The number of lines in Frege proofs with substitution. *Archive for Mathematical Logic, 37*, 15–19.
Urquhart, A. (1999). Beth's definability theorem in relevant logics. In E. Orłowska (Ed.), *Logic at Work: Essays dedicated to the Memory of Helena Rasiowa*. Heidelberg: Physica-Verlag.
Urquhart, A. (2001). Basic many-valued logic. In D. Gabbay, & F. Guenthner (Eds.), *Handbook of Philosophical Logic* (Vol, 2, 2nd Ed., pp. 249–295). Alphen aan den Rijn: Kluwer.
Urquhart, A. (2003). The theory of types. In N. Griffin (Ed.), *The Cambridge Companion to Russell* (pp. 286–309). Cambridge: Cambridge University Press.
Urquhart, A. (2005). The complexity of propositional proofs with the substitution rule. *Logic Journal of the IGPL, 13*, 287–291.
Urquhart, A. (2015). First degree formulas in quantified S5. *Australasian Journal of Logic, 12*, 204–210.

Urquhart, A. (2017). The geometry of relevant implication. *IFCoLog Journal of Logics and their Applications*, *4*(3), 591–604.
van Heijenoort, J. (1967). *From Frege to Gödel. A Source Book in Mathematical Logic, 1879–1931*. Cambridge: Harvard University Press.
Whitehead, A. N., & Russell, B. (1927). *Principia Mathematica* (Vol. 1, 2nd ed.). Cambridge: Cambridge University Press.
Wilde, O. (1891). *The Picture of Dorian Gray*. London: Ward Lock and Company.

CPSIA information can be obtained
at www.ICGtesting.com
Printed in the USA
LVHW060319180723
752364LV00035B/85